STUDENT'S SOLUTIONS MANUAL
JUDITH A. PENNA

ELEMENTARY AND INTERMEDIATE ALGEBRA
CONCEPTS AND APPLICATIONS
A COMBINED APPROACH

THIRD EDITION

Marvin L. Bittinger
Indiana University—Purdue University at Indianapolis

David J. Ellenbogen
Community College of Vermont

Barbara L. Johnson
Indiana University—Purdue University at Indianapolis

Addison
Wesley

Boston San Francisco New York
London Toronto Sydney Tokyo Singapore Madrid
Mexico City Munich Paris Cape Town Hong Kong Montreal

ISBN   0-201-64211-5

    6 7 8 9 10 BB 04

Contents

Chapter 1

Introduction to Algebraic Expressions

Exercise Set 1.1

1. Substitute 9 for a and multiply.

$3a = 3 \cdot 9 = 27$

2. 56

3. Substitute 2 for t and add.

$t + 6 = 2 + 6 = 8$

4. 4

5. $\dfrac{x+y}{4} = \dfrac{2+14}{4} = \dfrac{16}{4} = 4$

6. 5

7. $\dfrac{m-n}{2} = \dfrac{20-6}{2} = \dfrac{14}{2} = 7$

8. 3

9. $\dfrac{a}{b} = \dfrac{45}{9} = 5$

10. 6

11. $\dfrac{9m}{q} = \dfrac{9 \cdot 6}{18} = \dfrac{54}{18} = 3$

12. 3

13. $bh = (6 \text{ ft})(4 \text{ ft})$

$= (6)(4)(\text{ft})(\text{ft})$

$= 24 \text{ ft}^2$, or 24 square feet

14. 24 hr

15. $A = \dfrac{1}{2}bh$

$= \dfrac{1}{2}(5 \text{ cm})(6 \text{ cm})$

$= \dfrac{1}{2}(5)(6)(\text{cm})(\text{cm})$

$= \dfrac{5}{2} \cdot 6 \text{ cm}^2$

$= 15 \text{ cm}^2$, or 15 square centimeters

16. (a) 150 sec;

(b) 450 sec;

(c) 10 min

17. $\dfrac{h}{a} = \dfrac{10}{37}$, or about 0.270

18. 26 cm^2

19. Let a represent Jan's age. Then we have $j + 8$, or $8 + j$.

20. $4a$

21. $b + 6$, or $6 + b$

22. Let w represent Lou's weight; $w + 7$, or $7 + w$

23. $c - 9$

24. $d - 4$

25. $q + 6$, or $6 + q$

26. $z + 11$, or $11 + z$

27. Let s represent Phil's speed. Then we have $9s$, or $s9$.

28. $d + c$, or $c + d$

29. $y - x$

30. Let a represent Lorrie's age; $a - 2$

31. $x \div w$, or $\dfrac{x}{w}$

32. Let s and t represent the numbers; $s \div t$, or $\dfrac{s}{t}$

33. $n - m$

34. $q - p$

35. Let l and h represent the box's length and height, respectively. Then we have $l + h$, or $h + l$.

36. $d + f$, or $f + d$

37. $9 \cdot 2m$

38. Let p represent Paula's speed and w represent the wind speed; $p - 2w$

39. Let y represent "some number." Then we have $\dfrac{1}{4}y$, or $\dfrac{y}{4}$, or $y/4$, or $y \div 4$.

40. Let m and n represent the numbers; $\frac{1}{3}(m+n)$, or $\frac{m+n}{3}$

41. Let x represent the number of women attending. Then we have 64% of x, or $0.64x$.

42. Let y represent "a number;" 38% of y, or $0.38y$

43. $\$50 - x$

44. $65t$ mi

45.
$$\underset{\downarrow}{x + 17 = 32} \qquad \text{Writing the equation}$$
$$15 + 17 \;?\; 32 \qquad \text{Substituting 15 for } x$$
$$32 \;\big|\; 32 \qquad 32 = 32 \text{ is TRUE.}$$

Since the left-hand and right-hand sides are the same, 15 is a solution.

46. No

47.
$$\underset{\downarrow}{a - 28 = 75} \qquad \text{Writing the equation}$$
$$93 - 28 \;?\; 75 \qquad \text{Substituting 93 for } a$$
$$65 \;\big|\; 75 \qquad 65 = 75 \text{ is FALSE.}$$

Since the left-hand and right-hand sides are not the same, 93 is not a solution.

48. Yes

49.
$$\underset{\downarrow}{\dfrac{t}{7} = 9}$$
$$\dfrac{63}{7} \;?\; 9$$
$$9 \;\big|\; 9 \qquad 9 = 9 \text{ is TRUE.}$$

Since the left-hand and right-hand sides are the same, 63 is a solution.

50. No

51.
$$\underset{\downarrow}{\dfrac{108}{x} = 36}$$
$$\dfrac{108}{3} \;?\; 36$$
$$36 \;\big|\; 36 \qquad 36 = 36 \text{ is TRUE.}$$

Since the left-hand and right-hand sides are the same, 3 is a solution.

52. No

53. Let x represent the number.

What number added to 73 is 201?

Translating: $\quad x \quad + \quad 73 = 201$

$x + 73 = 201$

54. Let w represent the number; $7w = 2303$

55. Let y represent the number.

Rewording: 42 times what number is 2352?

Translating: 42 y $= 2352$

$42y = 2352$

56. Let x represent the number; $x + 345 = 987$

57. Let s represent the number of squares your opponent controls.

Rewording: The number of squares your opponent controls added to 35 is 64.

Translating: s $+$ $35 = 64$

$s + 35 = 64$

58. Let y represent the number of hours the carpenter worked; $25y = 53,400$

59. Let x represent the total amount of waste generated, in millions of tons.

Rewording: 27% of the total amount of waste is 56 million tons.

Translating: 27% $\cdot$ x $=$ 56

$27\% \cdot x = 56$, or $0.27x = 56$

60. Let m represent the length of the average commute in the West, in minutes; $m = 24.5 - 1.8$

61. Writing exercise

62. Writing exercise

63. Writing exercise

64. Writing exercise

65. Area of sign: $A = \dfrac{1}{2}(3 \text{ ft})(2.5 \text{ ft}) = 3.75 \text{ ft}^2$

Cost of sign: $\$90(3.75) = \$337.50$

66. 158.75 cm^2

67. When x is twice y, then y is one-half x, so
$$y = \frac{12}{2} = 6.$$
$$\frac{x-y}{3} = \frac{12-6}{3} = \frac{6}{3} = 2$$

68. 9

69. When a is twice b, then b is one-half a, so $b = \frac{16}{2} = 8$.
$$\frac{a+b}{4} = \frac{16+8}{4} = \frac{24}{4} = 6$$

70. 4

71. The next whole number is one more than $w + 3$:
$$w + 3 + 1 = w + 4$$

72. d

73. Let a and b represent the numbers. Then we have
$$\frac{1}{3} \cdot \frac{1}{2} \cdot ab.$$

74. $l + w + l + w$, or $2l + 2w$

75. $s + s + s + s$, or $4s$

76. $a + 9$

77. Writing exercise

Exercise Set 1.2

1. $x + 7$ Changing the order

2. $2 + a$

3. $c + ab$

4. $3y + x$

5. $3y + 9x$

6. $7b + 3a$

7. $5(1 + a)$

8. $9(5 + x)$

9. $a \cdot 2$ Changing the order

10. yx

11. ts

12. $x4$

13. $5 + ba$

14. $x + y3$

15. $(a + 1)5$

16. $(x + 5)9$

17. $a + (5 + b)$

18. $5 + (m + r)$

19. $(r + t) + 7$

20. $(x + 2) + y$

21. $ab + (c + d)$

22. $m + (np + r)$

23. $8(xy)$

24. $9(ab)$

25. $(2a)b$

26. $(9r)p$

27. $(3 \cdot 2)(a + b)$

28. $(5x)(2 + y)$

29. a) $r + (t + 6) = (t + 6) + r$ Using the commutative law
$$= (6 + t) + r \quad \text{Using the commutative law again}$$

     b) $r + (t + 6) = (t + 6) + r$ Using the commutative law
$$= t + (6 + r) \quad \text{Using the associative law}$$

Answers may vary.

30. Answers may vary; $v + (w + 5)$; $(v + 5) + w$

31. a) $(17a)b = b(17a)$ Using the commutative law
$$= b(a17) \quad \text{Using the commutative law again}$$

     b) $(17a)b = (a17)b$ Using the commutative law
$$= a(17b) \quad \text{Using the associative law}$$

Answers may vary.

32. Answers may vary; $3(yx)$; $(3x)y$

33. $\quad (5 + x) + 2$

$= (x + 5) + 2 \quad$ Commutative law

$= x + (5 + 2) \quad$ Associative law

$= x + 7 \qquad\quad$ Simplifying

34. $(2a)4 = 4(2a) \qquad$ Commutative law

$\qquad = (4 \cdot 2)a \quad$ Associative law

$\qquad = 8a \qquad\;$ Simplifying

35. $(m3)7 = m(3 \cdot 7) \quad$ Associative law

$\qquad = (3 \cdot 7)m \quad$ Commutative law

$\qquad = 21m \qquad$ Simplifying

36. $\quad 4 + (9 + x)$

$= (4 + 9) + x \quad$ Associative law

$= x + (4 + 9) \quad$ Commutative law

$= x + 13 \qquad\;$ Simplifying

37. $4(a + 3) = 4 \cdot a + 4 \cdot 3 = 4a + 12$

38. $3x + 15$

39. $6(1 + x) = 6 \cdot 1 + 6 \cdot x = 6 + 6x$

40. $6v + 24$

41. $3(x + 1) = 3 \cdot x + 3 \cdot 1 = 3x + 3$

42. $9x + 27$

43. $8(3 + y) = 8 \cdot 3 + 8 \cdot y = 24 + 8y$

44. $7s + 35$

45. $9(2x + 6) = 9 \cdot 2x + 9 \cdot 6 = 18x + 54$

46. $54m + 63$

47. $5(r + 2 + 3t) = 5 \cdot r + 5 \cdot 2 + 5 \cdot 3t = 5r + 10 + 15t$

48. $20x + 32 + 12p$

49. $(a + b)2 = a(2) + b(2) = 2a + 2b$

50. $7x + 14$

51. $(x + y + 2)5 = x(5) + y(5) + 2(5) = 5x + 5y + 10$

52. $12 + 6a + 6b$

53. $x + xyz + 19$

The terms are separated by plus signs. They are x, xyz, and 19.

54. $9, 17a, abc$

55. $2a + \dfrac{a}{b} + 5b$

The terms are separated by plus signs. They are $2a$, $\dfrac{a}{b}$, and $5b$.

56. $3xy,\ 20,\ \dfrac{4a}{b}$

57. $2a + 2b = 2(a + b) \qquad$ The common factor is 2.

Check: $2(a + b) = 2 \cdot a + 2 \cdot b = 2a + 2b$

58. $5(y + z)$

59. $\quad 7 + 7y = 7 \cdot 1 + 7 \cdot y \quad$ The common factor is 7.

$\qquad\; = 7(1 + y) \qquad$ Using the distributive law

Check: $7(1 + y) = 7 \cdot 1 + 7 \cdot y = 7 + 7y$

60. $113(1 + x)$

61. $18x + 3 = 3 \cdot 6x + 3 \cdot 1 = 3(6x + 1)$

Check: $3(6x + 1) = 3 \cdot 6x + 3 \cdot 1 = 18x + 3$

62. $5(4a + 1)$

63. $5x + 10 + 15y = 5 \cdot x + 5 \cdot 2 + 5 \cdot 3y = 5(x + 2 + 3y)$

Check: $5(x + 2 + 3y) = 5 \cdot x + 5 \cdot 2 + 5 \cdot 3y = 5x + 10 + 15y$

64. $3(1 + 9b + 2c)$

65. $12x + 9 = 3 \cdot 4x + 3 \cdot 3 = 3(4x + 3)$

Check: $3(4x + 3) = 3 \cdot 4x + 3 \cdot 3 = 12x + 9$

66. $6(x + 1)$

67. $3a + 9b = 3 \cdot a + 3 \cdot 3b = 3(a + 3b)$

Check: $3(a + 3b) = 3 \cdot a + 3 \cdot 3b = 3a + 9b$

68. $5(a + 3b)$

69. $44x + 11y + 22z = 11 \cdot 4x + 11 \cdot y + 11 \cdot 2z = 11(4x + y + 2z)$

Check: $11(4x + y + 2z) = 11 \cdot 4x + 11 \cdot y + 11 \cdot 2z = 44x + 11y + 22z$

70. $7(2a + 8b + 1)$

71. Writing exercise

72. Writing exercise

73. Let k represent Kara's salary. Then we have $2k$.

74. $\dfrac{1}{2}m$, or $\dfrac{m}{2}$, or $m/2$, or $m \div 2$

75. Writing exercise

76. Writing exercise

77. The expressions are equivalent.

$$8 + 4(a + b) = 8 + 4a + 4b = 4(2 + a + b)$$

78. The expressions are not equivalent.

Let $m = 1$. Then we have:

$$7 \div 3 \cdot 1 = \frac{7}{3} \cdot 1 = \frac{7}{3}, \text{ but}$$

$$1 \cdot 3 \div 7 = 3 \div 7 = \frac{3}{7}.$$

79. The expressions are equivalent.

$$(rt + st)5 = 5(rt + st) = 5 \cdot t(r + s) = 5t(r + s)$$

80. The expression are equivalent.

$$yax + ax = (y + 1)ax = (1 + y)ax = ax(1 + y) = xa(1 + y)$$

81. The expressions are not equivalent.

Let $x = 1$ and $y = 0$. Then we have:

$$30 \cdot 0 + 1 \cdot 15 = 0 + 15 = 15, \text{ but}$$

$$5[2(1 + 3 \cdot 0)] = 5[2(1)] = 5 \cdot 2 = 10.$$

82. The expressions are equivalent.

$$[c(2 + 3b)]5 = 5[c(2 + 3b)] = 5c(2 + 3b) = 10c + 15bc$$

83. Writing exercise

84. Writing exercise

Exercise Set 1.3

1. We write two factorizations of 50. There are other factorizations as well.

$$2 \cdot 25, \; 5 \cdot 10$$

List all of the factors of 50:

1, 2, 5, 10, 25, 50

2. $2 \cdot 35, \; 5 \cdot 14$; 1, 2, 5, 7, 10, 14, 35, 70

3. We write two factorizations of 42. There are other factorizations as well.

$$2 \cdot 21, \; 6 \cdot 7$$

List all of the factors of 42:

1, 2, 3, 6, 7, 14, 21, 42

4. $2 \cdot 30, \; 5 \cdot 12$; 1, 2, 3, 4, 5, 6, 10, 12, 15, 20, 30, 60

5. $26 = 2 \cdot 13$

6. $3 \cdot 5$

7. We begin factoring 30 in any way that we can and continue factoring until each factor is prime.

$$30 = 2 \cdot 15 = 2 \cdot 3 \cdot 5$$

8. $5 \cdot 11$

9. We begin by factoring 20 in any way that we can and continue factoring until each factor is prime.

$$20 = 4 \cdot 5 = 2 \cdot 2 \cdot 5$$

10. $2 \cdot 5 \cdot 5$

11. We begin by factoring 27 in any way that we can and continue factoring until each factor is prime.

$$27 = 3 \cdot 9 = 3 \cdot 3 \cdot 3$$

12. $2 \cdot 7 \cdot 7$

13. We begin by factoring 18 in any way that we can and continue factoring until each factor is prime.

$$18 = 2 \cdot 9 = 2 \cdot 3 \cdot 3$$

14. $2 \cdot 3 \cdot 3 \cdot 3$

15. We begin by factoring 40 in any way that we can and continue factoring until each factor is prime.

$$40 = 4 \cdot 10 = 2 \cdot 2 \cdot 2 \cdot 5$$

16. $2 \cdot 2 \cdot 2 \cdot 7$

17. 43 has exactly two different factors, 43 and 1. Thus, 43 is prime.

18. $2 \cdot 2 \cdot 2 \cdot 3 \cdot 5$

19. $210 = 2 \cdot 105 = 2 \cdot 3 \cdot 35 = 2 \cdot 3 \cdot 5 \cdot 7$

20. Prime

21. $115 = 5 \cdot 23$

22. $11 \cdot 13$

23. $\dfrac{10}{14} = \dfrac{2 \cdot 5}{2 \cdot 7}$ Factoring numerator and denominator

$$= \frac{2}{2} \cdot \frac{5}{7}$$ Rewriting as a product of two fractions

$$= 1 \cdot \frac{5}{7} \qquad \frac{2}{2} = 1$$

$$= \frac{5}{7}$$ Using the identity property of 1

24. $\dfrac{2}{3}$

25. $\dfrac{16}{56} = \dfrac{2 \cdot 8}{7 \cdot 8} = \dfrac{2}{7} \cdot \dfrac{8}{8} = \dfrac{2}{7} \cdot 1 = \dfrac{2}{7}$

26. $\dfrac{8}{3}$

27. $\dfrac{6}{48} = \dfrac{1 \cdot 6}{8 \cdot 6}$ Factoring and using the identity property of 1 to write 6 as $1 \cdot 6$

$= \dfrac{1}{8} \cdot \dfrac{6}{6}$

$= \dfrac{1}{8} \cdot 1 = \dfrac{1}{8}$

28. $\dfrac{6}{35}$

29. $\dfrac{49}{7} = \dfrac{7 \cdot 7}{1 \cdot 7} = \dfrac{7}{1} \cdot \dfrac{7}{7} = \dfrac{7}{1} \cdot 1 = 7$

30. 12

31. $\dfrac{19}{76} = \dfrac{1 \cdot 19}{4 \cdot 19}$ Factoring and using the identity property of 1 to write 19 as $1 \cdot 19$

$= \dfrac{1 \cdot \cancel{19}}{4 \cdot \cancel{19}}$ Removing a factor equal to 1: $\dfrac{19}{19} = 1$

$= \dfrac{1}{4}$

32. $\dfrac{1}{3}$

33. $\dfrac{150}{25} = \dfrac{6 \cdot 25}{1 \cdot 25}$ Factoring and using the identity property of 1 to write 25 as $1 \cdot 25$

$= \dfrac{6 \cdot \cancel{25}}{1 \cdot \cancel{25}}$ Removing a factor equal to 1: $\dfrac{25}{25} = 1$

$= \dfrac{6}{1}$

$= 6$ Simplifying

34. 5

35. $\dfrac{75}{80} = \dfrac{5 \cdot 15}{5 \cdot 16}$ Factoring the numerator and the denominator

$= \dfrac{\cancel{5} \cdot 15}{\cancel{5} \cdot 16}$ Removing a factor equal to 1: $\dfrac{5}{5} = 1$

$= \dfrac{15}{16}$

36. $\dfrac{21}{25}$

37. $\dfrac{120}{82} = \dfrac{2 \cdot 60}{2 \cdot 41}$ Factoring

$= \dfrac{\cancel{2} \cdot 60}{\cancel{2} \cdot 41}$ Removing a factor equal to 1: $\dfrac{2}{2} = 1$

$= \dfrac{60}{41}$

38. $\dfrac{5}{3}$

39. $\dfrac{210}{98} = \dfrac{2 \cdot 7 \cdot 15}{2 \cdot 7 \cdot 7}$ Factoring

$= \dfrac{\cancel{2} \cdot \cancel{7} \cdot 15}{\cancel{2} \cdot \cancel{7} \cdot 7}$ Removing a factor equal to 1: $\dfrac{2 \cdot 7}{2 \cdot 7} = 1$

$= \dfrac{15}{7}$

40. $\dfrac{2}{5}$

41. $\dfrac{1}{2} \cdot \dfrac{3}{7} = \dfrac{1 \cdot 3}{2 \cdot 7}$ Multiplying numerators and denominators

$= \dfrac{3}{14}$

42. $\dfrac{44}{25}$

43. $\dfrac{9}{2} \cdot \dfrac{3}{4} = \dfrac{9 \cdot 3}{2 \cdot 4} = \dfrac{27}{8}$

44. 1

45. $\dfrac{1}{8} + \dfrac{3}{8} = \dfrac{1 + 3}{8}$ Adding numerators; keeping the common denominator

$= \dfrac{4}{8}$

$= \dfrac{1 \cdot \cancel{4}}{2 \cdot \cancel{4}} = \dfrac{1}{2}$ Simplifying

46. $\dfrac{5}{8}$

47. $\dfrac{4}{9} + \dfrac{13}{18} = \dfrac{4}{9} \cdot \dfrac{2}{2} + \dfrac{13}{18}$ Using 18 as the common denominator

$= \dfrac{8}{18} + \dfrac{13}{18}$

$= \dfrac{21}{18}$

$= \dfrac{7 \cdot \cancel{3}}{6 \cdot \cancel{3}} = \dfrac{7}{6}$ Simplifying

48. $\dfrac{4}{3}$

49. $\dfrac{3}{a} \cdot \dfrac{b}{7} = \dfrac{3b}{7a}$ Multiplying numerators and denominators

50. $\dfrac{xy}{5z}$

51. $\dfrac{4}{a} + \dfrac{3}{a} = \dfrac{7}{a}$ Adding numerators; keeping the common denominator

52. $\dfrac{2}{a}$

53. $\dfrac{3}{10} + \dfrac{8}{15} = \dfrac{3}{10} \cdot \dfrac{3}{3} + \dfrac{8}{15} \cdot \dfrac{2}{2}$ Using 30 as the common denominator

$\phantom{\dfrac{3}{10} + \dfrac{8}{15}} = \dfrac{9}{30} + \dfrac{16}{30}$

$\phantom{\dfrac{3}{10} + \dfrac{8}{15}} = \dfrac{25}{30}$

$\phantom{\dfrac{3}{10} + \dfrac{8}{15}} = \dfrac{5 \cdot \cancel{5}}{6 \cdot \cancel{5}} = \dfrac{5}{6}$ Simplifying

54. $\dfrac{31}{24}$

55. $\dfrac{9}{7} - \dfrac{2}{7} = \dfrac{7}{7} = 1$

56. 2

57. $\dfrac{13}{18} - \dfrac{4}{9} = \dfrac{13}{18} - \dfrac{4}{9} \cdot \dfrac{2}{2}$ Using 18 as the common denominator

$\phantom{\dfrac{13}{18} - \dfrac{4}{9}} = \dfrac{13}{18} - \dfrac{8}{18}$

$\phantom{\dfrac{13}{18} - \dfrac{4}{9}} = \dfrac{5}{18}$

58. $\dfrac{31}{45}$

59. $\dfrac{5}{7} - \dfrac{5}{21} = \dfrac{5}{7} \cdot \dfrac{3}{3} - \dfrac{5}{21}$ Using 21 as the common denominator

$\phantom{\dfrac{5}{7} - \dfrac{5}{21}} = \dfrac{15}{21} - \dfrac{5}{21}$

$\phantom{\dfrac{5}{7} - \dfrac{5}{21}} = \dfrac{10}{21}$

60. 0

61. $\dfrac{7}{6} \div \dfrac{3}{5} = \dfrac{7}{6} \cdot \dfrac{5}{3}$ Multiplying by the reciprocal of the divisor

$\phantom{\dfrac{7}{6} \div \dfrac{3}{5}} = \dfrac{35}{18}$

62. $\dfrac{28}{15}$

63. $\dfrac{8}{9} \div \dfrac{4}{15} = \dfrac{8}{9} \cdot \dfrac{15}{4} = \dfrac{2 \cdot \cancel{4} \cdot \cancel{3} \cdot 5}{\cancel{3} \cdot 3 \cdot \cancel{4}} = \dfrac{10}{3}$

64. $\dfrac{1}{4}$

65. $12 \div \dfrac{3}{7} = \dfrac{12}{1} \cdot \dfrac{7}{3} = \dfrac{4 \cdot \cancel{3} \cdot 7}{1 \cdot \cancel{3}} = 28$

66. $\dfrac{1}{2}$

67. Note that we have a number divided by itself. Thus, the result is 1. We can also do this exercise as follows:

$\dfrac{7}{13} \div \dfrac{7}{13} = \dfrac{7}{13} \cdot \dfrac{13}{7} = \dfrac{7 \cdot 13}{7 \cdot 13} = 1$

68. $\dfrac{51}{20}$

69. $\dfrac{\frac{2}{7}}{\frac{5}{3}} = \dfrac{2}{7} \div \dfrac{5}{3} = \dfrac{2}{7} \cdot \dfrac{3}{5} = \dfrac{2 \cdot 3}{7 \cdot 5} = \dfrac{6}{35}$

70. $\dfrac{15}{8}$

71. $\dfrac{\frac{9}{1}}{\frac{1}{2}} = 9 \div \dfrac{1}{2} = \dfrac{9}{1} \cdot \dfrac{2}{1} = \dfrac{9 \cdot 2}{1 \cdot 1} = 18$

72. $\dfrac{7}{15}$

73. Writing exercise

74. Writing exercise

75. $5(x + 3) = 5(3 + x)$ Commutative law of addition
Answers may vary.

76. Answers may vary; $(a + b) + 7$, or $7 + (b + a)$

77. Writing exercise

78. Writing exercise

79. We need to find the smallest number that has both 6 and 8 as factors. Starting with 6 we list some numbers with a factor of 6, and starting with 8 we also list some numbers with a factor of 8. Then we find the first number that is on both lists.

6, 12, 18, 24, 30, 36, ...

8, 16, 24, 32, 40, 48, ...

Since 24 is the smallest number that is on both lists, the carton should be 24 in. long.

80.

Product	56	63	36	72	140	96	168
Factor	7	7	2	36	14	8	8
Factor	8	9	18	2	10	12	21
Sum	15	16	20	38	24	20	29

81. $\dfrac{16 \cdot 9 \cdot 4}{15 \cdot 8 \cdot 12} = \dfrac{\cancel{4} \cdot \cancel{4} \cdot \cancel{3} \cdot \cancel{3} \cdot \cancel{2} \cdot 2}{\cancel{3} \cdot 5 \cdot \cancel{2} \cdot \cancel{4} \cdot \cancel{3} \cdot \cancel{4}} = \dfrac{2}{5}$

82. 1

83. $\dfrac{27pqrs}{9prst} = \dfrac{3 \cdot \cancel{9} \cdot \cancel{p} \cdot q \cdot \cancel{r} \cdot \cancel{s}}{\cancel{9} \cdot \cancel{p} \cdot \cancel{r} \cdot \cancel{s} \cdot t} = \dfrac{3q}{t}$

84. $\dfrac{8}{3}$

85. $\dfrac{15 \cdot 4xy \cdot 9}{6 \cdot 25x \cdot 15y} = \dfrac{\cancel{15} \cdot \cancel{2} \cdot 2 \cdot \cancel{x} \cdot \cancel{y} \cdot \cancel{3} \cdot 3}{\cancel{2} \cdot \cancel{3} \cdot 25 \cdot \cancel{x} \cdot \cancel{15} \cdot \cancel{y}} = \dfrac{6}{25}$

86. $\dfrac{5}{2}$

87. $\dfrac{\frac{27ab}{15mn}}{\frac{18bc}{25np}} = \dfrac{27ab}{15mn} \div \dfrac{18bc}{25np} = \dfrac{27ab}{15mn} \cdot \dfrac{25np}{18bc} =$

$\dfrac{27ab \cdot 25np}{15mn \cdot 18bc} = \dfrac{\cancel{3} \cdot \cancel{9} \cdot a \cdot \cancel{b} \cdot \cancel{5} \cdot 5 \cdot \cancel{n} \cdot p}{\cancel{3} \cdot \cancel{5} \cdot m \cdot \cancel{n} \cdot 2 \cdot \cancel{9} \cdot \cancel{b} \cdot c} = \dfrac{5ap}{2mc}$

88. $\dfrac{2yc}{b}$

89. $A = lw = \left(\dfrac{4}{5}\text{ m}\right)\left(\dfrac{7}{9}\text{ m}\right)$

$= \left(\dfrac{4}{5}\right)\left(\dfrac{7}{9}\right)(\text{m})(\text{m})$

$= \dfrac{28}{45}\text{ m}^2, \text{ or } \dfrac{28}{45} \text{ square meters}$

90. $\dfrac{25}{28}\text{ m}^2$

91. $P = 4s = 4\left(3\dfrac{5}{9}\text{ m}\right) = 4 \cdot \dfrac{32}{9}\text{ m} = \dfrac{128}{9}\text{ m, or}$

$14\dfrac{2}{9}\text{ m}$

92. $\dfrac{142}{45}\text{ m, or } 3\dfrac{7}{45}\text{ m}$

93. Writing exercise

Exercise Set 1.4

1. The real number -19 corresponds to 19°F below zero, and the real number 59 corresponds to 59°F above zero.

2. $-2, 5$

3. The real number -150 corresponds to burning 150 calories, and the real number 65 corresponds to consuming 65 calories.

4. 1200, -800

5. The real number -1286 corresponds to 1286 ft below sea level. The real number 29,029 corresponds to 29,029 ft above sea level.

6. Jets: -34, Strikers: 34

7. The real number 750 corresponds to a \$750 deposit, and the real number -125 corresponds to a \$125 withdrawal.

8. 22, -9

9. The real numbers $20, -150$, and 300 correspond to the interception of the missile, the loss of the starship, and the capture of the base, respectively.

10. $-10, 235$

11. Since $\dfrac{10}{3} = 3\dfrac{1}{3}$, its graph is $\dfrac{1}{3}$ of a unit to the right of 3.

12.

13. The graph of -4.3 is $\dfrac{3}{10}$ of a unit to the left of -4.

14.

15.

16.

17. $\dfrac{7}{8}$ means $7 \div 8$, so we divide.

$$\begin{array}{r} 0.8\,7\,5 \\ 8\,\overline{\smash{)}\,7.0\,0\,0} \\ \underline{6\,4} \\ 6\,0 \\ \underline{5\,6} \\ 4\,0 \\ \underline{4\,0} \\ 0 \end{array}$$

We have $\dfrac{7}{8} = 0.875$.

18. -0.125

19. We first find decimal notation for $\frac{3}{4}$. Since $\frac{3}{4}$ means $3 \div 4$, we divide.

$$
\begin{array}{r}
0.7\,5 \\
4\,\overline{)\,3.0\,0} \\
2\,8 \\
\hline
2\,0 \\
2\,0 \\
\hline
0
\end{array}
$$

Thus, $\frac{3}{4} = 0.75$, so $-\frac{3}{4} = -0.75$.

20. $0.8\overline{3}$

21. $\frac{7}{6}$ means $7 \div 6$, so we divide.

$$
\begin{array}{r}
1.1\,6\,6 \\
6\,\overline{)\,7.0\,0\,0} \\
6 \\
\hline
1\,0 \\
6 \\
\hline
4\,0 \\
3\,6 \\
\hline
4\,0 \\
3\,6 \\
\hline
4
\end{array}
$$

We have $\frac{7}{6} = 1.1\overline{6}$.

22. $0.41\overline{6}$

23. $\frac{2}{3}$ means $2 \div 3$, so we divide.

$$
\begin{array}{r}
0.6\,6\,6\,\dots \\
3\,\overline{)\,2.0\,0\,0} \\
1\,8 \\
\hline
2\,0 \\
1\,8 \\
\hline
2\,0 \\
1\,8 \\
\hline
2
\end{array}
$$

We have $\frac{2}{3} = 0.\overline{6}$.

24. 0.25

25. We first find decimal notation for $\frac{1}{2}$. Since $\frac{1}{2}$ means $1 \div 2$, we divide.

$$
\begin{array}{r}
0.5 \\
2\,\overline{)\,1.0} \\
1\,0 \\
\hline
0
\end{array}
$$

Thus, $\frac{1}{2} = 0.5$, so $-\frac{1}{2} = -0.5$.

26. -0.375

27. Since the denominator is 100, we know that $\frac{13}{100} = 0.13$. We could also divide 13 by 100 to find this result.

28. -0.35

29. Since -8 is to the left of 2, we have $-8 < 2$.

30. $9 > 0$

31. Since 7 is to the right of 0, we have $7 > 0$.

32. $8 > -8$

33. Since -6 is to the left of 6, we have $-6 < 6$.

34. $0 > -7$

35. Since -8 is to the left of -5, we have $-8 < -5$.

36. $-4 < -3$

37. Since -5 is to the right of -11, we have $-5 > -11$.

38. $-3 > -4$

39. Since -12.5 is to the left of -9.4, we have $-12.5 < -9.4$.

40. $-10.3 > -14.5$

41. We convert to decimal notation. $\frac{5}{12} = 0.41\overline{6}$ and $\frac{11}{25} = 0.44$. Thus, $\frac{5}{12} < \frac{11}{25}$.

42. $-\frac{14}{17} < -\frac{27}{35}$

43. $-7 > x$ has the same meaning as $x < -7$.

44. $9 < a$

45. $-10 \leq y$ has the same meaning as $y \geq -10$.

46. $t \leq 12$

47. $-3 \geq -11$ is true, since $-3 > -11$ is true.

48. False

49. $0 \geq 8$ is false, since neither $0 > 8$ nor $0 = 8$ is true.

50. True

51. $-8 \leq -8$ is true because $-8 = -8$ is true.

52. True

53. $|-23| = 23$ since -23 is 23 units from 0.

54. 47

55. $|17| = 17$ since 17 is 17 units from 0.

56. 3.1

57. $|5.6| = 5.6$ since 5.6 is 5.6 units from 0.

58. $\dfrac{2}{5}$

59. $|329| = 329$ since 329 is 329 units from 0.

60. 456

61. $\left| -\dfrac{9}{7} \right| = \dfrac{9}{7}$ since $-\dfrac{9}{7}$ is $\dfrac{9}{7}$ units from 0.

62. 8.02

63. $|0| = 0$ since 0 is 0 units from itself.

64. 1.07

65. $|x| = |-8| = 8$

66. 5

67. $-83, -4.7, 0, \dfrac{5}{9}, 8.31, 62$

68. 62

69. $-83, 0, 62$

70. $\pi, \sqrt{17}$

71. All are real numbers.

72. 0, 62

73. Writing exercise

74. Writing exercise

75. $3xy = 3 \cdot 2 \cdot 7 = 42$

76. $5 + ab; \ ba + 5; \ 5 + ba$

77. Writing exercise

78. Writing exercise

79. Writing exercise

80. $-17, -12, 5, 13$

81. List the numbers as they occur on the number line, from left to right: $-23, -17, 0, 4$

82. $-\dfrac{4}{3}, \dfrac{4}{9}, \dfrac{4}{8}, \dfrac{4}{6}, \dfrac{4}{5}, \dfrac{4}{3}, \dfrac{4}{2}$

83. $-\dfrac{2}{3}, \dfrac{1}{2}, -\dfrac{3}{4}, -\dfrac{5}{6}, \dfrac{3}{8}, \dfrac{1}{6}$ can be written in decimal notation as $-0.66\overline{6}, 0.5, -0.75, -0.83\overline{3}, 0.375, 0.16\overline{6}$, respectively. Listing from least to greatest (in fractional form), we have
$$-\dfrac{5}{6}, -\dfrac{3}{4}, -\dfrac{2}{3}, \dfrac{1}{6}, \dfrac{3}{8}, \dfrac{1}{2}.$$

84. $|-5| > |-2|$

85. $|4| = 4$ and $|-7| = 7$, so $|4| < |-7|$.

86. $|-8| = |8|$

87. $|23| = 23$ and $|-23| = 23$, so $|23| = |-23|$.

88. $|-3| < |5|$

89. $|-19| = 19$ and $|-27| = 27$, so $|-19| < |-27|$.

90. $-7, 7$

91. x represents an integer whose distance from 0 is less than 3 units. Thus, $x = -2, -1, 0, 1, 2$.

92. $-4, -3, 3, 4$

93. $0.1\overline{1} = \dfrac{0.3\overline{3}}{3} = \dfrac{\frac{1}{3}}{3} = \dfrac{1}{3} \cdot \dfrac{1}{3} = \dfrac{1}{9}$

94. $\dfrac{3}{3}$

95. $5.5\overline{5} = 50(0.1\overline{1}) = 50 \cdot \dfrac{1}{9} = \dfrac{50}{9}$

(See Exercise 93.)

96. $\dfrac{70}{9}$

97. Writing exercise

Exercise Set 1.5

1. Start at 4. Move 7 units to the left.

$4 + (-7) = -3$

2. -3

3. Start at -5. Move 9 units to the right.

$-5 + 9 = 4$

4. 5

5. Start at 8. Move 8 units to the left.

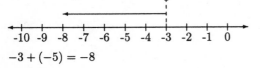

$8 + (-8) = 0$

6. 0

7. Start at -3. Move 5 units to the left.

$-3 + (-5) = -8$

8. -10

9. $-15 + 0$ One number is 0. The answer is the other number. $-15 + 0 = -15$

10. -6

11. $0 + (-8)$ One number is 0. The answer is the other number. $0 + (-8) = -8$

12. -2

13. $12 + (-12)$ The numbers have the same absolute value. The sum is 0. $12 + (-12) = 0$

14. 0

15. $-24 + (-17)$ Two negatives. Add the absolute values, getting 41. Make the answer negative. $-24 + (-17) = -41$

16. -42

17. $-15 + 15$ The numbers have the same absolute value. The sum is 0. $-15 + 15 = 0$

18. 0

19. $18 + (-11)$ The absolute values are 18 and 11. The difference is $18 - 11$, or 7. The positive number has the larger absolute value, so the answer is positive. $18 + (-11) = 7$

20. 3

21. $10 + (-12)$ The absolute values are 10 and 12. The difference is $12 - 10$, or 2. The negative number has the larger absolute value, so the answer is negative. $10 + (-12) = -2$

22. -4

23. $-3 + 14$ The absolute values are 3 and 14. The difference is $14 - 3$, or 11. The positive number has the larger absolute value, so the answer is positive. $-3 + 14 = 11$

24. 7

25. $-14 + (-19)$ Two negatives. Add the absolute values, getting 33. Make the answer negative. $-14 + (-19) = -33$

26. 2

27. $19 + (-19)$ The numbers has the same absolute value. The sum is 0. $19 + (-19) = 0$

28. -26

29. $23 + (-5)$ The absolute values are 23 and 5. The difference is $23 - 5$ or 18. The positive number has the larger absolute value, so the answer is positive. $23 + (-5) = 18$

30. -22

31. $-23 + (-9)$ Two negatives. Add the absolute values, getting 32. Make the answer negative. $-23 + (-9) = -32$

32. 32

33. $40 + (-40)$ The numbers have the same absolute value. The sum is 0. $40 + (-40) = 0$

34. 0

35. $85 + (-65)$ The absolute values are 85 and 65. The difference is $85 - 65$, or 20. The positive number has the larger absolute value, so the answer is positive. $85 + (-65) = 20$

36. 45

37. $-3.6 + 1.9$ The absolute values are 3.6 and 1.9. The difference is $3.6 - 1.9$, or 1.7. The negative number has the larger absolute value, so the answer is negative. $-3.6 + 1.9 = -1.7$

38. -1.8

39. $-5.4 + (-3.7)$ Two negatives. Add the absolute values, getting 9.1. Make the answer negative. $-5.4 + (-3.7) = -9.1$

40. -13.2

41. $\dfrac{-3}{5} + \dfrac{4}{5}$ The absolute values are $\dfrac{3}{5}$ and $\dfrac{4}{5}$. The difference is $\dfrac{4}{5} - \dfrac{3}{5}$, or $\dfrac{1}{5}$. The positive number has the larger absolute value, so the answer is positive.
$$\dfrac{-3}{5} + \dfrac{4}{5} = \dfrac{1}{5}$$

42. $\dfrac{1}{7}$

43. $\dfrac{-4}{7} + \dfrac{-2}{7}$ Two negatives. Add the absolute values, getting $\dfrac{6}{7}$. Make the answer negative.
$$\dfrac{-4}{7} + \dfrac{-2}{7} = \dfrac{-6}{7}$$

44. $\dfrac{-7}{9}$

45. $-\dfrac{2}{5} + \dfrac{1}{3}$ The absolute values are $\dfrac{2}{5}$ and $\dfrac{1}{3}$. The difference is $\dfrac{6}{15} - \dfrac{5}{15}$, or $\dfrac{1}{15}$. The negative number has the larger absolute value, so the answer is negative.
$$-\dfrac{2}{5} + \dfrac{1}{3} = -\dfrac{1}{15}$$

46. $\dfrac{5}{26}$

47. $\dfrac{-4}{9} + \dfrac{2}{3}$ The absolute values are $\dfrac{4}{9}$ and $\dfrac{2}{3}$. The difference is $\dfrac{6}{9} - \dfrac{4}{9}$, or $\dfrac{2}{9}$. The positive number has the larger absolute value, so the answer is positive.
$$\dfrac{-4}{9} + \dfrac{2}{3} = \dfrac{2}{9}$$

48. $\dfrac{1}{6}$

49.
$$35 + (-14) + (-19) + (-5)$$
$$= 35 + [(-14) + (-19) + (-5)] \quad \text{Using the associative law of addition}$$
$$= 35 + (-38) \qquad \text{Adding the negatives}$$
$$= -3 \qquad \text{Adding a positive and a negative}$$

50. -62

51. $-4.9 + 8.5 + 4.9 + (-8.5)$

Note that we have two pairs of numbers with different signs and the same absolute value: -4.9 and 4.9, 8.5 and -8.5. The sum of each pair is 0, so the result is $0 + 0$, or 0.

52. 37.9

53. Rewording:

Change from withdrawals	plus	change from additions
$\downarrow$	$\downarrow$	$\downarrow$

Translating: -5 $+$ 8

plus	change from drops	is	change in original size.
$\downarrow$	$\downarrow$	$\downarrow$	$\downarrow$
$+$	(-6)	$=$	change in original size

Since $-5 + 8 + (-6)$
$$= 3 + (-6)$$
$$= -3,$$
the class has 3 students less than the original class size.

54. Maya owed $69 at the end of August.

55. Rewording:

1998 loss	plus	1999 loss	plus
$\downarrow$	$\downarrow$	$\downarrow$	$\downarrow$

Translating: $-26,500$ $+$ $(-10,200)$ $+$

2000 profit	is	total profit or loss.
$\downarrow$	$\downarrow$	$\downarrow$
$32,400$	$=$	total profit or loss.

Since $-26,500 + (-10,200) + 32,400$
$$= -36,700 + 32,400$$
$$= -4300,$$
the loss was $4300.

56. The total gain was 22 yd.

57. Rewording:

Original balance	plus	change from writing first check	plus
$\downarrow$	$\downarrow$	$\downarrow$	$\downarrow$

Translating: 350 $+$ (-530) $+$

deposit	plus	change from writing second check	is	new balance.
$\downarrow$	$\downarrow$	$\downarrow$	$\downarrow$	$\downarrow$
75	$+$	(-90)	$=$	new balance

Since $350 + (-530) + (75) + (-90)$

$$= (350 + 75) + [-530 + (-90)]$$
$$= 425 + (-620)$$
$$= -195,$$

Leah's account is $195 overdrawn.

58. Lyle owes $85 on his credit card.

59. Rewording:

$$\underbrace{\text{First change}}_{\downarrow} \quad \text{plus} \atop \downarrow \quad \underbrace{\text{second change}}_{\downarrow} \quad \text{plus} \atop \downarrow$$

Translating: $\quad \dfrac{3}{16} \quad + \quad \left(-\dfrac{1}{2}\right) \quad +$

$$\underbrace{\text{third change}}_{\downarrow} \quad \text{is} \atop \downarrow \quad \underbrace{\text{total change.}}_{\downarrow}$$

$$\dfrac{1}{4} \quad = \text{total change.}$$

Since $\quad \dfrac{3}{16} + \left(-\dfrac{1}{2}\right) + \dfrac{1}{4}$

$$= \left(\dfrac{3}{16} + \dfrac{4}{16}\right) + \left(-\dfrac{8}{16}\right)$$
$$= \dfrac{7}{16} + \left(-\dfrac{8}{16}\right)$$
$$= \left(-\dfrac{1}{16}\right),$$

the value of the stock had fallen $\$\dfrac{1}{16}$ at the end of the day.

60. The elevation of the peak is 13,796 ft above sea level.

61. $\quad 7a + 5a = (7 + 5)a \quad$ Using the distributive law
$$= 12a$$

62. $11x$

63. $\quad -3x + 12x = (-3 + 12)x \quad$ Using the distributive law
$$= 9x$$

64. $-5m$

65. $5t + 8t = (5 + 8)t = 13t$

66. $14a$

67. $7m + (-9m) = [7 + (-9)]m = -2m$

68. 0

69. $-5a + (-2a) = [-5 + (-2)]a = -7a$

70. $-7n$

71. $\quad -3 + 8x + 4 + (-10x)$

$$= -3 + 4 + 8x + (-10x) \quad \text{Using the commutative law of addition}$$
$$= (-3 + 4) + [8 + (-10)]x \quad \text{Using the distributive law}$$
$$= 1 - 2x \quad \text{Adding}$$

72. $7a + 2$

73. $\text{Perimeter} = 8 + 5x + 9 + 7x$
$$= 8 + 9 + 5x + 7x$$
$$= (8 + 9) + (5 + 7)x$$
$$= 17 + 12x$$

74. $10a + 13$

75. $\text{Perimeter} = 9 + 6n + 7 + 8n + 4n$
$$= 9 + 7 + 6n + 8n + 4n$$
$$= (9 + 7) + (6 + 8 + 4)n$$
$$= 16 + 18n$$

76. $19n + 11$

77. Writing exercise

78. Writing exercise

79. $7(3z + y + 2) = 7 \cdot 3z + 7 \cdot y + 7 \cdot 2 = 21z + 7y + 14$

80. $\dfrac{28}{3}$

81. Writing exercise

82. Writing exercise

83. Starting with the final value, we "undo" the rise and drop in value by adding their opposites. The result is the original value.

Rewording: $\quad \underbrace{\text{Final value}}_{\downarrow} \quad \text{plus} \atop \downarrow \quad \underbrace{\text{opposite of rise}}_{\downarrow} \quad \text{plus} \atop \downarrow$

Translating: $\quad 64\dfrac{3}{8} \quad + \quad \left(-2\dfrac{3}{8}\right) \quad +$

$$\underbrace{\text{opposite of drop}}_{\downarrow} \quad \text{is original value.} \atop \downarrow \quad \underbrace{}_{\downarrow}$$

$$3\dfrac{1}{4} \quad = \text{original value.}$$

Since $64\dfrac{3}{8} + \left(-2\dfrac{3}{8}\right) + 3\dfrac{1}{4} = 62 + 3\dfrac{1}{4}$
$$= 65\dfrac{1}{4},$$

the stock's original value was 65\frac{1}{4}$.

84. $55.50

85. $4x + \underline{\quad} + (-9x) + (-2y)$

$= 4x + (-9x) + \underline{\quad} + (-2y)$

$= [4 + (-9)]x + \underline{\quad} + (-2y)$

$= -5x + \underline{\quad} + (-2y)$

This expression is equivalent to $-5x - 7y$, so the missing term is the term which yields $-7y$ when added to $-2y$. Since $-5y + (-2y) = -7y$, the missing term is $-5y$.

86. $-15b$

87. $3m + 2n + \underline{\quad} + (-2m)$

$= 2n + \underline{\quad} + (-2m) + 3m$

$= 2n + \underline{\quad} + (-2 + 3)m$

$= 2n + \underline{\quad} + m$

This expression is equivalent to $2n + (-6m)$, so the missing term is the term which yields $-6m$ when added to m. Since $-7m + m = -6m$, the missing term is $-7m$.

88. $-3y$

89. Note that, in order for the sum to be 0, the two missing terms must be the opposites of the given terms. Thus, the missing terms are $-7t$ and -23.

90. $\dfrac{7}{2}x$

91. $-3 + (-3) + 2 + (-2) + 1 = -5$

Since the total is 5 under par after the five rounds and $-5 = -1 + (-1) + (-1) + (-1) + (-1)$, the golfer was 1 under par on average.

Exercise Set 1.6

1. The opposite of 39 is -39 because $39 + (-39) = 0$.

2. 17

3. The opposite of -9 is 9 because $-9 + 9 = 0$.

4. $-\dfrac{7}{2}$

5. The opposite of -3.14 is 3.14 because $-3.14 + 3.14 = 0$.

6. -48.2

7. If $x = 23$, then $-x = -(23) = -23$. (The opposite of 23 is -23.)

8. 26

9. If $x = -\dfrac{14}{3}$, then $-x = -\left(-\dfrac{14}{3}\right) = \dfrac{14}{3}$.

$\left(\text{The opposite of} -\dfrac{14}{3} \text{ is } \dfrac{14}{3}.\right)$

10. $-\dfrac{1}{328}$

11. If $x = 0.101$, then $-x = -(0.101) = -0.101$. (The opposite of 0.101 is -0.101.)

12. 0

13. If $x = -72$, then $-(-x) = -(-72) = 72$ (The opposite of the opposite of 72 is 72.)

14. 29

15. If $x = -\dfrac{2}{5}$, then $-(-x) = -\left[-\left(-\dfrac{2}{5}\right)\right] = -\dfrac{2}{5}$.

$\left(\text{The opposite of the opposite of } -\dfrac{2}{5} \text{ is } -\dfrac{2}{5}.\right)$

16. -9.1

17. When we change the sign of -1 we obtain 1.

18. 7

19. When we change the sign of 7 we obtain -7.

20. -10

21. $-3 - 5$ is read "negative three minus five."

$-3 - 5 = -3 + (-5) = -8$

22. Negative four minus seven; $= -11$

23. $2 - (-9)$ is read "two minus negative nine."

$2 - (-9) = 2 + 9 = 11$

24. Five minus negative eight; 13

25. $4 - 6$ is read "four minus six."

$4 - 6 = 4 + (-6) = -2$

26. Nine minus twelve; -3

27. $-5 - (-7)$ is read "negative five minus negative seven."

$-5 - (-7) = -5 + 7 = 2$

28. Negative two minus negative five; 3

29. $6 - 8 = 6 + (-8) = -2$

30. -9

31. $0 - 5 = 0 + (-5) = -5$

32. -8

33. $3 - 9 = 3 + (-9) = -6$

34. -10

35. $0 - 10 = 0 + (-10) = -10$

36. -7

37. $-9 - (-3) = -9 + 3 = -6$

38. -4

39. Note that we are subtracting a number from itself. The result is 0. We could also do this exercise as follows:
$$-8 - (-8) = -8 + 8 = 0$$

40. 0

41. $14 - 19 = 14 + (-19) = -5$

42. -4

43. $30 - 40 = 30 + (-40) = -10$

44. -7

45. $-7 - (-9) = -7 + 9 = 2$

46. -5

47. $-9 - (-9) = -9 + 9 = 0$
(See Exercise 39.)

48. 0

49. $5 - 5 = 5 + (-5) = 0$
(See Exercise 39.)

50. 0

51. $4 - (-4) = 4 + 4 = 8$

52. 12

53. $-7 - 4 = -7 + (-4) = -11$

54. -14

55. $6 - (-10) = 6 + 10 = 16$

56. 15

57. $-14 - 2 = -14 + (-2) = -16$

58. -19

59. $-4 - (-3) = -4 + 3 = -1$

60. -1

61. $5 - (-6) = 5 + 6 = 11$

62. 17

63. $0 - 6 = 0 + (-6) = -6$

64. -5

65. $-3 - (-1) = -3 + 1 = -2$

66. -3

67. $-9 - 16 = -9 + (-16) = -25$

68. -21

69. $0 - (-1) = 0 + 1 = 1$

70. 5

71. $-9 - 0 = -9 + 0 = -9$

72. -8

73. $12 - (-5) = 12 + 5 = 17$

74. 10

75. $18 - 63 = 18 + (-63) = -45$

76. -23

77. $-18 - 63 = -18 + (-63) = -81$

78. -68

79. $-45 - 4 = -45 + (-4) = -49$

80. -58

81. $1.5 - 9.4 = 1.5 + (-9.4) = -7.9$

82. -5.5

83. $0.825 - 1 = 0.825 + (-1) = -0.175$

84. -0.928

85. $\dfrac{3}{7} - \dfrac{5}{7} = \dfrac{3}{7} + \left(-\dfrac{5}{7}\right) = -\dfrac{2}{7}$

86. $-\dfrac{7}{11}$

87. $\dfrac{-2}{9} - \dfrac{5}{9} = \dfrac{-2}{9} + \left(\dfrac{-5}{9}\right) = \dfrac{-7}{9}$, or $-\dfrac{7}{9}$

88. $-\dfrac{4}{5}$

89. $-\dfrac{2}{13} - \left(-\dfrac{5}{13} \right) = -\dfrac{2}{13} + \dfrac{5}{13} = \dfrac{3}{13}$

90. $\dfrac{5}{17}$

91. We subtract the smaller number from the larger.

Translate: $3.8 - (-5.2)$

Simplify: $3.8 - (-5.2) = 3.8 + 5.2 = 9$

92. $-2.1 - (-5.9);\ 3.8$

93. We subtract the smaller number from the larger.

Translate: $114 - (-79)$

Simplify: $114 - (-79) = 114 + 79 = 193$

94. $23 - (-17);\ 40$

95. $-21 - 37 = -21 + (-37) = -58$

96. -26

97. $9 - (-25) = 9 + 25 = 34$

98. 26

99. $25 - (-12) - 7 - (-2) + 9 = 25 + 12 + (-7) + 2 + 9 = 41$

100. -22

101. $-31 + (-28) - (-14) - 17 = (-31) + (-28) + 14 + (-17) = -62$

102. 22

103. $-34 - 28 + (-33) - 44 = (-34) + (-28) + (-33) + (-44) = -139$

104. 5

105. $-93 + (-84) - (-93) - (-84)$

Note that we are subtracting -93 from -93 and -84 from -84. Thus, the result will be 0. We could also do this exercise as follows:

$-93 + (-84) - (-93) - (-84) = -93 + (-84) + 93 + 84 = 0$

106. 4

107. $-7x - 4y = -7x + (-4y)$, so the terms are $-7x$ and $-4y$.

108. $7a,\ -9b$

109. $9 - 5t - 3st = 9 + (-5t) + (-3st)$, so the terms are $9,\ -5t,$ and $-3st$.

110. $-4,\ -3x,\ 2xy$

111. $\quad 4x - 7x$

$= 4x + (-7x) \qquad$ Adding the opposite

$= (4 + (-7))x \qquad$ Using the distributive law

$= -3x$

112. $-11a$

113. $\quad 7a - 12a + 4$

$= 7a + (-12a) + 4 \qquad$ Adding the opposite

$= (7 + (-12))a + 4 \qquad$ Using the distributive law

$= -5a + 4$

114. $-22x + 7$

115. $\quad -8n - 9 + n$

$= -8n + (-9) + n \qquad$ Adding the opposite

$= -8n + n + (-9) \qquad$ Using the commutative law of addition

$= -7n - 9 \qquad$ Adding like terms

116. $9n - 15$

117. $\quad 3x + 5 - 9x$

$= 3x + 5 + (-9x)$

$= 3x + (-9x) + 5$

$= -6x + 5$

118. $3a - 5$

119. $\quad 2 - 6t - 9 - 2t$

$= 2 + (-6t) + (-9) + (-2t)$

$= 2 + (-9) + (-6t) + (-2t)$

$= -7 - 8t$

120. $-2b - 12$

121. $\quad 5y + (-3x) - 9x + 1 - 2y + 8$

$= 5y + (-3x) + (-9x) + 1 + (-2y) + 8$

$= 5y + (-2y) + (-3x) + (-9x) + 1 + 8$

$= 3y - 12x + 9$

122. $46 + 3x + 6z$

123. $\quad 13x - (-2x) + 45 - (-21) - 7x$

$= 13x + 2x + 45 + 21 + (-7x)$

$= 13x + 2x + (-7x) + 45 + 21$

$= 8x + 66$

124. $6x + 39$

125. We subtract the lower temperature from the higher temperature:
$$44 - (-56) = 44 + 56 = 100$$
The temperature dropped 100°F.

126. $165

127. We subtract the lower elevation from the higher elevation:
$$29,028 - (-1312) = 30,340$$
The difference in elevation is 30,340 ft.

128. 14,494 ft

129. We subtract the lower elevation from the higher elevation:
$$-40 - (-156) = -40 + 156 = 116$$
Lake Assal is 116 m lower than the Valdes Peninsula.

130. 1767 m

131. Writing exercise

132. Writing exercise

133. Area $= lw = (36 \text{ ft})(12 \text{ ft}) = 432 \text{ ft}^2$

134. $2 \cdot 2 \cdot 2 \cdot 2 \cdot 2 \cdot 3 \cdot 3 \cdot 3$

135. Writing exercise

136. Writing exercise

137. True. For example, for $m = 5$ and $n = 3$, $5 > 3$ and $5 - 3 > 0$, or $2 > 0$. For $m = -4$ and $n = -9$, $-4 > -9$ and $-4 - (-9) > 0$, or $5 > 0$.

138. False. For example, let $m = -3$ and $n = -5$. Then $-3 > -5$, but $-3 + (-5) = -8 \not> 0$.

139. False. For example, let $m = 2$ and $n = -2$. Then 2 and -2 are opposites, but $2 - (-2) = 4 \neq 0$.

140. True. For example, for $m = 4$ and $n = -4$, $4 = -(-4)$ and $4 + (-4) = 0$; for $m = -3$ and $n = 3$, $-3 = -3$ and $-3 + 3 = 0$.

141. Writing exercise

142. Writing exercise

Exercise Set 1.7

1. $-4 \cdot 9 = -36$ Think: $4 \cdot 9 = 36$, make the answer negative.

2. -21

3. $-8 \cdot 7 = -56$ Think: $8 \cdot 7 = 56$, make the answer negative.

4. -18

5. $8 \cdot (-3) = -24$

6. -45

7. $-9 \cdot 8 = -72$

8. -30

9. $-6 \cdot (-7) = 42$ Multiplying absolute values; the answer is positive.

10. 10

11. $-5 \cdot (-9) = 45$ Multiplying absolute values; the answer is positive.

12. 18

13. $17 \cdot (-10) = -170$

14. 120

15. $-12 \cdot 12 = -144$

16. 195

17. $-25 \cdot (-48) = 1200$

18. -1677

19. $-3.5 \cdot (-28) = 98$

20. -203.7

21. $6 \cdot (-13) = -78$

22. -63

23. $-7 \cdot (-3.1) = 21.7$

24. 12.8

25. $\dfrac{2}{3} \cdot \left(-\dfrac{3}{5}\right) = -\left(\dfrac{2 \cdot 3}{3 \cdot 5}\right) = -\left(\dfrac{2}{5} \cdot \dfrac{3}{3}\right) = -\dfrac{2}{5}$

26. $-\dfrac{10}{21}$

27. $-\dfrac{3}{8} \cdot \left(-\dfrac{2}{9}\right) = \dfrac{\cancel{3} \cdot \cancel{2} \cdot 1}{4 \cdot \cancel{2} \cdot \cancel{3} \cdot 3} = \dfrac{1}{12}$

28. $\dfrac{1}{4}$

29. $(-5.3)(2.1) = -11.13$

30. -40.85

31. $-\dfrac{5}{9} \cdot \dfrac{3}{4} = -\dfrac{5 \cdot \cancel{3}}{\cancel{3} \cdot 3 \cdot 4} = -\dfrac{5}{12}$

32. -6

33. $\ 3 \cdot (-7) \cdot (-2) \cdot 6$

$= -21 \cdot (-12)$ Multiplying the first two numbers and the last two numbers

$= 252$

34. 756

35. 0, The product of 0 and any real number is 0.

36. 0

37. $-\dfrac{1}{3} \cdot \dfrac{1}{4} \cdot \left(-\dfrac{3}{7}\right) = -\dfrac{1}{12} \cdot \left(-\dfrac{3}{7}\right) = \dfrac{3}{12 \cdot 7} =$

$\dfrac{\cancel{3} \cdot 1}{\cancel{3} \cdot 4 \cdot 7} = \dfrac{1}{28}$

38. $\dfrac{3}{35}$

39. $-2 \cdot (-5) \cdot (-3) \cdot (-5) = 10 \cdot 15 = 150$

40. 30

41. 0, The product of 0 and any real number is 0.

42. 0

43. $(-8)(-9)(-10) = 72(-10) = -720$

44. 5040

45. $(-6)(-7)(-8)(-9)(-10) = 42 \cdot 72 \cdot (-10) = 3024 \cdot (-10) = -30,240$

46. $151,200$

47. $28 \div (-7) = -4$ Check: $-4 \cdot (-7) = 28$

48. -8

49. $\dfrac{36}{-9} = -4$ $-4 \cdot (-9) = 36$

50. -2

51. $\dfrac{-16}{8} = -2$ Check: $-2 \cdot 8 = -16$

52. 8

53. $\dfrac{-48}{-12} = 4$ Check: $4(-12) = -48$

54. 7

55. $\dfrac{-72}{9} = -8$ Check: $-8 \cdot 9 = -72$

56. -2

57. $-100 \div (-50) = 2$ Check: $2(-50) = -100$

58. -25

59. $-108 \div 9 = -12$ Check: $-12 \cdot 9 = -108$

60. $\dfrac{64}{7}$

61. $\dfrac{400}{-50} = -8$ Check: $-8 \cdot (-50) = 400$

62. $\dfrac{300}{13}$

63. Undefined

64. 0

65. $-4.8 \div 1.2 = -4$ Check: $-4(1.2) = -4.8$

66. -3

67. $\dfrac{0}{-9} = 0$

68. Undefined

69. Undefined

70. 0

71. $\dfrac{-8}{3} = \dfrac{8}{-3}$ and $\dfrac{-8}{3} = -\dfrac{8}{3}$

72. $\dfrac{12}{-7}, \ -\dfrac{12}{7}$

73. $\dfrac{29}{-35} = \dfrac{-29}{35}$ and $\dfrac{29}{-35} = -\dfrac{29}{35}$

74. $\dfrac{-9}{14}, \ -\dfrac{9}{14}$

75. $-\dfrac{7}{3} = \dfrac{-7}{3}$ and $-\dfrac{7}{3} = \dfrac{7}{-3}$

76. $\dfrac{-4}{15}, \ \dfrac{4}{-15}$

77. $\dfrac{-x}{2} = \dfrac{x}{-2}$ and $\dfrac{-x}{2} = -\dfrac{x}{2}$

78. $\dfrac{-9}{a}$, $-\dfrac{9}{a}$

79. The reciprocal of $\dfrac{4}{-5}$ is $\dfrac{-5}{4}$ $\left(\text{or equivalently, } -\dfrac{5}{4}\right)$ because $\dfrac{4}{-5} \cdot \dfrac{-5}{4} = 1$.

80. $-\dfrac{9}{2}$

81. The reciprocal of $-\dfrac{47}{13}$ is $-\dfrac{13}{47}$ because $-\dfrac{47}{13} \cdot \left(-\dfrac{13}{47}\right) = 1$.

82. $-\dfrac{12}{31}$

83. The reciprocal of -10 is $\dfrac{1}{-10}$ $\left(\text{or equivalently, } -\dfrac{1}{10}\right)$ because $-10\left(\dfrac{1}{-10}\right) = 1$.

84. $\dfrac{1}{13}$

85. The reciprocal of 4.3 is $\dfrac{1}{4.3}$ because $4.3\left(\dfrac{1}{4.3}\right) = 1$. Since $\dfrac{1}{4.3} = \dfrac{1}{4.3} \cdot \dfrac{10}{10} = \dfrac{10}{43}$, the reciprocal can also be expressed as $\dfrac{10}{43}$.

86. $\dfrac{1}{-8.5}$, or $-\dfrac{1}{8.5}$

87. The reciprocal of $\dfrac{-9}{4}$ is $\dfrac{4}{-9}$ $\left(\text{or equivalently, } -\dfrac{4}{9}\right)$ because $\dfrac{-9}{4} \cdot \dfrac{4}{-9} = 1$.

88. $\dfrac{11}{-6}$, or $-\dfrac{11}{6}$

89. The reciprocal of -1 is $\dfrac{1}{-1}$, or -1 because $(-1)(-1) = 1$.

90. $\dfrac{5}{3}$

91. $\left(\dfrac{-7}{4}\right)\left(-\dfrac{3}{5}\right)$
$= \left(-\dfrac{7}{4}\right)\left(-\dfrac{3}{5}\right)$ Rewriting $\dfrac{-7}{4}$ as $-\dfrac{7}{4}$
$= \dfrac{21}{20}$

92. $\dfrac{5}{18}$

93. $\left(\dfrac{-6}{5}\right)\left(\dfrac{2}{-11}\right)$
$= \left(\dfrac{-6}{5}\right)\left(\dfrac{-2}{11}\right)$ Rewriting $\dfrac{2}{-11}$ as $\dfrac{-2}{11}$
$= \dfrac{12}{55}$

94. $\dfrac{35}{12}$

95. $\dfrac{-3}{8} + \dfrac{-5}{8} = \dfrac{-8}{8} = -1$

96. $\dfrac{3}{5}$

97. $\left(\dfrac{-9}{5}\right)\left(\dfrac{5}{-9}\right)$
Note that this is the product of reciprocals. Thus, the result is 1.

98. $\dfrac{5}{28}$

99. $\left(-\dfrac{3}{11}\right) + \left(-\dfrac{6}{11}\right) = -\dfrac{9}{11}$

100. $-\dfrac{6}{7}$

101. $\dfrac{7}{8} \div \left(-\dfrac{1}{2}\right) = \dfrac{7}{8} \cdot \left(-\dfrac{2}{1}\right) = -\dfrac{14}{8} = -\dfrac{7 \cdot 2}{2 \cdot 4 \cdot 1} = -\dfrac{7}{4}$

102. $-\dfrac{9}{8}$

103. $\dfrac{9}{5} \cdot \dfrac{-20}{3} = \dfrac{9}{5}\left(-\dfrac{20}{3}\right) = -\dfrac{180}{15} = -\dfrac{3 \cdot 3 \cdot 4 \cdot 5}{5 \cdot 3 \cdot 1} = -12$

104. $-\dfrac{7}{36}$

105. $\left(-\dfrac{18}{7}\right) + \left(-\dfrac{3}{7}\right) = -\dfrac{21}{7} = -3$

106. -3

107. $-\dfrac{5}{9} \div \left(-\dfrac{5}{9}\right)$
Note that we have a number divided by itself. Thus, the result is 1.

108. $\dfrac{5}{3}$

109. $-44.1 \div (-6.3) = 7$ Do the long division. The answer is positive.

110. -2

111. $\frac{1}{9} - \frac{2}{9} = -\frac{1}{9}$

112. $-\frac{4}{7}$

113. $\frac{-3}{10} + \frac{2}{5} = \frac{-3}{10} + \frac{2}{5} \cdot \frac{2}{2} = \frac{-3}{10} + \frac{4}{10} = \frac{1}{10}$

114. $\frac{1}{9}$

115. $\frac{7}{10} \div \left(\frac{-3}{5}\right) = \frac{7}{10} \div \left(-\frac{3}{5}\right) = \frac{7}{10} \cdot \left(-\frac{5}{3}\right) = -\frac{35}{30} = -\frac{7 \cdot \cancel{5}}{2 \cdot \cancel{5} \cdot 3} = -\frac{7}{6}$

116. $-\frac{3}{2}$

117. $\frac{5}{7} - \frac{1}{-7} = \frac{5}{7} - \left(-\frac{1}{7}\right) = \frac{5}{7} + \frac{1}{7} = \frac{6}{7}$

118. $\frac{5}{9}$

119. $\frac{-4}{15} + \frac{2}{-3} = \frac{-4}{15} + \frac{-2}{3} = \frac{-4}{15} + \frac{-2}{3} \cdot \frac{5}{5} = \frac{-4}{15} + \frac{-10}{15} = \frac{-14}{15}$, or $-\frac{14}{15}$

120. $-\frac{1}{2}$

121. Writing exercise

122. Writing exercise

123. $\frac{264}{468} = \frac{\cancel{2} \cdot \cancel{2} \cdot 2 \cdot \cancel{3} \cdot 11}{\cancel{2} \cdot \cancel{2} \cdot \cancel{3} \cdot 3 \cdot 13} = \frac{22}{39}$

124. $12x - 2y - 9$

125. Writing exercise

126. Writing exercise

127. Consider the sum $2 + 3$. Its reciprocal is $\frac{1}{2+3}$, or $\frac{1}{5}$, but $\frac{1}{2} + \frac{1}{3} = \frac{5}{6}$.

128. $-1, 1$

129. When n is negative, $-n$ is positive, so $\frac{m}{-n}$ is the quotient of a negative and a positive number and, thus, is negative.

130. Positive

131. When n is negative, $-n$ is positive, so $\frac{-n}{m}$ is the quotient of a positive and a negative number and, thus, is negative. When m is negative, $-m$ is positive, so $-m \cdot \left(\frac{-n}{m}\right)$ is the product of a positive and a negative number and, thus, is negative.

132. Positive

133. $m + n$ is the sum of two negative numbers, so it is negative; $\frac{m}{n}$ is the quotient of two negative numbers, so it is positive. Then $(m + n) \cdot \frac{m}{n}$ is the product of a negative and a positive number and, thus, is negative.

134. Positive

135. a) m and n have different signs;

  b) either m or n is zero;

  c) m and n have the same sign

136. $a(-b) + ab = a[-b + b]$ Distributive law

       $= a(0)$ Law of opposites

       $= 0$ Multiplicative property of 0

Therefore, $a(-b)$ is the opposite of ab by the law of opposites.

137. Writing exercise

Exercise Set 1.8

1. $\underbrace{4 \cdot 4 \cdot 4}_{3 \text{ factors}} = 4^3$

2. 6^4

3. $\underbrace{x \cdot x \cdot x \cdot x \cdot x \cdot x \cdot x}_{7 \text{ factors}} = x^7$

4. y^6

5. $3t \cdot 3t \cdot 3t \cdot 3t \cdot 3t = (3t)^5$

6. $(5m)^5$

7. $2^4 = 2 \cdot 2 \cdot 2 \cdot 2 = 4 \cdot 4 = 16$

8. 125

9. $(-3)^2 = (-3)(-3) = 9$

10. 49

11. $-3^2 = -(3 \cdot 3) = -9$

12. -49

13. $4^3 = 4 \cdot 4 \cdot 4 = 16 \cdot 4 = 64$

14. 9

15. $(-5)^4 = (-5)(-5)(-5)(-5) = 25 \cdot 25 = 625$

16. 625

17. $7^1 = 7$ (1 factor)

18. -1

19. $(3t)^4 = (3t)(3t)(3t)(3t) =$
$3 \cdot 3 \cdot 3 \cdot 3 \cdot t \cdot t \cdot t \cdot t = 81t^4$

20. $25t^2$

21. $(-7x)^3 = (-7x)(-7x)(-7x) =$
$(-7)(-7)(-7)(x)(x)(x) = -343x^3$

22. $625x^4$

23. $\begin{aligned} 5 + 3 \cdot 7 &= 5 + 21 &&\text{Multiplying} \\ &= 26 &&\text{Adding} \end{aligned}$

24. -5

25. $\begin{aligned} 8 \cdot 7 + 6 \cdot 5 &= 56 + 30 &&\text{Multiplying} \\ &= 86 &&\text{Adding} \end{aligned}$

26. 51

27. $\begin{aligned} 19 - 5 \cdot 3 + 3 &= 19 - 15 + 3 &&\text{Multiplying} \\ &= 4 + 3 &&\text{Subtracting and add-} \\ &= 7 &&\text{ing from left to right} \end{aligned}$

28. 9

29. $\begin{aligned} 9 \div 3 + 16 \div 8 &= 3 + 2 &&\text{Dividing} \\ &= 5 &&\text{Adding} \end{aligned}$

30. 28

31. $84 \div 28 - 84 \div 28$

Note that we are subtracting a number, $84 \div 28$, from itself. Thus, the result is 0.

32. 21

33. $\begin{aligned} &\ \ \ 4 - 8 \div 2 + 3^2 \\ &= 4 - 8 \div 2 + 9 &&\text{Simplifying the exponential} \\ & &&\text{expression} \\ &= 4 - 4 + 9 &&\text{Dividing} \\ &= 0 + 9 &&\text{Subtracting and} \\ &= 9 &&\text{adding from left to right} \end{aligned}$

34. 298

35. $\begin{aligned} &\ \ \ 9 - 3^2 \div 9(-1) \\ &= 9 - 9 \div 9(-1) &&\text{Simplifying the exponential} \\ & &&\text{expression} \\ &= 9 - 1(-1) &&\text{Dividing and} \\ &= 9 + 1 &&\text{multiplying from left to right} \\ &= 10 &&\text{Adding} \end{aligned}$

36. 11

37. $\begin{aligned} (8 - 2 \cdot 3) - 9 &= (8 - 6) - 9 &&\text{Multiplying} \\ & &&\text{inside the parentheses} \\ &= 2 - 9 &&\text{Subtracting} \\ & &&\text{inside the parentheses} \\ &= -7 \end{aligned}$

38. -36

39. $(-24) \div (-3) \cdot \left(-\dfrac{1}{2}\right) = 8 \cdot \left(-\dfrac{1}{2}\right) = -\dfrac{8}{2} = -4$

40. 32

41. $\begin{aligned} &\ \ \ 13(-10)^2 + 45 \div (-5) \\ &= 13(100) + 45 \div (-5) &&\text{Simplifying the} \\ & &&\text{exponential expression} \\ &= 1300 + 45 \div (-5) &&\text{Multiplying and} \\ &= 1300 - 9 &&\text{dividing from left to right} \\ &= 1291 &&\text{Subtracting} \end{aligned}$

42. 13

43. $2^4 + 2^3 - 10 \div (-1)^4 = 16 + 8 - 10 \div 1 =$
$16 + 8 - 10 = 24 - 10 = 14$

44. 33

45. $5 + 3(2-9)^2 = 5 + 3(-7)^2 = 5 + 3 \cdot 49 = 5 + 147 = 152$

46. 13

47. $[2 \cdot (5 - 8)]^2 = [2 \cdot (-3)]^2 = (-6)^2 = 36$

48. 12

49. $\dfrac{7+2}{5^2 - 4^2} = \dfrac{9}{25 - 16} = \dfrac{9}{9} = 1$

50. 2

51. $8(-7) + |6(-5)| = -56 + |-30| = -56 + 30 = -26$

52. 49

53. $\dfrac{(-2)^3 + 4^2}{3 - 5^2 + 3 \cdot 6} = \dfrac{-8 + 16}{3 - 25 + 3 \cdot 6} = \dfrac{8}{3 - 25 + 18} =$
$\dfrac{8}{-22 + 18} = \dfrac{8}{-4} = -2$

54. -5

55. $\dfrac{27 - 2 \cdot 3^2}{8 \div 2^2 - (-2)^2} = \dfrac{27 - 2 \cdot 9}{8 \div 4 - 4} = \dfrac{27 - 18}{2 - 4} = \dfrac{9}{-2} = -\dfrac{9}{2}$

56. 5

57. $\begin{aligned} 7 - 5x &= 7 - 5 \cdot 3 && \text{Substituting 3 for } x \\ &= 7 - 15 && \text{Multiplying} \\ &= -8 && \text{Subtracting} \end{aligned}$

58. -7

59. $\begin{aligned} &\quad 24 \div t^3 \\ &= 24 \div (-2)^3 && \text{Substituting } -2 \text{ for } t \\ &= 24 \div (-8) && \text{Simplifying the exponential} \\ &&& \text{expression} \\ &= -3 && \text{Dividing} \end{aligned}$

60. 16

61. $\begin{aligned} 45 \div 3 \cdot a &= 45 \div 3 \cdot (-1) && \text{Substituting } -1 \text{ for} \\ &&& a \\ &= 15 \cdot (-1) && \text{Dividing} \\ &= -15 && \text{Multiplying} \end{aligned}$

62. -125

63. $\begin{aligned} &\quad 5x \div 15x^2 \\ &= 5 \cdot 3 \div 15(3)^2 && \text{Substituting 3 for } x \\ &= 5 \cdot 3 \div 15 \cdot 9 && \text{Simplifying the exponential} \\ &&& \text{expression} \\ &= 15 \div 15 \cdot 9 && \text{Multiplying and dividing} \\ &= 1 \cdot 9 && \text{in order from} \\ &= 9 && \text{left to right} \end{aligned}$

64. 8

65. $(12 \cdot 17) \div (17 \cdot 12)$

Since $12 \cdot 17$ and $17 \cdot 12$ are equivalent expressions, we have a number divided by itself so the result is 1.

66. 20

67. $-x^2 - 5x = -(-3)^2 - 5(-3) = -9 - 5(-3) = -9 + 15 = 6$

68. 24

69. $\dfrac{3a - 4a^2}{a^2 - 20} = \dfrac{3 \cdot 5 - 4(5)^2}{(5)^2 - 20} = \dfrac{3 \cdot 5 - 4 \cdot 25}{25 - 20} = \dfrac{15 - 100}{5} = \dfrac{-85}{5} = -17$

70. 0

71. $-(9x + 1) = -9x - 1$ Removing parentheses and changing the sign of each term

72. $-3x - 5$

73. $-(7 - 2x) = -7 + 2x$ Removing parentheses and changing the sign of each term

74. $-6x + 7$

75. $-(4a - 3b + 7c) = -4a + 3b - 7c$

76. $-5x + 2y + 3z$

77. $-(3x^2 + 5x - 1) = -3x^2 - 5x + 1$

78. $-8x^3 + 6x - 5$

79. $\begin{aligned} &\quad 5x - (2x + 7) \\ &= 5x - 2x - 7 && \text{Removing parentheses and} \\ &&& \text{changing the sign of each term} \\ &= 3x - 7 && \text{Collecting like terms} \end{aligned}$

80. $5y - 9$

81. $2a - (5a - 9) = 2a - 5a + 9 = -3a + 9$

82. $8n + 7$

83. $2x + 7x - (4x + 6) = 2x + 7x - 4x - 6 = 5x - 6$

84. $a - 7$

85. $9t - 5r - 2(3r + 6t) = 9t - 5r - 6r - 12t = -3t - 11r$

86. $-2m - 6n$

87. $\begin{aligned} &\quad 15x - y - 5(3x - 2y + 5z) \\ &= 15x - y - 15x + 10y - 25z && \text{Multiplying each} \\ &&& \text{term in parentheses by } -5 \\ &= 9y - 25z \end{aligned}$

88. $-16a + 27b - 32c$

89. $\begin{aligned} 3x^2 + 7 - (2x^2 + 5) &= 3x^2 + 7 - 2x^2 - 5 \\ &= x^2 + 2 \end{aligned}$

90. 0

91. $\begin{aligned} 5t^3 + t - 3(t + 2t^3) &= 5t^3 + t - 3t - 6t^3 \\ &= -t^3 - 2t \end{aligned}$

92. $2n^2 - n$

93. $\begin{aligned} &\quad 12a^2 - 3ab + 5b^2 - 5(-5a^2 + 4ab - 6b^2) \\ &= 12a^2 - 3ab + 5b^2 + 25a^2 - 20ab + 30b^2 \\ &= 37a^2 - 23ab + 35b^2 \end{aligned}$

94. $-20a^2 + 29ab + 48b^2$

95. $\quad -7t^3 - t^2 - 3(5t^3 - 3t)$
$= -7t^3 - t^2 - 15t^3 + 9t$
$= -22t^3 - t^2 + 9t$

96. $9t^4 - 45t^3 + 17t$

97. $\quad 5(2x - 7) - [4(2x - 3) + 2]$
$= 5(2x - 7) - [8x - 12 + 2]$
$= 5(2x - 7) - [8x - 10]$
$= 10x - 35 - 8x + 10$
$= 2x - 25$

98. $42x - 23$

99. Writing exercise

100. Writing exercise

101. Let x represent "a number." Then we have $2x + 9$.

102. Let x and y represent the numbers; $\frac{1}{2}(x + y)$.

103. Writing exercise

104. Writing exercise

105. $\quad 5t - \{7t - [4r - 3(t - 7)] + 6r\} - 4r$
$= 5t - \{7t - [4r - 3t + 21] + 6r\} - 4r$
$= 5t - \{7t - 4r + 3t - 21 + 6r\} - 4r$
$= 5t - \{10t + 2r - 21\} - 4r$
$= 5t - 10t - 2r + 21 - 4r$
$= -5t - 6r + 21$

106. $-4z$

107. $\quad \{x - [f - (f - x)] + [x - f]\} - 3x$
$= \{x - [f - f + x] + [x - f]\} - 3x$
$= \{x - [x] + [x - f]\} - 3x$
$= \{x - x + x - f\} - 3x$
$= x - f - 3x$
$= -2x - f$

108. Writing exercise

109. Writing exercise

110. True

111. False; let $m = 1$ and $n = 2$. Then $-2 + 1 = -(2 - 1) = -1$, but $-(2 + 1) = -3$.

112. True

113. False; let $m = 2$ and $n = 3$. Then $3(-3 - 2) = 3(-5) = -15$, but $-3^2 + 3 \cdot 2 = -9 + 6 = -3$.

114. False

115. True; $-m(-n + m) = mn - m^2 = m(n - m)$

116. True

117. $[x + 3(2 - 5x) \div 7 + x](x - 3)$
When $x = 3$, the factor $x - 3$ is 0, so the product is 0.

118. 1

119. $\quad 4 \cdot 20^3 + 17 \cdot 20^2 + 10 \cdot 20 + 0 \cdot 2$
$= 4 \cdot 8000 + 17 \cdot 400 + 10 \cdot 20 + 0 \cdot 2$
$= 32,000 + 6800 + 200 + 0$
$= 39,000$

120. 1, 5, 0

Chapter 2
Equations, Inequalities, and Problem Solving

Exercise Set 2.1

1. $x + 8 = 23$
$x + 8 - 8 = 23 - 8$ Subtracting 8 from both sides
$x = 15$ Simplifying

Check: $x + 8 = 23$
$15 + 8 \; ? \; 23$
$23 \mid 23$ TRUE

The solution is 15.

2. 3

3. $t + 9 = -4$
$t + 9 - 9 = -4 - 9$ Subtracting 9 from both sides
$t = -13$

Check: $t + 9 = -4$
$-13 + 9 \; ? \; -4$
$-4 \mid -4$ TRUE

The solution is -13.

4. 34

5. $y + 7 = -3$
$y + 7 - 7 = -3 - 7$
$y = -10$

Check: $y + 7 = -3$
$-10 + 7 \; ? \; -3$
$-3 \mid -3$ TRUE

The solution is -10.

6. -21

7. $-5 = x + 8$
$-5 - 8 = x + 8 - 8$
$-13 = x$

Check: $-5 = x + 8$
$-5 \; ? \; -13 + 8$
$-5 \mid -5$ TRUE

The solution is -13.

8. -31

9. $x - 9 = 6$
$x - 9 + 9 = 6 + 9$
$x = 15$

Check: $x - 9 = 6$
$15 - 9 \; ? \; 6$
$6 \mid 6$ TRUE

The solution is 15.

10. 13

11. $y - 6 = -14$
$y - 6 + 6 = -14 + 6$
$y = -8$

Check: $y - 6 = -14$
$-8 - 6 \; ? \; -14$
$-14 \mid -14$ TRUE

The solution is -8.

12. -15

13. $9 + t = 3$
$-9 + 9 + t = -9 + 3$
$t = -6$

Check: $9 + t = 3$
$9 - 6 \; ? \; 3$
$3 \mid 3$ TRUE

The solution is -6.

14. 18

15. $12 = -7 + y$
$7 + 12 = 7 + (-7) + y$
$19 = y$

Check: $12 = -7 + y$
$12 \; ? \; -7 + 19$
$12 \mid 12$ TRUE

The solution is 19.

16. 24

17. $-5 + t = -9$
$5 + (-5) + t = 5 + (-9)$
$t = -4$

Check: $-5 + t = -9$
$-5 + (-4) \; ? \; -9$
$-9 \mid -9$ TRUE

The solution is -4.

18. -15

19.
$$r + \frac{1}{3} = \frac{8}{3}$$
$$r + \frac{1}{3} - \frac{1}{3} = \frac{8}{3} - \frac{1}{3}$$
$$r = \frac{7}{3}$$

Check:
$$\frac{r + \frac{1}{3} = \frac{8}{3}}{\frac{7}{3} + \frac{1}{3} \;?\; \frac{8}{3}}$$
$$\frac{8}{3} \;\Big|\; \frac{8}{3} \quad \text{TRUE}$$

The solution is $\frac{7}{3}$.

20. $\frac{1}{4}$

21.
$$x + \frac{3}{5} = -\frac{7}{10}$$
$$x + \frac{3}{5} - \frac{3}{5} = -\frac{7}{10} - \frac{3}{5}$$
$$x = -\frac{7}{10} - \frac{3}{5} \cdot \frac{2}{2}$$
$$x = -\frac{7}{10} - \frac{6}{10}$$
$$x = -\frac{13}{10}$$

Check:
$$\frac{x + \frac{3}{5} = -\frac{7}{10}}{-\frac{13}{10} + \frac{3}{5} \;?\; -\frac{7}{10}}$$
$$-\frac{13}{10} + \frac{6}{10} \;\Big|$$
$$-\frac{7}{10} \;\Big|\; -\frac{7}{10} \quad \text{TRUE}$$

The solution is $-\frac{13}{10}$.

22. $-\frac{3}{2}$

23.
$$x - \frac{5}{6} = \frac{7}{8}$$
$$x - \frac{5}{6} + \frac{5}{6} = \frac{7}{8} + \frac{5}{6}$$
$$x = \frac{7}{8} \cdot \frac{3}{3} + \frac{5}{6} \cdot \frac{4}{4}$$
$$x = \frac{21}{24} + \frac{20}{24}$$
$$x = \frac{41}{24}$$

Check:
$$\frac{x - \frac{5}{6} = \frac{7}{8}}{\frac{41}{24} - \frac{5}{6} \;?\; \frac{7}{8}}$$
$$\frac{41}{24} - \frac{20}{24} \;\Big|\; \frac{21}{24}$$
$$\frac{21}{24} \;\Big|\; \frac{21}{24} \quad \text{TRUE}$$

The solution is $\frac{41}{24}$.

24. $\frac{19}{12}$

25.
$$-\frac{1}{5} + z = -\frac{1}{4}$$
$$\frac{1}{5} - \frac{1}{5} + z = \frac{1}{5} - \frac{1}{4}$$
$$z = \frac{1}{5} \cdot \frac{4}{4} - \frac{1}{4} \cdot \frac{5}{5}$$
$$z = \frac{4}{20} - \frac{5}{20}$$
$$z = -\frac{1}{20}$$

Check:
$$\frac{-\frac{1}{5} + z = -\frac{1}{4}}{-\frac{1}{5} + \left(-\frac{1}{20}\right) \;?\; -\frac{1}{4}}$$
$$-\frac{4}{20} + \left(-\frac{1}{20}\right) \;\Big|\; -\frac{5}{20}$$
$$-\frac{5}{20} \;\Big|\; -\frac{5}{20} \quad \text{TRUE}$$

The solution is $-\frac{1}{20}$.

26. $-\frac{5}{8}$

27.
$$m + 3.9 = 5.4$$
$$m + 3.9 - 3.9 = 5.4 - 3.9$$
$$m = 1.5$$

Check:
$$\frac{m + 3.9 = 5.4}{1.5 + 3.9 \;?\; 5.4}$$
$$5.4 \;\Big|\; 5.4 \quad \text{TRUE}$$

The solution is 1.5.

28. 3.4

29.
$$-9.7 = -4.7 + y$$
$$4.7 + (-9.7) = 4.7 + (-4.7) + y$$
$$-5 = y$$

Check: $\dfrac{-9.7 = -4.7 + y}{}$

$-9.7 \;?\; -4.7 + (-5)$

$-9.7 \;\big|\; -9.7$ TRUE

The solution is -5.

30. -10.6

31. $\quad 5x = 80$

$\dfrac{5x}{5} = \dfrac{80}{5}$ Dividing both sides by 5

$1 \cdot x = 16$ Simplifying

$x = 16$ Identity property of 1

Check: $\dfrac{5x = 80}{}$

$5 \cdot 16 \;?\; 80$

$80 \;\big|\; 80$ TRUE

The solution is 16.

32. 13

33. $\quad 9t = 36$

$\dfrac{9t}{9} = \dfrac{36}{9}$ Dividing both sides by 9

$1 \cdot t = 4$ Simplifying

$t = 4$ Identity property of 1

Check: $\dfrac{9t = 36}{}$

$9 \cdot 4 \;?\; 36$

$36 \;\big|\; 36$ TRUE

The solution is 4.

34. 12

35. $\quad 84 = 7x$

$\dfrac{84}{7} = \dfrac{7x}{7}$ Dividing both sides by 7

$12 = 1 \cdot x$

$12 = x$

Check: $\dfrac{84 = 7x}{}$

$84 \;?\; 7 \cdot 12$

$84 \;\big|\; 84$ TRUE

The solution is 12.

36. 8

37. $\quad\quad\quad -x = 23$

$-1 \cdot x = 23$

$-1 \cdot (-1 \cdot x) = -1 \cdot 23$

$1 \cdot x = -23$

$x = -23$

Check: $\dfrac{-x = 23}{}$

$-(-23) \;?\; 23$

$23 \;\big|\; 23$ TRUE

The solution is -23.

38. -100

39. $-t = -8$

The equation states that the opposite of t is the opposite of 8. Thus, $t = 8$. We could also do this exercise as follows.

$-t = -8$

$-1(-t) = -1(-8)$ Multiplying both sides by -1

$t = 8$

Check: $\dfrac{-t = -8}{}$

$-(8) \;?\; -8$

$-8 \;\big|\; -8$ TRUE

The solution is 8.

40. 68

41. $\quad 7x = -49$

$\dfrac{7x}{7} = \dfrac{-49}{7}$

$1 \cdot x = -7$

$x = -7$

Check: $\dfrac{7x = -49}{}$

$7(-7) \;?\; -49$

$-49 \;\big|\; -49$ TRUE

The solution is -7.

42. -4

43. $\quad -12x = 72$

$\dfrac{-12x}{-12} = \dfrac{72}{-12}$

$1 \cdot x = -6$

$x = -6$

Check: $\dfrac{-12x = 72}{}$

$-12(-6) \;?\; 72$

$72 \;\big|\; 72$ TRUE

The solution is -6.

44. -7

45. $\quad -3.4t = -20.4$

$\dfrac{-3.4t}{-3.4} = \dfrac{-20.4}{-3.4}$

$1 \cdot t = 6$

$t = 6$

Check: $\dfrac{-3.4t = -20.4}{}$

$-3.4(6) \;?\; -20.4$

$-20.4 \;\big|\; -20.4$ TRUE

The solution is 6.

46. 8

47.
$$\frac{a}{4} = 13$$
$$\frac{1}{4} \cdot a = 13$$
$$4 \cdot \frac{1}{4} \cdot a = 4 \cdot 13$$
$$a = 52$$

Check: $\dfrac{a}{4} = 13$

$$\begin{array}{c|c} \dfrac{52}{4} \ ? \ 13 & \\ \hline 13 & 13 \end{array} \quad \text{TRUE}$$

The solution is 52.

48. −88

49.
$$\frac{3}{4}x = 27$$
$$\frac{4}{3} \cdot \frac{3}{4}x = \frac{4}{3} \cdot 27$$
$$1 \cdot x = \frac{4 \cdot \cancel{3} \cdot 3 \cdot 3}{\cancel{3} \cdot 1}$$
$$x = 36$$

Check: $\dfrac{3}{4}x = 27$

$$\begin{array}{c|c} \dfrac{3}{4} \cdot 36 \ ? \ 27 & \\ \hline 27 & 27 \end{array} \quad \text{TRUE}$$

The solution is 36.

50. 20

51.
$$\frac{-t}{5} = 9$$
$$5 \cdot \frac{1}{5} \cdot (-t) = 5 \cdot 9$$
$$-t = 45$$
$$-1(-t) = -1 \cdot 45$$
$$t = -45$$

Check: $\dfrac{-t}{5} = 9$

$$\begin{array}{c|c} \dfrac{-(-45)}{5} \ ? \ 9 & \\ \dfrac{45}{5} & \\ \hline 9 & 9 \end{array} \quad \text{TRUE}$$

The solution is −45.

52. −54

53.
$$\frac{2}{7} = \frac{x}{3}$$
$$\frac{2}{7} = \frac{1}{3} \cdot x$$
$$3 \cdot \frac{2}{7} = 3 \cdot \frac{1}{3} \cdot x$$
$$\frac{6}{7} = x$$

Check: $\dfrac{2}{7} = \dfrac{x}{3}$

$$\begin{array}{c|c} \dfrac{2}{7} \ ? \ \dfrac{6/7}{3} & \\ & \dfrac{6}{7} \cdot \dfrac{1}{3} \\ & \dfrac{6}{21} \\ \hline \dfrac{2}{7} & \dfrac{2}{7} \end{array} \quad \text{TRUE}$$

The solution is $\dfrac{6}{7}$.

54. $\dfrac{5}{9}$

55. $-\dfrac{3}{5}r = -\dfrac{3}{5}$

The solution of the equation is the number that is multiplied by $-\dfrac{3}{5}$ to get $-\dfrac{3}{5}$. That number is 1. We could also do this exercise as follows:
$$-\frac{3}{5}r = -\frac{3}{5}$$
$$-\frac{5}{3} \cdot \left(-\frac{3}{5}r\right) = -\frac{5}{3}\left(-\frac{3}{5}\right)$$
$$r = 1$$

Check: $-\dfrac{3}{5}r = -\dfrac{3}{5}$

$$\begin{array}{c|c} -\dfrac{3}{5} \cdot 1 \ ? \ -\dfrac{3}{5} & \\ \hline -\dfrac{3}{5} & -\dfrac{3}{5} \end{array} \quad \text{TRUE}$$

The solution is 1.

56. $\dfrac{2}{3}$

57.
$$\frac{-3r}{2} = -\frac{27}{4}$$
$$-\frac{3}{2}r = -\frac{27}{4}$$
$$-\frac{2}{3} \cdot \left(-\frac{3}{2}r\right) = -\frac{2}{3} \cdot \left(-\frac{27}{4}\right)$$
$$r = \frac{\cancel{2} \cdot \cancel{3} \cdot 3 \cdot 3}{3 \cdot \cancel{2} \cdot 2}$$
$$r = \frac{9}{2}$$

Check:
$$\frac{-3r}{2} = -\frac{27}{4}$$

$$-\frac{3}{2} \cdot \frac{9}{2} \;?\; -\frac{27}{4}$$
$$-\frac{27}{4} \;\bigg|\; -\frac{27}{4} \qquad \text{TRUE}$$

The solution is $\frac{9}{2}$.

58. -1

59.
$$4.5 + t = -3.1$$
$$4.5 + t - 4.5 = -3.1 - 4.5$$
$$t = -7.6$$

The solution is -7.6.

60. 24

61.
$$-8.2x = 20.5$$
$$\frac{-8.2x}{-8.2} = \frac{20.5}{-8.2}$$
$$x = -2.5$$

The solution is -2.5.

62. -5.5

63.
$$12 = y + 29$$
$$12 - 29 = y + 29 - 29$$
$$-17 = y$$

The solution is -17.

64. -128

65.
$$a - \frac{1}{6} = -\frac{2}{3}$$
$$a - \frac{1}{6} + \frac{1}{6} = -\frac{2}{3} + \frac{1}{6}$$
$$a = -\frac{4}{6} + \frac{1}{6}$$
$$a = -\frac{3}{6}$$
$$a = -\frac{1}{2}$$

The solution is $-\frac{1}{2}$.

66. $-\dfrac{14}{9}$

67.
$$-24 = \frac{8x}{5}$$
$$-24 = \frac{8}{5}x$$
$$\frac{5}{8}(-24) = \frac{5}{8} \cdot \frac{8}{5}x$$
$$-\frac{5 \cdot \cancel{8} \cdot 3}{\cancel{8} \cdot 1} = x$$
$$-15 = x$$

The solution is -15.

68. $-\dfrac{1}{2}$

69.
$$-\frac{4}{3}t = -16$$
$$-\frac{3}{4}\left(-\frac{4}{3}t\right) = -\frac{3}{4}(-16)$$
$$t = \frac{3 \cdot \cancel{4} \cdot 4}{\cancel{4}}$$
$$t = 12$$

The solution is 12.

70. $-\dfrac{17}{35}$

71.
$$-483.297 = -794.053 + t$$
$$-483.297 + 794.053 = -794.053 + t + 794.053$$
$$310.756 = t \qquad \text{Using a calculator}$$

The solution is 310.756.

72. -8655

73. Writing exercise

74. Writing exercise

75.
$$9 - 2 \cdot 5^2 + 7$$
$$= 9 - 2 \cdot 25 + 7 \quad \text{Simplifying the exponential expression}$$
$$= 9 - 50 + 7 \quad \text{Multiplying}$$
$$= -41 + 7 \quad \text{Subtracting and}$$
$$= -34 \quad \text{adding from left to right}$$

76. 41

77.
$$16 \div (2 - 3 \cdot 2) + 5$$
$$= 16 \div (2 - 6) + 5 \quad \text{Simplifying inside}$$
$$= 16 \div (-4) + 5 \quad \text{the parentheses}$$
$$= -4 + 5 \quad \text{Dividing}$$
$$= 1 \quad \text{Adding}$$

78. -16

79. Writing exercise

80. Writing exercise

81. $2x = x + x$

$2x = 2x$ Adding on the right side

This is an identity.

82. Contradiction

83. $5x = 0$

$$\frac{5x}{5} = \frac{0}{5}$$

$x = 0$

The solution is 0.

84. Identity

85. $x + 8 = 3 + x + 7$

$x + 8 = 10 + x$ Adding on the right side

$x + 8 - x = 10 + x - x$

$8 = 10$

This is a contradiction.

86. 0

87. $2|x| = -14$

$$\frac{2|x|}{2} = -\frac{14}{2}$$

$|x| = -7$

Since the absolute value of a number is always non-negative, this is a contradiction.

88. $-2, 2$

89. $mx = 9.4m$

$$\frac{mx}{m} = \frac{9.4m}{m}$$

$x = 9.4$

The solution is 9.4.

90. 4

91.

$$\frac{7cx}{2a} = \frac{21}{a} \cdot c$$

$$\frac{7c}{2a} \cdot x = \frac{21}{a} \cdot c$$

$$\frac{2a}{7c} \cdot \frac{7c}{2a} \cdot x = \frac{2a}{7c} \cdot \frac{21}{a} \cdot \frac{c}{1}$$

$$x = \frac{2 \cdot \cancel{a} \cdot 3 \cdot 7 \cdot \cancel{c}}{7 \cdot \cancel{c} \cdot \cancel{a} \cdot 1}$$

$$x = 6$$

The solution is 6.

92. 2

93.

$$5a = ax - 3a$$

$$5a + 3a = ax - 3a + 3a$$

$$8a = ax$$

$$\frac{8a}{a} = \frac{ax}{a}$$

$$8 = x$$

The solution is 8.

94. $-13, 13$

95.

$$x - 4720 = 1634$$

$$x - 4720 + 4720 = 1634 + 4720$$

$$x = 6354$$

$$x + 4720 = 6354 + 4720$$

$$x + 4720 = 11,074$$

96. 250

97. Writing exercise

Exercise Set 2.2

1. $5x + 3 = 38$

$5x + 3 - 3 = 38 - 3$ Subtracting 3 from both sides

$5x = 35$ Simplifying

$\dfrac{5x}{5} = \dfrac{35}{5}$ Dividing both sides by 4

$x = 7$ Simplifying

Check: $\dfrac{5x + 3 = 38}{}$

$5 \cdot 7 + 3$? 38

$35 + 3$

38 | 38 TRUE

The solution is 7.

2. 8

3. $8x + 4 = 68$

$8x + 4 - 4 = 68 - 4$ Subtracting 4 from both sides

$8x = 64$ Simplifying

$\dfrac{8x}{8} = \dfrac{64}{8}$ Dividing both sides by 8

$x = 8$ Simplifying

Check: $\dfrac{8x + 4 = 68}{}$

$8 \cdot 8 + 4$? 68

$64 + 4$

68 | 68 TRUE

The solution is 8.

4. 9

5.
$$7t - 8 = 27$$
$$7t - 8 + 8 = 27 + 8 \qquad \text{Adding 8 to both sides}$$
$$7t = 35$$
$$\frac{7t}{7} = \frac{35}{7} \qquad \text{Dividing both sides by 7}$$
$$t = 5$$

Check:
$$\frac{7t - 8 = 27}{}$$
$$7 \cdot 5 - 8 \ ? \ 27$$
$$35 - 8 \ \Big|$$
$$27 \ \Big| \ 27 \qquad \text{TRUE}$$

The solution is 5.

6. 3

7.
$$3x - 9 = 33$$
$$3x - 9 + 9 = 33 + 9$$
$$3x = 42$$
$$\frac{3x}{3} = \frac{42}{3}$$
$$x = 14$$

Check:
$$\frac{3x - 9 = 33}{}$$
$$3 \cdot 14 - 9 \ ? \ 33$$
$$42 - 9 \ \Big|$$
$$33 \ \Big| \ 33 \qquad \text{TRUE}$$

The solution is 14.

8. 10

9.
$$8z + 2 = -54$$
$$8z + 2 - 2 = -54 - 2$$
$$8z = -56$$
$$\frac{8z}{8} = \frac{-56}{8}$$
$$z = -7$$

Check:
$$\frac{8z + 2 = -54}{}$$
$$8(-7) + 2 \ ? \ -54$$
$$-56 + 2 \ \Big|$$
$$-54 \ \Big| \ -54 \qquad \text{TRUE}$$

The solution is -7.

10. -6

11.
$$-39 = 1 + 8x$$
$$-39 - 1 = 1 + 8x - 1$$
$$-40 = 8x$$
$$\frac{-40}{8} = \frac{8x}{8}$$
$$-5 = x$$

Check:
$$\frac{-39 = 1 + 8x}{}$$
$$-39 \ ? \ 1 + 8(-5)$$
$$\Big| \ 1 - 40$$
$$-39 \ \Big| \ -39 \qquad \text{TRUE}$$

The solution is -5.

12. -11

13.
$$9 - 4x = 37$$
$$9 - 4x - 9 = 37 - 9$$
$$-4x = 28$$
$$\frac{-4x}{-4} = \frac{28}{-4}$$
$$x = -7$$

Check:
$$\frac{9 - 4x = 37}{}$$
$$9 - 4(-7) \ ? \ 37$$
$$9 + 28 \ \Big|$$
$$37 \ \Big| \ 37 \qquad \text{TRUE}$$

The solution is -7.

14. -24

15.
$$-7x - 24 = -129$$
$$-7x - 24 + 24 = -129 + 24$$
$$-7x = -105$$
$$\frac{-7x}{-7} = \frac{-105}{-7}$$
$$x = 15$$

Check:
$$\frac{-7x - 24 = -129}{}$$
$$-7 \cdot 15 - 24 \ ? \ -129$$
$$-105 - 24 \ \Big|$$
$$-129 \ \Big| \ -129 \qquad \text{TRUE}$$

The solution is 15.

16. 19

17.
$$48 = 5x + 7x$$
$$48 = 12x \qquad \text{Combining like terms}$$
$$\frac{48}{12} = \frac{12x}{12} \qquad \text{Dividing both sides by 12}$$
$$4 = x$$

Check:
$$\frac{48 = 5x + 7x}{}$$
$$48 \ ? \ 5 \cdot 4 + 7 \cdot 4$$
$$\Big| \ 20 + 28$$
$$48 \ \Big| \ 48 \qquad \text{TRUE}$$

The solution is 4.

18. 5

19.
$$27 - 6x = 99$$
$$27 - 6x - 27 = 99 - 27$$
$$-6x = 72$$
$$\frac{-6x}{-6} = \frac{72}{-6}$$
$$x = -12$$

Check: $\underline{\quad 27 - 6x = 99 \quad}$
$$27 - 6(-12) \ ? \ 99$$
$$27 + 72 \ \Big| $$
$$99 \ \Big| \ 99 \qquad \text{TRUE}$$

The solution is -12.

20. 3

21. $4x + 3x = 42$ Combining like terms
$$7x = 42$$
$$\frac{7x}{7} = \frac{42}{7}$$
$$x = 6$$

Check: $\underline{\quad 4x + 3x = 42 \quad}$
$$4 \cdot 6 + 3 \cdot 6 \ ? \ 42$$
$$24 + 18 \ \Big|$$
$$42 \ \Big| \ 42 \qquad \text{TRUE}$$

The solution is 6.

22. 4

23. $-2a + 5a = 24$
$$3a = 24$$
$$\frac{3a}{3} = \frac{24}{3}$$
$$a = 8$$

Check: $\underline{\quad -2a + 5a = 24 \quad}$
$$-2 \cdot 8 + 5 \cdot 8 \ ? \ 24$$
$$-16 + 40 \ \Big|$$
$$24 \ \Big| \ 24 \qquad \text{TRUE}$$

The solution is 8.

24. -3

25. $-7y - 8y = -15$
$$-15y = -15$$
$$\frac{-15y}{-15} = \frac{-15}{-15}$$
$$y = 1$$

Check: $\underline{\quad -7y - 8y = -15 \quad}$
$$-7 \cdot 1 - 8 \cdot 1 \ ? \ -15$$
$$-7 - 8 \ \Big|$$
$$-15 \ \Big| \ -15 \qquad \text{TRUE}$$

The solution is 1.

26. 4

27.
$$10.2y - 7.3y = -58$$
$$2.9y = -58$$
$$\frac{2.9y}{2.9} = \frac{-58}{2.9}$$
$$y = -\frac{58}{2.9}$$
$$y = -20$$

Check:

$\underline{\qquad 10.2y - 7.3y = -58 \qquad}$
$$10.2(-20) - 7.3(-20) \ ? \ -58$$
$$-204 + 146 \ \Big|$$
$$-58 \ \Big| \ -58 \qquad \text{TRUE}$$

The solution is -20.

28. -20

29.
$$x + \frac{1}{3}x = 8$$
$$\left(1 + \frac{1}{3}\right)x = 8$$
$$\frac{4}{3}x = 8$$
$$\frac{3}{4} \cdot \frac{4}{3}x = \frac{3}{4} \cdot 8$$
$$x = 6$$

Check: $\underline{\quad x + \frac{1}{3}x = 8 \quad}$
$$6 + \frac{1}{3} \cdot 6 \ ? \ 8$$
$$6 + 2 \ \Big|$$
$$8 \ \Big| \ 8 \qquad \text{TRUE}$$

The solution is 6.

30. 8

31. $9y - 35 = 4y$
$$9y = 4y + 35 \qquad \text{Adding 35 and simplifying}$$
$$9y - 4y = 35 \qquad \text{Subtracting } 4y \text{ and simplifying}$$
$$5y = 35 \qquad \text{Collecting like terms}$$
$$\frac{5y}{5} = \frac{35}{5} \qquad \text{Dividing both sides by 5}$$
$$y = 7$$

Check: $\underline{\quad 9y - 35 = 4y \quad}$
$$9 \cdot 7 - 35 \ ? \ 4 \cdot 7$$
$$63 - 35 \ \Big| \ 28$$
$$28 \ \Big| \ 28 \qquad \text{TRUE}$$

The solution is 7.

32. -3

33.
$$6x - 5 = 7 + 2x$$
$$6x - 5 - 2x = 7 + 2x - 2x \quad \text{Subtracting } 2x \text{ on both sides}$$
$$4x - 5 = 7 \quad \text{Simplifying}$$
$$4x - 5 + 5 = 7 + 5 \quad \text{Adding 5 on both sides}$$
$$4x = 12 \quad \text{Simplifying}$$
$$\frac{4x}{4} = \frac{12}{4} \quad \text{Dividing by 4 on both sides}$$
$$x = 3$$

Check:
$$\frac{6x - 5 = 7 + 2x}{}$$

| $6 \cdot 3 - 5$? $7 + 2 \cdot 3$ |
| $18 - 5$ \| $7 + 6$ |
| 13 \| 13 TRUE |

The solution is 3.

34. 5

35.
$$6x + 3 = 2x + 3$$
$$6x - 2x = 3 - 3$$
$$4x = 0$$
$$\frac{4x}{4} = \frac{0}{4}$$
$$x = 0$$

Check:
$$\frac{6x + 3 = 2x + 3}{}$$

| $6 \cdot 0 + 3$? $2 \cdot 0 + 3$ |
| $0 + 3$ \| $0 + 3$ |
| 3 \| 3 TRUE |

The solution is 0.

36. 4

37.
$$5 - 2x = 3x - 7x + 25$$
$$5 - 2x = -4x + 25$$
$$4x - 2x = 25 - 5$$
$$2x = 20$$
$$\frac{2x}{2} = \frac{20}{2}$$
$$x = 10$$

Check:
$$\frac{5 - 2x = 3x - 7x + 25}{}$$

| $5 - 2 \cdot 10$? $3 \cdot 10 - 7 \cdot 10 + 25$ |
$5 - 20$ | $30 - 70 + 25$
-15 | $-40 + 25$
-15 | -15 TRUE

The solution is 10.

38. 10

39.
$$7 + 3x - 6 = 3x + 5 - x$$
$$3x + 1 = 2x + 5 \quad \text{Combining like terms on each side}$$
$$3x - 2x = 5 - 1$$
$$x = 4$$

Check:
$$\frac{7 + 3x - 6 = 3x + 5 - x}{}$$

| $7 + 3 \cdot 4 - 6$? $3 \cdot 4 + 5 - 4$ |
$7 + 12 - 6$ | $12 + 5 - 4$
$19 - 6$ | $17 - 4$
13 | 13 TRUE

The solution is 4.

40. 0

41.
$$4y - 4 + y + 24 = 6y + 20 - 4y$$
$$5y + 20 = 2y + 20$$
$$5y - 2y = 20 - 20$$
$$3y = 0$$
$$y = 0$$

Check:
$$\frac{4y - 4 + y + 24 = 6y + 20 - 4y}{}$$

| $4 \cdot 0 - 4 + 0 + 24$? $6 \cdot 0 + 20 - 4 \cdot 0$ |
| $0 - 4 + 0 + 24$ \| $0 + 20 - 0$ |
| 20 \| 20 TRUE |

The solution is 0.

42. 7

43. $\dfrac{5}{4}x + \dfrac{1}{4}x = 2x + \dfrac{1}{2} + \dfrac{3}{4}x$

The number 4 is the least common denominator, so we multiply by 4 on both sides.

$$4\left(\frac{5}{4}x + \frac{1}{4}x\right) = 4\left(2x + \frac{1}{2} + \frac{3}{4}x\right)$$
$$4 \cdot \frac{5}{4}x + 4 \cdot \frac{1}{4}x = 4 \cdot 2x + 4 \cdot \frac{1}{2} + 4 \cdot \frac{3}{4}x$$
$$5x + x = 8x + 2 + 3x$$
$$6x = 11x + 2$$
$$6x - 11x = 2$$
$$-5x = 2$$
$$\frac{-5x}{-5} = \frac{2}{-5}$$
$$x = -\frac{2}{5}$$

Check:
$$\frac{\frac{5}{4}x + \frac{1}{4}x = 2x + \frac{1}{2} + \frac{3}{4}x}{}$$

| $\frac{5}{4}\left(-\frac{2}{5}\right) + \frac{1}{4}\left(-\frac{2}{5}\right)$? $2\left(-\frac{2}{5}\right) + \frac{1}{2} + \frac{3}{4}\left(-\frac{2}{5}\right)$ |
$-\frac{1}{2} - \frac{1}{10}$ | $-\frac{4}{5} + \frac{1}{2} - \frac{3}{10}$
$-\frac{5}{10} - \frac{1}{10}$ | $-\frac{8}{10} + \frac{5}{10} - \frac{3}{10}$
$-\frac{6}{10}$ | $-\frac{6}{10}$ TRUE

The solution is $-\dfrac{2}{5}$.

44. $\dfrac{1}{2}$

45. $\dfrac{2}{3} + \dfrac{1}{4}t = 6$

The number 12 is the least common denominator, so we multiply by 12 on both sides.

$$12\left(\dfrac{2}{3} + \dfrac{1}{4}t\right) = 12 \cdot 6$$

$$12 \cdot \dfrac{2}{3} + 12 \cdot \dfrac{1}{4}t = 72$$

$$8 + 3t = 72$$

$$3t = 72 - 8$$

$$3t = 64$$

$$t = \dfrac{64}{3}$$

Check:

$$\dfrac{2}{3} + \dfrac{1}{4}t = 6$$

$$\begin{array}{c|c} \dfrac{2}{3} + \dfrac{1}{4}\left(\dfrac{64}{3}\right) & ?\ 6 \\[2mm] \dfrac{2}{3} + \dfrac{16}{3} & \\[2mm] \dfrac{18}{3} & \\[2mm] 6 & 6 \quad \text{TRUE} \end{array}$$

The solution is $\dfrac{64}{3}$.

46. $-\dfrac{2}{3}$

47. $$\dfrac{2}{3} + 4t = 6t - \dfrac{2}{15}$$

The number 15 is the least common denominator, so we multiply by 15 on both sides.

$$15\left(\dfrac{2}{3} + 4t\right) = 15\left(6t - \dfrac{2}{15}\right)$$

$$15 \cdot \dfrac{2}{3} + 15 \cdot 4t = 15 \cdot 6t - 15 \cdot \dfrac{2}{15}$$

$$10 + 60t = 90t - 2$$

$$10 + 2 = 90t - 60t$$

$$12 = 30t$$

$$\dfrac{12}{30} = t$$

$$\dfrac{2}{5} = t$$

Check:

$$\dfrac{2}{3} + 4t = 6t - \dfrac{2}{15}$$

$$\begin{array}{c|c} \dfrac{2}{3} + 4 \cdot \dfrac{2}{5} & ?\ 6 \cdot \dfrac{2}{5} - \dfrac{2}{15} \\[2mm] \dfrac{2}{3} + \dfrac{8}{5} & \dfrac{12}{5} - \dfrac{2}{15} \\[2mm] \dfrac{10}{15} + \dfrac{24}{15} & \dfrac{36}{15} - \dfrac{2}{15} \\[2mm] \dfrac{34}{15} & \dfrac{34}{15} \quad \text{TRUE} \end{array}$$

The solution is $\dfrac{2}{5}$.

48. -3

49. $$\dfrac{1}{3}x + \dfrac{2}{5} = \dfrac{4}{15} + \dfrac{3}{5}x - \dfrac{2}{3}$$

The number 15 is the least common denominator, so we multiply by 15 on both sides.

$$15\left(\dfrac{1}{3}x + \dfrac{2}{5}\right) = 15\left(\dfrac{4}{15} + \dfrac{3}{5}x - \dfrac{2}{3}\right)$$

$$15 \cdot \dfrac{1}{3}x + 15 \cdot \dfrac{2}{5} = 15 \cdot \dfrac{4}{15} + 15 \cdot \dfrac{3}{5}x - 15 \cdot \dfrac{2}{3}$$

$$5x + 6 = 4 + 9x - 10$$

$$5x + 6 = -6 + 9x$$

$$5x - 9x = -6 - 6$$

$$-4x = -12$$

$$\dfrac{-4x}{-4} = \dfrac{-12}{-4}$$

$$x = 3$$

Check:

$$\dfrac{1}{3}x + \dfrac{2}{5} = \dfrac{4}{15} + \dfrac{3}{5}x - \dfrac{2}{3}$$

$$\begin{array}{c|c} \dfrac{1}{3} \cdot 3 + \dfrac{2}{5} & ?\ \dfrac{4}{15} + \dfrac{3}{5} \cdot 3 - \dfrac{2}{3} \\[2mm] 1 + \dfrac{2}{5} & \dfrac{4}{15} + \dfrac{9}{5} - \dfrac{2}{3} \\[2mm] \dfrac{5}{5} + \dfrac{2}{5} & \dfrac{4}{15} + \dfrac{27}{15} - \dfrac{10}{15} \\[2mm] \dfrac{7}{5} & \dfrac{21}{15} \\[2mm] \dfrac{7}{5} & \dfrac{7}{5} \quad \text{TRUE} \end{array}$$

The solution is 3.

50. -3

51.
$$2.1x + 45.2 = 3.2 - 8.4x$$
Greatest number of decimal places is 1
$$10(2.1x + 45.2) = 10(3.2 - 8.4x)$$
Multiplying by 10 to clear decimals
$$10(2.1x) + 10(45.2) = 10(3.2) - 10(8.4x)$$
$$21x + 452 = 32 - 84x$$
$$21x + 84x = 32 - 452$$
$$105x = -420$$
$$x = \frac{-420}{105}$$
$$x = -4$$

Check:
$$\frac{2.1x + 45.2 = 3.2 - 8.4x}{2.1(-4) + 45.2 \ ? \ 3.2 - 8.4(-4)}$$
$$\begin{array}{c|c} -8.4 + 45.2 & 3.2 + 33.6 \\ 36.8 & 36.8 \qquad \text{TRUE} \end{array}$$

The solution is -4.

52. $\dfrac{4}{5}$, or 0.8

53.
$$0.76 + 0.21t = 0.96t - 0.49$$
Greatest number of decimal places is 2
$$100(0.76 + 0.21t) = 100(0.96t - 0.49)$$
Multiplying by 100 to clear decimals
$$100(0.76) + 100(0.21t) = 100(0.96t) - 100(0.49)$$
$$76 + 21t = 96t - 49$$
$$76 + 49 = 96t - 21t$$
$$125 = 75t$$
$$\frac{125}{75} = t$$
$$\frac{5}{3} = t, \text{ or }$$
$$1.\overline{6} = t$$

The answer checks. The solution is $\dfrac{5}{3}$, or $1.\overline{6}$.

54. 1

55.
$$\frac{2}{5}x - \frac{3}{2}x = \frac{3}{4}x + 2$$
The least common denominator is 20.
$$20\left(\frac{2}{5}x - \frac{3}{2}x\right) = 20\left(\frac{3}{4}x + 2\right)$$
$$20 \cdot \frac{2}{5}x - 20 \cdot \frac{3}{2}x = 20 \cdot \frac{3}{4}x + 20 \cdot 2$$
$$8x - 30x = 15x + 40$$
$$-22x = 15x + 40$$
$$-22x - 15x = 40$$
$$-37x = 40$$
$$\frac{-37x}{-37} = \frac{40}{-37}$$
$$x = -\frac{40}{37}$$

Check:
$$\frac{2}{5}x - \frac{3}{2}x = \frac{3}{4}x + 2$$
$$\frac{2}{5}\left(-\frac{40}{37}\right) - \frac{3}{2}\left(-\frac{40}{37}\right) \ ? \ \frac{3}{4}\left(-\frac{40}{37}\right) + 2$$
$$\begin{array}{c|c} -\frac{16}{37} + \frac{60}{37} & -\frac{30}{37} + \frac{74}{37} \\ \frac{44}{37} & \frac{44}{37} \qquad \text{TRUE} \end{array}$$

The solution is $-\dfrac{40}{37}$.

56. $\dfrac{32}{7}$

57.
$$7(2a - 1) = 21$$
$$14a - 7 = 21 \qquad \text{Using the distributive law}$$
$$14a = 21 + 7 \quad \text{Adding 7}$$
$$14a = 28$$
$$a = 2 \qquad \text{Dividing by 14}$$

Check:
$$\frac{7(2a - 1) = 21}{7(2 \cdot 2 - 1) \ ? \ 21}$$
$$\begin{array}{c|c} 7(4 - 1) & \\ 7 \cdot 3 & \\ 21 & 21 \qquad \text{TRUE} \end{array}$$

The solution is 2.

58. $\dfrac{9}{2}$

59.
$$35 = 5(3x + 1)$$
$$35 = 15x + 5 \qquad \text{Using the distributive law}$$
$$35 - 5 = 15x$$
$$30 = 15x$$
$$2 = x$$

Check:
$$\frac{35 = 5(3x + 1)}{35 \ ? \ 5(3 \cdot 2 + 1)}$$
$$\begin{array}{c|c} & 5(6 + 1) \\ & 5 \cdot 7 \\ 35 & 35 \qquad \text{TRUE} \end{array}$$

The solution is 2.

60. 1

61.
$$2(3 + 4m) - 6 = 48$$
$$6 + 8m - 6 = 48$$
$$8m = 48 \quad \text{Combining like terms}$$
$$m = 6$$

Check: $2(3 + 4m) - 6 = 48$

$$\begin{array}{c|c} 2(3 + 4\cdot 6) - 6 \ ? \ 48 & \\ 2(3 + 24) - 6 & \\ 2\cdot 27 - 6 & \\ 54 - 6 & \\ 48 & 48 \quad \text{TRUE} \end{array}$$

The solution is 6.

62. 9

63. $7r - (2r + 8) = 32$
$7r - 2r - 8 = 32$
$5r - 8 = 32$ Combining like terms
$5r = 32 + 8$
$5r = 40$
$r = 8$

Check: $7r - (2r + 8) = 32$

$$\begin{array}{c|c} 7\cdot 8 - (2\cdot 8 + 8) \ ? \ 32 & \\ 56 - (16 + 8) & \\ 56 - 24 & \\ 32 & 32 \quad \text{TRUE} \end{array}$$

The solution is 8.

64. 8

65. $13 - 3(2x - 1) = 4$
$13 - 6x + 3 = 4$
$16 - 6x = 4$
$-6x = 4 - 16$
$-6x = -12$
$x = 2$

Check: $13 - 3(2x - 1) = 4$

$$\begin{array}{c|c} 13 - 3(2\cdot 2 - 1) \ ? \ 4 & \\ 13 - 3(4 - 1) & \\ 13 - 3\cdot 3 & \\ 13 - 9 & \\ 4 & 4 \quad \text{TRUE} \end{array}$$

The solution is 2.

66. 17

67. $3(t - 2) = 9(t + 2)$
$3t - 6 = 9t + 18$
$-6 - 18 = 9t - 3t$
$-24 = 6t$
$-4 = t$

Check: $3(t - 2) = 9(t + 2)$

$$\begin{array}{c|c} 3(-4 - 2) \ ? \ 9(-4 + 2) & \\ 3(-6) & 9(-2) \\ -18 & -18 \quad \text{TRUE} \end{array}$$

The solution is -4.

68. $-\dfrac{5}{3}$

69. $7(5x - 2) = 6(6x - 1)$
$35x - 14 = 36x - 6$
$-14 + 6 = 36x - 35x$
$-8 = x$

Check:

$$7(5x - 2) = 6(6x - 1)$$

$$\begin{array}{c|c} 7(5(-8) - 2) \ ? \ 6(6(-8) - 1) & \\ 7(-40 - 2) & 6(-48 - 1) \\ 7(-42) & 6(-49) \\ -294 & -294 \quad \text{TRUE} \end{array}$$

The solution is -8.

70. -12

71. $19 - (2x + 3) = 2(x + 3) + x$
$19 - 2x - 3 = 2x + 6 + x$
$16 - 2x = 3x + 6$
$16 - 6 = 3x + 2x$
$10 = 5x$
$2 = x$

Check: $19 - (2x + 3) = 2(x + 3) + x$

$$\begin{array}{c|c} 19 - (2\cdot 2 + 3) \ ? \ 2(2 + 3) + 2 & \\ 19 - (4 + 3) & 2\cdot 5 + 2 \\ 19 - 7 & 10 + 2 \\ 12 & 12 \quad \text{TRUE} \end{array}$$

The solution is 2.

72. 1

73. $\dfrac{1}{4}(3t - 4) = 5$

$4 \cdot \dfrac{1}{4}(3t - 4) = 4 \cdot 5$

$3t - 4 = 20$

$3t = 24$ Adding 4 to both sides

$t = 8$ Dividing both sides by 3

Check:

$$\dfrac{1}{4}(3t - 4) = 5$$

$$\begin{array}{c|c} \dfrac{1}{4}(3\cdot 8 - 4) \ ? \ 5 & \\ \dfrac{1}{4}(24 - 4) \ ? & \\ \dfrac{1}{4}\cdot 20 & \\ 5 & 5 \quad \text{TRUE} \end{array}$$

The solution is 8.

74. 11

75. $\dfrac{4}{3}(5x + 1) = 8$

$\dfrac{3}{4} \cdot \dfrac{4}{3}(5x + 1) = \dfrac{3}{4} \cdot 8$

$5x + 1 = 6$

$5x = 5$

$x = 1$

Check: $\dfrac{4}{3}(5x + 1) = 8$

$\dfrac{4}{3}(5 \cdot 1 + 1) \ ? \ 8$

$\dfrac{4}{3}(6) \ \bigg| \ 8$

$8 \ \bigg| \ 8$ TRUE

The solution is 1.

76. 6

77. $\dfrac{3}{2}(2x + 5) = -\dfrac{15}{2}$

$\dfrac{2}{3} \cdot \dfrac{3}{2}(2x + 5) = \dfrac{2}{3}\left(-\dfrac{15}{2}\right)$

$2x + 5 = -5$

$2x = -10$

$x = -5$

Check: $\dfrac{3}{2}(2x + 5) = -\dfrac{15}{2}$

$\dfrac{3}{2}(2(-5) + 5) \ ? \ -\dfrac{15}{2}$

$\dfrac{3}{2}(-10 + 5) \ \bigg|$

$\dfrac{3}{2}(-5) \ \bigg|$

$-\dfrac{15}{2} \ \bigg| \ -\dfrac{15}{2}$ TRUE

The solution is -5.

78. $\dfrac{16}{15}$

79. $\dfrac{3}{4}\left(3x - \dfrac{1}{2}\right) - \dfrac{2}{3} = \dfrac{1}{3}$

$\dfrac{9}{4}x - \dfrac{3}{8} - \dfrac{2}{3} = \dfrac{1}{3}$

Multiplying by the number 24 will clear all the fractions, so we multiply by 24 on both sides.

$24\left(\dfrac{9}{4}x - \dfrac{3}{8} - \dfrac{2}{3}\right) = 24 \cdot \dfrac{1}{3}$

$24 \cdot \dfrac{9}{4}x - 24 \cdot \dfrac{3}{8} - 24 \cdot \dfrac{2}{3} = 8$

$54x - 9 - 16 = 8$

$54x - 25 = 8$

$54x = 8 + 25$

$54x = 33$

$x = \dfrac{33}{54}$

$x = \dfrac{11}{18}$

The check is left to the student. The solution is $\dfrac{11}{18}$.

80. $-\dfrac{5}{32}$

81. $0.7(3x + 6) = 1.1 - (x + 2)$

$2.1x + 4.2 = 1.1 - x - 2$

$10(2.1x + 4.2) = 10(1.1 - x - 2)$ Clearing decimals

$21x + 42 = 11 - 10x - 20$

$21x + 42 = -10x - 9$

$21x + 10x = -9 - 42$

$31x = -51$

$x = -\dfrac{51}{31}$

The check is left to the student. The solution is $-\dfrac{51}{31}$.

82. $\dfrac{39}{14}$

83. $a + (a - 3) = (a + 2) - (a + 1)$

$a + a - 3 = a + 2 - a - 1$

$2a - 3 = 1$

$2a = 1 + 3$

$2a = 4$

$a = 2$

Check: $a + (a - 3) = (a + 2) - (a + 1)$

$2 + (2 - 3) \ ? \ (2 + 2) - (2 + 1)$

$2 - 1 \ \bigg| \ 4 - 3$

$1 \ \bigg| \ 1$ TRUE

The solution is 2.

84. -7.4

85. Writing exercise

86. Writing exercise

87. $3 - 5a = 3 - 5 \cdot 2 = 3 - 10 = -7$

88. 15

89. $7x - 2x = 7(-3) - 2(-3) = -21 + 6 = -15$

90. -28

91. Writing exercise

92. Writing exercise

93.
$$8.43x - 2.5(3.2 - 0.7x) = -3.455x + 9.04$$
$$8.43x - 8 + 1.75x = -3.455x + 9.04$$
$$10.18x - 8 = -3.455x + 9.04$$
$$10.18x + 3.455x = 9.04 + 8$$
$$13.635x = 17.04$$
$$x = 1.\overline{2497}$$
The solution is $1.\overline{2497}$.

94. 4.423346424

95.
$$-2[3(x - 2) + 4] = 4(5 - x) - 2x$$
$$-2[3x - 6 + 4] = 20 - 4x - 2x$$
$$-2[3x - 2] = 20 - 6x$$
$$-6x + 4 = 20 - 6x$$
$$4 = 20 \qquad \text{Adding } 6x \text{ to both sides}$$
This is contradiction.

96. $-\dfrac{7}{2}$

97.
$$3(x + 4) = 3(4 + x)$$
$$3x + 12 = 12 + 3x$$
$$3x + 12 - 12 = 12 + 3x - 12$$
$$3x = 3x$$
This is an identity.

98. Contradiction

99.
$$2x(x + 5) - 3(x^2 + 2x - 1) = 9 - 5x - x^2$$
$$2x^2 + 10x - 3x^2 - 6x + 3 = 9 - 5x - x^2$$
$$-x^2 + 4x + 3 = 9 - 5x - x^2$$
$$4x + 3 = 9 - 5x \quad \text{Adding } x^2$$
$$4x + 5x = 9 - 3$$
$$9x = 6$$
$$x = \frac{2}{3}$$
The solution is $\dfrac{2}{3}$.

100. -2

101.
$$9 - 3x = 2(5 - 2x) - (1 - 5x)$$
$$9 - 3x = 10 - 4x - 1 + 5x$$
$$9 - 3x = 9 + x$$
$$9 - 9 = x + 3x$$
$$0 = 4x$$
$$0 = x$$
The solution is 0.

102. Identity

103. $[7 - 2(8 \div (-2))]x = 0 \cdot$

Since $7 - 2(8 \div (-2)) \neq 0$ and the product on the left side of the equation is 0, then x must be 0.

104. $\dfrac{52}{45}$

105.
$$\frac{5x + 3}{4} + \frac{25}{12} = \frac{5 + 2x}{3}$$
$$12\left(\frac{5x + 3}{4} + \frac{25}{12}\right) = 12\left(\frac{5 + 2x}{3}\right)$$
$$12\left(\frac{5x + 3}{4}\right) + 12 \cdot \frac{25}{12} = 4(5 + 2x)$$
$$3(5x + 3) + 25 = 4(5 + 2x)$$
$$15x + 9 + 25 = 20 + 8x$$
$$15x + 34 = 20 + 8x$$
$$7x = -14$$
$$x = -2$$
The solution is -2.

Exercise Set 2.3

1. We substitute 10 for t and calculate M.
$$M = \frac{1}{5} \cdot 10 = 2$$
The storm is 2 miles away.

2. 3450 watts

3. We substitute 21,345 for n and calculate f.
$$f = \frac{21,345}{15} = 1423$$
There are 1423 full-time equivalent students.

4. 54 in^2

5. We substitute 84 for c and 8 for w and calculate D.
$$D = \frac{c}{w} = \frac{84}{8} = 10.5$$
The calorie density is 10.5 calories/oz.

6. $\frac{43}{3}$ m/cycle, or $14.\overline{3}$ m/cycle

7. Substitute 1 for t and calculate n.

$n = 0.5t^4 + 3.45t^3 - 96.65t^2 + 347.7t$

$= 0.5(1)^4 + 3.45(1)^3 - 96.65(1)^2 + 347.7(1)$

$= 0.5 + 3.45 - 96.65 + 347.7$

$= 255$

255 mg of ibuprofen remains in the bloodstream.

8. 42

9. $A = bh$

$\frac{A}{h} = \frac{bh}{h}$ Dividing both sides by h

$\frac{A}{h} = b$

10. $h = \frac{A}{b}$

11. $d = rt$

$\frac{d}{t} = \frac{rt}{t}$ Dividing both sides by t

$\frac{d}{t} = r$

12. $t = \frac{d}{r}$

13. $I = Prt$

$\frac{I}{rt} = \frac{Prt}{rt}$ Dividing both sides by rt

$\frac{I}{rt} = P$

14. $t = \frac{I}{Pr}$

15. $H = 65 - m$

$H + m = 65$ Adding m to both sides

$m = 65 - H$ Subtracting H from both sides

16. $h = d + 64$

17. $P = 2l + 2w$

$P - 2w = 2l + 2w - 2w$ Subtracting $2w$ from both sides

$P - 2w = 2l$

$\frac{P - 2w}{2} = \frac{2l}{2}$ Dividing both sides by 2

$\frac{P - 2w}{2} = l$, or

$\frac{P}{2} - w = l$

18. $w = \frac{P - 2l}{2}$, or $w = \frac{P}{2} - l$

19. $A = \pi r^2$

$\frac{A}{r^2} = \frac{\pi r^2}{r^2}$

$\frac{A}{r^2} = \pi$

20. $r^2 = \frac{A}{\pi}$

21. $A = \frac{1}{2}bh$

$2A = 2 \cdot \frac{1}{2}bh$ Multiplying both sides by 2

$2A = bh$

$\frac{2A}{b} = \frac{bh}{b}$ Dividing both sides by h

$\frac{2A}{b} = h$

22. $b = \frac{2A}{h}$

23. $E = mc^2$

$\frac{E}{c^2} = \frac{mc^2}{c^2}$ Dividing both sides by c^2

$\frac{E}{c^2} = m$

24. $c^2 = \frac{E}{m}$

25. $Q = \frac{c + d}{2}$

$2Q = 2 \cdot \frac{c + d}{2}$ Multiplying both sides by 2

$2Q = c + d$

$2Q - c = c + d - c$ Subtracting c from both sides

$2Q - c = d$

26. $p = 2Q + q$

27. $A = \frac{a + b + c}{3}$

$3A = 3 \cdot \frac{a + b + c}{3}$ Multiplying both sides by 3

$3A = a + b + c$

$3A - a - c = a + b + c - a - c$ Subtracting a and c from both sides

$3A - a - c = b$

28. $c = 3A - a - b$

29.
$$M = \frac{A}{s}$$

$s \cdot M = s \cdot \frac{A}{s}$ Multiplying both sides by s

$sM = A$

30. $b = \dfrac{Pc}{a}$

31.
$$A = at + bt$$

$A = t(a + b)$ Factoring

$\dfrac{A}{a+b} = t$ Dividing both sides by $a + b$

32. $x = \dfrac{S}{r + s}$

33.
$$A = \frac{1}{2}ah + \frac{1}{2}bh$$

$2A = 2\left(\dfrac{1}{2}ah + \dfrac{1}{2}bh\right)$

$2A = ah + bh$

$2A = h(a + b)$

$\dfrac{2A}{a+b} = h$

34. $P = \dfrac{A}{1 + rt}$

35.
$$R = r + \frac{400(W - L)}{N}$$

$N \cdot R = N\left(r + \dfrac{400(W - L)}{N}\right)$

Multiplying both sides by N

$NR = Nr + 400(W - L)$

$NR = Nr + 400W - 400L$

$NR + 400L = Nr + 400W$ Adding $400L$ to both sides

$400L = Nr + 400W - NR$ Adding $-NR$ to both sides

$L = \dfrac{Nr + 400W - NR}{400}$

36. $r^2 = \dfrac{360A}{\pi S}$

37. Writing exercise

38. Writing exercise

39. $0.79(38.4)0$

One factor is 0, so the product is 0.

40. 9.18

41.
$$20 \div (-4) \cdot 2 - 3$$

$= -5 \cdot 2 - 3$ Dividing and

$= -10 - 3$ multiplying from left to right

$= -13$ Subtracting

42. 65

43. Writing exercise

44. Writing exercise

45.
$$K = 19.18w + 7h - 9.52a + 92.4$$

$2627 = 19.18(82) + 7(185) - 9.52a + 92.4$

$2627 = 1572.76 + 1295 - 9.52a + 92.4$

$2627 = 2960.16 - 9.52a$

$-333.16 = -9.52a$

$35 \approx a$

The man is about 35 years old.

46. $T = t - \dfrac{h}{100},\ 0 \le h \le 12{,}000$

47.
$$c = \frac{w}{a} \cdot d$$

$ac = a \cdot \dfrac{w}{a} \cdot d$

$ac = wd$

$a = \dfrac{wd}{c}$

48. About 76.4 in.

49.
$$\frac{y}{z} \div \frac{z}{t} = 1$$

$\dfrac{y}{z} \cdot \dfrac{t}{z} = 1$

$\dfrac{yt}{z^2} = 1$

$\dfrac{z^2}{t} \cdot \dfrac{yt}{z^2} = \dfrac{z^2}{t} \cdot 1$

$y = \dfrac{z^2}{t}$

50. $c = \dfrac{d}{a - b}$

51.
$$qt = r(s + t)$$

$qt = rs + rt$

$qt - rt = rs$

$t(q - r) = rs$

$t = \dfrac{rs}{q - r}$

52. $a = \dfrac{c}{3 + b + d}$

53. We subtract the minimum output for a well-insulated house with a square feet from the minimum output for a poorly-insulated house with a square feet. Let S represent the number of BTU's saved.

$$S = 50a - 30a$$
$$S = 20a$$

54. $K = 917 + 13.2276w + 2.3622h - 6a$

55. $K = 19.18\left(\dfrac{w}{2.2046}\right) + 7\left(\dfrac{h}{0.3937}\right) - 9.52a + 92.4$

$K = 8.70w + 17.78h - 9.52a + 92.4$

Exercise Set 2.4

1. $82\% = 82 \times 0.01$ Replacing % by $\times 0.01$
       $= 0.82$

2. 0.49

3. $9\% = 9 \times 0.01$ Replacing % by $\times 0.01$
      $= 0.09$

4. 0.913

5. $43.7\% = 43.7 \times 0.01 = 0.437$

6. 0.02

7. $0.46\% = 0.46 \times 0.01 = 0.0046$

8. 0.048

9. 0.29

First move the decimal point   0.29.
two places to the right;   └─↑
then write a % symbol: 29%

10. 78%

11. 0.998

First move the decimal point   0.99.8
two places to the right;   └─↑
then write a % symbol: 99.8%

12. 35.8%

13. 1.92

First move the decimal point   1.92.
two places to the right;   └─↑
then write a % symbol: 192%

14. 139%

15. 2.1

First move the decimal point   2.10.
two places to the right;   └─↑
then write a % symbol: 210%

16. 920%

17. 0.0068

First move the decimal point   0.00.68
two places to the right;   └─↑
then write a % symbol: 0.68%

18. 0.95%

19. $\dfrac{3}{8}$ $\left(\text{Note: } \dfrac{3}{8} = 0.375\right)$

First move the decimal point   0.37.5
two places to the right;   └─↑
then write a % symbol: 37.5%

20. 75%

21. $\dfrac{7}{25}$ $\left(\text{Note: } \dfrac{7}{25} = 0.28\right)$

First move the decimal point   0.28.
two places to the right;   └─↑
then write a % symbol: 28%

22. 80%

23. $\dfrac{2}{3}$ $\left(\text{Note: } \dfrac{2}{3} = 0.66\overline{6}\right)$

First move the decimal point   0.66.$\overline{6}$
two places to the right;   └─↑
then write a % symbol: 66.$\overline{6}$%

Since $0.\overline{6} = \dfrac{2}{3}$, this can also be expressed as $66\dfrac{2}{3}\%$.

24. $83.\overline{3}\%$, or $83\dfrac{1}{3}\%$

25. *Translate*.

$$\underbrace{\text{What percent}}_{y} \text{ of } 68 \text{ is } 17?$$
$$y \quad\cdot\quad 68 \;=\; 17$$

We solve the equation and then convert to percent notation.

$$y \cdot 68 = 17$$
$$y = \dfrac{17}{68}$$
$$y = 0.25 = 25\%$$

The answer is 25%.

26. 26%

27. *Translate*.

$$\underbrace{\text{What percent}}_{y} \text{ of } 125 \text{ is } 30?$$
$$y \quad\cdot\quad 125 \;=\; 30$$

We solve the equation and then convert to percent notation.

$$y \cdot 125 = 30$$
$$y = \frac{30}{125}$$
$$y = 0.24 = 24\%$$

The answer is 24%.

28. 19%

29. *Translate*.

14 is 30% of what number?
$$14 = 30\% \cdot y$$

We solve the equation.

$$14 = 0.3y \quad (30\% = 0.3)$$
$$\frac{14}{0.3} = y$$
$$46.\overline{6} = y$$

The answer is $46.\overline{6}$, or $46\frac{2}{3}$, or $\frac{140}{3}$.

30. 225

31. *Translate*.

0.3 is 12% of what number?
$$0.3 = 12\% \cdot y$$

We solve the equation.

$$0.3 = 0.12y \quad (12\% = 0.12)$$
$$\frac{0.3}{0.12} = y$$
$$2.5 = y$$

The answer is 2.5.

32. 4

33. *Translate*.

What number is 35% of 240?
$$y = 35\% \cdot 240$$

We solve the equation.

$$y = 0.35 \cdot 240 \quad (35\% = 0.35)$$
$$y = 84 \qquad \text{Multiplying}$$

The answer is 84.

34. 10,000

35. *Translate*.

What percent of 60 is 75?
$$y \cdot 60 = 75$$

We solve the equation and then convert to percent notation.

$$y \cdot 60 = 75$$
$$y = \frac{75}{60}$$
$$y = 1.25 = 125\%$$

The answer is 125%.

36. 100%

37. *Translate*.

What is 2% of 40?
$$x = 2\% \cdot 40$$

We solve the equation.

$$x = 0.02 \cdot 40 \quad (2\% = 0.02)$$
$$x = 0.8 \qquad \text{Multiplying}$$

The answer is 0.8.

38. 0.8

39. Observe that 25 is half of 50. Thus, the answer is 0.5, or 50%. We could also do this exercise by translating to an equation.

Translate.

25 is what percent of 50?
$$25 = y \cdot 50$$

We solve the equation and convert to percent notation.

$$25 = y \cdot 50$$
$$\frac{25}{50} = y$$
$$0.5 = y, \text{ or } 50\% = y$$

The answer is 50%.

40. 400

41. Let I = the amount of interest Sarah will pay. Then we have:

I is 8% of $3500.
$$I = 0.08 \cdot \$3500$$
$$I = \$280$$

Sarah will pay $280 interest.

42. $168

43. Let p = the number of people who voted in the 2000 presidential election, in millions. Then we have:

48.62 is 48.36% of p.
$$48.62 = 0.4836 \cdot p$$

$$\frac{48.62}{0.4836} = p$$
$$100.54 \approx p$$

About 100.54 million, or 100,540,000, people voted in the 2000 presidential election.

44. About $11.9 billion

45. If $n =$ the number of women who had babies in good or excellent health, we have:

n is 8% of 300.
$\downarrow\downarrow$ $\downarrow$ $\downarrow$ $\downarrow$
$n = 0.08 \cdot 300$
$n = 24$

24 women had babies in good or excellent health.

46. 285

47. Let $a =$ the number of pounds of almonds the average American consumes each year. Then we have:

a is 25% of 2.25.
$\downarrow\downarrow$ $\downarrow$ $\downarrow$ $\downarrow$
$a = 0.25 \cdot 2.25$
$a = 0.5625$

The average American consumes 0.5625 lb of almonds each year.

48. 7410

49. Let $b =$ the number of bowlers you would expect to be left-handed. Then we have:

b is 17% of 160.
$\downarrow\downarrow$ $\downarrow$ $\downarrow$ $\downarrow$
$b = 0.17 \cdot 160$
$b \approx 27$

You would expect 27 bowlers to be left-handed.

50. 7%

51. Let $p =$ the percent that were correct. Then we have:

76 is what percent of 88?
$\downarrow$ $\downarrow$ $\downarrow$ $\downarrow$ $\downarrow$
$76 = p \cdot 88$

$$\frac{76}{88} = p$$
$$0.86\overline{4} = p, \text{ or}$$
$$86.\overline{4}\% = p$$

$86.\overline{4}\%$ of the items were correct.

52. 52%

53. When the sales tax is 5%, the total amount paid is 105% of the cost of the merchandise. Let $c =$ the cost of the merchandise. Then we have:

$37.80 is 105% of c.
$\downarrow$ $\downarrow$ $\downarrow$ $\downarrow\downarrow$
$37.80 = 1.05 \cdot c$

$$\frac{37.80}{1.05} = c$$
$$36 = c$$

The price of the merchandise was $36.

54. $940

55. When the sales tax is 5%, the total amount paid is 105% of the cost of the merchandise. Let $c =$ the amount the school group owes, or the cost of the software without tax. Then we have:

$157.41 is 106% of c.
$\downarrow$ $\downarrow$ $\downarrow$ $\downarrow\downarrow$
$157.41 = 1.06 \cdot c$

$$\frac{157.41}{1.06} = c$$
$$148.5 = c$$

The school group owes $148.50.

56. $138.95

57. A self-employed person must earn 120% as much as a non-self-employed person. Let $a =$ the amount Roy would need to earn, in dollars per hour, on his own for a comparable income. Then we have:

a is 120% of $15.
$\downarrow\downarrow$ $\downarrow$ $\downarrow$ $\downarrow$
$a = 1.2 \cdot 15$
$a = 18$

Roy would need to earn $18 per hour on his own.

58. $14.40 per hour

59. The number of calories in a serving of Light Style Bread is 85% of the number of calories in a serving of regular bread. Let $c =$ the number of calories in a serving of regular bread. Then we have:

140 calories is 85% of c.
$\downarrow$ $\downarrow$ $\downarrow$ $\downarrow\downarrow$
$140 = 0.85 \cdot c$

$$\frac{140}{0.85} = c$$
$$165 \approx c$$

There are about 165 calories in a serving of regular bread.

60. 58 calories

61. Writing exercise

62. Writing exercise

63. Let n represent "some number." Then we have $n + 5$, or $5 + n$.

64. Let w represent Tino's weight; $w - 4$

65. $8 \cdot 2a$, or $2a \cdot 8$.

66. Let m and n represent the numbers; $mn + 1$, or $1 + mn$

67. Writing exercise

68. Writing exercise

69. Let p = the population of Bardville. Then we have:

1332 is 15% of 48% of the population.

$$\downarrow \quad \downarrow \quad \downarrow \quad \downarrow \quad \downarrow \quad \downarrow \qquad \downarrow$$
$$1332 = 0.15 \cdot 0.48 \cdot \qquad p$$

$$\frac{1332}{0.15(0.48)} = p$$
$$18,500 = p$$

The population of Bardville is 18,500.

70. Rollie's: $12.83; Sound Warp: $12.97

71. The new price is 125% of the old price. Let p = the new price. Then we have:

p is 125% of $20,800.
$$\downarrow \downarrow \quad \downarrow \quad \downarrow \qquad \downarrow$$
$$p = 1.25 \cdot 20,800$$
$$p = 26,000$$

Now let x = the percent of the new price represented by the old price. We have:

$20,800 is what percent of $26,000.

$$\downarrow \qquad \downarrow \qquad \downarrow \qquad \downarrow \qquad \downarrow$$
$$20,800 = \qquad x \qquad \cdot \; 26,000$$

$$\frac{20,800}{26,000} = x$$
$$0.8 = x, \text{ or}$$
$$80\% = x$$

The old price is $100\% - 80\%$, or 20% lower than the new price.

72. $35.\overline{135}\%$, or $35\frac{5}{37}\%$

73. The number of births increased by $3.94 - 3.88$, or 0.06 million. Let p = the percent of increase. Then we have:

0.06 million is what percent of 3.88 million?

$$\downarrow \qquad \downarrow \qquad \downarrow \qquad \downarrow \qquad \downarrow$$
$$0.06 \quad = \quad p \quad \cdot \quad 3.88$$

$$\frac{0.06}{3.88} = p$$
$$0.0155 \approx p, \text{ or}$$
$$1.55\% \approx p$$

The number of births increased by about 1.55%.

74. Writing exercise

75. Writing exercise

Exercise Set 2.5

1. *Familiarize*. Let x = the number. Then "three less than twice a number" translates to $2x - 3$.

Translate.

Three less than twice a number is 19.

$$\underbrace{\qquad\qquad\qquad\qquad}\quad \downarrow \;\; \downarrow$$
$$2x - 3 \qquad\quad = \;\; 19$$

Carry out. We solve the equation.

$$2x - 3 = 19$$
$$2x = 22 \quad \text{Adding 3}$$
$$x = 11 \quad \text{Dividing by 2}$$

Check. Twice, or two times, 11 is 22. Three less than 22 is 19. The answer checks.

State. The number is 11.

2. 8

3. *Familiarize*. Let a = the number. Then "five times the sum of 3 and some number" translates to $5(a + 3)$.

Translate.

Five times the sum of 3 and some number is 70.

$$\underbrace{\qquad\qquad\qquad\qquad}\quad \downarrow \;\; \downarrow$$
$$5(a + 3) \qquad\quad = \;\; 70$$

Carry out. We solve the equation.

$$5(a + 3) = 70$$
$$5a + 15 = 70 \quad \text{Using the distributive law}$$
$$5a = 55 \quad \text{Subtracting 15}$$
$$a = 11 \quad \text{Dividing by 5}$$

Check. The sum of 3 and 11 is 14, and $5 \cdot 14 = 70$. The answer checks.

State. The number is 11.

4. 13

5. *Familiarize*. Let p = the regular price of the shoes. At 15% off, Amy paid 85% of the regular price.

Translate.

$63.75 is 85% of the regular price.

$$\downarrow \quad\;\; \downarrow \;\; \downarrow \;\; \downarrow \qquad\qquad \downarrow$$
$$63.75 = 0.85 \cdot \qquad\qquad p$$

Carry out. We solve the equation.

$$63.75 = 0.85p$$

$$\frac{63.75}{0.08} = p \qquad \text{Dividing both sides by 0.85}$$

$$75 = p$$

Check. 85% of $75, or 0.85($75), is $63.75. The answer checks.

State. The regular price was $75.

6. $90

7. *Familiarize*. Let $b =$ the price of the book itself. When the sales tax rate is 5%, the tax paid on the book is 5% of b, or $0.05b$.

Translate.

$$\underbrace{\text{Price of book}}_{b} \; \underbrace{\text{plus}}_{+} \; \underbrace{\text{sales tax}}_{0.05b} \; \underbrace{\text{is}}_{=} \; \underbrace{\$89.25.}_{89.25}$$

Carry out. We solve the equation.

$$b + 0.05b = 89.25$$

$$1.05b = 89.25$$

$$b = \frac{89.25}{1.05}$$

$$b = 85$$

Check. 5% of $85, or 0.05($85), is $4.25 and $85 + $4.25 is $89.25, the total cost. The answer checks.

State. The book itself cost $85.

8. $95

9. *Familiarize*. Let $d =$ Kouros' distance, in miles, from the start after 8 hr. Then the distance from the finish line is $2d$.

Translate.

$$\underbrace{\text{Distance from start}}_{d} \; \underbrace{\text{plus}}_{+} \; \underbrace{\text{distance from finish}}_{2d} \; \underbrace{\text{is}}_{=} \; \underbrace{188 \text{ mi.}}_{188}$$

Carry out. We solve the equation.

$$d + 2d = 188$$

$$3d = 188$$

$$d = \frac{188}{3}, \text{ or } 62\frac{2}{3}$$

Check. If Kouros is $\frac{188}{3}$ mi from the start, then he is $2 \cdot \frac{188}{3}$, or $\frac{376}{3}$ mi from the finish. Since $\frac{188}{3} + \frac{376}{3} = \frac{564}{3} = 188$, the total distance run, the answer checks.

State. Kouros had run $62\frac{2}{3}$ mi.

10. $699\frac{1}{3}$ mi

11. *Familiarize*. Let $x =$ the first page number. Then $x + 1 =$ the second page number, and $x + 2 =$ the third page number.

Translate.

$$\underbrace{\text{The sum of three consecutive page numbers}}_{x + (x + 1) + (x + 2)} \; \underbrace{\text{is}}_{=} \; \underbrace{60.}_{60}$$

Carry out. We solve the equation.

$$x + (x + 1) + (x + 2) = 60$$

$$3x + 3 = 60 \quad \text{Combining like terms}$$

$$3x = 57 \quad \text{Subtracting 3 from both sides}$$

$$x = 19 \quad \text{Dividing both sides by 3}$$

If x is 19, then $x + 1$ is 20 and $x + 2 = 21$.

Check. 19, 20, and 21 are consecutive integers, and $19 + 20 + 21 = 60$. The result checks.

State. The page numbers are 19, 20, and 21.

12. 32, 33, 34

13. *Familiarize*. Let $x =$ the smaller odd number. Then $x + 2 =$ the next odd number.

Translate. We reword the problem.

$$\underbrace{\text{Smaller odd number}}_{x} + \underbrace{\text{next odd number}}_{(x + 2)} \; \underbrace{\text{is}}_{=} \; \underbrace{60.}_{60}$$

Carry out. We solve the equation.

$$x + (x + 2) = 60$$

$$2x + 2 = 60 \quad \text{Combining like terms}$$

$$2x = 58 \quad \text{Subtracting 2 from both sides}$$

$$x = 29 \quad \text{Dividing both sides by 2}$$

If x is 29, then $x + 2$ is 31.

Check. 29 and 31 are consecutive odd integers, and their sum is 60. The answer checks.

State. The integers are 29 and 31.

14. 53, 55

15. *Familiarize*. Let $x =$ the first even integer. Then $x + 2 =$ the next even integer.

Translate.

$$\underbrace{\text{The sum of two consecutive even integers}}_{x + (x + 2)} \; \underbrace{\text{is}}_{=} \; \underbrace{126.}_{126}$$

Carry out. We solve the equation.

$$x + (x + 2) = 126$$
$$2x + 2 = 126 \quad \text{Combining like terms}$$
$$2x = 124 \quad \text{Subtracting 2 from both sides}$$
$$x = 62 \quad \text{Dividing both sides by 2}$$

If x is 62, then $x + 2$ is 64.

Check. 62 and 64 are consecutive even integers, and $62 + 64 = 126$. The result checks.

State. The numbers are 62 and 64.

16. 24, 26

17. *Familiarize.* Let $b =$ the bride's age. Then $b + 19 =$ the groom's age.

Translate.

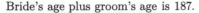

Bride's age plus groom's age is 187.
$$b \quad + \quad b + 19 \quad = \quad 187$$

Carry out. We solve the equation.

$$b + (b + 19) = 187$$
$$2b + 19 = 187$$
$$2b = 168$$
$$b = 84$$

If b is 84, then $b + 19$ is 103.

Check. 103 is 19 more than 84, and $84 + 103 = 187$. The answer checks.

State. The bride was 84 yr old, and the groom was 103 yr old.

18. Man: 97 yr; woman: 91 yr

19. *Familiarize.* We draw a picture. We let $x =$ the measure of the first angle. Then $3x =$ the measure of the second angle, and $x + 30 =$ the measure of the third angle.

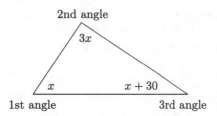

2nd angle
$3x$
x $x + 30$
1st angle   3rd angle

Recall that the measures of the angles of any triangle add up to 180°.

Translate.

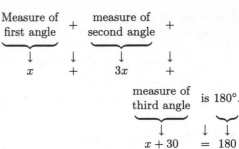

Measure of first angle + measure of second angle +
$$x \quad + \quad 3x \quad +$$

measure of third angle is 180°.
$$x + 30 \quad = \quad 180$$

Carry out. We solve the equation.

$$x + 3x + (x + 30) = 180$$
$$5x + 30 = 180$$
$$5x = 150$$
$$x = 30$$

Possible answers for the angle measures are as follows:

First angle: $x = 30°$

Second angle: $3x = 3(30)° = 90°$

Third angle: $x + 30° = 30° + 30° = 60°$

Check. Consider 30°, 90°, and 60°. The second angle is three times the first, and the third is 30° more than the first. The sum of the measures of the angles is 180°. These numbers check.

State. The measure of the first angle is 30°, the measure of the second angle is 90°, and the measure of the third angle is 60°.

20. 22.5°, 90°, 67.5°

21. *Familiarize.* Let $x =$ the measure of the first angle. Then $3x =$ the measure of the second angle, and $x + 3x + 10 = 4x + 10 =$ the measure of the third angle. Recall that the sum of the measures of the angles of a triangle is 180°.

Translate.

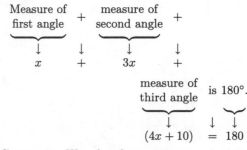

Measure of first angle + measure of second angle +
$$x \quad + \quad 3x \quad +$$

measure of third angle is 180°.
$$(4x + 10) \quad = \quad 180$$

Carry out. We solve the equation.

$$x + 3x + (4x + 10) = 180$$
$$8x + 10 = 180$$
$$8x = 170$$
$$x = 21.25$$

If x is 21.25, then $3x$ is 63.75, and $4x + 10$ is 95.

Check. Consider 21.25°, 63.75°, and 95°. The second is three times the first, and the third is 10° more than the sum of the other two. The sum of the measures of the angles is 180°. These numbers check.

State. The measure of the third angle is 95°.

22. 70°

23. **Familiarize**. The page numbers are consecutive integers. If we let p = the smaller number, then $p+1$ = the larger number.

Translate. We reword the problem.

$$\underbrace{\text{First integer}}_{x} + \underbrace{\text{Second integer}}_{(x+1)} = 385$$

Carry out. We solve the equation.

$x + (x + 1) = 385$

$2x + 1 = 385$ Combining like terms

$2x = 384$ Adding -1 on both sides

$x = 192$ Dividing on both sides by 2

Check. If $x = 192$, then $x+1 = 193$. These are consecutive integers, and $192 + 193 = 385$. The answer checks.

State. The page numbers are 192 and 193.

24. 140, 141

25. **Familiarize**. Let s = the length of the shortest side, in mm. Then $s + 2$ and $s + 4$ represent the lengths of the other two sides. The perimeter is the sum of the lengths of the sides.

Translate.

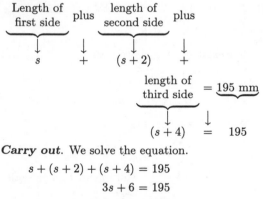

Carry out. We solve the equation.

$s + (s + 2) + (s + 4) = 195$

$3s + 6 = 195$

$3s = 189$

$s = 63$

If s is 63, then $s + 2$ is 65 and $s + 4$ is 67.

Check. The numbers 63, 65, and 67 are consecutive odd integers. Their sum is 195. These numbers check.

State. The lengths of the sides of the triangle are 63 mm, 65 mm, and 67 mm.

26. Width: 100 ft; length: 160 ft; area: 16,000 ft^2

27. **Familiarize**. We draw a picture. Let l = the length of the state, in miles. Then $l - 90$ = the width.

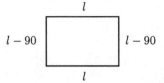

The perimeter is the sum of the lengths of the sides.

Translate. We use the definition of perimeter to write an equation.

$$\underbrace{\text{Width}}_{(l-90)} + \underbrace{\text{Width}}_{(l-90)} + \underbrace{\text{Length}}_{l} + \underbrace{\text{Length}}_{l} \text{ is } 1280.$$

$(l - 90) + (l - 90) + l + l = 1280$

Carry out. We solve the equation.

$(l - 90) + (l - 90) + l + l = 1280$

$4l - 180 = 1280$

$4l = 1460$

$l = 365$

Then $l - 90 = 275$.

Check. The width, 275 mi, is 90 mi less than the length, 365 mi. The perimeter is 275 mi + 275 mi + 365 mi + 365 mi, or 1280 mi. This checks.

State. The length is 365 mi, and the width is 275 mi.

28. Length: 27.9 cm; width: 21.6 cm

29. **Familiarize**. Let a = the amount Sarah invested. The investment grew by 28% of a, or $0.28a$.

Translate.

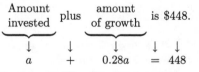

Carry out. We solve the equation.

$a + 0.28a = 448$

$1.28a = 448$

$a = 350$

Check. 28% of $350 is 0.28($350), or $98, and $350 + $98 = $448. The answer checks.

State. Sarah invested $350.

30. $6600

31. **Familiarize**. Let b = the balance in the account at the beginning of the month. The balance grew by 2% of b, or $0.02b$.

Translate.

$$\underbrace{\text{Original balance}}_{b} \underset{+}{\text{ plus }} \underbrace{\text{amount of growth}}_{0.02b} \underset{=}{\text{ is \$870.}} \underset{870}{}$$

Carry out. We solve the equation.

$$b + 0.02b = 870$$
$$1.02b = 870$$
$$b \approx \$852.94$$

Check. 2% of \$852.94 is 0.02(\$852.94), or \$17.06, and \$852.94 + \$17.06 = \$870. The answer checks.

State. The balance at the beginning of the month was \$852.94.

32. \$6540

33. *Familiarize.* The total cost is the initial charge plus the mileage charge. Let d = the distance, in miles, that Courtney can travel for \$12. The mileage charge is the cost per mile times the number of miles traveled or $0.75d$.

Translate.

$$\underbrace{\text{Initial charge}}_{3} \underset{+}{\text{ plus }} \underbrace{\text{mileage charge}}_{0.75d} \underset{=}{\text{ is \$12.}} \underset{12}{}$$

Carry out. We solve the equation.

$$3 + 0.75d = 12$$
$$0.75d = 9$$
$$d = 12$$

Check. A 12-mi taxi ride from the airport would cost \$3 + 12(\$0.75), or \$3 + \$9, or \$12. The answer checks.

State. Courtney can travel 12 mi from the airport for \$12.

34. 15 mi

35. *Familiarize.* The total cost is the daily charge plus the mileage charge. Let d = the distance that can be traveled, in miles, in one day for \$100. The mileage charge is the cost per mile times the number of miles traveled, or $0.39d$.

Translate.

$$\underbrace{\text{Daily rate}}_{49.95} \underset{+}{\text{ plus }} \underbrace{\text{mileage charge}}_{0.39d} \underset{=}{\text{ is \$100.}} \underset{100}{}$$

Carry out. We solve the equation.

$$49.95 + 0.39d = 100$$
$$0.39d = 50.05$$
$$d = 128.\overline{3}, \text{ or } 128\frac{1}{3}$$

Check. For a trip of $128\frac{1}{3}$ mi, the mileage charge is $\$0.39\left(128\frac{1}{3}\right)$, or \$50.05, and \$49.95+\$50.05 = \$100. The answer checks.

State. They can travel $128\frac{1}{3}$ mi in one day and stay within their budget.

36. 80 mi

37. *Familiarize.* Let x = the measure of one angle. Then $90 - x$ = the measure of its complement.

Translate.

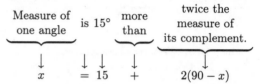

$$\underbrace{\text{Measure of one angle}}_{x} \underset{=}{\text{ is 15° }} \underset{15}{} \underset{+}{\text{ more than }} \underbrace{\text{twice the measure of its complement.}}_{2(90-x)}$$

Carry out. We solve the equation.

$$x = 15 + 2(90 - x)$$
$$x = 15 + 180 - 2x$$
$$x = 195 - 2x$$
$$3x = 195$$
$$x = 65$$

If x is 65, then $90 - x$ is 25.

Check. The sum of the angle measures is 90°. Also, 65° is 15° more than twice its complement, 25°. The answer checks.

State. The angle measures are 65° and 25°.

38. 105°, 75°

39. *Familiarize.* We will use the equation

$$T = \frac{1}{4}N + 40.$$

Translate. We substitute 80 for T.

$$80 = \frac{1}{4}N + 40$$

Carry out. We solve the equation.

$$80 = \frac{1}{4}N + 40$$
$$40 = \frac{1}{4}N$$
$$160 = N \qquad \text{Multiplying by 4 on both sides}$$

Check. When $N = 160$, we have $T = \frac{1}{4} \cdot 160 + 40 = 40 + 40 = 80$. The answer checks.

State. A cricket chirps 160 times per minute when the temperature is 80°F.

40. 2020

41. Writing exercise

42. Writing exercise

43. Since -9 is to the left of 5 on the number line, we have $-9 < 5$.

44. $1 < 3$

45. Since -4 is to the left of 7 on the number line, we have $-4 < 7$.

46. $-9 > -12$

47. Writing exercise

48. Writing exercise

49. **Familiarize**. Let $c =$ the amount the meal originally cost. The 15% tip is calculated on the original cost of the meal, so the tip is $0.15c$.

Translate.

$$\underbrace{\text{Original cost}}_{c} \quad \underset{+}{\text{plus}} \quad \underset{0.15c}{\text{tip}} \quad \underset{-}{\text{less}} \quad \underset{10}{\$10} \quad \underset{=}{\text{is}} \quad \underset{32.55}{\$32.55.}$$

Carry out. We solve the equation.
$$c + 0.15c - 10 = 32.55$$
$$1.15c - 10 = 32.55$$
$$1.15c = 42.55$$
$$c = 37$$

Check. If the meal originally cost \$37, the tip was 15% of \$37, or $0.15(\$37)$, or \$5.55. Since $\$37+\$5.55-\$10 = \$32.55$, the answer checks.

State. The meal originally cost \$37.

50. 19

51. **Familiarize**. Let $s =$ one score. Then four score $= 4s$ and four score and seven $= 4s + 7$.

Translate. We reword .

$$\underset{1776}{\underbrace{1776}} \quad \underset{+}{\text{plus}} \quad \underset{(4s+7)}{\underbrace{\text{four score and seven}}} \quad \underset{=}{\text{is}} \quad \underset{1863}{\underbrace{1863}}$$

Carry out. We solve the equation.
$$1776 + (4s + 7) = 1863$$
$$4s + 1783 = 1863$$
$$4s = 80$$
$$s = 20$$

Check. If a score is 20 years, then four score and seven represents 87 years. Adding 87 to 1776 we get 1863. This checks.

State. A score is 20.

52. 4, 16

53. **Familiarize**. We let $x =$ the length of the original rectangle. Then $\frac{3}{4}x =$ the width. We draw a picture of the enlarged rectangle. Each dimension is increased by 2 cm, so $x + 2 =$ the length of the enlarged rectangle and $\frac{3}{4}x + 2 =$ the width.

Translate. We use the perimeter of the enlarged rectangle to write an equation.

$$\underset{\left(\frac{3}{4}x+2\right)}{\text{Width}} + \underset{\left(\frac{3}{4}x+2\right)}{\text{Width}} + \underset{(x+2)}{\text{Length}} + \underset{(x+2)}{\text{Length}} \underset{=}{\text{is}}$$

$$\underset{50}{\text{Perimeter.}}$$

Carry out.
$$\left(\frac{3}{4}x+2\right)+\left(\frac{3}{4}x+2\right)+(x+2)+(x+2) = 50$$
$$\frac{7}{2}x + 8 = 50$$
$$2\left(\frac{7}{2}x + 8\right) = 2 \cdot 50$$
$$7x + 16 = 100$$
$$7x = 84$$
$$x = 12$$

Then $\frac{3}{4}x = \frac{3}{4}(12) = 9$.

Check. If the dimensions of the original rectangle are 12 cm and 9 cm, then the dimensions of the enlarged rectangle are 14 cm and 11 cm. The perimeter of the enlarged rectangle is $11+11+14+14 = 50$ cm. Also, 9 is $\frac{3}{4}$ of 12. These values check.

State. The length is 12 cm, and the width is 9 cm.

54. 87°, 89°, 91°, 93°

55. **Familiarize**. Let $x =$ the first even number. Then the next four even numbers are $x+2$, $x+4$, $x+6$, and $x+8$. The sum of the measures of the angles of an n-sided polygon is given by the formula $(n-2) \cdot 180°$. Thus, the sum of the measures of the angles of a pentagon is $(5-2) \cdot 180°$, or $3 \cdot 180°$, or $540°$.

Translate.

$$\underbrace{\text{The sum of the measures of the angles}} \quad \text{is } 540°.$$

$$x + (x+2) + (x+4) + (x+6) + (x+8) = 540$$

Carry out. We solve the equation.

$$x + (x+2) + (x+4) + (x+6) + (x+8) = 540$$
$$5x + 20 = 540$$
$$5x = 520$$
$$x = 104$$

If x is 104, then the other numbers are 106, 108, 110, and 112.

Check. The numbers 104, 106, 108, 110, and 112 are consecutive odd numbers. Their sum is 540. The answer checks.

State. The measures of the angles are 104°, 106°, 108°, 110°, and 112°.

56. 120

57. *Familiarize*. Let $p =$ the price before the two discounts. With the first 10% discount, the price becomes 90% of p, or $0.9p$. With the second 10% discount, the final price is 90% of $0.9p$, or $0.9(0.9p)$.

Translate.

$$\underbrace{\text{The final price}} \text{ is } \$77.75.$$

$$0.9(0.9p) = 77.75$$

Carry out. We solve the equation.

$$0.9(0.9p) = 77.75$$
$$p = \frac{77.75}{0.9(0.9)}$$
$$p \approx 95.99$$

Check. 90% of $95.99 is $86.39 and 90% of $86.39 is $77.75. The answer checks.

State. The price before the two discounts was $95.99.

58. 30

59. *Familiarize*. Let $n =$ the number of CD's purchased. Assume that two or more CD's were purchased. Then the first CD costs $8.49 and the total cost of the remaining $n-1$ CD's is $3.99(n-1)$. The shipping and handling costs are $2.47 for the first CD, $2.28 for the second, and a total of $1.99(n-2)$ for the remaining $n-2$ CD's. Then the total cost of the shipment is $\$8.49 + \$3.99(n-1) + \$2.47 + \$2.28 + \$1.99(n-2)$.

Translate.

$$\underbrace{\text{Total cost of shipment}} \quad \text{was } \$65.07.$$

$$8.49 + 3.99(n-1) + 2.47 + 2.28 + 1.99(n-2) = 65.07$$

Carry out. We solve the equation.

$$8.49 + 3.99(n-1) + 2.47 + 2.28 + 1.99(n-2) = 65.07$$

$$8.49 + 3.99n - 3.99 + 2.47 + 2.28 + 1.99n - 3.98 = 65.07$$

$$5.27 + 5.98n = 65.07$$
$$5.98n = 59.80$$
$$n = 10$$

Check. If 10 CD's are purchased, the total cost of the CD's is $\$8.49 + \$3.99(9) = \$44.40$. The total shipping and handling costs are $\$2.47 + \$2.28 + \$1.99(8) = \$20.67$. Then the total cost of the order is $\$44.40 + \$20.67 = \$65.07$.

State. There were 10 CD's in the shipment.

60. 76

61. *Familiarize*. At $0.30 per $\frac{1}{5}$ mile, the mileage charge can also be given as $5(\$0.30)$, or $1.50 per mile. Since it took 20 min to complete what is usually a 10-min drive, the taxi was stopped in traffic for $20 - 10$, or 10, min. Let $d =$ the distance, in miles, that Glenda traveled.

Translate.

$$\underbrace{\text{Initial charge}} \text{ plus } \underbrace{\text{mileage charge}} \text{ plus } \underbrace{\begin{array}{c}\text{charge for being stopped in traffic}\end{array}} \text{ is } \$13.$$

$$2 + 1.5d + 0.2(10) = 13$$

Carry out. We solve the equation.

$$2 + 1.5d + 0.2(10) = 13$$
$$2 + 1.5d + 2 = 13$$
$$1.5d + 4 = 13$$
$$1.5d = 9$$
$$d = 6$$

Check. The mileage charge for traveling 6 mi is $\$1.50(6) = \$9$. The charge for being stopped in traffic is $\$0.20(10) = \$2$. Since $\$2 + \$9 + \$2 = \$13$, the answer checks.

State. Glenda traveled 6 mi.

62. Writing exercise

63. Writing exercise

64. Width: 23.31 cm; length: 27.56 cm

65. *Familiarize*. Let s = the length of the first side, in cm. Then $s + 3.25$ = the length of the second side, and $(s + 3.25) + 4.35$, or $s + 7.6$ = the length of the third side.

Translate.

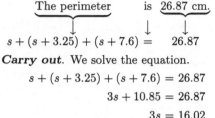

The perimeter $\quad$ is $\quad$ 26.87 cm.

$$s + (s + 3.25) + (s + 7.6) = 26.87$$

Carry out. We solve the equation.

$$s + (s + 3.25) + (s + 7.6) = 26.87$$
$$3s + 10.85 = 26.87$$
$$3s = 16.02$$
$$s = 5.34$$

If $s = 5.34$, then $s + 3.25 = 8.59$, and $s + 7.6 = 12.94$.

Check. Consider sides of 5.34 cm, 8.59 cm, and 12.94 cm. The second side is 3.25 cm longer than the first side, and the third side is 4.35 cm longer than the second side. The sum of the lengths of the sides is 26.87. The answer checks.

State. The lengths of the sides are 5.34 cm, 8.59 cm, and 12.94 cm.

Exercise Set 2.6

1. $x > -2$

a) Since $5 > -2$ is true, 5 is a solution.

b) Since $0 > -2$ is true, 0 is a solution.

c) Since $-1.9 > -2$ is true, -1.9 is a solution.

d) Since $-7.3 > -2$ is false, -7.3 is not a solution.

e) Since $1.6 > -2$ is true, 1.6 is a solution.

2. a) Yes, b) No, c) Yes, d) Yes, e) No

3. $x \geq 6$

a) Since $-6 \geq 6$ is false, -6 is not a solution.

b) Since $0 \geq 6$ is false, 0 is not a solution.

c) Since $6 \geq 6$ is true, 6 is a solution.

d) Since $6.01 \geq 6$ is true, 6.01 is a solution.

e) Since $-3\frac{1}{2} \geq 6$ is false, $-3\frac{1}{2}$ is not a solution.

4. a) Yes, b) Yes, c) Yes, d) No, e) Yes

5. The solutions of $x \leq 7$ are shown by shading the point 7 and all points to the left of 7. The closed circle at 7 indicates that 7 is part of the graph.

$$x \leq 7$$

6.

$$y < 2$$

7. The solutions of $t > -2$ are those numbers greater than -2. They are shown on the graph by shading all points to the right of -2. The open circle at -2 indicates that -2 is not part of the graph.

$$t > -2$$

8.

$$y > 4$$

9. The solutions of $1 \leq m$, or $m \geq 1$, are those numbers greater than or equal to 1. They are shown on the graph by shading the point 1 and all points to the right of 1. The closed circle at 1 indicates that 1 is part of the graph.

$$1 \leq m$$

10.

$$t \geq 0$$

11. In order to be a solution of the inequality $-3 < x \leq 5$, a number must be a solution of both $-3 < x$ and $x \leq 5$. The solution set is graphed as follows:

$$-3 < x \leq 5$$

The open circle at -3 means that -3 is not part of the graph. The closed circle at 5 means that 5 is part of the graph.

12.

$$-5 \leq x < 2$$

13. In order to be a solution of the inequality $0 < x < 3$, a number must be a solution of both $0 < x$ and $x < 3$. The solution set is graphed as follows:

$$0 < x < 3$$

The open circles at 0 and at 3 mean that 0 and 3 are not part of the graph.

14.

$$-5 \leq x \leq 0$$

15. All points to the right of -4 are shaded. The open circle at -4 indicates that -4 is not part of the graph. Using set-builder notation we have $\{x|x > -4\}$.

16. $\{x|x < 3\}$

17. The point 2 and all points to the left of 2 are shaded. Using set-builder notation we have $\{x|x \le 2\}$.

18. $\{x|x \ge -2\}$

19. All points to the left of -1 are shaded. The open circle at -1 indicates that -1 is not part of the graph. Using set-builder notation we have $\{x|x < -1\}$.

20. $\{x|x > 1\}$

21. The point 0 and all points to the right of 0 are shaded. Using set-builder notation we have $\{x|x \ge 0\}$.

22. $\{x|x \le 0\}$

23.
$$y + 2 > 9$$
$$y + 2 - 2 > 9 - 2 \qquad \text{Adding } -2 \text{ to both sides}$$
$$y > 7 \qquad \text{Simplifying}$$

The solution set is $\{y|y > 7\}$. The graph is as follows:

24. $\{y|y > 3\}$

25.
$$x + 8 \le -10$$
$$x + 8 - 8 \le -10 - 8 \qquad \text{Subtracting 8 from both sides}$$
$$x \le -18 \qquad \text{Simplifying}$$

The solution set is $\{x|x \le -18\}$. The graph is as follows:

26. $\{x|x \le -21\}$

27.
$$x - 3 < 7$$
$$x - 3 + 3 < 7 + 3$$
$$x < 10$$

The solution set is $\{x|x < 10\}$. The graph is as follows:

28. $\{x|x < 17\}$

29.
$$5 \le t + 8$$
$$5 - 8 \le t + 8 - 8$$
$$-3 \le t$$

The solution set is $\{t| -3 \le t\}$, or $\{t|t \ge -3\}$. The graph is as follows:

30. $\{t|t \ge -5\}$

31.
$$y - 7 > -12$$
$$y - 7 + 7 > -12 + 7$$
$$y > -5$$

The solution set is $\{y|y > -5\}$. The graph is as follows:

32. $\{y|y > -6\}$

33.
$$2x + 4 \le x + 9$$
$$2x + 4 - 4 \le x + 9 - 4 \qquad \text{Adding } -4$$
$$2x \le x + 5 \qquad \text{Simplifying}$$
$$2x - x \le x + 5 - x \qquad \text{Adding } -x$$
$$x \le 5 \qquad \text{Simplifying}$$

The solution set is $\{x|x \le 5\}$. The graph is as follows:

34. $\{x|x \le -3\}$

35.
$$5x - 6 \ge 4x - 1$$
$$5x - 6 + 6 \ge 4x - 1 + 6 \qquad \text{Adding 6 to both sides}$$
$$5x \ge 4x + 5$$
$$5x - 4x \ge 4x + 5 - 4x \qquad \text{Adding } -4x \text{ to both sides}$$
$$x \ge 5$$

The solution set is $\{x|x \ge 5\}$.

36. $\{x|x \geq 20\}$

37.
$$y + \frac{1}{3} \leq \frac{5}{6}$$
$$y + \frac{1}{3} - \frac{1}{3} \leq \frac{5}{6} - \frac{1}{3}$$
$$y \leq \frac{5}{6} - \frac{2}{6}$$
$$y \leq \frac{3}{6}$$
$$y \leq \frac{1}{2}$$

The solution set is $\left\{y \middle| y \leq \frac{1}{2}\right\}$.

38. $\left\{x \middle| x \leq \frac{1}{4}\right\}$

39.
$$t - \frac{1}{8} > \frac{1}{2}$$
$$t - \frac{1}{8} + \frac{1}{8} > \frac{1}{2} + \frac{1}{8}$$
$$t > \frac{4}{8} + \frac{1}{8}$$
$$t > \frac{5}{8}$$

The solution set is $\left\{t \middle| t > \frac{5}{8}\right\}$.

40. $\left\{y \middle| y > \frac{7}{12}\right\}$

41.
$$-9x + 17 > 17 - 8x$$
$$-9x + 17 - 17 > 17 - 8x - 17 \quad \text{Adding } -17$$
$$-9x > -8x$$
$$-9x + 9x > -8x + 9x \quad \text{Adding } 9x$$
$$0 > x$$

The solution set is $\{x|x < 0\}$.

42. $\{n|n < 0\}$

43. $-23 < -t$

The inequality states that the opposite of 23 is less than the opposite of t. Thus, t must be less than 23, so the solution set is $\{t|t < 23\}$. To solve this inequality using the addition principle, we would proceed as follows:
$$-23 < -t$$
$$t - 23 < 0 \quad \text{Adding } t \text{ to both sides}$$
$$t < 23 \quad \text{Adding 23 to both sides}$$
The solution set is $\{t|t < 23\}$.

44. $\{x|x < -19\}$

45.
$$5x < 35$$
$$\frac{1}{5} \cdot 5x < \frac{1}{5} \cdot 35 \quad \text{Multiplying by } \frac{1}{5}$$
$$x < 7$$

The solution set is $\{x|x < 7\}$. The graph is as follows:

46. $\{x|x \geq 4\}$

47.
$$9y \leq 81$$
$$\frac{1}{9} \cdot 9y \leq \frac{1}{9} \cdot 81 \quad \text{Multiplying by } \frac{1}{9}$$
$$y \leq 9$$

The solution set is $\{y|y \leq 9\}$. The graph is as follows:

48. $\{t|t < 35\}$

49.
$$-7x < 13$$
$$-\frac{1}{7} \cdot (-7x) > -\frac{1}{7} \cdot 13 \quad \text{Multiplying by } -\frac{1}{7}$$
$$\quad\quad\quad\text{The symbol has to be reversed.}$$
$$x > -\frac{13}{7} \quad \text{Simplifying}$$

The solution set is $\left\{x \middle| x > -\frac{13}{7}\right\}$.

50. $\left\{y \middle| y < \frac{17}{8}\right\}$

51.
$$-24 > 8t$$
$$-3 > t$$
The solution set is $\{t|t < -3\}$.

52. $\{x|x > 4\}$

53. $7y \geq -2$

$\dfrac{1}{7} \cdot 7y \geq \dfrac{1}{7}(-2)$ Multiplying by $\dfrac{1}{7}$

$y \geq -\dfrac{2}{7}$

The solution set is $\left\{ y \middle| y \geq -\dfrac{2}{7} \right\}$.

54. $\left\{ x \middle| x > -\dfrac{3}{5} \right\}$

55. $-2y \leq \dfrac{1}{5}$

$-\dfrac{1}{2} \cdot (-2y) \geq -\dfrac{1}{2} \cdot \dfrac{1}{5}$

$\underline{\qquad\qquad}$ The symbol has to be reversed.

$y \geq -\dfrac{1}{10}$

The solution set is $\left\{ y \middle| y \geq -\dfrac{1}{10} \right\}$.

56. $\left\{ x \middle| x \leq -\dfrac{1}{10} \right\}$

57. $-\dfrac{8}{5} > -2x$

$-\dfrac{1}{2} \cdot \left(-\dfrac{8}{5} \right) < -\dfrac{1}{2} \cdot (-2x)$

$\dfrac{8}{10} < x$

$\dfrac{4}{5} < x, \text{ or } x > \dfrac{4}{5}$

The solution set is $\left\{ x \middle| \dfrac{4}{5} < x \right\}$, or $\left\{ x \middle| x > \dfrac{4}{5} \right\}$.

58. $\left\{ y \middle| y < \dfrac{1}{16} \right\}$

59. $7 + 3x < 34$

$7 + 3x - 7 < 34 - 7$ Adding -7 to both sides

$3x < 27$ Simplifying

$x < 9$ Multiplying both sides

        by $\dfrac{1}{3}$

The solution set is $\{x|x < 9\}$.

60. $\{y|y < 8\}$

61. $6 + 5y \geq 26$

$6 + 5y - 6 \geq 26 - 6$ Adding -6

$5y \geq 20$

$y \geq 4$ Multiplying by $\dfrac{1}{5}$

The solution set is $\{y|y \geq 4\}$.

62. $\{x|x \geq 8\}$

63. $4t - 5 \leq 23$

$4t - 5 + 5 \leq 23 + 5$ Adding 5 to both sides

$4t \leq 28$

$\dfrac{1}{4} \cdot 4t \leq \dfrac{1}{4} \cdot 28$ Multiplying both sides

        by $\dfrac{1}{4}$

$x \leq 7$

The solution set is $\{x|x \leq 7\}$.

64. $\{y|y \leq 6\}$

65. $13x - 7 < -46$

$13x - 7 + 7 < -46 + 7$

$13x < -39$

$\dfrac{1}{13} \cdot 13x < \dfrac{1}{13} \cdot (-39)$

$x < -3$

The solution set is $\{x|x < -3\}$.

66. $\{y|y < -6\}$

67. $16 < 4 - 3y$

$16 - 4 < 4 - 3y - 4$ Adding -4 to both sides

$12 < -3y$

$-\dfrac{1}{3} \cdot 12 > -\dfrac{1}{3} \cdot (-3y)$ Multiplying by $-\dfrac{1}{3}$

$\underline{\qquad}$ The symbol has to be reversed.

$-4 > y$

The solution set is $\{y| -4 > y\}$, or $\{y|y < -4\}$.

68. $\{x| -2 > x\}$, or $\{x|x < -2\}$

69. $39 > 3 - 9x$

$39 - 3 > 3 - 9x - 3$ Adding -3

$36 > -9x$

$-\dfrac{1}{9} \cdot 36 < -\dfrac{1}{9} \cdot (-9x)$ Multiplying by $-\dfrac{1}{9}$

$\underline{\qquad}$ The symbol has to be reversed.

$-4 < x$

The solution set is $\{x| -4 < x\}$, or $\{x|x > -4\}$.

70. $\{y| -5 < y\}$, or $\{y|y > -5\}$

71.
$$5 - 6y > 25$$
$$-5 + 5 - 6y > -5 + 25$$
$$-6y > 20$$
$$-\frac{1}{6} \cdot (-6y) < -\frac{1}{6} \cdot 20$$

The symbol has to be reversed.

$$y < -\frac{20}{6}$$
$$y < -\frac{10}{3}$$

The solution set is $\left\{ y \middle| y < -\frac{10}{3} \right\}$.

72. $\{y | y < -3\}$

73.
$$-3 < 8x + 7 - 7x$$
$$-3 < x + 7 \qquad \text{Collecting like terms}$$
$$-3 - 7 < x + 7 - 7$$
$$-10 < x$$

The solution set is $\{x| -10 < x\}$, or $\{x | x > -10\}$.

74. $\{x| -13 < x\}$, or $\{x | x > -13\}$

75.
$$6 - 4y > 4 - 3y$$
$$6 - 4y + 4y > 4 - 3y + 4y \qquad \text{Adding } 4y$$
$$6 > 4 + y$$
$$-4 + 6 > -4 + 4 + y \qquad \text{Adding } -4$$
$$2 > y, \text{ or } y < 2$$

The solution set is $\{y | 2 > y\}$, or $\{y | y < 2\}$.

76. $\{y | 2 > y\}$, or $\{y | y < 2\}$

77.
$$7 - 9y \le 4 - 8y$$
$$7 - 9y + 9y \le 4 - 8y + 9y$$
$$7 \le 4 + y$$
$$-4 + 7 \le -4 + 4 + y$$
$$3 \le y, \text{ or } y \ge 3$$

The solution set is $\{y | 3 \le y\}$, or $\{y | y \ge 3\}$.

78. $\{y | 2 \le y\}$, or $\{y | y \ge 2\}$

79.
$$33 - 12x < 4x + 97$$
$$33 - 12x - 97 < 4x + 97 - 97$$
$$-64 - 12x < 4x$$
$$-64 - 12x + 12x < 4x + 12x$$
$$-64 < 16x$$
$$-4 < x$$

The solution set is $\{x| -4 < x\}$, or $\{x | x > -4\}$.

80. $\left\{ x \middle| x < \frac{9}{5} \right\}$

81.
$$2.1x + 43.2 > 1.2 - 8.4x$$
$$10(2.1x + 43.2) > 10(1.2 - 8.4x) \quad \text{Multiplying by}$$
$$\text{10 to clear decimals}$$
$$21x + 432 > 12 - 84x$$
$$21x + 84x > 12 - 432 \quad \text{Adding } 84x \text{ and}$$
$$-432$$
$$105x > -420$$
$$x > -4 \qquad \text{Multiplying by } \frac{1}{105}$$

The solution set is $\{x | x > -4\}$.

82. $\left\{ y \middle| y \le \frac{5}{3} \right\}$

83.
$$0.7n - 15 + n \ge 2n - 8 - 0.4n$$
$$1.7n - 15 \ge 1.6n - 8 \qquad \text{Collecting like terms}$$
$$10(1.7n - 15) \ge 10(1.6n - 8) \qquad \text{Multiplying by 10}$$
$$17n - 150 \ge 16n - 80$$
$$17n - 16n \ge -80 + 150 \quad \text{Adding } -16n \text{ and}$$
$$150$$
$$n \ge 70$$

The solution set is $\{n | n \ge 70\}$

84. $\{t | t > 1\}$

85.
$$\frac{x}{3} - 4 \le 1$$
$$3\left(\frac{x}{3} - 4\right) \le 3 \cdot 1 \qquad \text{Multiplying by 3 to}$$
$$\text{to clear the fraction}$$
$$x - 12 \le 3 \qquad \text{Simplifying}$$
$$x \le 15 \qquad \text{Adding 12}$$

The solution set is $\{x | x \le 15\}$.

86. $\{x | x > 2\}$

87.
$$3 < 5 - \frac{t}{7}$$
$$-2 < -\frac{t}{7}$$
$$-7(-2) > -7\left(-\frac{t}{7}\right)$$
$$14 > t$$

The solution set is $\{t | t < 14\}$.

88. $\{x | x > 35\}$

89.
$$4(2y - 3) < 36$$
$$8y - 12 < 36 \qquad \text{Removing parentheses}$$
$$8y < 48 \qquad \text{Adding 12}$$
$$y < 6 \qquad \text{Multiplying by } \frac{1}{8}$$

The solution set is $\{y | y < 6\}$.

90. $\{y | y > 5\}$

91. $3(t-2) \geq 9(t+2)$
$3t - 6 \geq 9t + 18$
$3t - 9t > 18 + 6$
$-6t \geq 24$
$t \leq -4$ Multiplying by $-\dfrac{1}{6}$ and reversing the symbol

The solution set is $\{t|t \leq -4\}$.

92. $\left\{ t \middle| t < -\dfrac{5}{3} \right\}$

93. $3(r-6) + 2 < 4(r+2) - 21$
$3r - 18 + 2 < 4r + 8 - 21$
$3r - 16 < 4r - 13$
$-16 + 13 < 4r - 3r$
$-3 < r$, or $r > -3$

The solution set is $\{r|r > -3\}$.

94. $\{t|t > -12\}$

95. $\dfrac{2}{3}(2x-1) \geq 10$

$\dfrac{3}{2} \cdot \dfrac{2}{3}(2x-1) \geq \dfrac{3}{2} \cdot 10$ Multiplying by $\dfrac{3}{2}$

$2x - 1 \geq 15$
$2x \geq 16$
$x \geq 8$

The solution set is $\{x|x \geq 8\}$.

96. $\{x|x \leq 7\}$

97. $\dfrac{3}{4}\left(3x - \dfrac{1}{2}\right) - \dfrac{2}{3} < \dfrac{1}{3}$

$\dfrac{3}{4}\left(3x - \dfrac{1}{2}\right) < 1$ Adding $\dfrac{2}{3}$

$\dfrac{9}{4}x - \dfrac{3}{8} < 1$ Removing parentheses

$8 \cdot \left(\dfrac{9}{4}x - \dfrac{3}{8}\right) < 8 \cdot 1$ Clearing fractions

$18x - 3 < 8$
$18x < 11$
$x < \dfrac{11}{18}$

The solution set is $\left\{ x \middle| x < \dfrac{11}{18} \right\}$.

98. $\left\{ x \middle| x > -\dfrac{5}{32} \right\}$

99. Writing exercise

100. Writing exercise

101. Let n represent "some number." Then we have $n+3$, or $3 + n$.

102. Let x and y represent the numbers; $2(x + y)$

103. Let x represent "a number." Then we have $2x - 3$.

104. Let y represent "a number;" $2y + 5$, or $5 + 2y$

105. Writing exercise

106. Writing exercise

107. $6[4 - 2(6 + 3t)] > 5[3(7 - t) - 4(8 + 2t)] - 20$
$6[4 - 12 - 6t] > 5[21 - 3t - 32 - 8t] - 20$
$6[-8 - 6t] > 5[-11 - 11t] - 20$
$-48 - 36t > -55 - 55t - 20$
$-48 - 36t > -75 - 55t$
$-36t + 55t > -75 + 48$
$19t > -27$
$t > -\dfrac{27}{19}$

The solution set is $\left\{ t \middle| t > -\dfrac{27}{19} \right\}$.

108. $\left\{ x \middle| x \leq \dfrac{5}{6} \right\}$

109. $-(x+5) \geq 4a - 5$
$-x - 5 \geq 4a - 5$
$-x \geq 4a - 5 + 5$
$-x \geq 4a$
$-1(-x) \leq -1 \cdot 4a$
$x \leq -4a$

The solution set is $\{x|x \leq -4a\}$.

110. $\{x|x > 7\}$

111. $y < ax + b$ Assume $a > 0$.
$y - b < ax$
$\dfrac{y - b}{a} < x$ Since $a > 0$, the inequality symbol stays the same.

The solution set is $\left\{ x \middle| x > \dfrac{y - b}{a} \right\}$.

112. $\left\{ x \middle| x < \dfrac{y - b}{a} \right\}$

113. $|x| < 3$

a) Since $|3.2| = 3.2$, and $3.2 < 3$ is false, 3.2 is not a solution.

b) Since $|-2| = 2$ and $2 < 3$ is true, -2 is a solution.

c) Since $|-3| = 3$ and $3 < 3$ is false, -3 is not a solution.

d) Since $|-2.9| = 2.9$ and $2.9 < 3$ is true, -2.9 is a solution.

e) Since $|3| = 3$ and $3 < 3$ is false, 3 is not a solution.

f) Since $|1.7| = 1.7$ and $1.7 < 3$ is true, 1.7 is a solution.

114.

115. $|x| > -3$

Since absolute value is always nonnegative, the absolute value of any real number will be greater than -3. Thus, the solution set is $\{x | x$ is a real number$\}$, or $(-\infty, \infty)$.

116. $\emptyset$

Exercise Set 2.7

1. Let n represent the number. Then we have
$$n \geq 7.$$

2. Let n represent the number; $n \geq 5$.

3. Let b represent the weight of the baby, in kilograms. Then we have
$$b > 2.$$

4. Let p represent the number of people who attended the concert; $75 < p < 100$.

5. Let s represent the average speed, in mph. Then we have
$$90 < s < 110.$$

6. Let n represent the number of people who attended the Million Man March; $n \geq 400,000$.

7. Let a represent the number of people who attended the Million Man March. Then we have
$$a \leq 1,200,000.$$

8. Let a represent the amount of acid, in liters; $a \leq 40$.

9. Let c represent the cost, per gallon, of gasoline. Then we have
$$c \geq \$1.50.$$

10. Let t represent the temperature; $t \leq -2$.

11. *Familiarize.* Let c = the number of copies Myra has made. The total cost of the copies is the setup fee of \$5 plus \$4 times the number of copies, or $\$4 \cdot c$.

Translate.

Setup fee	plus	copying cost	cannot exceed	\$65.
↓	↓	↓	↓	↓
5	+	$4c$	$\leq$	65

Carry out. We solve the inequality.
$$5 + 4c \leq 65$$
$$4c \leq 60$$
$$c \leq 15$$

Check. As a partial check, we show that Myra can have 15 copies made and not exceed her \$65 budget.
$$\$5 + \$4 \cdot 15 = 5 + 60 = \$65$$

State. Myra can have 15 or fewer copies made and stay within her budget.

12. 25 people

13. *Familiarize.* Let m represent the number of miles per day. Then the cost per day for those miles is $\$0.46m$. The total cost is the daily rate plus the daily mileage cost. The total cost cannot exceed \$200. In other words the total cost must be less than or equal to \$200, the daily budget.

Translate.

Daily rate	+	Mileage cost	$\leq$	Budget.
↓	↓	↓	↓	↓
42.95	+	$0.46m$	$\leq$	200

Carry out.
$$42.95 + 0.46m \leq 200$$
$$4295 + 46m \leq 20,000 \quad \text{Clearing decimals}$$
$$46m \leq 15,705$$
$$m \leq \frac{15,705}{46}$$
$$m \leq 341.4 \quad \text{Rounding to the nearest tenth}$$

Check. We can check to see if the solution set seems reasonable.

When $m = 342$, the total cost is
$$42.95 + 0.46(342), \text{ or } \$200.27.$$

When $m = 341.4$, the total cost is
$$42.95 + 0.46(341.4), \text{ or } \$199.99.$$

When $m = 341$, the total cost is
$$42.95 + 0.46(341), \text{ or } \$199.81.$$

From these calculations it would appear that $m \leq 341.4$ is the correct solution.

State. To stay within the budget, the number of miles the Letsons drive must not exceed 341.4.

14. 5 minutes or more

15. *Familiarize*. Let t = the number of hours the car is parked. Then $2t$ = the number of half-hours it is parked. The total parking cost is the initial $0.45 charge plus $0.25 per half hour, or $0.25 \cdot 2t$.

Translate.

Initial charge	plus	charge for time parked	is at least	$2.20.
↓	↓	↓	↓	↓
0.45	+	$0.25 \cdot 2t$	≥	2.20

Carry out. We solve the inequality.

$$0.45 + 0.25 \cdot 2t \geq 2.20$$
$$0.45 + 0.5t \geq 2.2$$
$$0.5t \geq 1.75$$
$$t \geq 3.5$$

Check. As a partial check, we can show that the parking charge for 3.5 hr is $2.20. Note that in 3.5 hr there are 2(3.5), or 7, half-hours.

$$\$0.45 + \$0.25(7) = \$0.45 + \$1.75 = \$2.20.$$

State. Laura's car is generally parked for 3.5 hr or more.

16. More than 2.5 hr

17. *Familiarize*. Let c = the number of courses for which Angelica registers. Her total tuition is the $35 registration fee plus $375 times the number of courses for which she registers, or $375 \cdot c$.

Translate.

Registration fee	plus	fee for courses	cannot exceed	$1000.
↓	↓	↓	↓	↓
35	+	$375 \cdot c$	≤	1000

Carry out. We solve the inequality.

$$35 + 375c \leq 1000$$
$$375c \leq 965$$
$$c \leq 2.57\overline{3}$$

Check. Although the solution set of the inequality is all numbers less than or equal to $2.57\overline{3}$, since c represents the number of courses for which Angelica registers, we round down to 2. If she registers for 2 courses, her tuition is $35 + $375 \cdot 2$, or $785 which does not exceed $1000. If she registers for 3 courses, her tuition is $35 + $375 \cdot 3$, or $1160 which exceeds $1000.

State. Angelica can register for at most 2 courses.

18. Mileages less than or equal to 525.8 mi

19. *Familiarize*. The average of the four scores is their sum divided by the number of tests, 4. We let s represent Nadia's score on the last test.

Translate. The average of the four scores is given by

$$\frac{82 + 76 + 78 + s}{4}.$$

Since this average must be at least 80, this means that it must be greater than or equal to 80. Thus, we can translate the problem to the inequality

$$\frac{82 + 76 + 78 + s}{4} \geq 80.$$

Carry out. We first multiply by 4 to clear the fraction.

$$4\left(\frac{82 + 76 + 78 + s}{4}\right) \geq 4 \cdot 80$$
$$82 + 76 + 78 + s \geq 320$$
$$236 + s \geq 320$$
$$s \geq 84$$

Check. As a partial check, we show that Nadia can get a score of 84 on the fourth test and have an average of at least 80:

$$\frac{82 + 76 + 78 + 84}{4} = \frac{320}{4} = 80.$$

State. Scores of 84 and higher will earn Nadia at least a B.

20. Scores greater than or equal to 97

21. *Familiarize*. Let s = the number of servings of fruits or vegetables Dale eats on Saturday.

Translate.

Average number of fruit or vegetable servings	is at least	5.
↓	↓	↓
$\dfrac{4 + 6 + 7 + 4 + 6 + 4 + s}{7}$	≥	5

Carry out. We first multiply by 7 to clear the fraction.

$$7\left(\frac{4 + 6 + 7 + 4 + 6 + 4 + s}{7}\right) \geq 7 \cdot 5$$
$$4 + 6 + 7 + 4 + 6 + 4 + s \geq 35$$
$$31 + s \geq 35$$
$$s \geq 4$$

Check. As a partial check, we show that Dale can eat 4 servings of fruits or vegetables on Saturday and average at least 5 servings per day for the week:

$$\frac{4 + 6 + 7 + 4 + 6 + 4 + 4}{7} = \frac{35}{7} = 5.$$

State. Dale should eat at least 4 servings of fruits or vegetables on Saturday.

22. 8 credits or more

23. *Familiarize*. Let m represent the number of minutes Monroe practices on the seventh day.

Translate.

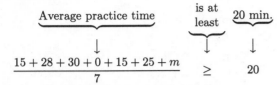

$$\frac{15 + 28 + 30 + 0 + 15 + 25 + m}{7} \geq 20$$

Carry out. We solve the inequality.

$$\frac{15 + 28 + 30 + 0 + 15 + 25 + m}{7} \geq 20$$
$$7\left(\frac{15 + 28 + 30 + 0 + 15 + 25 + m}{7}\right) \geq 7 \cdot 20$$
$$15 + 28 + 30 + 0 + 15 + 25 + m \geq 140$$
$$113 + m \geq 140$$
$$m \geq 27$$

Check. As a partial check, we show that if Monroe practices 27 min on the seventh day he meets expectations.

$$\frac{15 + 28 + 30 + 0 + 15 + 25 + 27}{7} = 20$$

State. Monroe must practice 27 min or more on the seventh day in order to meet expectations.

24. 21 calls or more

25. *Familiarize*. We first make a drawing. We let l represent the length, in feet.

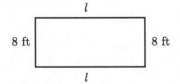

The perimeter is $P = 2l + 2w$, or $2l + 2 \cdot 8$, or $2l + 16$.

Translate. We translate to 2 inequalities.

$$2l + 16 \geq 200$$

The perimeter / is at least / 200 ft.

$$2l + 16 \leq 200$$

The perimeter / is at most / 200 ft.

Carry out. We solve each inequality.

$$2l + 16 \geq 200 \qquad 2l + 16 \leq 200$$
$$2l \geq 184 \qquad 2l \leq 184$$
$$l \geq 92 \qquad l \leq 92$$

Check. We check to see if the solutions seem reasonable.

When $l = 91$ ft, $P = 2 \cdot 91 + 16$, or 198 ft.

When $l = 92$ ft, $P = 2 \cdot 92 + 16$, or 200 ft.

When $l = 93$ ft, $P = 2 \cdot 93 + 16$, or 202 ft.

From these calculations, it appears that the solutions are correct.

State. Lengths greater than or equal to 92 ft will make the perimeter at least 200 ft. Lengths less than or equal to 92 ft will make the perimeter at most 200 ft.

26. Lengths greater than 6 cm

27. *Familiarize*. We first make a drawing. Let $w =$ the width, in feet. Then $2w =$ the length.

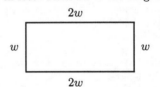

The perimeter is $P = 2l + 2w = 2 \cdot 2w + 2w = 4w + 2w = 6w$.

Translate.

$$6w \leq 70$$

The perimeter / cannot exceed / 70 ft.

Carry out. We solve the inequality.

$$6w \leq 70$$
$$w \leq \frac{35}{3}, \text{ or } 11\frac{2}{3}$$

Check. As a partial check we show that the perimeter is 70 ft when the width is $\frac{35}{3}$ ft and the length is $2 \cdot \frac{35}{3}$, or $\frac{70}{3}$ ft.

$$P = 2 \cdot \frac{70}{3} + 2 \cdot \frac{35}{3} = \frac{140}{3} + \frac{70}{3} = \frac{210}{3} = 70$$

State. Widths less than or equal to $11\frac{2}{3}$ ft will meet the given conditions.

28. George worked more than 12 hr; Joan worked more than 15 hr

29. *Familiarize*. Let $t =$ the number of 15-min units of time for a road call. Rick's Automotive charges $\$50 + \$15 \cdot t$ for a road call, and Twin City Repair charges $\$70 + \$10 \cdot t$.

Translate.

$$50 + 15t < 70 + 10t$$

Rick's charge / is less than / Twin City's charge.

Carry out. We solve the inequality.

$$50 + 15t < 70 + 10t$$
$$15t < 20 + 10t$$
$$5t < 20$$
$$t < 4$$

Check. We check to see if the solution seems reasonable. When $t = 3$, Rick's charges $50 + 15 \cdot 3$, or $95, and Twin City charges $70 + 10 \cdot 3$, or $100. When $t = 4$, Rick's charges $50 + 15 \cdot 4$, or $110, and Twin City charges $70 + 10 \cdot 4$, or $110. When $t = 5$, Rick's charges $50 + 15 \cdot 5$, or $125, and Twin City charges $70 + 10 \cdot 5$, or $120. From these calculations, it appears that the solution is correct.

State. It would be more economical to call Rick's for a service call of less than 4 15-min time units, or of less than 1 hr.

30. At most $49.02

31. *Familiarize*. We first make a drawing. We let l represent the length.

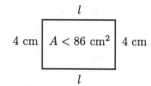

The area is the length times the width, or $4l$.

Translate.

$$\underbrace{\text{Area}}_{4l} \quad \underbrace{\text{is less than}}_{<} \quad \underbrace{86 \text{ cm}^2.}_{86}$$

Carry out.

$$4l < 86$$
$$l < 21.5$$

Check. We check to see if the solution seems reasonable.

When $l = 22$, the area is $22 \cdot 4$, or 88 cm^2.

When $l = 21.5$, the area is $21.5(4)$, or 86 cm^2.

When $l = 21$, the area is $21 \cdot 4$, or 84 cm^2.

From these calculations, it would appear that the solution is correct.

State. The area will be less than 86 cm^2 for lengths less than 21.5 cm.

32. Lengths greater than or equal to 16.5 yd

33. *Familiarize*. Let $v =$ the blue book value of the car. Since the car was repaired, we know that $8500 does not exceed $0.8v$ or, in other words, $0.8v$ is at least $8500.

Translate.

$$\underbrace{\text{80\% of the blue book value}}_{0.8v} \quad \underbrace{\text{is at least}}_{\geq} \quad \underbrace{\$8500.}_{8500}$$

Carry out.

$$0.8v \geq 8500$$
$$v \geq \frac{8500}{0.8}$$
$$v \geq 10,625$$

Check. As a partial check, we show that 80% of $10,625 is at least $8500:

$$0.8(\$10,625) = \$8500$$

State. The blue book value of the car was at least $10,625.

34. More than $16,800

35. *Familiarize*. We will use the formula $F = \frac{9}{5}C + 32$.

Translate.

$$\underbrace{\text{Fahrenheit temperature}}_{F} \quad \underbrace{\text{is above}}_{>} \quad \underbrace{98.6°.}_{98.6}$$

Substituting $\frac{9}{5}C + 32$ for F, we have

$$\frac{9}{5}C + 32 > 98.6.$$

Carry out. We solve the inequality.

$$\frac{9}{5}C + 32 > 98.6$$
$$\frac{9}{5}C > 66.6$$
$$C > \frac{333}{9}$$
$$C > 37$$

Check. We check to see if the solution seems reasonable.

When $C = 36$, $\frac{9}{5} \cdot 36 + 32 = 96.8$.

When $C = 37$, $\frac{9}{5} \cdot 37 + 32 = 98.6$.

When $C = 38$, $\frac{9}{5} \cdot 38 + 32 = 100.4$.

It would appear that the solution is correct, considering that rounding occurred.

State. The human body is feverish for Celsius temperatures greater than $37°$.

36. Temperatures less than 31.3°C

37. *Familiarize*. Let r = the amount of fat in a serving of the regular peanut butter, in grams. If reduced fat peanut butter has at least 25% less fat than regular peanut butter, then it has at most 75% as much fat as the regular peanut butter.

Translate.

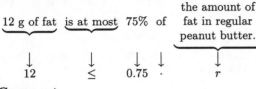

Carry out.

$$12 \leq 0.75r$$
$$16 \leq r$$

Check. As a partial check, we show that 12 g of fat does not exceed 75% of 16 g of fat:

$$0.75(16) = 12$$

State. Regular peanut butter contains at least 16 g of fat per serving.

38. They contain at least $6\frac{2}{3}$ g of fat.

39. *Familiarize*. Let d = the depth of the well, in feet. Then the cost on the pay-as-you-go plan is $500 + $8d$. The cost of the guaranteed-water plan is $4000. We want to find the values of d for which the pay-as-you-go plan costs less than the guaranteed-water plan.

Translate.

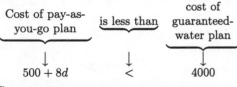

Carry out.

$$500 + 8d < 4000$$
$$8d < 3500$$
$$d < 437.5$$

Check. We check to see that the solution is reasonable.

When $d = 437$, $500 + $8 \cdot 437 = $3996 < 4000

When $d = 437.5$, $500 + $8(437.5) = 4000

When $d = 438$, $500 + $8(438) = $4004 > 4000

From these calculations, it appears that the solution is correct.

State. It would save a customer money to use the pay-as-you-go plan for a well of less than 437.5 ft.

40. 8 mi or more

41. *Familiarize*. $R = -0.012t + 20.8$

In the formula R represents the world record in the 200-m dash and t represents the years since 1920. When $t = 0$ (1920), the record was $-0.012(0) + 20.8$, or 20.8 sec. When $t = 2$ (1922), the record was $-0.012(2) + 20.8$, or 20.776 sec. For what values of t will $-0.012t + 20.8$ be less than 19.8?

Translate. The record is to be less than 19.8. We have the inequality

$$R < 19.8.$$

To find the t values which satisfy this condition we substitute $-0.012t + 20.8$ for R.

$$-0.012t + 20.8 < 19.8$$

Carry out.

$$-0.012t + 20.8 < 19.8$$
$$-0.012t < -1$$
$$t > \frac{-1}{-0.012}$$
$$t > 83.\overline{3}, \text{ or } 83\frac{1}{3}$$

Check. We check to see if the solution set we obtained seems reasonable.

When $t = 83\frac{1}{4}$, $R = -0.012(83.25) + 20.8 = 19.801$.

When $t = 83\frac{1}{3}$, $R = -0.012\left(\frac{250}{3}\right) + 20.8 = 19.8$.

When $t = 83\frac{1}{2}$, $R = -0.012(83.5) + 20.8 = 19.798$.

Since $r = 19.8$ when $t = 83\frac{1}{3}$ and R decreases as t increases, R will be less than 19.8 when t is greater than $83\frac{1}{3}$.

State. The world record will be less than 19.8 seconds when t is greater than $83\frac{1}{3}$ years (more than $83\frac{1}{3}$ years after 1920). This occurs in years after 2003.

42. Years after 2005

43. *Familiarize*. Let w = the number of weeks after July 1. After w weeks the water level has dropped $\frac{2}{3}w$ ft.

Translate.

Original depth	minus	drop in water level	does not exceed	21 ft.
↓	↓	↓	↓	↓
25	−	$\frac{2}{3}w$	≤	21

Carry out. We solve the inequality.

$$25 - \frac{2}{3}w \le 21$$

$$-\frac{2}{3}w \le -4$$

$$w \ge -\frac{3}{2}(-4)$$

$$w \ge 6$$

Check. As a partial check we show that the water level is 21 ft 6 weeks after July 1.

$$25 - \frac{2}{3} \cdot 6 = 25 - 4 = 21 \text{ ft}$$

Since the water level continues to drop during the weeks after July 1, the answer seems reasonable.

State. The water level will not exceed 21 ft for dates at least 6 weeks after July 1.

44. When the puppy is more than 18 weeks old

45. *Familiarize.* Let h = the height of the triangle, in ft. Recall that the formula for the area of a triangle with base b and height h is $A = \frac{1}{2}bh$.

Translate.

Area	is at least	3 ft².
↓	↓	↓
$\frac{1}{2}\left(1\frac{1}{2}\right)h$	$\ge$	3

Carry out. We solve the inequality.

$$\frac{1}{2}\left(1\frac{1}{2}\right)h \ge 3$$

$$\frac{1}{2} \cdot \frac{3}{2} \cdot h \ge 3$$

$$\frac{3}{4}h \ge 3$$

$$h \ge \frac{4}{3} \cdot 3$$

$$h \ge 3$$

Check. As a partial check, we show that the area of the triangle is 3 ft² when the height is 4 ft.

$$\frac{1}{2}\left(1\frac{1}{2}\right)(4) = \frac{1}{2} \cdot \frac{3}{2} \cdot \frac{4}{1} = 3$$

State. The height should be at least 4 ft.

46. Heights less than or equal to 3 ft

47. *Familiarize.* We will use the equation $y = 0.027x + 0.19$.

Translate.

The cost	is at most	$6.
↓	↓	↓
$0.027x + 0.19$	$\le$	6

Carry out. We solve the inequality.

$$0.027x + 0.19 \le 6$$

$$0.027x \le 5.81$$

$$x \le 215.2 \quad \text{Rounding to the nearest tenth}$$

Check. As a partial check, we show that the cost for driving 215.2 mi is $6.

$$0.027(215.2) + 0.19 \approx \$6$$

State. The cost will be at most $6 for mileages less than or equal to 215.2 mi.

48. 2001 and beyond

49. Writing exercise

50. Writing exercise

51. $\dfrac{9-5}{6-4} = \dfrac{4}{2} = 2$

52. $\dfrac{1}{2}$

53. $\dfrac{8-(-2)}{1-4} = \dfrac{10}{-3}$, or $-\dfrac{10}{3}$

54. $-\dfrac{1}{5}$

55. Writing exercise

56. Writing exercise

57. *Familiarize.* Let h = the number of hours the car has been parked. Then $h - 1$ = the number of hours after the first hour.

Translate.

Charge for first hour	plus	charge for additional hours	exceeds $16.50.	
↓	↓	↓	↓	↓
4.00	+	2.50(h − 1)	>	16.50

Carry out. We solve the inequality.

$$4.00 + 2.50(h-1) > 16.50$$

$$40 + 25(h-1) > 165 \quad \text{Multiplying by 10 to clear decimals}$$

$$40 + 25h - 25 > 165$$

$$25h + 15 > 165$$

$$25h > 150$$

$$h > 6$$

Check. We check to see if this solution seems reasonable.

When $h = 5$, $4.00 + 2.50(5 - 1) = 14.00$.

When $h = 6$, $4.00 + 2.50(6 - 1) = 16.50$.

When $h = 7$, $4.00 + 2.50(7 - 1) = 19.00$.

It appears that the solution is correct.

State. The charge exceeds $16.50 when the car has been parked for more than 6 hr.

58. Temperatures between $-15°C$ and $-9\frac{4}{9}°C$

59. Since $8^2 = 64$, the length of a side must be less than or equal to 8 cm (and greater than 0 cm, of course). We can also use the five-step problem-solving procedure.

Familiarize. Let s represent the length of a side of the square. The area s is the square of the length of a side, or s^2.

Translate.

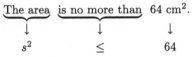

The area is no more than 64 cm².

$$s^2 \quad \leq \quad 64$$

Carry out.

$$s^2 \leq 64$$
$$s^2 - 64 \leq 0$$
$$(s + 8)(s - 8) \leq 0$$

We know that $(s+8)(s-8) = 0$ for $s = -8$ or $s = 8$. Now $(s + 8)(s - 8) < 0$ when the two factors have opposite signs. That is:

$s+8>0$ and $s-8<0$ or $s+8<0$ and $s-8>0$

$s>-8$ and $s<8$ or $s<-8$ and $s>8$

This can be expressed   This is not possible.

as $-8 < s < 8$.

Then $(s + 8)(s - 8) \leq 0$ for $-8 \leq s \leq 8$.

Check. Since the length of a side cannot be negative we only consider positive values of s, or $0 < s \leq 8$. We check to see if this solution seems reasonable.

When $s = 7$, the area is 7^2, or 49 cm².

When $s = 8$, the area is 8^2, or 64 cm².

When $s = 9$, the area is 9^2, or 81 cm².

From these calculations, it appears that the solution is correct.

State. Sides of length 8 cm or less will allow an area of no more than 64 cm². (Of course, the length of a side must be greater than 0 also.)

60. 47 and 49

61. Familiarize. Let f = the fat content of a serving of regular tortilla chips, in grams. A product that contains 60% less fat than another product has 40%

of the fat content of that product. If Reduced Fat Tortilla Pops cannot be labeled lowfat, then they contain at least 3 g of fat.

Translate.

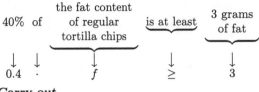

40% of	the fat content of regular tortilla chips	is at least	3 grams of fat
$0.4 \cdot$	f	$\geq$	3

Carry out.

$$0.4f \geq 3$$
$$f \geq 7.5$$

Check. As a partial check, we show that 40% of 7.5 g is not less than 3 g.

$$0.4(7.5) = 3$$

State. A serving of regular tortilla chips contains at least 7.5 g of fat.

62. Between 5 and 9 hr

63. Familiarize. Let p = the price of Neoma's tenth book. If the average price of each of the first 9 books is $12, then the total price of the 9 books is $9 \cdot \$12$, or $108. The average price of the first 10 books will be $\frac{\$108 + p}{10}$.

Translate.

The average price of 10 books is at least $15.

$$\frac{108 + p}{10} \quad \geq \quad 15$$

Carry out. We solve the inequality.

$$\frac{108 + p}{10} \geq 15$$
$$108 + p \geq 150$$
$$p \geq 42$$

Check. As a partial check, we show that the average price of the 10 books is $15 when the price of the tenth book is $42.

$$\frac{\$108 + \$42}{10} = \frac{\$150}{10} = \$15$$

State. Neoma's tenth book should cost at least $42 if she wants to select a $15 book for her free book.

64. Writing exercise

65. Let p = the total purchases for the year. Solving $10\%p > 25$, we get $p > 250$. Thus, when a customer's purchases are more than $250 for the year, the customer saves money by purchasing a card.

Chapter 3

Introduction to Graphing

Exercise Set 3.1

1. We go to the top of the bar that is above the body weight 100 lb. Then we move horizontally from the top of the bar to the vertical scale listing numbers of drinks. It appears that consuming approximately 2 drinks in one hour will give a 100 lb person a blood-alcohol level of 0.08%.

2. Approximately 3.5 drinks

3. From 4 on the vertical scale we move horizontally until we reach a bar whose top is above the horizontal line on which we are moving. The first such bar corresponds to a body weight of 220 lb. This means that for body weights represented by bars to the left of this one, consuming 4 drinks will yield a blood-alcohol level of 0.08%. The bar immediately to the left of the 220-pound bar represents 200 pounds. Thus, we can conclude an individual weighs more than 220 lb if 4 drinks are consumed in one hour without reaching a blood-alcohol level of 0.08%.

4. The individual weighs more than 240 lb.

5. *Familiarize*. Since there are 272 million Americans and about one-third of them live in the South, there are about $\frac{1}{3} \cdot 272$, or $\frac{272}{3}$ million Southerners. The pie chart indicates that 3% of Americans choose brown as their favorite color. Let $b =$ the number of Southerners, in millions, who choose brown as their favorite color.

   Translate. We reword and translate the problem.

   What is 3% of $\frac{272}{3}$ million?
   $\downarrow$ $\downarrow$ $\downarrow$ $\downarrow$ $\qquad$ $\downarrow$
   b $=$ 3% $\cdot$ $\qquad$ $\frac{272}{3}$

   Carry out. We solve the equation.

   $$b = 0.03 \cdot \frac{272}{3} = 2.72$$

   Check. We repeat the calculations. The answer checks.

   State. About 2.72 million, or 2,720,000 Southerners choose brown as their favorite color.

6. About 5,440,000

7. *Familiarize*. Since there are 272 million Americans and about one-eighth are senior citizens, there are about $\frac{1}{8} \cdot 272$ million, or 34 million senior citizens. The pie chart indicates that 4% of Americans choose black as their favorite color. Let $b =$ the number of senior citizens who choose black as their favorite color.

   Translate. We reword and translate the problem.

   What is 4% of 34 million?
   $\downarrow$ $\downarrow$ $\downarrow$ $\downarrow$ $\qquad$ $\downarrow$
   b $=$ 4% $\cdot$ $\quad$ 34, 000, 000

   Carry out. We solve the equation.

   $$b = 0.04 \cdot 34, 000, 000 = 1, 360, 000$$

   Check. We repeat the calculations. The answer checks.

   State. About 1,360,000 senior citizens choose black as their favorite color.

8. About 1,360,000

9. *Familiarize*. From the pie chart we see that 9.9% of solid waste is plastic. We let $x =$ the amount of plastic, in millions of tons, in the waste generated in 1998.

   Translate. We reword the problem.

   What is 9.9% of 210?
   $\downarrow$ $\downarrow$ $\downarrow$ $\downarrow$ $\downarrow$
   x $=$ 9.9% $\cdot$ $\;$ 210

   Carry out.

   $$x = 0.099 \cdot 210 = 20.79$$

   Check. We can repeat the calculation. The result checks.

   State. In 1998, about 20.79 million tons of waste was plastic.

10. About 1.7 lb

11. *Familiarize*. From the pie chart we see that 5.5% of solid waste is glass. From Exercise 9 we know that Americans generated 210 million tons of waste in 1998. Then the amount of this that is glass is

    0.055(210), or 11.55 million tons

    We let $x =$ the amount of glass, in millions of tons, that Americans recycled in 1998.

Translate. We reword the problem.

What is 26% of 11.55 million tons?

$$x = 26\% \cdot 11.55$$

Carry out.

$$x = 0.26(11.55) \approx 3.0$$

Check. We go over the calculations again. The result checks.

State. Americans recycled about 3.0 million tons of glass in 1998.

12. About 0.02 lb

13. Locate 1997 on the horizontal scale and then move up to the line that represents CD sales. Now move to the vertical axis and read that about 70% of recordings sold in 1997 were CDs.

14. About 20%

15. Locate 25% on the vertical scale, midway between 20% and 30%. Move right to the line representing cassette sales and then move down to the horizontal axis. We see that in 1995 approximately 25% of the recordings sold were cassettes.

16. 1998

17. The line slants upward most steeply from 1994 to 1995, so sales of CDs increased the most from 1994 to 1995.

18. 1996 to 1997

19. Starting at the origin:

(1,2) is 1 unit right and 2 units up;

(−2,3) is 2 units left and 3 units up;

(4, −1) is 4 units right and 1 unit down;

(−5, −3) is 5 units left and 3 units down;

(4,0) is 4 units right and 0 units up or down;

(0, −2) is 0 units right or left and 2 units down.

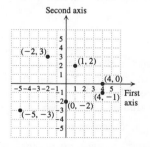

20.

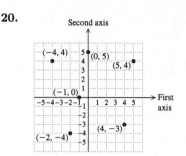

21. Starting at the origin:

(4,4) is 4 units right and 4 units up;

(−2, 4) is 2 units left and 4 units up;

(5, −3) is 5 units right and 3 units down;

(−5, −5) is 5 units left and 5 units down;

(0,4) is 0 units right or left and 4 units up;

(0, −4) is 0 units right or left and 4 units down;

(3,0) is 3 units right and 0 units up or down;

(−4, 0) is 4 units left and 0 units up or down.

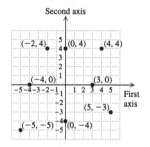

22.

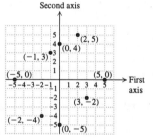

23.

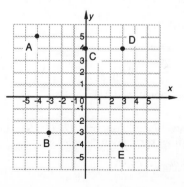

Point A is 4 units left and 5 units up. The coordinates of A are $(-4, 5)$.

Point B is 3 units left and 3 units down. The coordinates of B are $(-3, -3)$.

Point C is 0 units right or left and 4 units up. The coordinates of C are $(0,4)$.

Point D is 3 units right and 4 units up. The coordinates of D are $(3,4)$.

Point E is 3 units right and 4 units down. The coordinates of E are $(3, -4)$.

24. $A: (3,3)$, $B: (0,-4)$, $C: (-5,0)$, $D: (-1,-1)$, $E: (2,0)$

25.

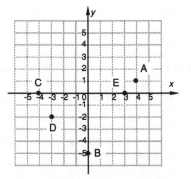

Point A is 4 units right and 1 unit up. The coordinates of A are $(4,1)$.

Point B is 0 units right or left and 5 units down. The coordinates of B are $(0, -5)$.

Point C is 4 units left and 0 units up or down. The coordinates of C are $(-4,0)$.

Point D is 3 units left and 2 units down. The coordinates of D are $(-3, -2)$.

Point E is 3 units right and 0 units up or down. The coordinates of E are $(3,0)$.

26. $A: (-5,1)$, $B: (0,5)$, $C: (5,3)$, $D: (0,-1)$, $E: (2,-4)$

27. Since the first coordinate is positive and the second coordinate negative, the point $(7, -2)$ is located in quadrant IV.

28. III

29. Since both coordinates are negative, the point $(-4, -3)$ is in quadrant III.

30. IV

31. Since both coordinates are positive, the point $(2,1)$ is in quadrant I.

32. II

33. Since the first coordinate is negative and the second coordinate is positive, the point $(-4.9, 8.3)$ is in quadrant II.

34. I

35. First coordinates are positive in the quadrants that lie to the right of the origin, or in quadrants I and IV.

36. III and IV

37. Points for which both coordinates are positive lie in quadrant I, and points for which both coordinates are negative life in quadrant III. Thus, both coordinates have the same sign in quadrants I and III.

38. II and IV

39. a) Draw a line segment connecting the points $(1991, 62.1)$ and $(1999, 49.6)$. Then locate 1995 on the horizontal axis and move up to the line. Now move horizontally to the vertical axis and estimate that the birth rate among teenagers in 1995 was approximately 56 births per 1000 females.

b) Extend the line segment on the graph until it is above 2003 on the horizontal axis. Then, from 2003, move up to the line and across to the vertical axis. We predict that the birth rate among teenagers in 2003 will be approximately 43 per 1000 females.

40. a) Approximately 23 million participants

b) Approximately 15 million participants

41. Draw a horizontal axis for the year and a vertical axis for the percentage of people age 26 to 34 who smoke. Number the axes with a scale that will permit us to view both the given data and the desired data. Then plot the points $(1985, 45.7)$ and $(1998, 32.5)$ and draw a line segment connecting them.

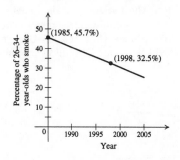

a) Locate 1990 on the horizontal scale and move up to the line. Then move horizontally to the vertical axis and estimate that approximately 40% of people age 26 to 34 smoked in 1990.

b) Extend the line segment on the graph until it is above 2003 on the horizontal axis. Then, from 2003, move up to the line and across to the vertical axis. We predict that in 2003 about 27.5% of people age 26 to 34 will smoke.

42.

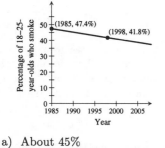

a) About 45%

b) About 40%

43. Draw a horizontal axis for the year and a vertical axis for U.S. college enrollment, in millions. Number the axes with a scale that will permit us to view both the given data and the desired data. Then plot the points $(1990, 60.3)$ and $(2000, 68.3)$ and draw a line segment connecting them.

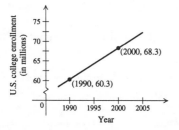

a) Locate 1996 on the horizontal axis and move up to the line. Then move horizontally to the vertical axis and estimate that U.S. college enrollment in 1996 was about 65 million students.

b) Extend the line segment on the graph until it is above 2005 on the horizontal axis. Then, from 2005, move up to the line and across to the vertical axis. We predict that in 2005 U.S. college enrollment will be about 72 million students.

44.

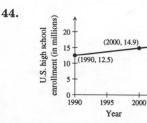

a) About 14 million students

b) About 16 million students

45. Draw a horizontal axis for the year and a vertical axis for the number of U.S. residents over the age of 65, in millions. Number the axes with a scale that will permit us to view both the given data and the desired data. Then plot the points $(1990, 31)$ and $(2000, 34.4)$ and draw a line segment connecting them.

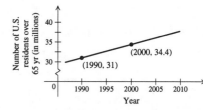

a) Since 1995 is midway between 1990 and 2000, it is reasonable to estimate that in 1995 the number of U.S. residents over the age of 65 will be about midway between 31 million and 34.4 million or approximately 33 million.

We could also use the graph to make this estimate. Locate 1995 on the horizontal axis and move up to the line. Then move horizontally to the vertical axis and estimate that in 1995 the number of U.S. residents over the age of 65 was about 33 million.

b) Extend the line segment on the graph until it is above 2010 on the horizontal axis. Then, from 2010, move up to the line and across to the vertical axis. We predict that in 2010 the number of U.S. residents over the age of 65 will be about 38 million.

46.

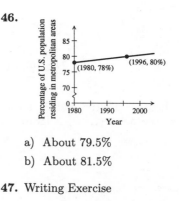

a) About 79.5%

b) About 81.5%

47. Writing Exercise

48. Writing Exercise

49. $4 \cdot 3 - 6 \cdot 5 = 12 - 30 = -18$

50. -31

51. $-\frac{1}{2}(-6) + 3 = 3 + 3 = 6$

52. 1

53.
$$3x - 2y = 6$$
$$-2y = -3x + 6 \quad \text{Adding } -3x \text{ to both sides}$$
$$-\frac{1}{2}(-2y) = -\frac{1}{2}(-3x + 6)$$
$$y = -\frac{1}{2}(-3x) - \frac{1}{2}(6)$$
$$y = \frac{3}{2}x - 3$$

54. $y = \frac{7}{4}x - \frac{7}{2}$

55. Writing Exercise

56. Writing Exercise

57. If the coordinates of a point are reciprocals of each other, they have the same sign. Thus, the point could be in quadrant I or quadrant III.

58. II or IV

59.

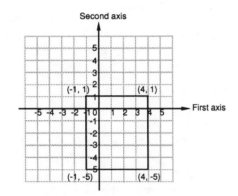

The coordinates of the fourth vertex are $(-1, -5)$.

60. $(5,2)$, $(-7,2)$, or $(3, -8)$

61. Answers may vary.

We select eight points such that the sum of the coordinates for each point is 7.

$$
\begin{array}{ll}
(0,7) & 0 + 7 = 7 \\
(1,6) & 1 + 6 = 7 \\
(2,5) & 2 + 5 = 7 \\
(3,4) & 3 + 4 = 7 \\
(4,3) & 4 + 3 = 7 \\
(5,2) & 5 + 2 = 7 \\
(6,1) & 6 + 1 = 7 \\
(7,0) & 7 + 0 = 7
\end{array}
$$

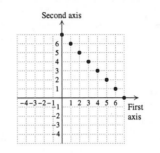

62. Answers may vary.

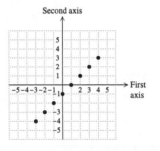

63. Plot the three given points and observe that the coordinates of the fourth vertex are $(5, 3)$

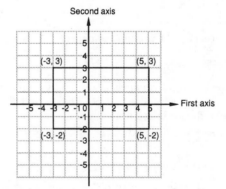

The length of the rectangle is 8 units, and the width is 5 units.

$$P = 2l + 2w$$
$$P = 2 \cdot 8 + 2 \cdot 5 = 16 + 10 = 26 \text{ units}$$

64. $\frac{65}{2}$ sq units

65. Latitude 32.5° North,

Longitude 64.5° West

66. Latitude 27° North,

Longitude 81° West

67. Writing Exercise

68. Writing Exercise

Exercise Set 3.2

1. We substitute 0 for x and 2 for y (alphabetical order of variables).

$$
\begin{array}{c|c}
\multicolumn{2}{c}{y = 7x + 1} \\
\hline
2 ~?~ 7 \cdot 0 + 1 & \\
& 0 + 1 \\
2 & 1 \qquad \text{FALSE}
\end{array}
$$

Since $2 = 1$ is false, the pair $(0, 2)$ is not a solution.

2. Yes

3. We substitute 4 for x and 2 for y.

$$
\begin{array}{c|c}
\multicolumn{2}{c}{3y + 2x = 12} \\
\hline
3 \cdot 2 + 2 \cdot 4 ~?~ 12 & \\
6 + 8 & \\
14 & 12 ~ \text{FALSE}
\end{array}
$$

Since $14 = 12$ is false, the pair $(4, 2)$ is not a solution.

4. No

5. We substitute 2 for a and -1 for b.

$$
\begin{array}{c|c}
\multicolumn{2}{c}{4a - 3b = 11} \\
\hline
4 \cdot 2 - 3(-1) ~?~ 11 & \\
8 + 3 & \\
11 & 11 ~ \text{TRUE}
\end{array}
$$

Since $11 = 11$ is true, the pair $(2, -1)$ is a solution.

6. Yes

7. To show that a pair is a solution, we substitute, replacing x with the first coordinate and y with the second coordinate in each pair.

$$
\begin{array}{c|c} \qquad
\begin{array}{c|c}
\multicolumn{2}{c}{y = x - 2} \\
\hline
1 ~?~ 3 - 2 & \\
1 & 1 \qquad \text{TRUE}
\end{array}
&
\begin{array}{c|c}
\multicolumn{2}{c}{y = x - 2} \\
\hline
-4 ~?~ -2 - 2 & \\
-4 & -4 \qquad \text{TRUE}
\end{array}
\end{array}
$$

In each case the substitution results in a true equation. Thus, $(3, 1)$ and $(-2, -4)$ are both solutions of $y = x - 2$. We graph these points and sketch the line passing through them.

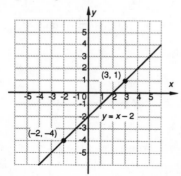

The line appears to pass through $(5, 3)$ also. We check to determine if $(5, 3)$ is a solution of $y = x - 2$.

$$
\begin{array}{c|c}
\multicolumn{2}{c}{y = x - 2} \\
\hline
3 ~?~ 5 - 2 & \\
3 & 3 \qquad \text{TRUE}
\end{array}
$$

Thus, $(5, 3)$ is another solution. There are other correct answers, including $(-3, -5)$, $(-1, -3)$, $(0, -2)$, $(1, -1)$, $(2, 0)$, and $(4, 2)$.

8.
$$
\begin{array}{c|c} \qquad
\begin{array}{c|c}
\multicolumn{2}{c}{y = x + 3} \\
\hline
2 ~?~ -1 + 3 & \\
2 & 2 \qquad \text{TRUE}
\end{array}
&
\begin{array}{c|c}
\multicolumn{2}{c}{y = x + 3} \\
\hline
7 ~?~ 4 + 3 & \\
7 & 7 \qquad \text{TRUE}
\end{array}
\end{array}
$$

$(0, 3)$; answers may vary

9. To show that a pair is a solution, we substitute, replacing x with the first coordinate and y with the second coordinate in each pair.

$$
\begin{array}{c|c} \qquad
\begin{array}{c|c}
\multicolumn{2}{c}{y = \frac{1}{2}x + 3} \\
\hline
5 ~?~ \frac{1}{2} \cdot 4 + 3 & \\
& 2 + 3 \\
5 & 5 \qquad \text{TRUE}
\end{array}
&
\begin{array}{c|c}
\multicolumn{2}{c}{y = \frac{1}{2}x + 3} \\
\hline
2 ~?~ \frac{1}{2}(-2) + 3 & \\
& -1 + 3 \\
2 & 2 \qquad \text{TRUE}
\end{array}
\end{array}
$$

In each case the substitution results in a true equation. Thus, $(4, 5)$ and $(-2, 2)$ are both solutions of $y = \frac{1}{2}x + 3$. We graph these points and sketch the line passing through them.

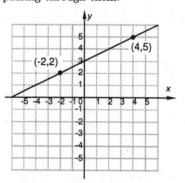

The line appears to pass through $(0, 3)$ also. We check to determine if $(0, 3)$ is a solution of $y = \frac{1}{2}x + 3$.

$$
\begin{array}{c|c}
\multicolumn{2}{c}{y = \frac{1}{2}x + 3} \\
\hline
3 ~?~ \frac{1}{2} \cdot 0 + 3 & \\
3 & 3 \qquad \text{TRUE}
\end{array}
$$

Thus, $(0, 3)$ is another solution. There are other correct answers, including $(-6, 0)$, $(-4, 1)$, $(2, 4)$, and $(6, 6)$.

10.

$$y = \frac{1}{2}x - 1$$
$$\frac{}{2 \ ? \ \frac{1}{2} \cdot 6 - 1}$$
$$3 - 1$$
$$2 \ \big| \ 2 \qquad \text{TRUE}$$

$$y = \frac{1}{2}x - 1$$
$$\frac{}{-1 \ ? \ \frac{1}{2} \cdot 0 - 1}$$
$$-1 \ \big| \ -1 \qquad \text{TRUE}$$

$(2, 0)$; answers may vary

11. To show that a pair is a solution, we substitute, replacing x with the first coordinate and y with the second coordinate in each pair.

$$y + 3x = 7$$
$$\frac{}{1 + 3 \cdot 2 \ ? \ 7}$$
$$1 + 6$$
$$7 \ \big| \ 7 \quad \text{TRUE}$$

$$y + 3x = 7$$
$$\frac{}{-5 + 3 \cdot 4 \ ? \ 7}$$
$$-5 + 12$$
$$7 \ \big| \ 7 \quad \text{TRUE}$$

In each case the substitution results in a true equation. Thus, $(2, 1)$ and $(4, -5)$ are both solutions of $y + 3x = 7$. We graph these points and sketch the line passing through them.

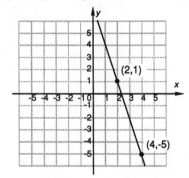

The line appears to pass through $(1, 4)$ also. We check to determine if $(1, 4)$ is a solution of $y + 3x = 7$.

$$y + 3x = 7$$
$$\frac{}{4 + 3 \cdot 1 \ ? \ 7}$$
$$4 + 3$$
$$7 \ \big| \ 7 \quad \text{TRUE}$$

Thus, $(1, 4)$ is another solution. There are other correct answers, including $(3, -2)$.

12.

$$2y + x = 5$$
$$\frac{}{2 \cdot 3 - 1 \ ? \ 5}$$
$$6 - 1$$
$$5 \ \big| \ 5 \quad \text{TRUE}$$

$$2y + x = 5$$
$$\frac{}{2(-1) + 7 \ ? \ 5}$$
$$-2 + 7$$
$$5 \ \big| \ 5 \quad \text{TRUE}$$

$(1, 2)$; answers may vary

13. To show that a pair is a solution, we substitute, replacing x with the first coordinate and y with the second coordinate in each pair.

$$4x - 2y = 10$$
$$\frac{}{4 \cdot 0 - 2(-5) \ ? \ 10}$$
$$10 \ \big| \ 10 \quad \text{TRUE}$$

$$4x - 2y = 10$$
$$\frac{}{4 \cdot 4 - 2 \cdot 3 \ ? \ 10}$$
$$16 - 6$$
$$10 \ \big| \ 10 \quad \text{TRUE}$$

In each case the substitution results in a true equation. Thus, $(0, -5)$ and $(4, 3)$ are both solutions of $4x - 2y = 10$. We graph these points and sketch the line passing through them.

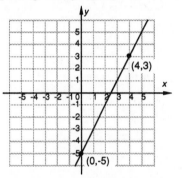

The line appears to pass through $(2, -1)$ also. We check to determine if $(2, -1)$ is a solution of $4x - 2y = 10$.

$$4x - 2y = 10$$
$$\frac{}{4 \cdot 2 - 2(-1) \ ? \ 10}$$
$$8 + 2$$
$$10 \ \big| \ 10 \quad \text{TRUE}$$

Thus, $(2, -1)$ is another solution. There are other correct answers, including $(1, -3)$, $(2, -1)$, $(3, 1)$, and $(5, 5)$.

14.

$$6x - 3y = 3$$
$$\frac{}{6 \cdot 1 - 3 \cdot 1 \ ? \ 3}$$
$$6 - 3$$
$$3 \ \big| \ 3 \quad \text{TRUE}$$

$$6x - 3y = 3$$
$$\frac{}{6(-1) - 3(-3) \ ? \ 3}$$
$$-6 + 9$$
$$3 \ \big| \ 3 \quad \text{TRUE}$$

$(0, -1)$; answers may vary

15. $y = x - 1$

The equation is equivalent to $y = x + (-1)$. The y-intercept is $(0, -1)$. We find two other pairs.

When $x = 3$, $y = 3 - 1 = 2$.

When $x = -5$, $y = -5 - 1 = -6$.

x	y
0	-1
3	2
-5	-6

Plot these points, draw the line they determine, and label the graph $y = x - 1$.

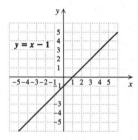

16.

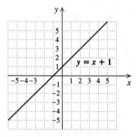

17. $y = x$

The equation is equivalent to $y = x + 0$. The y-intercept is $(0, 0)$. We find two other points.

When $x = -2$, $y = -2$.

When $x = 3$, $y = 3$.

x	y
0	0
-2	-2
3	3

Plot these points, draw the line they determine, and label the graph $y = x$.

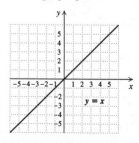

18.

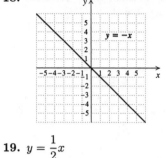

19. $y = \dfrac{1}{2}x$

The equation is equivalent to $y = \dfrac{1}{2}x + 0$. The y-intercept is $(0, 0)$. We find two other points.

When $x = -4$, $y = \dfrac{1}{2}(-4) = -2$.

When $x = 4$, $y = \dfrac{1}{2} \cdot 4 = 2$.

x	y
0	0
-4	-2
4	2

Plot these points, draw the line they determine, and label the graph $y = \dfrac{1}{2}x$.

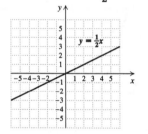

20.

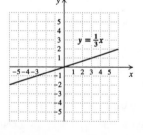

21. $y = x + 2$

The equation is in the form $y = mx + b$. The y-intercept is $(0, 2)$. We find two other points.

When $x = -2$, $y = -2 + 2 = 0$.

When $x = 3$, $y = 3 + 2 = 5$.

x	y
0	2
-2	0
3	5

Plot these points, draw the line they determine, and label the graph $y = x + 2$.

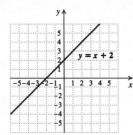

22.

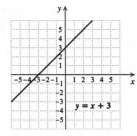

23. $y = 3x - 2 = 3x + (-2)$

The y-intercept is $(0, -2)$. We find two other points.

When $x = -2$, $y = 3(-2) + 2 = -6 + 2 = -4$.

When $x = 1$, $y = 3 \cdot 1 + 2 = 3 + 2 = 5$.

x	y
0	-2
-2	-4
1	5

Plot these points, draw the line they determine, and label the graph $y = 3x + 2$.

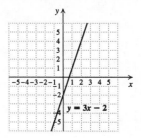

24.

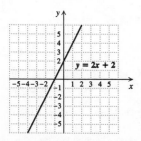

25. $y = \dfrac{1}{2}x + 1$

The y-intercept is $(0, 1)$. We find two other points using multiples of 2 for x to avoid fractions.

When $x = -4$, $y = \dfrac{1}{2}(-4) + 1 = -2 + 1 = -1$.

When $x = 4$, $y = \dfrac{1}{2} \cdot 4 + 1 = 2 + 1 = 3$.

x	y
0	1
-4	-1
4	3

Plot these points, draw the line they determine, and label the graph $y = \dfrac{1}{2}x + 1$.

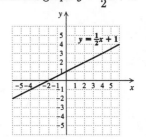

26.

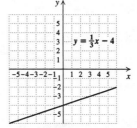

27. $x + y = -5$

$$y = -x - 5$$
$$y = -x + (-5)$$

The y-intercept is $(0, -5)$. We find two other points.

When $x = -4$, $y = -(-4) - 5 = 4 - 5 = -1$.

When $x = -1$, $y = -(-1) - 5 = 1 - 5 = -4$.

x	y
0	-5
-4	-1
-1	-4

Plot these points, draw the line they determine, and label the graph $x + y = -5$.

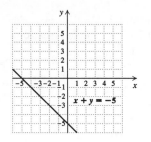

28.

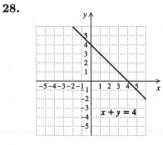

29. $y = \dfrac{5}{3}x - 2 = \dfrac{5}{3}x + (-2)$

The y-intercept is $(0, -2)$. We find two other points using multiples of 3 for x to avoid fractions.

When $x = -3$, $y = \dfrac{5}{3}(-3) - 2 = -5 - 2 = -7$.

When $x = 3$, $y = \dfrac{5}{3} \cdot 3 - 2 = 5 - 2 = 3$.

x	y
0	-2
-3	-7
3	3

Plot these points, draw the line they determine, and label the graph $y = \dfrac{5}{3}x - 2$.

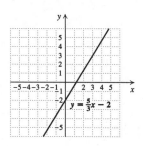

30.

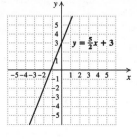

31. $x + 2y = 8$

$\qquad 2y = -x + 8$

$\qquad y = -\dfrac{1}{2}x + 4$

The y-intercept is $(0, 4)$. We find two other points using multiples of 2 for x to avoid fractions.

When $x = -2$, $y = -\dfrac{1}{2}(-2) + 4 = 1 + 4 = 5$.

When $x = 4$, $y = -\dfrac{1}{2} \cdot 4 + 4 = -2 + 4 = 2$.

x	y
0	4
-2	5
4	2

Plot these points, draw the line they determine, and label the graph $x + 2y = 8$.

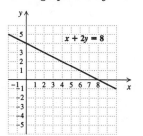

32.

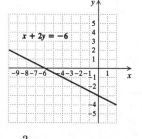

33. $y = \dfrac{3}{2}x + 1$

The y-intercept is $(0, 1)$. We find two other points using multiples of 2 for x to avoid fractions.

When $x = -4$, $y = \dfrac{3}{2}(-4) + 1 = -6 + 1 = -5$.

When $x = 2$, $y = \dfrac{3}{2} \cdot 2 + 1 = 3 + 1 = 4$.

x	y
0	1
−4	−5
2	4

Plot these points, draw the line they determine, and label the graph $y = \dfrac{3}{2}x + 1$.

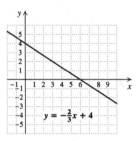

34.

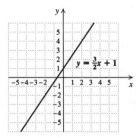

35. $6x - 3y = 9$

$$-3y = -6x + 9$$
$$y = 2x - 3$$
$$y = 2x + (-3)$$

The y-intercept is $(0, -3)$. We find two other points.

When $x = -1$, $y = 2(-1) - 3 = -2 - 3 = -5$.

When $x = 3$, $y = 2 \cdot 3 - 3 = 6 - 3 = 3$.

x	y
0	−3
−1	−5
3	3

Plot these points, draw the line they determine, and label the graph $6x - 3y = 9$.

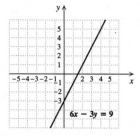

36.

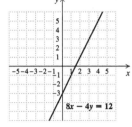

37. $8y + 2x = -4$

$$8y = -2x - 4$$
$$y = -\frac{1}{4}x - \frac{1}{2}$$
$$y = -\frac{1}{4}x + \left(-\frac{1}{2}\right)$$

The y-intercept is $\left(0, -\dfrac{1}{2}\right)$. We find two other points.

When $x = -2$, $y = -\dfrac{1}{4}(-2) - \dfrac{1}{2} = \dfrac{1}{2} - \dfrac{1}{2} = 0$.

When $x = 2$, $y = -\dfrac{1}{4} \cdot 2 - \dfrac{1}{2} = -\dfrac{1}{2} - \dfrac{1}{2} = -1$.

x	y
0	$-\dfrac{1}{2}$
−2	0
2	−1

Plot these points, draw the line they determine, and label the graph $8y + 2x = -4$.

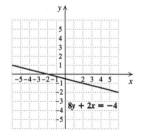

38.

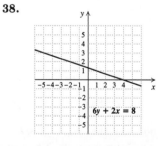

39. We graph $w = \dfrac{1}{2}t + 5$. Since the number of gallons of bottled water consumed cannot be negative in this application, we select only nonnegative values for t.

If $t = 0$, $w = \dfrac{1}{2} \cdot 0 + 5 = 5$.

If $t = 4$, $w = \dfrac{1}{2} \cdot 4 + 5 = 2 + 5 = 7$.

If $t = 10$, $w = \dfrac{1}{2} \cdot 10 + 5 = 5 + 5 = 10$.

t	w
0	5
4	7
10	10

We plot the points and draw the graph.

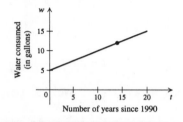

Number of years since 1990

To predict the number of gallons consumed per person in 2004 we find the second coordinate associated with 14. (2004 is 14 years after 1990.) Locate the point on the line that is above 14 and then find the value on the vertical axis that corresponds to that point. That value is 12, so we predict that 12 gallons of bottled water will be consumed per person in 2004.

40.

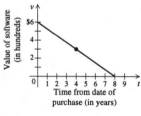

Time from date of purchase (in years)

$300

41. We graph $t + w = 15$, or $w = -t + 15$. Since time cannot be negative in this application, we select only nonnegative values for t.

If $t = 0$, $w = -0 + 15 = 15$.

If $t = 2$, $w = -2 + 15 = 13$.

If $t = 5$, $w = -5 + 15 = 10$.

t	w
0	15
2	13
5	10

We plot the points and draw the graph. Since the likelihood of death cannot be negative, the graph stops at the horizontal axis.

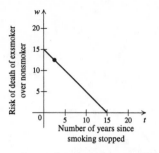

Number of years since smoking stopped

To estimate how much more likely it is for Sandy to die from lung cancer than Polly, we find the second coordinate associated with $2\dfrac{1}{2}$. Locate the point on the line that is above $2\dfrac{1}{2}$ and then find the value on the vertical axis that corresponds to that point. That value is about $12\dfrac{1}{2}$, so it is $12\dfrac{1}{2}$ times more likely for Sandy to die from lung cancer than Polly.

42.

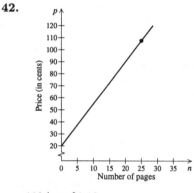

Number of pages

110¢, or $1.10

43. We graph $T = \dfrac{6}{5}c + 1$. Since the number of credits cannot be negative, we select only nonnegative values for c.

If $c = 5$, $T = \dfrac{6}{5} \cdot 5 + 1 = 6 + 1 = 7$.

If $c = 10$, $T = \dfrac{6}{5} \cdot 10 + 1 = 12 + 1 = 13$.

If $c = 15$, $T = \dfrac{6}{5} \cdot 15 + 1 = 18 + 1 = 19$.

c	T
5	7
10	13
15	19

We plot the points and draw the graph.

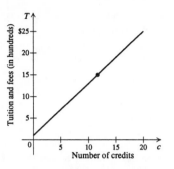

Four three-credit courses total $4 \cdot 3$, or 12, credits. To estimate the cost of tuition and fees for a student who is registered for 12 credits, we find the second coordinate associated with 12. Locate the point on the line that is above 12 and then find the value on the vertical axis that corresponds to that point. That value is about 15, so tuition and fees will cost about $1500.

44.

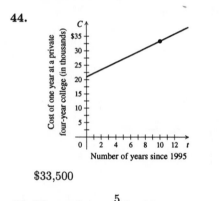

$33,500

45. We graph $n = \dfrac{5}{2}d + 20$.

When $d = 0$, $n = \dfrac{5}{2} \cdot 0 + 20 = 20$.

When $d = 2$, $n = \dfrac{5}{2} \cdot 2 + 20 = 25$.

When $d = 8$, $n = \dfrac{5}{2} \cdot 8 + 20 = 40$.

d	n
0	20
2	25
8	40

We plot the points and draw the graph.

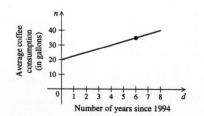

To estimate coffee consumption in 2000, we first note that 2000 is 6 years after 1994. Then we find the second coordinate associated with 6. Locate the point on the line that is above 6 and find the value on the vertical axis that corresponds to that point. That value is about 35, so 35 gal of coffee were consumed by the average U.S. consumer in 2000.

46.

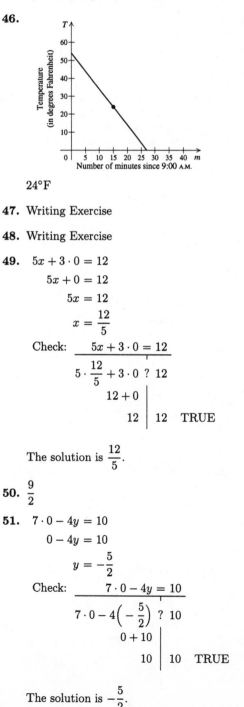

$24°F$

47. Writing Exercise

48. Writing Exercise

49.
$$5x + 3 \cdot 0 = 12$$
$$5x + 0 = 12$$
$$5x = 12$$
$$x = \frac{12}{5}$$

Check:
$$\frac{5x + 3 \cdot 0 = 12}{\begin{array}{c} 5 \cdot \dfrac{12}{5} + 3 \cdot 0 \ ? \ 12 \\ 12 + 0 \ \Big| \\ 12 \ \Big|\ 12 \quad \text{TRUE} \end{array}}$$

The solution is $\dfrac{12}{5}$.

50. $\dfrac{9}{2}$

51.
$$7 \cdot 0 - 4y = 10$$
$$0 - 4y = 10$$
$$y = -\frac{5}{2}$$

Check:
$$\frac{7 \cdot 0 - 4y = 10}{\begin{array}{c} 7 \cdot 0 - 4\left(-\dfrac{5}{2}\right) \ ? \ 10 \\ 0 + 10 \ \Big| \\ 10 \ \Big|\ 10 \quad \text{TRUE} \end{array}}$$

The solution is $-\dfrac{5}{2}$.

52. $p = \dfrac{w}{q+1}$

53. $\quad Ax + By = C$
$\qquad\quad By = C - Ax \qquad$ Subtracting Ax

$\qquad\qquad y = \dfrac{C - Ax}{B} \qquad$ Dividing by B

54. $Q = 2A - T$

55. Writing Exercise

56. Writing Exercise

57. Let s represent the gear that Lauren uses on the southbound portion of her ride and n represent the gear she uses on the northbound portion. Then we have $s + n = 18$. We graph this equation, using only positive integer values for s and n.

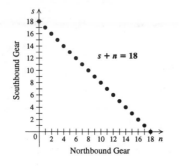

58. $x + y = 5$, or $y = -x + 5$

59. Note that the sum of the coordinates of each point on the graph is 2. Thus, we have $x + y = 2$, or $y = -x + 2$.

60. $y = x + 2$

61. Note that when $x = 0$, $y = -5$ and when $y = 0$, $x = 3$. An equation that fits this situation is $5x - 3y = 15$, or $y = \dfrac{5}{3}x - 5$.

62.

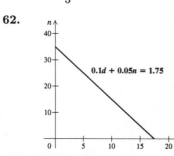

5 dimes, 25 nickels; 10 dimes, 15 nickels; 12 dimes, 11 nickels

63. The equation is $25d + 5l = 225$.

Since the number of dinners cannot be negative, we choose only nonnegative values of d when graphing the equation. The graph stops at the horizontal axis since the number of lunches cannot be negative.

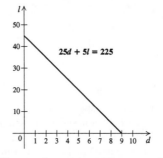

We see that three points on the graph are $(1, 40)$, $(5, 20)$, and $(8, 5)$. Thus, three combinations of dinners and lunches that total \$225 are

$\qquad$ 1 dinner, 40 lunches,

$\qquad$ 5 dinners, 20 lunches,

$\qquad$ 8 dinners, 5 lunches.

64.

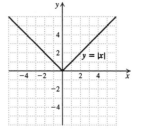

65. $y = -|x|$

x	y
-3	-3
-2	-2
-1	-1
0	0
1	-1
2	-2
3	-3

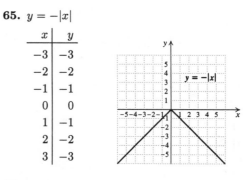

66.

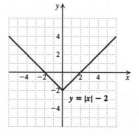

67. $y = -|x| + 2$

x	y
-3	-1
-2	0
-1	1
0	2
1	1
2	0
3	-1

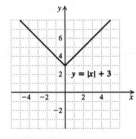

68.

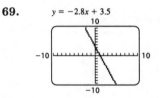

69. $y = -2.8x + 3.5$

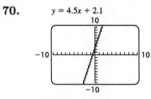

70. $y = 4.5x + 2.1$

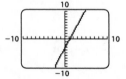

71. $y = 2.8x - 3.5$

72. $y = -4.5x - 2.1$

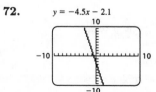

73. $y = x^2 + 4x + 1$

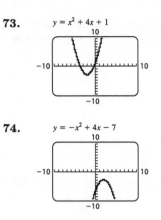

74. $y = -x^2 + 4x - 7$

75. Writing Exercise

Exercise Set 3.3

1. (a) The graph crosses the y-axis at $(0, 5)$, so the y-intercept is $(0, 5)$.

   (b) The graph crosses the x-axis at $(2, 0)$, so the x-intercept is $(2, 0)$.

2. (a) $(0, 3)$; (b) $(4, 0)$

3. (a) The graph crosses the y-axis at $(0, -4)$, so the y-intercept is $(0, -4)$.

   (b) The graph crosses the x-axis at $(3, 0)$, so the x-intercept is $(3, 0)$.

4. (a) $(0, 5)$; (b) $(-3, 0)$

5. (a) The graph crosses the y-axis at $(0, -2)$, so the y-intercept is $(0, -2)$.

   (b) The graph crosses the x-axis at $(-3, 0)$ and also at $(3, 0)$, so the x-intercepts are $(-3, 0)$ and $(3, 0)$.

6. (a) $(0, 1)$; (b) $(-3, 0)$

7. (a) The graph crosses the y-axis at $(0, 4)$, so the y-intercept is $(0, 4)$.

   (b) The graph crosses the x-axis at $(-3, 0)$, $(3, 0)$, and $(5, 0)$. Each of these points is an x-intercept.

8. (a) $(0, -3)$; (b) $(-2, 0)$, $(2, 0)$, $(5, 0)$

9. $5x + 3y = 15$

   (a) To find the y-intercept, let $x = 0$. This is the same as ignoring the x-term and then solving.

$$3y = 15$$
$$y = 5$$

   The y-intercept is $(0, 5)$.

(b) To find the x-intercept, let $y = 0$. This is the same as ignoring the y-term and then solving.
$$5x = 15$$
$$x = 3$$
The x-intercept is $(3, 0)$.

10. (a) $(0, 10)$

(b) $(4, 0)$

11. $7x - 2y = 28$

(a) To find the y-intercept, let $x = 0$. This is the same as ignoring the x-term and then solving.
$$-2y = 28$$
$$y = -14$$
The y-intercept is $(0, -14)$.

(b) To find the x-intercept, let $y = 0$. This is the same as ignoring the y-term and then solving.
$$7x = 28$$
$$x = 4$$
The x-intercept is $(4, 0)$.

12. (a) $(0, -8)$

(b) $(6, 0)$.

13. $-4x + 3y = 10$

(a) To find the y-intercept, let $x = 0$. This is the same as ignoring the x-term and then solving.
$$3y = 10$$
$$y = \frac{10}{3}$$
The y-intercept is $\left(0, \frac{10}{3}\right)$.

(b) To find the x-intercept, let $y = 0$. This is the same as ignoring the y-term and then solving.
$$-4x = 10$$
$$x = -\frac{5}{2}$$
The x-intercept is $\left(-\frac{5}{2}, 0\right)$.

14. (a) $\left(0, \frac{7}{3}\right)$

(b) $\left(-\frac{7}{2}, 0\right)$

15. $y = 9$

Observe that this is the equation of a horizontal line 9 units above the x-axis. Thus, (a) the y-intercept is $(0, 9)$ and (b) there is no x-intercept.

16. (a) None

(b) $(8, 0)$

17. $x + 2y = 6$

Find the y-intercept:
$$2y = 6 \quad \text{Ignoring the } x\text{-term}$$
$$y = 3$$
The y-intercept is $(0, 3)$.

Find the x-intercept:
$$x = 6 \qquad \text{Ignoring the } y\text{-term}$$
The x-intercept is $(6, 0)$.

To find a third point we replace x with 2 and solve for y.
$$2 + 2y = 6$$
$$2y = 4$$
$$y = 2$$

The point $(2, 2)$ appears to line up with the intercepts, so we draw the graph.

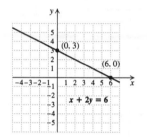

18.

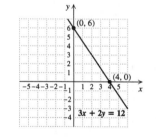

19. $6x + 9y = 36$

Find the y-intercept:
$$9y = 36 \quad \text{Ignoring the } x\text{-term}$$
$$y = 4$$
The y-intercept is $(0, 4)$.

Find the x-intercept:
$$6x = 36 \quad \text{Ignoring the } y\text{-term}$$
$$x = 6$$
The x-intercept is $(6, 0)$.

To find a third point we replace x with -3 and solve for y.
$$6(-3) + 9y = 36$$
$$-18 + 9y = 36$$
$$9y = 54$$
$$y = 6$$

The point $(-3, 6)$ appears to line up with the intercepts, so we draw the graph.

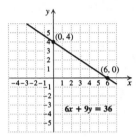

20.

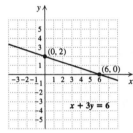

21. $-x + 3y = 9$

Find the y-intercept:

$$3y = 9 \quad \text{Ignoring the } x\text{-term}$$
$$y = 3$$

The y-intercept is $(0, 3)$.

Find the x-intercept:

$$-x = 9 \quad \text{Ignoring the } y\text{-term}$$
$$x = -9$$

The x-intercept is $(-9, 0)$.

To find a third point we replace x with 3 and solve for y.

$$-3 + 3y = 9$$
$$3y = 12$$
$$y = 4$$

The point $(3, 4)$ appears to line up with the intercepts, so we draw the graph.

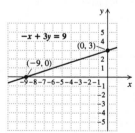

22.

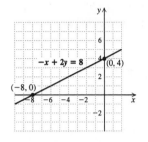

23. $2x - y = 8$

Find the y-intercept:

$$-y = 8 \quad \text{Ignoring the } x\text{-term}$$
$$y = -8$$

The y-intercept is $(0, -8)$.

Find the x-intercept:

$$2x = 8 \quad \text{Ignoring the } y\text{-term}$$
$$x = 4$$

The x-intercept is $(4, 0)$.

To find a third point we replace x with 2 and solve for y.

$$2 \cdot 2 - y = 8$$
$$4 - y = 8$$
$$-y = 4$$
$$y = -4$$

The point $(2, -4)$ appears to line up with the intercepts, so we draw the graph.

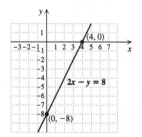

24.

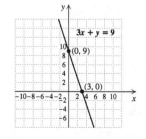

25. $y = -3x + 6$

Find the y-intercept:

$$y = 6 \quad \text{Ignoring the } x\text{-term}$$

The y-intercept is $(0, 6)$.

Find the x−intercept:

$$0 = -3x + 6 \quad \text{Replacing } y \text{ with } 0$$
$$3x = 6$$
$$x = 2$$

The x-intercept is $(2, 0)$.

To find a third point we replace x with 3 and find y.

$$y = -3 \cdot 3 + 6 = -9 + 6 = -3$$

The point $(3, -3)$ appears to line up with the intercepts, so we draw the graph.

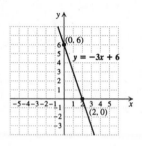

26.

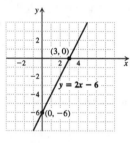

27. $5x - 10 = 5y$

We can leave the equation in the given form or rewrite it in the form $Ax + By = C$. We will use the given form.

Find the y-intercept:

$$-10 = 5y \quad \text{Ignoring the } x\text{-term}$$
$$-2 = y$$

The y-intercept is $(0, -2)$.

To find the x-intercept, let $y = 0$.

$$5x - 10 = 5 \cdot 0$$
$$5x - 10 = 0$$
$$5x = 10$$
$$x = 2$$

The x-intercept is $(2, 0)$.

To find a third point we replace x with 5 and solve for y.

$$5 \cdot 5 - 10 = 5y$$
$$25 - 10 = 5y$$
$$15 = 5y$$
$$3 = y$$

The point $(5, 3)$ appears to line up with the intercepts, so we draw the graph.

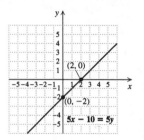

28.

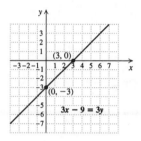

29. $2x - 5y = 10$

Find the y-intercept:

$$-5y = 10 \quad \text{Ignoring the } x\text{-term}$$
$$y = -2$$

The y-intercept is $(0, -2)$.

Find the x-intercept:

$$2x = 10 \quad \text{Ignoring the } y\text{-term}$$
$$x = 5$$

The x-intercept is $(5, 0)$.

To find a third point we replace x with -5 and solve for y.

$$2(-5) - 5y = 10$$
$$-10 - 5y = 10$$
$$-5y = 20$$
$$y = -4$$

The point $(-5, -4)$ appears to line up with the intercepts, so we draw the graph.

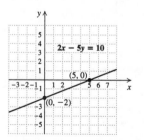

30.

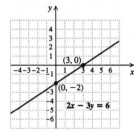

31. $6x + 2y = 12$

Find the y-intercept:

$$2y = 12 \quad \text{Ignoring the } x\text{-term}$$
$$y = 6$$

The y-intercept is $(0, 6)$.

Find the x-intercept:

$$6x = 12 \quad \text{Ignoring the } y\text{-term}$$
$$x = 2$$

The x-intercept is $(2, 0)$.

To find a third point we replace x with 3 and solve for y.

$$6 \cdot 3 + 2y = 12$$
$$18 + 2y = 12$$
$$2y = -6$$
$$y = -3$$

The point $(3, -3)$ appears to line up with the intercepts, so we draw the graph.

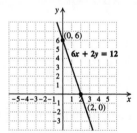

32.

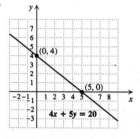

33. $x - 1 = y$

We can leave the equation in the given form or rewrite it in the form $Ax + By = C$. We will use the given form.

Find the y-intercept:

$$-1 = y \quad \text{Ignoring the } x\text{-term}$$

The y-intercept is $(0, -1)$.

To find the x-intercept, let $y = 0$.

$$x - 1 = 0$$
$$x = 1$$

The x-intercept is $(1, 0)$.

To find a third point we replace x with -3 and solve for y.

$$-3 - 1 = y$$
$$-4 = y$$

The point $(-3, -4)$ appears to line up with the intercepts, so we draw the graph.

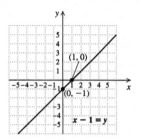

34.

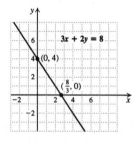

35. $2x - 6y = 18$

Find the y-intercept:

$$-6y = 18 \quad \text{Ignoring the } x\text{-term}$$
$$y = -3$$

The y-intercept is $(0, -3)$.

Find the x-intercept:

$$2x = 18 \quad \text{Ignoring the } y\text{-term}$$
$$x = 9$$

The x-intercept is $(9, 0)$.

To find a third point we replace x with 3 and solve for y.

$$2 \cdot 3 - 6y = 18$$
$$6 - 6y = 18$$
$$-6y = 12$$
$$y = -2$$

The point $(3, -2)$ appears to line up with the intercepts, so we draw the graph.

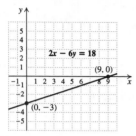

36.

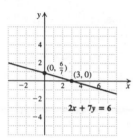

37. $4x - 3y = 12$

Find the y-intercept:

$$-3y = 12 \quad \text{Ignoring the } x\text{-term}$$
$$y = -4$$

The y-intercept is $(0, -4)$.

Find the x-intercept:

$$4x = 12 \quad \text{Ignoring the } y\text{-term}$$
$$x = 3$$

The x-intercept is $(3, 0)$.

To find a third point we replace x with 6 and solve for y.

$$4 \cdot 6 - 3y = 12$$
$$24 - 3y = 12$$
$$-3y = -12$$
$$y = 4$$

The point $(6, 4)$ appears to line up with the intercepts, so we draw the graph.

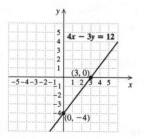

38.

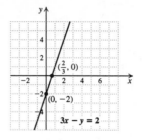

39. $-3x = 6y - 2$

We can leave the equation in the given form or rewrite it in the form $Ax + By = C$. We will use the given form.

To find the y-intercept, let $x = 0$.

$$-3 \cdot 0 = 6y - 2$$
$$0 = 6y - 2$$
$$2 = 6y$$
$$\frac{1}{3} = y$$

The y-intercept is $\left(0, \frac{1}{3}\right)$.

Find the x-intercept:

$$-3x = -2 \quad \text{Ignoring the } y\text{-term}$$
$$x = \frac{2}{3}$$

The x-intercept is $\left(\frac{2}{3}, 0\right)$.

To find a third point we replace x with -4 and solve for y.

$$-3(-4) = 6y - 2$$
$$12 = 6y - 2$$
$$14 = 6y$$
$$\frac{7}{3} = y$$

The point $\left(-4, \frac{7}{3}\right)$ appears to line up with the intercepts, so we draw the graph.

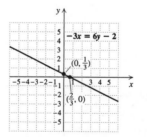

40.

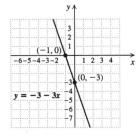

41. $3 = 2x - 5y$

Find the y-intercept:

$$3 = -5y \quad \text{Ignoring the } x\text{-term}$$
$$-\frac{3}{5} = y$$

The y-intercept is $\left(0, -\frac{3}{5}\right)$.

Find the x-intercept:

$$3 = 2x \quad \text{Ignoring the } y\text{-term}$$
$$\frac{3}{2} = x$$

The x-intercept is $\left(\frac{3}{2}, 0\right)$.

To find a third point we replace x with -1 and solve for y.

$$3 = 2(-1) - 5y$$
$$3 = -2 - 5y$$
$$5 = -5y$$
$$-1 = y$$

The point $(-1, -1)$ appears to line up with the intercepts, so we draw the graph.

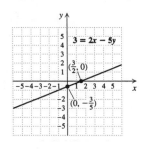

42.

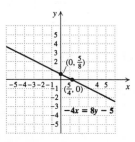

43. $x + 2y = 0$

Find the y-intercept:

$$2y = 0 \quad \text{Ignoring the } x\text{-term}$$
$$y = 0$$

The y-intercept is $(0, 0)$. Note that this is also the x-intercept.

In order to graph the line, we will find a second point. When $x = 4$, $4 + 2y = 0$

$$2y = -4$$
$$y = -2.$$

Thus, a second point is $(4, -2)$.

To find a third point we replace x with -2 and solve for y.

$$-2 + 2y = 0$$
$$2y = 2$$
$$y = 1$$

The point $(-2, 1)$ appears to line up with the other two points, so we draw the graph.

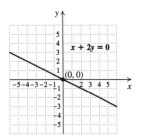

44.

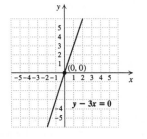

45. $y = 5$

Any ordered pair $(x, 5)$ is a solution. The variable y must be 5, but the x variable can be any number we choose. A few solutions are listed below. Plot these points and draw the line.

x	y
-3	5
0	5
2	5

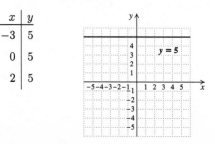

46.

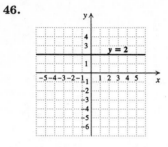

47. $x = 4$

Any ordered pair $(4, y)$ is a solution. The variable x must be 4, but the y variable can be any number we choose. A few solutions are listed below. Plot these points and draw the line.

x	y
4	-2
4	0
4	4

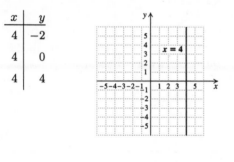

48.

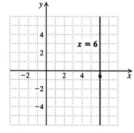

49. $y = -2$

Any ordered pair $(x, -2)$ is a solution. The variable y must be -2, but the x variable can be any number we choose. A few solutions are listed below. Plot these points and draw the line.

x	y
-3	-2
0	-2
4	-2

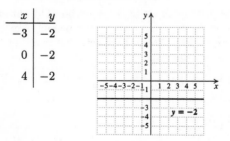

50.

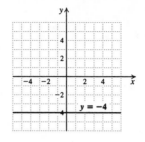

51. $x = -1$

Any ordered pair $(-1, y)$ is a solution. The variable x must be -1, but the y variable can be any number we choose. A few solutions are listed below. Plot these points and draw the line.

x	y
-1	-3
-1	0
-1	2

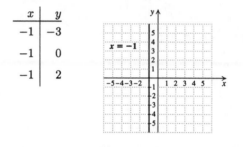

52.

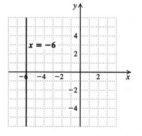

53. $y = 7$

Any ordered pair $(x, 7)$ is a solution. A few solutions are listed below. Plot these points and draw the line.

x	y
-3	7
0	7
4	7

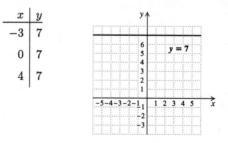

54.

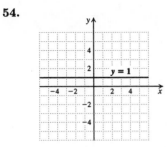

55. $x = 1$

Any ordered pair $(1, y)$ is a solution. A few solutions are listed below. Plot these points and draw the line.

x	y
1	-1
1	4
1	5

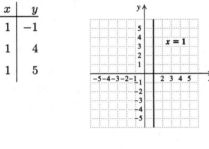

56.

57. $y = 0$

Any ordered pair $(x, 0)$ is a solution. A few solutions are listed below. Plot these points and draw the line.

x	y
-4	0
0	0
2	0

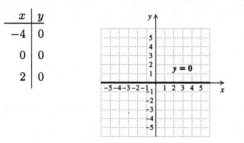

58.

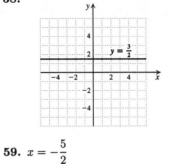

59. $x = -\dfrac{5}{2}$

Any ordered pair $\left(-\dfrac{5}{2}, y \right)$ is a solution. A few solutions are listed below. Plot these points and draw the line.

x	y
$-\dfrac{5}{2}$	-3
$-\dfrac{5}{2}$	0
$-\dfrac{5}{2}$	5

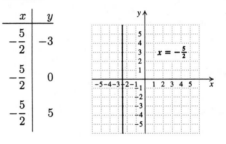

60.

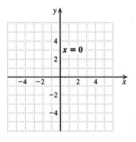

61. $-5y = 15$

Observe that $-5y = 15$ is equivalent to $y = -3$. Thus, the graph is a horizontal line 3 units below the x-axis.

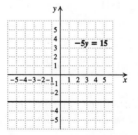

62.

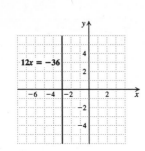

63. $35 + 7y = 0$

$$7y = -35$$

$$y = -5$$

The graph is a horizontal line 5 units below the x-axis.

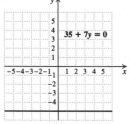

64.

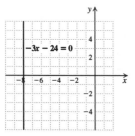

65. Note that every point on the horizontal line passing through $(0, -1)$ has -1 as the y-coordinate. Thus, the equation of the line is $y = -1$.

66. $x = -1$

67. Note that every point on the vertical line passing through $(4, 0)$ has 4 as the x-coordinate. Thus, the equation of the line is $x = 4$.

68. $y = -5$

69. Note that every point on the horizontal line passing through $(0, 0)$ has 0 as the y-coordinate. Thus, the equation of the line is $y = 0$.

70. $x = 0$

71. Writing Exercise

72. Writing Exercise

73. $d - 7$

74. $w + 5$, or $5 + w$

75. Let x represent "a number." Then we have $2 + x$, or $x + 2$.

76. Let y represent "a number;" $3y$

77. Let x and y represent the numbers. Then we have $2(x + y)$.

78. Let m and n represent the numbers; $\frac{1}{2}(m + n)$

79. Writing Exercise

80. Writing Exercise

81. The x-axis is a horizontal line, so it is of the form $y = b$. All points on the x-axis are of the form $(x, 0)$, so b must be 0 and the equation is $y = 0$.

82. $y = 5$

83. A line parallel to the y-axis has an equation of the form $x = a$. Since the x-coordinate of one point on the line is -2, then $a = -2$ and the equation is $x = -2$.

84. $(6, 6)$

85. Since the x-coordinate of the point of intersection must be -3 and y must equal x, the point of intersection is $(-3, -3)$.

86. $y = -\frac{5}{2}x + 5$, or $5x + 2y = 10$

87. The y-intercept is $(0, 5)$, so we have $y = mx + 5$. Another point on the line is $(-3, 0)$ so we have

$$0 = m(-3) + 5$$

$$-5 = -3m$$

$$\frac{5}{3} = m$$

The equation is $y = \frac{5}{3}x + 5$, or $5x - 3y = -15$.

88. 12

89. Substitute 0 for x and -8 for y.

$$4 \cdot 0 = C - 3(-8)$$

$$0 = C + 24$$

$$-24 = C$$

90. Writing Exercise

91. Find the y-intercept:

$$2y = 50 \quad \text{Covering the } x\text{-term}$$

$$y = 25$$

The y-intercept is $(0, 25)$.

Find the x-intercept:

$$3x = 50 \qquad \text{Covering the } y\text{-term}$$

$$x = \frac{50}{3} = 16.\overline{6}$$

The x-intercept is $\left(\frac{50}{3}, 0\right)$, or $(16.\overline{6}, 0)$.

92. $\left(0, -\frac{80}{7}\right)$, or $(0, -11.\overline{428571})$; $(40, 0)$

93. From the equation we see that the y-intercept is $(0, -9)$.

To find the x-intercept, let $y = 0$.

$$0 = 0.2x - 9$$

$$9 = 0.2x$$

$$45 = x$$

The x-intercept is $(45, 0)$.

94. $(0, -15)$; $\left(\frac{150}{13}, 0\right)$, or $(11.\overline{538461}, 0)$

95. Find the y-intercept.

$$-20y = 1 \qquad \text{Covering the } x\text{-term}$$

$$y = -\frac{1}{20}, \text{ or } -0.05$$

The y-intercept is $\left(0, -\frac{1}{20}\right)$, or $(0, -0.05)$.

Find the x-intercept:

$$25x = 1 \qquad \text{Covering the } y\text{-term}$$

$$x = \frac{1}{25}, \text{ or } 0.04$$

The x-intercept is $\left(\frac{1}{25}, 0\right)$, or $(0.04, 0)$.

96. $\left(0, \frac{1}{25}\right)$, or $(0, 0.04)$; $\left(\frac{1}{50}, 0\right)$, or $(0.02, 0)$

Exercise Set 3.4

1. a) We divide the number of miles traveled by the number of gallons of gas used for that amount of driving.

Rate, in miles per gallon

$$= \frac{14,014 \text{ mi} - 13,741 \text{ mi}}{13 \text{ gal}}$$

$$= \frac{273 \text{ mi}}{13 \text{ gal}}$$

$$= 21 \text{ mi/gal}$$

$$= 21 \text{ miles per gallon}$$

b) We divide the cost of the rental by the number of days. From July 1 to July 4 is $4 - 1$, or 3 days.

Average cost, in dollars per day

$$= \frac{118 \text{ dollars}}{3 \text{ days}}$$

$$\approx 39.33 \text{ dollars/day}$$

$$\approx \$39.33 \text{ per day}$$

c) We divide the number of miles traveled by the number of days. In part (a) we found that the van was driven 273 mi, and in part (b) we found that it was rented for 3 days.

Rate, in miles per day

$$= \frac{273 \text{ mi}}{3 \text{ days}}$$

$$= 91 \text{ mi/day}$$

$$= 91 \text{ mi per day}$$

2. a) 16.5 mi per gal

b) $46 per day

c) 115.5 mi per day

d) 40¢ per mi

3. a) From 2:00 to 5:00 is $5 - 2$, or 3 hr.

Average speed, in miles per hour

$$= \frac{18 \text{ mi}}{3 \text{ hr}}$$

$$= 6 \text{ mph}$$

b) From part (a) we know that the bike was rented for 3 hr.

$$\text{Rate, in dollars per hour} = \frac{\$10.50}{3 \text{ hr}}$$

$$= \$3.50 \text{ per hr}$$

c) $\quad \text{Rate, in dollars per mile} = \dfrac{\$10.50}{18 \text{ mi}}$

$$\approx \$0.58 \text{ per mile}$$

4. a) 7 mph

b) $6 per hr

c) $0.86 per mi

5. a) It is 3 hr from 9:00 A.M. to noon and 5 more hours from noon to 5:00 P.M., so the typist worked $3 + 5$, or 8 hr.

$$\text{Rate, in dollars per hour} = \frac{\$128}{8 \text{ hr}}$$
$$= \$16 \text{ per hr}$$

b) The number of pages typed is $48 - 12$, or 36.

In part (a) we found that the typist worked 8 hr.

$$\text{Rate, in pages per hour} = \frac{36 \text{ pages}}{8 \text{ hr}}$$
$$= 4.5 \text{ pages per hr}$$

c) In part (b) we found that 36 pages were typed.

$$\text{Rate, in dollars per page} = \frac{\$128}{36 \text{ pages}}$$
$$\approx \$3.56 \text{ per page}$$

6. a) $15 per hr

b) 5.25 pages per hr

c) $2.86 per page

7. The tuition increased $1318-$1239, or $79, in 1998−1996 or 2 yr.

$$\text{Rate of increase} = \frac{\text{Change in tuition}}{\text{Change in time}}$$
$$= \frac{\$79}{2 \text{ yr}}$$
$$= \$39.50 \text{ per yr}$$

8. $170.67 per yr

9. a) The elevator traveled $34-5$, or 29 floors in 2:40 − 2:38, or 2 min.

$$\text{Average rate of travel} = \frac{29 \text{ floors}}{2 \text{ min}}$$
$$= 14.5 \text{ floors per min}$$

b) In part (a) we found that the elevator traveled 29 floors in 2 min. Note that $2 \text{ min} = 2 \times 1 \text{ min} = 2 \times 60 \text{ sec} = 120 \text{ sec}$.

$$\text{Average rate of travel} = \frac{120 \text{ sec}}{29 \text{ floors}}$$
$$\approx 4.14 \text{ sec per floor}$$

10. a) 1 driveway per hour

b) 1 hr per driveway

11. a) Krakauer ascended $29,028 \text{ ft} - 27,600 \text{ ft}$, or 1428 ft. From 7:00 A.M. to noon it is $5 \text{ hr} = 5 \times 1 \text{ hr} = 5 \times 60 \text{ min} = 300 \text{ min}$. From noon to 1:25 P.M. is another 1 hr, 25 min, or $1 \text{ hr} + 25 \text{ min} = 60 \text{ min} + 25 \text{ min} = 85 \text{ min}$. The total time of the ascent is $300 \text{ min} + 85 \text{ min}$, or 385 min.

$$\text{Rate, in feet per minute} = \frac{1428 \text{ ft}}{385 \text{ min}}$$
$$\approx 3.71 \text{ ft per min}$$

b) We use the information found in part (a).

$$\text{Rate, in minutes per foot} = \frac{385 \text{ min}}{1428 \text{ ft}}$$
$$\approx 0.27 \text{ min per ft}$$

12. a) 9.38 ft per min

b) 0.11 min per ft

13. The rate is given in millions of crimes per year, so we list number of crimes, in millions, on the vertical axis and years on the horizontal axis. If we count by 10's of millions on the vertical axis we can easily reach 37 million without needed a terribly large graph. We plot the point (1996, 37 million). Then, to display the rate of growth, we move from that point to a point that represents 2.5 million fewer crimes 1 year later. The coordinates of this point are 1996+1, 37−2.5 million), or (1997, 34.5 million). Finally, we draw a line through the two points.

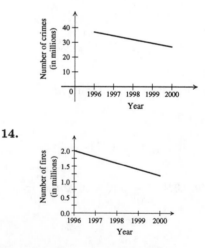

14.

15. The rate is given in miles per hour, so we list the number of miles traveled on the vertical axis and the time of day on the horizontal axis. If we count by 100's of miles on the vertical axis we can easily reach 230 without needing a terribly large graph. We plot the point (3:00, 230). Then to display the rate of travel, we move from that point to a point that represents 90 more miles traveled 1 hour later. The coordinates of this point are (3:00 + 1 hr, 230 + 90), or (4:00, 320). Finally, we draw a line through the two points.

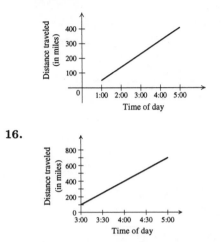

16.

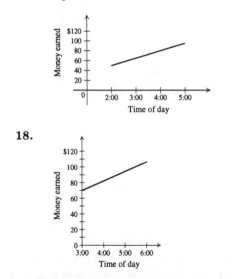

17. The rate is given in dollars per hour so we list money earned on the vertical axis and the time of day on the horizontal axis. We can count by $20 on the vertical axis and reach $50 without needing a terribly large graph. Next we plot the point (2:00 P.M., $50). To display the rate we move from that point to a point that represents $15 more 1 hour later. The coordinates of this point are (2 + 1, $50 + $15), or (3:00 P.M., $65). Finally, we draw a line through the two points.

18.

19. The rate is given in cost per minute so we list the amount of the telephone bill on the vertical axis and the number of additional minutes on the horizontal axis. We begin with $7.50 on the vertical axis and count by $0.50. A jagged line at the base of the axis indicates that we are not showing amounts smaller

than $7.50. We begin with 0 additional minutes on the horizontal axis and plot the point (0, $7.50). We move from there to a point that represents $0.10 more 1 minute later. The coordinates of this point are (0+1 min, $7.50+$0.10), or (1 min, $7.60). Then we draw a line through the two points.

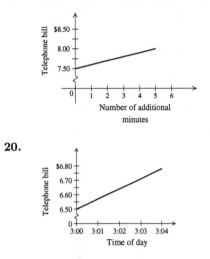

20.

21. The points (2:00, 7 haircuts) and (5:00, 12 haircuts) are on the graph. This tells us that in the 3 hr between 2:00 and 5:00 there were $12 - 7 = 5$ haircuts completed. The rate is

$$\frac{5 \text{ haircuts}}{3 \text{ hr}} = \frac{5}{3}, \text{ or } 1\frac{2}{3} \text{ haircuts per hour.}$$

22. 4 manicures per hour

23. The points (12:00, 100 mi) and (2:00, 250 mi) are on the graph. This tells us that in the 2 hr between 12:00 and 2:00 the train traveled $250 - 100 = 150$ mi. The rate is

$$\frac{150 \text{ mi}}{2 \text{hr}} = 75 \text{ mi per hr.}$$

24. 87.5 mi per hr

25. The points (15 min, 150¢) and (30 min, 300¢) are on the graph. This tells us that in $30 - 15 = 15$ min the cost of the call increased $300¢ - 150¢ = 150¢$. The rate is

$$\frac{150¢}{15 \text{ min}} = 10¢ \text{ per min.}$$

26. 7¢ per min

27. The points (2 yr, $2000) and (4 yr, $1000) are on the graph. This tells us that in $4 - 2 = 2$ yr the value of the copier changes $1000 - 2000 = -1000$. The rate is

$$\frac{-\$1000}{2 \text{ yr}} = -\$500 \text{ per yr.}$$

This means that the value of the copier is decreasing at a rate of $500 per yr.

28. −$0.15 billion per yr

29. The points (50 mi, 2 gal) and (200 mi, 8 gal) are on the graph. This tells us that when driven $200 - 50 = 150$ mi the vehicle consumed $8 - 2 = 6$ gal of gas. The rate is

$$\frac{6 \text{ gal}}{150 \text{ mi}} = 0.04 \text{ gal per mi.}$$

30. $0.08\overline{3}$ gal per mi

31. Writing Exercise

32. Writing Exercise

33. $-2 - (-7) = -2 + 7 = 5$

34. -6

35. $\dfrac{5 - (-4)}{-2 - 7} = \dfrac{9}{-9} = -1$

36. $-\dfrac{4}{3}$

37. $\dfrac{-4 - 8}{7 - (-2)} = \dfrac{-12}{9} = -\dfrac{4}{3}$

38. $-\dfrac{4}{5}$

39. Writing Exercise

40. Writing Exercise

41. Let $t =$ flight time and $a =$ altitude. While the plane is climbing at a rate of 6500 ft/min, the equation $a = 6500t$ describes the situation. Solving $34,000 = 6500t$, we find that the cruising altitude of 34,000 ft is reached after about 5.23 min. Thus we graph $a = 6500t$ for $0 \leq t \leq 5.23$.

The plane cruises at 34,000 ft for 3 min, so we graph $a = 34,000$ for $5.23 < t \leq 8.23$. After 8.23 min the plane descends at a rate of 3500 ft/min and lands. The equation $a = 34,000 - 3500(t - 8.23)$, or $a = -3500t + 62,805$, describes this situation. Solving $0 = -3500t + 62,805$, we find that the plane lands after about 17.94 min. Thus we graph $a = -3500t + 62,805$ for $8.23 < t \leq 17.94$. The entire graph is show below.

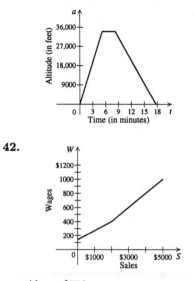

42.

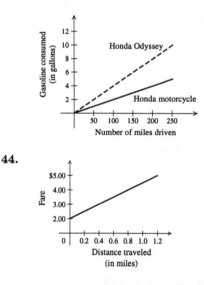

About $550

43. We begin with the graph in Exercise 29 showing the gas consumption of a Honda Odyssey. For each point (x, y) on this graph we can plot a point $(2x, y)$ on the graph that represents the gas consumption of the motorcycle.

44.

45. Penny walks forward at a rate of $\dfrac{24 \text{ ft}}{3 \text{ sec}}$, or 8 ft per sec. In addition, the boat is traveling at a rate of 5 ft per sec. Thus, with respect to land, Penny is traveling at a rate of $8 + 5$, or 13 ft per sec.

46. 0.45 min per mi

47. First we find Annette's speed in minutes per kilometer.

$$\text{Speed} = \frac{15.5 \text{ min}}{7 \text{ km} - 4 \text{ km}} = \frac{15.5 \text{ min}}{3 \text{ km}}$$

Now we convert min/km to min/mi.

$$\frac{15.5}{3} \frac{\text{min}}{\text{km}} \approx \frac{15.5}{3} \frac{\text{min}}{\text{km}} \cdot \frac{1 \text{ km}}{0.621 \text{ min}} \approx \frac{15.5}{1.863} \frac{\text{min}}{\text{mi}}$$

At a rate of $\dfrac{15.5}{1.863} \dfrac{\text{min}}{\text{mi}}$, to run a 5-mi race it would take $\dfrac{15.5}{1.863} \dfrac{\text{min}}{\text{mi}} \cdot 5 \text{ mi} \approx 41.6 \text{ min}$.

(Answers may vary slightly depending on the conversion factor used.)

48. 51.8 min

(Answers may vary slightly depending on the conversion factor used.)

49. First we find Ryan's rate. Then we double it to find Marcy's rate. Note that 50 minutes $= \dfrac{50}{60} \text{ hr} = \dfrac{5}{6} \text{ hr}$.

$$\begin{aligned}
\text{Ryan's rate} &= \frac{\text{change in number of bushels picked}}{\text{corresponding change in time}} \\[2mm]
&= \frac{5\frac{1}{2} - 4 \text{ bushels}}{\frac{5}{6} \text{ hr}} \\[2mm]
&= \frac{1\frac{1}{2} \text{ bushels}}{\frac{5}{6} \text{ hr}} \\[2mm]
&= \frac{3}{2} \cdot \frac{6}{5} \frac{\text{bushels}}{\text{hr}} \\[2mm]
&= \frac{9}{5} \text{ bushels per hour, or} \\[1mm]
&\quad 1.8 \text{ bushels per hour}
\end{aligned}$$

Then Marcy's rate is $2(1.8) = 3.6$ bushels per hour.

50. 27 candles per hour

Exercise Set 3.5

1. The rate can be found using the coordinates of any two points on the line. We use $(2, 30)$ and $(6, 90)$.

$$\begin{aligned}
\text{Rate} &= \frac{\text{change in number of calories burned}}{\text{corresponding change in time}} \\[2mm]
&= \frac{90 - 30 \text{ calories}}{6 - 2 \text{ min}} \\[2mm]
&= \frac{60 \text{ calories}}{4 \text{ min}} \\[2mm]
&= 15 \text{ calories per min}
\end{aligned}$$

2. 2.5 million people per year

3. The rate can be found using the coordinates of any two points on the line. We use $(35, 490)$ and $(45, 500)$, where 35 and 45 are in \$1000's.

$$\begin{aligned}
\text{Rate} &= \frac{\text{change in score}}{\text{corresponding change in income}} \\[2mm]
&= \frac{500 - 490 \text{ points}}{45 - 35} \\[2mm]
&= \frac{10 \text{ points}}{10} \\[2mm]
&= 1 \text{ point per \$1000 income}
\end{aligned}$$

4. $1\dfrac{1}{3}$ points per \$1000 income

5. The rate can be found using the coordinates of any two points on the line. We use $(1993, 20)$ and $(1997, 17)$.

$$\begin{aligned}
\text{Rate} &= \frac{\text{change in percent}}{\text{corresponding change in time}} \\[2mm]
&= \frac{17\% - 20\%}{1997 - 1993} \\[2mm]
&= \frac{-3\%}{4 \text{ yr}} \\[2mm]
&= -\frac{3}{4}\% \text{ per yr, or } -0.75\% \text{ per yr}
\end{aligned}$$

6. -0.4% per yr, or $-\dfrac{2}{5}\%$ per yr

7. We can use any two points on the line, such as $(0, 1)$ and $(4, 4)$.

$$\begin{aligned}
m &= \frac{\text{change in } y}{\text{change in } x} \\[2mm]
&= \frac{4 - 1}{4 - 0} = \frac{3}{4}
\end{aligned}$$

8. $\dfrac{2}{3}$

9. We can use any two points on the line, such as $(1, 0)$ and $(3, 3)$.

$$\begin{aligned}
m &= \frac{\text{change in } y}{\text{change in } x} \\[2mm]
&= \frac{3 - 0}{3 - 1} = \frac{3}{2}
\end{aligned}$$

10. $\dfrac{1}{3}$

11. We can use any two points on the line, such as $(-3, -4)$ and $(0, -3)$.

$$\begin{aligned}
m &= \frac{\text{change in } y}{\text{change in } x} \\[2mm]
&= \frac{-3 - (-4)}{0 - (-3)} = \frac{1}{3}
\end{aligned}$$

12. 3

13. We can use any two points on the line, such as $(0, 2)$ and $(2, 0)$.

$$m = \frac{\text{change in } y}{\text{change in } x}$$
$$= \frac{2 - 0}{0 - 2} = \frac{2}{-2} = -1$$

14. $-\dfrac{1}{2}$

15. This is the graph of a horizontal line. Thus, the slope is 0.

16. $-\dfrac{3}{2}$

17. We can use any two points on the line, such as $(0, 2)$ and $(3, 1)$.

$$m = \frac{\text{change in } y}{\text{change in } x}$$
$$= \frac{1 - 2}{3 - 0} = -\frac{1}{3}$$

18. -2

19. This is the graph of a vertical line. Thus, the slope is undefined.

20. Undefined

21. We can use any two points on the line, such as $(-2, 3)$ and $(2, 2)$.

$$m = \frac{\text{change in } y}{\text{change in } x}$$
$$= \frac{2 - 3}{2 - (-2)} = -\frac{1}{4}$$

22. 0

23. We can use any two points on the line, such as $(-2, -3)$ and $(2, 3)$.

$$m = \frac{\text{change in } y}{\text{change in } x}$$
$$= \frac{3 - (-3)}{2 - (-2)} = \frac{6}{4} = \frac{3}{2}$$

24. $-\dfrac{2}{3}$

25. This is the graph of a horizontal line, so the slope is 0.

26. 5

27. We can use any two points on the line, such as $(-3, 5)$ and $(0, -4)$.

$$m = \frac{\text{change in } y}{\text{change in } x}$$
$$= \frac{-4 - 5}{0 - (-3)} = \frac{-9}{3} = -3$$

28. 0

29. $(1, 2)$ and $(5, 8)$

$$m = \frac{8 - 2}{5 - 1} = \frac{6}{4} = \frac{3}{2}$$

30. 2

31. $(-2, 4)$ and $(3, 0)$

$$m = \frac{4 - 0}{-2 - 3} = \frac{4}{-5} = -\frac{4}{5}$$

32. $-\dfrac{5}{6}$

33. $(-4, 0)$ and $(5, 7)$

$$m = \frac{7 - 0}{5 - (-4)} = \frac{7}{9}$$

34. $\dfrac{2}{3}$

35. $(0, 8)$ and $(-3, 10)$

$$m = \frac{8 - 10}{0 - (-3)} = \frac{8 - 10}{0 + 3} = \frac{-2}{3} = -\frac{2}{3}$$

36. $-\dfrac{1}{2}$

37. $(-2, 3)$ and $(-6, 5)$

$$m = \frac{5 - 3}{-6 - (-2)} = \frac{2}{-6 + 2} = \frac{2}{-4} = -\frac{1}{2}$$

38. $-\dfrac{11}{8}$

39. $\left(-2, \dfrac{1}{2}\right)$ and $\left(-5, \dfrac{1}{2}\right)$

Observe that the points have the same y-coordinate. Thus, they lie on a horizontal line and its slope is 0. We could also compute the slope.

$$m = \frac{\dfrac{1}{2} - \dfrac{1}{2}}{-2 - (-5)} = \frac{\dfrac{1}{2} - \dfrac{1}{2}}{-2 + 5} = \frac{0}{3} = 0$$

40. $\dfrac{4}{7}$

41. $(3, 4)$ and $(9, -7)$

$$m = \frac{-7 - 4}{9 - 3} = \frac{-11}{6} = -\frac{11}{6}$$

42. Undefined

43. $(6, -4)$ and $(6, 5)$

Observe that the points have the same x-coordinate. Thus, they lie on a vertical line and its slope is undefined. We could also compute the slope.

$$m = \frac{-4 - 5}{6 - 6} = \frac{-9}{0}, \text{ undefined}$$

44. 0

45. The line $x = -3$ is a vertical line. The slope is undefined.

46. Undefined

47. The line $y = 4$ is a horizontal line. A horizontal line has slope 0.

48. 0

49. The line $x = 9$ is a vertical line. The slope is undefined.

50. Undefined

51. The line $y = -9$ is a horizontal line. A horizontal line has slope 0.

52. 0

53. The grade is expressed as a percent.

$$m = \frac{106}{1325} = 0.08 = 8\%$$

54. $0\frac{1}{20}$, or 0.05

55. The grade is expressed as a percent.

$$m = \frac{1}{12} = 0.08\overline{3} = 8.\overline{3}\%$$

56. 7%

57. $m = \frac{2.4}{8.2} = \frac{12}{41}$, or about 29%

58. 0.08, or 8%

59. Longs Peak rises $14,255 - 9600 = 4655$ ft.

$$m = \frac{4655}{15,840} \approx 0.29 \approx 29\%$$

60. About 64%

61. Writing Exercise

62. Writing Exercise

63. $ax + by = c$

$by = c - ax$ Adding $-ax$ to both sides

$y = \dfrac{c - ax}{b}$ Dividing both sides by b

64. $r = \dfrac{p + mn}{x}$

65. $ax - by = c$

$-by = c - ax$ Adding $-ax$ to both sides

$y = \dfrac{c - ax}{-b}$ Dividing both sides by $-b$

We could also express this result as $y = \dfrac{ax - c}{b}$.

66. $t = \dfrac{q - rs}{n}$

67. $\dfrac{2}{3}x - 5 = \dfrac{2}{3} \cdot 12 - 5$ Substituting

$= 8 - 5$

$= 3$

68. 2

69. Writing Exercise

70. Writing Exercise

71. If the line passes through $(4, -7)$ and never enters the first quadrant, then it slants down from left to right or is horizontal. This means that its slope is not positive ($m \leq 0$). The line will slant most steeply if it passes through $(0, 0)$. In this case, $m = \dfrac{-7 - 0}{4 - 0} = -\dfrac{7}{4}$. Thus, the numbers the line could have for its slope are $\left\{ m \middle| -\dfrac{7}{4} \leq m \leq 0 \right\}$.

72. $\left\{ m \middle| m \geq \dfrac{5}{2} \right\}$

73. $x + y = 18$

$y = 18 - x$

The slope is $\dfrac{y}{x}$, or $\dfrac{18 - x}{x}$.

74. $\dfrac{1}{2}$

75. Let $t =$ the number of units each tick mark on the horizontal axis represents. Note that the graph drops 1 unit for every 6 tick marks of horizontal change. Then we have:

$$\frac{-1}{6t} = -\frac{2}{3}$$

$$-1 = -4t$$

$$\frac{1}{4} = t$$

Each tick mark on the horizontal axis represents $\frac{1}{4}$ unit.

Exercise Set 3.6

1. Slope $\frac{2}{5}$; y-intercept $(0, 1)$

We plot $(0, 1)$ and from there move up 2 units and right 5 units. This locates the point $(5, 3)$. We plot $(5, 3)$ and draw a line passing through $(0, 1)$ and $(5, 3)$.

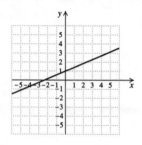

2.

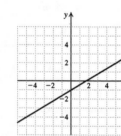

3. Slope $\frac{5}{3}$; y-intercept $(0, -2)$

We plot $(0, -2)$ and from there move up 5 units and right 3 units. This locates the point $(3, 3)$. We plot $(3, 3)$ and draw a line passing through $(0, -2)$ and $(3, 3)$.

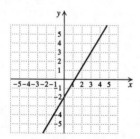

4.

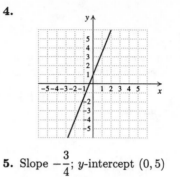

5. Slope $-\frac{3}{4}$; y-intercept $(0, 5)$

We plot $(0, 5)$. We can think of the slope as $\frac{-3}{4}$, so from $(0, 5)$ we move down 3 units and right 4 units. This locates the point $(4, 2)$. We plot $(4, 2)$ and draw a line passing through $(0, 5)$ and $(4, 2)$.

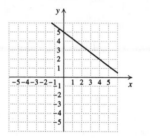

6.

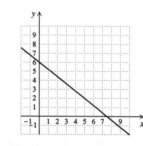

7. Slope 2; y-intercept $(0, -4)$

We plot $(0, -4)$. We can think of the slope as $\frac{2}{1}$, so from $(0, -4)$ we move up 2 units and right 1 unit. This locates the point $(1, -2)$. We plot $(1, -2)$ and draw a line passing through $(0, -4)$ and $(1, -2)$.

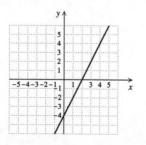

8.

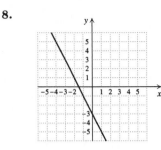

9. Slope -3; y-intercept $(0, 2)$

We plot $(0, 2)$. We can think of the slope as $\dfrac{-3}{1}$, so from $(0, 2)$ we move down 3 units and right 1 unit. This locates the point $(1, -1)$. We plot $(1, -1)$ and draw a line passing through $(0, 2)$ and $(1, -1)$.

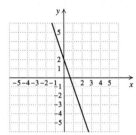

10.

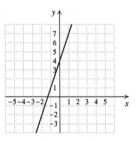

11. We read the slope and y-intercept from the equation.

$$y = \frac{3}{7}x + 5$$

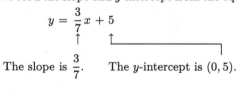

The slope is $\dfrac{3}{7}$.   The y-intercept is $(0, 5)$.

12. $-\dfrac{3}{8}$, $(0, 6)$

13. We read the slope and y-intercept from the equation.

$$y = -\frac{5}{6}x + 2$$

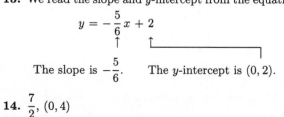

The slope is $-\dfrac{5}{6}$.   The y-intercept is $(0, 2)$.

14. $\dfrac{7}{2}$, $(0, 4)$

15. $y = \dfrac{9}{4}x - 7$

$$y = \frac{9}{4}x + (-7)$$

The slope is $\dfrac{9}{4}$, and the y-intercept is $(0, -7)$.

16. $\dfrac{2}{9}$; $(0, -1)$

17. $y = -\dfrac{2}{5}x$

$$y = -\frac{2}{5}x + 0$$

The slope is $-\dfrac{2}{5}$, and the y-intercept is $(0, 0)$.

18. $\dfrac{4}{3}$; $(0, 0)$

19. We solve for y to rewrite the equation in the form $y = mx + b$.

$$-2x + y = 4$$
$$y = 2x + 4$$

The slope is 2, and the y-intercept is $(0, 4)$.

20. 5; $(0, 5)$

21. $3x - 4y = 12$

$$-4y = -3x + 12$$
$$y = -\frac{1}{4}(-3x + 12)$$
$$y = \frac{3}{4}x - 3$$

The slope is $\dfrac{3}{4}$, and the y-intercept is $(0, -3)$.

22. $\dfrac{3}{2}$; $(0, -9)$

23. $x - 5y = -8$

$$-5y = -x - 8$$
$$y = -\frac{1}{5}(-x - 8)$$
$$y = \frac{1}{5}x + \frac{8}{5}$$

The slope is $\dfrac{1}{5}$, and the y-intercept is $\left(0, \dfrac{8}{5}\right)$.

24. $\dfrac{1}{6}$; $\left(0, -\dfrac{3}{2}\right)$

25. Observe that this is the equation of a horizontal line that lies 4 units above the x-axis. Thus, the slope is 0, and the y-intercept is $(0, 4)$. We could also write the equation in slope-intercept form.

$y = 4$

$y = 0x + 4$

The slope is 0, and the y-intercept is $(0, 4)$.

26. 0; $(0, 8)$

27. We use the slope-intercept equation, substituting 3 for m and 7 for b:

$$y = mx + b$$

$$y = 3x + 7$$

28. $y = -4x - 2$

29. We use the slope-intercept equation, substituting $\frac{7}{8}$ for m and -1 for b:

$$y = mx + b$$

$$y = \frac{7}{8}x - 1$$

30. $y = \frac{5}{7}x + 4$

31. We use the slope-intercept equation, substituting $-\frac{5}{3}$ for m and -8 for b:

$$y = mx + b$$

$$y = -\frac{5}{3}x - 8$$

32. $y = \frac{3}{4}x + 23$

33. Since the slope is 0, we know that the line is horizontal. Its y-intercept is $(0, 3)$, so the equation of the line must be $y = 3$.

We could also use the slope-intercept equation, substituting 0 for m and 3 for b.

$$y = mx + b$$

$$y = 0 \cdot x + 3$$

$$y = 3$$

34. $y = 7x$

35. $y = \frac{3}{5}x + 2$

First we plot the y-intercept $(0, 2)$. We can start at the y-intercept and use the slope, $\frac{3}{5}$, to find another point. We move up 3 units and right 5 units to get a new point $(5, 5)$. Thinking of the slope as $\frac{-3}{-5}$ we can start at $(0, 2)$ and move down 3 units and left 5 units to get another point $(-5, -1)$.

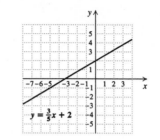

36.

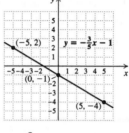

37. $y = -\frac{3}{5}x + 1$

First we plot the y-intercept $(0, 1)$. We can start at the y-intercept and, thinking of the slope as $\frac{-3}{5}$, find another point by moving down 3 units and right 5 units to the point $(5, -2)$. Thinking of the slope as $\frac{3}{-5}$ we can start at $(0, 1)$ and move up 3 units and left 5 units to get another point $(-5, 4)$.

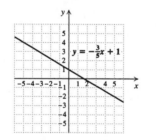

38.

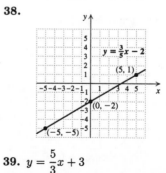

39. $y = \frac{5}{3}x + 3$

First we plot the y-intercept $(0, 3)$. We can start at the y-intercept and use the slope, $\frac{5}{3}$, to find another point. We move up 5 units and right 3 units to get a new point $(3, 8)$. Thinking of the slope as $\frac{-5}{-3}$ we can start at $(0, 3)$ and move down 5 units and left 3 units to get another point $(-3, -2)$.

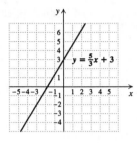

40.

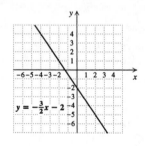

41. $y = -\frac{3}{2}x - 2$

First we plot the y-intercept $(0, -2)$. We can start at the y-intercept and, thinking of the slope as $\frac{-3}{2}$, find another point by moving down 3 units and right 2 units to the point $(2, -5)$. Thinking of the slope as $\frac{3}{-2}$ we can start at $(0, -2)$ and move up 3 units and left 2 units to get another point $(-2, 1)$.

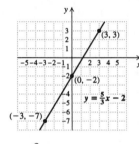

42.

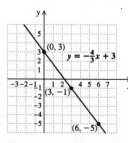

43. We first rewrite the equation in slope-intercept form.
$$2x + y = 1$$
$$y = -2x + 1$$

Now we plot the y-intercept $(0, 1)$. We can start at the y-intercept and, thinking of the slope as $\frac{-2}{1}$, find another point by moving down 2 units and right 1 unit to the point $(1, -1)$. In a similar manner, we can move from the point $(1, -1)$ to find a third point $(2, -3)$.

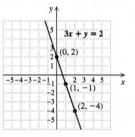

44.

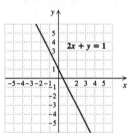

45. We first rewrite the equation in slope-intercept form.
$$3x - y = 4$$
$$-y = -3x + 4$$
$$y = 3x - 4 \quad \text{Multiplying by } -1$$

Now we plot the y-intercept $(0, -4)$. We can start at the y-intercept and, thinking of the slope as $\frac{3}{1}$, find another point by moving up 3 units and right 1 unit to the point $(1, -1)$. In a similar manner, we can move from the point $(1, -1)$ to find a third point $(2, 2)$.

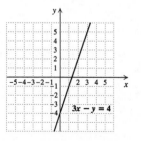

46.

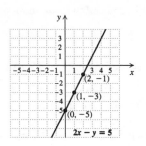

47. We first rewrite the equation in slope-intercept form.
$$2x + 3y = 9$$
$$3y = -2x + 9$$
$$y = \frac{1}{3}(-2x + 9)$$
$$y = -\frac{2}{3}x + 3$$

Now we plot the y-intercept $(0, 3)$. We can start at the y-intercept and, thinking of the slope as $\frac{-2}{3}$, find another point by moving down 2 units and right 3 units to the point $(3, 1)$. Thinking of the slope as $\frac{2}{-3}$ we can start at $(0, 3)$ and move up 2 units and left 3 units to get another point $(-3, 5)$.

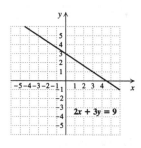

48.

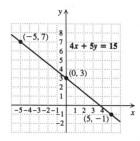

49. We first rewrite the equation in slope-intercept form.
$$x - 4y = 12$$
$$-4y = -x + 12$$
$$y = -\frac{1}{4}(-x + 12)$$
$$y = \frac{1}{4}x - 3$$

Now we plot the y-intercept $(0, -3)$. We can start at the y-intercept and use the slope, $\frac{1}{4}$, to find another point. We move up 1 unit and right 4 units to the point $(4, -2)$. Thinking of the slope as $\frac{-1}{-4}$ we can start at $(0, -3)$ and move down 1 unit and left 4 units to get another point $(-4, -4)$.

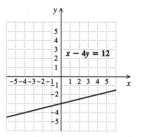

50.

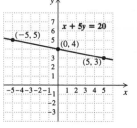

51. Two points on the graph are $(0, 9)$ and $(7, 16)$, so we see that the y-intercept will be $(0, 9)$. Now we find the slope:
$$m = \frac{16 - 9}{7 - 0} = \frac{7}{7} = 1$$
Then the equation is $y = x + 9$.

To graph the equation we first plot $(0, 9)$. We can think of the slope as $\frac{1}{1}$, so from the y-intercept we move up 1 unit and right 1 unit to the point $(1, 10)$. We plot $(1, 10)$ and draw the line passing through $(0, 9)$ and $(1, 10)$.

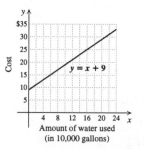

Since the slope is 1, the rate is \$1 per 10,000 gallons.

52.

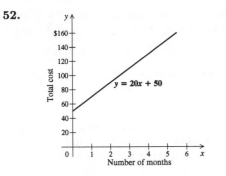

$20 per month

53. Two points on the graph are $(0, 16)$ and $(1, 16 + 1.5)$, or $(1, 17.5)$, so the y-intercept will be $(0, 16)$. Now we find the slope:
$$m = \frac{17.5 - 16}{1 - 0} = \frac{1.5}{1} = 1.5$$
Then the equation is $y = 1.5x + 16$.

54. $y = 0.07x + 4.95$

55. $y = \frac{2}{3}x + 7$: The slope is $\frac{2}{3}$, and the y-intercept is $(0, 7)$.

$y = \frac{2}{3}x - 5$: The slope is $\frac{2}{3}$, and the y-intercept is $(0, -5)$.

Since both lines have slope $\frac{2}{3}$ but different y-intercepts, their graphs are parallel.

56. No

57. The equation $y = 2x - 5$ represents a line with slope 2 and y-intercept $(0, -5)$. We rewrite the second equation in slope-intercept form.
$$4x + 2y = 9$$
$$2y = -4x + 9$$
$$y = \frac{1}{2}(-4x + 9)$$
$$y = -2x + \frac{9}{2}$$

The slope is -2 and the y-intercept is $\left(0, \frac{9}{2}\right)$. Since the lines have different slopes, their graphs are not parallel.

58. Yes

59. Rewrite each equation in slope-intercept form.
$$3x + 4y = 8$$
$$4y = -3x + 8$$
$$y = \frac{1}{4}(-3x + 8)$$
$$y = -\frac{3}{4}x + 2$$

The slope is $-\frac{3}{4}$, and the y-intercept is $(0, 2)$.
$$7 - 12y = 9x$$
$$-12y = 9x - 7$$
$$y = -\frac{1}{12}(9x - 7)$$
$$y = -\frac{3}{4}x + \frac{7}{12}$$

The slope is $-\frac{3}{4}$, and the y-intercept is $\left(0, \frac{7}{12}\right)$.

Since both lines have slope $-\frac{3}{4}$ but different y-intercepts, their graphs are parallel.

60. No

61. $y = 4x - 5$,
$\quad 4y = 8 - x$

The first equation is in slope-intercept form. It represents a line with slope 4. Now we rewrite the second equation in slope-intercept form.
$$4y = 8 - x$$
$$y = \frac{1}{4}(8 - x)$$
$$y = 2 - \frac{1}{4}x$$
$$y = -\frac{1}{4}x + 2$$

The slope of the line is $-\frac{1}{4}$.

Since $4\left(-\frac{1}{4}\right) = -1$, the equations represent perpendicular lines.

62. No

63. $y - 2y = 5$,
$\quad 2x + 4y = 8$

We write each equation in slope-intercept form.
$$x - 2y = 5$$
$$-2y = -x + 5$$
$$y = -\frac{1}{2}(-x + 5)$$
$$y = \frac{1}{2}x - \frac{5}{2}$$

The slope is $\frac{1}{2}$.

$$2x + 4y = 8$$
$$4y = -2x + 8$$
$$y = \frac{1}{4}(-2x + 8)$$
$$y = -\frac{1}{2}x + 2$$

The slope is $-\frac{1}{2}$.

Since $\frac{1}{2}\left(-\frac{1}{2}\right) = -\frac{1}{4} \neq -1$, the equations do not represent perpendicular lines.

64. Yes

65. $2x + 3y = 1$,
$3x - 2y = 1$

We write each equation in slope-intercept form.

$$2x + 3y = 1$$
$$3y = -2x + 1$$
$$y = \frac{1}{3}(-2x + 1)$$
$$y = -\frac{2}{3}x + \frac{1}{3}$$

The slope is $-\frac{2}{3}$.

$$3x - 2y = 1$$
$$-2y = -3x + 1$$
$$y = -\frac{1}{2}(-3x + 1)$$
$$y = \frac{3}{2}x - \frac{1}{2}$$

The slope is $\frac{3}{2}$.

Since $-\frac{2}{3}\left(\frac{3}{2}\right) = -1$, the equations represent perpendicular lines.

66. No

67. The slope of the line represented by $y = 5x - 7$ is 5. Then a line parallel to the graph of $y = 5x - 7$ has slope 5 also. Since the y-intercept is $(0, 11)$, the desired equation is $y = 5x + 11$.

68. $y = 2x - 3$

69. First find the slope of the line represented by $2x + y = 0$.

$$2x + y = 0$$
$$y = -2x$$

The slope is -2. Then the slope of a line perpendicular to the graph of $2x + y = 0$ is the negative reciprocal of -2, or $\frac{1}{2}$. Since the y-intercept is $(0, 0)$, the desired equation is $y = \frac{1}{2}x + 0$, or $y = \frac{1}{2}x$.

70. $y = -3x + 5$

71. The slope of the line represented by $y = x$ is 1. Then a line parallel to this line also has slope 1. Since the y-intercept is $(0, 3)$, the desired equation is $y = 1 \cdot x + 3$, or $y = x + 3$.

72. $y = -x$

73. First find the slope of the line represented by $x + y = 3$.

$$x + y = 3$$
$$y = -x + 3, \text{ or } y = -1 \cdot x + 3$$

The slope is -1. Then the slope of a line perpendicular to this line is the negative reciprocal is -1, or 1. Since the y-intercept is -4, the desired equation is $y = 1 \cdot x - 4$, or $y = x - 4$.

74. $y = -\frac{3}{2}x - 1$

75. Writing Exercise

76. Writing Exercise

77. $y - k = m(x - h)$
$\quad y = m(x - h) + k \quad$ Adding k to both sides

78. $y = -2(x + 4) + 9$

79. $-5 - (-7) = -5 + 7 = 2$

80. 16

81. $-3 - 6 = -3 + (-6) = -9$

82. -10

83. Writing Exercise

84. Writing Exercise

85. See the answer section in the text.

86. $y = -\frac{5}{2}x - \frac{10}{7}$

87. Rewrite each equation in slope-intercept form.

$$2x - 6y = 10$$
$$-6y = -2x + 10$$
$$y = \frac{1}{3}x - \frac{5}{3}$$

The slope of the line is $\frac{1}{3}$.

$$9x + 6y = 18$$
$$6y = -9x + 18$$
$$y = -\frac{3}{2}x + 3$$

The y-intercept of the line is $(0, 3)$.

The equation of the line is $y = \frac{1}{3}x + 3$.

88. $y = \frac{3}{2}x - 2$

89. Rewrite $2x + 5y = 6$ in slope-intercept form.
$$2x + 5y = 6$$
$$5y = -2x + 6$$
$$y = \frac{1}{5}(-2x + 6)$$
$$y = -\frac{2}{5}x + \frac{6}{5}$$

The slope is $-\frac{2}{5}$.

The slope of a line perpendicular to this line is a number m such that
$$-\frac{2}{5}m = -1, \text{ or}$$
$$m = \frac{5}{2}.$$

We graph the line whose equation we want to find. First we plot the given point $(2, 6)$. Now think of the slope as $\frac{-5}{-2}$. From the point $(2, 6)$ go down 5 units and left 2 units to the point $(0, 1)$. Plot this point and draw the graph.

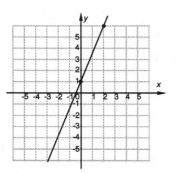

We see that the y-intercept is $(0, 1)$, so the desired equation is $y = \frac{5}{2}x + 1$.

90. Writing Exercise

Exercise Set 3.7

1. $y - y_1 = m(x - x_1)$

We substitute 6 for m, 2 for x_1, and 7 for y_1.
$$y - 7 = 6(x - 2)$$

2. $y - 5 = 4(x - 3)$

3. $y - y_1 = m(x - x_1)$

We substitute $\frac{3}{5}$ for m, 9 for x_1, and 2 for y_1.
$$y - 2 = \frac{3}{5}(x - 9)$$

4. $y - 1 = \frac{2}{3}(x - 4)$

5. $y - y_1 = m(x - x_1)$

We substitute -4 for m, 3 for x_1, and 1 for y_1.
$$y - 1 = -4(x - 3)$$

6. $y - 2 = -5(x - 6)$

7. $y - y_1 = m(x - x_1)$

We substitute $\frac{3}{2}$ for m, 5 for x_1, and -4 for y_1.
$$y - (-4) = \frac{3}{2}(x - 5)$$

8. $y - (-1) = \frac{4}{3}(x - 7)$

9. $y - y_1 = m(x - x_1)$

We substitute $\frac{5}{4}$ for m, -2 for x_1, and 6 for y_1.
$$y - 6 = \frac{5}{4}(x - (-2))$$

10. $y - 4 = \frac{7}{2}(x - (-3))$

11. $y - y_1 = m(x - x_1)$

We substitute -2 for m, -4 for x_1, and -1 for y_1.
$$y - (-1) = -2(x - (-4))$$

12. $y - (-5) = -3(x - (-2))$

13. $y - y_1 = m(x - x_1)$

We substitute 1 for m, -2 for x_1, and 8 for y_1.
$$y - 8 = 1(x - (-2))$$

14. $y - 6 = -1(x - (-3))$

15. First we write the equation in point-slope form.
$$y - y_1 = m(x - x_1)$$
$$y - 7 = 2(x - 5) \quad \text{Substituting}$$

Next we find an equivalent equation of the form $y = mx + b$.
$$y - 7 = 2(x - 5)$$
$$y - 7 = 2x - 10$$
$$y = 2x - 3$$

16. $y = 3x - 16$

17. First we write the equation in point-slope form.
$$y - y_1 = m(x - x_1)$$
$$y - (-2) = \frac{7}{4}(x - 4) \quad \text{Substituting}$$

Next we find an equivalent equation of the form $y = mx + b$.
$$y - (-2) = \frac{7}{4}(x - 4)$$
$$y + 2 = \frac{7}{4}x - 7$$
$$y = \frac{7}{4}x - 9$$

18. $y = \frac{8}{3}x - 12$

19. First we write the equation in point-slope form.
$$y - y_1 = m(x - x_1)$$
$$y - (-5) = -3(x - 1)$$

Next we find an equivalent equation of the form $y = mx + b$.
$$y - (-5) = -3(x - 1)$$
$$y + 5 = -3x + 3$$
$$y = -3x - 2$$

20. $y = -2x + 5$

21. First we write the equation in point-slope form.
$$y - y_1 = m(x - x_1)$$
$$y - (-1) = -4(x - (-2))$$

Next we find an equivalent equation of the form $y = mx + b$.
$$y - (-1) = -4(x - (-2))$$
$$y + 1 = -4(x + 2)$$
$$y + 1 = -4x - 8$$
$$y = -4x - 9$$

22. $y = -5x - 9$

23. First we write the equation in point-slope form.
$$y - y_1 = m(x - x_1)$$
$$y - 5 = \frac{2}{3}(x - 6)$$

Next we find an equivalent equation of the form $y = mx + b$.

$$y - 5 = \frac{2}{3}(x - 6)$$
$$y - 5 = \frac{2}{3}x - 4$$
$$y = \frac{2}{3}x + 1$$

24. $y = \frac{3}{2}x + 1$

25. First we write the equation in point-slope form.
$$y - y_1 = m(x - x_1)$$
$$y - 2 = -\frac{5}{6}(x - 3)$$

Next we find an equivalent equation of the form $y = mx + b$.
$$y - 2 = -\frac{5}{6}(x - 3)$$
$$y - 2 = -\frac{5}{6}x + \frac{5}{2}$$
$$y = -\frac{5}{6}x + \frac{9}{2}$$

26. $y = -\frac{3}{4}x + \frac{13}{2}$

27. We plot $(1, 2)$, move up 4 and to the right 3 to $(4, 6)$ and draw the line.

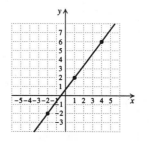

28.

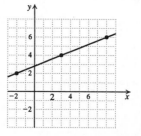

29. We plot $(2, 5)$, move down 3 and to the right 4 to $(6, 2)$ $\left(\text{since} -\frac{3}{4} = \frac{-3}{4}\right)$, and draw the line. We could also think of $-\frac{3}{4}$ and $\frac{3}{-4}$ and move up 3 and to the left 4 from the point $(2, 5)$ to $(-2, 8)$.

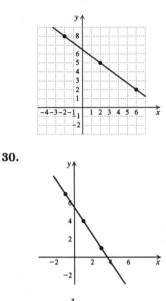

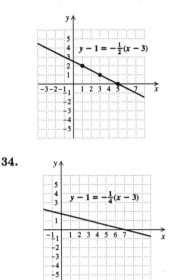

30.

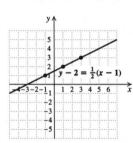

34.

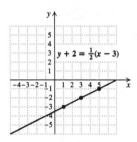

31. $y - 2 = \dfrac{1}{2}(x - 1)$ Point-slope form

The line has slope $\dfrac{1}{2}$ and passes through $(1, 2)$. We plot $(1, 2)$ and then find a second point by moving up 1 unit and right 2 units to $(3, 3)$. We draw the line through these points.

35. $y + 2 = \dfrac{1}{2}(x - 3)$, or $y - (-2) = \dfrac{1}{2}(x - 3)$

The line has slope $\dfrac{1}{2}$ and passes through $(3, -2)$. We plot $(3, -2)$ and then find a second point by moving up 1 unit and right 2 units to $(5, -1)$. We draw the line through these points.

32.

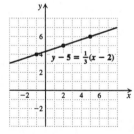

36.

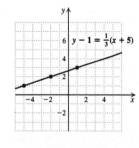

33. $y - 1 = -\dfrac{1}{2}(x - 3)$ Point-slope form

The line has slope $-\dfrac{1}{2}$, or $\dfrac{1}{-2}$ passes through $(3, 1)$. We plot $(3, 1)$ and then find a second point by moving up 1 unit and left 2 units to $(1, 2)$. We draw the line through these points.

37. $y + 4 = 3(x + 1)$, or $y - (-4) = 3(x - (-1))$

The line has slope 3, or $\dfrac{3}{1}$, and passes through $(-1, -4)$. We plot $(-1, -4)$ and then find a second point by moving up 3 units and right 1 unit to $(0, -1)$. We draw the line through these points.

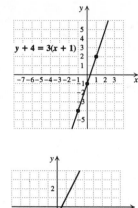

38.

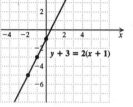

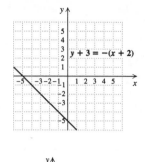

42.

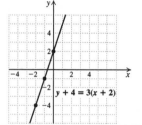

39. $y - 4 = -2(x + 1)$, or $y - 4 = -2(x - (-1))$

The line has slope -2, or $\dfrac{-2}{1}$, and passes through $(-1, 4)$. We plot $(-1, 4)$ and then find a second point by moving down 2 units and right 1 unit to $(0, 2)$. We draw the line through these points.

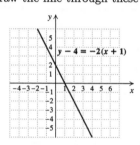

40.

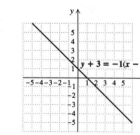

41. $y + 3 = -(x + 2)$, or $y - (-3) = -1(x - (-2))$

The line has slope -1, or $\dfrac{-1}{1}$, and passes through $(-2, -3)$. We plot $(-2, -3)$ and then find a second point by moving down 1 unit and right 1 unit to $(-1, -4)$. We draw the line through these points.

43. $y + 1 = -\dfrac{3}{5}(x + 2)$, or $y - (-1) = -\dfrac{3}{5}(x - (-2))$

The line has slope $-\dfrac{3}{5}$, or $\dfrac{-3}{5}$ and passes through $(-2, -1)$. We plot $(-2, -1)$ and then find a second point by moving down 3 units and right 5 units to $(3, -4)$, and draw the line.

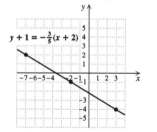

44.

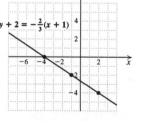

45. $y - 1 = -\dfrac{7}{2}(x + 5)$, or $y - 1 = -\dfrac{7}{2}(x - (-5))$

The line has slope $-\dfrac{7}{2}$, or $\dfrac{-7}{2}$ and passes through $(-5, 1)$. We plot $(-5, 1)$ and then find a second point by moving down 7 units and right 2 units to $(-3, -6)$, and draw the line.

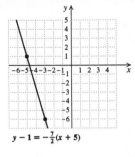

$$y - 1 = -\frac{7}{2}(x + 5)$$

46.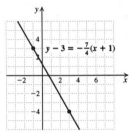

$$y - 3 = -\frac{7}{4}(x + 1)$$

47. Writing Exercise

48. Writing Exercise

49. $(-5)^3 = (-5)(-5)(-5) = -125$

50. 64

51. $3 \cdot 2^4 - 5 \cdot 2^3$

   $= 3 \cdot 16 - 5 \cdot 8$ Evaluating the exponential
                        expressions

   $= 48 - 40$ Multiplying

   $= 8$ Subtracting

52. 24

53. $(-2)^3(-3)^2 = -8 \cdot 9 = -72$

54. -4

55. Writing Exercise

56. Writing Exercise

57. $y - 3 = 0(x - 52)$

Observe that the slope is 0. Then this is the equation of a horizontal line that passes through $(52, 3)$. Thus, its graph is a horizontal line 3 units above the x-axis.

$$y - 3 = 0(x - 52)$$

58.

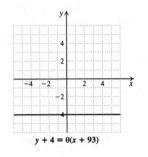

$$y + 4 = 0(x + 93)$$

59. First find the slope of the line passing through $(1, 2)$ and $(3, 7)$.

$$m = \frac{7 - 2}{3 - 1} = \frac{5}{2}$$

Then write an equation of the line containing $(1, 2)$ and having slope $\frac{5}{2}$.

$$y - 2 = \frac{5}{2}(x - 1)$$

We can also write an equation of the line containing $(3, 7)$ and having slope $\frac{5}{2}$.

$$y - 7 = \frac{5}{2}(x - 3)$$

60. $y - 1 = \frac{1}{2}(x - 3);\ y - 3 = \frac{1}{2}(x - 7)$

61. First find the slope of the line passing through $(-1, 2)$ and $(3, 8)$.

$$m = \frac{2 - 8}{-1 - 3} = \frac{-6}{-4} = \frac{3}{2}$$

Then write an equation of the line containing $(-1, 2)$ and having slope $\frac{3}{2}$.

$$y - 2 = \frac{3}{2}(x - (-1))$$

We can also write an equation of the line containing $(3, 8)$ and having slope $\frac{3}{2}$.

$$y - 8 = \frac{3}{2}(x - 3)$$

62. $y - 1 = \frac{2}{7}(x - (-3));\ y - 3 = \frac{2}{7}(x - 4)$

63. First find the slope of the line passing through $(-3, 8)$ and $(1, -2)$.

$$m = \frac{8 - (-2)}{-3 - 1} = \frac{10}{-4} = -\frac{5}{2}$$

Then write an equation of the line containing $(-3, 8)$ and having slope $-\frac{5}{2}$.

$$y - 8 = -\frac{5}{2}(x - (-3))$$

We can also write an equation of the line containing $(1, -2)$ and having slope $-\frac{5}{2}$.

$$y - (-2) = -\frac{5}{2}(x - 1)$$

64. $y - 7 = -\frac{5}{3}(x - (-2)); \ y - (-3) = -\frac{5}{3}(x - 4)$

65. First we find the slope of the line using any two points on the line. We will use $(3, -3)$ and $(4, -1)$.

$$m = \frac{-3 - (-1)}{3 - 4} = \frac{-2}{-1} = 2$$

Then we write an equation of the line in point-slope form using either of the points above.

$$y - (-3) = 2(x - 3)$$

Finally, we find an equivalent equation in slope-intercept form.

$$y - (-3) = 2(x - 3)$$
$$y + 3 = 2x - 6$$
$$y = 2x - 9$$

66. $y = -3x + 7$

67. First we find the slope of the line using any two points on the line. We will use $(2, 5)$ and $(5, 1)$.

$$m = \frac{5 - 1}{2 - 5} = \frac{4}{-3} = -\frac{4}{3}$$

Then we write an equation of the line in point-slope form using either of the points above.

$$y - 5 = -\frac{4}{3}(x - 2)$$

Finally, we find an equivalent equation in slope-intercept form.

$$y - 5 = -\frac{4}{3}(x - 2)$$
$$y - 5 = -\frac{4}{3}x + \frac{8}{3}$$
$$y = -\frac{4}{3}x + \frac{23}{3}$$

68. $y = \frac{5}{3}x + \frac{28}{3}$

69. $(1, 5)$ and $(4, 2)$

First we find the slope.

$$m = \frac{5 - 2}{1 - 4} = \frac{3}{-3} = -1$$

Then we write an equation of the line in point-slope form using either of the points above.

$$y - 5 = -1(x - 1)$$

Finally, we find an equivalent equation in slope-intercept form.

$$y - 5 = -1(x - 1)$$
$$y - 5 = -x + 1$$
$$y = -x + 6$$

70. $y = x + 4$

71. $(-3, 1)$ and $(3, 5)$

First we find the slope.

$$m = \frac{1 - 5}{-3 - 3} = \frac{-4}{-6} = \frac{2}{3}$$

Then we write an equation of the line in point-slope form using either of the points above.

$$y - 5 = \frac{2}{3}(x - 3)$$

Finally, we find an equivalent equation in slope-intercept form.

$$y - 5 = \frac{2}{3}(x - 3)$$
$$y - 5 = \frac{2}{3}x - 2$$
$$y = \frac{2}{3}x + 3$$

72. $y = \frac{1}{2}x + 4$

73. $(5, 0)$ and $(0, -2)$

First we find the slope.

$$m = \frac{0 - (-2)}{5 - 0} = \frac{2}{5}$$

Then we write an equation of the line in point-slope form using either of the points above.

$$y - 0 = \frac{2}{5}(x - 5)$$

Finally, we find an equivalent equation in slope-intercept form.

$$y - 0 = \frac{2}{5}(x - 5)$$
$$y = \frac{2}{5}x - 2$$

74. $y = \frac{3}{2}x + 3$

75. $(-2, -4)$ and $(2, -1)$

First we find the slope.

$$m = \frac{-4 - (-1)}{-2 - 2} = \frac{-4 + 1}{-2 - 2} = \frac{-3}{-4} = \frac{3}{4}$$

Then we write an equation of the line in point-slope form using either of the points above.

$$y - (-4) = \frac{3}{4}(x - (-2))$$

Finally, we find an equivalent equation in slope-intercept form.

$$y - (-4) = \frac{3}{4}(x - (-2))$$

$$y + 4 = \frac{3}{4}(x + 2)$$

$$y + 4 = \frac{3}{4}x + \frac{3}{2}$$

$$y = \frac{3}{4}x - \frac{5}{2}$$

76. $y = -4x - 7$

77. First find the slope of $2x + 3y = 11$.

$$2x + 3y = 11$$

$$3y = -2x + 11$$

$$y = -\frac{2}{3}x + \frac{11}{3}$$

The slope is $-\frac{2}{3}$.

Then write a point-slope equation of the line containing $(-4, 7)$ and having slope $-\frac{2}{3}$.

$$y - 7 = -\frac{2}{3}(x - (-4))$$

78. $y - (-1) = -\frac{5}{4}(x - 3)$

79. The slope of $y = 3 - 4x$ is -4. We are given the y-intercept of the line, so we use slope-intercept form. The equation is $y = -4x + 7$.

80. $y = \frac{1}{5}x - 2$

81. First find the slope of the line passing through $(2, 7)$ and $(-1, -3)$.

$$m = \frac{-3 - 7}{-1 - 2} = \frac{-10}{-3} = \frac{10}{3}$$

Now find an equation of the line containing the point $(-1, 5)$ and having slope $-\frac{3}{10}$.

$$y - 5 = -\frac{3}{10}(x - (-1))$$

$$y - 5 = -\frac{3}{10}(x + 1)$$

$$y - 5 = -\frac{3}{10}x - \frac{3}{10}$$

$$y = -\frac{3}{10}x + \frac{47}{10}$$

82. $y = \frac{1}{2}x + 1$

83. Writing exercise

Chapter 4
Polynomials

1. $r^4 \cdot r^6 = r^{4+6} = r^{10}$

2. 8^7

3. $9^5 \cdot 9^3 = 9^{5+3} = 9^8$

4. n^{23}

5. $a^6 \cdot a = a^6 \cdot a^1 = a^{6+1} = a^7$

6. y^{16}

7. $5^7 \cdot 5^8 = 5^{7+8} = 5^{15}$

8. t^{16}

9. $(3y)^4(3y)^8 = (3y)^{4+8} = (3y)^{12}$

10. $(2t)^{25}$

11. $(5t)(5t)^6 = (5t)^1(5t)^6 = (5t)^{1+6} = (5t)^7$

12. $8x$

13. $(a^2b^7)(a^3b^2) = a^2b^7a^3b^2$ Using an associative law
$= a^2a^3b^7b^2$ Using a commutative law
$= a^5b^9$ Adding exponents

14. $(m-3)^9$

15. $(x+1)^5(x+1)^7 = (x+1)^{5+7} = (x+1)^{12}$

16. $a^{12}b^4$

17. $r^3 \cdot r^7 \cdot r^0 = r^{3+7+0} = r^{10}$

18. s^{11}

19. $(xy^4)(xy)^3 = (xy^4)(x^3y^3)$
$= x \cdot x^3 \cdot y^4 \cdot y^3$
$= x^{1+3}y^{4+3}$
$= x^4y^7$

20. a^7b^5

21. $\dfrac{7^5}{7^2} = 7^{5-2} = 7^3$ Subtracting exponents

22. 4^4

23. $\dfrac{x^{15}}{x^3} = x^{15-3} = x^{12}$ Subtracting exponents

24. a^8

25. $\dfrac{t^5}{t} = \dfrac{t^5}{t^1} = t^{5-1} = t^4$

26. x^6

27. $\dfrac{(5a)^7}{(5a)^6} = (5a)^{7-6} = (5a)^1 = 5a$

28. $3m$

29. $\dfrac{(x+y)^8}{(x+y)^8}$

Observe that we have an expression divided by itself. Thus, the result is 1.

We could also do this exercise as follows:
$\dfrac{(x+y)^8}{(x+y)^8} = (x+y)^{8-8} = (x+y)^0 = 1$

30. $a-b$

31. $\dfrac{18m^5}{6m^2} = \dfrac{18}{6}m^{5-2} = 3m^3$

32. $5n^4$

33. $\dfrac{a^9b^7}{a^2b} = \dfrac{a^9}{a^2} \cdot \dfrac{b^7}{b^1} = a^{9-2}b^{7-1} = a^7b^6$

34. r^8s^6

35. $\dfrac{m^9n^8}{m^0n^4} = \dfrac{m^9}{m^0} \cdot \dfrac{n^8}{n^4} = m^{9-0}n^{8-4} = m^9n^4$

36. a^8b^{12}

37. When $x = 13$, $x^0 = 13^0 = 1$. (Any nonzero number raised to the 0 power is 1.)

38. 1

39. When $x = -4$, $5x^0 = 5(-4)^0 = 5 \cdot 1 = 5$.

40. 7

41. $8^0 + 5^0 = 1 + 1 = 2$

42. 1

43. $(-3)^1 - (-3)^0 = -3 - 1 = -4$

44. 5

45. $(x^4)^7 = x^{4\cdot 7} = x^{28}$ Multiplying exponents

46. a^{24}

47. $(5^8)^2 = 5^{8\cdot 2} = 5^{16}$ Multiplying exponents

48. 2^{15}, or 32,768

49. $(m^7)^5 = m^{7\cdot 5} = m^{35}$

50. n^{18}

51. $(t^{20})^4 = t^{20\cdot 4} = t^{80}$

52. t^{27}

53. $(7x)^2 = 7^2 \cdot x^2 = 49x^2$

54. $25a^2$

55. $(-2a)^3 = (-2)^3 a^3 = -8a^3$

56. $-27x^3$

57. $(4m^3)^2 = 4^2(m^3)^2 = 16m^6$

58. $25n^8$

59. $(a^2b)^7 = (a^2)^7(b^7) = a^{14}b^7$

60. x^9y^{36}

61. $(x^3y)^2(x^2y^5) = (x^3)^2y^2x^2y^5 = x^6y^2x^2y^5 = x^8y^7$

62. $a^{14}b^{11}$

63. $(2x^5)^3(3x^4) = 2^3(x^5)^3(3x^4) = 8x^{15} \cdot 3x^4 = 24x^{19}$

64. $50x^{13}$

65. $\left(\dfrac{a}{4}\right)^3 = \dfrac{a^3}{4^3} = \dfrac{a^3}{64}$ Raising the numerator and the denominator to the third power

66. $\dfrac{81}{x^4}$

67. $\left(\dfrac{7}{5a}\right)^2 = \dfrac{7^2}{(5a)^2} = \dfrac{49}{5^2a^2} = \dfrac{49}{25a^2}$

68. $\dfrac{125x^3}{8}$

69. $\left(\dfrac{a^4}{b^3}\right)^5 = \dfrac{(a^4)^5}{(b^3)^5} = \dfrac{a^{20}}{b^{15}}$

70. $\dfrac{x^{35}}{y^{14}}$

71. $\left(\dfrac{y^3}{2}\right)^2 = \dfrac{(y^3)^2}{2^2} = \dfrac{y^6}{4}$

72. $\dfrac{a^{15}}{8}$

73. $\left(\dfrac{x^2y}{z^3}\right)^4 = \dfrac{(x^2y)^4}{(z^3)^4} = \dfrac{(x^2)^4(y^4)}{z^{12}} = \dfrac{x^8y^4}{z^{12}}$

74. $\dfrac{x^{15}}{y^{10}z^5}$

75. $\left(\dfrac{a^3}{-2b^5}\right)^4 = \dfrac{(a^3)^4}{(-2b^5)^4} = \dfrac{a^{12}}{(-2)^4(b^5)^4} = \dfrac{a^{12}}{16b^{20}}$

76. $\dfrac{x^{20}}{81y^{12}}$

77. $\left(\dfrac{5x^7y}{2z^4}\right)^3 = \dfrac{(5x^7y)^3}{(2z^4)^3} = \dfrac{5^3(x^7)^3y^3}{2^3(z^4)^3} = \dfrac{125x^{21}y^3}{8z^{12}}$

78. $\dfrac{64a^6b^3}{27c^{21}}$

79. $\left(\dfrac{4x^3y^5}{3z^7}\right)^0$

Observe that for $x \neq 0$, $y \neq 0$, and $z \neq 0$, we have a nonzero number raised to the 0 power. Thus, the result is 1.

80. 1

81. Writing exercise

82. Writing exercise

83. $3s - 3r + 3t = 3 \cdot s - 3 \cdot r + 3 \cdot t = 3(s - r + t)$

84. $-7(x - y + z)$

85. $9x + 2y - x - 2y = 9x - x + 2y - 2y = (9-1)x + (2-2)y = 8x + 0y = 8x$

86. $-3a - 6b$

87. $2y + 3x$

88. $5z + 2xy$

89. Writing exercise

90. Writing exercise

91. Writing exercise

92. Writing exercise

93. Choose any number except 0.

For example, let $a = 1$. Then $(a + 5)^2 = (1 + 5)^2 = 6^2 = 36$, but $a^2 + 5^2 = 1^2 + 5^2 = 1 + 25 = 26$.

94. Choose any number except 0. For example, let $x = 1$.

$$3x^2 = 3 \cdot 1^2 = 3 \cdot 1 = 3, \text{ but}$$

$$(3x)^2 = (3 \cdot 1)^2 = 3^2 = 9.$$

95. Choose any number except $\frac{7}{6}$. For example let $a = 0$.

Then $\frac{0 + 7}{7} = \frac{7}{7} = 1$, but $a = 0$.

96. Choose any number except 0 or 1. For example, let $t = -1$. Then $\frac{t^6}{t^2} = \frac{(-1)^6}{(-1)^2} = \frac{1}{1} = 1$, but $t^3 = (-1)^3 = -1$.

97. $a^{10k} \div a^{2k} = a^{10k - 2k} = a^{8k}$

98. y^{6x}

99. $\dfrac{\left(\frac{1}{2}\right)^3 \left(\frac{2}{3}\right)^4}{\left(\frac{5}{6}\right)^3} = \dfrac{\frac{1}{8} \cdot \frac{16}{81}}{\frac{125}{216}} = \dfrac{1}{8} \cdot \dfrac{16}{81} \cdot \dfrac{216}{125} =$

$\dfrac{1 \cdot 2 \cdot \not{8} \cdot \not{27} \cdot 8}{\not{8} \cdot 3 \cdot \not{27} \cdot 125} = \dfrac{16}{375}$

100. x^t

101. $\dfrac{t^{26}}{t^x} = t^x$

$t^{26 - x} = t^x$

$26 - x = x \quad$ Equating exponents

$26 = 2x$

$13 = x$

The solution is 13.

102. $3^5 > 3^4$

103. Since the bases are the same, the expression with the larger exponent is larger. Thus, $4^2 < 4^3$.

104. $4^3 < 5^3$

105. $4^3 = 64$, $3^4 = 81$, so $4^3 < 3^4$.

106. $9^7 > 3^{13}$

107. $25^8 = (5^2)^8 = 5^{16}$

$125^5 = (5^3)^5 = 5^{15}$

$5^{16} > 5^{15}$, or $25^8 > 125^5$.

108. 16,000; 16,384; 384

109. $2^{22} = 2^{10} \cdot 2^{10} \cdot 2^2 \approx 10^3 \cdot 10^3 \cdot 4 \approx 1000 \cdot 1000 \cdot 4 \approx 4,000,000$

Using a calculator, we find that $2^{22} = 4,194,304$. The difference between the exact value and the approximation is $4,194,304 - 4,000,000$, or $194,304$.

110. 64,000,000; 67,108,864; 3,108,864

111. $2^{31} = 2^{10} \cdot 2^{10} \cdot 2^{10} \cdot 2 \approx 10^3 \cdot 10^3 \cdot 10^3 \cdot 2 \approx 1000 \cdot 1000 \cdot 1000 \cdot 2 = 2,000,000,000$

Using a calculator, we find that $2^{31} = 2,147,483,648$. The difference between the exact value and the approximation is $2,147,483,648 - 2,000,000,000 = 147,483,648$.

112. 57,344 bytes

113. $64 \text{ K} = 64 \times 1 \times 2^{10}$ bytes $= 65,536$ bytes

Exercise Set 4.2

1. $7x^4 + x^3 - 5x + 8 = 7x^4 + x^3 + (-5x) + 8$

The terms are $7x^4$, x^3, $-5x$, and 8.

2. $5a^3$, $4a^2$, $-a$, -7

3. $-t^4 + 7t^3 - 3t^2 + 6 = -t^4 + 7t^3 + (-3t^2) + 6$

The terms are $-t^4$, $7t^3$, $-3t^2$, and 6.

4. n^5, $-4n^3$, $2n$, -8

5. $4x^5 + 7x$

Term	Coefficient	Degree
$4x^5$	4	5
$7x$	7	1

6.

Term	Coefficient	Degree
$9a^3$	9	3
$-4a^2$	-4	2

7. $9t^2 - 3t + 4$

Term	Coefficient	Degree
$9t^2$	9	2
$-3t$	-3	1
4	4	0

8.

Term	Coefficient	Degree
$7x^4$	7	4
$5x$	5	1
-3	-3	0

9. $7a^4 + 9a + a^3$

Term	Coefficient	Degree
$7a^4$	7	4
$9a$	9	1
a^3	1	3

10.

Term	Coefficient	Degree
$6t^5$	6	5
$-3t^2$	-3	2
$-t$	-1	1

11. $x^4 - x^3 + 4x - 3$

Term	Coefficient	Degree
x^4	1	4
$-x^3$	-1	3
$4x$	4	1
-3	-3	0

12.

Term	Coefficient	Degree
$3a^4$	3	4
$-a^3$	-1	3
a	1	1
-9	-9	0

13. $2a^3 + 7a^5 + a^2$

a)

Term	$2a^3$	$7a^5$	a^2
Degree	3	5	2

b) The term of highest degree is $7a^5$. This is the leading term. Then the leading coefficient is 7.

c) Since the term of highest degree is $7a^5$, the degree of the polynomial is 5.

14. a)

Term	$5x$	$-9x^2$	$3x^6$
Degree	1	2	6

b) $3x^6$; 3

c) 6

15. $2t + 3 + 4t^2$

a)

Term	$2t$	3	$4t^2$
Degree	1	0	2

b) The term of highest degree is $4t^2$. This is the leading term. Then the leading coefficient is 4.

c) Since the term of highest degree is $4t^2$, the degree of the polynomial is 2.

16. a)

Term	$3a^2$	-7	$2a^4$
Degree	2	0	4

b) $2a^4$; 2

c) 4

17. $9x^4 + x^2 + x^7 + 4$

a)

Term	$9x^4$	x^2	x^7	4
Degree	4	2	7	0

b) The term of highest degree is x^7. This is the leading term. Then the leading coefficient is 1.

c) Since the term of highest degree is x^7, the degree of the polynomial is 7.

18. a)

Term	8	$6x^2$	$-3x$	$-x^5$
Degree	0	2	1	5

b) $-x^5$; -1

c) 5

19. $9a - a^4 + 3 + 2a^3$

a)

Term	$9a$	$-a^4$	3	$2a^3$
Degree	1	4	0	3

b) The term of highest degree is $-a^4$. This is the leading term. Then the leading coefficient is -1.

c) Since the term of highest degree is $-a^4$, the degree of the polynomial is 4.

20. a)

Term	$-x$	$2x^5$	$-5x^2$	x^6
Degree	1	5	2	6

b) x^6; 1

c) 6

21. $7x^2 + 8x^5 - 4x^3 + 6 - \dfrac{1}{2}x^4$

Term	Coefficient	Degree of Term	Degree of Polynomial
$8x^5$	8	5	
$-\dfrac{1}{2}x^4$	$-\dfrac{1}{2}$	4	
$-4x^3$	-4	3	5
$7x^2$	7	2	
6	6	0	

22.

Term	Coefficient	Degree of Term	Degree of Polynomial
$-3x^4$	-3	4	
$6x^3$	6	3	
$-2x^2$	-2	2	4
$8x$	8	1	
7	7	0	

23. Three monomials are added, so $x^2 - 23x + 17$ is a trinomial.

24. Monomial

25. The polynomial $x^3 - 7x^2 + 2x - 4$ is none of these because it is composed of four monomials.

26. Binomial

27. Two monomials are added, so $8t^2 + 5t$ is a binomial.

28. Trinomial

29. The polynomial 17 is a monomial because it is the product of a constant and a variable raised to a whole number power. (In this case the variable is raised to the power 0.)

30. None of these

31. $7x^2 + 3x + 4x^2 = (7+4)x^2 + 3x = 11x^2 + 3x$

32. $7a^2 + 8a$

33. $3a^4 - 2a + 2a + a^4 = (3+1)a^4 + (-2+2)a = 4a^4 + 0a = 4a^4$

34. $7b^5$

35. $2x^2 - 6x + 3x + 4x^2 = (2+4)x^2 + (-6+3)x = 6x^2 - 3x$

36. $4x^4 - 9x$

37. $9x^3 + 2x - 4x^3 + 5 - 3x = (9-4)x^3 + (2-3)x + 5 = 5x^3 - x + 5$

38. x^4

39. $10x^2 + 2x^3 - 3x^3 - 4x^2 - 6x^2 - x^4 = -x^4 + (2-3)x^3 + (10-4-6)x^2 = -x^4 - x^3$

40. $-x^6 + 10x^5$

41. $\dfrac{1}{5}x^4 + 7 - 2x^2 + 3 - \dfrac{2}{15}x^4 + 2x^2 =$

$\left(\dfrac{1}{5} - \dfrac{2}{15}\right)x^4 + (-2+2)x^2 + (7+3) =$

$\left(\dfrac{3}{15} - \dfrac{2}{15}\right)x^4 + 0x^2 + 10 = \dfrac{1}{15}x^4 + 10$

42. $-\dfrac{1}{6}x^3 + 4x^2 - 3$

43. $5.9x^2 - 2.1x + 6 + 3.4x - 2.5x^2 - 0.5 = (5.9 - 2.5)x^2 + (-2.1 + 3.4)x + (6 - 0.5) = 3.4x^2 + 1.3x + 5.5$

44. $9.3x^3 - 8.4x - 1.4$

45. $6t - 9t^3 + 8t^4 + 4t + 2t^4 + 7t - 3t^3 = (8+2)t^4 + (-9-3)t^3 + (6+4+7)t = 10t^4 - 12t^3 + 17t$

46. $6b^3 + 3b^2 + b$

47. $-7x + 5 = -7 \cdot 3 + 5$
$= -21 + 5$
$= -16$

48. -6

49. $2x^2 - 3x + 7 = 2 \cdot 3^2 - 3 \cdot 3 + 7$
$= 2 \cdot 9 - 3 \cdot 3 + 7$
$= 18 - 9 + 7$
$= 16$

50. 27

51. $5x + 7 = 5(-2) + 7$
$= -10 + 7$
$= -3$

52. 13

53. $x^2 - 3x + 1 = (-2)^2 - 3(-2) + 1$
$= 4 - 3(-2) + 1$
$= 4 + 6 + 1$
$= 11$

54. -15

55. $\quad -3x^3 + 7x^2 - 4x - 5$
$= -3(-2)^3 + 7(-2)^2 - 4(-2) - 5$
$= -3(-8) + 7 \cdot 4 - 4(-2) - 5$
$= 24 + 28 + 8 - 5$
$= 55$

56. -5

57. Locate 10 on the horizontal axis. From there move vertically to the graph and then horizontally to the M-axis. This locates an M-value of about 9. Thus, about 9 words were memorized in 10 minutes.

58. About 17

59. Locate 8 on the horizontal axis. From there move vertically to the graph and then horizontally to the M-axis. This locates an M-value of about 6. Thus, the value of $-0.001t^3 + 0.1t^2$ for $t = 8$ is approximately 6.

60. About 13

61. Locate 13 on the horizontal axis. It is halfway between 12 and 14. From there move vertically to the graph and then horizontally to the M-axis. This locates an M-value of about 15. Thus, the value of $-0.001t^3 + 0.1t^2$ when t is 13 is approximately 15.

62. About 4.5

63. $11.12t^2 = 11.12(10)^2 = 11.12(100) = 1112$

A skydiver has fallen approximately 1112 ft 10 seconds after jumping from a plane.

64. 3091 ft

65. $0.4r^2 - 40r + 1039 = 0.4(18)^2 - 40(18) + 1039 =$
$0.4(324) - 720 + 1039 = 129.6 - 720 + 1039 =$
448.6

There are approximately 449 accidents daily involving an 18-year-old driver.

66. 399

67. Evaluate the polynomial for $x = 40$:
$250x - 0.5x^2 = 250(40) - 0.5(40)^2 =$
$10,000 - 800 = 9200$
The total revenue is \$9200.

68. \$13,200

69. Evaluate the polynomial for $x = 200$:
$4000 + 0.6x^2 = 4000 + 0.6(200)^2 =$
$4000 + 0.6(40,000) = 4000 + 24,000 = 28,000$
The total cost is \$28,000.

70. \$58,000

71. $2\pi r = 2(3.14)(10)$ Substituting 3.14 for π
and 10 for r
$\qquad = 62.8$
The circumference is 62.8 cm.

72. 31.4 ft

73. $\pi r^2 = 3.14(7)^2$ Substituting 3.14 for π
and 7 for r
$\qquad = 3.14(49)$
$\qquad = 153.86$
The area is 153.86 m^2.

74. 113.04 ft^2

75. Writing exercise

76. Writing exercise

77. $-19 + 24$ A negative and a positive number. We subtract the absolute values: $24 - 19 = 5$. The positive number has the larger absolute value so the answer is positive.
$-19 + 24 = 5$

78. -9

79. $5x + 15 = 5 \cdot x + 5 \cdot 3 = 5(x + 3)$

80. $7(a - 3)$

81. *Familiarize.* Let $x =$ the cost per mile of gasoline in dollars. Then the total cost of the gasoline for the year was $14,800x$.

Translate.

$$\underbrace{\text{Cost of insurance}} + \underbrace{\text{cost of registration and oil}} + \underbrace{\text{cost of gasoline}} = \$2011.$$
$$\quad\downarrow \qquad\quad \downarrow \qquad\quad \downarrow \qquad\qquad \downarrow \quad\; \downarrow \qquad \downarrow \quad \downarrow$$
$$\;\, 972 \quad\;\; + \qquad 114 \qquad + \; 14,800x \; = \; 2011$$

Carry out. We solve the equation.
$$972 + 114 + 14,800x = 2011$$
$$1086 + 14,800x = 2011$$
$$14,800x = 925$$
$$x = 0.0625$$

Check. If gasoline cost \$0.0625 per mile, then the total cost of the gasoline was $14,800(\$0.0625)$, or \$925. Then the total auto expense was $\$972 + \$114 + \$925$, or \$2011. The answer checks.

State. Gasoline cost \$0.0625, or 6.25¢ per mile.

82. 274 and 275

83. Writing exercise

84. Writing exercise

85. Answers may vary. Use an ax^5-term, where a is an integer, and 3 other terms with different degrees, each less than degree 5, and integer coefficients. Three answers are $-6x^5 + 14x^4 - x^2 + 11$, $x^5 - 8x^3 + 3x + 1$, and $23x^5 + 2x^4 - x^2 + 5x$.

86. Answers may vary; $0.2y^4 - y + \dfrac{5}{2}$

87. $(5m^5)^2 = 5^2 m^{5 \cdot 2} = 25m^{10}$

The degree is 10.

88. Answers may vary; $9y^4$, $-\dfrac{3}{2}y^4$, $4.2y^4$

89.
$$\frac{9}{2}x^8 + \frac{1}{9}x^2 + \frac{1}{2}x^9 + \frac{9}{2}x + \frac{9}{2}x^9 + \frac{8}{9}x^2 +$$
$$\frac{1}{2}x - \frac{1}{2}x^8$$
$$= \left(\frac{1}{2} + \frac{9}{2}\right)x^9 + \left(\frac{9}{2} - \frac{1}{2}\right)x^8 + \left(\frac{1}{9} + \frac{8}{9}\right)x^2 +$$
$$\left(\frac{9}{2} + \frac{1}{2}\right)x$$
$$= \frac{10}{2}x^9 + \frac{8}{2}x^8 + \frac{9}{9}x^2 + \frac{10}{2}x$$
$$= 5x^9 + 4x^8 + x^2 + 5x$$

90. $3x^6$

91. Let $c =$ the coefficient of x^3. Solve:
$$\begin{aligned}
c + (c - 3) + 3(c - 3) + (c + 2) &= -4 \\
c + c - 3 + 3c - 9 + c + 2 &= -4 \\
6c - 10 &= -4 \\
6c &= 6 \\
c &= 1
\end{aligned}$$

Coefficient of x^3, c: 1

Coefficient of x^2, $c - 3$: $1 - 3$, or -2

Coefficient of x, $3(c - 3)$: $3(1 - 3)$, or -6

Coefficient remaining (constant term), $c + 2$: $1 + 2$, or 3

The polynomial is $x^3 - 2x^2 - 6x + 3$.

92.

d	$-0.0064d^2 + 0.8d + 2$
0	2
30	20.24
60	26.96
90	22.16
120	5.84

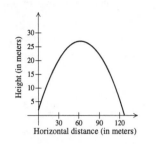

93. We first find q, the quiz average, and t, the test average.
$$q = \frac{60 + 85 + 72 + 91}{4} = \frac{308}{4} = 77$$
$$t = \frac{89 + 93 + 90}{3} = \frac{272}{3} \approx 90.7$$

Now we substitute in the polynomial.
$$\begin{aligned}
A &= 0.3q + 0.4t + 0.2f + 0.1h \\
&= 0.3(77) + 0.4(90.7) + 0.2(84) + 0.1(88) \\
&= 23.1 + 36.28 + 16.8 + 8.8 \\
&= 84.98 \\
&\approx 85.0
\end{aligned}$$

94. 84.1

95. Using a calculator, evaluate $0.4r^2 - 40r + 1039$ for $r = 10, 20, 30, 40, 50, 60,$ and 70 and list the values in a table.

Age	Average number of accidents per day
r	$0.4r^2 - 40r + 1039$
10	679
20	399
30	199
40	79
50	39
60	79
70	199

The numbers in the chart increase both below and above age 50. We would assume the number of accidents is the smallest near age 50. Now we evaluate for 49 and 51.

49	39.4
50	39
51	39.4

Again the numbers increase below and above 50.

We conclude that the smallest number of daily accidents occurs at age 50.

96.

t	$-t^2 + 10t - 18$
3	3
4	6
5	7
6	6
7	3

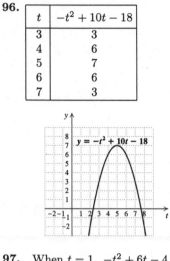

97. When $t = 1$, $-t^2 + 6t - 4 = -1^2 + 6 \cdot 1 - 4 =$
$-1 + 6 - 4 = 1$.
When $t = 2$, $-t^2 + 6t - 4 = -2^2 + 6 \cdot 2 - 4 =$
$-4 + 12 - 4 = 4$.
When $t = 3$, $-t^2 + 6t - 4 = -3^2 + 6 \cdot 3 - 4 =$
$-9 + 18 - 4 = 5$.
When $t = 4$, $-t^2 + 6t - 4 = -4^2 + 6 \cdot 4 - 4 =$
$-16 + 24 - 4 = 4$.
When $t = 5$, $-t^2 + 6t - 4 = -5^2 + 6 \cdot 5 - 4 =$
$-25 + 30 - 4 = 1$.

We complete the table. Then we plot the points and connect them with a smooth curve.

t	$-t^2 + 6t - 4$
1	1
2	4
3	5
4	4
5	1

Exercise Set 4.3

1. $(2x + 3) + (-7x + 6) = (2 - 7)x + (3 + 6) = -5x + 9$

2. $-4x + 5$

3. $(-6x + 2) + (x^2 + x - 3) =$
$x^2 + (-6 + 1)x + (2 - 3) = x^2 - 5x - 1$

4. $x^2 + 3x - 5$

5. $(7t^2 - 3t + 6) + (2t^2 + 8t - 9) =$
$(7 + 2)t^2 + (-3 + 8)t + (6 - 9) = 9t^2 + 5t - 3$

6. $15a^2 + a - 6$

7. $(2m^3 - 4m^2 + m - 7) + (4m^3 + 7m^2 - 4m - 2) =$
$(2 + 4)m^3 + (-4 + 7)m^2 + (1 - 4)m + (-7 - 2) =$
$6m^3 + 3m^2 - 3m - 9$

8. $7n^3 - 5n^2 + 7n - 7$

9. $(3 + 6a + 7a^2 + 8a^3) + (4 + 7a - a^2 + 6a^3) =$
$(3 + 4) + (6 + 7)a + (7 - 1)a^2 + (8 + 6)a^3 =$
$7 + 13a + 6a^2 + 14a^3$

10. $9 + 5t + t^2 + 2t^3$

11. $(9x^8 - 7x^4 + 2x^2 + 5) + (8x^7 + 4x^4 - 2x) =$
$9x^8 + 8x^7 + (-7 + 4)x^4 + 2x^2 - 2x + 5 =$
$9x^8 + 8x^7 - 3x^4 + 2x^2 - 2x + 5$

12. $4x^5 + 9x^2 + 1$

13. $\left(\frac{1}{4}x^4 + \frac{2}{3}x^3 + \frac{5}{8}x^2 + 7\right) + \left(-\frac{3}{4}x^4 + \frac{3}{8}x^2 - 7\right) =$
$\left(\frac{1}{4} - \frac{3}{4}\right)x^4 + \frac{2}{3}x^3 + \left(\frac{5}{8} + \frac{3}{8}\right)x^2 + (7 - 7) =$
$-\frac{2}{4}x^4 + \frac{2}{3}x^3 + \frac{8}{8}x^2 + 0 =$
$-\frac{1}{2}x^4 + \frac{2}{3}x^3 + x^2$

14. $\frac{2}{15}x^9 - \frac{2}{5}x^5 + \frac{1}{4}x^4 - \frac{1}{2}x^2 + 7$

15. $(5.3t^2 - 6.4t - 9.1) + (4.2t^3 - 1.8t^2 + 7.3) =$
$4.2t^3 + (5.3 - 1.8)t^2 - 6.4t + (-9.1 + 7.3) =$
$4.2t^3 + 3.5t^2 - 6.4t - 1.8$

16. $4.9a^3 + 5.3a^2 - 8.8a + 4.6$

17. $-3x^4 + 6x^2 + 2x - 1$
$\underline{ - 3x^2 + 2x + 1}$
$-3x^4 + 3x^2 + 4x + 0$
$-3x^4 + 3x^2 + 4x$

18. $-4x^3 + 4x^2 + 6x$

19. Rewrite the problem so the coefficients of like terms have the same number of decimal places.

$$
\begin{array}{l}
0.15x^4 + 0.10x^3 - 0.90x^2 \\
 - 0.01x^3 + 0.01x^2 + x \\
1.25x^4 + 0.11x^2 + 0.01 \\
 0.27x^3 + 0.99 \\
\underline{-0.35x^4 + 15.00x^2 - 0.03} \\
1.05x^4 + 0.36x^3 + 14.22x^2 + x + 0.97
\end{array}
$$

20. $1.3x^4 + 0.35x^3 + 9.53x^2 + 2x + 0.96$

21. Two forms of the opposite of $-t^3 + 4t^2 - 9$ are

i) $-(-t^3 + 4t^2 - 9)$ and

ii) $t^3 - 4t^2 + 9$. (Changing the sign of every term)

22. $-(-4x^3 - 5x^2 + 2x)$, $4x^3 + 5x^2 - 2x$

23. Two forms for the opposite of $12x^4 - 3x^3 + 3$ are

i) $-(12x^4 - 3x^3 + 3)$ and

ii) $-12x^4 + 3x^3 - 3$. (Changing the sign of every term)

24. $-(5a^3 + 2a - 17)$, $-5a^3 - 2a + 17$

25. We change the sign of every term inside parentheses.
$-(8x - 9) = -8x + 9$

26. $6x - 5$

27. We change the sign of every term inside parentheses.
$-(3a^4 - 5a^2 + 9) = -3a^4 + 5a^2 - 9$

28. $6a^3 - 2a^2 + 7$

29. We change the sign of every term inside parentheses.
$-\left(-4x^4 + 6x^2 + \dfrac{3}{4}x - 8\right) = 4x^4 - 6x^2 - \dfrac{3}{4}x + 8$

30. $5x^4 - 4x^3 + x^2 - 0.9$

31. $(7x + 4) - (-2x + 1)$
$= 7x + 4 + 2x - 1$ $\quad$ Changing the sign of every term inside parentheses
$= 9x + 3$

32. $7x + 2$

33. $(-5t + 4) - (t^2 + 2t - 1) = -5t + 4 - t^2 - 2t + 1 = -t^2 - 7t + 5$

34. $-2a^2 - 7a + 6$

35. $(6x^4 + 3x^3 - 1) - (4x^2 - 3x + 3)$
$= 6x^4 + 3x^3 - 1 - 4x^2 + 3x - 3$
$= 6x^4 + 3x^3 - 4x^2 + 3x - 4$

36. $-3x^3 + x^2 + 2x - 3$

37. $(1.2x^3 + 4.5x^2 - 3.8x) - (-3.4x^3 - 4.7x^2 + 23)$
$= 1.2x^3 + 4.5x^2 - 3.8x + 3.4x^3 + 4.7x^2 - 23$
$= 4.6x^3 + 9.2x^2 - 3.8x - 23$

38. $-1.8x^4 - 0.6x^2 - 1.8x + 4.6$

39. $(7x^3 - 2x^2 + 6) - (7x^3 - 2x^2 + 6)$

Observe that we are subtracting the polynomial $7x^3 - 2x^2 + 6$ from itself. The result is 0.

40. x

41. $(6 + 5a + 3a^2 - a^3) - (2 + 3a - 4a^2 + 2a^3) =$
$6 + 5a + 3a^2 - a^3 - 2 - 3a + 4a^2 - 2a^3 =$
$4 + 2a + 7a^2 - 3a^3$

42. $6 - t - t^2 - 3t^3$

43. $\dfrac{5}{8}x^3 - \dfrac{1}{4}x - \dfrac{1}{3} - \left(-\dfrac{1}{8}x^3 + \dfrac{1}{4}x - \dfrac{1}{3}\right)$
$= \dfrac{5}{8}x^3 - \dfrac{1}{4}x - \dfrac{1}{3} + \dfrac{1}{8}x^3 - \dfrac{1}{4}x + \dfrac{1}{3}$
$= \dfrac{6}{8}x^3 - \dfrac{2}{4}x$
$= \dfrac{3}{4}x^3 - \dfrac{1}{2}x$

44. $\dfrac{3}{5}x^3 - \dfrac{307}{1000}$

45. $(0.07t^3 - 0.03t^2 + 0.01t) - (0.02t^3 + 0.04t^2 - 1) =$
$0.07t^3 - 0.03t^2 + 0.01t - 0.02t^3 - 0.04t^2 + 1 =$
$0.05t^3 - 0.07t^2 + 0.01t + 1$

46. $-0.7a^4 + 0.9a^3 + 0.5a - 4.9$

47. $x^2 + 5x + 6$
$\underline{-(x^2 + 2x + 1)}$

$x^2 + 5x + 6$ $\quad$ Changing signs and
$\underline{-x^2 - 2x - 1}$ $\quad$ removing parentheses
$ 3x + 5$ $\quad$ Adding

48. $2x^2 + 6$

49.
$$5x^4 + 6x^3 - 9x^2$$
$$-(-6x^4 - 6x^3 + x^2)$$

$\begin{array}{ll} 5x^4 + 6x^3 - 9x^2 & \text{Changing signs and} \\ \underline{6x^4 + 6x^3 - x^2} & \text{removing parentheses} \\ 11x^4 + 12x^3 - 10x^2 & \text{Adding} \end{array}$

50. $-2x^4 - 8x^3 - x^2$

51. a)

Familiarize. The area of a rectangle is the product of the length and the width.

Translate. The sum of the areas is found as follows:

$$\begin{array}{ccccccc} \text{Area} & & \text{Area} & & \text{Area} & & \text{Area} \\ \text{of } A & + & \text{of } B & + & \text{of } C & + & \text{of } D \\ = 3x \cdot x & + & x \cdot x & + & 4 \cdot x & + & x \cdot x \end{array}$$

Carry out. We collect like terms.

$$3x^2 + x^2 + 4x + x^2 = 5x^2 + 4x$$

Check. We can go over our calculations. We can also assign some value to x, say 2, and carry out the computation of the area in two ways.

$$\begin{array}{l} \text{Sum of areas: } 3 \cdot 2 \cdot 2 + 2 \cdot 2 + 4 \cdot 2 + 2 \cdot 2 = \\ \qquad 12 + 4 + 8 + 4 = 28 \end{array}$$

Substituting in the polynomial:
$$5(2)^2 + 4 \cdot 2 = 20 + 8 = 28$$

Since the results are the same, our solution is probably correct.

State. A polynomial for the sum of the areas is $5x^2 + 4x$.

b)   For $x = 5$: $5x^2 + 4x = 5 \cdot 5^2 + 4 \cdot 5 = $
$$5 \cdot 25 + 4 \cdot 5 = 125 + 20 = 145$$

When $x = 5$, the sum of the areas is 145 square units.

For $x = 7$: $5x^2 + 4x = 5 \cdot 7^2 + 4 \cdot 7 = $
$$5 \cdot 49 + 4 \cdot 7 = 245 + 28 = 273$$

When $x = 7$, the sum of the areas is 273 square units.

52. a) $r^2\pi + 13\pi$

b) 38π; $= 140.69\pi$

53.

Familiarize. The perimeter is the sum of the lengths of the sides.

Translate. The sum of the lengths is found as follows:

$$3y + 7y + (2y + 3) + 5 + 7 + 2y + 7 + 3$$

Carry out. We collect like terms.

$$(3 + 7 + 2 + 2)y + (3 + 5 + 7 + 7 + 3) = 14y + 25$$

Check. We can go over our calculations. We can also assign some value to y, say 3, and carry out the computation of the perimeter in two ways.

Sum of lengths: $3 \cdot 3 + 7 \cdot 3 + (2 \cdot 3 + 3) + 5 + 7 + 2 \cdot 3 + 7 + 3 = $
$$9 + 21 + 9 + 5 + 7 + 6 + 7 + 3 = 67$$

Substituting in the polynomial:
$$14 \cdot 3 + 25 = 42 + 25 = 67$$

Since the results are the same, our solution is probably correct.

State. A polynomial for the perimeter of the figure is $14y + 25$.

54. $11\frac{1}{2}a + 12$, or $\frac{23}{2}a + 12$

55.

The length and width of the figure can be expressed as $r + 11$ and $r + 9$, respectively. The area of this figure (a rectangle) is the product of the length and width. An algebraic expression for the area is $(r + 11) \cdot (r + 9)$.

The algebraic expressions $9r + 99 + r^2 + 11r$ and $(r + 11) \cdot (r + 9)$ represent the same area.

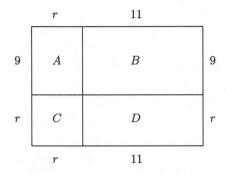

The area of the figure can be found by adding the areas of the four rectangles A, B, C, and D. The area of a rectangle is the product of the length and the width.

$$\begin{array}{ccccccc} \text{Area} & + & \text{Area} & + & \text{Area} & + & \text{Area} \\ \text{of } A & & \text{of } B & & \text{of } C & & \text{of } D \\ = 9 \cdot r & + & 11 \cdot 9 & + & r \cdot r & + & 11 \cdot r \\ = 9r & + & 99 & + & r^2 & + & 11r \end{array}$$

An algebraic expression for the area of the figure is $9r + 99 + r^2 + 11r$.

56. $(t + 5) \cdot (t + 3)$; $t^2 + 5t + 3t + 15$

57.

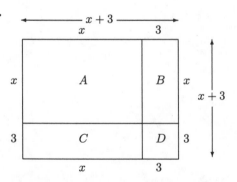

The length and width of the figure can each be expressed as $x + 3$. The area can be expressed as $(x + 3) \cdot (x + 3)$, or $(x + 3)^2$. Another way to express the area is to find an expression for the sum of the areas of the four rectangles A, B, C, and D. The area of each rectangle is the product of its length and width.

$$\begin{array}{ccccccc} \text{Area} & + & \text{Area} & + & \text{Area} & + & \text{Area} \\ \text{of } A & & \text{of } B & & \text{of } C & & \text{of } D \\ = x \cdot x & + & 3 \cdot x & + & 3 \cdot x & + & 3 \cdot 3 \\ = x^2 & + & 3x & + & 3x & + & 9 \end{array}$$

The algebraic expressions $(x+3)^2$ and $x^2+3x+3x+9$ represent the same area.

$$(x + 3)^2 = x^2 + 3x + 3x + 9$$

58. $(x + 10) \cdot (x + 8)$; $8x + 80 + x^2 + 10x$

59.

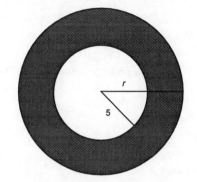

Familiarize. Recall that the area of a circle is the product of π and the square of the radius, r^2.

$$A = \pi r^2$$

Translate.

$$\begin{array}{ccccc} \text{Area of circle} & - & \text{Area of circle} & = & \text{Shaded} \\ \text{with radius } r & & \text{with radius 5} & & \text{area} \\ \pi \cdot r^2 & - & \pi \cdot 5^2 & = & \text{Shaded area} \end{array}$$

Carry out. We simplify the expression.

$$\pi \cdot r^2 - \pi \cdot 5^2 = \pi r^2 - 25\pi$$

Check. We can go over our calculations. We can also assign some value to r, say 7, and carry out the computation in two ways.

Difference of areas: $\pi \cdot 7^2 - \pi \cdot 5^2 = 49\pi - 25\pi = 24\pi$

Substituting in the polynomial: $\pi \cdot 7^2 - 25\pi = 49\pi - 25\pi = 24\pi$

Since the results are the same, our solution is probably correct.

State. A polynomial for the shaded area is $\pi r^2 - 25\pi$.

60. $m^2 - 40$

61. *Familiarize*. We label the figure with additional information.

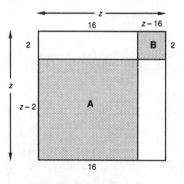

Translate.

Area of shaded sections $=$ Area of A $+$ Area of B

Area of shaded sections $= 16(z - 2) + 2(z - 16)$

Carry out. We simplify the expression.

$16(z - 2) + 2(z - 16) = 16z - 32 + 2z - 32 = 18z - 64$

Check. We can go over the calculations. We can also assign some value to z, say 30, and carry out the computation in two ways.

Sum of areas:

$16 \cdot 28 + 2 \cdot 14 = 448 + 28 = 476$

Substituting in the polynomial:

$18 \cdot 30 - 64 = 540 - 64 = 476$

Since the results are the same, our solution is probably correct.

State. A polynomial for the shaded area is $18z - 64$.

62. $\pi r^2 - 49$

63.

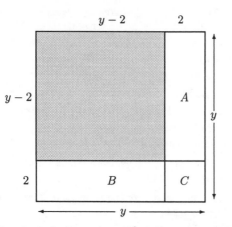

The shaded area is $(y - 2)^2$. We find it as follows:

$$\begin{array}{lcl} \text{Shaded} \\ \text{area} \end{array} = \begin{array}{c} \text{Area of} \\ \text{square} \end{array} - \begin{array}{c} \text{Area} \\ \text{of } A \end{array} - \begin{array}{c} \text{Area} \\ \text{of } B \end{array} - \begin{array}{c} \text{Area} \\ \text{of } C \end{array}$$

$(y - 2)^2 = \quad y^2 \quad - 2(y - 2) - 2(y - 2) - \; 2 \cdot 2$

$(y - 2)^2 = y^2 - 2y + 4 - 2y + 4 - 4$

$(y - 2)^2 = y^2 - 4y + 4$

64. $100 - 40x + 4x^2$

65. Writing exercise

66. Writing exercise

67. $5(4 + 3) - 5 \cdot 4 - 5 \cdot 3$

$= 5 \cdot 7 - 5 \cdot 4 - 5 \cdot 3$ Adding inside the parentheses

$= 35 - 20 - 15$ Multiplying

$= 0$ Subtracting

68. 0

69. $2(5t + 7) + 3t = 10t + 14 + 3t = 13t + 14$

70. $14t - 15$

71. $2(x + 3) > 5(x - 3) + 7$

$2x + 6 > 5x - 15 + 7$ Removing parentheses

$2x + 6 > 5x - 8$ Collecting like terms

$2x + 14 > 5x$ Adding 8 to both sides

$14 > 3x$ Adding $-2x$ to both sides

$\dfrac{14}{3} > x$ Dividing both sides by 3

The solution set is $\left\{ x \left| \dfrac{14}{3} > x \right. \right\}$, or $\left\{ x \left| x < \dfrac{14}{3} \right. \right\}$.

72. $\{x | x \leq 12\}$

73. Writing exercise

74. Writing exercise

75. $(6t^2 - 7t) + (3t^2 - 4t + 5) - (9t - 6)$

$= 6t^2 - 7t + 3t^2 - 4t + 5 - 9t + 6$

$= 9t^2 - 20t + 11$

76. $5x^2 - 9x - 1$

77. $(-8y^2 - 4) - (3y + 6) - (2y^2 - y)$

$= -8y^2 - 4 - 3y - 6 - 2y^2 + y$

$= -10y^2 - 2y - 10$

78. $4x^3 - 5x^2 + 6$

79. $(-y^4 - 7y^3 + y^2) + (-2y^4 + 5y - 2) - (-6y^3 + y^2)$

$= -y^4 - 7y^3 + y^2 - 2y^4 + 5y - 2 + 6y^3 - y^2$

$= -3y^4 - y^3 + 5y - 2$

80. $2 + x + 2x^2 + 4x^3$

81. $(345.099x^3 - 6.178x) - (94.508x^3 - 8.99x)$

$= 345.099x^3 - 6.178x - 94.508x^3 + 8.99x$

$= 250.591x^3 + 2.812x$

82. $36x + 2x^2$

83. *Familiarize.* The surface area is $2lw + 2lh + 2wh$, where $l =$ length, $w =$ width, and $h =$ height of the rectangular solid. Here we have $l = 3$, $w = w$, and $h = 7$.

Translate. We substitute in the formula above.

$2 \cdot 3 \cdot w + 2 \cdot 3 \cdot 7 + 2 \cdot w \cdot 7$

Carry out. We simplify the expression.

$$2 \cdot 3 \cdot w + 2 \cdot 3 \cdot 7 + 2 \cdot w \cdot 7$$
$$= 6w + 42 + 14w$$
$$= 20w + 42$$

Check. We can go over the calculations. We can also assign some value to w, say 6, and carry out the computation in two ways.

Using the formula: $2 \cdot 3 \cdot 6 + 2 \cdot 3 \cdot 7 + 2 \cdot 6 \cdot 7 = 36 + 42 + 84 = 162$

Substituting in the polynomial: $20 \cdot 6 + 42 = 120 + 42 = 162$

Since the results are the same, our solution is probably correct.

State. A polynomial for the surface area is $20w+42$.

84. $22a + 56$

85. ***Familiarize***. The surface area is $2lw + 2lh + 2wh$, where $l = $ length, $w = $ width, and $h = $ height of the rectangular solid. Here we have $l = x$, $w = x$, and $h = 5$.

Translate. We substitute in the formula above.

$$2 \cdot x \cdot x + 2 \cdot x \cdot 5 + 2 \cdot x \cdot 5$$

Carry out. We simplify the expression.

$$2 \cdot x \cdot x + 2 \cdot x \cdot 5 + 2 \cdot x \cdot 5$$
$$= 2x^2 + 10x + 10x$$
$$= 2x^2 + 20x$$

Check. We can go over the calculations. We can also assign some value to x, say 3, and carry out the computation in two ways.

Using the formula: $2 \cdot 3 \cdot 3 + 2 \cdot 3 \cdot 5 + 2 \cdot 3 \cdot 5 = 18 + 30 + 30 = 78$

Substituting in the polynomial: $2 \cdot 3^2 + 20 \cdot 3 = 2 \cdot 9 + 60 = 18 + 60 = 78$

Since the results are the same, our solution is probably correct.

State. A polynomial for the surface area is $2x^2+20x$.

86. a) $P = -x^2 + 280x - 5000$

   b) \$10,375

   c) \$13,000

87. Writing exercise

1. $(5x^4)6 = (5 \cdot 6)x^4 = 30x^4$

2. $28x^3$

3. $(-x^2)(-x) = (-1 \cdot x^2)(-1 \cdot x) = (-1)(-1)(x^2 \cdot x) = x^3$

4. $-x^7$

5. $(-x^5)(x^3) = (-1 \cdot x^5)(1x^3) = (-1)(1)(x^5 \cdot x^3) = -x^8$

6. x^8

7. $(7t^5)(4t^3) = (7 \cdot 4)(t^5 \cdot t^3) = 28t^8$

8. $30a^4$

9. $(-0.1x^6)(0.2x^4) = (-0.1)(0.2)(x^6 \cdot x^4) = -0.02x^{10}$

10. $-0.12x^9$

11. $\left(-\dfrac{1}{5}x^3\right)\left(-\dfrac{1}{3}x\right) = \left(-\dfrac{1}{5}\right)\left(-\dfrac{1}{3}\right)(x^3 \cdot x) = \dfrac{1}{15}x^4$

12. $-\dfrac{1}{20}x^{12}$

13. $19t^2 \cdot 0 = 0$ Any number multiplied by 0 is 0.

14. $5n^3$

15. $7x^2(-2x^3)(2x^6) = 7(-2)(2)(x^2 \cdot x^3 \cdot x^6) = -28x^{11}$

16. $72y^{10}$

17. $3x(-x + 5) = 3x(-x) + 3x(5)$
$$= -3x^2 + 15x$$

18. $8x^2 - 12x$

19. $4x(x + 1) = 4x(x) + 4x(1)$
$$= 4x^2 + 4x$$

20. $3x^2 + 6x$

21. $(a + 9)3a = a \cdot 3a + 9 \cdot 3a = 3a^2 + 27a$

22. $4a^2 - 28a$

23. $x^2(x^3 + 1) = x^2(x^3) + x^2(1)$
$$= x^5 + x^2$$

24. $-2x^5 + 2x^3$

25. $3x(2x^2 - 6x + 1) = 3x(2x^2) + 3x(-6x) + 3x(1)$
$$= 6x^3 - 18x^2 + 3x$$

26. $-8x^4 + 24x^3 + 20x^2 - 4x$

27. $5t^2(3t + 6) = 5t^2(3t) + 5t^2(6) = 15t^3 + 30t^2$

28. $14t^3 + 7t^2$

29. $-6x^2(x^2 + x) = -6x^2(x^2) - 6x^2(x)$
$$= -6x^4 - 6x^3$$

30. $-4x^4 + 4x^3$

31. $\dfrac{2}{3}a^4\left(6a^5 - 12a^3 - \dfrac{5}{8}\right)$
$$= \frac{2}{3}a^4(6a^5) - \frac{2}{3}a^4(12a^3) - \frac{2}{3}a^4\left(\frac{5}{8}\right)$$
$$= \frac{12}{3}a^9 - \frac{24}{3}a^7 - \frac{10}{24}a^4$$
$$= 4a^9 - 8a^7 - \frac{5}{12}a^4$$

32. $6t^{11} - 9t^9 + \dfrac{9}{7}t^5$

33. $(x + 6)(x + 3) = (x + 6)x + (x + 6)3$
$$= x \cdot x + 6 \cdot x + x \cdot 3 + 6 \cdot 3$$
$$= x^2 + 6x + 3x + 18$$
$$= x^2 + 9x + 18$$

34. $x^2 + 7x + 10$

35. $(x + 5)(x - 2) = (x + 5)x + (x + 5)(-2)$
$$= x \cdot x + 5 \cdot x + x(-2) + 5(-2)$$
$$= x^2 + 5x - 2x - 10$$
$$= x^2 + 3x - 10$$

36. $x^2 + 4x - 12$

37. $(a - 6)(a - 7) = (a - 6)a + (a - 6)(-7)$
$$= a \cdot a - 6 \cdot a + a(-7) + (-6)(-7)$$
$$= a^2 - 6a - 7a + 42$$
$$= a^2 - 13a + 42$$

38. $a^2 - 12a - 32$

39. $(x + 3)(x - 3) = (x + 3)x + (x + 3)(-3)$
$$= x \cdot x + 3 \cdot x + x(-3) + 3(-3)$$
$$= x^2 + 3x - 3x - 9$$
$$= x^2 - 9$$

40. $x^2 - 36$

41. $(5 - x)(5 - 2x) = (5 - x)5 + (5 - x)(-2x)$
$$= 5 \cdot 5 - x \cdot 5 + 5(-2x) - x(-2x)$$
$$= 25 - 5x - 10x + 2x^2$$
$$= 25 - 15x + 2x^2$$

42. $18 + 12x + 2x^2$

43. $\left(t + \dfrac{3}{2}\right)\left(t + \dfrac{4}{3}\right) = \left(t + \dfrac{3}{2}\right)t + \left(t + \dfrac{3}{2}\right)\left(\dfrac{4}{3}\right)$
$$= t \cdot t + \frac{3}{2} \cdot t + t \cdot \frac{4}{3} + \frac{3}{2} \cdot \frac{4}{3}$$
$$= t^2 + \frac{3}{2}t + \frac{4}{3}t + 2$$
$$= t^2 + \frac{9}{6}t + \frac{8}{6}t + 2$$
$$= t^2 + \frac{17}{6}t + 2$$

44. $a^2 + \dfrac{21}{10}a - 1$

45. $\left(\dfrac{1}{4}a + 2\right)\left(\dfrac{3}{4}a - 1\right)$
$$= \left(\frac{1}{4}a + 2\right)\left(\frac{3}{4}a\right) + \left(\frac{1}{4}a + 2\right)(-1)$$
$$= \frac{1}{4}a\left(\frac{3}{4}a\right) + 2 \cdot \frac{3}{4}a + \frac{1}{4}a(-1) + 2(-1)$$
$$= \frac{3}{16}a^2 + \frac{3}{2}a - \frac{1}{4}a - 2$$
$$= \frac{3}{16}a^2 + \frac{6}{4}a - \frac{1}{4}a - 2$$
$$= \frac{3}{16}a^2 + \frac{5}{4}a - 2$$

46. $\dfrac{6}{25}t^2 - \dfrac{1}{5}t - 1$

47. Illustrate $x(x + 5)$ as the area of a rectangle with width x and length $x + 5$.

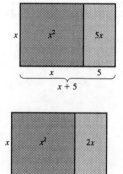

48.

49. Illustrate $(x + 1)(x + 2)$ as the area of a rectangle with width $x + 1$ and length $x + 2$.

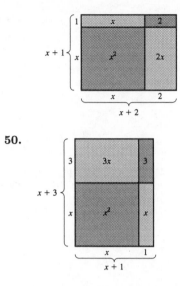

50.

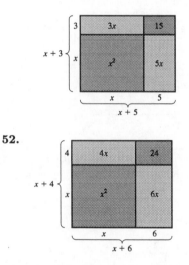

51. Illustrate $(x + 5)(x + 3)$ as the area of a rectangle with length $x + 5$ and width $x + 3$.

52.

53. Illustrate $(3x + 2)(3x + 2)$ as the area of a square with sides of length $3x + 2$.

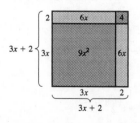

54.

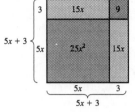

55.
$$(x^2 - x + 5)(x + 1)$$
$$= (x^2 - x + 5)x + (x^2 - x + 5)1$$
$$= x^3 - x^2 + 5x + x^2 - x + 5$$
$$= x^3 + 4x + 5$$

A partial check can be made by selecting a convenient replacement for x, say 1, and comparing the values of the original expression and the result.

$$(1^2 - 1 + 5)(1 + 1) \qquad 1^3 + 4 \cdot 1 + 5$$
$$= (1 - 1 + 5)(1 + 1) \qquad = 1 + 4 + 5$$
$$= 5 \cdot 2 \qquad\qquad = 10$$
$$= 10$$

Since the value of both expressions is 10, the multiplication is very likely correct.

56. $x^3 + 3x^2 - 5x - 14$

57.
$$(2a + 5)(a^2 - 3a + 2)$$
$$= (2a + 5)a^2 - (2a + 5)(3a) + (2a + 5)2$$
$$= 2a \cdot a^2 + 5 \cdot a^2 - 2a \cdot 3a - 5 \cdot 3a + 2a \cdot 2 + 5 \cdot 2$$
$$= 2a^3 + 5a^2 - 6a^2 - 15a + 4a + 10$$
$$= 2a^3 - a^2 - 11a + 10$$

A partial check can be made as in Exercise 55.

58. $3t^3 - 11t^2 - 17t + 4$

59.
$$(y^2 - 7)(2y^3 + y + 1)$$
$$= (y^2 - 7)(2y^3) + (y^2 - 7)y + (y^2 - 7)(1)$$
$$= y^2 \cdot 2y^3 - 7 \cdot 2y^3 + y^2 \cdot y - 7 \cdot y + y^2 \cdot 1 - 7 \cdot 1$$
$$= 2y^5 - 14y^3 + y^3 - 7y + y^2 - 7$$
$$= 2y^5 - 13y^3 + y^2 - 7y - 7$$

A partial check can be made as in Exercise 55.

60. $5a^5 + 17a^3 - a^2 - 12a - 4$

61.
$$(5x^3 - 7x^2 + 1)(x - 3x^2)$$
$$= (5x^3 - 7x^2 + 1)x - (5x^3 - 7x^2 + 1)(3x^2)$$
$$= 5x^3 \cdot x - 7x^2 \cdot x + 1 \cdot x - 5x^3 \cdot 3x^2 + 7x^2 \cdot 3x^2 - 1 \cdot 3x^2$$
$$= 5x^4 - 7x^3 + x - 15x^5 + 21x^4 - 3x^2$$
$$= -15x^5 + 26x^4 - 7x^3 - 3x^2 + x$$

A partial check can be made in Exercise 55.

62. $8x^5 - 6x^3 - 6x^2 - 5x - 3$

63.

$$
\begin{array}{ll}
\quad\; x^2 - 3x + 2 & \text{Line up like terms} \\
\quad\; x^2 + x + 1 & \text{in columns} \\
\hline
\quad\; x^2 - 3x + 2 & \text{Multiplying by 1} \\
\; x^3 - 3x^2 + 2x & \text{Multiplying by } x \\
x^4 - 3x^3 + 2x^2 & \text{Multiplying by } x^2 \\
\hline
x^4 - 2x^3 \qquad\; - x + 2
\end{array}
$$

A partial check can be made as in Exercise 55.

64. $x^4 + 4x^3 - 3x^2 + 16x - 3$

65.

$$
\begin{array}{ll}
\quad\; 2t^2 - 5t - 4 & \\
\quad\; 3t^2 - t + 1 & \\
\hline
\quad\; 2t^2 - 5t - 4 & \text{Multiplying by 1} \\
\; - 2t^3 + 5t^2 + 4t & \text{Multiplying by } -t \\
6t^4 - 15t^3 - 12t^2 & \text{Multiplying by } 3t^2 \\
\hline
6t^4 - 17t^3 - 5t^2 - t - 4
\end{array}
$$

A partial check can be made as in Exercise 55.

66. $10t^4 + 3t^3 - 14t^2 + 4t - 3$

67. We will multiply horizontally while still aligning like terms.

$$(x+1)(x^3 + 7x^2 + 5x + 4)$$

$$
\begin{array}{ll}
= x^4 + 7x^3 + 5x^2 + 4x & \text{Multiplying by } x \\
\;\; + x^3 + 7x^2 + 5x + 4 & \text{Multiplying by 1} \\
\hline
= x^4 + 8x^3 + 12x^2 + 9x + 4
\end{array}
$$

A partial check can be made as in Exercise 55.

68. $x^4 + 7x^3 + 19x^2 + 21x + 6$

69. We will multiply horizontally while still aligning like terms.

$$\left(x - \frac{1}{2}\right)\left(2x^3 - 4x^2 + 3x - \frac{2}{5}\right)$$

$$
\begin{array}{l}
= 2x^4 - 4x^3 + 3x^2 - \dfrac{2}{5}x \\[2mm]
\quad\;\; - x^3 + 2x^2 - \dfrac{3}{2}x + \dfrac{1}{5} \\[2mm]
\hline
2x^4 - 5x^3 + 5x^2 - \dfrac{19}{10}x + \dfrac{1}{5}
\end{array}
$$

A partial check can be made as in Exercise 55.

70. $6x^4 - 10x^3 - 9x^2 - \dfrac{7}{6}x + \dfrac{1}{6}$

71. Writing exercise

72. Writing exercise

73. $5 - 3 \cdot 2 + 7 = 5 - 6 + 7 = -1 + 7 = 6$

74. 31

75.
$$
\begin{aligned}
& (8-2)(8+2) + 2^2 - 8^2 \\
&= 6 \cdot 10 + 2^2 - 8^2 \\
&= 6 \cdot 10 + 4 - 64 \\
&= 60 + 4 - 64 \\
&= 64 - 64 \\
&= 0
\end{aligned}
$$

76. 0

77. Writing exercise

78. Writing exercise

79. The shaded area is the area of the large rectangle, $6y(14y - 5)$ less the area of the unshaded rectangle, $3y(3y + 5)$. We have:

$$
\begin{aligned}
& 6y(14y - 5) - 3y(3y + 5) \\
&= 84y^2 - 30y - 9y^2 - 15y \\
&= 75y^2 - 45y
\end{aligned}
$$

80. $78t^2 + 40t$

81. Let $n =$ the missing number. Label the figure with the known areas.

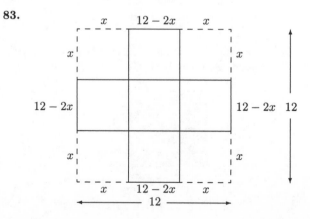

Then the area of the figure is $x^2 + 2x + nx + 2n$. This is equivalent to $x^2 + 7x + 10$, so we have $2x + nx = 7x$ and $2n = 10$. Solving either equation for n, we find that the missing number is 5.

82. 5

83.

The dimensions, in inches, of the box are $12 - 2x$ by $12 - 2x$ by x. The volume is the product of the dimensions (volume = length × width × height):

$$
\begin{aligned}
\text{Volume} &= (12 - 2x)(12 - 2x)x \\
&= (144 - 48x + 4x^2)x \\
&= 144x - 48x^2 + 4x^3 \text{ in}^3, \text{ or} \\
&\quad 4x^3 - 48x^2 + 144x \text{ in}^3
\end{aligned}
$$

The outside surface area is the sum of the area of the bottom and the areas of the four sides. The dimensions, in inches, of the bottom are $12 - 2x$ by $12 - 2x$, and the dimensions, in inches, of each side are x by $12 - 2x$.

$$
\begin{aligned}
\frac{\text{Surface}}{\text{area}} &= \frac{\text{Area of bottom} +}{4 \cdot \text{Area of each side}} \\
&= (12 - 2x)(12 - 2x) + 4 \cdot x(12 - 2x) \\
&= 144 - 24x - 24x + 4x^2 + 48x - 8x^2 \\
&= 144 - 48x + 4x^2 + 48x - 8x^2 \\
&= 144 - 4x^2 \text{ in}^2, \text{ or } -4x^2 + 144 \text{ in}^2
\end{aligned}
$$

84. $x^3 - 5x^2 + 8x - 4 \text{ cm}^3$

85. We have a rectangular solid with dimensions x m by x m by $x+2$ m with a rectangular solid piece with dimensions 6 m by 5 m by 7 m cut out of it.

$$
\begin{aligned}
\text{Volume} &= \frac{\text{Volume of}}{\text{large solid}} - \frac{\text{Volume of}}{\text{small solid}} \\
&= (x \text{ m})(x \text{ m})(x + 2 \text{ m}) - (6 \text{ m})(5 \text{ m})(7 \text{ m}) \\
&= x^2(x + 2) \text{ m}^3 - 210 \text{ m}^3 \\
&= x^3 + 2x^2 - 210 \text{ m}^3
\end{aligned}
$$

86. $x^3 + 6x^2 + 12x + 8 \text{ cm}^3$

87. Let x = the width of the garden. Then $2x$ = the length of the garden.

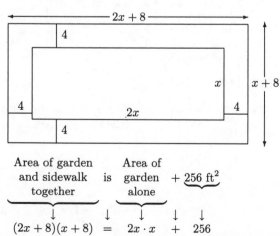

Area of garden and sidewalk together   is   Area of garden alone   + $\underbrace{256 \text{ ft}^2}$

$$(2x + 8)(x + 8) = 2x \cdot x + 256$$

$$2x^2 + 24x + 64 = 2x^2 + 256$$
$$24x = 192$$
$$x = 8$$

The dimensions are 8 ft by 16 ft.

88. $2x^2 + 18x + 36$

89. $(x - 2)(x - 7) - (x - 7)(x - 2)$

First observe that, by the commutative law of multiplication, $(x - 2)(x - 7)$ and $(x - 7)(x - 2)$ are equivalent expressions. Then when we subtract $(x - 7)(x - 2)$ from $(x - 2)(x - 7)$, the result is 0.

90. $16x + 16$

91.
$$
\begin{aligned}
&(x - a)(x - b) \cdots (x - x)(x - y)(x - z) \\
&= (x - a)(x - b) \cdots 0 \cdot (x - y)(x - z) \\
&= 0
\end{aligned}
$$

92. ▮

Exercise Set 4.5

1. $(x + 4)(x^2 + 3)$
$$
\begin{array}{cccc}
\text{F} & \text{O} & \text{I} & \text{L}
\end{array}
$$
$$
\begin{aligned}
&= x \cdot x^2 + x \cdot 3 + 4 \cdot x^2 + 4 \cdot 3 \\
&= x^3 + 3x + 4x^2 + 12, \text{ or } x^3 + 4x^2 + 3x + 12
\end{aligned}
$$

2. $x^3 - x^2 - 3x + 3$

3. $(x^3 + 6)(x + 2)$
$$
\begin{array}{cccc}
\text{F} & \text{O} & \text{I} & \text{L}
\end{array}
$$
$$
\begin{aligned}
&= x^3 \cdot x + x^3 \cdot 2 + 6 \cdot x + 6 \cdot 2 \\
&= x^4 + 2x^3 + 6x + 12
\end{aligned}
$$

4. $x^5 + 12x^4 + 2x + 24$

5. $(y + 2)(y - 3)$
$$
\begin{array}{cccc}
\text{F} & \text{O} & \text{I} & \text{L}
\end{array}
$$
$$
\begin{aligned}
&= y \cdot y + y \cdot (-3) + 2 \cdot y + 2 \cdot (-3) \\
&= y^2 - 3y + 2y - 6 \\
&= y^2 - y - 6
\end{aligned}
$$

6. $a^2 + 4a + 4$

7. $(3x + 2)(3x + 5)$
$$
\begin{array}{cccc}
\text{F} & \text{O} & \text{I} & \text{L}
\end{array}
$$
$$
\begin{aligned}
&= 3x \cdot 3x + 3x \cdot 5 + 2 \cdot 3x + 2 \cdot 5 \\
&= 9x^2 + 15x + 6x + 10 \\
&= 9x^2 + 21x + 10
\end{aligned}
$$

8. $8x^2 + 30x + 7$

9. $(5x - 6)(x + 2)$

      F   O   I   L

$= 5x \cdot x + 5x \cdot 2 + (-6) \cdot x + (-6) \cdot 2$

$= 5x^2 + 10x - 6x - 12$

$= 5x^2 + 4x - 12$

10. $t^2 - 81$

11. $(1 + 3t)(2 - 3t)$

     F   O   I   L

$= 1 \cdot 2 + 1(-3t) + 3t \cdot 2 + 3t(-3t)$

$= 2 - 3t + 6t - 9t^2$

$= 2 + 3t - 9t^2$

12. $14 + 19a - 3a^2$

13. $(2x - 7)(x - 1)$

     F   O   I   L

$= 2x \cdot x + 2x \cdot (-1) + (-7) \cdot x + (-7) \cdot (-1)$

$= 2x^2 - 2x - 7x + 7$

$= 2x^2 - 9x + 7$

14. $6x^2 - x - 1$

15. $\left(p - \frac{1}{4}\right)\left(p + \frac{1}{4}\right)$

     F   O   I   L

$= p \cdot p + p \cdot \frac{1}{4} + \left(-\frac{1}{4}\right) \cdot p + \left(-\frac{1}{4}\right) \cdot \frac{1}{4}$

$= p^2 + \frac{1}{4}p - \frac{1}{4}p - \frac{1}{16}$

$= p^2 - \frac{1}{16}$

16. $q^2 + \frac{3}{2}q + \frac{9}{16}$

17. $(x - 0.1)(x + 0.1)$

     F   O   I   L

$= x \cdot x + x \cdot (0.1) + (-0.1) \cdot x + (-0.1)(0.1)$

$= x^2 + 0.1x - 0.1x - 0.01$

$= x^2 - 0.01$

18. $x^2 - 0.1x - 0.12$

19. $(2x^2 + 6)(x + 1)$

     F   O   I   L

$= 2x^3 + 2x^2 + 6x + 6$

20. $4x^3 - 2x^2 + 6x - 3$

21. $(-2x + 1)(x + 6)$

     F   O   I   L

$= -2x^2 - 12x + x + 6$

$= -2x^2 - 11x + 6$

22. $-2x^2 + 13x - 20$

23. $(a + 9)(a + 9)$

     F   O   I   L

$= a^2 + 9a + 9a + 81$

$= a^2 + 18a + 81$

24. $4y^2 + 28y + 49$

25. $(1 + 3t)(1 - 5t)$

     F   O   I   L

$= 1 - 5t + 3t - 15t^2$

$= 1 - 2t - 15t^2$

26. $1 + 2t - 3t^2 - 6t^3$

27. $(x^2 + 3)(x^3 - 1)$

     F   O   I   L

$= x^5 - x^2 + 3x^3 - 3$, or $x^5 + 3x^3 - x^2 - 3$

28. $2x^5 + x^4 - 6x - 3$

29. $(3x^2 - 2)(x^4 - 2)$

     F   O   I   L

$= 3x^6 - 6x^2 - 2x^4 + 4$, or $3x^6 - 2x^4 - 6x^2 + 4$

30. $x^{20} - 9$

31. $(2t^3 + 5)(2t^3 + 3)$

     F   O   I   L

$= 4t^6 + 6t^3 + 10t^3 + 15$

$= 4t^6 + 16t^3 + 15$

32. $10t^4 + 17t^2 + 3$

33. $(8x^3 + 5)(x^2 + 2)$

     F   O   I   L

$= 8x^5 + 16x^3 + 5x^2 + 10$

34. $20 - 10x - 8x^2 + 4x^3$

35. $(4x^2 + 3)(x - 3)$

     F   O   I   L

$= 4x^3 - 12x^2 + 3x - 9$

36. $14x^2 - 53x + 14$

37. $(x + 8)(x - 8)$ Product of sum and difference of the same two terms

$= x^2 - 8^2$

$= x^2 - 64$

38. $x^2 - 1$

39. $(2x + 1)(2x - 1)$ Product of sum and difference of the same two terms

$= (2x)^2 - 1^2$

$= 4x^2 - 1$

40. $x^4 - 1$

41. $\quad (5m-2)(5m+2)$ Product of sum and difference of the same two terms

$\quad = (5m)^2 - 2^2$

$\quad = 25m^2 - 4$

42. $9x^8 - 4$

43. $\quad (2x^2+3)(2x^2-3)$ Product of sum and difference of the same two terms

$\quad = (2x^2)^2 - 3^2$

$\quad = 4x^4 - 9$

44. $36x^{10} - 25$

45. $\quad (3x^4-1)(3x^4+1)$

$\quad = (3x^4)^2 - 1^2$

$\quad = 9x^8 - 1$

46. $t^4 - 0.04$

47. $\quad (x^4+7)(x^4-7)$

$\quad = (x^4)^2 - 7^2$

$\quad = x^8 - 49$

48. $t^6 - 16$

49. $\quad \left(t - \dfrac{3}{4}\right)\left(t + \dfrac{3}{4}\right)$

$\quad = t^2 - \left(\dfrac{3}{4}\right)^2$

$\quad = t^2 - \dfrac{9}{16}$

50. $m^2 - \dfrac{4}{9}$

51. $\quad (x+2)^2$

$\quad = x^2 + 2 \cdot x \cdot 2 + 2^2$ Square of a binomial

$\quad = x^2 + 4x + 4$

52. $4x^2 - 4x + 1$

53. $\quad (3x^5+1)^2$ Square of a binomial

$\quad = (3x^5)^2 + 2 \cdot 3x^5 \cdot 1 + 1^2$

$\quad = 9x^{10} + 6x^5 + 1$

54. $16x^6 + 8x^3 + 1$

55. $\quad \left(a - \dfrac{2}{5}\right)^2$ Square of a binomial

$\quad = a^2 - 2 \cdot a \cdot \dfrac{2}{5} + \left(\dfrac{2}{5}\right)^2$

$\quad = a^2 - \dfrac{4}{5}a + \dfrac{4}{25}$

56. $t^2 - \dfrac{2}{5}t + \dfrac{1}{25}$

57. $\quad = (t^3+3)^2$ Square of a binomial

$\quad = (t^3)^2 + 2 \cdot t^3 \cdot 3 + 3^2$

$\quad = t^6 + 6t^3 + 9$

58. $a^8 + 4a^4 + 4$

59. $\quad (2-3x^4)^2 = 2^2 - 2 \cdot 2 \cdot 3x^4 + (3x^4)^2$

$\quad\quad\quad\quad\quad = 4 - 12x^4 + 9x^8$

60. $25 - 20t^3 + 4t^6$

61. $\quad (5+6t^2)^2 = 5^2 + 2 \cdot 5 \cdot 6t^2 + (6t^2)^2$

$\quad\quad\quad\quad\quad = 25 + 60t^2 + 36t^4$

62. $9p^4 - 6p^3 + p^2$

63. $\quad (7x-0.3)^2 = (7x)^2 - 2(7x)(0.3) + (0.3)^2$

$\quad\quad\quad\quad\quad = 49x^2 - 4.2x + 0.09$

64. $16a^2 - 4.8a + 0.36$

65. $\quad 5a^3(2a^2-1)$

$\quad = 5a^3 \cdot 2a^2 - 5a^3 \cdot 1$ Multiplying each term of the binomial by the monomial

$\quad = 10a^5 - 5a^3$

66. $18x^5 - 45x^3$

67. $\quad (a-3)(a^2+2a-4)$

$\quad = a^3 + 2a^2 - 4a$ Multiplying horizontally

$\quad \underline{\quad\quad -3a^2 - 6a + 12}$ and aligning like terms

$\quad = a^3 - a^2 - 10a + 12$

68. $x^4 + x^3 - 6x^2 - 5x + 5$

69. $\quad (3-2x^3)^2$

$\quad = 3^2 - 2 \cdot 3 \cdot 2x^3 + (2x^3)^2$ Squaring a binomial

$\quad = 9 - 12x^3 + 4x^6$

70. $x^2 - 8x^4 + 16x^6$

71. $\quad 4x(x^2+6x-3)$

$\quad = 4x \cdot x^2 + 4x \cdot 6x + 4x(-3)$ Multiplying each term of the trinomial by the monomial

$\quad = 4x^3 + 24x^2 - 12x$

72. $-8x^6 + 48x^3 + 72x$

73. $\quad (-t^3+1)^2$

$\quad = (-t^3)^2 + 2(-t)^3(1) + 1^2$ Squaring a binomial

$\quad = t^6 - 2t^3 + 1$

74. $x^4 - 2x^2 + 1$

75. $3t^2(5t^3 - t^2 + t)$

$= 3t^2 \cdot 5t^3 + 3t^2(-t^2) + 3t^2 \cdot t$ Multiplying each
                                 term of the trinomial
                                 by the monomial

$= 15t^5 - 3t^4 + 3t^3$

76. $-5x^5 - 40x^4 + 45x^3$

77. $(6x^4 - 3)^2$ Squaring a binomial

$= (6x^4)^2 - 2 \cdot 6x^4 \cdot 3 + 3^2$

$= 36x^8 - 36x^4 + 9$

78. $64a^2 + 80a^3 + 25$

79. $(3x + 2)(4x^2 + 5)$ Product of two
                               binomials; use FOIL

$= 3x \cdot 4x^2 + 3x \cdot 5 + 2 \cdot 4x^2 + 2 \cdot 5$

$= 12x^3 + 15x + 8x^2 + 10,$ or
    $12x^3 + 8x^2 + 15x + 10$

80. $6x^4 - 3x^2 - 63$

81. $(5 - 6x^4)^2$ Squaring a binomial

$= 5^2 - 2 \cdot 5 \cdot 6x^4 + (6x^4)^2$

$= 25 - 60x^4 + 36x^8$

82. $9 - 24t^5 + 16t^{10}$

83. $(a+1)(a^2 - a + 1)$

$= a^3 - a^2 + a$ Multiplying horizontally
       $a^2 - a + 1$ and aligning like terms
          ‾‾‾‾‾‾‾‾‾‾‾‾‾‾
    $a^3 \quad\quad\quad + 1$

84. $x^3 - 125$

85.

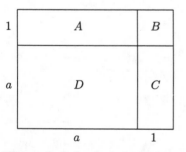

We can find the shaded area in two ways.

Method 1: The figure is a square with side $a + 1$, so
the area is $(a + 1)^2 = a^2 + 2a + 1$.

Method 2: We add the areas of A, B, C, and D.

$1 \cdot a + 1 \cdot 1 + 1 \cdot a + a \cdot a = a + 1 + a + a^2 =$
$a^2 + 2a + 1$.

Either way we find that the total shaded area is
$a^2 + 2a + 1$.

86. $x^2 + 6x + 9$

87.

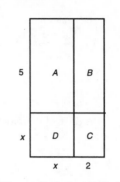

We can find the shaded area in two ways.

Method 1: The figure is a rectangle with dimensions
$x + 5$ by $x + 2$, so the area is

$(x + 5)(x + 2) = x^2 + 2x + 5x + 10 = x^2 + 7x + 10.$

Method 2: We add the areas of A, B, C, and D.

$5 \cdot x + 2 \cdot 5 + 2 \cdot x + x \cdot x = 5x + 10 + 2x + x^2 = x^2 + 7x + 10.$

Either way, we find that the area is $x^2 + 7x + 10$.

88. $t^2 + 7t + 12$

89.

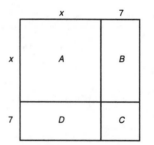

We can find the shaded area in two ways.

Method 1: The figure is a square with side $x + 7$, so
the area is $(x + 7)^2 = x^2 + 14x + 49$.

Method 2: We add the areas of A, B, C, and D.

$x \cdot x + x \cdot 7 + 7 \cdot 7 + 7 \cdot x = x^2 + 7x + 49 + 7x = x^2 + 14x + 49.$

Either way, we find that the total shaded area is
$x^2 + 14x + 49$.

90. $a^2 + 10a + 25$

91.

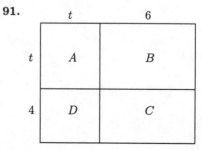

We can find the shaded area in two ways.

Method 1: The figure is a rectangle with dimensions $t + 6$ by $t + 4$, so the area is $(t + 6)(t + 4) =$ $t^2 + 4t + 6t + 24 = t^2 + 10t + 24$.

Method 2: We add the areas of A, B, C, and D.

$t \cdot t + t \cdot 6 + 6 \cdot 4 + 4 \cdot t = t^2 + 6t + 24 + 4t = t^2 + 10t + 24$.

Either way, we find that the total shaded area is $t^2 + 10t + 24$.

92. $x^2 + 10x + 21$

93.

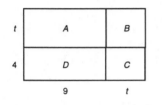

We can find the shaded area in two ways.

Method 1: The figure is a rectangle with dimensions $t + 9$ by $t + 4$, so the area is

$(t + 9)(t + 4) = t^2 + 4t + 9t + 36 = t^2 + 13t + 36$

Method 2: We add the areas of A, B, C, and D.

$9 \cdot t + t \cdot t + 4 \cdot t + 4 \cdot 9 = 9t + t^2 + 4t + 36 = t^2 + 13t + 36$.

Either way, we find that the total shaded area is $t^2 + 13t + 36$.

94. $a^2 + 8a + 7$

95.

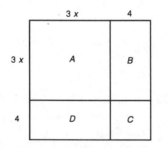

We can find the shaded area in two ways.

Method 1: The figure is a square with side $3x + 4$, so the area is $(3x + 4)^2 = 9x^2 + 24x + 16$.

Method 2: We add the areas of A, B, C, and D.

$3x \cdot 3x + 3x \cdot 4 + 4 \cdot 4 + 3x \cdot 4 = 9x^2 + 12x + 16 + 12x = 9x^2 + 24x + 16$.

Either way, we find that the total shaded area is $9x^2 + 24x + 16$.

96. $25t^2 + 20t + 4$

97. We draw a square with side $x + 5$.

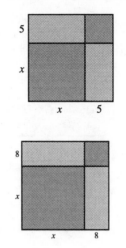

98.

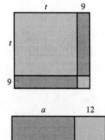

99. We draw a square with side $t + 9$.

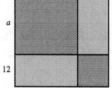

100.

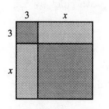

101. We draw a square with side $3 + x$.

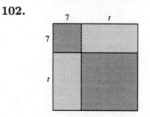

102.

103. Writing exercise

104. Writing exercise

105. *Familiarize*. Let t = the number of watts used by the television set. Then $10t$ = the number of watts used by the lamps, and $40t$ = the number of watts used by the air conditioner.

Translate.

Lamp watts	+	Air conditioner watts	+	Television watts	=	Total watts
↓ ↓		↓		↓		↓ ↓
$10t$ +		$40t$	+	t	=	2550

Solve. We solve the equation.

$$10t + 40t + t = 2550$$
$$51t = 2550$$
$$t = 50$$

The possible solution is:

Television, t: 50 watts

Lamps, $10t$: $10 \cdot 50$, or 500 watts

Air conditioner, $40t$: $40 \cdot 50$, or 2000 watts

Check. The number of watts used by the lamps, 500, is 10 times 50, the number used by the television. The number of watts used by the air conditioner, 2000, is 40 times 50, the number used by the television. Also, $50 + 500 + 2000 = 2550$, the total wattage used.

State. The television uses 50 watts, the lamps use 500 watts, and the air conditioner uses 2000 watts.

106. II

107. $5xy = 8$

$y = \dfrac{8}{5x}$ Dividing both sides by $5x$

108. $a = \dfrac{c}{3b}$

109. $ax - b = c$

$ax = b + c$ Adding b to both sides

$x = \dfrac{b+c}{a}$ Dividing both sides by a

110. $t = \dfrac{u - r}{s}$

111. Writing exercise

112. Writing exercise

113. $(4x^2 + 9)(2x + 3)(2x - 3)$
$= (4x^2 + 9)(4x^2 - 9)$
$= 16x^4 - 81$

114. $81a^4 - 1$

115. $(3t - 2)^2(3t + 2)^2$
$= [(3t - 2)(3t + 2)]^2$
$= (9t^2 - 4)^2$
$= 81t^4 - 72t^2 + 16$

116. $625a^4 - 50a^2 + 1$

117. $(t^3 - 1)^4(t^3 + 1)^4$
$= [(t^3 - 1)(t^3 + 1)]^4$
$= (t^6 - 1)^4$
$= [(t^6 - 1)^2]^2$
$= (t^{12} - 2t^6 + 1)^2$
$= (t^{12} - 2t^6 + 1)(t^{12} - 2t^6 + 1)$
$= t^{24} - 2t^{18} + t^{12} - 2t^{18} + 4t^{12} - 2t^6 +$
$\quad t^{12} - 2t^6 + 1$
$= t^{24} - 4t^{18} + 6t^{12} - 4t^6 + 1$

118. $1050.4081x^2 + 348.0834x + 28.8369$

119. $18 \times 22 = (20 - 2)(20 + 2) = 20^2 - 2^2 =$
$400 - 4 = 396$

120. 9951

121. $(x + 2)(x - 5) = (x + 1)(x - 3)$
$x^2 - 5x + 2x - 10 = x^2 - 3x + x - 3$
$x^2 - 3x - 10 = x^2 - 2x - 3$
$-3x - 10 = -2x - 3$ Adding $-x^2$
$-3x + 2x = 10 - 3$ Adding $2x$ and 10
$-x = 7$
$x = -7$

The solution is -7.

122. 0

123. If l = the length, then $l + 1$ = the height, and $l - 1$ = the width. Recall that the volume of a rectangular solid is given by length $\times$ width $\times$ height.

Volume = $l(l - 1)(l + 1) = l(l^2 - 1) = l^3 - l$

124. $w^3 + 3w^2 + 2w$

125.

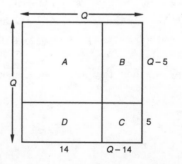

The dimensions of the shaded area, B, are $Q - 14$ by $Q - 5$, so one expression is $(Q - 14)(Q - 5)$.

To find another expression we find the area of regions B and C together and subtract the area of region C. The region consisting of B and C together has dimensions Q by $Q - 14$, so its area is $Q(Q - 14)$. Region C has dimensions 5 by $Q - 14$, so its area is $5(Q - 14)$. Then another expression for the shaded area, B, is $Q(Q - 14) - 5(Q - 14)$.

It is possible to find other equivalent expressions also.

126. $F^2 - (F - 7)(F - 17)$; $24F - 119$; other equivalent expressions are possible.

127.

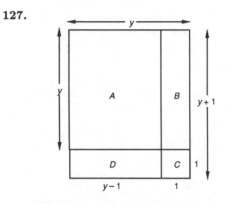

The dimensions of the shaded area, regions A and D together, are $y+1$ by $y-1$ so the area is $(y+1)(y-1)$.

To find another expression we add the areas of regions A and D. The dimensions of region A are y by $y - 1$, and the dimensions of region D are $y - 1$ by 1, so the sum of the areas is $y(y - 1) + (y - 1)(1)$, or $y(y - 1) + y - 1$.

It is possible to find other equivalent expressions also.

128. 10, 11, 12

129.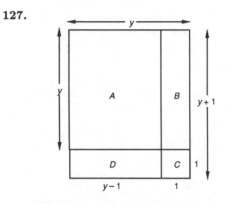

Exercise Set 4.6

1. We replace x by 5 and y by -2.
$$x^2 - 3y^2 + 2xy = 5^2 - 3(-2)^2 + 2 \cdot 5(-2) =$$
$$25 - 12 - 20 = -7.$$

2. 85

3. We replace x by 2, y by -3, and z by -4.
$$xyz^2 - z = 2(-3)(-4)^2 - (-4) = -96 + 4 = -92$$

4. 14

5. Evaluate the polynomial for $h = 160$ and $A = 50$.
$$0.041h - 0.018A - 2.69$$
$$= 0.041(160) - 0.018(50) - 2.69$$
$$= 6.56 - 0.9 - 2.69$$
$$= 2.97$$

The woman's lung capacity is 2.97 liters.

6. 3.715 liters

7. Evaluate the polynomial for $h = 300$, $v = 40$, and $t = 2$.
$$h + vt - 4.9t^2$$
$$= 300 + 40 \cdot 2 - 4.9(2)^2$$
$$= 300 + 80 - 19.6$$
$$= 360.4$$

The rocket will be 360.4 m above the ground 2 seconds after blast off.

8. 250.9 m

9. Evaluate the polynomial for $h = 7\frac{1}{2}$, or $\frac{15}{2}$, $r = 1\frac{1}{4}$, or $\frac{5}{4}$, and $\pi \approx 3.14$.
$$2\pi rh + \pi r^2 \approx 2(3.14)\left(\frac{5}{4}\right)\left(\frac{15}{2}\right) + (3.14)\left(\frac{5}{4}\right)^2$$
$$\approx 2(3.14)\left(\frac{5}{4}\right)\left(\frac{15}{2}\right) + (3.14)\left(\frac{25}{16}\right)$$
$$\approx 58.875 + 4.90625$$
$$\approx 63.78125$$

The surface area is about 63.78125 in^2.

10. 20.60625 in^2

11. $x^3y - 2xy + 3x^2 - 5$

Term	Coefficient	Degree	
x^3y	1	4	(Think: $x^3y = x^3y^1$)
$-2xy$	-2	2	(Think: $-2xy = -2x^1y^1$)
$3x^2$	3	2	
-5	-5	0	(Think: $-5 = -5x^0$)

The degree of the polynomial is the degree of the term of highest degree. The term of highest degree is x^3y. Its degree is 4, so the degree of the polynomial is 4.

12. Coefficients: 1, -1, 9, 7

Degrees: 3, 2, 3, 0; 3

13. $17x^2y^3 - 3x^3yz - 7$

Term	Coefficient	Degree	
$17x^2y^3$	17	5	
$-3x^3yz$	-3	5	(Think: $-3x^3yz = -3x^3y^1z^1$)
-7	-7	0	(Think: $-7 = -7x^0$)

The terms of highest degree are $17x^2y^3$ and $-3x^3yz$. Each has degree 5. The degree of the polynomial is 5.

14. Coefficients: 6, -1, 8, -1

Degrees: 0, 2, 4, 5; 5

15. $7a + b - 4a - 3b = (7 - 4)a + (1 - 3)b = 3a - 2b$

16. $3r - 3s$

17. $3x^2y - 2xy^2 + x^2 + 5x$

There are <u>no</u> like terms, so none of the terms can be collected.

18. $m^3 + 2m^2n - 3m^2 + 3mn^2$

19. $\quad 2u^2v - 3uv^2 + 6u^2v - 2uv^2 + 7u^2$
$= (2 + 6)u^2v + (-3 - 2)uv^2 + 7u^2$
$= 8u^2v - 5uv^2 + 7u^2$

20. $-2x^2 - 4xy + 3y^2$

21. $\quad 5a^2c - 2ab^2 + a^2b - 3ab^2 + a^2c - 2ab^2$
$= (5 + 1)a^2c + (-2 - 3 - 2)ab^2 + a^2b$
$= 6a^2c - 7ab^2 + a^2b$

22. $2s^2t - 6r^2t - st^2$

23. $\quad (4x^2 - xy + y^2) + (-x^2 - 3xy + 2y^2)$
$= (4 - 1)x^2 + (-1 - 3)xy + (1 + 2)y^2$
$= 3x^2 - 4xy + 3y^2$

24. $-3r^3 + 2rs - 9s^2$

25. $\quad (3a^4 - 5ab + 6ab^2) - (9a^4 + 3ab - ab^2)$
$= 3a^4 - 5ab + 6ab^2 - 9a^4 - 3ab + ab^2$
$\qquad\qquad\qquad$ Adding the opposite
$= (3 - 9)a^4 + (-5 - 3)ab + (6 + 1)ab^2$
$= -6a^4 - 8ab + 7ab^2$

26. $-5r^2t - 6rt + 6rt^2$

27. $(5r^2 - 4rt + t^2) + (-6r^2 - 5rt - t^2) + (-5r^2 + 4rt - t^2)$

Observe that the polynomials $5r^2 - 4rt + t^2$ and $-5r^2 + 4rt - t^2$ are opposites. Thus, their sum is 0 and the sum in the exercise is the remaining polynomial, $-6r^2 - 5rt - t^2$.

28. $2x^2 - 3xy + y^2$

29. $\quad (x^3 - y^3) - (-2x^3 + x^2y - xy^2 + 2y^3)$
$= x^3 - y^3 + 2x^3 - x^2y + xy^2 - 2y^3$
$= 3x^3 - 3y^3 - x^2y + xy^2,$ or
$\quad 3x^3 - x^2y + xy^2 - 3y^3$

30. $6a^3 - 2a^2b + ab^2 - 2b^3$

31. $\quad (2y^4x^2 - 5y^3x) + (5y^4x^2 - y^3x) + (3y^4x^2 - 2y^3x)$
$= (2 + 5 + 3)y^4x^2 + (-5 - 1 - 2)y^3x$
$= 10y^4x^2 - 8y^3x$

32. $15a^2b - 4ab$

33. $\quad (4x + 5y) + (-5x + 6y) - (7x + 3y)$
$= 4x + 5y - 5x + 6y - 7x - 3y$
$= (4 - 5 - 7)x + (5 + 6 - 3)y$
$= -8x + 8y$

34. $-5b$

35. $\qquad\qquad\qquad\qquad\quad$ F $\quad$ O $\quad$ I $\quad$ L
$(3z - u)(2z + 3u) = 6z^2 + 9zu - 2uz - 3u^2$
$\qquad\qquad\qquad\quad = 6z^2 + 7zu - 3u^2$

36. $10x^2 - 13xy - 3y^2$

37. $\qquad\qquad\qquad\qquad$ F $\quad$ O $\quad$ I $\quad$ L
$(xy + 7)(xy - 4) = x^2y^2 - 4xy + 7xy - 28 - 28$
$\qquad\qquad\qquad\quad = x^2y^2 + 3xy - 28$

38. $a^2b^2 - 2ab - 15$

39. $\quad (2a - b)(2a + b) \quad [(A + B)(A - B) = A^2 - B^2]$
$= 4a^2 - b^2$

40. $a^2 - 9b^2$

41. $\qquad\qquad\qquad\qquad$ F $\quad$ O $\quad$ I $\quad$ L
$(5rt - 2)(3rt + 1) = 15r^2t^2 + 5rt - 6rt - 2$
$\qquad\qquad\qquad\quad = 15r^2t^2 - rt - 2$

42. $12x^2y^2 + 2xy - 2$

43. $\quad (m^3n + 8)(m^3n - 6)$
$\qquad\quad$ F $\qquad$ O $\qquad$ I $\qquad$ L
$= m^6n^2 - 6m^3n + 8m^3n - 48$
$= m^6n^2 + 2m^3n - 48$

44. $12 - c^2d^2 - c^4d^4$

45. $\quad (6x - 2y)(5x - 3y)$
$\qquad\quad$ F $\qquad$ O $\qquad$ I $\qquad$ L
$= 30x^2 - 18xy - 10xy + 6y^2$
$= 30x^2 - 28xy + 6y^2$

46. $35a^2 - 2ab - 24b^2$

47. $(pq + 0.2)(0.4pq - 0.1)$
    F   O   I   L
$= 0.4p^2q^2 - 0.1pq + 0.08pq - 0.02$
$= 0.4p^2q^2 - 0.02pq - 0.02$

48. $0.2a^2b^2 + 0.18ab - 0.18$

49. $(x + h)^2$
$= x^2 + 2xh + h^2 \quad [(A + B)^2 = A^2 + 2AB + B^2]$

50. $r^2 + 2rt + t^2$

51. $(4a + 5b)^2$
$= 16a^2 + 40ab + 25b^2 \quad [(A+B)^2 = A^2 + 2AB + B^2]$

52. $9x^2 + 12xy + 4y^2$

53. $(c^2 - d)(c^2 + d) = (c^2)^2 - d^2$
$= c^4 - d^2$

54. $p^6 - 25q^2$

55. $(ab + cd^2)(ab - cd^2) = (ab)^2 - (cd^2)^2$
$= a^2b^2 - c^2d^4$

56. $x^2y^2 - p^2q^2$

57. $(a + b - c)(a + b + c)$
$= [(a + b) - c][(a + b) + c]$
$= (a + b)^2 - c^2$
$= a^2 + 2ab + b^2 - c^2$

58. $x^2 + 2xy + y^2 - z^2$

59. $[a + b + c][a - (b + c)]$
$= [a + (b + c)][a - (b + c)]$
$= a^2 - (b + c)^2$
$= a^2 - (b^2 + 2bc + c^2)$
$= a^2 - b^2 - 2bc - c^2$

60. $a^2 - b^2 - 2bc - c^2$

61. The figure is a rectangle with dimensions $a + b$ by $a + c$. Its area is $(a + b)(a + c) = a^2 + ac + ab + bc$.

62. $x^2 + 2xy + y^2$

63. The figure is a parallelogram with base $x + z$ and height $x - z$. Thus the area is $(x+z)(x-z) = x^2 - z^2$.

64. $\frac{1}{2}a^2b^2 - 2$

65. The figure is a square with side $x + y + z$. Thus the area is
$$(x + y + z)^2$$
$$= [(x + y) + z]^2$$
$$= (x + y)^2 + 2(x + y)(z) + z^2$$
$$= x^2 + 2xy + y^2 + 2xz + 2yz + z^2.$$

66. $a^2 + 2ac + c^2 + ad + cd + ab + bc + bd$

67. The figure is a triangle with base $x + 2y$ and height $x - y$. Thus the area is $\frac{1}{2}(x + 2y)(x - y) = \frac{1}{2}(x^2 + xy - 2y^2) = \frac{1}{2}x^2 + \frac{1}{2}xy - y^2$.

68. $m^2 - n^2$

69. We draw a rectangle with dimensions $r + s$ by $u + v$.

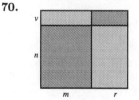

70.

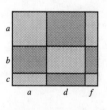

71. We draw a rectangle with dimensions $a + b + c$ by $a + d + f$.

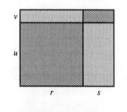

72.

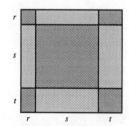

73. Writing exercise

74. Writing exercise

75. $5 + \dfrac{7 + 4 + 2 \cdot 5}{3}$

$= 5 + \dfrac{7 + 4 + 10}{3}$ Multiplying

$= 5 + \dfrac{21}{3}$ Adding in the numerator

$= 5 + 7$ Dividing

$= 12$ Adding

76. 5

77. $(4 + 3 \cdot 5 + 8) \div 3 \cdot 3$

$= (4 + 15 + 8) \div 3 \cdot 3$ Multiplying inside the parentheses

$= 27 \div 3 \cdot 3$ Adding

$= 9 \cdot 3$ Dividing

$= 27$ Multiplying

78. 36

79. $[3 \cdot 5 - 4 \cdot 2 + 7(-3)] \div (-2)$

$= (15 - 8 - 21) \div (-2)$ Multiplying

$= -14 \div (-2)$ Subtracting

$= 7$ Dividing

80. 5

81. Writing exercise

82. Writing exercise

83. The unshaded region is a circle with radius $a - b$. Then the shaded area is the area of a circle with radius a less the area of a circle with radius $a - b$. Thus, we have:

Shaded area $= \pi a^2 - \pi(a - b)^2$

$= \pi a^2 - \pi(a^2 - 2ab + b^2)$

$= \pi a^2 - \pi a^2 + 2\pi ab - \pi b^2$

$= 2\pi ab - \pi b^2$

84. $4xy - 4y^2$

85. The shaded area is the area of a square with side a less the areas of 4 squares with side b. Thus, the shaded area is $a^2 - 4 \cdot b^2$, or $a^2 - 4b^2$.

86. $\pi x^2 + 2xy$

87. a) The figure is a square with side A, so its area is $A \cdot A$, or A^2. The unshaded square has side B, so its area is $B \cdot B$, or B^2. Then the shaded area is $A^2 - B^2$.

b) In the upper left-hand corner we have a square with side $A - B$, so its area is $(A - B)(A - B)$, or $A^2 - 2AB + B^2$. The rectangles in the upper right-hand and lower left-hand corners have dimensions $A - B$ by B, so each has area $(A - B)(B)$, or $AB - B^2$. The sum of these three areas is $A^2 - 2AB + B^2 + AB - B^2 + AB - B^2$, or $A^2 - B^2$.

88. $2\pi nh + 2\pi mh + 2\pi n^2 - 2\pi m^2$

89. The surface area of the solid consists of the surface area of a rectangular solid with dimensions x by x by h less the areas of 2 circles with radius r plus the lateral surface area of a right circular cylinder with radius r and height h. Thus, we have

$2x^2 + 2xh + 2xh - 2\pi r^2 + 2\pi rh$, or

$2x^2 + 4xh - 2\pi r^2 + 2\pi rh$.

90. Writing exercise

91. $(x + a)(x - b)(x - a)(x + b)$

$= [(x + a)(x - a)][(x - b)(x + b)]$

$= (x^2 - a^2)(x^2 - b^2)$

$= x^4 - b^2 x^2 - a^2 x^2 + a^2 b^2$

92. $P + 2Pr + Pr^2$

93. Replace t with 2 and multiply.

$P(1 - r)^2$

$= P(1 - 2r + r^2)$

$= P - 2Pr + Pr^2$

94. $15,638.03

95. Substitute $90,000 for P, 12.5% or 0.125 for r, and 4 for t.

$P(1 - r)^t$

$= \$90,000(1 - 0.125)^4$

$\approx \$52,756.35$

Exercise Set 4.7

1. $\dfrac{40x^5 - 16x}{8} = \dfrac{40x^5}{8} - \dfrac{16x}{8}$

$= \dfrac{40}{8}x^5 - \dfrac{16}{8}x$ Dividing coefficients

$= 5x^5 - 2x$

To check, we multiply the quotient by 8:

$(5x^5 - 2x)8 = 40x^5 - 16x$

The answer checks.

2. $2a^4 - \dfrac{1}{2}a^2$

3. $\dfrac{u - 2u^2 + u^7}{u}$

$= \dfrac{u}{u} - \dfrac{2u^2}{u} + \dfrac{u^7}{u}$

$= 1 - 2u + u^6$

Check: We multiply.

$$
\begin{array}{r}
1 - 2u + u^6 \\
\underline{\hspace{2cm} u} \\
u - 2u^2 + u^7
\end{array}
$$

4. $50x^4 - 7x^3 + x$

5. $(15t^3 - 24t^2 + 6t) \div (3t)$

$= \dfrac{15t^3 - 24t^2 + 6t}{3t}$

$= \dfrac{15t^3}{3t} - \dfrac{24t^2}{3t} + \dfrac{6t}{3t}$

$= 5t^2 - 8t + 2$

Check: We multiply.

$$
\begin{array}{r}
5t^2 - 8t + 2 \\
\underline{\hspace{2.5cm} 3t} \\
15t^3 - 24t^2 + 6t
\end{array}
$$

6. $4t^2 - 3t + 6$

7. $(25x^6 - 20x^4 - 5x^2) \div (-5x^2)$

$= \dfrac{25x^6 - 20x^4 - 5x^2}{-5x^2}$

$= \dfrac{25x^6}{-5x^2} - \dfrac{20x^4}{-5x^2} - \dfrac{5x^2}{-5x^2}$

$= -5x^4 - (-4x^2) - (-1)$

$= -5x^4 + 4x^2 + 1$

Check: We multiply.

$$
\begin{array}{r}
-5x^4 + 4x^2 + 1 \\
\underline{\hspace{3cm} -5x^2} \\
25x^6 - 20x^4 - 5x^2
\end{array}
$$

8. $-2x^4 - 4x^3 + 1$

9. $(24t^5 - 40t^4 + 6t^3) \div (4t^3)$

$= \dfrac{24t^5 - 40t^4 + 6t^3}{4t^3}$

$= \dfrac{24t^5}{4t^3} - \dfrac{40t^4}{4t^3} + \dfrac{6t^3}{4t^3}$

$= 6t^2 - 10t + \dfrac{3}{2}$

Check: We multiply.

$$
\begin{array}{r}
6t^2 - 10t + \dfrac{3}{2} \\
\underline{\hspace{3cm} 4t^3} \\
24t^5 - 40t^4 + 6t^3
\end{array}
$$

10. $2t^3 - 3t^2 - \dfrac{1}{3}$

11. $\dfrac{6x^2 - 10x + 1}{2}$

$= \dfrac{6x^2}{2} - \dfrac{10x}{2} + \dfrac{1}{2}$

$= 3x^2 - 5x + \dfrac{1}{2}$

Check: We multiply.

$$
\begin{array}{r}
3x^2 - 5x + \dfrac{1}{2} \\
\underline{\hspace{3cm} 2} \\
6x^2 - 10x + 1
\end{array}
$$

12. $3x^2 + x - \dfrac{2}{3}$

13. $\dfrac{4x^3 + 6x^2 + 4x}{2x^2}$

$= \dfrac{4x^3}{2x^2} + \dfrac{6x^2}{2x^2} + \dfrac{4x}{2x^2}$

$= 2x + 3 + \dfrac{2}{x}$

Check: We multiply.

$$
\begin{array}{r}
2x + 3 + \dfrac{2}{x} \\
\underline{\hspace{3cm} 2x^2} \\
4x^3 + 6x^2 + 4x
\end{array}
$$

14. $2x^2 + 3x + \dfrac{1}{x}$

15. $\dfrac{9r^2s^2 + 3r^2s - 6rs^2}{-3rs}$

$= \dfrac{9r^2s^2}{-3rs} + \dfrac{3r^2s}{-3rs} - \dfrac{6rs^2}{-3rs}$

$= -3rs - r + 2s$

Check: We multiply.

$$
\begin{array}{r}
-3rs - r + 2s \\
\underline{\hspace{3cm} -3rs} \\
9r^2s^2 + 3r^2s - 6rs^2
\end{array}
$$

16. $1 - 2x^2y + 3x^4y^5$

17.

$$\begin{array}{r} x+6 \\ x-2\,\overline{\smash{\big)}\,x^2+4x-12} \\ \underline{x^2-2x} \\ 6x-12 \\ \underline{6x-12} \\ 0 \end{array}$$

$\leftarrow (x^2+4x)-(x^2-2x)=6x$

$\leftarrow (6x-12)-(6x-12)=0$

The answer is $x+6$.

18. $x-2$

19.

$$\begin{array}{r} t-5 \\ t-5\,\overline{\smash{\big)}\,t^2-10t-20} \\ \underline{t^2-5t} \\ -5t-20 \\ \underline{-5t+25} \\ -45 \end{array}$$

$\leftarrow (t^2-10t)-(t^2-5t)= -5t$

$\leftarrow (-5t-20)-(-5t+25)= -45$

The answer is $t-5+\dfrac{-45}{t-5}$, or $t-5-\dfrac{45}{t-5}$.

20. $t+4+\dfrac{-31}{t+4}$

21.

$$\begin{array}{r} 2x-1 \\ x+6\,\overline{\smash{\big)}\,2x^2+11x-5} \\ \underline{2x^2+12x} \\ -x-5 \\ \underline{-x-6} \\ 1 \end{array}$$

$\leftarrow (2x^2+11x)-(2x^2+12x)= -x$

$\leftarrow (-x-5)-(-x-6)=1$

The answer is $2x-1+\dfrac{1}{x+6}$.

22. $3x+4+\dfrac{-5}{x-2}$

23.

$$\begin{array}{r} a^2-2a+4 \\ a+2\,\overline{\smash{\big)}\,a^3+0a^2+0a+8} \\ \underline{a^3+2a^2} \\ -2a^2+0a \\ \underline{-2a^2-4a} \\ 4a+8 \\ \underline{4a+8} \\ 0 \end{array}$$

$\leftarrow$Writing in the missing terms

$\leftarrow a^3-(a^3+2a^2)=-2a^2$

$\leftarrow -2a^2-(-2a^2-4a)=4a$

$\leftarrow (4a+8)-(4a+8)=0$

The answer is a^2-2a+4.

24. t^2-3t+9

25.

$$\begin{array}{r} t+4 \\ t-4\,\overline{\smash{\big)}\,t^2+0t-15} \\ \underline{t^2-4t} \\ 4t-15 \\ \underline{4t-16} \\ 1 \end{array}$$

$\leftarrow$Writing in the missing term

$\leftarrow t^2-(t^2-4t)=4t$

$\leftarrow (4t-15)-(4t-16)=1$

The answer is $t+4+\dfrac{1}{t-4}$.

26. $a+5+\dfrac{2}{a-5}$

27.

$$\begin{array}{r} x+4 \\ 3x-1\,\overline{\smash{\big)}\,3x^2+11x-4} \\ \underline{3x^2-x} \\ 12x-4 \\ \underline{12x-4} \\ 0 \end{array}$$

$\leftarrow (3x^2+11x)-(3x^2-x)=12x$

$\leftarrow (12x-4)-(12x-4)=0$

The answer is $x+4$.

28. $2x+3$

29.

$$\begin{array}{r} 3a+1 \\ 2a+5\,\overline{\smash{\big)}\,6a^2+17a+8} \\ \underline{6a^2+15a} \\ 2a+8 \\ \underline{2a+5} \\ 3 \end{array}$$

$\leftarrow (6a^2+17a)-(6a^2+15a)=2a$

$\leftarrow (2a+8)-(2a+5)=3$

The answer is $3a+1+\dfrac{3}{2a+5}$.

30. $5a+2+\dfrac{3}{2a+3}$

31.

$$\begin{array}{r} t^2-3t+1 \\ 2t-3\,\overline{\smash{\big)}\,2t^3-9t^2+11t-3} \\ \underline{2t^3-3t^2} \\ -6t^2+11t \\ \underline{-6t^2+9t} \\ 2t-3 \\ \underline{2t-3} \\ 0 \end{array}$$

$\leftarrow (2t^3-9t^2)-(2t^3-3t^2)=-6t^2$

$\leftarrow (-6t^2+11t)-(-6t^2+9t)=2t$

$\leftarrow (2t-3)-(2t-3)=0$

The answer is t^2-3t+1.

32. $2t^2-7t+4$

33.

$$\begin{array}{r} t^2-2t+3 \\ t+1\,\overline{\smash{\big)}\,t^3-t^2+t-1} \\ \underline{t^3+t^2} \\ -2t^2+t \\ \underline{-2t^2-2t} \\ 3t-1 \\ \underline{3t+3} \\ -4 \end{array}$$

$\leftarrow (t^3-t^2)-(t^3+t^2)=-2t^2$

$\leftarrow (-2t^2+t)-(-2t^2-2t)=3t$

The answer is $t^2-2t+3+\dfrac{-4}{t+1}$, or

$t^2-2t+3-\dfrac{4}{t+1}$.

34. x^2+1

35.

$$t^2 + 5 \overline{\smash)t^4 + 0t^3 + 4t^2 + 3t - 6}$$ ←Writing in the

$$\underline{t^4 + 5t^2}$$ missing term

$$-t^2 + 3t - 6 \leftarrow (t^4 + 4t^2)-$$

$$\underline{-t^2 - 5} \quad (t^4 + 5t^2) = -t^2$$

$$3t - 1 \leftarrow (-t^2 + 3t - 6)-$$

$$(-t^2 - 5) = 3t - 1$$

The answer is $t^2 - 1 + \dfrac{3t - 1}{t^2 + 5}$.

36. $t^2 + 1 + \dfrac{4t - 2}{t^2 - 3}$

37.

$$2x^2 - 3 \overline{\smash)4x^4 + 0x^3 - 4x^2 - x - 3}$$ ←Writing in the

$$\underline{4x^4 - 6x^2}$$ missing term

$$2x^2 - x - 3 \leftarrow (4x^4 - 4x^2)-$$

$$(4x^4 - 6x^2) = 2x^2$$

$$\underline{2x^2 - 3}$$

$$-x \leftarrow (2x^2 - x - 3)-$$

$$(2x^2 - 3) = -x$$

The answer is $2x^2 + 1 + \dfrac{-x}{2x^2 - 3}$, or

$2x^2 + 1 - \dfrac{x}{2x^2 - 3}$.

38. $3x^2 - 3 + \dfrac{x - 1}{2x^2 + 1}$

39. Writing exercise

40. Writing exercise

41. $-4 + (-13)$ Two negative numbers. Add the absolute values, 4 and 13, to get 17. Make the answer negative.

     $-4 + (-13) = -17$

42. -23

43. $-9 - (-7) = -9 + 7 = -2$

44. 5

45. *Familiarize.* Let $w =$ the width. Then $w + 15 =$ the length. We draw a picture.

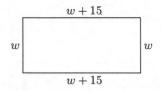

We will use the fact that the perimeter is 640 ft to find w (the width). Then we can find $w + 15$ (the length) and multiply the length and the width to find the area.

Translate.

Width+Width+ Length + Length =Perimeter

$w \quad + \quad w \quad +(w + 15)+(w + 15)= \quad 640$

Carry out.

$$w + w + (w + 15) + (w + 15) = 640$$
$$4w + 30 = 640$$
$$4w = 610$$
$$w = 152.5$$

If the width is 152.5, then the length is $152.5 + 15$, or 167.5.

Check. The length, 167.5 ft, is 15 ft greater than the width, 152.5 ft. The perimeter is $152.5 + 152.5 + 167.5 + 167.5$, or 640 ft. The answer checks.

State. The length is 167.5 ft.

46. 2

47. Graph: $3x - 2y = 12$.

We will graph the equation using intercepts. To find the y-intercept, we let $x = 0$.

$$-2y = 12 \quad \text{Ignoring the } x\text{-term}$$
$$y = -6$$

The y-intercept is $(0, -6)$.

To find the x-intercept, we let $y = 0$.

$$3x = 12 \quad \text{Ignoring the } y\text{-term}$$
$$x = 4$$

The x-intercept is $(4, 0)$.

To find a third point, replace x with -2 and solve for y:

$$3(-2) - 2y = 12$$
$$-6 - 2y = 12$$
$$-2y = 18$$
$$y = -9$$

The point $(-2, -9)$ appears to line up with the intercepts, so we draw the graph.

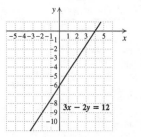

48.

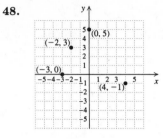

49. Writing exercise

50. Writing exercise

51. $(10x^{9k} - 32x^{6k} + 28x^{3k}) \div (2x^{3k})$

$= \dfrac{10x^{9k} - 32x^{6k} + 28x^{3k}}{2x^{3k}}$

$= \dfrac{10x^{9k}}{2x^{3k}} - \dfrac{32x^{6k}}{2x^{3k}} + \dfrac{28x^{3k}}{2x^{3k}}$

$= 5x^{9k-3k} - 16x^{6k-3k} + 14x^{3k-3k}$

$= 5x^{6k} - 16x^{3k} + 14$

52. $15a^{6k} + 10a^{4k} - 20a^{2k}$

53.
$$
\begin{array}{r}
3t^{2h} + 2t^h - 5 \\
2t^h + 3 \,\overline{\smash{\big)}\, 6t^{3h} + 13t^{2h} - 4t^h - 15} \\
\underline{6t^{3h} + 9t^{2h}} \\
4t^{2h} - 4t^h \\
\underline{4t^{2h} + 6t^h} \\
-10t^h - 15 \\
\underline{-10t^h - 15} \\
0
\end{array}
$$

The answer is $3t^{2h} + 2t^h - 5$.

54. $x^3 - ax^2 + a^2x - a^3 + \dfrac{a^2 + a^4}{x + a}$

55.
$$
\begin{array}{r}
a + 3 \\
5a^2 - 7a - 2 \,\overline{\smash{\big)}\, 5a^3 + 8a^2 - 23a - 1} \\
\underline{5a^3 - 7a^2 - 2a} \\
15a^2 - 21a - 1 \\
\underline{15a^2 - 21a - 6} \\
5
\end{array}
$$

The answer is $a + 3 + \dfrac{5}{5a^2 - 7a - 2}$.

56. $5y + 2 + \dfrac{-10y + 11}{3y^2 - 5y - 2}$

57. $(4x^5 - 14x^3 - x^2 + 3) +$
$ (2x^5 + 3x^4 + x^3 - 3x^2 + 5x)$
$= 6x^5 + 3x^4 - 13x^3 - 4x^2 + 5x + 3$

$$
\begin{array}{r}
2x^2 + x - 3 \\
3x^3 - 2x - 1 \,\overline{\smash{\big)}\, 6x^5 + 3x^4 - 13x^3 - 4x^2 + 5x + 3} \\
\underline{6x^5 - 4x^3 - 2x^2} \\
3x^4 - 9x^3 - 2x^2 + 5x \\
\underline{3x^4 - 2x^2 - x} \\
-9x^3 + 6x + 3 \\
\underline{-9x^3 + 6x + 3} \\
0
\end{array}
$$

The answer is $2x^2 + x - 3$.

58. $5x^5 + 5x^4 - 8x^2 - 8x + 2$

59.
$$
\begin{array}{r}
x - 3 \\
x - 1 \,\overline{\smash{\big)}\, x^2 - 4x + c} \\
\underline{x^2 - x} \\
-3x + c \\
\underline{-3x + 3} \\
c - 3
\end{array}
$$

We set the remainder equal to 0.

$c - 3 = 0$

$c = 3$

Thus, c must be 3.

60. -2

61.
$$
\begin{array}{r}
c^2x + (2c + c^2) \\
x - 1 \,\overline{\smash{\big)}\, c^2x^2 + 2cx + 1} \\
\underline{c^2x^2 - c^2x} \\
(2c + c^2)x + 1 \\
\underline{(2c + c^2)x - (2c + c^2)} \\
1 + (2c + c^2)
\end{array}
$$

We set the remainder equal to 0.

$c^2 + 2c + 1 = 0$

$(c + 1)^2 = 0$

$c + 1 = 0 \quad or \quad c + 1 = 0$

$c = -1 \quad or c = -1$

Thus, c must be -1.

Exercise Set 4.8

1. $5^{-2} = \dfrac{1}{5^2} = \dfrac{1}{25}$

2. $\dfrac{1}{2^4} = \dfrac{1}{16}$

3. $10^{-4} = \dfrac{1}{10^4} = \dfrac{1}{10,000}$

4. $\dfrac{1}{5^3} = \dfrac{1}{125}$

5. $(-2)^{-6} = \dfrac{1}{(-2)^6} = \dfrac{1}{64}$

6. $\dfrac{1}{(-3)^4} = \dfrac{1}{81}$

7. $x^{-8} = \dfrac{1}{x^8}$

8. $\dfrac{1}{t^5}$

9. $xy^{-2} = x \cdot \dfrac{1}{y^2} = \dfrac{x}{y^2}$

10. $\dfrac{b}{a^3}$

11. $r^{-5}t = \dfrac{1}{r^5} \cdot t = \dfrac{t}{r^5}$

12. $\dfrac{x}{y^9}$

13. $\dfrac{1}{t^{-7}} = t^7$

14. z^9

15. $\dfrac{1}{h^{-8}} = h^8$

16. a^{12}

17. $7^{-1} = \dfrac{1}{7^1} = \dfrac{1}{7}$

18. $\dfrac{1}{3}$

19. $\left(\dfrac{2}{5}\right)^{-2} = \left(\dfrac{5}{2}\right)^2 = \dfrac{5^2}{2^2} = \dfrac{25}{4}$

20. $\left(\dfrac{4}{3}\right)^2 = \dfrac{16}{9}$

21. $\left(\dfrac{a}{2}\right)^{-3} = \left(\dfrac{2}{a}\right)^3 = \dfrac{2^3}{a^3} = \dfrac{8}{a^3}$

22. $\left(\dfrac{3}{x}\right)^4 = \dfrac{81}{x^4}$

23. $\left(\dfrac{s}{t}\right)^{-7} = \left(\dfrac{t}{s}\right)^7 = \dfrac{t^7}{s^7}$

24. $\left(\dfrac{v}{r}\right)^5 = \dfrac{v^5}{r^5}$

25. $\dfrac{1}{7^2} = 7^{-2}$

26. 5^{-2}

27. $\dfrac{1}{t^6} = t^{-6}$

28. y^{-2}

29. $\dfrac{1}{a^4} = a^{-4}$

30. t^{-5}

31. $\dfrac{1}{p^8} = p^{-8}$

32. m^{-12}

33. $\dfrac{1}{5} = \dfrac{1}{5^1} = 5^{-1}$

34. 8^{-1}

35. $\dfrac{1}{t} = \dfrac{1}{t^1} = t^{-1}$

36. m^{-1}

37. $2^{-5} \cdot 2^8 = 2^{-5+8} = 2^3$, or 8

38. 5

39. $x^{-2} \cdot x^{-7} = x^{-2+(-7)} = x^{-9} = \dfrac{1}{x^9}$

40. $\dfrac{1}{x^{11}}$

41. $t^{-3} \cdot t = t^{-3} \cdot t^1 = t^{-3+1} = t^{-2} = \dfrac{1}{t^2}$

42. $\dfrac{1}{y^4}$

43. $(a^{-2})^9 = a^{-2 \cdot 9} = a^{-18} = \dfrac{1}{a^{18}}$

44. $\dfrac{1}{x^{30}}$

45. $(t^{-3})^{-6} = t^{-3(-6)} = t^{18}$

46. a^{28}

47. $(t^4)^{-3} = t^{4(-3)} = t^{-12} = \dfrac{1}{t^{12}}$

48. $\dfrac{1}{t^{10}}$

49. $(x^{-2})^{-4} = x^{-2(-4)} = x^8$

50. t^{30}

51. $(ab)^{-3} = \dfrac{1}{(ab)^3} = \dfrac{1}{a^3b^3}$

52. $\dfrac{1}{x^6 y^6}$

53. $(mn)^{-7} = \dfrac{1}{(mn)^7} = \dfrac{1}{m^7 n^7}$

54. $\dfrac{1}{a^9 b^9}$

55. $(3x^{-4})^2 = 3^2 (x^{-4})^2 = 9x^{-8} = \dfrac{9}{x^8}$

56. $\dfrac{8}{a^{15}}$

57. $(5r^{-4} t^3)^2 = 5^2 (r^{-4})^2 (t^3)^2 = 25 r^{-8} t^6 = \dfrac{25 t^6}{r^8}$

58. $\dfrac{64 x^{15}}{y^{18}}$

59. $\dfrac{t^7}{t^{-3}} = t^{7-(-3)} = t^{10}$

60. x^9

61. $\dfrac{y^{-7}}{y^{-3}} = y^{-7-(-3)} = y^{-4} = \dfrac{1}{y^4}$

62. $\dfrac{1}{z^4}$

63. $\dfrac{y^{-4}}{y^{-9}} = y^{-4-(-9)} = y^5$

64. a^4

65. $\dfrac{x^6}{x} = \dfrac{x^6}{x^1} = x^{6-1} = x^5$

66. x^2

67. $\dfrac{a^{-7}}{b^{-9}} = \dfrac{b^9}{a^7}$

68. $\dfrac{y^{10}}{x^6}$

69. $\dfrac{t^{-7}}{t^{-7}}$

Note that we have an expression divided by itself. Thus, the result is 1. We could also find this result as follows:
$$\dfrac{t^{-7}}{t^{-7}} = t^{-7-(-7)} = t^0 = 1.$$

70. $\dfrac{b^7}{a^5}$

71. $\dfrac{3x^{-5}}{y^{-6} z^{-2}} = \dfrac{3 y^6 z^2}{x^5}$

72. $\dfrac{4 b^5 c^7}{a^6}$

73. $\dfrac{3 t^4}{s^{-2} u^{-4}} = 3 s^2 t^4 u^4$

74. $\dfrac{5 y^3}{x^8 z^2}$

75. $(x^4 y^5)^{-3} = (x^4)^{-3} (y^5)^{-3} = x^{-12} y^{-15} = \dfrac{1}{x^{12} y^{15}}$

76. $\dfrac{1}{t^{20} x^{12}}$

77. $(x^{-6} y^{-2})^{-4} = (x^{-6})^{-4} (y^{-2})^{-4} = x^{24} y^8$

78. $x^{10} y^{35}$

79. $(a^{-5} b^7 c^{-2})(a^{-3} b^{-2} c^6) = a^{-5+(-3)} b^{7+(-2)} c^{-2+6} = a^{-8} b^5 c^4 = \dfrac{b^5 c^4}{a^8}$

80. $\dfrac{z^4}{x y^6}$

81. $\left(\dfrac{a^4}{3}\right)^{-2} = \left(\dfrac{3}{a^4}\right)^2 = \dfrac{3^2}{(a^4)^2} = \dfrac{9}{a^8}$

82. $\dfrac{4}{y^4}$

83. $\left(\dfrac{7}{x^{-3}}\right)^2 = (7 x^3)^2 = 7^2 (x^3)^2 = 49 x^6$

84. $27 a^6$

85. $\left(\dfrac{m^{-1}}{n^{-4}}\right)^3 = \dfrac{(m^{-1})^3}{(n^{-4})^3} = \dfrac{m^{-3}}{n^{-12}} = \dfrac{n^{12}}{m^3}$

86. $x^6 y^3 z^{15}$

87. $\left(\dfrac{2a^2}{3b^4}\right)^{-3} = \left(\dfrac{3b^4}{2a^2}\right)^3 = \dfrac{(3b^4)^3}{(2a^2)^3} = \dfrac{3^3 (b^4)^3}{2^3 (a^2)^3} = \dfrac{27 b^{12}}{8 a^6}$

88. $\dfrac{c^5 d^{15}}{a^{10} b^5}$

89. $\left(\dfrac{5 x^{-2}}{3 y^{-2} z}\right)^0$

Any nonzero expression raised to the 0 power is equal to 1. Thus, the answer is 1.

90. $\dfrac{4 a^3 c^3}{5 b^2}$

91. 7.12×10^4

Since the exponent is positive, the decimal point will move to the right.

7.1200.   The decimal point moves right 4 places.

$7.12 \times 10^4 = 71,200$

92. 892

93. 8.92×10^{-3}

Since the exponent is negative, the decimal point will move to the left.

.008.92   The decimal point moves left 3 places.

$8.92 \times 10^{-3} = 0.00892$

94. 0.000726

95. 9.04×10^8

Since the exponent is positive, the decimal point will move to the right.

9.04000000.

 8 places

$9.04 \times 10^8 = 904,000,000$

96. 13,500,000

97. 2.764×10^{-10}

Since the exponent is negative, the decimal point will move to the left.

0.0000000002.764

 10 places

$2.764 \times 10^{-10} = 0.0000000002764$

98. 0.009043

99. 4.209×10^9

Since the exponent is positive, the decimal point will move to the right.

4.2090000.

 7 places

$4.209 \times 10^7 = 42,090,000$

100. 502,900,000

101. $490,000 = 4.9 \times 10^m$

To write 4.9 as 490,000 we move the decimal point 5 places to the right. Thus, m is 5 and

$490,000 = 4.9 \times 10^5$.

102. 7.15×10^4

103. $0.00583 = 5.83 \times 10^m$

To write 5.83 as 0.00583 we move the decimal point 3 places to the left. Thus, m is -3 and

$0.00583 = 5.83 \times 10^{-3}$.

104. 8.14×10^{-2}

105. $78,000,000,000 = 7.8 \times 10^m$

To write 7.8 as 78,000,000,000 we move the decimal point 10 places to the right. Thus, m is 10 and

$78,000,000,000 = 7.8 \times 10^{10}$.

106. 3.7×10^{12}

107. $907,000,000,000,000,000 = 9.07 \times 10^m$

To write 9.07 as 907,000,000,000,000,000 we move the decimal point 17 places to the right. Thus, m is 17 and

$907,000,000,000,000,000 = 9.07 \times 10^{17}$.

108. 1.68×10^{14}

109. $0.000000527 = 5.27 \times 10^m$

To write 5.27 as 0.000000527 we move the decimal point 7 places to the left. Thus, m is -7 and

$0.000000527 = 5.27 \times 10^{-7}$.

110. 6.48×10^{-9}

111. $0.000000018 = 1.8 \times 10^m$

To write 1.8 as 0.000000018 we move the decimal point 8 places to the left. Thus, m is -8 and

$0.000000018 = 1.8 \times 10^{-8}$.

112. 2×10^{-11}

113. $1,094,000,000,000,000 = 1.094 \times 10^m$

To write 1.094 as 1,094,000,000,000,000 we move the decimal point 15 places to the right. Thus, m is 15 and

$1,094,000,000,000,000 = 1.094 \times 10^{15}$.

114. 1.0302×10^{18}

115. $(4 \times 10^7)(2 \times 10^5) = (4 \cdot 2) \times (10^7 \cdot 10^5)$

$ = 8 \times 10^{7+5}$ Adding exponents

$ = 8 \times 10^{12}$

116. 6.46×10^5

117. $(3.8 \times 10^9)(6.5 \times 10^{-2}) = (3.8 \cdot 6.5) \times (10^9 \cdot 10^{-2})$
$$= 24.7 \times 10^7$$

The answer is not yet in scientific notation since 24.7 is not a number between 1 and 10. We convert to scientific notation.
$$24.7 \times 10^7 = (2.47 \times 10) \times 10^7 = 2.47 \times 10^8$$

118. 6.106×10^{-11}

119. $(8.7 \times 10^{-12})(4.5 \times 10^{-5})$
$$= (8.7 \cdot 4.5) \times (10^{-12} \cdot 10^{-5})$$
$$= 39.15 \times 10^{-17}$$

The answer is not yet in scientific notation since 39.15 is not a number between 1 and 10. We convert to scientific notation.
$$39.15 \times 10^{-17} = (3.915 \times 10) \times 10^{-17} = 3.915 \times 10^{-16}$$

120. 2.914×10^{-6}

121. $\dfrac{8.5 \times 10^8}{3.4 \times 10^{-5}} = \dfrac{8.5}{3.4} \times \dfrac{10^8}{10^{-5}}$
$$= 2.5 \times 10^{8-(-5)}$$
$$= 2.5 \times 10^{13}$$

122. 2.24×10^{-7}

123. $(3.0 \times 10^6) \div (6.0 \times 10^9) = \dfrac{3.0 \times 10^6}{6.0 \times 10^9}$
$$= \dfrac{3.0}{6.0} \times \dfrac{10^6}{10^9}$$
$$= 0.5 \times 10^{6-9}$$
$$= 0.5 \times 10^{-3}$$

The answer is not yet in scientific notation because 0.5 is not between 1 and 10. We convert to scientific notation.
$$0.5 \times 10^{-3} = (5.0 \times 10^{-1}) \times 10^{-3} =$$
$$5.0 \times 10^{-4}$$

124. 9.375×10^2

125. $\dfrac{7.5 \times 10^{-9}}{2.5 \times 10^{12}} = \dfrac{7.5}{2.5} \times \dfrac{10^{-9}}{10^{12}}$
$$= 3.0 \times 10^{-9-12}$$
$$= 3.0 \times 10^{-21}$$

126. 5×10^{-24}

127. Writing exercise

128. Writing exercise

129. $(3 - 8)(9 - 12)$
$$= (-5)(-3) \qquad \text{Subtracting}$$
$$= 15 \qquad\qquad \text{Multiplying}$$

130. 49

131. $7 \cdot 2 + 8^2$
$$= 7 \cdot 2 + 64 \quad \text{Evaluating the exponential expression}$$
$$= 14 + 64 \quad \text{Multiplying}$$
$$= 78 \qquad\quad \text{Adding}$$

132. 6

133. To plot $(-3, 2)$, we start at the origin and move 3 units to the left and then 2 units up. To plot $(4, -1)$, we start at the origin and move 4 units to the right and then 1 unit down. To plot $(5, 3)$, we start at the origin and move 5 units to the right and 3 units up. To plot $(-5, -2)$, we start at the origin and move 5 units to the left and then 2 units down.

134. $t = \dfrac{r - cx}{b}$

135. Writing exercise

136. Writing exercise

137. $\dfrac{4.2 \times 10^8 [(2.5 \times 10^{-5}) \div (5.0 \times 10^{-9})]}{3.0 \times 10^{-12}}$
$$= \dfrac{4.2 \times 10^8 [0.5 \times 10^4]}{3.0 \times 10^{-12}}$$
$$= \dfrac{2.1 \times 10^{12}}{3.0 \times 10^{-12}}$$
$$= 0.7 \times 10^{24}$$
$$= (7 \times 10^{-1}) \times 10^{24}$$
$$= 7 \times 10^{23}$$

138. 8×10^5

139. $\dfrac{1}{2.5 \times 10^9} = \dfrac{1}{2.5} \times \dfrac{1}{10^9} = 0.4 \times 10^{-9} =$
$$(4 \times 10^{-1}) \times 10^{-9} = 4 \times 10^{-10}$$

140. 2^{-12}

141. $81^3 \cdot 27 \div 9^2 = (3^4)^3 \cdot 3^3 \div (3^2)^2 = 3^{12} \cdot 3^3 \div 3^4 =$
$3^{15} \div 3^4 = 3^{11}$

142. 7

143. $\dfrac{125^{-4}(25^2)^4}{125} = \dfrac{(5^3)^{-4}((5^2)^2)^4}{5^3} =$

$\dfrac{5^{-12}(5^4)^4}{5^3} = \dfrac{5^{-12} \cdot 5^{16}}{5^3} = \dfrac{5^4}{5^3} = 5^1 = 5$

144. 9

145. a) False; let $x = 2$, $y = 3$, $m = 4$, and $n = 2$:
$$2^4 \cdot 3^2 = 16 \cdot 9 = 144, \text{ but}$$
$$(2 \cdot 3)^{4 \cdot 2} = 6^8 = 1,679,616$$

b) False; let $x = 3$, $y = 4$, and $m = 2$:
$$3^2 \cdot 4^2 = 9 \cdot 16 = 144, \text{ but}$$
$$(3 \cdot 4)^{2 \cdot 2} = 12^4 = 20,736$$

c) False; let $x = 5$, $y = 3$, and $m = 2$:
$$(5 - 3)^2 = 2^2 = 4, \text{ but}$$
$$5^2 - 3^2 = 25 - 9 = 16$$

146. $4.894179894 \times 10^{26}$

147.
$$\frac{5.8 \times 10^{17}}{(4.0 \times 10^{-13})(2.3 \times 10^4)}$$
$$= \frac{5.8}{(4.0 \cdot 2.3)} \times \frac{10^{17}}{(10^{-13} \cdot 10^4)}$$
$$\approx 0.6304347826 \times 10^{17-(-13)-4}$$
$$\approx (6.304347826 \times 10^{-1}) \times 10^{26}$$
$$\approx 6.304347826 \times 10^{25}$$

148. 3.12×10^{43}

149.
$$\frac{(2.5 \times 10^{-8})(6.1 \times 10^{-11})}{1.28 \times 10^{-3}}$$
$$= \frac{(2.5 \cdot 6.1)}{1.28} \times \frac{(10^{-8} \cdot 10^{-11})}{10^{-3}}$$
$$= 11.9140625 \times 10^{-8+(-11)-(-3)}$$
$$= 11.9140625 \times 10^{-16}$$
$$= (1.19140625 \times 10) \times 10^{-16}$$
$$= 1.19140625 \times 10^{-15}$$

150. $\$5.09425 \times 10^{12}$

151. *Familiarize*. Express 1 billion and 2500 in scientific notation:
$$1 \text{ billion} = 1,000,000,000 = 10^9$$
$$2500 = 2.5 \times 10^3$$

Let b = the number of bytes in the network.

Translate. We reword the problem.

What is 2500 times 1 gigabyte?

$b = 2.5 \times 10^3 \times 10^9$

Carry out. We do the computation.
$$b = (2.5 \times 10^3) \times 10^9$$
$$b = 2.5 \times (10^3 \times 10^9)$$
$$b = 2.5 \times 10^{12}$$

Check. We review the computation. Also, the answer seems reasonable since it is larger than 1 billion.

State. There are 2.5×10^{12} bytes in the network.

152. 1.325×10^{14} cubic feet

153. *Familiarize*. We must express both dimensions using the same units. Let's choose centimeters. First, convert 1.5 m to centimeters and express the result in scientific notation.

$1.5 \text{ m} = 1.5 \times 1 \text{ m} = 1.5 \times 100 \text{ cm} = 1.5 \times 10^2 \text{ cm}$

Let l represent how many times the DNA is longer than it is wide.

Translate. We reword the problem.

The length is how many times the width.

$1.5 \times 10^2 = l \cdot 1.3 \times 10^{-10}$

Carry out. We solve the equation.
$$1.5 \times 10^2 = l \cdot 1.3 \times 10^{-10}$$
$$\frac{1.5 \times 10^2}{1.3 \times 10^{-10}} = l$$
$$1.15385 \times 10^{12} \approx l$$

Check. Since $(1.15385 \times 10^{12}) \times (1.3 \times 10^{-10}) = 1.498705 \times 10^2 \approx 1.5 \times 10^2$, the answer checks.

State. A strand of DNA is about 1.15385×10^{12} times longer than it is wide.

154. 2×10^{14} gal

Chapter 5

Polynomials and Factoring

Exercise Set 5.1

1. Answers may vary. $10x^3 = (5x)(2x^2) = (10x^2)(x) = (-2)(-5x^3)$

2. Answers may vary. $(6x)(x^2)$; $(3x^2)(2x)$; $(2x^2)(3x)$

3. Answers may vary. $-15a^4 = (-15)(a^4) = (-5a)(3a^3) = (-3a^2)(5a^2)$

4. Answers may vary. $(-8t)(t^4)$; $(-2t^2)(4t^3)$; $(-4t^3)(2t^2)$

5. Answers may vary. $26x^5 = (2x^4)(13x) = (2x^3)(13x^2) = (-x^2)(-26x^3)$

6. Answers may vary. $(5x^2)(5x^2)$; $(x^3)(25x)$; $(-5x)(-5x^3)$

7. $\begin{aligned}x^2 + 8x &= x \cdot x + x \cdot 8 \\ &= x(x+8)\end{aligned}$

8. $x(x+6)$

9. $\begin{aligned}10t^2 - 5t &= 5t \cdot 2t - 5t \cdot 1 \\ &= 5t(2t-1)\end{aligned}$

10. $5a(a-3)$

11. $\begin{aligned}x^3 + 6x^2 &= x^2 \cdot x + x^2 \cdot 6 \\ &= x^2(x+6)\end{aligned}$

12. $x^2(4x^2+1)$

13. $\begin{aligned}8x^4 - 24x^2 &= 8x^2 \cdot x^2 - 8x^2 \cdot 3 \\ &= 8x^2(x^2-3)\end{aligned}$

14. $5x^3(x^2+2)$

15. $\begin{aligned}2x^2 + 2x - 8 &= 2 \cdot x^2 + 2 \cdot x - 2 \cdot 4 \\ &= 2(x^2+x-4)\end{aligned}$

16. $3(2x^2+x-5)$

17. $\begin{aligned}7a^6 - 10a^4 - 14a^2 &= a^2 \cdot 7a^4 - a^2 \cdot 10a^2 - a^2 \cdot 14 \\ &= a^2(7a^4 - 10a^2 - 14)\end{aligned}$

18. $t^3(10t^2 - 15t + 9)$

19. $\begin{aligned}&2x^8 + 4x^6 - 8x^4 + 10x^2 \\ &= 2x^2 \cdot x^6 + 2x^2 \cdot 2x^4 - 2x^2 \cdot 4x^2 + 2x^2 \cdot 5 \\ &= 2x^2(x^6 + 2x^4 - 4x^2 + 5)\end{aligned}$

20. $5(x^4 - 3x^3 - 5x - 2)$

21. $\begin{aligned}&x^5y^5 + x^4y^3 + x^3y^3 - x^2y^2 \\ &= x^2y^2 \cdot x^3y^3 + x^2y^2 \cdot x^2y + x^2y^2 \cdot xy - x^2y^2 \cdot 1 \\ &= x^2y^2(x^3y^3 + x^2y + xy - 1)\end{aligned}$

22. $x^3y^3(x^6y^3 - x^4y^2 + xy + 1)$

23. $\begin{aligned}&5a^3b^4 + 10a^2b^3 - 15a^3b^2 \\ &= 5a^2b^2 \cdot ab^2 + 5a^2b^2 \cdot 2b - 5a^2b^2 \cdot 3a \\ &= 5a^2b^2(ab^2 + 2b - 3a)\end{aligned}$

24. $7r^3t^4(3r^2 - 2rt^2 + 3t^2)$

25. $\begin{aligned}&y(y-2) + 7(y-2) \\ &= (y-2)(y+7) \quad \text{Factoring out the common binomial factor } y-2\end{aligned}$

26. $(b+5)(b+3)$

27. $\begin{aligned}&x^2(x+3) - 7(x+3) \\ &= (x+3)(x^2-7) \quad \text{Factoring out the common binomial factor } x+3\end{aligned}$

28. $(2z+9)(3z^2+1)$

29. $\begin{aligned}y^2(y+8) + (y+8) &= y^2(y+8) + 1(y+8) \\ &= (y+8)(y^2+1) \quad \text{Factoring out the common factor}\end{aligned}$

30. $(x-7)(x^2-3)$

31. $\begin{aligned}&x^3 + 3x^2 + 4x + 12 \\ &= (x^3 + 3x^2) + (4x + 12) \\ &= x^2(x+3) + 4(x+3) \quad \text{Factoring each binomial} \\ &= (x+3)(x^2+4) \quad \text{Factoring out the common factor } x+3\end{aligned}$

32. $(2z+1)(3z^2+1)$

33. $\qquad 3a^3 + 9a^2 + 2a + 6$

$= (3a^3 + 9a^2) + (2a + 6)$

$= 3a^2(a + 3) + 2(a + 3)$ $\quad$ Factoring each binomial

$= (a + 3)(3a^2 + 2)$ $\quad$ Factoring out the common factor $a + 3$

34. $(3a + 2)(a^2 + 2)$

35. $\qquad 9x^3 - 12x^2 + 3x - 4$

$= 3x^2(3x - 4) + 1(3x - 4)$

$= (3x - 4)(3x^2 + 1)$

36. $(2x - 5)(5x^2 + 2)$

37. $\qquad 4t^3 - 20t^2 + 3t - 15$

$= 4t^2(t - 5) + 3(t - 5)$

$= (t - 5)(4t^2 + 3)$

38. $(3a - 4)(2a^2 + 3)$

39. $\qquad 7x^3 + 2x^2 - 14x - 4$

$= x^2(7x + 2) - 2(7x + 2)$

$= (7x + 2)(x^2 - 2)$

40. $(5x + 4)(x^2 - 2)$

41. $\qquad 6a^3 - 7a^2 + 6a - 7$

$= a^2(6a - 7) + 1(6a - 7)$

$= (6a - 7)(a^2 + 1)$

42. $(7t - 5)(t^2 + 1)$

43. $x^3 + 8x^2 - 3x - 24 = x^2(x + 8) - 3(x + 8)$

$= (x + 8)(x^2 - 3)$

44. $(x + 7)(x^2 - 2)$

45. $2x^3 + 12x^2 - 5x - 30 = 2x^2(x + 6) - 5(x + 6)$

$= (x + 6)(2x^2 - 5)$

46. $(x + 5)(3x^2 - 5)$

47. $w^3 - 7w^2 + 4w - 28 = w^2(w - 7) + 4(w - 7)$

$= (w - 7)(w^2 + 4)$

48. Cannot be factored by grouping

49. We try factoring by grouping.

$x^3 - x^2 - 2x + 5 = x^2(x - 1) - 1(2x - 5)$, or

$x^3 - 2x - x^2 + 5 = x(x^2 - 2) - (x^2 - 5)$

Because we cannot find a common binomial factor, this polynomial cannot be factored using factoring by grouping.

50. $(y + 8)(y^2 - 2)$

51. $2x^3 - 8x^2 - 9x + 36 = 2x^2(x - 4) - 9(x - 4)$

$= (x - 4)(2x^2 - 9)$

52. $(5g - 1)(4g^2 - 5)$

53. Writing exercise

54. Writing exercise

55. $\qquad (x + 3)(x + 5)$

$\qquad$ F $\qquad$ O $\qquad$ I $\qquad$ L

$= x \cdot x + x \cdot 5 + 3 \cdot x + 3 \cdot 5$

$= x^2 + 5x + 3x + 15$

$= x^2 + 8x + 15$

56. $x^2 + 9x + 14$

57. $\qquad (a - 7)(a + 3)$

$\qquad$ F $\qquad$ O $\qquad$ I $\qquad$ L

$= a \cdot a + a \cdot 3 - 7 \cdot a - 7 \cdot 3$

$= a^2 + 3a - 7a - 21$

$= a^2 - 4a - 21$

58. $a^2 - 3a - 40$

59. $\qquad (2x + 5)(3x - 4)$

$\qquad$ F $\qquad$ O $\qquad$ I $\qquad$ L

$= 2x \cdot 3x - 2x \cdot 4 + 5 \cdot 3x - 5 \cdot 4$

$= 6x^2 - 8x + 15x - 20$

$= 6x^2 + 7x - 20$

60. $12t^2 - 13t - 14$

61. $\qquad (3t - 5)^2$

$= (3t)^2 - 2 \cdot 3t \cdot 5 + 5^2$

$\qquad [(A - B)^2 = A^2 - 2AB + B^2]$

$= 9t^2 - 30t + 25$

62. $4t^2 - 36t + 81$

63. Writing exercise

64. Writing exercise

65. $4x^5 + 6x^3 + 6x^2 + 9 = 2x^3(2x^2 + 3) + 3(2x^2 + 3)$

$= (2x^2 + 3)(2x^3 + 3)$

66. $(x^2 + 1)(x^4 + 1)$

67. $x^{12} + x^7 + x^5 + 1 = x^7(x^5 + 1) + (x^5 + 1)$

$= (x^5 + 1)(x^7 + 1)$

68. Cannot be factored by grouping

69.
$$5x^5 - 5x^4 + x^3 - x^2 + 3x - 3$$
$$= 5x^4(x-1) + x^2(x-1) + 3(x-1)$$
$$= (x-1)(5x^4 + x^2 + 3)$$

We could also do this exercise as follows:
$$5x^5 - 5x^4 + x^3 - x^2 + 3x - 3$$
$$= (5x^5 + x^3 + 3x) - (5x^4 + x^2 + 3)$$
$$= x(5x^4 + x^2 + 3) - 1(5x^4 + x^2 + 3)$$
$$= (5x^4 + x^2 + 3)(x-1)$$

70. $(x^2 + 2x + 3)(a + 1)$

71. Answers may vary. $3x^4y^3 - 9x^3y^3 + 27x^2y^4$

Exercise Set 5.2

1. $x^2 + 6x + 5$

Since the constant term and the coefficient of the middle term are both positive, we look for a factorization of 5 in which both factors are positive. The only possible positive factors are 1 and 5. Their sum is 6, so these are the numbers we want.

$x^2 + 6x + 5 = (x+1)(x+5)$.

2. $(x + 1)(x + 6)$

3. $x^2 + 7x + 10$

Since the constant term is positive and the coefficient of the middle term is positive, we look for a factorization of 10 in which both factors are positive. Their sum must be 7.

Pairs of factors	Sums of factors
1, 10	11
2, 5	7

The numbers we want are 2 and 5.

$x^2 + 7x + 10 = (x+2)(x+5)$

4. $(x + 3)(x + 4)$

5. $y^2 + 11y + 28$

Since the constant term is positive and the coefficient of the middle term is positive, we look for a factorization of 28 in which both factors are positive. Their sum must be 11.

Pairs of factors	Sums of factors
1, 28	29
2, 14	16
4, 7	11

The numbers we want are 4 and 7.

$y^2 + 11y + 28 = (y+4)(y+7)$

6. $(x - 3)(x - 3)$, or $(x - 3)^2$

7. $a^2 + 11a + 30$

Since the constant term is positive and the coefficient of the middle term is positive, we look for a factorization of 30 in which both factors are positive. Their sum must be 11.

Pairs of factors	Sums of factors
1, 30	31
2, 15	17
3, 10	13
5, 6	11

The numbers we want are 5 and 6.

$a^2 + 11a + 30 = (a+5)(a+6)$.

8. $(x + 2)(x + 7)$

9. $x^2 - 5x + 4$

Since the constant term is positive and the coefficient of the middle term is negative, we look for a factorization of 4 in which both factors are negative. Their sum must be -5.

Pairs of factors	Sums of factors
$-1, -4$	-5
$-2, -2$	-4

The numbers we want are -1 and -4.

$x^2 - 5x + 4 = (x-1)(x-4)$.

10. $(b + 1)(b + 4)$

11. $z^2 - 8z + 7$

Since the constant term is positive and the coefficient of the middle term is negative, we look for a factorization of 7 in which both factors are negative. Their sum must be -8. The only possible negative factors are -1 and -7. Their sum is -8, so these are the numbers we want.

$z^2 - 8z + 7 = (z-1)(z-7)$

12. $(a + 2)(a - 6)$

13. $x^2 - 8x + 15$

Since the constant term is positive and the coefficient of the middle term is negative, we look for a factorization of 15 in which both factors are negative. Their sum must be -8.

Pairs of factors	Sums of factors
$-1, -15$	-16
$-3, -5$	-8

The numbers we want are -3 and -5.

$x^2 - 8x + 15 = (x-3)(x-5)$.

14. $(d-2)(d-5)$

15. $y^2 - 11y + 10$

Since the constant term is positive and the coefficient of the middle term is negative, we look for a factorization of 10 in which both factors are negative. Their sum must be -11.

Pairs of factors	Sums of factors
-1, -10	-11
-2, -5	-7

The numbers we want are -1 and -10.
$y^2 - 11y + 10 = (y-1)(y-10)$.

16. $(x+3)(x-5)$

17. $x^2 + x - 42$

Since the constant term is negative, we look for a factorization of -42 in which one factor is positive and one factor is negative. Their sum must be 1, the coefficient of the middle term, so the positive factor must have the larger absolute value. Thus we consider only pairs of factors in which the positive factor has the larger absolute value.

Pairs of factors	Sums of factors
-1, 42	41
-2, 21	19
-3, 14	11
-6, 7	1

The numbers we need are -6 and 7.
$x^2 + x - 42 = (x-6)(x+7)$.

18. $(x-3)(x+5)$

19. $2x^2 - 14x - 36 = 2(x^2 - 7x - 18)$

After factoring out the common factor, 2, we consider $x^2 - 7x - 18$. Since the constant term is negative, we look for a factorization of -18 in which one factor is positive and one factor is negative. Their sum must be -7, the coefficient of the middle term, so the negative factor must have the larger absolute value. Thus we consider only pairs of factors in which the negative factor has the larger absolute value.

Pairs of factors	Sums of factors
1, -18	-17
2, -9	-7
3, -6	-3

The numbers we want are 2 and -9. The factorization of $x^2 - 7x - 18$ is $(x+2)(x-9)$. We must not forget the common factor, 2. The factorization of $2x^2 - 14x - 36$ is $2(x+2)(x-9)$.

20. $3(y+4)(y-7)$

21. $x^3 - 6x^2 - 16x = x(x^2 - 6x - 16)$

After factoring out the common factor, x, we consider $x^2 - 6x - 16$. Since the constant term is negative, we look for a factorization of -16 in which one factor is positive and one factor is negative. Their sum must be -6, the coefficient of the middle term, so the negative factor must have the larger absolute value. Thus we consider only pairs of factors in which the negative factor has the large absolute value.

Pairs of factors	Sums of factors
1, -16	-15
2, -8	-6

The numbers we want are 2 and -8.
Then $x^2 - 6x - 16 = (x+2)(x-8)$, so $x^3 - 6x^2 - 16x = x(x+2)(x-8)$.

22. $x(x+6)(x-7)$

23. $y^2 + 4y - 45$

The constant term, 45, must be expressed as the product of a negative number and a positive number. Since the sum of those two numbers must be positive, the positive number must have the greater absolute value.

Pairs of factors	Sums of factors
-1, 45	44
-3, 15	12
-5, 9	4

The numbers we need are -5 and 9.
$y^2 + 4y - 45 = (y-5)(y+9)$

24. $(x-5)(x+12)$

25. $-2x - 99 + x^2 = x^2 - 2x - 99$

Since the constant term is negative, we look for a factorization of -99 in which one factor is positive and one factor is negative. Their sum must be -2, the coefficient of the middle term, so the negative factor must have the larger absolute value. Thus we consider only pairs of factors in which the negative factor has the larger absolute value.

Pairs of factors	Sums of factors
1, -99	-98
3, -33	-30
9, -11	-2

The numbers we want are 9 and -11.
$-2x - 99 + x^2 = (x+9)(x-11)$

26. $(x-6)(x+12)$

27. $c^4 + c^3 - 56c^2 = c^2(c^2 + c - 56)$

After factoring out the common factor, c^2, we consider $c^2 + c - 56$. Since the constant term is negative, we look for a factorization of -56 in which one factor is positive and one factor is negative. Their sum must be 1, so the positive factor must have the larger absolute value. Thus we consider only pairs of factors in which the positive factor has the larger absolute value.

Pairs of factors	Sums of factors
-1, 56	55
-2, 28	26
-4, 14	10
-7, 8	1

The numbers we want are -7 and 8. The factorization of $c^2 + c - 56$ is $(c-7)(c+8)$, so $c^4 + c^3 - 56c^2 = c^2(c-7)(c+8)$.

28. $5(b-3)(b+8)$

29. $2a^2 - 4a - 70 = 2(a^2 - 2a - 35)$

After factoring out the common factor, 2, we consider $a^2 - 2a - 35$. Since the constant term is negative, we look for a factorization of -35 in which one factor is positive and one factor is negative. Their sum must be -2, so the negative factor must have the large absolute value. Thus we consider only pairs of factors in which the negative factor has the larger absolute value.

Pairs of factors	Sums of factors
1, -35	-34
5, -7	-2

The numbers we want are 5 and -7. The factorization of $a^2 - 2a - 35$ is $(a+5)(a-7)$, so $2a^2 - 4a - 70 = 2(a+5)(a-7)$.

30. $x^3(x-2)(x+1)$

31. $x^2 + x + 1$

Since the constant term and the coefficient of the middle term are both positive, we look for a factorization of 1 in which both factors are positive. Their sum must be 1. The only possible pair of factors is 1 and 1, but their sum is not 1. Thus, this polynomial is not factorable into polynomials with integer coefficients. It is prime.

32. Prime

33. $7 - 2p + p^2 = p^2 - 2p + 7$

Since the constant term is positive and the coefficient of the middle term is negative, we look for a factorization of 7 in which both factors are negative. Their sum must be -2. The only possible pair of factors is -1 and -7, but their sum is not -2. Thus, this polynomial is not factorable into polynomials with integer coefficients. It is prime.

34. Prime

35. $x^2 + 20x + 100$

We look for two factors, both positive, whose product is 100 and whose sum is 20.

They are 10 and 10: $10 \cdot 10 = 100$ and $10 + 10 = 20$.

$x^2 + 20x + 100 = (x+10)(x+10)$, or $(x+10)^2$.

36. $(x+9)(x+11)$

37. $3x^3 - 63x^2 - 300x = 3x(x^2 - 21x - 100)$

After factoring out the common factor, $3x$, we consider $x^2 - 21x - 100$. We look for two factors, one positive and one negative, whose product is -100 and whose sum is -21.

They are 4 and -25: $4 \cdot (-25) = -100$ and $4 + (-25) = -21$.

$x^2 - 21x - 100 = (x+4)(x-25)$, so $3x^3 - 63x^2 - 300x = 3x(x+4)(x-25)$.

38. $2x(x-8)(x-12)$

39. $x^2 - 21x - 72$

We look for two factors, both negative, whose product is -72 and whose sum is -21. They are 3 and -24.

$x^2 - 21x - 72 = (x+3)(x-24)$

40. $4(x+5)^2$

41. $x^2 - 25x + 144$

We look for two factors, both negative, whose product is 144 and whose sum is -25. They are -9 and -16.

$x^2 - 25x + 144 = (x-9)(x-16)$

42. $(y-9)(y-12)$

43. $a^4 + a^3 - 132a^2 = a^2(a^2 + a - 132)$

After factoring out the common factor, a^2, we consider $a^2 + a - 132$. We look for two factors, one positive and one negative, whose product is -132 and whose sum is 1. They are -11 and 12.

$a^2 + a - 132 = (a-11)(a+12)$, so $a^4 + a^3 - 132a^2 = a^2(a-11)(a+12)$.

44. $a^4(a-6)(a+15)$

45. $x^2 - \dfrac{2}{5}x + \dfrac{1}{25}$

We look for two factors, both negative, whose product is $\dfrac{1}{25}$ and whose sum is $-\dfrac{2}{5}$. They are $-\dfrac{1}{5}$ and $-\dfrac{1}{5}$.

$x^2 - \dfrac{2}{5}x + \dfrac{1}{25} = \left(x - \dfrac{1}{5}\right)\left(x - \dfrac{1}{5}\right)$, or $\left(x - \dfrac{1}{5}\right)^2$

46. $\left(t + \dfrac{1}{3}\right)\left(t + \dfrac{1}{3}\right)$, or $\left(t + \dfrac{1}{3}\right)^2$

47. $27 + 12y + y^2 = y^2 + 12y + 27$

We look for two factors, both positive, whose product is 27 and whose sum is 12. They are 3 and 9.

$27 + 12y + y^2 = (y+3)(y+9)$, or $(3+y)(9+y)$

48. $(5+x)(10+x)$

49. $t^2 - 0.3t - 0.10$

We look for two factors, one positive and one negative, whose product is -0.10 and whose sum is -0.3. They are 0.2 and -0.5.

$t^2 - 0.3t - 0.10 = (t + 0.2)(t - 0.5)$

50. $(y + 0.2)(y - 0.4)$

51. $p^2 + 3pq - 10q^2 = p^2 + 3qp - 10q^2$

Think of $3q$ as a "coefficient" of p. Then we look for factors of $-10q^2$ whose sum is $3q$. They are $5q$ and $-2q$.

$p^2 + 3pq - 10q^2 = (p + 5q)(p - 2q)$.

52. $(a - 3b)(a + b)$

53. $m^2 + 5mn + 5n^2 = m^2 + 5nm + 5n^2$

We look for factors of $5n^2$ whose sum is $5n$. The only reasonable possibilities are shown below.

Pairs of factors	Sums of factors
$5n, \quad n$	$6n$
$-5n, \; -n$	$-6n$

There are no factors whose sum is $5n$. Thus, the polynomial is not factorable into polynomials with integer coefficients. It is prime.

54. $(x - 8y)(x - 3y)$

55. $s^2 - 2st - 15t^2 = s^2 - 2ts - 15t^2$

We look for factors of $-15t^2$ whose sum is $-2t$. They are $-5t$ and $3t$.

$s^2 - 2st - 15t^2 = (s - 5t)(s + 3t)$

56. $(b + 10c)(b - 2c)$

57. $6a^{10} - 30a^9 - 84a^8 = 6a^8(a^2 - 5a - 14)$

After factoring out the common factor, $6a^8$, we consider $a^2 - 5a - 14$. We look for two factors, one positive and one negative, whose product is -14 and whose sum is -5. They are 2 and -7.

$a^2 - 5a - 14 = (a+2)(a-7)$, so $6a^{10} - 30a^9 - 84a^8 = 6a^8(a+2)(a-7)$.

58. $7x^7(x+1)(x-5)$

59. Writing exercise

60. Writing exercise

61. $3x - 8 = 0$

$\qquad 3x = 8 \qquad$ Adding 8 on both sides

$\qquad x = \dfrac{8}{3} \qquad$ Dividing by 3 on both sides

The solution is $\dfrac{8}{3}$.

62. $-\dfrac{7}{2}$

63. $(x + 6)(3x + 4)$

$= 3x^2 + 4x + 18x + 24 \qquad$ Using FOIL

$= 3x^2 + 22x + 24$

64. $49w^2 + 84w + 36$

65. *Familiarize.* Let $n =$ the number of people arrested the year before.

Translate. We reword the problem.

$$\underbrace{\begin{matrix}\text{Number}\\ \text{arrested the}\\ \text{year before}\end{matrix}}_{n} \underbrace{\text{less}}_{-} \ \underbrace{1.2\%}_{1.2\%} \ \underbrace{\text{of}}_{\cdot} \ \underbrace{\begin{matrix}\text{that}\\ \text{number}\end{matrix}}_{n} \ \underbrace{\text{is}}_{=} \ \underbrace{29,090.}_{29,090}$$

Carry out. We solve the equation.

$n - 1.2\% \cdot n = 29,090$

$1 \cdot n - 0.012n = 29,090$

$\qquad 0.988n = 29,090$

$\qquad\qquad n \approx 29,443 \qquad$ Rounding

Check. 1.2% of 29,443 is $0.012(29,443) \approx 353$ and $29,443 - 353 = 29,090$. The answer checks.

State. Approximately 29,443 people were arrested the year before.

66. $100°$, $25°$, $55°$

67. Writing exercise

68. Writing exercise

69. $a^2 + ba - 50$

We look for all pairs of integer factors whose product is -50. The sum of each pair is represented by b.

Pairs of factors whose product is -50	Sums of factors
$-1, \quad 50$	49
$1, \quad -50$	-49
$-2, \quad 25$	23
$2, \quad -25$	-23
$-5, \quad 10$	5
$5, \quad -10$	-5

The polynomial $a^2 + ba - 50$ can be factored if b is 49, -49, 23, -23, 5, or -5.

70. $51, -51, 27, -27, 15, -15$

71.
$$\begin{aligned}
30 + 7x - x^2 &= -1(-30) - 1(-7x) - 1 \cdot x^2 \\
&= -1(-30 - 7x + x^2) \\
&= -1(x^2 - 7x - 30)
\end{aligned}$$

We look for factors of -30 whose sum is -7. The numbers we want are -10 and 3.

$30 + 7x - x^2 = -1(x-10)(x+3)$, or $-1(-10+x)(3+x)$

72. $-1(x-9)(x+5)$, or $-1(-9+x)(5+x)$

73.
$$\begin{aligned}
24 - 10a - a^2 &= -1(-24) - 1(10a) - 1 \cdot a^2 \\
&= -1(-24 + 10a + a^2) \\
&= -1(a^2 + 10a - 24)
\end{aligned}$$

We look for factors of -24 whose sum is 10. The numbers we want are 12 and -2.

$24 - 10a - a^2 = -1(a+12)(a-2)$, or $-1(12+a)(-2+a)$

74. $-1(a+12)(a-3)$, or $-1(12+a)(-3+a)$

75.
$$\begin{aligned}
84 - 8t - t^2 &= -1(-84) - 1(8t) - 1 \cdot t^2 \\
&= -1(-84 + 8t + t^2) \\
&= -1(t^2 + 8t - 84)
\end{aligned}$$

We look for factors of -84 whose sum is 8. the numbers we want are 14 and -6.

$84 - 8t - t^2 = -1(t+14)(t-6)$, or $-1(14+t)(-6+t)$

76. $-1(t+12)(t-6)$, or $-1(12+t)(-6+t)$

77. $x^2 + \dfrac{1}{4}x - \dfrac{1}{8}$

We look for two factors, one positive and one negative, whose product is $-\dfrac{1}{8}$ and whose sum is $\dfrac{1}{4}$. They are $\dfrac{1}{2}$ and $-\dfrac{1}{4}$.

$$x^2 + \frac{1}{4}x - \frac{1}{8} = \left(x + \frac{1}{2}\right)\left(x - \frac{1}{4}\right)$$

78. $\left(x + \dfrac{3}{4}\right)\left(x - \dfrac{1}{4}\right)$

79. $\dfrac{1}{3}a^3 - \dfrac{1}{3}a^2 - 2a = \dfrac{1}{3}a(a^2 - a - 6)$

After factoring out the common factor, $\dfrac{1}{3}a$, we consider $a^2 - a - 6$. We look for two factors, one positive and one negative, whose product is -6 and whose sum is -1. They are 2 and -3.

$a^2 - a - 6 = (a+2)(a-3)$, so
$\dfrac{1}{3}a^3 - \dfrac{1}{3}a^2 - 2a = \dfrac{1}{3}a(a+2)(a-3)$.

80. $a^5(a-5)\left(a + \dfrac{5}{7}\right)$

81. $x^{2m} + 11x^m + 28 = (x^m)^2 + 11x^m + 28$

We look for numbers p and q such that $x^{2m} + 11x^m + 28 = (x^m + p)(x^m + q)$. We find two factors, both positive, whose product is 28 and whose sum is 11. They are 4 and 7.

$x^{2m} + 11x^m + 28 = (x^m + 4)(x^m + 7)$

82. $(t^n - 2)(t^n - 5)$

83.
$$\begin{aligned}
&(a+1)x^2 + (a+1)3x + (a+1)2 \\
&= (a+1)(x^2 + 3x + 2)
\end{aligned}$$

After factoring out the common factor $a+1$, we consider $x^2 + 3x + 2$. We look for two factors, whose product is 2 and whose sum is 3. They are 1 and 2.

$x^2 + 3x + 2 = (x+1)(x+2)$, so
$(a+1)x^2 + (a+1)3x + (a+1)2 =$
$(a+1)(x+1)(x+2)$.

84. $(a-5)(x+9)(x-1)$

85. We first label the drawing with additional information.

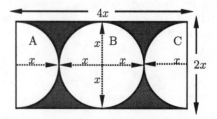

$4x$ represents the length of the rectangle and $2x$ the width. The area of the rectangle is $4x \cdot 2x$, or $8x^2$.

The area of semicircle A is $\dfrac{1}{2}\pi x^2$.

The area of circle B is πx^2.

The area of semicircle C is $\frac{1}{2}\pi x^2$.

$$\begin{array}{l}\text{Area of}\\\text{shaded}\\\text{region}\end{array} = \begin{array}{l}\text{Area of}\\\text{rectangle}\end{array} - \begin{array}{l}\text{Area}\\\text{of}\\A\end{array} - \begin{array}{l}\text{Area}\\\text{of}\\B\end{array} - \begin{array}{l}\text{Area}\\\text{of}\\C\end{array}$$

$$\begin{array}{l}\text{Area of}\\\text{shaded}\\\text{region}\end{array} = \quad 8x^2 \quad - \frac{1}{2}\pi x^2 - \pi x^2 - \frac{1}{2}\pi x^2$$

$$= 8x^2 - 2\pi x^2$$

$$= 2x^2(4-\pi)$$

The shaded area can be represented by $2x^2(4-\pi)$.

86. $x^2(\pi - 1)$

87. $6x^2 + 36x + 54 = 6(x^2 + 6x + 9) = 6(x+3)(x+3) = 6(x+3)^2$

Since the surface area of a cube with sides is given by $6s^2$, we know that this cube has side $x + 3$. The volume of a cube with side s is given by s^3, so the volume of this cube is $(x+3)^3$, or $x^3 + 9x^2 + 27x + 27$.

88. Writing exercise

Exercise Set 5.3

1. $2x^2 + 7x - 4$

(1) There is no common factor (other than 1 or -1).

(2) Because $2x^2$ can be factored as $2x \cdot x$, we have this possibility:

$$(2x + \quad)(x + \quad)$$

(3) There are 3 pairs of factors of -4 and they can be listed two ways:

$$-4,1 \quad 4,-1 \quad 2,-2$$

$$\text{and} \quad 1,-4 \quad -1,4 \quad -2,2$$

(4) Look for Outer and Inner products resulting from steps (2) and (3) for which the sum is $7x$. We can immediately reject all possibilities in which a factor has a common factor, such as $(2x - 4)$ or $(2x + 2)$, because we determined at the outset that there is no common factor other than 1 and -1. We try some possibilities:

$$(2x+1)(x-4) = 2x^2 - 7x - 4$$

$$(2x-1)(x+4) = 2x^2 + 7x - 4$$

The factorization is $(2x - 1)(x + 4)$.

2. $(3x + 4)(x - 1)$

3. $3t^2 + 4t - 15$

(1) There is no common factor (other than 1 or -1).

(2) Because $3t^2$ can be factored as $3t \cdot t$, we have this possibility:

$$(3t + \quad)(t + \quad)$$

(3) There are 4 pairs of factors of -15 and they can be listed two ways:

$$-15,1 \quad 15,-1 \quad -5,3 \quad 5,-3$$

$$\text{and} \quad 1,-15 \quad -1,15 \quad 3,-5 \quad -3,5$$

(4) Look for Outer and Inner products resulting from steps (2) and (3) for which the sum is $4t$. We can immediately reject all possibilities in which either factor has a common factor, such as $(3t - 15)$ or $(3t + 3)$, because at the outset we determined that there is no common factor other than 1 or -1. We try some possibilities:

$$(3t+1)(t-15) = 3t^2 - 44t - 15$$

$$(3t-5)(t+3) = 3t^2 + 4t - 15$$

The factorization is $(3t - 5)(t + 3)$.

4. $(5t - 9)(t + 2)$

5. $6x^2 - 23x + 7$

(1) There is no common factor (other than 1 or -1).

(2) Because $6x^2$ can be factored as $6x \cdot x$ or $3x \cdot 2x$, we have these possibilities:

$$(6x + \quad)(x + \quad) \text{ and } (3x + \quad)(2x + \quad)$$

(3) There are 2 pairs of factors of 7 and they can be listed two ways:

$$7,1 \quad -7,-1$$

$$\text{and} \quad 1,7 \quad -1,-7$$

(4) Look for Outer and Inner products resulting from steps (2) and (3) for which the sum is $-23x$. Since the sign of the middle term is negative and the sign of the last term is positive, the factors of 7 must both be negative. We try some possibilities:

$$(6x-7)(x-1) = 6x^2 - 13x + 7$$

$$(3x-7)(2x-1) = 6x^2 - 17x + 7$$

$$(6x-1)(x-7) = 6x^2 - 43x + 7$$

$$(3x-1)(2x-7) = 6x^2 - 23x + 7$$

The factorization is $(3x - 1)(2x - 7)$.

6. $(3x - 2)(2x - 3)$

7. $7x^2 + 15x + 2$

(1) There is no common factor (other than 1 or -1).

(2) Because $7x^2$ can be factored as $7x \cdot x$, we have this possibility:

$$(7x + \quad)(x + \quad)$$

(3) There are 2 pairs of factors of 2 and they can be listed two ways:

$$2, 1 \quad -2, -1$$
$$\text{and} \quad 1, 2 \quad -1, -2$$

(4) Look for Outer and Inner products resulting from steps (2) and (3) for which the sum is $15x$. Since all coefficients are positive, we need consider only positive factors of 2. We try some possibilities:

$$(7x + 2)(x + 1) = 7x^2 + 9x + 2$$
$$(7x + 1)(x + 2) = 7x^2 + 15x + 2$$

The factorization is $(7x + 1)(x + 2)$.

8. $(3x + 1)(x + 1)$

9. $9a^2 - 6a - 8$

(1) There is no common factor (other than 1 or -1).

(2) Because $9a^2$ can be factored as $9a \cdot a$ and $3a \cdot 3a$, we have these possibilities:

$$(9a + \quad)(a + \quad) \text{ and } (3a + \quad)(3a + \quad)$$

(3) There are 4 pairs of factors of -8 and they can be listed two ways:

$$-8, 1 \quad 8, -1 \quad -4, 2 \quad 4, -2$$
$$\text{and} \quad 1, -8 \quad -1, 8 \quad 2, -4 \quad -2, 4$$

(4) Look for Outer and Inner products resulting from steps (2) and (3) for which the sum is $-6a$. We try some possibilities:

$$(9a - 8)(a + 1) = 9a^2 + a - 8$$
$$(9a - 4)(a + 2) = 9a^2 + 14a - 8$$
$$(3a + 8)(3a - 1) = 9a^2 + 21a - 8$$
$$(3a - 4)(3a + 2) = 9a^2 - 6a - 8$$

The factorization is $(3a - 4)(3a + 2)$.

10. $(2a - 5)(2a + 3)$

11. $3x^2 - 5x - 2$

(1) There is no common factor (other than 1 or -1).

(2) Because $3x^2$ can be factored as $3x \cdot x$, we have this possibility:

$$(3x + \quad)(x + \quad)$$

(3) There are 2 pairs of factors of -2 and they can be listed two ways:

$$-2, 1 \quad 2, -1$$
$$\text{and} \quad 1, -2 \quad -1, 2$$

(4) Look for Outer and Inner products resulting from steps (2) and (3) for which the sum is $-5x$. We try some possibilities:

$$(3x - 2)(x + 1) = 3x^2 + x - 2$$
$$(3x + 2)(x - 1) = 3x^2 - x - 2$$
$$(3x + 1)(x - 2) = 3x^2 - 5x - 2$$

The factorization is $(3x + 1)(x - 2)$.

12. $(5x + 2)(3x - 5)$

13. $12t^2 - 6t - 6$

(1) We factor out the common factor, 6:

$$6(2t^2 - t - 1).$$

Then we factor the trinomial $2t^2 - t - 1$.

(2) Because $2t^2$ can be factored as $2t \cdot t$, we have this possibility:

$$(2t + \quad)(t + \quad)$$

(3) There are 2 pairs of factors of -1. In this case they can be listed in only one way:

$$-1, 1 \quad 1, -1$$

(4) Look for Outer and Inner products resulting from steps (2) and (3) for which the sum is $-t$. We try some possibilities:

$$(2t - 1)(t + 1) = 2t^2 + t - 1$$
$$(2t + 1)(t - 1) = 2t^2 - t - 1$$

The factorization of $2t^2 - t - 1$ is $(2t + 1)(t - 1)$. We must include the common factor in order to get a factorization of the original trinomial.

$$12t^2 - 6t - 6 = 6(2t + 1)(t - 1)$$

14. $2(3t - 2)(3t + 8)$

15. $18t^2 + 3t - 10$

(1) There is no common factor (other than 1 or -1).

(2) Because $18t^2$ can be factored as $18t \cdot t$, $9t \cdot 2t$, and $6t \cdot 3t$, we have these possibilities:

$$(18t + \quad)(t + \quad) \text{ and } (9t + \quad)(2t + \quad) \text{ and }$$
$$(6t + \quad)(3t + \quad)$$

(3) There are 4 pairs of factors of -10 and they can be listed two ways:

$$-10, 1 \quad 10, -1 \quad -5, 2 \quad 5, -2$$
$$\text{and} \quad 1, -10 \quad -1, 10 \quad 2, -5 \quad -2, 5$$

(4) We can immediately reject all possibilities in which either factor has a common factor, such as $(18t - 10)$ or $(2t + 2)$, because we determined at the outset that there is no common factor other than 1 or -1. We try some possibilities:

$$(18t - 5)(t + 2) = 18t^2 + 31t - 10$$
$$(9t - 5)(t + 2) = 9t^2 + 13t - 10$$
$$(6t - 5)(3t + 2) = 18t^2 - 3t - 10$$
$$(6t + 5)(3t - 2) = 18t^2 + 3t - 10$$

The factorization is $(6t + 5)(3t - 2)$.

16. $(2t + 1)(t + 2)$

17. $15x^2 + 19x + 6$

(1) There is no common factor (other than 1 or -1).

(2) Because $15x^2$ can be factored as $15x \cdot x$ and $5x \cdot 3x$, we have these possibilities:

$$(15x + \quad)(x + \quad) \text{ and } (5x + \quad)(3x + \quad)$$

(3) Since all coefficients are positive, we need consider only positive pairs of factors of 6. There are 2 such pairs and they can be listed two ways:

$$6, 1 \quad 3, 2$$
$$\text{and} \quad 1, 6 \quad 2, 3$$

(4) We can immediately reject all possibilities in which either factor has a common factor, such as $(15x + 6)$ or $(3x + 3)$, because we determined at the outset that there is no common factor other than 1 or -1. We try some possibilities:

$$(15x + 2)(x + 3) = 15x^2 + 47x + 6$$
$$(5x + 6)(3x + 1) = 15x^2 + 23x + 6$$
$$(5x + 3)(3x + 2) = 15x^2 + 19x + 6$$

The factorization is $(5x + 3)(3x + 2)$.

18. $(4x - 5)(3x - 4)$

19. $35x^2 + 34x + 8$

(1) There is no common factor (other than 1 or -1).

(2) Because $35x^2$ can be factored as $35x \cdot x$ or $7x \cdot 5x$, we have these possibilities:

$$(35x + \quad)(x + \quad) \text{ and } (7x + \quad)(5x + \quad)$$

(3) Since all coefficients are positive, we need consider only positive pairs of factors of 8. There are 2 such pairs and they can be listed two ways:

$$8, 1 \quad 4, 2$$
$$\text{and} \quad 1, 8 \quad 2, 4$$

(4) We try some possibilities:

$$(35x + 8)(x + 1) = 35x^2 + 43x + 8$$
$$(7x + 8)(5x + 1) = 35x^2 + 47x + 8$$
$$(7x + 4)(5x + 2) = 35x^2 + 34x + 8$$

The factorization is $(7x + 4)(5x + 2)$.

20. $2(7x - 1)(2x + 3)$

21. $4 + 6t^2 - 13t = 6t^2 - 13t + 4$

(1) There is no common factor (other than 1 or -1).

(2) Because $6t^2$ can be factored as $6t \cdot t$ or $3t \cdot 2t$, we have these possibilities:

$$(6t + \quad)(t + \quad) \text{ and } (3t + \quad)(2t + \quad)$$

(3) Since the sign of the middle term is negative but the sign of the last term is positive, we need to consider only negative factors of 4. There is only 1 such pair and it can be listed two ways:

$$-4, -1 \text{ and } -1, -4$$

(4) We can immediately reject all possibilities in which either factor has a common factor, such as $(6t - 4)$ or $(2t - 4)$, because we determined at the outset that there is no common factor other than 1 or -1. We try some possibilities:

$$(6t - 1)(t - 4) = 6t^2 - 25t + 4$$
$$(3t - 4)(2t - 1) = 6t^2 - 11t + 4$$

These are the only possibilities that do not contain a common factor. Since neither is the desired factorization, we must conclude that $4 + 6t^2 - 13t$ is prime.

22. $(2t - 3)(4t - 3)$

23. $25x^2 + 40x + 16$

(1) There is no common factor (other than 1 or -1).

(2) Because $25x^2$ can be factored as $25x \cdot x$ or $5x \cdot 5x$, we have these possibilities:

$$(25x + \quad)(x + \quad) \text{ and } (5x + \quad)(5x + \quad)$$

(3) Since all coefficients are positive, we need consider only positive pairs of factors of 16. There are 3 such pairs and two of them can be listed two ways:

$$16, 1 \quad 8, 2 \quad 4, 4$$
$$\text{and} \quad 1, 16 \quad 2, 8$$

(4) We try some possibilities:

$$(25x + 16)(x + 1) = 25x^2 + 41x + 16$$
$$(5x + 8)(5x + 2) = 25x^2 + 50x + 16$$
$$(5x + 4)(5x + 4) = 25x^2 + 40x + 16$$

The factorization is $(5x + 4)(5x + 4)$, or $(5x + 4)^2$.

24. $(7t + 3)(7t + 3)$, or $(7t + 3)^2$

25. $16a^2 + 78a + 27$

(1) There is no common factor (other than 1 or -1).

(2) Because $16a^2$ can be factored as $16a \cdot a$, $8a \cdot 2a$, or $4a \cdot 4a$, we have these possibilities:

$$(16a + \quad)(a + \quad) \text{ and } (8a + \quad)(2a + \quad) \text{ and } (4a + \quad)(4a + \quad)$$

(3) Since all coefficients are positive, we need consider only positive pairs of factors of 27. There are 2 such pairs and two of them can be listed two ways:

$$27, 1 \quad 3, 9$$
$$\text{and} \quad 1, 27 \quad 9, 3$$

(4) We try some possibilities:

$$(16a + 27)(a + 1) = 16a^2 + 43a + 27$$
$$(8a + 3)(2a + 9) = 16a^2 + 78a + 27$$

The factorization is $(8a + 3)(2a + 9)$.

26. $(24x - 1)(x + 2)$

27. $18t^2 + 24t - 10$

(1) Factor out the common factor, 2:

$2(9t^2 + 12t - 5)$

Then we factor the trinomial $9t^2 + 12t - 5$.

(2) Because $9t^2$ can be factored as $9t \cdot t$ or $3t \cdot 3t$, we have these possibilities:

$(9t +\quad)(t +\quad)$ and $(3t +\quad)(3t +\quad)$

(3) There are 2 pairs of factors of -5 and they can be listed two ways:

$$-5, 1 \quad 5, -1$$
$$\text{and} \quad 1, -5 \quad -1, 5$$

(4) We try some possibilities:

$$(9t - 5)(t + 1) = 9t^2 + 4t - 5$$
$$(9t + 1)(t - 5) = 9t^2 - 44t - 5$$
$$(3t + 1)(3t - 5) = 9t^2 - 12t - 5$$
$$(3t - 1)(3t + 5) = 9t^2 + 12t - 5$$

The factorization of $9t^2 + 12t - 5$ is $(3t - 1)(3t + 5)$. We must include the common factor in order to get a factorization of the original trinomial.

$18t^2 + 24t - 10 = 2(3t - 1)(3t + 5)$

28. $(7x + 4)(5x - 11)$

29. $2x^2 - 15 - x = 2x^2 - x - 15$

(1) There is no common factor (other than 1 or -1).

(2) Because $2x^2$ can be factored as $2x \cdot x$ we have this possibility:

$(2x +\quad)(x +\quad)$

(3) There are 4 pairs of factors of -15 and they can be listed two ways:

$$-15, 1 \quad 15, -1 \quad -5, 3 \quad 5, -3$$
$$\text{and} \quad 1, -15 \quad -1, 15 \quad 3, -5 \quad -3, 5$$

(4) We try some possibilities:

$$(2x - 15)(x + 1) = 2x^2 - 13x - 15$$
$$(2x - 5)(x + 3) = 2x^2 + x - 15$$
$$(2x + 5)(x - 3) = 2x^2 - x - 15$$

The factorization is $(2x + 5)(x - 3)$.

30. Prime

31. $6x^2 + 33x + 15$

(1) Factor out the common factor, 3:

$3(2x^2 + 11x + 5)$

Then we factor the trinomial $2x^2 + 11x + 5$.

(2) Because $2x^2$ can be factored as $2x \cdot x$ we have this possibility:

$(2x +\quad)(x +\quad)$

(3) Since all coefficients are positive, we need consider only positive pairs of factors of 5. There is one such pair and it can be listed two ways:

$5, 1 \quad \text{and} \quad 1, 5$

(4) We try some possibilities:

$$(2x + 5)(x + 1) = 2x^2 + 7x + 5$$
$$(2x + 1)(x + 5) = 2x^2 + 11x + 5$$

The factorization of $2x^2 + 11x + 5$ is $(2x + 1)(x + 5)$. We must include the common factor in order to get a factorization of the original trinomial.

$6x^2 + 33x + 15 = 3(2x + 1)(x + 5)$

32. $4(3x - 2)(x + 3)$

33. $20x^2 - 25x + 5$

(1) Factor out the common factor, 5:

$5(4x^2 - 5x + 1)$

Then we factor the trinomial $4x^2 - 5x + 1$.

(2) Because $4x^2$ can be factored as $4x \cdot x$ or $2x \cdot 2x$, we have these possibilities:

$(4x +\quad)(x +\quad)$ and $(2x +\quad)(2x +\quad)$

(3) Since the sign of the middle term is negative but the sign of the last term is positive, we need to consider only negative factors of 1. There is only 1 such pair, $-1, -1$.

(4) We try the possibilities:

$$(4x - 1)(x - 1) = 4x^2 - 5x + 1$$

The factorization of $4x^2 - 5x + 1$ is $(4x - 1)(x - 1)$. We must include the common factor in order to get a factorization of the original trinomial.

$20x^2 - 25x + 5 = 5(4x - 1)(x - 1)$

34. $6(5x - 9)(x + 1)$

35. $12x^2 + 68x - 24$

(1) Factor out the common factor, 4:

$4(3x^2 + 17x - 6)$

Then we factor the trinomial $3x^2 + 17x - 6$.

(2) Because $3x^2$ can be factored as $3x \cdot x$ we have this possibility:

$(3x +\quad)(x +\quad)$

(3) There are 4 pairs of factors of -6 and they can be listed two ways:

$$6, -1 \quad -6, 1 \quad 3, -2 \quad -3, 2$$
$$\text{and} \quad -1, 6 \quad 1, -6 \quad -2, 3 \quad 2, -3$$

(4) We can immediately reject all possibilities in which either factor has a common factor, such as $(3x + 6)$ or $(3x - 3)$, because we determined at the outset that there is no common factor other than 1 or -1. We try some possibilities:

$$(3x - 1)(x + 6) = 3x^2 + 17x - 6$$

The factorization of $3x^2 + 17x - 6$ is $(3x - 1)(x + 6)$. We must include the common factor in order to get a factorization of the original trinomial.

$$12x^2 + 68x - 24 = 4(3x - 1)(x + 6)$$

36. $3(2x + 5)(x + 1)$

37. $4x + 1 + 3x^2 = 3x^2 + 4x + 1$

(1) There is no common factor (other than 1 or -1).

(2) Because $3x^2$ can be factored as $3x \cdot x$ we have this possibility:

$$(3x + \quad)(x + \quad)$$

(3) Since all coefficients are positive, we need consider only positive pairs of factors of 1. There is one such pair: 1,1.

(4) We try the possible factorization:

$$(3x + 1)(x + 1) = 3x^2 + 4x + 1$$

The factorization is $(3x + 1)(x + 1)$.

38. $3(3x - 1)(2x + 3)$

39. $y^2 + 4y - 2y - 8 = y(y + 4) - 2(y + 4)$
$$= (y + 4)(y - 2)$$

40. $(x + 5)(x - 2)$

41. $8t^2 - 6t - 28t + 21 = 2t(4t - 3) - 7(4t - 3)$
$$= (4t - 3)(2t - 7)$$

42. $(7t - 8)(5t + 3)$

43. $6x^2 + 4x + 9x + 6 = 2x(3x + 2) + 3(3x + 2)$
$$= (3x + 2)(2x + 3)$$

44. $(3x - 2)(x + 1)$

45. $2t^2 + 6t - t - 3 = 2t(t + 3) - 1(t + 3)$
$$= (t + 3)(2t - 1)$$

46. $(t + 2)(5t - 1)$

47. $3a^2 - 12a - a + 4 = 3a(a - 4) - 1(a - 4)$
$$= (a - 4)(3a - 1)$$

48. $(a - 5)(2a - 1)$

49. $9t^2 + 14t + 5$

(1) First note that there is no common factor (other than 1 or -1).

(2) Multiply the leading coefficient, 9, and the constant, 5:

$$9 \cdot 5 = 45$$

(3) We look for factors of 45 that add to 14. Since all coefficients are positive, we need to consider only positive factors.

Pairs of factors	Sums of factors
1, 45	46
3, 15	18
5, 9	14

The numbers we need are 5 and 9.

(4) Rewrite the middle term:

$$14t = 5t + 9t$$

(5) Factor by grouping:

$$9t^2 + 14t + 5 = 9t^2 + 5t + 9t + 5$$
$$= t(9t + 5) + 1(9t + 5)$$
$$= (9t + 5)(t + 1)$$

50. $(t + 1)(16t + 7)$

51. $16x^2 + 32x + 7$

(1) First note that there is no common factor (other than 1 or -1).

(2) Multiply the leading coefficient, 16, and the constant, 7:

$$16 \cdot 7 = 112$$

(3) We look for factors of 112 that add to 32. Since all coefficients are positive, we need to consider only positive factors.

Pairs of factors	Sums of factors
1, 112	113
2, 56	58
4, 28	32
7, 16	23
8, 14	22

The numbers we need are 4 and 28.

(4) Rewrite the middle term:

$$32x = 4x + 28x$$

(5) Factor by grouping:

$$16x^2 + 32x + 7 = 16x^2 + 4x + 28x + 7$$
$$= 4x(4x + 1) + 7(4x + 1)$$
$$= (4x + 1)(4x + 7)$$

52. $(3x + 5)(3x + 1)$

53. $10a^2 + 25a - 15$

(1) Factor out the largest common factor, 5:

$$10a^2 + 25a - 15 = 5(2a^2 + 5a - 3)$$

(2) To factor $2a^2 + 5a - 3$ by grouping we first multiply the leading coefficient, 2, and the constant, -3:

$$2(-3) = -6$$

(3) We look for factors of -6 that add to 5.

Pairs of factors	Sums of factors
$-1, 6$	5
$-6, 1$	-5
$-2, 3$	1
$2, -3$	-1

The numbers we need are -1 and 6.

(4) Rewrite the middle term:

$$5a = -a + 6a$$

(5) Factor by grouping:

$$\begin{aligned} 2a^2 + 5a - 3 &= 2a^2 - a + 6a - 3 \\ &= a(2a - 1) + 3(2a - 1) \\ &= (2a - 1)(a + 3) \end{aligned}$$

The factorization of $2a^2 + 5a - 3$ is $(2a - 1)(a + 3)$. We must include the common factor in order to get a factorization of the original trinomial:

$$10a^2 + 25a - 15 = 5(2a - 1)(a + 3)$$

54. $(2a - 3)(5a + 6)$

55. $2x^2 - 6x - 14$

(1) Factor out the largest common factor, 2.

$$2x^2 + 6x - 14 = 2(x^2 + 3x - 7)$$

To factor the trinomial $x^2 + 3x - 7$ we must find a pair of factors of -7 whose sum is 3. Since there is no such pair, we conclude that $x^2 + 3x - 7$ is prime. Thus, we have $2x^2 + 6x - 14 = 2(x^2 + 3x - 7)$.

56. $7(x - 2)(2x - 1)$

57. $18x^3 + 21x^2 - 9x$

(1) Factor out the largest common factor, $3x$:

$$18x^3 + 21x^2 - 9x = 3x(6x^2 + 7x - 3)$$

(2) To factor $6x^2 + 7x - 3$ by grouping we first multiply the leading coefficient, 6, and the constant, -3:

$$6(-3) = -18$$

(3) We look for factors of -18 that add to 7.

Pairs of factors	Sums of factors
$-1, 18$	17
$1, -18$	-17
$-2, 9$	7
$2, -9$	-7
$-3, 6$	3
$3, -6$	-3

The numbers we need are -2 and 9.

(4) Rewrite the middle term:

$$7x = -2x + 9x$$

(5) Factor by grouping:

$$\begin{aligned} 6x^2 + 7x - 3 &= 6x^2 - 2x + 9x - 3 \\ &= 2x(3x - 1) + 3(3x - 1) \\ &= (3x - 1)(2x + 3) \end{aligned}$$

The factorization of $6x^2 + 7x - 3$ is $(3x - 1)(2x + 3)$. We must include the common factor in order to get a factorization of the original trinomial:

$$18x^3 + 21x^2 - 9x = 3x(3x - 1)(2x + 3)$$

58. $2x(3x - 5)(x + 1)$

59. $89x + 64 + 25x^2 = 25x^2 + 89x + 64$

(1) First note that there is no common factor (other than 1 or -1).

(2) Multiply the leading coefficient, 25, and the constant, 64:

$$25 \cdot 64 = 1600$$

(3) We look for factors of 1600 that add to 89. Since all coefficients are positive, we need to consider only positive factors. The numbers we need are 25 and 64.

(4) Rewrite the middle term:

$$89x = 25x + 64x$$

(5) Factor by grouping:

$$\begin{aligned} 25x^2 + 89x + 64 &= 25x^2 + 25x + 64x + 64 \\ &= 25x(x + 1) + 64(x + 1) \\ &= (x + 1)(25x + 64) \end{aligned}$$

60. Prime

61. $168x^3 + 45x^2 + 3x$

(1) Factor out the largest common factor, $3x$:

$$168x^3 + 45x^2 + 3x = 3x(56x^2 + 15x + 1)$$

(2) To factor $56x^2 + 15x + 1$ we first multiply the leading coefficient, 56, and the constant, 1:

$$56 \cdot 1 = 56$$

(3) We look for factors of 56 that add to 15. Since all coefficients are positive, we need to consider only positive factors. The numbers we need are 7 and 8.

(4) Rewrite the middle term:

$$15x = 7x + 8x$$

(5) Factor by grouping:

$$56x^2 + 15x + 1 = 56x^2 + 7x + 8x + 1$$
$$= 7x(8x + 1) + 1(8x + 1)$$
$$= (8x + 1)(7x + 1)$$

The factorization of $56x^2 + 15x + 1$ is $(8x+1)(7x+1)$. We must include the common factor in order to get a factorization of the original trinomial:

$$168x^3 + 45x^2 + 3x = 3x(8x + 1)(7x + 1)$$

62. $24x^3(3x - 2)(2x - 1)$

63. $14t^4 - 19t^3 - 3t^2$

(1) Factor out the largest common factor, t^2:

$$14t^4 - 19t^3 - 3t^2 = t^2(14t^2 - 19t - 3)$$

(2) To factor $14t^2 - 19t - 3$ we first multiply the leading coefficient, 14, and the constant, -3:

$$14(-3) = -42$$

(3) We look for factors of -42 that add to -19. The numbers we need are -21 and 2.

(4) Rewrite the middle term:

$$-19t = -21t + 2t$$

(5) Factor by grouping:

$$14t^2 - 19t - 3 = 14t^2 - 21t + 2t - 3$$
$$= 7t(2t - 3) + 1(2t - 3)$$
$$= (2t - 3)(7t + 1)$$

The factorization of $14t^2 - 19t - 3$ is $(2t-3)(7t+1)$. We must include the common factor in order to get a factorization of the original trinomial:

$$14t^4 - 19t^3 - 3t^2 = t^2(2t - 3)(7t + 1)$$

64. $2a^2(5a - 2)(7a - 4)$

65. $3x + 45x^2 - 18 = 45x^2 + 3x - 18$

(1) Factor out the largest common factor, 3:

$$45x^2 + 3x - 18 = 3(15x^2 + x - 6)$$

(2) To factor $15x^2 + x - 6$ we first multiply the leading coefficient, 15, and the constant, -6:

$$15(-6) = -90$$

(3) We look for factors of -90 that add to 1. The numbers we need are 10 and -9.

(4) Rewrite the middle term:

$$x = 10x - 9x$$

(5) Factor by grouping:

$$15x^2 + x - 6 = 15x^2 + 10x - 9x - 6$$
$$= 5x(3x + 2) - 3(3x + 2)$$
$$= (3x + 2)(5x - 3)$$

The factorization of $15x^2 + x - 6$ is $(3x+2)(5x-3)$. We must include the common factor in order to get a factorization of the original trinomial:

$$3x + 45x^2 - 18 = 3(3x + 2)(5x - 3)$$

66. $2(4x - 5)(3x + 4)$

67. $9a^2 + 18ab + 8b^2$

(1) First note that there is no common factor (other than 1 or -1).

(2) Multiply the leading coefficient, 9, and the constant, 8:

$$9 \cdot 8 = 72$$

(3) We look for factors of 72 that add to 18. The numbers we need are 6 and 12.

(4) Rewrite the middle term:

$$18ab = 6ab + 12ab$$

(5) Factor by grouping:

$$9a^2 + 18ab + 8b^2 = 9a^2 + 6ab + 12ab + 8b^2$$
$$= 3a(3a + 2b) + 4b(3a + 2b)$$
$$= (3a + 2b)(3a + 4b)$$

68. $(p - 6q)(3p + 2q)$

69. $35p^2 + 34pq + 8q^2$

(1) First note that there is no common factor (other than 1 or -1).

(2) Multiply the leading coefficient, 35, and the constant, 8:

$$35 \cdot 8 = 280$$

(3) We look for factors of 280 that add to 34. The numbers we need are 14 and 20.

(4) Rewrite the middle term:

$$34pq = 14pq + 20pq$$

(5) Factor by grouping:

$$35p^2 + 34pq + 8q^2 = 35p^2 + 14pq + 20pq + 8q^2$$
$$= 7p(5p + 2q) + 4q(5p + 2q)$$
$$= (5p + 2q)(7p + 4q)$$

70. $2(s + t)(5s - 3t)$

71. $18x^2 - 6xy - 24y^2$

(1) Factor out the largest common factor, 6:

$$18x^2 - 6xy - 24y^2 = 6(3x^2 - xy - 4y^2)$$

(2) To factor $3x^2 - xy - 4y^2$, we first multiply the leading coefficient, 3, and the constant, -4:

$3(-4) = -12$

(3) We look for factors of -12 that add to -1. The numbers we need are -4 and 3.

(4) Rewrite the middle term:

$-xy = -4xy + 3xy$

(5) Factor by grouping:

$$3x^2 - xy - 4y^2 = 3x^2 - 4xy + 3xy - 4y^2$$
$$= x(3x - 4y) + y(3x - 4y)$$
$$= (3x - 4y)(x + y)$$

The factorization of $3x^2 - xy - 4y^2$ is $(3x-4y)(x+y)$. We must include the common factor in order to get a factorization of the original trinomial:

$18x^2 - 6xy - 24y^2 = 6(3x - 4y)(x + y)$

72. $3(5a + 2b)(2a + 5b)$

73. $24a^2 - 34ab + 12b^2$

(1) Factor out the largest common factor, 2:

$24a^2 - 34ab + 12b^2 = 2(12a^2 - 17ab + 6b^2)$

(2) To factor $12a^2 - 17ab + 6b^2$, we first multiply the leading coefficient, 12, and the constant, 6:

$12 \cdot 6 = 72$

(3) We look for factors of 72 that add to -17. The numbers we need are -8 and -9.

(4) Rewrite the middle term:

$-17ab = -8ab - 9ab$

(5) Factor by grouping:

$$12a^2 - 17ab + 6b^2 = 12a^2 - 8ab - 9ab + 6b^2$$
$$= 4a(3a - 2b) - 3b(3a - 2b)$$
$$= (3a - 2b)(4a - 3b)$$

The factorization of $12a^2 - 17ab + 6b^2$ is $(3a - 2b)(4a - 3b)$. We must include the common factor in order to get a factorization of the original trinomial:

$24a^2 - 34ab + 12b^2 = 2(3a - 2b)(4a - 3b)$

74. $5(a + b)(3a - 4b)$

75. $35x^2 + 34x^3 + 8x^4 = 8x^4 + 34x^3 + 35x^2$

(1) Factor out the largest common factor, x^2:

$x^2(8x^2 + 34x + 35)$

(2) To factor $8x^2 + 34x + 35$ by grouping we first multiply the leading coefficient, 8, and the constant, 35:

$8 \cdot 35 = 280$

(3) We look for factors of 280 that add to 34. The numbers we need are 14 and 20.

(4) Rewrite the middle term:

$34x = 14x + 20x$

(5) Factor by grouping:

$$8x^2 + 34x + 35 = 8x^2 + 14x + 20x + 35$$
$$= 2x(4x + 7) + 5(4x + 7)$$
$$= (4x + 7)(2x + 5)$$

The factorization of $8x^2 + 34x + 35$ is $(4x+7)(2x+5)$. We must include the common factor in order to get a factorization of the original trinomial:

$35x^2 + 34x^3 + 8x^4 = x^2(4x + 7)(2x + 5)$

76. $x^2(2x + 3)(7x - 1)$

77. $18a^7 + 8a^6 + 9a^8 = 9a^8 + 18a^7 + 8a^6$

(1) Factor out the largest common factor, a^6:

$9a^8 + 18a^7 + 8a^6 = a^6(9a^2 + 18a + 8)$

(2) To factor $9a^2 + 18a + 8$ we first multiply the leading coefficient, 9, and the constant, 8:

$9 \cdot 8 = 72$

(3) Look for factors of 72 that add to 18. The numbers we need are 6 and 12.

(4) Rewrite the middle term:

$18a = 6a + 12a$

(5) Factor by grouping:

$$9a^2 + 18a + 8 = 9a^2 + 6a + 12a + 8$$
$$= 3a(3a + 2) + 4(3a + 2)$$
$$= (3a + 2)(3a + 4)$$

The factorization of $9a^2 + 18a + 8$ is $(3a+2)(3a+4)$. We must include the common factor in order to get a factorization of the original trinomial:

$18a^7 + 8a^6 + 9a^8 = a^6(3a + 2)(3a + 4)$

78. $a^7(5a + 4)(5a + 4)$, or $a^7(5a + 4)^2$

79. Writing exercise

80. Writing exercise

81. *Familiarize.* We will use the formula $C = 2\pi r$, where C is circumference and r is radius, to find the radius in kilometers. Then we will multiply that number by 0.62 to find the radius in miles.

Translate.

$$\underbrace{\text{Circumference}}_{40,000} = \underbrace{2 \cdot \pi \cdot \text{radius}}_{2(3.14)r}$$

Carry out. First we solve the equation.

$$40,000 \approx 2(3.14)r$$
$$40,000 \approx 6.28r$$
$$6369 \approx r$$

Then we multiply to find the radius in miles:

$$6369(0.62) \approx 3949$$

Check. If $r = 6369$, then $2\pi r = 2(3.14)(6369) \approx 40,000$. We should also recheck the multiplication we did to find the radius in miles. Both values check.

State. The radius of the earth is about 6369 km or 3949 mi. (These values may differ slightly if a different approximation is used for π.)

82. $40°$

83. $\quad (3x+1)^2 = (3x)^2 + 2 \cdot 3x \cdot 1 + 1^2$
$$[(A+B)^2 = A^2 + 2AB + B^2]$$
$$= 9x^2 + 6x + 1$$

84. $25x^2 - 20x + 4$

85. $\quad (4t-5)^2 = (4t)^2 - 2 \cdot 4t \cdot 5 + 5^2$
$$[(A-B)^2 = A^2 - 2AB + B^2]$$
$$= 16t^2 - 40t + 25$$

86. $49a^2 + 14a + 1$

87. $\quad (5x-2)(5x+2) = (5x)^2 - 2^2$
$$[(A+B)(A-B) = A^2 - B^2]$$
$$= 25x^2 - 4$$

88. $4x^2 - 9$

89. $\quad (2t+7)(2t-7) = (2t)^2 - 7^2$
$$[(A+B)(A-B) = A^2 - B^2]$$
$$= 4t^2 - 49$$

90. $16a^2 - 49$

91. Writing exercise

92. Writing exercise

93. $9a^2b^2 - 15ab - 2$

(1) There is no common factor (other than 1 or -1).

(2) Because $9a^2b^2$ can be factored as $9ab \cdot ab$ or $3ab \cdot 3ab$, we have these possibilities:

$$(9ab + \quad)(ab + \quad) \text{ and } (3ab + \quad)(3ab + \quad)$$

(3) There are 2 pairs of factors of -2 and they can be listed two ways:

$$-2, 1 \quad 2, -1$$
$$\text{and} \quad 1, -2 \quad -1, 2$$

(4) We try some possibilities:

$$(9ab - 2)(ab + 1) = 9a^2b^2 + 7ab - 2$$
$$(9ab + 2)(ab - 1) = 9a^2b^2 - 7ab - 2$$
$$(9ab + 1)(ab - 2) = 9a^2b^2 - 17ab - 2$$
$$(9ab - 1)(ab + 2) = 9a^2b^2 + 17ab - 2$$
$$(3ab - 2)(3ab + 1) = 9a^2b^2 - 3ab - 2$$
$$(3ab + 2)(3ab - 1) = 9a^2b^2 + 3ab - 2$$

Since none of the possibilities is the correct factorization, we conclude that $9a^2b^2 - 15ab - 2$ is prime.

94. $(3xy + 2)(6xy - 5)$

95. $8x^2y^3 + 10xy^2 + 2y$

(1) We factor out the common factor, $2y$:

$$2y(4x^2y^2 + 5xy + 1)$$

Then we factor the trinomial $4x^2y^2 + 5xy + 1$.

(2) Because $4x^2y^2$ can be factored as $4xy \cdot xy$ or $2xy \cdot 2xy$, we have these possibilities:

$$(4xy + \quad)(xy + \quad) \text{ and } (2xy + \quad)(2xy + \quad)$$

(3) Since all coefficients are positive, we need consider only positive pairs of factors of 1. The only such pair is 1, 1.

(4) We try some possibilities:

$$(4xy + 1)(xy + 1) = 4x^2y^2 + 5xy + 1$$

The factorization of $4x^2y^2 + 5xy + 1$ is $(4xy + 1)(xy + 1)$. We must include the common factor in order to get a factorization of the original trinomial.

$$8x^2y^3 + 10xy^2 + 2y = 2y(4xy + 1)(xy + 1)$$

96. Prime

97. $9t^{10} + 12t^5 + 4 = 9(t^5)^2 + 12t^5 + 4$

(1) There is no common factor (other than 1 or -1).

(2) Because $9t^{10}$ can be factored as $9t^5 \cdot t^5$ or $3t^5 \cdot 3t^5$, we have these possibilities:

$$(9t^5 + \quad)(t^5 + \quad) \text{ and } (3t^5 + \quad)(3t^5 + \quad)$$

(3) Since all coefficients are positive, we need consider only positive pairs of factors of 4. There are two such pairs and one of them can be listed two ways:

$$4, 1 \quad 2, 2$$
$$\text{and} \quad 1, 4$$

(4) We try some possibilities:

$$(9t^5 + 4)(t^5 + 1) = 9t^{10} + 13t^5 + 4$$
$$(3t^5 + 2)(3t^5 + 2) = 9t^{10} + 12t^5 + 4$$

The factorization is $(3t^5 + 2)(3t^5 + 2)$, or $(3t^5 + 2)^2$.

98. $(4t^5 - 1)^2$

99. $-15x^{2m} + 26x^m - 8 = -(15x^{2m} - 26x^m + 8)$

We will factor $15x^{2m} - 26x^m - 8$.

(1) We have factored -1 out of the original trinomial in order to work with a trinomial that has a positive leading term.

(2) Multiply the leading coefficient, 15, and the constant, 8:

$$15 \cdot 8 = 120$$

(3) We look for factors of 120 that add to -26. The numbers we need are -6 and -20.

(4) Rewrite the middle term.

$$-26x^m = -6x^m - 20x^m$$

(5) Factor by grouping:

$$15x^{2m} - 26x^m + 8 = 15x^{2m} - 6x^m - 20x^m + 8$$
$$= 3x^m(5x^m - 2) - 4(5x^m - 2)$$
$$= (5x^m - 2)(3x^m - 4)$$

The factorization of $15x^{2m} - 26x^m + 8$ is $(5x^m - 2)(3x^m - 4)$. We must include the common factor in order to get a factorization of the original trinomial:

$$-15x^{2m} + 26x^m - 8 = -(5x^m - 2)(3x^m - 4)$$

100. $(10x^n + 3)(2x^n + 1)$

101. $a^{2n+1} - 2a^{n+1} + a$

(1) Factor out the largest common factor, a:

$$a^{2n+1} - 2a^{n+1} + a = a(a^{2n} - 2a^n + 1)$$

(2) Multiply the leading coefficient, 1, and the constant, 1:

$$1 \cdot 1 = 1$$

(3) Look for factors of 1 that add to -2. The numbers we need are -1 and -1.

(4) Rewrite the middle term.

$$-2a^n = -a^n - a^n$$

(5) Factor by grouping:

$$a^{2n} - 2a^n + 1 = a^{2n} - a^n - a^n + 1$$
$$= a^n(a^n - 1) - 1(a^n - 1)$$
$$= (a^n - 1)(a^n - 1), \text{ or } (a^n - 1)^2$$

The factorization of $a^{2n} - 2a^n + 1$ is $(a^n - 1)^2$. We must include the common factor in order to get a factorization of the original trinomial:

$$a^{2n+1} - 2a^{n+1} + a = a(a^n - 1)^2$$

102. $(3a^{3n} + 1)(a^{3n} - 1)$

103.
$$3(a + 1)^{n+1}(a + 3)^2 - 5(a + 1)^n(a + 3)^3$$
$$= (a + 1)^n(a + 3)^2[3(a + 1) - 5(a + 3)]$$

Removing the common factors

$$= (a + 1)^n(a + 3)^2[3a + 3 - 5a - 15] \text{ Simplify-}$$
$$= (a + 1)^n(a + 3)^2(-2a - 12) \quad \text{ing inside the brackets}$$
$$= (a + 1)^n(a + 3)^2(-2)(a + 6) \text{ Removing the common factor}$$
$$= -2(a + 1)^n(a + 3)^2(a + 6) \quad \text{Rearranging}$$

104. $[7(t - 3)^n - 2][(t - 3)^n + 1]$

Exercise Set 5.4

1. $x^2 - 18x + 81$

(1) Two terms, x^2 and 81, are squares.

(2) There is no minus sign before x^2 or 81.

(3) Twice the product of the square roots, $2 \cdot x \cdot 9$, is $18x$, the opposite of the remaining term, $-18x$.

Thus, $x^2 - 18x + 81$ is a perfect-square trinomial.

2. Yes

3. $x^2 + 16x - 64$

(1) Two terms, x^2 and 64, are squares.

(2) There is a minus sign before 64, so $x^2 + 16x - 64$ is not a perfect-square trinomial.

4. No

5. $x^2 - 3x + 9$

(1) Two terms, x^2 and 9, are squares.

(2) There is no minus sign before x^2 or 9.

(3) Twice the product of the square roots, $2 \cdot x \cdot 3$, is $6x$. This is neither the remaining term nor its opposite, so $x^2 - 3x + 9$ is not a perfect-square trinomial.

6. No

7. $9x^2 - 36x + 24$

(1) Only one term, $9x^2$, is a square. Thus, $9x^2 - 36x + 24$ is not a perfect-square trinomial.

8. No

9.
$$x^2 - 16x + 64$$
$$= x^2 - 2 \cdot x \cdot 8 + 8^2 = (x - 8)^2$$
$$\uparrow \quad \uparrow \quad \uparrow \quad \uparrow \quad \uparrow$$
$$= A^2 - 2 \quad A \quad B + B^2 = (A - B)^2$$

10. $(x - 7)^2$

11. $\quad x^2 + 14x + 49$

$\qquad = x^2 + 2 \cdot x \cdot 7 + 7^2 = (x+7)^2$

$\qquad \quad \uparrow \quad \uparrow \ \uparrow \ \uparrow \qquad \uparrow$

$\qquad = A^2 + 2 \ \ A \ \ B + B^2 = (A+B)^2$

12. $(x+8)^2$

13. $\quad 3x^2 - 6x + 3 = 3(x^2 - 2x + 1)$

$\qquad\qquad\qquad\ = 3(x^2 - 2 \cdot x \cdot 1 + 1^2)$

$\qquad\qquad\qquad\ = 3(x-1)^2$

14. $5(x-1)^2$

15. $\quad 4 + 4x + x^2 = 2^2 + 2 \cdot 2 \cdot x + x^2$

$\qquad\qquad\qquad\ = (2+x)^2, \text{ or } (x+2)^2$

16. $(x-2)^2$

17. $\quad 18x^2 - 12x + 2 = 2(9x^2 - 6x + 1)$

$\qquad\qquad\qquad\quad = 2[(3x)^2 - 2 \cdot 3x \cdot 1 + 1^2]$

$\qquad\qquad\qquad\quad = 2(3x-1)^2$

18. $(5x+1)^2$

19. $\quad 49 + 56y + 16y^2 = 16y^2 + 56y + 49$

$\qquad\qquad\qquad\qquad = (4y)^2 + 2 \cdot 4y \cdot 7 + 7^2$

$\qquad\qquad\qquad\qquad = (4y+7)^2$

$\quad$ We could also factor as follows:

$\qquad 49 + 56y + 16y^2 = 7^2 + 2 \cdot 7 \cdot 4y + (4y)^2$

$\qquad\qquad\qquad\qquad = (7+4y)^2$

20. $3(4m+5)^2$

21. $\quad x^5 - 18x^4 + 81x^3 = x^3(x^2 - 18x + 81)$

$\qquad\qquad\qquad\qquad = x^3(x^2 - 2 \cdot x \cdot 9 + 9^2)$

$\qquad\qquad\qquad\qquad = x^3(x-9)^2$

22. $2(x-10)^2$

23. $\quad 2x^3 - 4x^2 + 2x = 2x(x^2 - 2x + 1)$

$\qquad\qquad\qquad\quad = 2x(x^2 - 2 \cdot x \cdot 1 + 1^2)$

$\qquad\qquad\qquad\quad = 2x(x-1)^2$

24. $x(x+12)^2$

25. $\quad 20x^2 + 100x + 125 = 5(4x^2 + 20x + 25)$

$\qquad\qquad\qquad\qquad = 5[(2x)^2 + 2 \cdot 2x \cdot 5 + 5^2]$

$\qquad\qquad\qquad\qquad = 5(2x+5)^2$

26. $3(2x+3)^2$

27. $\quad 49 - 42x + 9x^2 = 7^2 - 2 \cdot 7 \cdot 3x + (3x)^2 = (7-3x)^2,$

$\qquad$ or $(3x-7)^2$

28. $(8-7x)^2$, or $(7x-8)^2$

29. $\quad 16x^2 + 24x + 9 = (4x)^2 + 2 \cdot 4x \cdot 3 + 3^2 =$

$\qquad (4x+3)^2$

30. $2(a+7)^2$

31. $\quad 2 + 20x + 50x^2 = 2(1 + 10x + 25x^2)$

$\qquad\qquad\qquad\quad = 2[1^2 + 2 \cdot 1 \cdot 5x + (5x)^2]$

$\qquad\qquad\qquad\quad = 2(1+5x)^2, \text{ or } 2(5x+1)^2$

32. $(3x+5)^2$

33. $\quad 4p^2 + 12pq + 9q^2 = (2p)^2 + 2 \cdot 2p \cdot 3q + (3q)^2$

$\qquad\qquad\qquad\qquad = (2p+3q)^2$

34. $(5m+2n)^2$

35. $a^2 - 12ab + 49b^2$

$\quad$ This is not a perfect square trinomial because $-2 \cdot a \cdot 7b = -14ab \neq -12ab$. Nor can it be factored using the methods of Sections 5.2 and 5.3. Thus, it is prime.

36. Prime

37. $\quad 64m^2 + 16mn + n^2 = (8m)^2 + 2 \cdot 8m \cdot n + n^2$

$\qquad\qquad\qquad\qquad = (8m+n)^2$

38. $(9p-q)^2$

39. $\quad 32s^2 - 80st + 50t^2 = 2(16s^2 - 40st + 25t^2)$

$\qquad\qquad\qquad\qquad = 2[(4s)^2 - 2 \cdot 4s \cdot 5t + (5t)^2]$

$\qquad\qquad\qquad\qquad = 2(4s-5t)^2$

40. $4(3a+4b)^2$

41. $x^2 - 100$

$\quad$ (1) The first expression is a square: x^2

$\qquad$ The second expression is a square: $100 = 10^2$

$\quad$ (2) The terms have different signs.

$\quad$ Thus, $x^2 - 100$ is a difference of squares, $x^2 - 10^2$.

42. Yes

43. $x^2 + 36$

$\quad$ (1) The first expression is a square: x^2

$\qquad$ The second expression is a square: $36 = 6^2$

$\quad$ (2) The terms do not have different signs.

$\quad$ Thus, $x^2 + 36$ is not a difference of squares.

44. No

45. $9t^2 - 32$

    (1) The expression 32 is not a square.

    Thus, $9t^2 - 32$ is not a difference of squares.

46. No

47. $-25 + 4t^2$

    (1) The expressions 25 and $4t^2$ are squares:
        $25 = 5^2$ and $4t^2 = (2t)^2$.

    (2) The terms have different signs.

    Thus, $-25 + 4t^2$ is a difference of squares, $(2t)^2 - 5^2$.

48. Yes

49. $y^2 - 4 = y^2 - 2^2 = (y + 2)(y - 2)$

50. $(x + 6)(x - 6)$

51. $p^2 - 9 = p^2 - 3^2 = (p + 3)(p - 3)$

52. Prime

53. $-49 + t^2 = t^2 - 49 = t^2 - 7^2 = (t + 7)(t - 7)$, or $(7 + t)(-7 + t)$

54. $(m + 8)(m - 8)$, or $(8 + m)(-8 + m)$

55. $6a^2 - 54 = 6(a^2 - 9) = 6(a^2 - 3^2) = 6(a + 3)(a - 3)$

56. $(x - 4)^2$

57. $49x^2 - 14x + 1 = (7x)^2 - 2 \cdot 7x \cdot 1 + 1^2 = (7x - 1)^2$

58. $3(t + 2)(t - 2)$

59. $200 - 2t^2 = 2(100 - t^2) = 2(10^2 - t^2) = 2(10 + t)(10 - t)$

60. $2(7 + 2w)(7 - 2w)$

61. $80a^2 - 45 = 5(16a^2 - 9) = 5[(4a^2) - 3^2] = 5(4a + 3)(4a - 3)$

62. $(5x + 2)(5x - 2)$

63. $5t^2 - 80 = 5(t^2 - 16) = 5(t^2 - 4^2) = 5(t + 4)(t - 4)$

64. $4(t + 4)(t - 4)$

65. $8x^2 - 98 = 2(4x^2 - 49) = 2[(2x)^2 - 7^2] = 2(2x + 7)(2x - 7)$

66. $6(2x + 3)(2x - 3)$

67. $36x - 49x^3 = x(36 - 49x^2) = x[6^2 - (7x)^2] = x(6 + 7x)(6 - 7x)$

68. $x(4 + 9x)(4 - 9x)$

69. $49a^4 - 20$

    There is no common factor (other than 1 or -1). Since 20 is not a square, this is not a difference of squares. Thus, the polynomial is prime.

70. $(5a^2 + 3)(5a^2 - 3)$

71. $\quad t^4 - 1$

$= (t^2)^2 - 1^2$

$= (t^2 + 1)(t^2 - 1)$

$= (t^2 + 1)(t + 1)(t - 1)$ Factoring further; $t^2 - 1$ is a difference of squares

72. $(x^2 + 4)(x + 2)(x - 2)$

73. $\quad 3x^3 - 24x^2 + 48x = 3x(x^2 - 8x + 16)$

$= 3x(x^2 - 2 \cdot x \cdot 4 + 4^2)$

$= 3x(x - 4)^2$

74. $2a^2(a - 9)^2$

75. $48t^2 - 27 = 3(16t^2 - 9)$

$= 3[(4t)^2 - 3^2]$

$= 3(4t + 3)(4t - 3)$

76. $5(5t + 3)(5t - 3)$

77. $\quad a^8 - 2a^7 + a^6 = a^6(a^2 - 2a + 1)$

$= a^6(a^2 - 2 \cdot a \cdot 1 + 1^2)$

$= a^6(a - 1)^2$

78. $x^6(x - 4)^2$

79. $7a^2 - 7b^2 = 7(a^2 - b^2)$

$= 7(a + b)(a - b)$

80. $6(p + q)(p - q)$

81. $25x^2 - 4y^2 = (5x)^2 - (2y)^2$

$= (5x + 2y)(5x - 2y)$

82. $(4a + 3b)(4a - 3b)$

83. $1 - a^4b^4 = 1^2 - (a^2b^2)^2$

$= (1 + a^2b^2)(1 - a^2b^2)$

$= (1 + a^2b^2)[1^2 - (ab)^2]$

$= (1 + a^2b^2)(1 + ab)(1 - ab)$

84. $3(5 + m^2n^2)(5 - m^2n^2)$

85. $18t^2 - 8s^2 = 2(9t^2 - 4s^2)$
$$= 2[(3t)^2 - (2s)^2]$$
$$= 2(3t + 2s)(3t - 2s)$$

86. $(7x + 4y)(7x - 4y)$

87. Writing exercise

88. Writing exercise

89. *Familiarize.* Let a = the amount of oxygen, in liters, that can be dissolved in 100 L of water at $20°$ C.

Translate. We reword the problem.

$\underbrace{5 \text{ L}}$ is 1.6 times $\underbrace{\text{amount } a.}$
$\quad \downarrow \quad \downarrow \quad \downarrow \quad \downarrow \qquad \downarrow$
$\quad 5 \quad = 1.6 \quad \cdot \qquad a$

Carry out. We solve the equation.
$$5 = 1.6a$$
$$3.125 = a \qquad \text{Dividing both sides by 1.6}$$

Check. Since 1.6 times 3.125 is 5, the answer checks.

State. 3.125 L of oxygen can be dissolved in 100 L of water at $20°$ C.

90. Scores of 77 or higher

91. $(x^3y^5)(x^9y^7) = x^{3+9}y^{5+7} = x^{12}y^{12}$

92. $25a^4b^6$

93. Graph: $y = \frac{3}{2}x - 3$

Because the equation is in the form $y = mx + b$, we know the y-intercept is $(0, -3)$. We find two other points on the line, substituting multiples of 2 for x to avoid fractions.

When $x = -2$, $y = \frac{3}{2}(-2) - 3 = -3 - 3 = -6$.

When $x = 4$, $y = \frac{3}{2} \cdot 4 - 3 = 6 - 3 = 3$.

x	y
0	-3
-2	-6
4	3

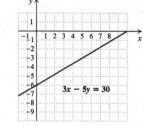

94.

95. Writing exercise

96. Writing exercise

97. $x^8 - 2^8 = (x^4 + 2^4)(x^4 - 2^4)$
$$= (x^4 + 2^4)(x^2 + 2^2)(x^2 - 2^2)$$
$$= (x^4 + 2^4)(x^2 + 2^2)(x + 2)(x - 2), \text{ or}$$
$$(x^4 + 16)(x^2 + 4)(x + 2)(x - 2)$$

98. $3\left(x + \frac{1}{3}\right)\left(x - \frac{1}{3}\right);\ \frac{1}{3}(3x + 1)(3x - 1)$

99. $18x^3 - \frac{8}{25}x = 2x\left(9x^2 - \frac{4}{25}\right) =$
$$2x\left(3x + \frac{2}{5}\right)\left(3x - \frac{2}{5}\right)$$

100. $p(0.7 + p)(0.7 - p)$

101. $0.64x^2 - 1.21 = (0.8x)^2 - (1.1)^2 =$
$$(0.8x + 1.1)(0.8x - 1.1)$$

102. $x(x + 6)(x^2 + 6x + 18)$

103. $(y - 5)^4 - z^8$
$$= [(y - 5)^2 + z^4][(y - 5)^2 - z^4]$$
$$= [(y - 5)^2 + z^4][y - 5 + z^2][y - 5 - z^2]$$
$$= (y^2 - 10y + 25 + z^4)(y - 5 + z^2)(y - 5 - z^2)$$

104. $\left(x + \frac{1}{x}\right)\left(x - \frac{1}{x}\right)$

105. $a^{2n} - 49b^{2n} = (a^n)^2 - (7b^n)^2 =$
$$(a^n + 7b^n)(a^n - 7b^n)$$

106. $(9 + b^{2k})(3 + b^k)(3 - b^k)$

107. $x^4 - 8x^2 - 9 = (x^2 - 9)(x^2 + 1)$
$$= (x + 3)(x - 3)(x^2 + 1)$$

108. $(3b^n + 2)^2$

109. $16x^4 - 96x^2 + 144 = 16(x^4 - 6x^2 + 9)$
$$= 16(x^2 - 3)^2$$

110. $(y + 4)^2$

111. $49(x + 1)^2 - 42(x + 1) + 9 = [7(x + 1) - 3]^2 =$
$(7x + 7 - 3)^2 = (7x + 4)^2$

112. $(3x - 7)^2(3x + 7)$

113. $\quad x^2(x + 1)^2 - (x^2 + 1)^2$
$= x^2(x^2 + 2x + 1) - (x^4 + 2x^2 + 1)$
$= x^4 + 2x^3 + x^2 - x^4 - 2x^2 - 1$
$= 2x^3 + x^2 - 2x^2 - 1$
$= (2x^3 - 2x^2) + (x^2 - 1)$
$= 2x^3 - x^2 - 1$

114. $(a + 4)(a - 2)$

115. $\quad y^2 + 6y + 9 - x^2 - 8x - 16$
$= (y^2 + 6y + 9) - (x^2 + 8x + 16)$
$= (y + 3)^2 - (x + 4)^2$
$= [(y + 3) + (x + 4)][(y + 3) - (x + 4)]$
$= (y + 3 + x + 4)(y + 3 - x - 4)$
$= (y + x + 7)(y - x - 1)$

116. 9

117. For $c = a^2$, $2 \cdot a \cdot 3 = 24$. Then $a = 4$, so $c = 4^2 = 16$.

118. 0, 2

119. $\quad (x + 1)^2 - x^2$
$= [(x + 1) + x][(x + 1) - x]$
$= 2x + 1$

Exercise Set 5.5

---

1. $t^3 + 8 = t^3 + 2^3$
$= (t + 2)(t^2 - t \cdot 2 + 2^2)$
$= (t + 2)(t^2 - 2t + 4)$

2. $(p + 3)(p^2 - 3p + 9)$

3. $a^3 - 64 = a^3 - 4^3$
$= (a - 4)(a^2 + a \cdot 4 + 4^2)$
$= (a - 4)(a^2 + 4a + 16)$

4. $(w - 1)(w^2 + w + 1)$

5. $z^3 + 125 = z^3 + 5^3$
$= (z + 5)(z^2 - x \cdot 5 + 5^2)$
$= (z + 5)(x^2 - 5z + 25)$

6. $(x + 1)(x^2 - x + 1)$

7. $8a^3 - 1 = (2a)^3 - 1^3$
$= (2a - 1)[(2a)^2 + 2a \cdot 1 + 1^2]$
$= (2a - 1)(4a^2 + 2a + 1)$

8. $(3x - 1)(9x^2 + 3x + 1)$

9. $y^3 - 27 = y^3 - 3^3$
$= (y - 3)(y^3 + y \cdot 3 + 3^2)$
$= (y - 3)(y^2 + 3y + 9)$

10. $(p - 2)(p^2 + 2p + 4)$

11. $64 + 125x^3 = 4^3 + (5x)^3$
$= (4 + 5x)[4^2 - 4 \cdot 5x + (5x)^2]$
$= (4 + 5x)(16 - 20x + 25x^2)$

12. $(2 + 3b)(4 - 6b + 9b^2)$

13. $125p^3 - 1 = (5p)^3 - 1^3$
$= (5p - 1)[(5p)^2 + 5p \cdot 1 + 1^2]$
$= (5p - 1)(25p^2 + 5p + 1)$

14. $(4w - 1)(16w^2 + 4w + 1)$

15. $27m^3 + 64 = (3m)^3 + 4^3$
$= (3m + 4)[(3m)^2 - 3m \cdot 4 + 4^2]$
$= (3m + 4)(9m^2 - 12m + 16)$

16. $(2t + 3)(4t^2 - 6t + 9)$

17. $p^3 - q^3 = (p - q)(p^2 + pq + q^2)$

18. $(a + b)(a^2 - ab + b^2)$

19. $x^3 + \dfrac{1}{8} = x^3 + \left(\dfrac{1}{2}\right)^3$
$= \left(x + \dfrac{1}{2}\right)\left[x^2 - x \cdot \dfrac{1}{2} + \left(\dfrac{1}{2}\right)^2\right]$
$= \left(x + \dfrac{1}{2}\right)\left(x^2 - \dfrac{1}{2}x + \dfrac{1}{4}\right)$

20. $\left(y + \dfrac{1}{3}\right)\left(y^2 - \dfrac{1}{3}y + \dfrac{1}{9}\right)$

21. $2y^3 - 128 = 2(y^3 - 64)$
$= 2(y^3 - 4^3)$
$= 2(y - 4)(y^2 + 4 \cdot y + 4^2)$
$= 2(y - 4)(y^2 + 4y + 16)$

22. $3(z - 1)(z^2 + z + 1)$

23. $24a^3 + 3 = 3(8a^3 + 1)$
$$= 3[(2a)^3 + 1^3]$$
$$= 3(2a + 1)[(2a)^2 - 2a \cdot 1 + 1^2]$$
$$= 3(2a + 1)(4a^2 - 2a + 1)$$

24. $2(3x + 1)(9x^2 - 3x + 1)$

25. $rs^3 - 64r = r(s^3 - 64)$
$$= r(s^3 - 4^3)$$
$$= r(s - 4)(s^2 + s \cdot 4 + 4^2)$$
$$= r(s - 4)(s^2 + 4s + 16)$$

26. $a(b + 5)(b^2 - 5b + 25)$

27. $5x^3 + 40z^3 = 5(x^3 + 8z^3)$
$$= 5[x^3 + (2z)^3]$$
$$= 5(x + 2z)[x^2 - x \cdot 2z + (2z)^2]$$
$$= 5(x + 2z)(x^2 - 2xz + 4z^2)$$

28. $2(y - 3z)(y^2 + 3yz + 9z^2)$

29. $x^3 + 0.001 = x^3 + (0.1)^3$
$$= (x + 0.1)[x^2 - x(0.1) + (0.1)^2]$$
$$= (x + 0.1)(x^2 - 0.1x + 0.01)$$

30. $(y + 0.5)(y^2 - 0.5y + 0.25)$

31. $3z^5 - 3z^2 = 3z^2(z^3 - 1)$
$$= 3z^2(z^3 - 1^3)$$
$$= 3z^2(z - 1)(z^2 + z \cdot 1 + 1^2)$$
$$= 3z^2(z - 1)(z^2 + z + 1)$$

32. $2y(y - 4)(y^2 + 4y + 16)$

33. $t^6 + 1 = (t^2)^3 + 1^3$
$$= (t^2 + 1)[(t^2)^2 - t^2 \cdot 1 + 1^2]$$
$$= (t^2 + 1)(t^4 - t^2 + 1)$$

34. $(z + 1)(z^2 - z + 1)(z - 1)(z^2 + z + 1)$

35. $p^6 - q^6$
$$= (p^3)^2 - (q^3)^2$$
$$= (p^3 + q^3)(p^3 - q^3)$$
$$= (p + q)(p^2 - pq + q^2)(p - q)(p^2 + pq + q^2)$$

36. $(x + 2y)(x^2 - 2xy + 4y^2)(x - 2y)(x^2 + 2xy + 4y^2)$

37. Writing exercise

38. Writing exercise

39. $(3x + 5)(3x - 5) = (3x)^2 - 5^2 = 9x^2 - 25$

40. $9x^2 + 30x + 25$

41. $(x - 7)(x + 4) = x^2 + 4x - 7x - 28 = x^2 - 3x - 28$

42. $x^3 + 1$

43. **Familiarize.** Let $n =$ the number of people who had high-speed Internet access at work in 2000.

Translate.

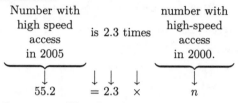

Carry out. We solve the equation.
$$55.2 = 2.3n$$
$$24 = n \qquad \text{Dividing both sides by 2.3}$$

Check. Since $2.3(24) = 55.2$, the answer checks.

State. In 2000, 24 million people had high-speed Internet access at work.

44. 32.8 million barrels per day

45. Writing exercise

46. Writing exercise

47. $125c^6 + 8d^6$
$$= (5c^2)^3 + (2d^2)^3$$
$$= (5c^2 + 2d^2)[(5c^2)^2 - 5c^2 \cdot 2d^2 + (2d^2)^2]$$
$$= (5c^2 + 2d^2)(25c^4 - 10c^2d^2 + 4d^4)$$

48. $8(2x^2 + t^2)(4x^4 - 2x^2t^2 + t^4)$

49. $3x^{3a} - 24y^{3b}$
$$= 3(x^{3a} - 8y^{3b})$$
$$= 3[(x^a)^3 - (2y^b)^3]$$
$$= 3(x^a - 2y^b)[(x^a)^2 + x^a \cdot 2y^b + (2y^b)^2]$$
$$= 3(x^a - 2y^b)(x^{2a} + 2x^ay^b + 4y^{2b})$$

50. $\left(\dfrac{2}{3}x - \dfrac{1}{4}y\right)\left(\dfrac{4}{9}x^2 + \dfrac{1}{6}xy + \dfrac{1}{16}y^2\right)$

51. $\dfrac{1}{24}x^3y^3 + \dfrac{1}{3}z^3$
$$= \dfrac{1}{3}\left(\dfrac{1}{8}x^3y^3 + z^3\right)$$
$$= \dfrac{1}{3}\left[\left(\dfrac{1}{2}xy\right)^3 + z^3\right]$$
$$= \dfrac{1}{3}\left(\dfrac{1}{2}xy + z\right)\left[\left(\dfrac{1}{2}xy\right)^2 - \dfrac{1}{2}xy \cdot z + z^2\right]$$
$$= \dfrac{1}{3}\left(\dfrac{1}{2}xy + 1\right)\left(\dfrac{1}{4}x^2y^2 - \dfrac{1}{2}xyz + z^2\right)$$

52. $\frac{1}{2}\left(\frac{1}{2}x^a + y^{2a}z^{3b}\right)\left(\frac{1}{4}x^{2a} - \frac{1}{2}x^a y^{2a}z^{3b} + y^{4a}z^{6b}\right)$

53. $(x+5)^3 + (x-5)^3$

$= (x{+}5{+}x{-}5)[(x{+}5)^2 - (x{+}5)(x{-}5) + (x{-}5)^2]$

$= 2x(x^2{+}10x{+}25 - (x^2{-}25) + x^2{-}10x{+}25)$

$= 2x(x^2{+}10x{+}25 - x^2{+}25 + x^2{-}10x{+}25)$

$= 2x(x^2 + 75), \text{ or } 2x^3 + 150x$

54. $-y(3x^2 + 3xy + y^2)$

55. $t^4 - 8t^3 - t + 8$

$= t^4 - t - 8t^3 + 8$

$= t(t^3 - 1) - 8(t^3 - 1)$

$= (t^3 - 1)(t - 8)$

$= (t - 1)(t^2 + t + 1)(t - 8)$

56. $5\left(xy^2 - \frac{1}{2}\right)\left(x^2 y^4 + \frac{1}{2}xy^2 + \frac{1}{4}\right)$

Exercise Set 5.6

---

1. $5x^2 - 45$

$= 5(x^2 - 9)$ 5 is a common factor.

$= 5(x + 3)(x - 3)$ Factoring the difference of squares

2. $10(a + 8)(a - 8)$

3. $a^2 + 25 + 10a$

$= a^2 + 10a + 25$ Perfect-square trinomial

$= (a + 5)^2$

4. $(y - 7)^2$

5. $8t^2 - 18t - 5$

There is no common factor (other than 1). This polynomial has three terms, but it is not a perfect-square trinomial. Multiply the leading coefficient and the constant, 8 and -5: $8(-5) = -40$. Try to factor -40 so that the sum of the factors is -18. The numbers we want are -20 and 2: $-20 \cdot 2 = -40$ and $-20 + 2 = -18$. Split the middle term and factor by grouping.

$8t^2 - 18t - 5 = 8t^2 - 20t + 2t - 5$

$\qquad\qquad\qquad = 4t(2t - 5) + 1(2t - 5)$

$\qquad\qquad\qquad = (2t - 5)(4t + 1)$

6. $(2t + 3)(t + 4)$

7. $x^3 - 24x^2 + 144x$

$= x(x^2 - 24x + 144)$ x is a common factor.

$= x(x^2 - 2 \cdot x \cdot 12 + 12^2)$ Perfect-square trinomial

$= x(x - 12)^2$

8. $x(x - 9)^2$

9. $x^3 + 3x^2 - 4x - 12$

$= x^2(x + 3) - 4(x + 3)$ Factoring by grouping

$= (x + 3)(x^2 - 4)$

$= (x + 3)(x + 2)(x - 2)$ Factoring the difference of squares

10. $(x + 5)(x - 5)^2$

11. $98t^2 - 18$

$= 2(49t^2 - 9)$ 2 is a common factor.

$= 2[(7t)^2 - 3^2]$ Difference of squares

$= 2(7t + 3)(7t - 3)$

12. $3t(3t + 1)(3t - 1)$

13. $20x^3 - 4x^2 - 72x$

$= 4x(5x^2 - x - 18)$ $4x$ is a common factor.

$= 4x(5x + 9)(x - 2)$ Factoring the trinomial using trial and error

14. $3x(x + 3)(3x - 5)$

15. $x^2 + 4$

The polynomial has no common factor and is not a difference of squares. It is prime.

16. Prime

17. $a^4 + 8a^2 + 8a^3 + 64a$

$= a(a^3 + 8a + 8a^2 + 64)$ a is a common factor.

$= a[a(a^2 + 8) + 8(a^2 + 8)]$ Factoring by grouping

$= a(a^2 + 8)(a + 8)$

18. $t(t^2 + 7)(t - 3)$

19. $x^5 - 14x^4 + 49x^3$

$= x^3(x^2 - 14x + 49)$ x^3 is a common factor.

$= x^3(x^2 - 2 \cdot x \cdot 7 + 7^2)$ Trinomial square

$= x^3(x - 7)^2$

20. $2x^4(x + 2)^2$

21. $20 - 6x - 2x^2$

$= -2x^2 - 6x + 20$ Rewriting

$= -2(x^2 + 3x - 10)$ -2 is a common factor.

$= -2(x + 5)(x - 2)$ Using trial and error

We could also express this result as $2(5 + x)(2 - x)$.

22. $-3(2x - 5)(x + 3)$, or $3(5 - 2x)(3 + x)$

23. $t^2 - 7t - 6$

There is no common factor (other than 1). This is not a trinomial square, because only t^2 is a square. We try factoring by trial and error. We look for two factors whose product is -6 and whose sum is -7. There are none. The polynomial cannot be factored. It is prime.

24. Prime

25. $4x^4 - 64$

$= 4(x^4 - 16)$ 4 is a common factor.

$= 4[(x^2)^2 - 4^2]$ Difference of squares

$= 4(x^2 + 4)(x^2 - 4)$ Difference of squares

$= 4(x^2 + 4)(x + 2)(x - 2)$

26. $5x(x^2 + 4)(x + 2)(x - 2)$

27. $9 + t^8$

There is no common factor (other than 1). Although both 9 and t^8 are squares, this is not a difference of squares because the terms do not have different signs. This polynomial cannot be factored. It is prime.

28. $(t^2 + 3)(t^2 - 3)$

29. $x^5 - 4x^4 + 3x^3$

$= x^3(x^2 - 4x + 3)$ x^3 is a common factor.

$= x^3(x - 3)(x - 1)$ Factoring the trinomial using trial and error

30. $x^4(x^2 - 2x + 7)$

31. $x^3 - y^3$ Difference of cubes

$= (x - y)(x^2 + xy + y^2)$

32. $(2t + 1)(4t^2 - 2t + 1)$

33. $12n^2 + 24n^3 = 12n^2(1 + 2n)$

34. $a(x^2 + y^2)$

35. $ab^2 - a^2b = ab(b - a)$

36. $9mn(4 - mn)$

37. $2\pi rh + 2\pi r^2 = 2\pi r(h + r)$

38. $2\pi r(2r + 1)$

39. $(a + b)(x - 3) + (a + b)(x + 4)$

$= (a + b)[(x - 3) + (x + 4)]$ $(a + b)$ is a common factor.

$= (a + b)(2x + 1)$

40. $(a^3 + b)(5c - 1)$

41. $n^2 + 2n + np + 2p$

$= n(n + 2) + p(n + 2)$ Factoring by grouping

$= (n + 2)(n + p)$

42. $(x + 1)(x + y)$

43. $2x^2 - 4x + xz - 2z$

$= 2x(x - 2) + z(x - 2)$ Factoring by grouping

$= (x - 2)(2x + z)$

44. $(a - 3)(a + y)$

45. $x^2 + y^2 + 2xy$

$= x^2 + 2xy + y^2$ Perfect-square trinomial

$= (x + y)^2$

46. $(3x - 2y)(x + 5y)$

47. $9c^2 - 6cd + d^2$

$= (3c)^2 - 2 \cdot 3c \cdot d + d^2$ Perfect-square trinomial

$= (3c - d)^2$

48. $(2b - a)^2$, or $(a - 2b)^2$

49. $7p^4 - 7q^4$

$= 7(p^4 - q^4)$ 7 is a common factor.

$= 7(p^2 + q^2)(p^2 - q^2)$ Difference of squares

$= 7(p^2 + q^2)(p + q)(p - q)$ Difference of squares

50. $(a^2b^2 + 4)(ab + 2)(ab - 2)$

51. $25z^2 + 10zy + y^2$

$= (5z)^2 + 2 \cdot 5z \cdot y + y^2$ Perfect-square trinomial

$= (5z + y)^2$

52. $(2xy + 3z)^2$

53. $m^6 - 1$

$= (m^3)^2 - 1^2$ Difference of squares

$= (m^3 + 1)(m^3 - 1)$ Sum of cubes; difference of cubes

$= (m + 1)(m^2 - m + 1)(m - 1)(m^2 + m + 1)$

54. $(2t+1)(4t^2-2t+1)(2t-1)(4t^2+2t+1)$

55. $a^2+ab+2b^2$

There is no common factor (other than 1). This is not a perfect-square trinomial because only a^2 is a square. We try factoring by trial and error. We look for two factors whose product is 2 and whose sum is 1. There are none. This polynomial cannot be factored. It is prime.

56. $p(4pq-q^2+4p^2)$

57. $2mn-360n^2+m^2$

$=m^2+2mn-360n^2$ Rewriting

$=(m+20n)(m-18n)$ Using trial and error

58. $(3b-a)(b+6a)$

59. $m^2n^2-4mn-32$

$=(mn-8)(mn+4)$ Using trial and error

60. $(xy+2)(xy+6)$

61. $a^5b^2+3a^4b-10a^3$

$=a^3(a^2b^2+3ab-10)$ a^3 is a common factor.

$=a^3(ab+5)(ab-2)$ Using trial and error

62. $(pq+1)(pq+6)$

63. $54a^4+16ab^3$

$=2a(27a^3+8b^3)$ $2a$ is a common factor.

$=2a[(3a)^3+(2b)^3]$ Sum of cubes

$=2a(3a+2b)[(3a)^2-3a\cdot 2b+(2b)^2]$

$=2a(3a+2b)(9a^2-6ab+4b^2)$

64. $2y(3x-5y)(9x^2+15xy+25y^2)$

65. $2s^6t^2+10s^3t^3+12t^4$

$=2t^2(s^6+5s^3t+6t^2)$ $2t^2$ is a common factor.

$=2t^2(s^3+3t)(s^3+2t)$ Using trial and error

66. $x^4(x+2y)(x-y)$

67. $a^2+2a^2bc+a^2b^2c^2$

$=a^2(1+2bc+b^2c^2)$ a^2 is a common factor.

$=a^2[1^2+2\cdot 1\cdot bc+(bc)^2]$ Perfect-square trinomial

$=a^2(1+bc)^2$

68. $\left(6a-\dfrac{5}{4}\right)^2$

69. $\dfrac{1}{81}x^2-\dfrac{8}{27}x+\dfrac{16}{9}$

$=\left(\dfrac{1}{9}x\right)^2-2\cdot\dfrac{1}{9}x\cdot\dfrac{4}{3}+\left(\dfrac{4}{3}\right)^2$ Perfect-square trinomial

$=\left(\dfrac{1}{9}x-\dfrac{4}{3}\right)^2$

If we had factored out $\dfrac{1}{9}$ at the outset, the final result would have been $\dfrac{1}{9}\left(\dfrac{1}{3}x-4\right)^2$.

70. $\left(\dfrac{1}{2}a+\dfrac{1}{3}b\right)^2$

71. $1-16x^{12}y^{12}$

$=(1+4x^6y^6)(1-4x^6y^6)$ Difference of squares

$=(1+4x^6y^6)(1+2x^3y^3)(1-2x^3y^3)$ Difference of squares

72. $a(b^2+9a^2)(b+3a)(b-3a)$

73. Writing exercise

74. Writing exercise

75.

$$\begin{array}{c|l}
\multicolumn{2}{l}{y=-4x+7} \\ \hline
11 \;\;? & -4(-1)+7 \\
 & 4+7 \\
11 & 11 \qquad\qquad \text{TRUE}
\end{array}$$

Since $11=11$ is true, $(-1,11)$ is a solution.

$$\begin{array}{c|l}
\multicolumn{2}{l}{y=-4x+7} \\ \hline
7 \;\;? & -4\cdot 0+7 \\
 & 0+7 \\
7 & 7 \qquad\qquad \text{TRUE}
\end{array}$$

Since $7=7$ is true, $(0,7)$ is a solution.

$$\begin{array}{c|l}
\multicolumn{2}{l}{y=-4x+7} \\ \hline
-5 \;\;? & -4\cdot 3+7 \\
 & -12+7 \\
-5 & -5 \qquad\qquad \text{TRUE}
\end{array}$$

Since $-5=-5$ is true, $(3,-5)$ is a solution.

76. $\dfrac{4}{5}$

77. $3x+7=0$

$3x=-7$ Subtracting 7 from both sides

$x=-\dfrac{7}{3}$ Dividing both sides by 3

The solution is $-\dfrac{7}{3}$.

78. $-\dfrac{9}{2}$

79. $4x - 9 = 0$

$\qquad 4x = 9 \qquad$ Adding 9 to both sides

$\qquad x = \dfrac{9}{4} \qquad$ Dividing both sides by 4

The solution is $\dfrac{9}{4}$.

80.

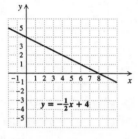

$y = -\frac{1}{2}x + 4$

81. Writing exercise

82. Writing exercise

83. $\quad -(x^5 + 7x^3 - 18x)$

$= -x(x^4 + 7x^2 - 18)$

$= -x(x^2 + 9)(x^2 - 2)$

84. $(a - 2)(a + 3)(a - 3)$

85. $\quad 3a^4 - 15a^2 + 12$

$= 3(a^4 - 5a^2 + 4)$

$= 3(a^2 - 1)(a^2 - 4)$

$= 3(a + 1)(a - 1)(a + 2)(a - 2)$

86. $(x^2 + 2)(x + 3)(x - 3)$

87. $\quad y^2(y + 1) - 4y(y + 1) - 21(y + 1)$

$= (y + 1)(y^2 - 4y - 21)$

$= (y + 1)(y - 7)(y + 3)$

88. $(y - 1)^3$

89. $\quad 6(x - 1)^2 + 7y(x - 1) - 3y^2$

$= [2(x - 1) + 3y][3(x - 1) - y]$

$= (2x + 3y - 2)(3x - y - 3)$

90. $(y + 4 + x)^2$

91. $\quad 2(a + 3)^4 - (a + 3)^3(b - 2) - (a + 3)^2(b - 2)^2$

$= (a + 3)^2[2(a + 3)^2 - (a + 3)(b - 2) - (b - 2)^2]$

$= (a + 3)^2[2(a + 3) + (b - 2)][(a + 3) - (b - 2)]$

$= (a + 3)^2(2a + 6 + b - 2)(a + 3 - b + 2)$

$= (a + 3)^2(2a + b + 4)(a - b + 5)$

92. $(t - 1)^3(5t - s - 4)(t - s)$

Exercise Set 5.7

1. $(x + 5)(x + 6) = 0$

We use the principle of zero products.

$\qquad x + 5 = 0 \quad or \quad x + 6 = 0$

$\qquad x = -5 \quad or \qquad x = -6$

Check:

For -5:

$$\begin{array}{c|c} (x + 5)(x + 6) = 0 & \\ \hline (-5 + 5)(-5 + 6) \; ? \; 0 & \\ 0 \cdot 1 & \\ 0 & 0 \quad \text{TRUE} \end{array}$$

For -6:

$$\begin{array}{c|c} (x + 5)(x + 6) = 0 & \\ \hline (-6 + 5)(-6 + 6) \; ? \; 0 & \\ -1 \cdot 0 & \\ 0 & 0 \quad \text{TRUE} \end{array}$$

The solutions are -5 and -6.

2. $-1,\ -2$

3. $(x - 3)(x + 7) = 0$

$\qquad x - 3 = 0 \quad or \quad x + 7 = 0$

$\qquad x = 3 \quad or \qquad x = -7$

Check:

For 3:

$$\begin{array}{c|c} (x - 3)(x + 5) = 0 & \\ \hline (3 - 3)(3 + 5) \; ? \; 0 & \\ 0 \cdot 8 & \\ 0 & 0 \quad \text{TRUE} \end{array}$$

For -7:

$$\begin{array}{c|c} (x - 3)(x + 7) = 0 & \\ \hline (-7 - 3)(-7 + 7) \; ? \; 0 & \\ -10 \cdot 0 & \\ 0 & 0 \quad \text{TRUE} \end{array}$$

The solutions are 3 and -7.

4. $-9,\ 3$

5. $(2x - 9)(x + 4) = 0$

$\qquad 2x - 9 = 0 \quad or \quad x + 4 = 0$

$\qquad 2x = 9 \quad or \qquad x = -4$

$\qquad x = \dfrac{9}{2} \quad or \qquad x = -4$

The solutions are $\dfrac{9}{2}$ and -4.

6. $\dfrac{5}{3}$, -1

7. $(10x - 9)(4x + 7) = 0$

$10x - 9 = 0 \quad$ or $\quad 4x + 7 = 0$

$10x = 9 \quad$ or $\quad 4x = -7$

$x = \dfrac{9}{10} \quad$ or $\quad x = -\dfrac{7}{4}$

The solutions are $\dfrac{9}{10}$ and $-\dfrac{7}{4}$.

8. $\dfrac{7}{2}$, $-\dfrac{4}{3}$

9. $x(x + 6) = 0$

$x = 0 \quad$ or $\quad x + 6 = 0$

$x = 0 \quad$ or $\quad x = -6$

The solutions are 0 and -6.

10. 0, -9

11. $\left(\dfrac{2}{3}x - \dfrac{12}{11}\right)\left(\dfrac{7}{4}x - \dfrac{1}{12}\right) = 0$

$\dfrac{2}{3}x - \dfrac{12}{11} = 0 \quad$ or $\quad \dfrac{7}{4}x - \dfrac{1}{12} = 0$

$\dfrac{2}{3}x = \dfrac{12}{11} \quad$ or $\quad \dfrac{7}{4}x = \dfrac{1}{12}$

$x = \dfrac{3}{2} \cdot \dfrac{12}{11} \quad$ or $\quad x = \dfrac{4}{7} \cdot \dfrac{1}{12}$

$x = \dfrac{18}{11} \quad$ or $\quad x = \dfrac{1}{21}$

The solutions are $\dfrac{18}{11}$ and $\dfrac{1}{21}$.

12. $-\dfrac{1}{10}$, $\dfrac{1}{27}$

13. $5x(2x + 9) = 0$

$5x = 0 \quad$ or $\quad 2x + 9 = 0$

$x = 0 \quad$ or $\quad 2x = -9$

$x = 0 \quad$ or $\quad x = -\dfrac{9}{2}$

The solutions are 0 and $-\dfrac{9}{2}$.

14. 0, $-\dfrac{5}{4}$

15. $(20x - 0.4x)(7 - 0.1x) = 0$

$20 - 0.4x = 0 \quad$ or $\quad 7 - 0.1x = 0$

$-0.4x = -20 \quad$ or $\quad -0.1x = -7$

$x = 50 \quad$ or $\quad x = 70$

The solutions are 50 and 70.

16. $3.\overline{3}$, or $\dfrac{10}{3}$; 20

17. $(3x - 2)(x + 5)(x - 1) = 0$

$3x - 2 = 0 \quad$ or $\quad x + 5 = 0 \quad$ or $\quad x - 1 = 0$

$3x = 2 \quad$ or $\quad x = -5 \quad$ or $\quad x = 1$

$x = \dfrac{2}{3} \quad$ or $\quad x = -5 \quad$ or $\quad x = 1$

The solutions are $\dfrac{2}{3}$, -5, and 1.

18. -3, $-\dfrac{1}{2}$, 5

19. $\quad x^2 + 7x + 6 = 0$

$(x + 6)(x + 1) = 0 \quad$ Factoring

$x + 6 = 0 \quad$ or $\quad x + 1 = 0 \quad$ Using the principle of zero products

$x = -6 \quad$ or $\quad x = -1$

The solutions are -6 and -1.

20. 1, 5

21. $\quad x^2 - 4x - 21 = 0$

$(x + 3)(x - 7) = 0 \quad$ Factoring

$x + 3 = 0 \quad$ or $\quad x - 7 = 0 \quad$ Using the principle of zero products

$x = -3 \quad$ or $\quad x = 7$

The solutions are -3 and 7.

22. -2, 9

23. $\quad x^2 - 6x = 0$

$x(x - 6) = 0$

$x = 0 \quad$ or $\quad x - 6 = 0$

$x = 0 \quad$ or $\quad x = 6$

The solutions are 0 and 6.

24. -8, 0

25. $\quad x^3 - 3x^2 + 2x = 0$

$x(x^2 - 3x + 2) = 0$

$x(x - 1)(x - 2) = 0$

$x = 0 \quad$ or $\quad x - 1 = 0 \quad$ or $\quad x - 2 = 0$

$x = 0 \quad$ or $\quad x = 1 \quad$ or $\quad x = 2$

The solutions are 0, 1, and 2.

26. -1, 0, 7

27. $\quad\quad\quad 9x^2 = 4$

$9x^2 - 4 = 0$

$(3x + 2)(3x - 2) = 0$

$$3x + 2 = 0 \quad or \quad 3x - 2 = 0$$
$$3x = -2 \quad or \quad 3x = 2$$
$$x = -\frac{2}{3} \quad or \quad x = \frac{2}{3}$$

The solutions are $-\frac{2}{3}$ and $\frac{2}{3}$.

28. $-\frac{7}{2}, \frac{7}{2}$

29. $\quad 0 = 25 + x^2 + 10x$

$\quad 0 = x^2 + 10x + 25$ Writing in descending
                                     order

$\quad 0 = (x + 5)(x + 5)$

$\quad x + 5 = 0 \quad or \quad x + 5 = 0$

$\quad\quad x = -5 \quad or \quad\quad x = -5$

The solution is -5.

30. -3

31. $\quad\quad 1 + x^2 = 2x$

$x^2 - 2x + 1 = 0$

$(x - 1)(x - 1) = 0$

$x - 1 = 0 \quad or \quad x - 1 = 0$

$\quad x = 1 \quad or \quad\quad x = 1$

The solution is 1.

32. 4

33. $\quad\quad 8x^2 = 5x$

$8x^2 - 5x = 0$

$x(8x - 5) = 0$

$x = 0 \quad or \quad 8x - 5 = 0$

$x = 0 \quad or \quad\quad 8x = 5$

$x = 0 \quad or \quad\quad x = \frac{5}{8}$

The solutions are 0 and $\frac{5}{8}$.

34. $0, \frac{7}{3}$

35. $\quad\quad 3x^2 - 7x = 20$

$3x^2 - 7x - 20 = 0$

$(3x + 5)(x - 4) = 0$

$3x + 5 = 0 \quad or \quad x - 4 = 0$

$\quad 3x = -5 \quad or \quad\quad x = 4$

$\quad\quad x = -\frac{5}{3} \quad or \quad\quad x = 4$

The solutions are $-\frac{5}{3}$ and 4.

36. $-1, \frac{5}{3}$

37. $\quad\quad 2y^2 + 12y = -10$

$2y^2 + 12y + 10 = 0$

$2(y^2 + 6y + 5) = 0$

$2(y + 5)(y + 1) = 0$

$y + 5 = 0 \quad or \quad y + 1 = 0$

$\quad y = -5 \quad or \quad\quad y = -1$

The solutions are -5 and -1.

38. $-\frac{1}{4}, \frac{2}{3}$

39. $\quad (x - 7)(x + 1) = -16$

$x^2 - 6x - 7 = -16$

$x^2 - 6x + 9 = 0$

$(x - 3)(x - 3) = 0$

$x - 3 = 0 \quad or \quad x - 3 = 0$

$\quad x = 3 \quad or \quad\quad x = 3$

The solution is 3.

40. 1, 4

41. $\quad\quad y(3y + 1) = 2$

$\quad 3y^2 + y = 2$

$3y^2 + y - 2 = 0$

$(3y - 2)(y + 1) = 0$

$3y - 2 = 0 \quad or \quad y + 1 = 0$

$\quad 3y = 2 \quad or \quad\quad y = -1$

$\quad\quad y = \frac{2}{3} \quad or \quad\quad y = -1$

The solutions are $\frac{2}{3}$ and -1.

42. $-2, 7$

43. $\quad\quad 81x^2 - 5 = 20$

$81x^2 - 25 = 0$

$(9x + 5)(9x - 5) = 0$

$9x + 5 = 0 \quad or \quad 9x - 5 = 0$

$\quad 9x = -5 \quad or \quad\quad 9x = 5$

$\quad\quad x = -\frac{5}{9} \quad or \quad\quad x = \frac{5}{9}$

The solutions are $-\frac{5}{9}$ and $\frac{5}{9}$.

44. $-\frac{7}{6}, \frac{7}{6}$

45. $(x-1)(5x+4) = 2$

$\quad 5x^2 - x - 4 = 2$

$\quad 5x^2 - x - 6 = 0$

$\quad (5x-6)(x+1) = 0$

$\quad 5x - 6 = 0 \quad or \quad x + 1 = 0$

$\quad\quad 5x = 6 \quad or \quad\quad x = -1$

$\quad\quad x = \dfrac{6}{5} \quad or \quad\quad x = -1$

The solutions are $\dfrac{6}{5}$ and -1.

46. $-4, -\dfrac{2}{3}$

47. $\quad\quad x^2 - 2x = 18 + 5x$

$\quad x^2 - 7x - 18 = 0 \quad\quad$ Subtracting 18 and $5x$

$\quad (x-9)(x+2) = 0$

$\quad x - 9 = 0 \quad or \quad x + 2 = 0$

$\quad\quad x = 9 \quad or \quad\quad x = -2$

The solutions are 9 and -2.

48. $-3, 1$

49. $\quad\quad x^2(2x-1) = 3x$

$\quad\quad 2x^3 - x^2 = 3x$

$\quad 2x^3 - x^2 - 3x = 0$

$\quad x(2x^2 - x - 3) = 0$

$\quad x(2x-3)(x+1) = 0$

$\quad x = 0 \quad or \quad 2x - 3 = 0 \quad or \quad x + 1 = 0$

$\quad x = 0 \quad or \quad\quad 2x = 3 \quad or \quad\quad x = -1$

$\quad x = 0 \quad or \quad\quad x = \dfrac{3}{2} \quad or \quad\quad x = -1$

The solutions are 0, $\dfrac{3}{2}$, and -1.

50. $-2, 0, \dfrac{5}{3}$

51. The solutions of the equation are the first coordinates of the x-intercepts of the graph. From the graph we see that the x-intercepts are $(-1, 0)$ and $(4, 0)$, so the solutions of the equation are -1 and 4.

52. $-3, 2$

53. The solutions of the equation are the first coordinates of the x-intercepts of the graph. From the graph we see that the x-intercepts are $(-1, 0)$ and $(3, 0)$, so the solutions of the equation are -1 and 3.

54. $-3, 2$

55. We let $y = 0$ and solve for x.

$\quad 0 = x^2 + 3x - 4$

$\quad 0 = (x+4)(x-1)$

$\quad x + 4 = 0 \quad or \quad x - 1 = 0$

$\quad\quad x = -4 \quad or \quad\quad x = 1$

The x-intercepts are $(-4, 0)$ and $(1, 0)$.

56. $(-2, 0), (3, 0)$

57. We let $y = 0$ and solve for x.

$\quad 0 = x^2 - 2x - 15$

$\quad 0 = (x-5)(x+3)$

$\quad x - 5 = 0 \quad or \quad x + 3 = 0$

$\quad\quad x = 5 \quad or \quad\quad x = -3$

The x-intercepts are $(5, 0)$ and $(-3, 0)$.

58. $(-4, 0), (2, 0)$

59. We let $y = 0$ and solve for x

$\quad 0 = 2x^2 + x - 10$

$\quad 0 = (2x+5)(x-2)$

$\quad 2x + 5 = 0 \quad or \quad x - 2 = 0$

$\quad\quad 2x = -5 \quad or \quad\quad x = 2$

$\quad\quad x = -\dfrac{5}{2} \quad or \quad\quad x = 2$

The x-intercepts are $\left(-\dfrac{5}{2}, 0\right)$ and $(2, 0)$.

60. $(-3, 0), \left(\dfrac{3}{2}, 0\right)$

61. Writing exercise

62. Writing exercise

63. $(a+b)^2$

64. $a^2 + b^2$

65. Let x represent the smaller integer; $x + (x+1)$

66. Let x represent the number; $2x + 5 < 19$

67. Let x represent the number; $\dfrac{1}{2}x - 7 > 24$

68. Let n represent the number; $n - 3 \geq 34$

69. Writing exercise

70. Writing exercise

71. $(2x - 5)(x + 7)(3x + 8) = 0$

$2x - 5 = 0 \quad or \quad x + 7 = 0 \quad or \quad 3x + 8 = 0$

$2x = 5 \quad or \qquad x = -7 \quad or \qquad 3x = -8$

$x = \dfrac{5}{2} \quad or \qquad x = -7 \quad or \qquad x = -\dfrac{8}{3}$

The solutions are $\dfrac{5}{2}$, -7, and $-\dfrac{8}{3}$.

72. $-\dfrac{9}{4}, \dfrac{2}{3}, -\dfrac{1}{5}$

73. a)
$\qquad x = -4 \quad or \qquad x = 5$

$\qquad x + 4 = 0 \quad or \quad x - 5 = 0$

$(x + 4)(x - 5) = 0 \qquad$ Principle of zero products

$x^2 - x - 20 = 0 \qquad$ Multiplying

b)
$\qquad x = -1 \quad or \qquad x = 7$

$\qquad x + 1 = 0 \quad or \quad x - 7 = 0$

$(x + 1)(x - 7) = 0$

$x^2 - 6x + -7 = 0$

c)
$\qquad x = \dfrac{1}{4} \quad or \qquad x = 3$

$\qquad x - \dfrac{1}{4} = 0 \quad or \quad x - 3 = 0$

$\left(x - \dfrac{1}{4}\right)(x - 3) = 0$

$x^2 - \dfrac{13}{4}x + \dfrac{3}{4} = 0$

$4\left(x^2 - \dfrac{13}{4}x + \dfrac{3}{4}\right) = 4 \cdot 0 \quad$ Multiplying both sides by 4

$4x^2 - 13x + 3 = 0$

d)
$\qquad x = \dfrac{1}{2} \quad or \qquad x = \dfrac{1}{3}$

$\qquad x - \dfrac{1}{2} = 0 \quad or \quad x - \dfrac{1}{3} = 0$

$\left(x - \dfrac{1}{2}\right)\left(x - \dfrac{1}{3}\right) = 0$

$x^2 - \dfrac{5}{6}x + \dfrac{1}{6} = 0$

$6x^2 - 5x + 1 = 0 \quad$ Multiplying by 6

e)
$\qquad x = \dfrac{2}{3} \quad or \qquad x = \dfrac{3}{4}$

$\qquad x - \dfrac{2}{3} = 0 \quad or \quad x - \dfrac{3}{4} = 0$

$\left(x - \dfrac{2}{3}\right)\left(x - \dfrac{3}{4}\right) = 0$

$x^2 - \dfrac{17}{12}x + \dfrac{1}{2} = 0$

$12x^2 - 17x + 6 = 0 \quad$ Multiplying by 12

f)
$\qquad x = -1 \quad or \qquad x = 2 \ or \qquad x = 3$

$x + 1 = 0 \quad or \quad x - 2 = 0 \ or \quad x - 3 = 0$

$(x + 1)(x - 2)(x - 3) = 0$

$(x^2 - x - 2)(x - 3) = 0$

$x^3 - 4x^2 + x + 6 = 0$

74. 4

75.
$\qquad a(9 + a) = 4(2a + 5)$

$\qquad 9a + a^2 = 8a + 20$

$a^2 + a - 20 = 0 \qquad$ Subtracting $8a$ and 20

$(a + 5)(a - 4) = 0$

$a + 5 = 0 \quad or \quad a - 4 = 0$

$\qquad a = -5 \quad or \qquad a = 4$

The solutions are -5 and 4.

76. 3, 5

77.
$\qquad x^2 - \dfrac{9}{25} = 0$

$\left(x - \dfrac{3}{5}\right)\left(x + \dfrac{3}{5}\right) = 0$

$x - \dfrac{3}{5} = 0 \quad or \quad x + \dfrac{3}{5} = 0$

$\qquad x = \dfrac{3}{5} \quad or \qquad x = -\dfrac{3}{5}$

The solutions are $\dfrac{3}{5}$ and $-\dfrac{3}{5}$.

78. $-\dfrac{5}{6}, \dfrac{5}{6}$

79. $(t + 1)^2 = 9$

Observe that $t + 1$ is a number which yields 9 when it is squared. Thus, we have

$t + 1 = -3 \quad or \quad t + 1 = 3$

$\qquad t = -4 \quad or \qquad t = 2$

The solutions are -4 and 2.

We could also do this exercise as follows:

$\qquad (t + 1)^2 = 9$

$\qquad t^2 + 2t + 1 = 9$

$\qquad t^2 + 2t - 8 = 0$

$\qquad (t + 4)(t - 2) = 0$

$t + 4 = 0 \quad or \quad t - 2 = 0$

$\qquad t = -4 \quad or \qquad t = 2$

Again we see that the solutions are -4 and 2.

80. $-\dfrac{5}{9}, \dfrac{5}{9}$

81. a) $2(x^2 + 10x - 2) = 2 \cdot 0$ Multiplying (a) by 2

$\qquad 2x^2 + 20x - 4 = 0$

(a) and $2x^2 + 20x - 4 = 0$ are equivalent.

b) $(x - 6)(x + 3) = x^2 - 3x - 18$ Multiplying

(b) and $x^2 - 3x - 18 = 0$ are equivalent.

c) $5x^2 - 5 = 5(x^2 - 1) = 5(x + 1)(x - 1) =$
$(x + 1)5(x - 1) = (x + 1)(5x - 5)$

(c) and $(x + 1)(5x - 5) = 0$ are equivalent.

d) $2(2x - 5)(x + 4) = 2 \cdot 0$ Multiplying (d) by 2

$\qquad 2(x + 4)(2x - 5) = 0$

$\qquad (2x + 8)(2x - 5) = 0$

(d) and $(2x + 8)(2x - 5) = 0$ are equivalent.

e) $4(x^2 + 2x + 9) = 4 \cdot 0$ Multiplying (e) by 4

$\qquad 4x^2 + 8x + 36 = 0$

(e) and $4x^2 + 8x + 36 = 0$ are equivalent.

f) $3(3x^2 - 4x + 8) = 3 \cdot 0$ Multiplying (f) by 3

$\qquad 9x^2 - 12x + 24 = 0$

(f) and $9x^2 - 12x + 24 = 0$ are equivalent.

82. Writing exercise

83. Writing exercise

84. $-3.45, 1.65$

85. $2.33, 6.77$

86. $-0.25, 0.88$

87. $-4.59, -9.15$

88. $4.55, -3.23$

89. $0, 2.74$

90. $-3.76, 0$

Exercise Set 5.8

1. *Familiarize*. Let $x =$ the number (or numbers).

***Translate*.** We reword the problem.

$$\underbrace{\text{The square of a number}}_{x^2} \quad \underset{-}{\text{minus}} \quad \underbrace{\text{the number}}_{x} \quad \underset{= \; 6}{\text{is 6.}}$$

***Carry out*.** We solve the equation.

$$x^2 - x = 6$$
$$x^2 - x - 6 = 0$$
$$(x - 3)(x + 2) = 0$$

$$x - 3 = 0 \quad or \quad x + 2 = 0$$
$$x = 3 \quad or \qquad x = -2$$

***Check*.** For 3: The square of 3 is 3^2, or 9, and $9 - 3 = 6$.

For -2: The square of -2 is $(-2)^2$, or 4, and $4 - (-2) = 4 + 2 = 6$. Both numbers check.

***State*.** The numbers are 3 and -2.

2. $-1, 2$

3. *Familiarize*. Let $x =$ the length of the shorter leg, in cm. Then $x + 3 =$ the length of the longer leg.

***Translate*.** we use the Pythagorean theorem.

$$a^2 + b^2 = c^2$$
$$x^2 + (x + 3)^2 = 15^2$$

***Carry out*.** We solve the equation.

$$x^2 + (x + 3)^2 = 15^2$$
$$x^2 + x^2 + 6x + 9 = 225$$
$$2x^2 + 6x + 9 = 225$$
$$2x^2 + 6x - 216 = 0$$
$$2(x^2 + 3x - 108) = 0$$
$$2(x + 12)(x - 9) = 0$$

$$x + 12 = 0 \quad or \quad x - 9 = 0$$
$$x = -12 \, or \qquad x = 9$$

***Check*.** The number -12 cannot be the length of a side because it is negative. When $x = 9$, then $x + 3 = 12$, and $9^2 + 12^2 = 81 + 144 = 225 = 15^2$, so the number 9 checks.

***State*.** The lengths of the sides are 9 cm, 12 cm, and 15 cm.

4. 6 cm, 8 cm, 10 cm

5. *Familiarize*. The page numbers on facing pages are consecutive integers. Let $x =$ the smaller integer. Then $x + 1 =$ the larger integer.

***Translate*.** We reword the problem.

$$\underbrace{\text{Smaller integer}}_{x} \quad \underset{\cdot}{\text{times}} \quad \underbrace{\text{larger integer}}_{(x + 1)} \quad \underset{= \; 110}{\text{is 110.}}$$

***Carry out*.** We solve the equation.

$$x(x + 1) = 110$$
$$x^2 + x = 110$$
$$x^2 + x - 110 = 0$$
$$(x + 11)(x - 10) = 0$$
$$x + 11 = 0 \quad or \quad x - 10 = 0$$
$$x = -11 \quad or \quad\quad x = 10$$

Check. The solutions of the equation are -11 and 10. Since a page number cannot be negative, -11 cannot be a solution of the original problem. We only need to check 10. When $x = 10$, then $x + 1 = 11$, and $10 \cdot 11 = 110$. This checks.

State. The page numbers are 10 and 11.

6. 14, 15

7. Familiarize. Let $x =$ the smaller odd integer. Then $x + 2 =$ the larger odd integer.

Translate. We reword the problem.

$$
\underbrace{\text{Smaller odd integer}}_{x} \quad \overset{\text{times}}{\cdot} \quad \underbrace{\text{larger odd integer}}_{(x + 2)} \quad \overset{\text{is 255.}}{= 255}
$$

Carry out.

$$x(x + 2) = 255$$
$$x^2 + 2x = 255$$
$$x^2 + 2x - 255 = 0$$
$$(x + 17)(x - 15) = 0$$
$$x + 17 = 0 \quad or \quad x - 15 = 0$$
$$x = -17 \quad or \quad\quad x = 15$$

Check. The solutions of the equation are -17 and 15. When x is -17, then $x + 2$ is -15 and $-17(-15) = 225$. The numbers -17 and -15 are consecutive odd integers which are solutions of the problem. When x is 15, then $x + 2 = 17$ and $15 \cdot 17 = 255$. The numbers 15 and 17 are also consecutive odd integers which are solutions of the problem.

State. We have two solutions, each of which consists of a pair of numbers: -17 and -15 or 15 and 17.

8. 14 and 16; -16 and -14

9. Familiarize. Let $w =$ the width of the frame, in inches. Then $2w =$ the length of the frame. Recall that the area of a rectangle is Length $\cdot$ Width.

Translate.

$$
\underbrace{\text{The area of the rectangle}}_{2w \cdot w} \quad \overset{\text{is}}{=} \quad \underbrace{288 \text{ in}^2.}_{288}
$$

Carry out. We solve the equation.

$$2w \cdot w = 288$$
$$2w^2 = 288$$
$$2w^2 - 288 = 0$$
$$2(w^2 - 144) = 0$$
$$2(w + 12)(w - 12) = 0$$
$$w + 12 = 0 \quad or \quad w - 12 = 0$$
$$w = -12 \quad or \quad\quad w = 12$$

Check. Since the width must be positive, -12 cannot be a solution. If the width is 12 in., then the length is $2 \cdot 12$ in, or 24 in., and the area is 12 in. $\cdot$ 24 in. $= 288$ in^2. Thus, 12 checks.

State. The frame is 12 in. wide and 24 in. long.

10. Length: 12 ft; width: 2 ft

11. Familiarize. We make a drawing. Let $w =$ the width, in cm. Then $2w + 2 =$ the length, in cm.

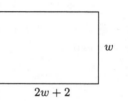

Recall that the area of a rectangle is length times width.

Translate. We reword the problem.

$$
\begin{array}{ccccc}
\text{Length} & \text{times} & \text{width} & \text{is} & \underline{144 \text{ cm}^2.} \\
\downarrow & \downarrow & \downarrow & \downarrow & \downarrow \\
(2w + 2) & \cdot & w & = & 144
\end{array}
$$

Carry out. We solve the equation.

$$(2w + 2)w = 144$$
$$2w^2 + 2w = 144$$
$$2w^2 + 2w - 144 = 0$$
$$2(w^2 + w - 72) = 0$$
$$2(w + 9)(w - 8) = 0$$
$$w + 9 = 0 \quad or \quad w - 8 = 0$$
$$w = -9 \quad or \quad\quad w = 8$$

Check. Since the width must be positive, -9 cannot be a solution. If the width is 8 cm, then the length is $2 \cdot 8 + 2$, or 18 cm, and the area is $8 \cdot 18$, or 144 cm^2. Thus, 8 checks.

State. The width is 8 cm, and the length is 18 cm.

12. Length: 12 m; width: 8 m

13. Familiarize. We make a drawing. Let $l =$ the length of the cable, in ft.

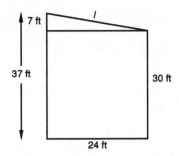

Note that we have a right triangle with hypotenuse l and legs of 24 ft and $37 - 30$, or 7 ft.

Translate. We use the Pythagorean theorem.

$$a^2 + b^2 = c^2$$
$$7^2 + 24^2 = l^2 \quad \text{Substituting}$$

Carry out.

$$7^2 + 24^2 = l^2$$
$$49 + 576 = l^2$$
$$625 = l^2$$
$$0 = l^2 - 625$$
$$0 = (l + 25)(l - 25)$$
$$l + 25 = 0 \quad or \quad l - 25 = 0$$
$$l = -25 \quad or \quad l = 25$$

Check. The integer -25 cannot be the length of the cable, because it is negative. When $l = 25$, we have $7^2 + 24^2 = 25^2$. This checks.

State. The cable is 25 ft long.

14. 8000 ft

15. Familiarize. Using the labels shown on the drawing in the text, we let h = the height, in cm, and $h + 10$ = the base, in cm. Recall that the formula for the area of a triangle is $\frac{1}{2} \cdot$ (base) $\cdot$ (height).

Translate.

$$\frac{1}{2} \text{ times base times height is } \underbrace{28 \text{ cm}^2.}$$
$$\downarrow \quad \downarrow \quad \downarrow \quad \downarrow \quad \downarrow \quad \downarrow$$
$$\frac{1}{2} \cdot (h + 10) \cdot h = 28$$

Carry out.

$$\frac{1}{2}(h + 10)h = 28$$
$$(h + 10)h = 56 \quad \text{Multiplying by 2}$$
$$h^2 + 10h = 56$$
$$h^2 + 10h - 56 = 0$$
$$(h + 14)(h - 4) = 0$$
$$h + 14 = 0 \quad or \quad h - 4 = 0$$
$$h = -14 \quad or \quad h = 4$$

Check. Since the height of the triangle must be positive, -14 cannot be a solution. If the height is 4 cm, then the base is $4 + 10$, or 14 cm, and the area is $\frac{1}{2} \cdot 14 \cdot 4$, or 28 cm^2. Thus, 4 checks.

State. The height of the triangle is 4 cm, and the base is 14 cm.

16. Base: 10 cm; height; 7 cm

17. Familiarize. Using the labels show on the drawing in the text, we let x = the length of the foot of the sail, in ft, and $x + 5$ = the height of the sail, in ft. Recall that the formula for the area of a triangle is $\frac{1}{2} \cdot$ (base) $\cdot$ (height).

Translate.

$$\frac{1}{2} \text{ times base times height is } \underbrace{42 \text{ ft}^2.}$$
$$\downarrow \quad \downarrow \quad \downarrow \quad \downarrow \quad \downarrow \quad \downarrow$$
$$\frac{1}{2} \cdot x \cdot (x + 5) = 42$$

Carry out.

$$\frac{1}{2}x(x + 5) = 42$$
$$x(x + 5) = 84 \quad \text{Multiplying by 2}$$
$$x^2 + 5x = 84$$
$$x^2 + 5x - 84 = 0$$
$$(x + 12)(x - 7) = 0$$
$$x + 12 = 0 \quad or \quad x - 7 = 0$$
$$x = -12 \quad or \quad x = 7$$

Check. The solutions of the equation are -12 and 7. The length of the base of a triangle cannot be negative, so -12 cannot be a solution. Suppose the length of the foot of the sail is 7 ft. Then the height is $7 + 5$, or 12 ft, and the area is $\frac{1}{2} \cdot 7 \cdot 12$, or 42 ft^2. These numbers check.

State. The length of the foot of the sail is 7 ft, and the height is 12 ft.

18. Base: 8 m; height: 16 m

19. Familiarize. We will use the formula $n^2 - n = N$.

Translate. Substitute 20 for n.

$$20^2 - 20 = N$$

Carry out. We do the computation on the left.

$$20^2 - 20 = N$$
$$400 - 20 = N$$
$$380 = N$$

Check. We can recheck the computation or we can solve $n^2 - n = 380$. The answer checks.

State. 380 games will be played.

20. 182

21. *Familiarize*. We will use the formula $n^2 - n = N$.

Translate. Substitute 132 for N.

$$n^2 - n = 132$$

Carry out.

$$n^2 - n = 132$$
$$n^2 - n - 132 = 0$$
$$(n - 12)(n + 11) = 0$$
$$n - 12 = 0 \quad \text{or} \quad n + 11 = 0$$
$$n = 12 \quad \text{or} \qquad n = -11$$

Check. The solutions of the equation are 12 and -11. Since the number of teams cannot be negative, -11 cannot be a solution. But 12 checks since $12^2 - 12 = 144 - 12 = 132$.

State. There are 12 teams in the league.

22. 10

23. *Familiarize*. Let h = the vertical height to which each brace reaches, in feet. We have a right triangle with hypotenuse 15 ft and legs 12 ft and h.

Translate. We use the Pythagorean theorem.

$$a^2 + b^2 = c^2$$
$$12^2 + h^2 = 15^2$$

Carry out. We solve the equation.

$$12^2 + h^2 = 15^2$$
$$144 + h^2 = 225$$
$$h^2 - 81 = 0$$
$$(h + 9)(h - 9) = 0$$
$$h + 9 = 0 \quad \text{or} \quad h - 9 = 0$$
$$h = -9 \quad \text{or} \qquad h = 9$$

Check. Since the vertical height must be positive, -9 cannot be a solution. If the height is 9 ft, then we have $12^2 + 9^2 = 144 + 81 = 225 = 15^2$. The number 9 checks.

State. Each brace reaches 9 ft vertically.

24. 24 ft

25. *Familiarize*. We will use the formula
$$N = \frac{1}{2}(n^2 - n).$$

Translate. Substitute 15 for n.

$$N = \frac{1}{2}(15^2 - 15)$$

Carry out. We do the computation on the right.

$$N = \frac{1}{2}(15^2 - 15)$$

$$N = \frac{1}{2}(225 - 15)$$

$$N = \frac{1}{2}(210)$$

$$N = 105$$

Check. We can recheck the computation, or we can solve the equation $105 = \frac{1}{2}(n^2 - n)$. The answer checks.

State. 105 handshakes are possible.

26. 435

27. *Familiarize*. We will use the formula $N = \frac{1}{2}(n^2 - n)$, since "high fives" can be substituted for handshakes.

Translate. Substitute 66 for N.

$$66 = \frac{1}{2}(n^2 - n)$$

Carry out.

$$66 = \frac{1}{2}(n^2 - n)$$
$$132 = n^2 - n \qquad \text{Multiplying by 2}$$
$$0 = n^2 - n - 132$$
$$0 = (n - 12)(n + 11)$$
$$n - 12 = 0 \quad or \quad n + 11 = 0$$
$$n = 12 \; or \qquad n = -11$$

Check. The solutions of the equation are 12 and -11. Since the number of people cannot be negative, -11 cannot be a solution. However, 12 checks since $\frac{1}{2}(12^2 - 12) = \frac{1}{2}(144 - 12) = \frac{1}{2}(132) = 66$.

State. 12 people were on the team.

28. 20

29. *Familiarize*. We label the drawing. Let x = the length of a side of the dining room, in ft. Then the dining room has dimensions x by x and the kitchen has dimensions x by 10. The entire rectangular space has dimension x by $x + 10$. Recall that we multiply these dimensions to find the area of the rectangle.

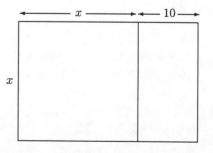

Translate.

$$\underbrace{\text{The area of the rectangular space}}_{x(x+10)} \quad \text{is} \quad \underbrace{264 \text{ ft}^2}_{264}.$$

Carry out. We solve the equation.

$$x(x+10) = 264$$
$$x^2 + 10x = 264$$
$$x^2 + 10x - 264 = 0$$
$$(x+22)(x-12) = 0$$
$$x + 22 = 0 \quad \text{or} \quad x - 12 = 0$$
$$x = -22 \quad \text{or} \quad x = 12$$

Check. Since the length of a side of the dining room must be positive, -22 cannot be a solution. If x is 12 ft, then $x + 10$ is 22 ft, and the area of the space is $12 \cdot 22$, or 264 ft^2. The number 12 checks.

State. The dining room is 12 ft by 12 ft, and the kitchen is 12 ft by 10 ft.

30. 4 m

31. *Familiarize*. We will use the formula $h = 48t - 16t^2$.

Translate. Substitute $\frac{1}{2}$ for t.

$$h = 48 \cdot \frac{1}{2} - 16\left(\frac{1}{2}\right)^2$$

Carry out. We do the computation on the right.

$$h = 48 \cdot \frac{1}{2} - 16\left(\frac{1}{2}\right)^2$$
$$h = 48 \cdot \frac{1}{2} - 16 \cdot \frac{1}{4}$$
$$h = 24 - 4$$
$$h = 20$$

Check. We can recheck the computation, or we can solve the equation $20 = 48t - 16t^2$. The answer checks.

State. The rocket is 20 ft high $\frac{1}{2}$ sec after it is launched.

32. 36 ft

33. *Familiarize*. We will use the formula $h = 48t - 16t^2$.

Translate. Substitute 32 for h.

$$32 = 48t - 16t^2$$

Carry out. We solve the equation.

$$32 = 48t - 16t^2$$
$$0 = -16t^2 + 48t - 32$$
$$0 = -16(t^2 - 3t + 2)$$
$$0 = -16(t-1)(t-2)$$

$$t - 1 = 0 \quad \text{or} \quad t - 2 = 0$$
$$t = 1 \quad \text{or} \quad t = 2$$

Check. When $t = 1$, $h = 48 \cdot 1 - 16 \cdot 1^2 = 48 - 16 = 32$. When $t = 2$, $h = 48 \cdot 2 - 16 \cdot 2^2 = 96 - 64 = 32$. Both numbers check.

State. The rocket will be exactly 32 ft above the ground at 1 sec and at 2 sec after it is launched.

34. 3 sec after launch

35. Writing exercise

36. Writing exercise

37. $-\dfrac{2}{3} \cdot \dfrac{4}{7} = -\dfrac{2 \cdot 4}{3 \cdot 7} = -\dfrac{8}{21}$

38. $-\dfrac{8}{45}$

39. $\dfrac{5}{6}\left(\dfrac{-7}{9}\right) = \dfrac{5(-7)}{6 \cdot 9} = \dfrac{-35}{54}$, or $-\dfrac{35}{54}$

40. $-\dfrac{5}{16}$

41. $-\dfrac{2}{3} + \dfrac{4}{7} = -\dfrac{2}{3} \cdot \dfrac{7}{7} + \dfrac{4}{7} \cdot \dfrac{3}{3}$

$$= -\dfrac{14}{21} + \dfrac{12}{21}$$
$$= -\dfrac{2}{21}$$

42. $-\dfrac{26}{45}$

43. $\dfrac{5}{6} + \dfrac{-7}{9} = \dfrac{5}{6} \cdot \dfrac{3}{3} + \dfrac{-7}{9} \cdot \dfrac{2}{2}$

$$= \dfrac{15}{18} + \dfrac{-14}{18}$$
$$= \dfrac{1}{18}$$

44. $-\dfrac{11}{24}$

45. Writing exercise

46. Writing exercise

47. *Familiarize*. First we can use the Pythagorean theorem to find x, in ft. Then the height of the telephone pole is $x + 5$.

Translate. We use the Pythagorean theorem.

$$a^2 + b^2 = c^2$$
$$\left(\frac{1}{2}x + 1\right)^2 + x^2 = 34^2$$

Carry out. We solve the equation.

$$\left(\frac{1}{2}x + 1\right)^2 + x^2 = 34^2$$

$$\frac{1}{4}x^2 + x + 1 + x^2 = 1156$$

$$x^2 + 4x + 4 + 4x^2 = 4624 \qquad \text{Multiplying by 4}$$

$$5x^2 + 4 + 4 = 4624$$

$$5x^2 + 4x - 4620 = 0$$

$$(5x + 154)(x - 30) = 0$$

$$5x + 154 = 0 \qquad or \quad x - 30 = 0$$

$$5x = -154 \quad or \qquad x = 30$$

$$x = -30.8 \quad or \qquad x = 30$$

Check. Since the length x must be positive, -30.8 cannot be a solution. If x is 30 ft, then $\frac{1}{2}x + 1$ is $\frac{1}{2} \cdot 30 + 1$, or 16 ft. Since $16^2 + 30^2 = 1156 = 34^2$, the number 30 checks. When x is 30 ft, then $x + 5$ is 35 ft.

State. The height of the telephone pole is 35 ft.

48. $1200

49. *Familiarize*. From the drawing in the text we see that the length of each half of the roof is 32 ft. Next we need to find the width w of each half of the roof, in ft. Then we will find the area of the roof and determine how many squares of shingles are needed. We make a drawing.

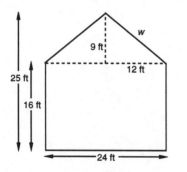

Translate. We use the Pythagorean theorem to find w.

$$a^2 + b^2 = c^2$$

$$9^2 + 12^2 = w^2 \qquad \text{Substituting}$$

Carry out.

$$9^2 + 12^2 = w^2$$

$$81 + 144 = w^2$$

$$225 = w^2$$

$$0 = w^2 - 225$$

$$0 = (w + 15)(w - 15)$$

$$w + 15 = 0 \qquad or \quad w - 15 = 0$$

$$w = -15 \quad or \qquad w = 15$$

Since the width of the roof cannot be negative, we use $w = 15$ ft. The roof consists of two rectangles, each of which has dimensions 15 ft by 32 ft. We find the area of the roof:

$$2 \cdot 32 \cdot 15 = 960$$

Since a square of shingles covers 100 ft^2, we divide 960 by 100 to find the number of squares needed: $960 \div 100 = 9.6$. Assuming it is not possible to buy a fraction of a square, we round up, finding that 10 squares would be needed.

Check. Recheck the calculations.

State. 10 squares of shingles will be needed.

50. 39 cm

51. *Familiarize*. Let $y =$ the ten's digit. Then $y + 4 =$ the one's digit and $10y + y + 4$, or $11y + 4$, represents the number.

Translate.

$$\underbrace{\text{The number}}_{} \text{ plus } \underbrace{\text{the product of the digits}}_{} \text{ is } 58.$$

$$\downarrow \qquad \downarrow \qquad \downarrow \qquad \downarrow \quad \downarrow$$

$$11y + 4 \quad + \quad y(y + 4) \quad = \quad 58$$

Carry out. We solve the equation.

$$11y + 4 + y(y + 4) = 58$$

$$11y + 4 + y^2 + 4y = 58$$

$$y^2 + 15y + 4 = 58$$

$$y^2 + 15y - 54 = 0$$

$$(y + 18)(y - 3) = 0$$

$$y + 18 = 0 \qquad or \quad y - 3 = 0$$

$$y = -18 \quad or \qquad y = 3$$

Check. Since -18 cannot be a digit of the number, we only need to check 3. When $y = 3$, then $y + 4 = 7$ and the number is 37. We see that $37 + 3 \cdot 7 = 37 + 21$, or 58. The result checks.

State. The number is 37.

52. 5 ft

53. *Familiarize*. Let $w =$ the width of the piece of cardboard, in cm. Then $2w =$ the length, in cm. The length and width of the base of the box are $2x - 8$ and $x - 8$, respectively, and its height is 4.

Recall that the formula for the volume of a rectangular solid is given by length $\cdot$ width $\cdot$ height.

Translate.

$$\underbrace{\text{The volume}}_{} \text{ is } \underbrace{616 \text{ cm}^3}_{}.$$

$$\downarrow \qquad \downarrow \qquad \downarrow$$

$$(2w - 8)(w - 8)(4) = \qquad 616$$

Carry out. We solve the equation.

$$(2w - 8)(w - 8)(4) = 616$$
$$(2w^2 - 24w + 64)(4) = 616$$
$$8w^2 - 96 + 256 = 616$$
$$8w^2 - 96w - 360 = 0$$
$$8(w^2 - 12w - 45) = 0$$
$$w^2 - 12w - 45 = 0 \qquad \text{Dividing by 8}$$
$$(w - 15)(w + 3) = 0$$
$$w - 15 = 0 \quad or \quad w + 3 = 0$$
$$w = 15 \quad or \qquad w = -3$$

Check. The width cannot be negative, so we only need to check 15. When $w = 15$, then $2w = 30$ and the dimensions of the box are $30 - 8$ by $15 - 8$ by 4, or 22 by 7 by 4. The volume is $22 \cdot 7 \cdot 4$, or 616.

State. The original dimension of the cardboard are 15 cm by 30 cm.

54. 7 m

55. *Familiarize*. We make a drawing. Let $x =$ the depth of the gutter, in inches.

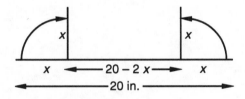

The cross-section has dimensions $20 - 2x$ by x.

Translate.

$$\underbrace{\text{Area of cross-section}}_{(20 - 2x)(x)} \quad \underset{=}{\text{is}} \quad \underbrace{48 \text{ in}^2}_{48}.$$

Carry out.

$$(20 - 2x)(x) = 48$$
$$20x - 2x^2 = 48$$
$$-2x^2 + 20x - 48 = 0$$
$$-2(x^2 - 10x + 24) = 0$$
$$-2(x - 4)(x - 6) = 0$$
$$x - 4 = 0 \quad or \quad x - 6 = 0$$
$$x = 4 \quad or \qquad x = 6$$

Check. If the depth of the gutter is 4 in., then the cross-section has dimension $20 - 2 \cdot 4$, or 12 in. by 4 in., and its area is $12 \cdot 4 = 48 \text{ in}^2$. If the depth of the gutter is 6 in., then the cross-section has dimensions $20 - 2 \cdot 6$, or 8 in. by 6 in., and its area is $8 \cdot 6$, or 48 in^2. Both answers check.

State. The gutter is 4 in. deep or 6 in. deep.

56. 100 cm^2; 225 cm^2

57. The circle has radius x, so its area is πx^2. The length s of a side of the square is the hypotenuse of a right triangle with legs of length x. Thus, we have $s^2 = x^2 + x^2 = 2x^2$. Note that s^2 is the area of the square. The shaded area is the area of the circle less the area of the square. Thus, the shaded area is

$$\pi x^2 - 2x^2, \text{ or } (\pi - 2)x^2.$$

Chapter 6

Rational Expressions and Equations

Exercise Set 6.1

1. $\dfrac{25}{-7x}$

We find the real number(s) that make the denominator 0. To do so we set the denominator equal to 0 and solve for x:

$$-7x = 0$$
$$x = 0$$

The expression is undefined for $x = 0$.

2. 0

3. $\dfrac{t-3}{t+8}$

Set the denominator equal to 0 and solve for t:

$$t + 8 = 0$$
$$t = -8$$

The expression is undefined for $t = -8$.

4. -7

5. $\dfrac{a-4}{3a-12}$

Set the denominator equal to 0 and solve for a:

$$3a - 12 = 0$$
$$3a = 12$$
$$a = 4$$

The expression is undefined for $a = 4$.

6. 3

7. $\dfrac{x^2 - 16}{x^2 - 3x - 28}$

Set the denominator equal to 0 and solve for x:

$$x^2 - 3x - 28 = 0$$
$$(x - 7)(x + 4) = 0$$
$$x - 7 = 0 \quad or \quad x + 4 = 0$$
$$x = 7 \quad or \qquad x = -4$$

The expression is undefined for $x = 7$ and $x = -4$.

8. 2, 5

9. $\dfrac{m^3 - 2m}{m^2 - 25}$

Set the denominator equal to 0 and solve for m:

$$m^2 - 25 = 0$$
$$(m + 5)(m - 5) = 0$$
$$m + 5 = 0 \quad or \quad m - 5 = 0$$
$$m = -5 \quad or \qquad m = 5$$

The expression is undefined for $m = -5$ and $m = 5$.

10. $-7, 7$

11. $\dfrac{60a^2b}{40ab^3}$

$\displaystyle = \frac{3a \cdot 20ab}{2b^2 \cdot 20ab}$ Factoring the numerator and denominator. Note the common factor of $20ab$.

$\displaystyle = \frac{3a}{2b^2} \cdot \frac{20ab}{20ab}$ Rewriting as a product of two rational expressions

$\displaystyle = \frac{3a}{2b^2} \cdot 1 \qquad \frac{20ab}{20ab} = 1$

$\displaystyle = \frac{3a}{2b^2}$ Removing the factor 1

12. $\dfrac{5y}{x^2}$

13. $\dfrac{35x^2y}{14x^3y^5} = \dfrac{5 \cdot 7x^2y}{2xy^4 \cdot 7x^2y}$

$\displaystyle = \frac{5}{2xy^4} \cdot \frac{7x^2y}{7x^2y}$

$\displaystyle = \frac{5}{2xy^4} \cdot 1$

$\displaystyle = \frac{5}{2xy^4}$

14. $\dfrac{2a^2b^5}{3}$

15. $\dfrac{9x + 15}{12x + 20} = \dfrac{3(3x + 5)}{4(3x + 5)}$

$\displaystyle = \frac{3}{4} \cdot \frac{3x + 5}{3x + 5}$

$\displaystyle = \frac{3}{4} \cdot 1$

$\displaystyle = \frac{3}{4}$

16. $\dfrac{7}{5}$

17. $\dfrac{a^2-9}{a^2+4a+3} = \dfrac{(a+3)(a-3)}{(a+3)(a+1)}$

$= \dfrac{a+3}{a+3}\cdot\dfrac{a-3}{a+1}$

$= 1\cdot\dfrac{a-3}{a+1}$

$= \dfrac{a-3}{a+1}$

18. $\dfrac{a+2}{a-3}$

19. $\dfrac{36x^6}{24x^9} = \dfrac{3\cdot 12x^6}{2x^3\cdot 12x^6}$

$= \dfrac{3}{2x^3}\cdot\dfrac{12x^6}{12x^6}$

$= \dfrac{3}{2x^3}\cdot 1$

$= \dfrac{3}{2x^3}$

Check: Let $x = 1$.

$\dfrac{36x^6}{24x^9} = \dfrac{36\cdot 1^6}{24\cdot 1^9} = \dfrac{36}{24} = \dfrac{3}{2}$

$\dfrac{3}{2x^3} = \dfrac{3}{2\cdot 1^3} = \dfrac{3}{2}$

The answer is probably correct.

20. $\dfrac{3a^2}{2}$

21. $\dfrac{-2y+6}{-8y} = \dfrac{-2(y-3)}{-2\cdot 4y}$

$= \dfrac{-2}{-2}\cdot\dfrac{y-3}{4y}$

$= 1\cdot\dfrac{y-3}{4y}$

$= \dfrac{y-3}{4y}$

Check: Let $x = 2$.

$\dfrac{-2y+6}{-8y} = \dfrac{-2\cdot 2+6}{-8\cdot 2} = \dfrac{2}{-16} = -\dfrac{1}{8}$

$\dfrac{y-3}{4y} = \dfrac{2-3}{4\cdot 2} = \dfrac{-1}{8} = -\dfrac{1}{8}$

The answer is probably correct.

22. $\dfrac{2(x-3)}{3x}$

23. $\dfrac{6a^2-3a}{7a^2-7a} = \dfrac{3a(2a-1)}{7a(a-1)}$

$= \dfrac{a}{a}\cdot\dfrac{3(2a-1)}{7(a-1)}$

$= 1\cdot\dfrac{3(2a-1)}{7(a-1)}$

$= \dfrac{3(2a-1)}{7(a-1)}$

Check: Let $a = 2$.

$\dfrac{6a^2-3a}{7a^2-7a} = \dfrac{6\cdot 2^2-3\cdot 2}{7\cdot 2^2-7\cdot 2} = \dfrac{18}{14} = \dfrac{9}{7}$

$\dfrac{3(2a-1)}{7(a-1)} = \dfrac{3(2\cdot 2-1)}{7(2-1)} = \dfrac{3\cdot 3}{7\cdot 1} = \dfrac{9}{7}$

The answer is probably correct.

24. $\dfrac{m+1}{2m+3}$

25. $\dfrac{t^2-16}{t^2+t-20} = \dfrac{(t+4)(t-4)}{(t+5)(t-4)}$

$= \dfrac{t+4}{t+5}\cdot\dfrac{t-4}{t-4}$

$= \dfrac{t+4}{t+5}\cdot 1$

$= \dfrac{t+4}{t+5}$

Check: Let $t = 1$.

$\dfrac{t^2-16}{t^2+t-20} = \dfrac{1^2-16}{1^2+1-20} = \dfrac{-15}{-18} = \dfrac{5}{6}$

$\dfrac{t+4}{t+5} = \dfrac{1+4}{1+5} = \dfrac{5}{6}$

The answer is probably correct.

26. $\dfrac{a-2}{a+3}$

27. $\dfrac{3a^2+9a-12}{6a^2-30a+24} = \dfrac{3(a^2+3a-4)}{6(a^2-5a+4)}$

$= \dfrac{3(a+4)(a-1)}{3\cdot 2(a-4)(a-1)}$

$= \dfrac{3(a-1)}{3(a-1)}\cdot\dfrac{a+4}{2(a-4)}$

$= 1\cdot\dfrac{a+4}{2(a-4)}$

$= \dfrac{a+4}{2(a-4)}$

Check: Let $a = 2$.

$\dfrac{3a^2+9a-12}{6a^2-30a+24} = \dfrac{3\cdot 2^2+9\cdot 2-12}{6\cdot 2^2-30\cdot 2+24} = \dfrac{18}{-12} = -\dfrac{3}{2}$

$\dfrac{a+4}{2(a-4)} = \dfrac{2+4}{2(2-4)} = \dfrac{6}{-4} = -\dfrac{3}{2}$

The answer is probably correct.

28. $\dfrac{t-2}{2(t+4)}$

29. $\dfrac{x^2 + 8x + 16}{x^2 - 16} = \dfrac{(x+4)(x+4)}{(x+4)(x-4)}$

$\qquad\qquad\quad = \dfrac{x+4}{x+4} \cdot \dfrac{x+4}{x-4}$

$\qquad\qquad\quad = 1 \cdot \dfrac{x+4}{x-4}$

$\qquad\qquad\quad = \dfrac{x+4}{x-4}$

Check: Let $x = 1$.

$\dfrac{x^2 + 8x + 16}{x^2 - 16} = \dfrac{1^2 + 8 \cdot 1 + 16}{1^2 - 16} = \dfrac{25}{-15} = -\dfrac{5}{3}$

$\dfrac{x+4}{x-4} = \dfrac{1+4}{1-4} = \dfrac{5}{-3} = -\dfrac{5}{3}$

The answer is probably correct.

30. $\dfrac{x+5}{x-5}$

31. $\dfrac{t^2 - 1}{t+1} = \dfrac{(t+1)(t-1)}{t+1}$

$\qquad\quad = \dfrac{t+1}{t+1} \cdot \dfrac{t-1}{1}$

$\qquad\quad = 1 \cdot \dfrac{t-1}{1}$

$\qquad\quad = t - 1$

Check: Let $t = 2$.

$\dfrac{t^2 - 1}{t+1} = \dfrac{2^2 - 1}{2+1} = \dfrac{3}{3} = 1$

$t - 1 = 2 - 1 = 1$

The answer is probably correct.

32. $a + 1$

33. $\dfrac{y^2 + 4}{y+2}$ cannot be simplified.

Neither the numerator nor the denominator can be factored.

34. $\dfrac{x^2 + 1}{x+1}$

35. $\dfrac{5x^2 - 20}{10x^2 - 40} = \dfrac{5(x^2 - 4)}{10(x^2 - 4)}$

$\qquad\qquad\quad = \dfrac{1 \cdot \cancel{5} \cdot \cancel{(x^2 - 4)}}{2 \cdot \cancel{5} \cdot \cancel{(x^2 - 4)}}$

$\qquad\qquad\quad = \dfrac{1}{2}$

Check: Let $x = 1$.

$\dfrac{5x^2 - 20}{10x^2 - 40} = \dfrac{5 \cdot 1^2 - 20}{10 \cdot 1^2 - 40} = \dfrac{-15}{-30} = \dfrac{1}{2}$

$\dfrac{1}{2} = \dfrac{1}{2}$

The answer is probably correct.

36. $\dfrac{3}{2}$

37. $\dfrac{5y + 5}{y^2 + 7y + 6} = \dfrac{5(y+1)}{(y+1)(y+6)}$

$\qquad\qquad\qquad = \dfrac{y+1}{y+1} \cdot \dfrac{5}{y+6}$

$\qquad\qquad\qquad = 1 \cdot \dfrac{5}{y+6}$

$\qquad\qquad\qquad = \dfrac{5}{y+6}$

Check: Let $x = 1$.

$\dfrac{5y + 5}{y^2 + 7y + 6} = \dfrac{5 \cdot 1 + 5}{1^2 + 7 \cdot 1 + 6} = \dfrac{10}{14} = \dfrac{5}{7}$

$\dfrac{5}{y+6} = \dfrac{5}{1+6} = \dfrac{5}{7}$

The answer is probably correct.

38. $\dfrac{6}{t-3}$

39. $\dfrac{y^2 + 3y - 18}{y^2 + 2y - 15} = \dfrac{(y+6)(y-3)}{(y+5)(y-3)}$

$\qquad\qquad\qquad\quad = \dfrac{y+6}{y+5} \cdot \dfrac{y-3}{y-3}$

$\qquad\qquad\qquad\quad = \dfrac{y+6}{y+5} \cdot 1$

$\qquad\qquad\qquad\quad = \dfrac{y+6}{y+5}$

Check: Let $y = 1$.

$\dfrac{y^2 + 3y - 18}{y^2 + 2y - 15} = \dfrac{1^2 + 3 \cdot 1 - 18}{1^2 + 2 \cdot 1 - 15} = \dfrac{-14}{-12} = \dfrac{7}{6}$

$\dfrac{y+6}{y+5} = \dfrac{1+6}{1+5} = \dfrac{7}{6}$

The answer is probably correct.

40. $\dfrac{a+3}{a+4}$

41. $\dfrac{(a-3)^2}{a^2 - 9} = \dfrac{(a-3)(a-3)}{(a+3)(a-3)}$

$\qquad\qquad\quad = \dfrac{a-3}{a+3} \cdot \dfrac{a-3}{a-3}$

$\qquad\qquad\quad = \dfrac{a-3}{a+3} \cdot 1$

$\qquad\qquad\quad = \dfrac{a-3}{a+3}$

Check: Let $a = 2$.

$\dfrac{(a-3)^2}{a^2 - 9} = \dfrac{(2-3)^2}{2^2 - 9} = \dfrac{1}{-5} = -\dfrac{1}{5}$

$\dfrac{a-3}{a+3} = \dfrac{2-3}{2+3} = \dfrac{-1}{5} = -\dfrac{1}{5}$

The answer is probably correct.

42. $\dfrac{t-2}{t+2}$

43. $\dfrac{x-8}{8-x} = \dfrac{x-8}{-(x-8)}$

$\qquad = \dfrac{1}{-1} \cdot \dfrac{x-8}{x-8}$

$\qquad = \dfrac{1}{-1} \cdot 1$

$\qquad = -1$

Check: Let $x = 2$.

$\dfrac{x-8}{8-x} = \dfrac{2-8}{8-2} = \dfrac{-6}{6} = -1$

The answer is probably correct.

44. -1

45. $\dfrac{7t-14}{2-t} = \dfrac{7(t-2)}{-(t-2)}$

$\qquad = \dfrac{7}{-1} \cdot \dfrac{t-2}{t-2}$

$\qquad = \dfrac{7}{-1} \cdot 1$

$\qquad = -7$

Check: Let $t = 1$.

$\dfrac{7t-14}{2-t} = \dfrac{7 \cdot 1 - 14}{2-1} = \dfrac{-7}{1} = -7$

The answer is probably correct.

46. -4

47. $\dfrac{a-b}{3b-3a} = \dfrac{a-b}{-3(a-b)}$

$\qquad = \dfrac{1}{-3} \cdot \dfrac{a-b}{a-b}$

$\qquad = \dfrac{1}{-3} \cdot 1$

$\qquad = -\dfrac{1}{3}$

Check: Let $a = 2$ and $b = 1$.

$\dfrac{a-b}{3b-3a} = \dfrac{2-1}{3 \cdot 1 - 3 \cdot 2} = \dfrac{1}{-3} = -\dfrac{1}{3}$

The answer is probably correct.

48. $-\dfrac{1}{2}$

49. $\dfrac{3x^2 - 3y^2}{2y^2 - 2x^2} = \dfrac{3(x^2 - y^2)}{2(y^2 - x^2)}$

$\qquad = \dfrac{3(x^2 - y^2)}{2(-1)(x^2 - y^2)}$

$\qquad = \dfrac{3}{2(-1)} \cdot \dfrac{x^2 - y^2}{x^2 - y^2}$

$\qquad = \dfrac{3}{2(-1)} \cdot 1$

$\qquad = -\dfrac{3}{2}$

Check: Let $x = 1$ and $y = 2$.

$\dfrac{3x^2 - 3y^2}{2y^2 - 2x^2} = \dfrac{3 \cdot 1^2 - 3 \cdot 2^2}{2 \cdot 2^2 - 2 \cdot 1^2} = \dfrac{-9}{6} = -\dfrac{3}{2}$

$-\dfrac{3}{2} = -\dfrac{3}{2}$

The answer is probably correct.

50. $-\dfrac{7}{3}$

51. $\dfrac{7s^2 - 28t^2}{28t^2 - 7s^2}$

Note that the numerator and denominator are opposites. Thus, we have an expression divided by its opposite, so the result is -1.

52. -1

53. Writing exercise

54. Writing exercise

55. $-\dfrac{2}{3} \cdot \dfrac{6}{7} = -\dfrac{2 \cdot 6}{3 \cdot 7}$

$\qquad = -\dfrac{2 \cdot 2 \cdot \cancel{3}}{\cancel{3} \cdot 7}$

$\qquad = -\dfrac{4}{7}$

56. $-\dfrac{10}{33}$

57. $\dfrac{5}{8} \div \left(-\dfrac{1}{6}\right) = \dfrac{5}{8} \cdot (-6)$

$\qquad = -\dfrac{5 \cdot 6}{8}$

$\qquad = -\dfrac{5 \cdot \cancel{2} \cdot 3}{\cancel{2} \cdot 4}$

$\qquad = -\dfrac{15}{4}$

58. $-\dfrac{21}{16}$

59. $\dfrac{7}{9} - \dfrac{2}{3} \cdot \dfrac{6}{7} = \dfrac{7}{9} - \dfrac{4}{7} = \dfrac{7}{9} \cdot \dfrac{7}{7} - \dfrac{4}{7} \cdot \dfrac{9}{2} =$

$\dfrac{49}{63} - \dfrac{36}{63} = \dfrac{13}{63}$

60. $\dfrac{5}{48}$

61. Writing exercise

62. Writing exercise

63. $\dfrac{x^4 - y^4}{(y-x)^4} = \dfrac{(x^2 + y^2)(x^2 - y^2)}{[-(x-y)]^4}$

$\qquad = \dfrac{(x^2 + y^2)(x+y)(x-y)}{(-1)^4(x-y)(x-y)^3}$

$\qquad = \dfrac{x^2 + y^2)(x+y)}{(x-y)^3}$

64. $-(2y + x)$

65. $\dfrac{(x-1)(x^4 - 1)(x^2 - 1)}{(x^2 + 1)(x-1)^2(x^4 - 2x^2 + 1)} =$

$\dfrac{(x-1)(x^4 - 1)(x^2 - 1)}{(x^2 + 1)(x-1)^2(x^2 - 1)^2} =$

$\dfrac{(x-1)(x^2 + 1)(x^2 - 1)(x+1)(x-1)}{(x^2 + 1)(x-1)(x-1)(x^2 - 1)(x^2 - 1)} =$

$\dfrac{(x-1)(x^2+1)(x^2-1)(x+1)(x-1)\cdot 1}{(x^2+1)\,(x-1)(x-1)(x^2-1)(x+1)\,(x-1)} =$

$\dfrac{1}{x-1}$

66. $\dfrac{x^3 + 4}{(x^3 + 2)(x^2 + 2)}$

67. $\dfrac{10t^4 - 8t^3 + 15t - 12}{8 - 10t + 12t^2 - 15t^3}$

$= \dfrac{2t^3(5t-4) + 3(5t-4)}{2(4-5t) + 3t^2(4-5t)}$

$= \dfrac{(5t-4)(2t^3 + 3)}{(4-5t)(2 + 3t^2)}$

$= \dfrac{(5t-4)(2t^3 + 3)}{(-1)(5t-4)(2 + 3t^2)}$

$= -\dfrac{2t^3 + 3}{2 + 3t^2}$, or $\dfrac{-2t^3 - 3}{2 + 3t^2}$, or $\dfrac{2t^3 + 3}{-2 - 3t^2}$

68. $\dfrac{(t-1)(t-9)^2}{(t^2 + 9)(t+1)}$

69. $\dfrac{(t+2)^3(t^2 + 2t + 1)(t+1)}{(t+1)^3(t^2 + 4t + 4)(t+2)} =$

$\dfrac{(t+2)^3(t+1)^2(t+1)}{(t+1)^3(t+2)^2(t+2)} = \dfrac{(t+2)^3(t+1)^3}{(t+1)^3(t+2)^3} = 1$

70. $\dfrac{(x-y)^3}{(x+y)^2(x-5y)}$

71. Writing exercise

Exercise Set 6.2

1. $\dfrac{9x}{4} \cdot \dfrac{x-5}{2x+1} = \dfrac{9x(x-5)}{4(2x+1)}$

2. $\dfrac{3x(5x+2)}{4(x-1)}$

3. $\dfrac{a-4}{a+6} \cdot \dfrac{a+2}{a+6} = \dfrac{(a-4)(a+2)}{(a+6)(a+6)}$, or $\dfrac{(a-4)(a+2)}{(a+6)^2}$

4. $\dfrac{(a+3)^2}{(a+6)(a-1)}$

5. $\dfrac{2x+3}{4} \cdot \dfrac{x+1}{x-5} = \dfrac{(2x+3)(x+1)}{4(x-5)}$

6. $\dfrac{4(x+2)}{(3x-4)(5x+6)}$

7. $\dfrac{a-5}{a^2+1} \cdot \dfrac{a+2}{a^2-1} = \dfrac{(a-5)(a+2)}{(a^2+1)(a^2-1)}$

8. $\dfrac{(t+3)^2}{(t^2-2)(t^2-4)}$

9. $\dfrac{x+4}{2+x} \cdot \dfrac{x-1}{x+1} = \dfrac{(x+4)(x-1)}{(2+x)(x+1)}$

10. $\dfrac{(m+4)(2+m)}{(m+8)(m+5)}$

11. $\qquad \dfrac{5a^4}{6a} \cdot \dfrac{2}{a}$

$= \dfrac{5a^4 \cdot 2}{6a \cdot a}$ $\quad$ Multiplying the numerators and the denominators

$= \dfrac{5 \cdot a \cdot a \cdot a \cdot a \cdot 2}{2 \cdot 3 \cdot a \cdot a}$ $\quad$ Factoring the numerator and the denominator

$= \dfrac{5 \cdot a \cdot a \cdot a \cdot a \cdot 2}{2 \cdot 3 \cdot a \cdot a}$ $\quad$ Removing a factor equal to 1

$= \dfrac{5a^2}{3}$ $\quad$ Simplifying

12. $\dfrac{6}{5t^6}$

13. $\dfrac{3c}{d^2} \cdot \dfrac{8d}{6c^3}$

$= \dfrac{3c \cdot 8d}{d^2 \cdot 6c^3}$ Multiplying the numerators and
the denominators

$= \dfrac{3 \cdot c \cdot 2 \cdot 4 \cdot d}{d \cdot d \cdot 3 \cdot 2 \cdot c \cdot c \cdot c}$ Factoring the numerator and the denominator

$= \dfrac{\cancel{3} \cdot \cancel{c} \cdot \cancel{2} \cdot 4 \cdot \cancel{d}}{\cancel{d} \cdot d \cdot \cancel{3} \cdot \cancel{2} \cdot \cancel{c} \cdot c \cdot c}$

$= \dfrac{4}{dc^2}$

14. $\dfrac{6x}{y^2}$

15. $\dfrac{x^2 - 3x - 10}{(x-2)^2} \cdot \dfrac{x-2}{x-5} = \dfrac{(x^2 - 3x - 10)(x-2)}{(x-2)^2(x-5)}$

$\quad = \dfrac{(x-5)(x+2)(x-2)}{(x-2)(x-2)(x-5)}$

$\quad = \dfrac{\cancel{(x-5)}(x+2)\cancel{(x-2)}}{\cancel{(x-2)}(x-2)\cancel{(x-5)}}$

$\quad = \dfrac{x+2}{x-2}$

16. $\dfrac{t-3}{t+2}$

17. $\dfrac{a^2 + 25}{a^2 - 4a + 3} \cdot \dfrac{a-5}{a+5} = \dfrac{(a^2 + 25)(a-5)}{(a^2 - 4a + 3)(a+5)}$

$\quad = \dfrac{(a^2 + 25)(a-5)}{(a-3)(a-1)(a+5)}$

(No simplification is possible.)

18. $\dfrac{(x+3)(x+4)(x+1)}{(x^2+9)(x+9)}$

19. $\dfrac{a^2 - 9}{a^2} \cdot \dfrac{5a}{a^2 + a - 12} = \dfrac{(a+3)(a-3) \cdot 5 \cdot a}{a \cdot a(a+4)(a-3)}$

$\quad = \dfrac{(a+3)\cancel{(a-3)} \cdot 5 \cdot \cancel{a}}{\cancel{a} \cdot a(a+4)\cancel{(a-3)}}$

$\quad = \dfrac{5(a+3)}{a(a+4)}$

20. $\dfrac{x^2(x-1)}{5}$

21. $\dfrac{4a^2}{3a^2 - 12a + 12} \cdot \dfrac{3a-6}{2a}$

$= \dfrac{4a^2(3a-6)}{(3a^2 - 12a + 12)2a}$

$= \dfrac{2 \cdot 2 \cdot a \cdot a \cdot 3 \cdot (a-2)}{3 \cdot (a-2) \cdot (a-2) \cdot 2 \cdot a}$

$= \dfrac{\cancel{2} \cdot 2 \cdot \cancel{a} \cdot a \cdot \cancel{3} \cdot \cancel{(a-2)}}{\cancel{3} \cdot \cancel{(a-2)} \cdot (a-2) \cdot \cancel{2} \cdot \cancel{a}}$

$= \dfrac{2a}{a-2}$

22. $\dfrac{10(v-2)}{v-1}$

23. $\dfrac{t^2 + 2t - 3}{t^2 + 4t - 5} \cdot \dfrac{t^2 - 3t - 10}{t^2 + 5t + 6}$

$= \dfrac{(t^2 + 2t - 3)(t^2 - 3t - 10)}{(t^2 + 4t - 5)(t^2 + 5t + 6)}$

$= \dfrac{(t+3)(t-1)(t-5)(t+2)}{(t+5)(t-1)(t+3)(t+2)}$

$= \dfrac{\cancel{(t+3)}\cancel{(t-1)}(t-5)\cancel{(t+2)}}{(t+5)\cancel{(t-1)}\cancel{(t+3)}\cancel{(t+2)}}$

$= \dfrac{t-5}{t+5}$

24. $\dfrac{x+4}{x-4}$

25. $\dfrac{5a^2 - 180}{10a^2 - 10} \cdot \dfrac{20a + 20}{2a - 12}$

$= \dfrac{(5a^2 - 180)(20a + 20)}{(10a^2 - 10)(2a - 12)}$

$= \dfrac{5(a+6)(a-6)(2)(10)(a+1)}{10(a+1)(a-1)(2)(a-6)}$

$= \dfrac{5(a+6)\cancel{(a-6)}\cancel{(2)}\cancel{(10)}\cancel{(a+1)}}{\cancel{10}\cancel{(a+1)}(a-1)\cancel{(2)}\cancel{(a-6)}}$

$= \dfrac{5(a+6)}{a-1}$

26. $\dfrac{t+7}{4(t-1)}$

27. $\dfrac{x^2 + 4x + 4}{(x-1)^2} \cdot \dfrac{x^2 - 2x + 1}{(x+2)^2} = \dfrac{(x+2)^2(x-1)^2}{(x-1)^2(x+2)^2} = 1$

28. $\dfrac{1}{x+2}$

29. $\dfrac{t^2+8t+16}{(t+4)^3} \cdot \dfrac{(t+2)^3}{t^2+4t+4} = \dfrac{(t+4)^2(t+2)^3}{(t+4)^3(t+2)^2}$

$ = \dfrac{(t+4)^2(t+2)^2(t+2)}{(t+4)^2(t+4)(t+2)^2}$

$ = \dfrac{(t+4)^2(t+2)^2}{(t+4)^2(t+2)^2} \cdot \dfrac{t+2}{t+4}$

$ = 1 \cdot \dfrac{t+2}{t+4}$

$ = \dfrac{t+2}{t+4}$

30. $\dfrac{y-1}{y-2}$

31. The reciprocal of $\dfrac{3x}{7}$ is $\dfrac{7}{3x}$ because $\dfrac{3x}{7} \cdot \dfrac{7}{3x} = 1.$

32. $\dfrac{x^2+4}{3-x}$

33. The reciprocal of $a^3 - 8a$ is $\dfrac{1}{a^3-8a}$ because

$\dfrac{a^3-8a}{1} \cdot \dfrac{1}{a^3-8a} = 1.$

34. $\dfrac{a^2-b^2}{7}$

35. The reciprocal of $\dfrac{x^2+2x-5}{x^2-4x+7}$ is $\dfrac{x^2-4x+7}{x^2+2x-5}$ because

$\dfrac{x^2+2x-5}{x^2-4x+7} \cdot \dfrac{x^2-4x+7}{x^2+2x-5} = 1.$

36. $\dfrac{x^2+7xy-y^2}{x^2-3xy+y^2}$

37. $\dfrac{3}{8} \div \dfrac{5}{2}$

$ = \dfrac{3}{8} \cdot \dfrac{2}{5}$ Multiplying by the reciprocal of the divisor

$ = \dfrac{3 \cdot 2}{8 \cdot 5}$

$ = \dfrac{3 \cdot 2}{2 \cdot 4 \cdot 5}$ Factoring the denominator

$ = \dfrac{2}{2} \cdot \dfrac{3}{4 \cdot 5}$ Factoring the fractional expression

$ = \dfrac{3}{20}$ Simplifying

38. $\dfrac{35}{18}$

39. $\dfrac{x}{4} \div \dfrac{5}{x}$

$ = \dfrac{x}{4} \cdot \dfrac{x}{5}$ Multiplying by the reciprocal of the divisor

$ = \dfrac{x \cdot x}{4 \cdot 5}$

$ = \dfrac{x^2}{20}$

40. $\dfrac{60}{x^2}$

41. $\dfrac{a^5}{b^4} \div \dfrac{a^2}{b} = \dfrac{a^5}{b^4} \cdot \dfrac{b}{a^2}$

$ = \dfrac{a^5 \cdot b}{b^4 \cdot a^2}$

$ = \dfrac{a^2 \cdot a^3 \cdot b}{b \cdot b^3 \cdot a^2}$

$ = \dfrac{a^2 b}{a^2 b} \cdot \dfrac{a^3}{b^3}$

$ = \dfrac{a^3}{b^3}$

42. $\dfrac{x^3}{y}$

43. $\dfrac{y+5}{4} \div \dfrac{y}{2} = \dfrac{y+5}{4} \cdot \dfrac{2}{y}$

$ = \dfrac{(y+5)(2)}{4 \cdot y}$

$ = \dfrac{(y+5)(\cancel{2})}{\cancel{2} \cdot 2y}$

$ = \dfrac{y+5}{2y}$

44. $\dfrac{(a+2)(a+3)}{(a-3)(a-1)}$

45. $\dfrac{4y-8}{y+2} \div \dfrac{y-2}{y^2-4} = \dfrac{4y-8}{y+2} \cdot \dfrac{y^2-4}{y-2}$

$ = \dfrac{(4y-8)(y^2-4)}{(y+2)(y-2)}$

$ = \dfrac{4(\cancel{y-2})(\cancel{y+2})(y-2)}{(\cancel{y+2})(\cancel{y-2})(1)}$

$ = 4(y-2)$

46. $\dfrac{(x-1)^2}{x}$

47. $\dfrac{a}{a-b} \div \dfrac{b}{b-a} = \dfrac{a}{a-b} \cdot \dfrac{b-a}{b}$

$\qquad\qquad = \dfrac{a(b-a)}{(a-b)(b)}$

$\qquad\qquad = \dfrac{a(-1)(a-b)}{(a-b)(b)}$

$\qquad\qquad = \dfrac{-a}{b} = -\dfrac{a}{b}$

48. $-\dfrac{1}{2}$

49. $(y^2-9) \div \dfrac{y^2-2y-3}{y^2+1} = \dfrac{(y^2-9)}{1} \cdot \dfrac{y^2+1}{y^2-2y-3}$

$\qquad\qquad = \dfrac{(y^2-9)(y^2+1)}{y^2-2y-3}$

$\qquad\qquad = \dfrac{(y+3)(y-3)(y^2+1)}{(y-3)(y+1)}$

$\qquad\qquad = \dfrac{(y+3)(y-3)(y^2+1)}{(y-3)(y+1)}$

$\qquad\qquad = \dfrac{(y+3)(y^2+1)}{y+1}$

50. $\dfrac{(x-6)(x+6)}{x-1}$

51. $\dfrac{5x-5}{16} \div \dfrac{x-1}{6} = \dfrac{5x-5}{16} \cdot \dfrac{6}{x-1}$

$\qquad\qquad = \dfrac{(5x-5)\cdot 6}{16(x-1)}$

$\qquad\qquad = \dfrac{5(x-1)\cdot 2\cdot 3}{2\cdot 8(x-1)}$

$\qquad\qquad = \dfrac{5(x-1)\cdot 2\cdot 3}{2\cdot 8(x-1)}$

$\qquad\qquad = \dfrac{15}{8}$

52. $\dfrac{2}{5}$

53. $\dfrac{-6+3x}{5} \div \dfrac{4x-8}{25} = \dfrac{-6+3x}{5} \cdot \dfrac{25}{4x-8}$

$\qquad\qquad = \dfrac{(-6+3x)\cdot 25}{5(4x-8)}$

$\qquad\qquad = \dfrac{3(x-2)\cdot 5\cdot 5}{5\cdot 4(x-2)}$

$\qquad\qquad = \dfrac{3(x-2)\cdot 5\cdot 5}{5\cdot 4(x-2)}$

$\qquad\qquad = \dfrac{15}{4}$

54. 1

55. $\dfrac{a+2}{a-1} \div \dfrac{3a+6}{a-5} = \dfrac{a+2}{a-1} \cdot \dfrac{a-5}{3a+6}$

$\qquad\qquad = \dfrac{(a+2)(a-5)}{(a-1)(3a+6)}$

$\qquad\qquad = \dfrac{(a+2)(a-5)}{(a-1)\cdot 3\cdot (a+2)}$

$\qquad\qquad = \dfrac{(a+2)(a-5)}{(a-1)\cdot 3\cdot (a+2)}$

$\qquad\qquad = \dfrac{a-5}{3(a-1)}$

56. $\dfrac{t+1}{4(t+2)}$

57. $(2x-1) \div \dfrac{2x^2-11x+5}{4x^2-1}$

$\qquad = \dfrac{2x-1}{1} \cdot \dfrac{4x^2-1}{2x^2-11x+5}$

$\qquad = \dfrac{(2x-1)(4x^2-1)}{1\cdot (2x^2-11x+5)}$

$\qquad = \dfrac{(2x-1)(2x+1)(2x-1)}{1\cdot (2x-1)(x-5)}$

$\qquad = \dfrac{(2x-1)(2x+1)(2x-1)}{1\cdot (2x-1)(x-5)}$

$\qquad = \dfrac{(2x-1)(2x+1)}{x-5}$

58. $\dfrac{(a+7)(a+1)}{3a-7}$

59. $\dfrac{x-5}{x+5} \div \dfrac{2x^2-50}{x^2+25} = \dfrac{x-5}{x+5} \cdot \dfrac{x^2+25}{2x^2-50}$

$\qquad\qquad = \dfrac{(x-5)(x^2+25)}{(x+5)(2)(x^2-25)}$

$\qquad\qquad = \dfrac{(x-5)(x^2+25)}{(x+5)(2)(x+5)(x-5)}$

$\qquad\qquad = \dfrac{x^2+25}{2(x+5)^2}$

60. $\dfrac{3(x-3)^2}{x^2+1}$

61. $\dfrac{a^2-10a+25}{a^2+7a+12} \div \dfrac{a^2-a-20}{a^2+6a+9}$

$\qquad = \dfrac{a^2-10a+25}{a^2+7a+12} \cdot \dfrac{a^2+6a+9}{a^2-a-20}$

$\qquad = \dfrac{(a-5)(a-5)(a+3)(a+3)}{(a+3)(a+4)(a-5)(a+4)}$

$\qquad = \dfrac{(a-5)(a+3)}{(a+4)^2}$

62. $\dfrac{(a+1)^2(a-6)}{(a-1)^2(a+4)}$

63. $\quad \dfrac{c^2+10c+21}{c^2-2c-15} \div (c^2+2c-35)$

$\quad = \dfrac{c^2+10c+21}{c^2-2c-25} \cdot \dfrac{1}{c^2+2c-35}$

$\quad = \dfrac{(c^2+10c+21)\cdot 1}{(c^2-2c-15)(c^2+2c-35)}$

$\quad = \dfrac{(c+7)(c+3)}{(c-5)(c+3)(c+7)(c-5)}$

$\quad = \dfrac{(c+7)(c+3)}{(c+7)(c+3)} \cdot \dfrac{1}{(c-5)(c-5)}$

$\quad = \dfrac{1}{(c-5)^2}$

64. $\dfrac{1}{1+2z-z^2}$

65. $\quad \dfrac{x-y}{x^2+2xy+y^2} \div \dfrac{x^2-y^2}{x^2-5xy+4y^2}$

$\quad = \dfrac{x-y}{x^2+2xy+y^2} \cdot \dfrac{x^2-5xy+4y^2}{x^2-y^2}$

$\quad = \dfrac{(x-y)(x-y)(x-4y)}{(x+y)(x+y)(x+y)(x-y)}$

$\quad = \dfrac{(x-y)(x-4y)}{(x+y)^3}$

66. $\dfrac{a+b}{(a-2b)^2}$

67. Writing exercise

68. Writing exercise

69. $\quad \dfrac{3}{4}+\dfrac{5}{6} = \dfrac{3}{4}\cdot\dfrac{3}{3}+\dfrac{5}{6}\cdot\dfrac{2}{2}$

$\quad = \dfrac{9}{12}+\dfrac{10}{12}$

$\quad = \dfrac{19}{12}$

70. $\dfrac{41}{24}$

71. $\quad \dfrac{2}{9}-\dfrac{1}{6} = \dfrac{2}{9}\cdot\dfrac{2}{2}-\dfrac{1}{6}\cdot\dfrac{3}{3}$

$\quad = \dfrac{4}{18}-\dfrac{3}{18}$

$\quad = \dfrac{1}{18}$

72. $-\dfrac{1}{6}$

73. $\dfrac{2}{5}-\left(\dfrac{3}{2}\right)^2 = \dfrac{2}{5}-\dfrac{9}{4} = \dfrac{8}{20}-\dfrac{45}{20} = -\dfrac{37}{20}$

74. $\dfrac{49}{45}$

75. Writing exercise

76. Writing exercise

77. $\dfrac{3x-y}{2x+y} \div \dfrac{3x-y}{2x+y}$

We have the rational expression $\dfrac{3x-y}{2x+y}$ divided by itself. Thus, the result is 1.

78. $\dfrac{a}{(c-3d)(2a+5b)}$

79. $\quad (x-2a) \div \dfrac{a^2x^2-4a^4}{a^2x+2a^3} = \dfrac{x-2a}{1} \cdot \dfrac{a^2x+2a^3}{a^2x^2-4a^4}$

$\quad\quad = \dfrac{(x-2a)(a^2)(x+2a)}{a^2(x+2a)(x-2a)}$

$\quad\quad = 1$

80. $\dfrac{1}{b^3(a-3b)}$

81. $\quad \dfrac{3x^2-2xy-y^2}{x^2-y^2} \div (3x^2+4xy+y^2)^2 =$

$\quad \dfrac{3x^2-2xy-y^2}{x^2-y^2} \cdot \dfrac{1}{(3x^2+4xy+y^2)^2} =$

$\quad \dfrac{(3x+y)(x-y)\cdot 1}{(x+y)(x-y)(3x+y)(3x+y)(x+y)(x+y)} =$

$\quad \dfrac{1}{(x+y)^3(3x+y)}$

82. $\dfrac{x^3y^2}{4}$

83. $\dfrac{a^2-3b}{a^2+2b} \cdot \dfrac{a^2-2b}{a^2+3b} \cdot \dfrac{a^2+2b}{a^2-3b}$

Note that $\dfrac{a^2-3b}{a^2+2b} \cdot \dfrac{a^2+2b}{a^2-3b}$ is the product of reciprocals and thus is equal to 1. Then the product in the original exercise is the remaining factor, $\dfrac{a^2-2b}{a^2+3b}$.

84. $\dfrac{(z+4)^3}{3(z-4)^2}$

85. $\dfrac{x^2 - x + xy - y}{x^2 + 6x - 7} \div \dfrac{x^2 + 2xy + y^2}{4x + 4y}$

$= \dfrac{x^2 - x + xy - y}{x^2 + 6x - 7} \cdot \dfrac{4x + 4y}{x^2 + 2xy + y^2}$

$= \dfrac{x(x - 1) + y(x - 1)}{x^2 + 6x - 7} \cdot \dfrac{4x + 4y}{x^2 + 2xy + y^2}$

$= \dfrac{(x - 1)(x + y) \cdot 4(x + y)}{(x + 7)(x - 1)(x + y)(x + y)}$

$= \dfrac{(x - 1)(x + y)(x + y)}{(x - 1)(x + y)(x + y)} \cdot \dfrac{4}{x + 7}$

$= \dfrac{4}{x + 7}$

86. $\dfrac{x(x^2 + 1)}{3(x + y - 1)}$

87. $\dfrac{(t + 2)^3}{(t + 1)^3} \div \dfrac{t^2 + 4t + 4}{t^2 + 2t + 1} \cdot \dfrac{t + 1}{t + 2}$

$= \dfrac{(t + 2)^3}{(t + 1)^3} \cdot \dfrac{t^2 + 2t + 1}{t^2 + 4t + 4} \cdot \dfrac{t + 1}{t + 2}$

$= \dfrac{(t + 2)(t + 2)(t + 2)(t + 1)(t + 1)(t + 1)}{(t + 1)(t + 1)(t + 1)(t + 2)(t + 2)(t + 2)}$

$= 1$

88. $\dfrac{3(y + 2)^3}{y(y - 1)}$

89. $\dfrac{6y - 4x}{(2x + 3y)^2} \cdot \dfrac{2x - 3y}{x^2 - 9y^2} \div \dfrac{4x^2 - 12xy + 9y^2}{9y^2 + 12xy + 4x^2}$

$= \dfrac{6y - 4x}{(2x + 3y)^2} \cdot \dfrac{2x - 3y}{x^2 - 9y^2} \cdot \dfrac{9y^2 + 12xy + 4x^2}{4x^2 - 12xy + 9y^2}$

$= \dfrac{2(3y - 2x)(2x - 3y)(3y + 2x)(3y + 2x)}{(2x + 3y)(2x + 3y)(x + 3y)(x - 3y)(2x - 3y)(2x - 3y)}$

$= \dfrac{2(-1)(2x - 3y)(2x - 3y)(2x + 3y)(2x + 3y)}{(2x + 3y)(2x + 3y)(x + 3y)(x - 3y)(2x - 3y)(2x - 3y)}$

$= \dfrac{-2}{(x + 3y)(x - 3y)}, \text{ or } -\dfrac{2}{(x + 3y)(x - 3y)}$

90. $\dfrac{a - 3b}{c}$

Exercise Set 6.3

1. $\dfrac{3}{x} + \dfrac{9}{x} = \dfrac{12}{x}$ Adding numerators

2. $\dfrac{13}{a^2}$

3. $\dfrac{x}{15} + \dfrac{2x + 5}{15} = \dfrac{3x + 5}{15}$ Adding numerators

4. $\dfrac{4a - 4}{7}$

5. $\dfrac{4}{a + 3} + \dfrac{5}{a + 3} = \dfrac{9}{a + 3}$

6. $\dfrac{13}{x + 2}$

7. $\dfrac{9}{a + 2} - \dfrac{3}{a + 2} = \dfrac{6}{a + 2}$ Subtracting numerators

8. $\dfrac{6}{x + 7}$

9. $\dfrac{3y + 8}{2y} - \dfrac{y + 1}{2y}$

$= \dfrac{3y + 8 - (y + 1)}{2y}$

$= \dfrac{3y + 8 - y - 1}{2y}$ Removing parentheses

$= \dfrac{2y + 7}{2y}$

10. $\dfrac{t + 4}{4t}$

11. $\dfrac{7x + 8}{x + 1} + \dfrac{4x + 3}{x + 1}$

$= \dfrac{11x + 11}{x + 1}$ Adding numerators

$= \dfrac{11(x + 1)}{x + 1}$ Factoring

$= \dfrac{11(x + 1)}{x + 1}$ Removing a factor equal to 1

$= 11$

12. 5

13. $\dfrac{7x + 8}{x + 1} - \dfrac{4x + 3}{x + 1} = \dfrac{7x + 8 - (4x + 3)}{x + 1}$

$= \dfrac{7x + 8 - 4x - 3}{x + 1}$

$= \dfrac{3x + 5}{x + 1}$

14. $\dfrac{a + 6}{a + 4}$

15. $\dfrac{a^2}{a - 4} + \dfrac{a - 20}{a - 4} = \dfrac{a^2 + a - 20}{a - 4}$

$= \dfrac{(a + 5)(a - 4)}{a - 4}$

$= \dfrac{(a + 5)(a - 4)}{a - 4}$

$= a + 5$

16. $x + 2$

17. $\dfrac{x^2}{x-2} - \dfrac{6x-8}{x-2} = \dfrac{x^2 - (6x-8)}{x-2}$

$\qquad\qquad = \dfrac{x^2 - 6x + 8}{x - 2}$

$\qquad\qquad = \dfrac{(x-4)(x-2)}{x-2}$

$\qquad\qquad = \dfrac{(x-4)(\cancel{x-2})}{\cancel{x-2}}$

$\qquad\qquad = x - 4$

18. $a - 5$

19. $\dfrac{t^2 - 5t}{t - 1} + \dfrac{5t - t^2}{t - 1}$

Note that the numerators are opposites, so their sum is 0. Then we have $\dfrac{0}{t-1}$, or 0.

20. $y + 6$

21. $\dfrac{x-4}{x^2 + 5x + 6} + \dfrac{7}{x^2 + 5x + 6} = \dfrac{x+3}{x^2 + 5x + 6}$

$\qquad\qquad = \dfrac{x+3}{(x+3)(x+2)}$

$\qquad\qquad = \dfrac{\cancel{x+3}}{(\cancel{x+3})(x+2)}$

$\qquad\qquad = \dfrac{1}{x+2}$

22. $\dfrac{1}{x-1}$

23. $\dfrac{3a^2 + 14}{a^2 + 5a - 6} - \dfrac{13a}{a^2 + 5a - 6} = \dfrac{3a^2 - 13a + 14}{a^2 + 5a - 6}$

$\qquad\qquad = \dfrac{(3a-7)(a-2)}{(a+6)(a-1)}$

(No simplification is possible.)

24. $\dfrac{2a - 5}{a - 4}$

25. $\dfrac{t^2 - 3t}{t^2 + 6t + 9} + \dfrac{2t - 12}{t^2 + 6t + 9} = \dfrac{t^2 - t - 12}{t^2 + 6t + 9}$

$\qquad\qquad = \dfrac{(t-4)(t+3)}{(t+3)^2}$

$\qquad\qquad = \dfrac{(t-4)(\cancel{t+3})}{(t+3)(\cancel{t+3})}$

$\qquad\qquad = \dfrac{t-4}{t+3}$

26. $\dfrac{y-5}{y+4}$

27. $\dfrac{2x^2 + x}{x^2 - 8x + 12} - \dfrac{x^2 - 2x + 10}{x^2 - 8x + 12}$

$\quad = \dfrac{2x^2 + x - (x^2 - 2x + 10)}{x^2 - 8x + 12}$

$\quad = \dfrac{2x^2 + x - x^2 + 2x - 10}{x^2 - 8x + 12}$

$\quad = \dfrac{x^2 + 3x - 10}{x^2 - 8x + 12}$

$\quad = \dfrac{(x+5)(x-2)}{(x-6)(x-2)}$

$\quad = \dfrac{(x+5)(\cancel{x-2})}{(x-6)(\cancel{x-2})}$

$\quad = \dfrac{x+5}{x-6}$

28. 0

29. $\dfrac{3 - 2x}{x^2 - 6x + 8} + \dfrac{7 - 3x}{x^2 - 6x + 8}$

$\quad = \dfrac{10 - 5x}{x^2 - 6x + 8}$

$\quad = \dfrac{5(2 - x)}{(x-4)(x-2)}$

$\quad = \dfrac{5(-1)(x-2)}{(x-4)(x-2)}$

$\quad = \dfrac{5(-1)(\cancel{x-2})}{(x-4)(\cancel{x-2})}$

$\quad = \dfrac{-5}{x-4}$, or $-\dfrac{5}{x-4}$, or $\dfrac{5}{4-x}$

30. $-\dfrac{5}{t-4}$, or $\dfrac{5}{4-t}$

31. $\dfrac{x-7}{x^2 + 3x - 4} - \dfrac{2x-3}{x^2 + 3x - 4}$

$\quad = \dfrac{x - 7 - (2x - 3)}{x^2 + 3x - 4}$

$\quad = \dfrac{x - 7 - 2x + 3}{x^2 + 3x - 4}$

$\quad = \dfrac{-x - 4}{x^2 + 3x - 4}$

$\quad = \dfrac{-(x+4)}{(x+4)(x-1)}$

$\quad = \dfrac{-1(\cancel{x+4})}{(\cancel{x+4})(x-1)}$

$\quad = \dfrac{-1}{x-1}$, or $-\dfrac{1}{x-1}$, or $\dfrac{1}{1-x}$

32. $-\dfrac{4}{x-1}$, or $\dfrac{4}{1-x}$

33. $15 = 3 \cdot 5$

$27 = 3 \cdot 3 \cdot 3$

LCM $= 3 \cdot 3 \cdot 3 \cdot 5$, or 135

34. 30

35. $8 = 2 \cdot 2 \cdot 2$

$9 = 3 \cdot 3$

LCM $= 2 \cdot 2 \cdot 2 \cdot 3 \cdot 3$, or 72

36. 60

37. $6 = 2 \cdot 3$

$9 = 3 \cdot 3$

$21 = 3 \cdot 7$

LCM $= 2 \cdot 3 \cdot 3 \cdot 7$, or 126

38. 360

39. $12x^2 = 2 \cdot 2 \cdot 3 \cdot x \cdot x$

$6x^3 = 2 \cdot 3 \cdot x \cdot x \cdot x$

LCM $= 2 \cdot 2 \cdot 3 \cdot x \cdot x \cdot x$, or $12x^3$

40. $10t^4$

41. $15a^4b^7 = 3 \cdot 5 \cdot a \cdot a \cdot a \cdot a \cdot b \cdot b \cdot b \cdot b \cdot b \cdot b \cdot b$

$10a^2b^8 = 2 \cdot 5 \cdot a \cdot a \cdot b \cdot b \cdot b \cdot b \cdot b \cdot b \cdot b \cdot b$

LCM $= 2 \cdot 3 \cdot 5 \cdot a \cdot a \cdot a \cdot a \cdot b \cdot b \cdot b \cdot b \cdot b \cdot b \cdot b \cdot b$,

or $30a^4b^8$

42. $18a^5b^7$

43. $2(y - 3) = 2 \cdot (y - 3)$

$6(y - 3) = 2 \cdot 3 \cdot (y - 3)$

LCM $= 2 \cdot 3 \cdot (y - 3)$, or $6(y - 3)$

44. $8(x - 1)$

45. $x^2 - 4 = (x + 2)(x - 2)$

$x^2 + 5x + 6 = (x + 3)(x + 2)$

LCM $= (x + 2)(x - 2)(x + 3)$

46. $(x + 2)(x + 1)(x - 2)$

47. $t^3 + 4t^2 + 4t = t(t^2 + 4t + 4) = t(t + 2)(t + 2)$

$t^2 - 4t = t(t - 4)$

LCM $= t(t + 2)(t + 2)(t - 4) = t(t + 2)^2(t - 4)$

48. $y^2(y + 1)(y - 1)$

49. $10x^2y = 2 \cdot 5 \cdot x \cdot x \cdot y$

$6y^2z = 2 \cdot 3 \cdot y \cdot y \cdot z$

$5xz^3 = 5 \cdot x \cdot z \cdot z \cdot z$

LCM $= 2 \cdot 3 \cdot 5 \cdot x \cdot x \cdot y \cdot y \cdot z \cdot z \cdot z = 30x^2y^2z^3$

50. $24x^3y^5z^2$

51. $a + 1 = a + 1$

$(a - 1)^2 = (a - 1)(a - 1)$

$a^2 - 1 = (a + 1)(a - 1)$

LCM $= (a + 1)(a - 1)(a - 1) = (a + 1)(a - 1)^2$

52. $(x + 3)(x - 3)^2$

53. $m^2 - 5m + 6 = (m - 3)(m - 2)$

$m^2 - 4m + 4 = (m - 2)(m - 2)$

LCM $= (m - 3)(m - 2)(m - 2) = (m - 3)(m - 2)^2$

54. $(2x + 1)(x + 2)(x - 1)$

55. $t - 3, t + 3, (t^2 - 9)^2$

Note that $(t^2 - 9)^2 = [(t + 3)(t - 3)]^2$, so this expression is a multiple of each of the other expressions. Thus, the LCM is $(t^2 - 9)^2$.

56. $(a^2 - 10a + 25)^2$

57. $6x^3 - 24x^2 + 18x = 6x(x^2 - 4x + 3) =$

$2 \cdot 3 \cdot x(x - 1)(x - 3)$

$4x^5 - 24x^4 + 20x^3 = 4x^3(x^2 - 6x + 5) =$

$2 \cdot 2 \cdot x \cdot x \cdot x(x - 1)(x - 5)$

LCM $= 2 \cdot 2 \cdot 3 \cdot x \cdot x \cdot x(x - 1)(x - 3)(x - 5) =$

$12x^3(x - 1)(x - 3)(x - 5)$

58. $18x^3(x - 2)^2(x + 1)$

59. $6x^5 = 2 \cdot 3 \cdot x \cdot x \cdot x \cdot x \cdot x$

$12x^3 = 2 \cdot 2 \cdot 3 \cdot x \cdot x \cdot x$

The LCD is $2 \cdot 2 \cdot 3 \cdot x \cdot x \cdot x \cdot x \cdot x$, or $12x^5$.

The factor of the LCD that is missing from the first denominator is 2. We multiply by 1 using $2/2$:

$$\frac{5}{6x^5} \cdot \frac{2}{2} = \frac{10}{12x^5}$$

The second denominator is missing two factors of x, or x^2. We multiply by 1 using x^2/x^2:

$$\frac{y}{12x^3} \cdot \frac{x^2}{x^2} = \frac{x^2y}{12x^5}$$

60. $\dfrac{3a^3}{10a^6}; \dfrac{2b}{10a^6}$

61. $2a^2b = 2 \cdot a \cdot a \cdot b$

$8ab^2 = 2 \cdot 2 \cdot 2 \cdot a \cdot b \cdot b$

The LCD is $2 \cdot 2 \cdot 2 \cdot a \cdot a \cdot b \cdot b$, or $8a^2b^2$.

We multiply the first expression by $\dfrac{4b}{4b}$ to obtain the LCD:

$$\frac{3}{2a^2b} \cdot \frac{4b}{4b} = \frac{12b}{8a^2b^2}$$

We multiply the second expression by a/a to obtain the LCD:

$$\frac{7}{8ab^2} \cdot \frac{a}{a} = \frac{7a}{8a^2b^2}$$

62. $\dfrac{21y}{9x^4y^3}; \ \dfrac{4x^3}{9x^4y^3}$

63. The LCD is $(x+2)(x-2)(x+3)$. (See Exercise 45.)

$$\frac{2x}{x^2-4} = \frac{2x}{(x+2)(x-2)} \cdot \frac{x+3}{x+3}$$

$$= \frac{(2x)(x+3)}{(x+2)(x-2)(x+3)}$$

$$\frac{4x}{x^2+5x+6} = \frac{4x}{(x+3)(x+2)} \cdot \frac{x-2}{x-2}$$

$$= \frac{4x(x-2)}{(x+3)(x+2)(x-2)}$$

64. $\dfrac{5x(x+8)}{(x+3)(x-3)(x+8)}; \ \dfrac{2x(x-3)}{(x+3)(x+8)(x-3)}$

65. Writing exercise

66. Writing exercise

67. $\dfrac{7}{-9} = -\dfrac{7}{9} = \dfrac{-7}{9}$

68. $\dfrac{-3}{2}, \ \dfrac{3}{-2}$

69. $\dfrac{5}{18} - \dfrac{7}{12} = \dfrac{5}{18} \cdot \dfrac{2}{2} - \dfrac{7}{12} \cdot \dfrac{3}{3}$

$$= \frac{10}{36} - \frac{21}{36}$$

$$= -\frac{11}{36}$$

70. $-\dfrac{7}{60}$

71. The shaded area has dimensions $x-6$ by $x-3$. Then the area is $(x-6)(x-3)$, or $x^2 - 9x + 18$.

72. $s^2 - \pi r^2$

73. Writing exercise

74. Writing exercise

75. $\dfrac{6x-1}{x-1} + \dfrac{3(2x+5)}{x-1} + \dfrac{3(2x-3)}{x-1}$

$$= \frac{6x-1+6x+15+6x-9}{x-1}$$

$$= \frac{18x+5}{x-1}$$

76. $\dfrac{30}{(x-3)(x+4)}$

77. $\dfrac{x^2}{3x^2-5x-2} - \dfrac{2x}{3x+1} \cdot \dfrac{1}{x-2}$

$$= \frac{x^2}{(3x+1)(x-2)} - \frac{2x}{(3x+1)(x-2)}$$

$$= \frac{x^2-2x}{(3x+1)(x-2)}$$

$$= \frac{x(x-2)}{(3x+1)(x-2)}$$

$$= \frac{x}{3x+1}$$

78. 0

79. The smallest number of strands that can be used is the LCM of 10 and 3.

$10 = 2 \cdot 5$

$3 = 3$

$\text{LCM} = 2 \cdot 5 \cdot 3 = 30$

80. 24

81. If the number of strands must also be a multiple of 4, we find the smallest multiple of 30 that is also a multiple of 4.

$1 \cdot 30 = 30$, not a multiple of 4

$2 \cdot 30 = 60 = 15 \cdot 4$, a multiple of 4

The smallest number of strands that can be used is 60.

82. 1440

83. $8x^2 - 8 = 8(x^2-1) = 2 \cdot 2 \cdot 2(x+1)(x-1)$

$6x^2 - 12x + 6 = 6(x^2-2x+1) = 2 \cdot 3(x-1)(x-1)$

$10 - 10x = 10(1-x) = 2 \cdot 5(1-x)$

Note that $x-1$ and $1-x$ and opposites.

$\text{LCM} = 2 \cdot 2 \cdot 2 \cdot 3 \cdot 5(x+1)(x-1)(x-1) = 120(x+1)(x-1)^2$

$\Big($We could also express the LCM as

$120(x+1)(x-1)(1-x)$. It is not necessary to include both a factor and its opposite in the LCM since $\dfrac{a}{-b} = \dfrac{-a}{b} = -\dfrac{a}{b}.\Big)$

84. $(3x + 4)(3x - 4)^2(2x - 3)$

85. The time it takes Kim and Jed to meet again at the starting place is the LCM of the times it takes them to complete one round of the course.

$6 = 2 \cdot 3$

$8 = 2 \cdot 2 \cdot 2$

$\text{LCM} = 2 \cdot 2 \cdot 2 \cdot 3, \text{ or } 24$

It takes 24 min.

86. 7:55 A.M.

87. The number of years after 2002 in which all three appliances will need to be replaced at once is the LCM of the average numbers of years each will last.

$10 = 2 \cdot 5$

$14 = 2 \cdot 7$

$20 = 2 \cdot 2 \cdot 5$

$\text{LCM} = 2 \cdot 2 \cdot 5 \cdot 7 = 140$

All three appliances will need to be replaced 140 years after 2002, or in 2142.

88. Writing exercise

89. Writing exercise

Exercise Set 6.4

1. $\dfrac{3}{x} + \dfrac{7}{x^2} = \dfrac{3}{x} + \dfrac{7}{x \cdot x}$ $\text{LCD} = x \cdot x, \text{ or } x^2$

$\qquad = \dfrac{3}{x} \cdot \dfrac{x}{x} + \dfrac{7}{x \cdot x}$

$\qquad = \dfrac{3x + 7}{x^2}$

2. $\dfrac{5x + 6}{x^2}$

3. $\left. \begin{array}{l} 6r = 2 \cdot 3 \cdot r \\ 8r = 2 \cdot 2 \cdot 2 \cdot r \end{array} \right\} \text{LCD} = 2 \cdot 2 \cdot 2 \cdot 3 \cdot r, \text{ or } 24r$

$\dfrac{1}{6r} - \dfrac{3}{8r} = \dfrac{1}{6r} \cdot \dfrac{4}{4} - \dfrac{3}{8r} \cdot \dfrac{3}{3}$

$\qquad = \dfrac{4 - 9}{24r}$

$\qquad = \dfrac{-5}{24r}, \text{ or } -\dfrac{5}{24r}$

4. $-\dfrac{13}{18t}$

5. $\left. \begin{array}{l} xy^2 = x \cdot y \cdot y \\ x^2 y = x \cdot x \cdot y \end{array} \right\} \text{LCD} = x \cdot x \cdot y \cdot y, \text{ or } x^2 y^2$

$\dfrac{4}{xy^2} + \dfrac{2}{x^2 y} = \dfrac{4}{xy^2} \cdot \dfrac{x}{x} + \dfrac{2}{x^2 y} \cdot \dfrac{y}{y}$

$\qquad = \dfrac{4x + 2y}{x^2 y^2}$

6. $\dfrac{2d^2 + 7c}{c^2 d^3}$

7. $\left. \begin{array}{l} 9t^3 = 3 \cdot 3 \cdot t \cdot t \cdot t \\ 6t^2 = 2 \cdot 3 \cdot t \cdot t \end{array} \right\} \text{LCD} = 2 \cdot 3 \cdot 3 \cdot t \cdot t \cdot t, \text{ or } 18t^3$

$\dfrac{8}{9t^3} - \dfrac{5}{6t^2} = \dfrac{8}{9t^3} \cdot \dfrac{2}{2} - \dfrac{5}{6t^2} \cdot \dfrac{3t}{3t}$

$\qquad = \dfrac{16 - 15t}{18t^3}$

8. $\dfrac{-2xy - 18}{3x^2 y^3}$

9. $\text{LCD} = 24$ (See Example 1.)

$\dfrac{x + 5}{8} + \dfrac{x - 3}{12} = \dfrac{x + 5}{8} \cdot \dfrac{3}{3} + \dfrac{x - 3}{12} \cdot \dfrac{2}{2}$

$\qquad = \dfrac{3(x + 5)}{24} + \dfrac{2(x - 3)}{24}$

$\qquad = \dfrac{3x + 15}{24} + \dfrac{2x - 6}{24}$

$\qquad = \dfrac{5x + 9}{24}$ Adding numerators

10. $\dfrac{5x + 7}{18}$

11. $\left. \begin{array}{l} 2 = 2 \\ 4 = 2 \cdot 2 \end{array} \right\} \text{LCD} = 4$

$\dfrac{a + 2}{2} - \dfrac{a - 4}{4} = \dfrac{a + 2}{2} \cdot \dfrac{2}{2} - \dfrac{a - 4}{4}$

$\qquad = \dfrac{2a + 4}{4} - \dfrac{a - 4}{4}$

$\qquad = \dfrac{2a + 4 - (a - 4)}{4}$

$\qquad = \dfrac{2a + 4 - a + 4}{4}$

$\qquad = \dfrac{a + 8}{4}$

12. $\dfrac{-x - 4}{6}$

13. $\left. \begin{array}{l} 3a^2 = 3 \cdot a \cdot a \\ 9a = 3 \cdot 3 \cdot a \end{array} \right\} \text{LCD} = 3 \cdot 3 \cdot a \cdot a, \text{ or } 9a^2$

$$\frac{2a-1}{3a^2} + \frac{5a+1}{9a} = \frac{2a-1}{3a^2} \cdot \frac{3}{3} + \frac{5a+1}{9a} \cdot \frac{a}{a}$$

$$= \frac{6a-3}{9a^2} + \frac{5a^2+a}{9a^2}$$

$$= \frac{5a^2+7a-3}{9a^2}$$

14. $\dfrac{a^2+16a+16}{16a^2}$

15. $\left.\begin{array}{l} 4x = 4 \cdot x \\ x = x \end{array}\right\}$ LCD $= 4x$

$$\frac{x-1}{4x} - \frac{2x+3}{x} = \frac{x-1}{4x} - \frac{2x+3}{x} \cdot \frac{4}{4}$$

$$= \frac{x-1}{4x} - \frac{8x+12}{4x}$$

$$= \frac{x-1-(8x+12)}{4x}$$

$$= \frac{x-1-8x-12}{4x}$$

$$= \frac{-7x-13}{4x}$$

16. $\dfrac{7z-12}{12z}$

17. $\left.\begin{array}{l} c^2 d = c \cdot c \cdot d \\ cd^2 = c \cdot d \cdot d \end{array}\right\}$ LCD $= c \cdot c \cdot d \cdot d$, or $c^2 d^2$

$$\frac{2c-d}{c^2 d} + \frac{c+d}{cd^2} = \frac{2c-d}{c^2 d} \cdot \frac{d}{d} + \frac{c+d}{cd^2} \cdot \frac{c}{c}$$

$$= \frac{d(2c-d) + c(c+d)}{c^2 d^2}$$

$$= \frac{2cd - d^2 + c^2 + cd}{c^2 d^2}$$

$$= \frac{c^2 + 3cd - d^2}{c^2 d^2}$$

18. $\dfrac{x^2 + 4xy + y^2}{x^2 y^2}$

19. $\left.\begin{array}{l} 2x^2 y = 2 \cdot x \cdot x \cdot y \\ xy^2 = x \cdot y \cdot y \end{array}\right\}$ LCD $= 2 \cdot x \cdot x \cdot y \cdot y$, or $2x^2 y^2$

$$\frac{5x+3y}{2x^2 y} - \frac{3x+4y}{xy^2} = \frac{5x+3y}{2x^2 y} \cdot \frac{y}{y} - \frac{3x+4y}{xy^2} \cdot \frac{2x}{2x}$$

$$= \frac{5xy + 3y^2}{2x^2 y^2} - \frac{6x^2 + 8xy}{2x^2 y^2}$$

$$= \frac{5xy + 3y^2 - (6x^2 + 8xy)}{2x^2 y^2}$$

$$= \frac{5xy + 3y^2 - 6x^2 - 8xy}{2x^2 y^2}$$

$$= \frac{3y^2 - 3xy - 6x^2}{2x^2 y^2}$$

(Although $3y^2 - 3xy - 6x^2$ can be factored, doing so will not enable us to simplify the result further.)

20. $\dfrac{4x^2 - 13xt + 9t^2}{3x^2 t^2}$

21. The denominators cannot be factored, so the LCD is their product, $(x-1)(x+1)$.

$$\frac{5}{x-1} + \frac{5}{x+1} = \frac{5}{x-1} \cdot \frac{x+1}{x+1} + \frac{5}{x+1} \cdot \frac{x-1}{x-1}$$

$$= \frac{5(x+1) + 5(x-1)}{(x-1)(x+1)}$$

$$= \frac{5x+5+5x-5}{(x-1)(x+1)}$$

$$= \frac{10x}{(x-1)(x+1)}$$

22. $\dfrac{6x}{(x-2)(x+2)}$

23. The denominators cannot be factored, so the LCD is their product, $(z-1)(z+1)$.

$$\frac{4}{z-1} - \frac{2}{z+1} = \frac{4}{z-1} \cdot \frac{z+1}{z+1} - \frac{2}{z+1} \cdot \frac{z-1}{z-1}$$

$$= \frac{4z+4}{(z-1)(z+1)} - \frac{2z-2}{(z-1)(z+1)}$$

$$= \frac{4z+4-(2z-2)}{(z-1)(z+1)}$$

$$= \frac{4z+4-2z+2}{(z-1)(z+1)}$$

$$= \frac{2z+6}{(z-1)(z+1)}$$

(Although $2z + 6$ can be factored, doing so will not enable us to simplify the result further.)

24. $\dfrac{2x-40}{(x+5)(x-5)}$

25. $\left.\begin{array}{l} x+5 = x+5 \\ 4x = 4 \cdot x \end{array}\right\}$ LCD $= 4x(x+5)$

$$\frac{2}{x+5} + \frac{3}{4x} = \frac{2}{x+5} \cdot \frac{4x}{4x} + \frac{3}{4x} \cdot \frac{x+5}{x+5}$$

$$= \frac{2 \cdot 4x + 3(x+5)}{4x(x+5)}$$

$$= \frac{8x + 3x + 15}{4x(x+5)}$$

$$= \frac{11x + 15}{4x(x+5)}$$

26. $\dfrac{11x+2}{3x(x+1)}$

27. $3t^2 - 15t = 3t(t-5)$
$2t - 10 = 2(t-5)$ $\Big\}$ LCD $= 6t(t-5)$

$$\frac{8}{3t(t-5)} - \frac{3}{2(t-5)}$$
$$= \frac{8}{3t(t-5)} \cdot \frac{2}{2} - \frac{3}{2(t-5)} \cdot \frac{3t}{3t}$$
$$= \frac{16}{6t(t-5)} - \frac{9t}{6t(t-5)}$$
$$= \frac{16 - 9t}{6t(t-5)}$$

28. $\dfrac{3 - 5t}{2t(t-1)}$

29. $\dfrac{4x}{x^2 - 25} + \dfrac{x}{x+5}$

$$= \frac{4x}{(x+5)(x-5)} + \frac{x}{x+5} \quad \text{LCD} = (x+5)(x-5)$$
$$= \frac{4x + x(x-5)}{(x+5)(x-5)}$$
$$= \frac{4x + x^2 - 5x}{(x+5)(x-5)}$$
$$= \frac{x^2 - x}{(x+5)(x-5)}$$

(Although $x^2 - x$ can be factored, doing so will not enable us to simplify the result further.)

30. $\dfrac{x^2 + 6x}{(x+4)(x-4)}$

31. $\dfrac{t}{t-3} - \dfrac{5}{4t-12}$

$$= \frac{t}{t-3} - \frac{5}{4(t-3)} \quad \text{LCD} = 4(t-3)$$
$$= \frac{t}{t-3} \cdot \frac{4}{4} - \frac{5}{4(t-3)}$$
$$= \frac{4t - 5}{4(t-3)}$$

32. $\dfrac{16}{3(z+4)}$

33. $\dfrac{2}{x+3} + \dfrac{4}{(x+3)^2} \quad \text{LCD} = (x+3)^2$

$$= \frac{2}{x+3} \cdot \frac{x+3}{x+3} + \frac{4}{(x+3)^2}$$
$$= \frac{2(x+3) + 4}{(x+3)^2}$$
$$= \frac{2x + 6 + 4}{(x+3)^2}$$
$$= \frac{2x + 10}{(x+3)^2}$$

(Although $2x + 10$ can be factored, doing so will not enable us to simplify the result further.)

34. $\dfrac{3x - 1}{(x-1)^2}$

35. $\dfrac{3}{x+2} - \dfrac{8}{x^2 - 4}$

$$= \frac{3}{x+2} - \frac{8}{(x+2)(x-2)} \quad \text{LCD} = (x+2)(x-2)$$
$$= \frac{3}{x+2} \cdot \frac{x-2}{x-2} - \frac{8}{(x+2)(x-2)}$$
$$= \frac{3(x-2) - 8}{(x+2)(x-2)}$$
$$= \frac{3x - 6 - 8}{(x+2)(x-2)}$$
$$= \frac{3x - 14}{(x+2)(x-2)}$$

36. $\dfrac{-t - 9}{(t+3)(t-3)}$

37. $\dfrac{3a}{4a - 20} + \dfrac{9a}{6a - 30}$

$$= \frac{3a}{2 \cdot 2(a-5)} + \frac{9a}{2 \cdot 3(a-5)}$$
$$\text{LCD} = 2 \cdot 2 \cdot 3(a-5)$$
$$= \frac{3a}{2 \cdot 2(a-5)} \cdot \frac{3}{3} + \frac{9a}{2 \cdot 3(a-5)} \cdot \frac{2}{2}$$
$$= \frac{9a + 18a}{2 \cdot 2 \cdot 3(a-5)}$$
$$= \frac{27a}{2 \cdot 2 \cdot 3(a-5)}$$
$$= \frac{\cancel{3} \cdot 9 \cdot a}{2 \cdot 3 \cdot \cancel{3}(a-5)}$$
$$= \frac{9a}{4(a-5)}$$

38. $\dfrac{11a}{10(a-2)}$

39. $\dfrac{x}{x-5} + \dfrac{x}{5-x} = \dfrac{x}{x-5} + \dfrac{x}{5-x} \cdot \dfrac{-1}{-1}$

$$= \frac{x}{x-5} + \frac{-x}{x-5}$$
$$= 0$$

40. $\dfrac{2x^2 + 8x + 16}{x(x+4)}$

41.

$$\frac{7}{a^2 + a - 2} + \frac{5}{a^2 - 4a + 3}$$

$$= \frac{7}{(a+2)(a-1)} + \frac{5}{(a-3)(a-1)}$$

$$\text{LCD} = (a+2)(a-1)(a-3)$$

$$= \frac{7}{(a+2)(a-1)} \cdot \frac{a-3}{a-3} + \frac{5}{(a-3)(a-1)} \cdot \frac{a+2}{a+2}$$

$$= \frac{7(a-3) + 5(a+2)}{(a+2)(a-1)(a-3)}$$

$$= \frac{7a - 21 + 5a + 10}{(a+2)(a-1)(a-3)}$$

$$= \frac{12a - 11}{(a+2)(a-1)(a-3)}$$

42. $\dfrac{x^2 + 5x + 1}{(x+1)^2(x+4)}$

43.

$$\frac{x}{x^2 + 9x + 20} - \frac{4}{x^2 + 7x + 12}$$

$$= \frac{x}{(x+4)(x+5)} - \frac{4}{(x+3)(x+4)}$$

$$\text{LCD} = (x+3)(x+4)(x+5)$$

$$= \frac{x}{(x+4)(x+5)} \cdot \frac{x+3}{x+3} - \frac{4}{(x+3)(x+4)} \cdot \frac{x+5}{x+5}$$

$$= \frac{x(x+3) - 4(x+5)}{(x+3)(x+4)(x+5)}$$

$$= \frac{x^2 + 3x - 4x - 20}{(x+3)(x+4)(x+5)}$$

$$= \frac{x^2 - x - 20}{(x+3)(x+4)(x+5)}$$

$$= \frac{(x+4)(x-5)}{(x+3)(x+4)(x+5)}$$

$$= \frac{x-5}{(x+3)(x+5)}$$

44. $\dfrac{x-3}{(x+3)(x+1)}$

45.

$$\frac{3z}{z^2 - 4x + 4} + \frac{10}{z^2 + z - 6}$$

$$= \frac{3z}{(z-2)^2} + \frac{10}{(z-2)(z+3)},$$

$$\text{LCD} = (z-2)^2(z+3)$$

$$= \frac{3z}{(z-2)^2} \cdot \frac{z+3}{z+3} + \frac{10}{(z-2)(z+3)} \cdot \frac{z-2}{z-2}$$

$$= \frac{3z(z+3) + 10(z-2)}{(x-2)^2(z+3)}$$

$$= \frac{3z^2 + 9z + 10z - 20}{(z-2)^2(z+3)}$$

$$= \frac{3z^2 + 19z - 20}{(z-2)^2(z+3)}$$

46. $\dfrac{5x + 12}{(x+3)(x-3)(x+2)}$

47. $\dfrac{-5}{x^2 + 17x + 16} - \dfrac{0}{x^2 + 9x + 8}$

Note that $\dfrac{0}{x^2 + 9x + 8} = 0$, so the difference is

$\dfrac{-5}{x^2 + 17x + 16}$.

48. $\dfrac{x^2 + 5x - 8}{(x+7)(x+8)(x+6)}$

49.

$$\frac{2x}{5} - \frac{x-3}{-5} = \frac{2x}{5} - \frac{x-3}{-5} \cdot \frac{-1}{-1}$$

$$= \frac{2x}{5} - \frac{3-x}{5}$$

$$= \frac{2x - (3-x)}{5}$$

$$= \frac{2x - 3 + x}{5}$$

$$= \frac{3x - 3}{5}$$

(Although $3x - 3$ can be factored, doing so will not enable us to simplify the result further.)

50. $\dfrac{4x - 5}{4}$

51.

$$\frac{y^2}{y-3} + \frac{9}{3-y} = \frac{y^2}{y-3} + \frac{9}{3-y} \cdot \frac{-1}{-1}$$

$$= \frac{y^2}{y-3} + \frac{-9}{-3+y}$$

$$= \frac{y^2 - 9}{y-3}$$

$$= \frac{(y+3)(y-3)}{y-3}$$

$$= y + 3$$

52. $t + 2$

53.

$$\frac{b-7}{b^2 - 16} + \frac{7-b}{16 - b^2} = \frac{b-7}{b^2 - 16} + \frac{7-b}{16 - b^2} \cdot \frac{-1}{-1}$$

$$= \frac{b-7}{b^2 - 16} + \frac{b-7}{b^2 - 16}$$

$$= \frac{2b - 14}{b^2 - 16}$$

(Although both $2b - 14$ and $b^2 - 16$ can be factored, doing so will not enable us to simplify the result further.)

54. 0

55.
$$\frac{y+2}{y-7} + \frac{3-y}{49-y^2}$$

$$= \frac{y+2}{y-7} + \frac{3-y}{(7+y)(7-y)}$$

$$= \frac{y+2}{y-7} + \frac{3-y}{(7+y)(7-y)} \cdot \frac{-1}{-1}$$

$$= \frac{y+2}{y-7} + \frac{y-3}{(y+7)(y-7)} \quad \text{LCD} = (y+7)(y-7)$$

$$= \frac{y+2}{y-7} \cdot \frac{y+7}{y+7} + \frac{y-3}{(y+7)(y-7)}$$

$$= \frac{y^2 + 9y + 14 + y - 3}{(y+7)(y-7)}$$

$$= \frac{y^2 + 10y + 11}{(y+7)(y-7)}$$

56. $\dfrac{p^2 + 7p + 1}{(p+5)(p-5)}$

57.
$$\frac{5x}{x^2-9} - \frac{4}{3-x}$$

$$= \frac{5x}{(x+3)(x-3)} - \frac{4}{3-x}$$

$$= \frac{5x}{(x+3)(x-3)} - \frac{4}{3-x} \cdot \frac{-1}{-1}$$

$$= \frac{5x}{(x+3)(x-3)} - \frac{-4}{x-3} \quad \text{LCD} = (x+3)(x-3)$$

$$= \frac{5x}{(x+3)(x-3)} - \frac{-4}{x-3} \cdot \frac{x+3}{x+3}$$

$$= \frac{5x - (-4)(x+3)}{(x+3)(x-3)}$$

$$= \frac{5x + 4x + 12}{(x+3)(x-3)}$$

$$= \frac{9x + 12}{(x+3)(x-3)}$$

(Although $9x + 12$ can be factored, doing so will not enable us to simplify the result further.)

58. $\dfrac{13x+20}{(4+x)(4-x)}$, or $\dfrac{-13x-20}{(x+4)(x-4)}$

59.
$$\frac{3x+2}{3x+6} + \frac{x}{4-x^2}$$

$$= \frac{3x+2}{3(x+2)} + \frac{x}{(2+x)(2-x)}$$

$$\qquad\qquad \text{LCD} = 3(x+2)(2-x)$$

$$= \frac{3x+2}{3(x+2)} \cdot \frac{2-x}{2-x} + \frac{x}{(2+x)(2-x)} \cdot \frac{3}{3}$$

$$= \frac{(3x+2)(2-x) + x \cdot 3}{3(x+2)(2-x)}$$

$$= \frac{-3x^2 + 4x + 4 + 3x}{3(x+2)(2-x)}$$

$$= \frac{-3x^2 + 7x + 4}{3(x+2)(2-x)}, \text{ or}$$

$$\frac{3x^2 - 7x - 4}{3(x+2)(x-2)}$$

60. $\dfrac{-a-2}{(a+1)(a-1)}$, or $\dfrac{a+2}{(1+a)(1-a)}$

61.
$$\frac{4-a^2}{a^2-9} - \frac{a-2}{3-a}$$

$$= \frac{4-a^2}{(a+3)(a-3)} - \frac{a-2}{3-a}$$

$$= \frac{4-a^2}{(a+3)(a-3)} - \frac{a-2}{3-a} \cdot \frac{-1}{-1}$$

$$= \frac{4-a^2}{(a+3)(a-3)} - \frac{2-a}{a-3} \quad \text{LCD} = (a+3)(a-3)$$

$$= \frac{4-a^2}{(a+3)(a-3)} - \frac{2-a}{a-3} \cdot \frac{a+3}{a+3}$$

$$= \frac{4-a^2 - (2a+6-a^2-3a)}{(a+3)(a-3)}$$

$$= \frac{4-a^2 - 2a - 6 + a^2 + 3a}{(a+3)(a-3)}$$

$$= \frac{a-2}{(a+3)(a-3)}$$

62. $\dfrac{10x+6y}{(x+y)(x-y)}$

63.

$$\frac{x-3}{2-x} - \frac{x+3}{x+2} + \frac{x+6}{4-x^2}$$

$$= \frac{x-3}{2-x} - \frac{x+3}{x+2} + \frac{x+6}{(2+x)(2-x)}$$

$$\text{LCD} = (2+x)(2-x)$$

$$= \frac{x-3}{2-x} \cdot \frac{2+x}{2+x} - \frac{x+3}{x+2} \cdot \frac{2-x}{2-x} + \frac{x+6}{(2+x)(2-x)}$$

$$= \frac{(x-3)(2+x) - (x+3)(2-x) + (x+6)}{(2+x)(2-x)}$$

$$= \frac{x^2 - x - 6 - (-x^2 - x + 6) + x + 6}{(2+x)(2-x)}$$

$$= \frac{x^2 - x - 6 + x^2 + x - 6 + x + 6}{(2+x)(2-x)}$$

$$= \frac{2x^2 + x - 6}{(2+x)(2-x)}$$

$$= \frac{(2x-3)(x+2)}{(2+x)(2-x)}$$

$$= \frac{2x-3}{2-x}$$

64. $\dfrac{-2t^2 + 2t + 11}{(t+1)(t-1)}$

65.

$$\frac{x+5}{x+3} + \frac{x+7}{x+2} - \frac{7x+19}{(x+3)(x+2)}$$

$$\text{LCD is } (x+3)(x+2)$$

$$= \frac{x+5}{x+3} \cdot \frac{x+2}{x+2} + \frac{x+7}{x+2} \cdot \frac{x+3}{x+3} - \frac{7x+19}{(x+3)(x+2)}$$

$$= \frac{(x+5)(x+2) + (x+7)(x+3) - (7x+19)}{(x+3)(x+2)}$$

$$= \frac{x^2 + 7x + 10 + x^2 + 10x + 21 - 7x - 19}{(x+3)(x+2)}$$

$$= \frac{2x^2 + 10x + 12}{(x+3)(x+2)}$$

$$= \frac{2(x^2 + 5x + 6)}{(x+3)(x+2)}$$

$$= \frac{2(x+3)(x+2)}{(x+3)(x+2)}$$

$$= 2$$

66. 3

67.

$$\frac{t}{s+t} - \frac{t}{s-t} \qquad \text{LCD} = (s+t)(s-t)$$

$$= \frac{t}{s+t} \cdot \frac{s-t}{s-t} - \frac{t}{s-t} \cdot \frac{s+t}{s+t}$$

$$= \frac{t(s-t) - t(s+t)}{(s+t)(s-t)}$$

$$= \frac{st - t^2 - st - t^2}{(s+t)(s-t)}$$

$$= \frac{-2t^2}{(s+t)(s-t)}$$

68. $\dfrac{a^2 + 2ab - b^2}{(b-a)(b+a)}$

69.

$$\frac{1}{x+y} + \frac{1}{x-y} - \frac{2x}{x^2 - y^2}$$

$$\text{LCD} = (x+y)(x-y)$$

$$= \frac{1}{x+y} \cdot \frac{x-y}{x-y} + \frac{1}{x-y} \cdot \frac{x+y}{x+y} - \frac{2x}{(x+y)(x-y)}$$

$$= \frac{(x-y) + (x+y) - 2x}{(x+y)(x-y)}$$

$$= 0$$

70. $\dfrac{2}{r+s}$

71. Writing exercise

72. Writing exercise

73.

$$-\frac{3}{7} \div \frac{6}{13} = -\frac{3}{7} \cdot \frac{13}{6}$$

$$= -\frac{3 \cdot 13}{7 \cdot 2 \cdot 3}$$

$$= -\frac{13}{14}$$

74. $-\dfrac{5}{9}$

75.

$$\frac{\frac{2}{9}}{\frac{5}{3}} = \frac{2}{9} \div \frac{5}{3}$$

$$= \frac{2}{9} \cdot \frac{3}{5}$$

$$= \frac{2 \cdot 3}{3 \cdot 3 \cdot 5}$$

$$= \frac{2}{15}$$

76. $\dfrac{7}{6}$

77. Graph: $y = -\dfrac{1}{2}x - 5$

Since the equation is in the form $y = mx + b$, we know the y-intercept is $(0, -5)$. We find two other solutions, substituting multiples of 2 for x to avoid fractions.

When $x = -2$, $y = -\dfrac{1}{2}(-2) - 5 = 1 - 5 = -4$.

When $x = -4$, $y = -\dfrac{1}{2}(-4) - 5 = 2 - 5 = -3$.

x	y
0	-5
-2	-4
-4	-3

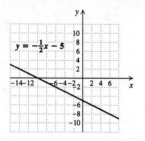

78.

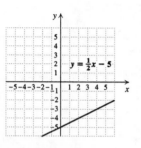

79. Writing exercise

80. Writing exercise

81. $P = 2\left(\dfrac{3}{x+4}\right) + 2\left(\dfrac{2}{x-5}\right)$

$= \dfrac{6}{x+4} + \dfrac{4}{x-5}$ LCD $= (x+4)(x-5)$

$= \dfrac{6}{x+4} \cdot \dfrac{x-5}{x-5} + \dfrac{4}{x-5} \cdot \dfrac{x+4}{x+4}$

$= \dfrac{6x-30+4x+16}{(x+4)(x-5)}$

$= \dfrac{10x-14}{(x+4)(x-5)}$, or $\dfrac{10x-14}{x^2-x-20}$

$A = \left(\dfrac{3}{x+4}\right)\left(\dfrac{2}{x-5}\right) = \dfrac{6}{(x+4)(x-5)}$, or

$\dfrac{6}{x^2-x-20}$

82. $\dfrac{4x^2+18x}{(x+4)(x+5)}$; $\dfrac{x^2}{(x+4)(x+5)}$

83. $\dfrac{2x+11}{x-3} \cdot \dfrac{3}{x+4} + \dfrac{2x+1}{4+x} \cdot \dfrac{3}{3-x}$

$= \dfrac{6x+33}{(x-3)(x+4)} + \dfrac{6x+3}{(4+x)(3-x)}$

$= \dfrac{6x+33}{(x-3)(x+4)} + \dfrac{6x+3}{(4+x)(3-x)} \cdot \dfrac{-1}{-1}$

$= \dfrac{6x+33}{(x-3)(x+4)} + \dfrac{-6x-3}{(x+4)(x-3)}$

$= \dfrac{6x+33-6x-3}{(x-3)(x+4)}$

$= \dfrac{30}{(x-3)(x+4)}$

84. $\dfrac{x}{3x+1}$

85. $\left(\dfrac{x}{x+7} - \dfrac{3}{x+2}\right)\left(\dfrac{x}{x+7} + \dfrac{3}{x+2}\right)$

$= \dfrac{x^2}{(x+7)^2} - \dfrac{9}{(x+2)^2}$ LCD $= (x+7)^2(x+2)^2$

$= \dfrac{x^2}{(x+7)^2} \cdot \dfrac{(x+2)^2}{(x+2)^2} - \dfrac{9}{(x+2)^2} \cdot \dfrac{(x+7)^2}{(x+7)^2}$

$= \dfrac{x^2(x+2)^2 - 9(x+7)^2}{(x+7)^2(x+2)^2}$

$= \dfrac{x^2(x^2+4x+4) - 9(x^2+14x+49)}{(x+7)^2(x+2)^2}$

$= \dfrac{x^4+4x^3+4x^2-9x^2-126x-441}{(x+7)^2(x+2)^2}$

$= \dfrac{x^4+4x^3-5x^2-126x-441}{(x+7)^2(x+2)^2}$

86. $\dfrac{-3xy-3a+6x}{(y-3)^2(a+2x)(a-2x)}$

87. $\dfrac{2x^2+5x-3}{2x^2-9x+9} + \dfrac{x+1}{3-2x} + \dfrac{4x^2+8x+3}{x-3} \cdot \dfrac{x+3}{9-4x^2}$

$= \dfrac{2x^2+5x-3}{(2x-3)(x-3)} + \dfrac{x+1}{3-2x} +$

$\dfrac{(4x^2+8x+3)(x+3)}{(x-3)(3+2x)(3-2x)}$

$= \dfrac{2x^2+5x-3}{(2x-3)(x-3)} \cdot \dfrac{-1}{-1} + \dfrac{x+1}{3-2x} +$

$\dfrac{4x^3+20x^2+27x+9}{(x-3)(3+2x)(3-2x)}$

$= \dfrac{-2x^2-5x+3}{(3-2x)(x-3)} + \dfrac{x+1}{3-2x} + \dfrac{4x^3+20x^2+27x+9}{(x-3)(3+2x)(3-2x)}$

LCD $= (x-3)(3+2x)(3-2x)$

$= \dfrac{-2x^2-5x+3}{(3-2x)(x-3)} \cdot \dfrac{3+2x}{3+2x} + \dfrac{x+1}{3-2x} \cdot \dfrac{(x-3)(3+2x)}{(x-3)(3+2x)} +$

$\dfrac{4x^3+20x^2+27x+9}{(x-3)(3+2x)(3-2x)}$

$= [(-4x^3-16x^2-9x+9+2x^3-x^2-12x-9+$

$4x^3+20x^2+27x+9)]/$

$[(x-3)(3+2x)(3-2x)]$

$= \dfrac{2x^3+3x^2+6x+9}{(x-3)(3+2x)(3-2x)}$

$= \dfrac{x^2(2x+3)+3(2x+3)}{(x-3)(3+2x)(3-2x)}$

$= \dfrac{(2x+3)(x^2+3)}{(x-3)(3+2x)(3-2x)}$

$= \dfrac{x^2+3}{(x-3)(3-2x)}$, or $\dfrac{-x^2-3}{(x-3)(2x-3)}$

88. $\dfrac{5(a^2+2ab-b^2)}{(a-b)(3a+b)(3a-b)}$

89. Answers may vary. $\dfrac{a}{a-b}+\dfrac{3b}{b-a}$

Exercise Set 6.5

---

1. $\dfrac{1+\dfrac{1}{2}}{1+\dfrac{1}{4}}$ LCD is 4

$=\dfrac{1+\dfrac{1}{2}}{1+\dfrac{1}{4}}\cdot\dfrac{4}{4}$ Multiplying by $\dfrac{4}{4}$

$=\dfrac{\left(1+\dfrac{1}{2}\right)4}{\left(1+\dfrac{1}{4}\right)4}$ Multiplying numerator and denominator by 4

$=\dfrac{1\cdot4+\dfrac{1}{2}\cdot4}{1\cdot4+\dfrac{1}{4}\cdot4}$

$=\dfrac{4+2}{4+1}$

$=\dfrac{6}{5}$

2. $\dfrac{7}{6}$

3. $\dfrac{1+\dfrac{1}{3}}{5-\dfrac{5}{27}}$

$=\dfrac{1\cdot\dfrac{3}{3}+\dfrac{1}{3}}{5\cdot\dfrac{27}{27}-\dfrac{5}{27}}$ Getting a common denominator in numerator and in denominator

$=\dfrac{\dfrac{3}{3}+\dfrac{1}{3}}{\dfrac{135}{27}-\dfrac{5}{27}}$

$=\dfrac{\dfrac{4}{3}}{\dfrac{130}{27}}$ Adding in the numerator; subtracting in the denominator

$=\dfrac{4}{3}\cdot\dfrac{27}{130}$ Multiplying by the reciprocal of the divisor

$=\dfrac{2\cdot2\cdot3\cdot9}{3\cdot2\cdot65}$

$=\dfrac{\cancel{2}\cdot2\cdot\cancel{3}\cdot9}{\cancel{3}\cdot\cancel{2}\cdot65}$

$=\dfrac{18}{65}$

4. 8

5. $\dfrac{\dfrac{s}{3}+s}{\dfrac{3}{s}+s}$ LCD is $3s$

$=\dfrac{\dfrac{s}{3}+s}{\dfrac{3}{s}+s}\cdot\dfrac{3s}{3s}$

$=\dfrac{\left(\dfrac{s}{3}+s\right)(3s)}{\left(\dfrac{3}{s}+s\right)(3s)}$

$=\dfrac{\dfrac{s}{3}\cdot3s+s\cdot3s}{\dfrac{3}{s}\cdot3s+s\cdot3s}$

$=\dfrac{s^2+3s^2}{9+3s^2}$

$=\dfrac{4s^2}{9+3s^2}$

6. $\dfrac{1-5x}{1+3x}$

7. $\dfrac{\dfrac{2}{x}}{\dfrac{3}{x}+\dfrac{1}{x^2}}$ LCD is x^2

$=\dfrac{\dfrac{2}{x}}{\dfrac{3}{x}+\dfrac{1}{x^2}}\cdot\dfrac{x^2}{x^2}$

$=\dfrac{\dfrac{2}{x}\cdot x^2}{\left(\dfrac{3}{x}+\dfrac{1}{x^2}\right)x^2}$

$=\dfrac{2x}{\dfrac{3}{x}\cdot x^2+\dfrac{1}{x^2}\cdot x^2}$

$=\dfrac{2x}{3x+1}$

8. $\dfrac{4x-1}{2x}$

9. $\dfrac{\dfrac{2a-5}{3a}}{\dfrac{a-1}{6a}}$

$=\dfrac{2a-5}{3a}\cdot\dfrac{6a}{a-1}$ Multiplying by the reciprocal of the divisor

$=\dfrac{(2a-5)\cdot 2\cdot 3a}{3a\cdot(a-1)}$

$=\dfrac{(2a-5)\cdot 2\cdot 3\!\!\!/a}{3\!\!\!/a\cdot(a-1)}$

$=\dfrac{2(2a-5)}{a-1}$

$=\dfrac{4a-10}{a-1}$

10. $\dfrac{3a+12}{a^2-2a}$

11. $\dfrac{\dfrac{x}{4}-\dfrac{4}{x}}{\dfrac{1}{4}+\dfrac{1}{x}}$ LCD is $4x$

$=\dfrac{\dfrac{x}{4}-\dfrac{4}{x}}{\dfrac{1}{4}+\dfrac{1}{x}}\cdot\dfrac{4x}{4x}$

$=\dfrac{\dfrac{x}{4}\cdot 4x-\dfrac{4}{x}\cdot 4x}{\dfrac{1}{4}\cdot 4x+\dfrac{1}{x}\cdot 4x}$

$=\dfrac{x^2-16}{x+4}$

$=\dfrac{(x+4)(x-4)}{x+4}$

$=\dfrac{(x\!\!\!/+4)(x-4)}{(x\!\!\!/+4)\cdot 1}$

$=x-4$

12. $\dfrac{24+3x}{x^2-24}$

13. $\dfrac{\dfrac{1}{5}+\dfrac{1}{x}}{\dfrac{5+x}{5}}$ LCD is $5x$

$=\dfrac{\dfrac{1}{5}+\dfrac{1}{x}}{\dfrac{5+x}{5}}\cdot\dfrac{5x}{5x}$

$=\dfrac{\dfrac{1}{5}\cdot 5x+\dfrac{1}{x}\cdot 5x}{\left(\dfrac{5+x}{5}\right)(5x)}$

$=\dfrac{x+5}{x(5+x)}$

$=\dfrac{(x+5)\cdot 1}{x(5+x)}$ $(5+x=x+5)$

$=\dfrac{1}{x}$

14. $-\dfrac{1}{a}$

15. $\dfrac{\dfrac{1}{t^2}+1}{\dfrac{1}{t}-1}$ LCD is t^2

$=\dfrac{\dfrac{1}{t^2}+1}{\dfrac{1}{t}-1}\cdot\dfrac{t^2}{t^2}$

$=\dfrac{\dfrac{1}{t^2}\cdot t^2+1\cdot t^2}{\dfrac{1}{t}\cdot t^2-1\cdot t^2}$

$=\dfrac{1+t^2}{t-t^2}$

(Although the denominator can be factored, doing so will not enable us to simplify further.)

16. $\dfrac{2x^2+x}{2x^2-1}$

17. $\dfrac{\dfrac{x^2}{x^2-y^2}}{\dfrac{x}{x+y}}$

$=\dfrac{x^2}{x^2-y^2}\cdot\dfrac{x+y}{x}$ Multiplying by the reciprocal of the divisor

$=\dfrac{x^2(x+y)}{(x^2-y^2)(x)}$

$=\dfrac{x\cdot x\cdot(x+y)}{(x+y)(x-y)(x)}$

$=\dfrac{x\!\!\!/\cdot x\cdot(x\!\!\!+\!\!\!y)}{(x\!\!\!+\!\!\!y)(x-y)(x\!\!\!/)}$

$=\dfrac{x}{x-y}$

18. $\dfrac{a^2 + 3a}{2}$

19. $\dfrac{\dfrac{2}{a} + \dfrac{4}{a^2}}{\dfrac{5}{a^3} - \dfrac{3}{a}}$ LCD is a^3

$= \dfrac{\dfrac{2}{a} + \dfrac{4}{a^2}}{\dfrac{5}{a^3} - \dfrac{3}{a}} \cdot \dfrac{a^3}{a^3}$

$= \dfrac{\dfrac{2}{a} \cdot a^3 + \dfrac{4}{a^2} \cdot a^3}{\dfrac{5}{a^3} \cdot a^3 - \dfrac{3}{a} \cdot a^3}$

$= \dfrac{2a^2 + 4a}{5 - 3a^2}$

(Although the numerator can be factored, doing so will not enable us to simplify further.)

20. $\dfrac{5 - x}{2x^2 + 3x}$

21. $\dfrac{\dfrac{2}{7a^4} - \dfrac{1}{14a}}{\dfrac{3}{5a^2} + \dfrac{2}{15a}} = \dfrac{\dfrac{2}{7a^4} \cdot \dfrac{2}{2} - \dfrac{1}{14a} \cdot \dfrac{a^3}{a^3}}{\dfrac{3}{5a^2} \cdot \dfrac{3}{3} + \dfrac{2}{15a} \cdot \dfrac{a}{a}}$

$= \dfrac{\dfrac{4 - a^3}{14a^4}}{\dfrac{9 + 2a}{15a^2}}$

$= \dfrac{4 - a^3}{14a^4} \cdot \dfrac{15a^2}{9 + 2a}$

$= \dfrac{15 \cdot a^2(4 - a^3)}{14a^2 \cdot a^2(9 + 2a)}$

$= \dfrac{15(4 - a^3)}{14a^2(9 + 2a)}$, or $\dfrac{60 - 15a^3}{126a^2 + 28a^3}$

22. $\dfrac{10 - 3x^2}{12x^2 + 6}$

23. $\dfrac{\dfrac{x}{5y^3} + \dfrac{3}{10y}}{\dfrac{3}{10y} + \dfrac{x}{5y^3}}$

Observe that, by the commutative law of addition, the numerator and denominator are equivalent, so the result is 1.

24. $\dfrac{3a + 8b}{15b^2 - 2}$

25. $\dfrac{\dfrac{5}{ab^4} + \dfrac{2}{a^3b}}{\dfrac{5}{a^3b} - \dfrac{3}{ab}} = \dfrac{\dfrac{5}{ab^4} \cdot \dfrac{a^2}{a^2} + \dfrac{2}{a^3b} \cdot \dfrac{b^3}{b^3}}{\dfrac{5}{a^3b} - \dfrac{3}{ab} \cdot \dfrac{a^2}{a^2}}$

$= \dfrac{\dfrac{5a^2 + 2b^3}{a^3b^4}}{\dfrac{5 - 3a^2}{a^3b}}$

$= \dfrac{5a^2 + 2b^3}{a^3b^4} \cdot \dfrac{a^3b}{5 - 3a^2}$

$= \dfrac{a^3b(5a^2 + 2b^3)}{a^3b \cdot b^3(5 - 3a^2)}$

$= \dfrac{5a^2 + 2b^3}{b^3(5 - 3a^2)}$, or $\dfrac{5a^2 + 2b^3}{5b^3 - 3a^2b^3}$

26. 1

27. $\dfrac{2 - \dfrac{3}{x^2}}{2 + \dfrac{3}{x^4}} = \dfrac{2 - \dfrac{3}{x^2}}{2 + \dfrac{3}{x^4}} \cdot \dfrac{x^4}{x^4}$

$= \dfrac{2 \cdot x^4 - \dfrac{3}{x^2} \cdot x^4}{2 \cdot x^4 + \dfrac{3}{x^4} \cdot x^4}$

$= \dfrac{2x^4 - 3x^2}{2x^4 + 3}$

28. $\dfrac{3a^4 - 2}{2a^4 + 3a}$

29. $\dfrac{t - \dfrac{2}{t}}{t + \dfrac{5}{t}} = \dfrac{t \cdot \dfrac{t}{t} - \dfrac{2}{t}}{t \cdot \dfrac{t}{t} + \dfrac{5}{t}}$

$= \dfrac{\dfrac{t^2 - 2}{t}}{\dfrac{t^2 + 5}{t}}$

$= \dfrac{t^2 - 2}{t} \cdot \dfrac{t}{t^2 + 5}$

$= \dfrac{t(t^2 - 2)}{t(t^2 + 5)}$

$= \dfrac{t^2 - 2}{t^2 + 5}$

30. $\dfrac{x^2 + 3}{x^2 - 2}$

31. $\dfrac{3+\dfrac{4}{ab^3}}{\dfrac{3+a}{a^2b}} = \dfrac{3+\dfrac{4}{ab^3}}{\dfrac{3+a}{a^2b}} \cdot \dfrac{a^2b^3}{a^2b^3}$

$\quad = \dfrac{3 \cdot a^2b^3 + \dfrac{4}{ab^3} \cdot a^2b^3}{\dfrac{3+a}{a^2b} \cdot a^2b^3}$

$\quad = \dfrac{3a^2b^3 + 4a}{b^2(3+a)}, \text{ or } \dfrac{3a^2b^3 + 4a}{3b^2 + ab^2}$

32. $\dfrac{5x^3y + 3x}{3+x}$

33. $\dfrac{\dfrac{x+5}{x^2}}{\dfrac{2}{x} - \dfrac{3}{x^2}} = \dfrac{\dfrac{x+5}{x^2}}{\dfrac{2}{x} \cdot \dfrac{x}{x} - \dfrac{3}{x^2}}$

$\quad = \dfrac{\dfrac{x+5}{x^2}}{\dfrac{2x-3}{x^2}}$

$\quad = \dfrac{x+5}{x^2} \cdot \dfrac{x^2}{2x-3}$

$\quad = \dfrac{x^2(x+5)}{x^2(2x-3)}$

$\quad = \dfrac{x+5}{2x-3}$

34. $\dfrac{a+6}{2a+3a^2}$

35. $\dfrac{x-3+\dfrac{2}{x}}{x-4+\dfrac{3}{x}} = \dfrac{x \cdot \dfrac{x}{x} - 3 \cdot \dfrac{x}{x} + \dfrac{2}{x}}{x \cdot \dfrac{x}{x} - 4 \cdot \dfrac{x}{x} + \dfrac{3}{x}}$

$\quad = \dfrac{\dfrac{x^2 - 3x + 2}{x}}{\dfrac{x^2 - 4x + 3}{x}}$

$\quad = \dfrac{x^2 - 3x + 2}{x} \cdot \dfrac{x}{x^2 - 4x + 3}$

$\quad = \dfrac{(x-2)(x-1)}{x} \cdot \dfrac{x}{(x-3)(x-1)}$

$\quad = \dfrac{x(x-1)}{x(x-1)} \cdot \dfrac{x-2}{x-3}$

$\quad = \dfrac{x-2}{x-3}$

36. $\dfrac{x-1}{x-4}$

37. Writing exercise

38. Writing exercise

39. $3x - 5 + 2(4x - 1) = 12x - 3$

$\quad 3x - 5 + 8x - 2 = 12x - 3$

$\quad\quad\quad 11x - 7 = 12x - 3$

$\quad\quad\quad\quad\quad -7 = x - 3$

$\quad\quad\quad\quad\quad -4 = x$

The solution is -4.

40. -4

41. $\dfrac{3}{4}x - \dfrac{5}{8} = \dfrac{3}{8}x + \dfrac{7}{4}$ LCD is 8

$8\left(\dfrac{3}{4}x - \dfrac{5}{8}\right) = 8\left(\dfrac{3}{8}x + \dfrac{7}{4}\right)$

$8 \cdot \dfrac{3}{4}x - 8 \cdot \dfrac{5}{8} = 8 \cdot \dfrac{3}{8}x + 8 \cdot \dfrac{7}{4}$

$\quad\quad 6x - 5 = 3x + 14$

$\quad\quad 3x - 5 = 14$

$\quad\quad\quad 3x = 19$

$\quad\quad\quad\quad x = \dfrac{19}{3}$

The solution is $\dfrac{19}{3}$.

42. $-\dfrac{14}{27}$

43. $x^2 - 7x - 30 = 0$

$\quad (x - 10)(x + 3) = 0$

$\quad x - 10 = 0 \quad or \quad x + 3 = 0$

$\quad\quad x = 10 \ or \quad\quad x = -3$

The solutions are 10 and -3.

44. $-10, 2$

45. Writing exercise

46. Writing exercise

47.
$$\dfrac{\dfrac{x-5}{x-6}}{\dfrac{x-7}{x-8}}$$

This expression is undefined for any value of x that makes a denominator 0. We see that $x-6=0$ when $x=6$, $x-7=0$ when $x=7$, and $x-8=0$ when $x=8$, so the expression is undefined for the x-values 6, 7, and 8.

48. $-2, -3, -4$

49.
$$\dfrac{\dfrac{2x+3}{5x+4}}{\dfrac{3}{7}-\dfrac{2x}{9}}$$

This expression is undefined for any value of x that makes a denominator 0. First we find the value of x for which $5x+4=0$.

$$5x+4=0$$
$$5x=-4$$
$$x=-\dfrac{4}{5}$$

Then we find the value of x for which $\dfrac{3}{7}-\dfrac{2x}{9}=0$:

$$\dfrac{3}{7}-\dfrac{2x}{9}=0$$
$$63\left(\dfrac{3}{7}-\dfrac{2x}{9}\right)=63\cdot 0$$
$$63\cdot\dfrac{3}{7}-63\cdot\dfrac{2x}{9}=0$$
$$27-14x=0$$
$$27=14x$$
$$\dfrac{27}{14}=x$$

The expression is undefined for the x-values $-\dfrac{4}{5}$ and $\dfrac{27}{14}$.

50. $\dfrac{25}{24}, \dfrac{7}{2}$

51.
$$\dfrac{P\left(1+\dfrac{i}{12}\right)^2}{\dfrac{\left(1+\dfrac{1}{12}\right)^2-1}{\dfrac{i}{12}}}=\dfrac{P\left(1+\dfrac{i}{6}+\dfrac{i^2}{144}\right)}{\dfrac{\left(1+\dfrac{i}{6}+\dfrac{i^2}{144}\right)-1}{\dfrac{i}{12}}}$$

$$=\dfrac{P\left(1+\dfrac{i}{6}+\dfrac{i^2}{144}\right)}{\dfrac{\dfrac{i}{6}+\dfrac{i^2}{144}}{\dfrac{i}{12}}}$$

$$=\dfrac{P\left(1+\dfrac{i}{6}+\dfrac{i^2}{144}\right)}{\left(\dfrac{i}{6}+\dfrac{i^2}{144}\right)\left(\dfrac{12}{i}\right)}$$

$$=\dfrac{P\left(1+\dfrac{i}{6}+\dfrac{i^2}{144}\right)}{2+\dfrac{i}{12}}$$

$$=\dfrac{P\left(1+\dfrac{i}{6}+\dfrac{i^2}{144}\right)}{2+\dfrac{i}{12}}\cdot\dfrac{144}{144}$$

$$=\dfrac{144P\left(1+\dfrac{i}{6}+\dfrac{i^2}{144}\right)}{144\left(2+\dfrac{i}{12}\right)}$$

$$=\dfrac{P(144+24i+i^2)}{288+12i}$$

$$=\dfrac{P(12+i)^2}{12(24+i)}, \text{ or}$$

$$\dfrac{P(i+12)^2}{12(i+24)}$$

52. $\dfrac{(x-1)(3x-2)}{5x-3}$

53.
$$\dfrac{\dfrac{5}{x+2} - \dfrac{3}{x-2}}{\dfrac{x}{x-1} + \dfrac{x}{x+1}} = \dfrac{\dfrac{5}{x+2} \cdot \dfrac{x-2}{x-2} - \dfrac{3}{x-2} \cdot \dfrac{x+2}{x+2}}{\dfrac{x}{x-1} \cdot \dfrac{x+1}{x+1} + \dfrac{x}{x+1} \cdot \dfrac{x-1}{x-1}}$$

$$= \dfrac{\dfrac{5(x-2) - 3(x+2)}{(x+2)(x-2)}}{\dfrac{x(x+1) + x(x-1)}{(x+1)(x-1)}}$$

$$= \dfrac{\dfrac{5x - 10 - 3x - 6}{(x+2)(x-2)}}{\dfrac{x^2 + x + x^2 - x}{(x+1)(x-1)}}$$

$$= \dfrac{\dfrac{2x - 16}{(x+2)(x-2)}}{\dfrac{2x^2}{(x+1)(x-1)}}$$

$$= \dfrac{2x - 16}{(x+2)(x-2)} \cdot \dfrac{(x+1)(x-1)}{2x^2}$$

$$= \dfrac{\cancel{2}(x-8)(x+1)(x-1)}{\cancel{2} \cdot x^2(x+2)(x-2)}$$

$$= \dfrac{(x-8)(x+1)(x-1)}{x^2(x+2)(x-2)}$$

54. $\dfrac{x^2 + 5x + 5}{-x^2 + 10}$

55. $\left[\dfrac{\dfrac{x-1}{x-1} - 1}{\dfrac{x+1}{x-1} + 1} \right]^5$

Consider the numerator of the complex rational expression:
$$\dfrac{x-1}{x-1} - 1 = 1 - 1 = 0$$

Since the denominator, $\dfrac{x+1}{x-1} + 1$ is not equal to 0, the simplified form of the original expression is 0.

56. $\dfrac{3x+2}{2x+1}$

57.
$$\dfrac{\dfrac{z}{1 - \dfrac{z}{2+2z}} - 2z}{\dfrac{2z}{5z - 2} - 3} = \dfrac{\dfrac{z}{\dfrac{2+2z - z}{2+2z}} - 2z}{\dfrac{2z - 15z + 6}{5z - 2}}$$

$$= \dfrac{\dfrac{z}{\dfrac{2+z}{2+2z}} - 2z}{\dfrac{-13z + 6}{5z - 2}}$$

$$= \dfrac{z \cdot \dfrac{2+2z}{2+z} - 2z}{\dfrac{-13z + 6}{5z - 2}}$$

$$= \dfrac{\dfrac{z(2+2z) - 2z(2+z)}{2+z}}{\dfrac{-13z + 6}{5z - 2}}$$

$$= \dfrac{\dfrac{2z + 2z^2 - 4z - 2z^2}{2+z}}{\dfrac{-13z + 6}{5z - 2}}$$

$$= \dfrac{\dfrac{-2z}{2+z}}{\dfrac{-13z + 6}{5z - 2}}$$

$$= \dfrac{-2z}{2+z} \cdot \dfrac{5z - 2}{-13z + 6}$$

$$= \dfrac{-2z(5z - 2)}{(2+z)(-13z + 6)}, \text{ or}$$

$$\dfrac{2z(5z - 2)}{(2+z)(13z - 6)}$$

58. Writing exercise

59.

Exercise Set 6.6

1. Because no variable appears in a denominator, no restrictions exist.
$$\dfrac{5}{8} - \dfrac{4}{5} = \dfrac{x}{20}, \ \text{LCD} = 40$$

$$40\left(\dfrac{5}{8} - \dfrac{4}{5}\right) = 40 \cdot \dfrac{x}{20}$$

$$40 \cdot \dfrac{5}{8} - 40 \cdot \dfrac{4}{5} = 40 \cdot \dfrac{x}{20}$$

$$25 - 32 = 2x$$

$$-7 = 2x$$

$$-\dfrac{7}{2} = x$$

Check:

$$\frac{5}{8} - \frac{4}{5} = \frac{x}{20}$$

$$\frac{5}{8} - \frac{4}{5} \ ? \ \frac{-\frac{7}{2}}{20}$$

$$\frac{25}{40} - \frac{32}{40} \ \bigg| \ -\frac{7}{2} \cdot \frac{1}{20}$$

$$-\frac{7}{40} \ \bigg| \ -\frac{7}{40} \quad \text{TRUE}$$

This checks, so the solution is $-\frac{7}{2}$.

2. $\frac{6}{5}$

3. Note that x cannot be 0.

$$\frac{1}{3} + \frac{5}{6} = \frac{1}{x}, \ \text{LCD} = 6x$$

$$6x\left(\frac{1}{3} + \frac{5}{6}\right) = 6x \cdot \frac{1}{x}$$

$$6x \cdot \frac{1}{3} + 6x \cdot \frac{5}{6} = 6x \cdot \frac{1}{x}$$

$$2x + 5x = 6$$

$$7x = 6$$

$$x = \frac{6}{7}$$

Check:

$$\frac{1}{3} + \frac{5}{6} = \frac{1}{x}$$

$$\frac{1}{3} + \frac{5}{6} \ ? \ \frac{1}{\frac{6}{7}}$$

$$\frac{2}{6} + \frac{5}{6} \ \bigg| \ 1 \cdot \frac{7}{6}$$

$$\frac{7}{6} \ \bigg| \ \frac{7}{6} \quad \text{TRUE}$$

This checks, so the solution is $\frac{6}{7}$.

4. $\frac{40}{29}$

5. Note that t cannot be 0.

$$\frac{1}{6} + \frac{1}{8} = \frac{1}{t}, \ \text{LCD} = 24t$$

$$24t\left(\frac{1}{6} + \frac{1}{8}\right) = 24t \cdot \frac{1}{t}$$

$$24t \cdot \frac{1}{6} + 24t \cdot \frac{1}{8} = 24t \cdot \frac{1}{t}$$

$$4t + 3t = 24$$

$$7t = 24$$

$$t = \frac{24}{7}$$

Check:

$$\frac{1}{6} + \frac{1}{8} = \frac{1}{t}$$

$$\frac{1}{6} + \frac{1}{8} \ ? \ \frac{1}{\frac{24}{7}}$$

$$\frac{4}{24} + \frac{3}{24} \ \bigg| \ 1 \cdot \frac{7}{24}$$

$$\frac{7}{24} \ \bigg| \ \frac{7}{24} \quad \text{TRUE}$$

This checks, so the solution is $\frac{24}{7}$.

6. $\frac{40}{9}$

7. Note that x cannot be 0.

$$x + \frac{5}{x} = -6, \ \text{LCD} = x$$

$$x\left(x + \frac{5}{x}\right) = -6 \cdot x$$

$$x \cdot x + x \cdot \frac{5}{x} = -6 \cdot x$$

$$x^2 + 5 = -6x$$

$$x^2 + 6x + 5 = 0$$

$$(x + 5)(x + 1) = 0$$

$$x + 5 = 0 \quad or \quad x + 1 = 0$$

$$x = -5 \quad or \quad x = -1$$

Check:

$$x + \frac{5}{x} = -6 \qquad\qquad x + \frac{5}{x} = -6$$

$$-5 + \frac{5}{-5} \ ? \ -6 \qquad -1 + \frac{5}{-1} \ ? \ -6$$

$$-5 - 1 \ \bigg| \qquad\qquad -1 - 5 \ \bigg|$$

$$-6 \ \bigg| \ -6 \ \text{TRUE} \qquad -6 \ \bigg| \ -6 \ \text{TRUE}$$

Both of these check, so the two solutions are -5 and -1.

8. $-6, -1$

9. Note that x cannot be 0.

$$\frac{x}{6} - \frac{6}{x} = 0, \ \text{LCD} = 6x$$

$$6x\left(\frac{x}{6} - \frac{6}{x}\right) = 6x \cdot 0$$

$$6x \cdot \frac{x}{6} - 6x \cdot \frac{6}{x} = 6x \cdot 0$$

$$x^2 - 36 = 0$$

$$(x + 6)(x - 6) = 0$$

$x + 6 = 0 \quad or \quad x - 6 = 0$

$x = -6 \quad or \quad x = 6$

Check:

$$\dfrac{x}{6} - \dfrac{6}{x} = 0$$

$$\dfrac{-6}{6} - \dfrac{6}{-6} \;?\; 0$$

$$-1 + 1$$

$$0 \;\Big|\; 0 \quad \text{TRUE}$$

$$\dfrac{x}{6} - \dfrac{6}{x} = 0$$

$$\dfrac{6}{6} - \dfrac{6}{6} \;?\; 0$$

$$1 - 1$$

$$0 \;\Big|\; 0 \quad \text{TRUE}$$

Both of these check, so the two solutions are -6 and 6.

10. $-7, 7$

11. Note that x cannot be 0.

$$\dfrac{5}{x} = \dfrac{6}{x} - \dfrac{1}{3}, \text{ LCD} = 3x$$

$$3x \cdot \dfrac{5}{x} = 3x\left(\dfrac{6}{x} - \dfrac{1}{3}\right)$$

$$3x \cdot \dfrac{5}{x} = 3x \cdot \dfrac{6}{x} - 3x \cdot \dfrac{1}{3}$$

$$15 = 18 - x$$

$$-3 = -x$$

$$3 = x$$

Check:

$$\dfrac{5}{x} = \dfrac{6}{x} - \dfrac{1}{3}$$

$$\dfrac{5}{3} \;?\; \dfrac{6}{3} - \dfrac{1}{3}$$

$$\dfrac{5}{3} \;\Big|\; \dfrac{5}{3} \quad \text{TRUE}$$

This checks, so the solution is 3.

12. 2

13. Note that t cannot be 0.

$$\dfrac{5}{3t} + \dfrac{3}{t} = 1, \text{ LCD} = 3t$$

$$3t\left(\dfrac{5}{3t} + \dfrac{3}{t}\right) = 3t \cdot 1$$

$$3t \cdot \dfrac{5}{3t} + 3t \cdot \dfrac{3}{t} = 3t \cdot 1$$

$$5 + 9 = 3t$$

$$14 = 3t$$

$$\dfrac{14}{3} = t$$

Check:

$$\dfrac{5}{3t} + \dfrac{3}{t} = 1$$

$$\dfrac{5}{3 \cdot \frac{14}{3}} + \dfrac{3}{\frac{14}{3}} \;?\; 1$$

$$\dfrac{5}{14} + \dfrac{9}{14}$$

$$\dfrac{14}{14}$$

$$1 \;\Big|\; 1 \quad \text{TRUE}$$

This checks, so the solution is $\dfrac{14}{3}$.

14. $\dfrac{23}{4}$

15. To avoid division by 0, we must have $x + 3 \neq 0$, or $x \neq -3$.

$$\dfrac{x - 8}{x + 3} = \dfrac{1}{4}, \text{ LCD} = 4(x + 3)$$

$$4(x + 3) \cdot \dfrac{x - 8}{x + 3} = 4(x + 3) \cdot \dfrac{1}{4}$$

$$4(x - 8) = x + 3$$

$$4x - 32 = x + 3$$

$$3x = 35$$

$$x = \dfrac{35}{3}$$

Check:

$$\dfrac{x - 8}{x + 3} = \dfrac{1}{4}$$

$$\dfrac{\frac{35}{3} - 8}{\frac{35}{3} + 3} \;?\; \dfrac{1}{4}$$

$$\dfrac{\frac{35}{3} - \frac{24}{3}}{\frac{35}{3} + \frac{9}{3}}$$

$$\dfrac{\frac{11}{3}}{\frac{44}{3}}$$

$$\dfrac{11}{3} \cdot \dfrac{3}{44}$$

$$\dfrac{1}{4} \;\Big|\; \dfrac{1}{4} \quad \text{TRUE}$$

This checks, so the solution is $\dfrac{35}{3}$.

16. $\dfrac{47}{5}$

17. To avoid division by 0, we must have $x + 1 \neq 0$ and $x - 2 \neq 0$, or $x \neq -1$ and $x \neq 2$.

$$\frac{2}{x+1} = \frac{1}{x-2},$$
$$\text{LCD} = (x+1)(x-2)$$
$$(x+1)(x-2) \cdot \frac{2}{x+1} = (x+1)(x-2) \cdot \frac{1}{x-2}$$
$$2(x-2) = x+1$$
$$2x - 4 = x + 1$$
$$x = 5$$

This checks, so the solution is 5.

18. $-\dfrac{13}{2}$

19. Because no variable appears in a denominator, no restrictions exist.

$$\frac{a}{6} - \frac{a}{10} = \frac{1}{6}, \text{ LCD} = 30$$
$$30\left(\frac{a}{6} - \frac{a}{10}\right) = 30 \cdot \frac{1}{6}$$
$$30 \cdot \frac{a}{6} - 30 \cdot \frac{a}{10} = 30 \cdot \frac{1}{6}$$
$$5a - 3a = 5$$
$$2a = 5$$
$$a = \frac{5}{2}$$

This checks, so the solution is $\dfrac{5}{2}$.

20. 3

21. Because no variable appears in a denominator, no restrictions exist.

$$\frac{x+1}{3} - 1 = \frac{x-1}{2}, \text{ LCD} = 6$$
$$6\left(\frac{x+1}{3} - 1\right) = 6 \cdot \frac{x-1}{2}$$
$$6 \cdot \frac{x+1}{3} - 6 \cdot 1 = 6 \cdot \frac{x-1}{2}$$
$$2(x+1) - 6 = 3(x-1)$$
$$2x + 2 - 6 = 3x - 3$$
$$2x - 4 = 3x - 3$$
$$-1 = x$$

This checks, so the solution is -1.

22. -2

23. To avoid division by 0, we must have $t - 5 \neq 0$, or $t \neq 5$.

$$\frac{4}{t-5} = \frac{t-1}{t-5}, \text{ LCD} = t-5$$
$$(t-5) \cdot \frac{4}{t-5} = (t-5) \cdot \frac{t-1}{t-5}$$
$$4 = t - 1$$
$$5 = t$$

Because of the restriction $t \neq 5$, the number 5 must be rejected as a solution. The equation has no solution.

24. No solution

25. To avoid division by 0, we must have $x + 4 \neq 0$ and $x \neq 0$, or $x \neq -4$ and $x \neq 0$.

$$\frac{3}{x+4} = \frac{5}{x}, \text{ LCD} = x(x+4)$$
$$x(x+4) \cdot \frac{3}{x+4} = x(x+4) \cdot \frac{5}{x}$$
$$3x = 5(x+4)$$
$$3x = 5x + 20$$
$$-2x = 20$$
$$x = -10$$

This checks, so the solution is -10.

26. $-\dfrac{21}{5}$

27. To avoid division by 0, we must have $a - 1 \neq 0$ and $a - 2 \neq 0$, or $a \neq 1$ and $a \neq 2$.

$$\frac{a-4}{a-1} = \frac{a+2}{a-2}, \text{ LCD} = (a-1)(a-2)$$
$$(a-1)(a-2) \cdot \frac{a-4}{a-1} = (a-1)(a-2) \cdot \frac{a+2}{a-2}$$
$$(a-2)(a-4) = (a-1)(a+2)$$
$$a^2 - 6a + 8 = a^2 + a - 2$$
$$-6a + 8 = a - 2$$
$$10 = 7a$$
$$\frac{10}{7} = a$$

This checks, so the solution is $\dfrac{10}{7}$.

28. $\dfrac{5}{3}$

29. To avoid division by 0, we must have $x - 3 \neq 0$ and $x + 3 \neq 0$, or $x \neq 3$ and $x \neq -3$.

$$\frac{4}{x-3} + \frac{2x}{x^2-9} = \frac{1}{x+3},$$
$$\text{LCD} = (x-3)(x+3)$$
$$(x-3)(x+3)\left(\frac{4}{x-3} + \frac{2x}{(x+3)(x-3)}\right) =$$
$$(x-3)(x+3)\cdot\frac{1}{x+3}$$
$$4(x+3) + 2x = x-3$$
$$4x+12+2x = x-3$$
$$6x+12 = x-3$$
$$5x = -15$$
$$x = -3$$

Because of the restriction of $x \neq -3$, we must reject the number -3 as a solution. The equation has no solution.

30. No solution.

31. To avoid division by 0, we must have $y - 3 \neq 0$ and $y + 3 \neq 0$, or $y \neq 3$ and $y \neq -3$.

$$\frac{5}{y-3} - \frac{30}{y^2-9} = 1$$
$$\frac{5}{y-3} - \frac{30}{(y+3)(y-3)} = 1,$$
$$\text{LCD} = (y-3)(y+3)$$
$$(y-3)(y+3)\left(\frac{5}{y-3} - \frac{30}{(y+3)(y-3)}\right) =$$
$$(y-3)(y+3)\cdot 1$$
$$5(y+3) - 30 = (y+3)(y-3)$$
$$5y+15-30 = y^2-9$$
$$0 = y^2 - 5y + 6$$
$$0 = (y-3)(y-2)$$

$$y - 3 = 0 \;\;or\;\; y - 2 = 0$$
$$y = 3 \;\;or\;\;\;\;\;\; y = 2$$

Because of the restriction $y \neq 3$, we must reject the number 3 as a solution. The number 2 checks, so it is the solution.

32. $\dfrac{1}{2}$

33. To avoid division by 0, we must have $8 - a \neq 0$ (or equivalently $a - 8 \neq 0$), or $a \neq 8$.

$$\frac{4}{8-a} = \frac{4-a}{a-8}$$
$$\frac{-1}{-1}\cdot\frac{4}{8-a} = \frac{4-a}{a-8}$$
$$\frac{-4}{a-8} = \frac{4-a}{a-8}, \;\text{LCD} = a-8$$
$$(a-8)\cdot\frac{-4}{a-8} = (a-8)\cdot\frac{4-a}{a-8}$$
$$-4 = 4-a$$
$$-8 = -a$$
$$8 = a$$

Because of the restriction $a \neq 8$, we must reject the number 8 as a solution. The equation has no solution.

34. -13

35. $\dfrac{-2}{x+2} = \dfrac{x}{x+2}$

To avoid division by 0, we must have $x + 2 \neq 0$, or $x \neq -2$. Now observe that the denominators are the same, so the numerators must be the same. Thus, we have $-2 = x$, but because of the restriction $x \neq -2$ this cannot be a solution. The equation has no solution.

36. No solution

37. Writing exercise

38. Writing exercise

39. **Familiarize.** Let $x = $ the first odd integer. Then $x + 2 = $ the next odd integer.

Translate.

$$\underbrace{\text{The sum of two consecutive odd integers}}_{\displaystyle\downarrow \atop \displaystyle x+(x+2)} \;\; \overset{\text{is}}{\underset{=}{\downarrow}} \;\; \overset{276.}{\underset{276}{\downarrow}}$$

Carry out. We solve the equation.
$$x + (x+2) = 276$$
$$2x + 2 = 276$$
$$2x = 274$$
$$x = 137$$

When $x = 137$, then $x + 2 = 137 + 2 = 139$.

Check. The numbers 137 and 139 are consecutive odd integers and $137 + 139 = 276$. These numbers check.

State. The integers are 137 and 139.

40. 14 yd

41. **Familiarize.** Let $b = $ the base of the triangle, in cm. Then $b + 3 = $ the height. Recall that the area of a triangle is given by $\dfrac{1}{2} \times$ base $\times$ height.

Translate.

$$\underbrace{\text{The area of the triangle}}_{\displaystyle\downarrow \atop \displaystyle \frac{1}{2}\cdot b\cdot(b+3)} \;\; \overset{\text{is}}{\underset{=}{\downarrow}} \;\; \overset{54\text{ cm}^2.}{\underset{54}{\downarrow}}$$

Carry out. We solve the equation.

$$\frac{1}{2}b(b+3) = 54$$

$$2 \cdot \frac{1}{2}b(b+3) = 2 \cdot 54$$

$$b(b+3) = 108$$

$$b^2 + 3b = 108$$

$$b^2 + 3b - 108 = 0$$

$$(b-9)(b+12) = 0$$

$$b - 9 = 0 \ \ or \ \ b + 12 = 0$$

$$b = 9 \ \ or \ \ \ \ \ \ b = -12$$

Check. The length of the base cannot be negative so we need to check only 9. If the base is 9 cm, then the height is $9+3$, or 12 cm, and the area is $\frac{1}{2} \cdot 9 \cdot 12$, or 54 cm^2. The answer checks.

State. The base measures 9 cm, and the height measures 12 cm.

42. $-8, -6$; 6, and 8

43. To find the rate, in centimeters per day, we divide the amount of growth by the number of days. From June 9 to June 24 is $24 - 9 = 15$ days.

$$\text{Rate, in cm per day} = \frac{0.9 \text{ cm}}{15 \text{ days}}$$
$$= 0.06 \text{ cm/day}$$
$$= 0.06 \text{ cm per day}$$

44. 0.28 in. per day

45. Writing exercise

46. Writing exercise

47. To avoid division by 0, we must have $x - 3 \neq 0$, or $x \neq 3$.

$$1 + \frac{x-1}{x-3} = \frac{2}{x-3} - x, \ \text{LCD} = x - 3$$

$$(x-3)\left(1 + \frac{x-1}{x-3}\right) = (x-3)\left(\frac{2}{x-3} - x\right)$$

$$(x-3) \cdot 1 + (x-3) \cdot \frac{x-1}{x-3} = (x-3) \cdot \frac{2}{x-3} - (x-3)x$$

$$x - 3 + x - 1 = 2 - x^2 + 3x$$

$$2x - 4 = 2 - x^2 + 3x$$

$$x^2 - x - 6 = 0$$

$$(x-3)(x+2) = 0$$

$$x - 3 = 0 \ \ or \ \ x + 2 = 0$$

$$x = 3 \ \ or \ \ \ \ \ \ x = -2$$

Because of the restriction $x \neq 3$, we must reject the number 3 as a solution. The number -2 checks, so it is the solution.

48. 7

49. To avoid division by 0, we must have $x + 4 \neq 0$ and $x - 1 \neq 0$ and $x + 2 \neq 0$, or $x \neq -4$ and $x \neq 1$ and $x \neq -2$.

$$\frac{x}{x^2 + 3x - 4} + \frac{x}{x^2 + 6x + 8} =$$

$$\frac{2x}{x^2 + x - 2} - \frac{1}{x^2 + 6x + 8}$$

$$\frac{x}{(x+4)(x-1)} + \frac{x}{(x+2)(x+4)} =$$

$$\frac{2x}{(x+2)(x-1)} - \frac{1}{(x+2)(x+4)},$$

$$\text{LCD} = (x+4)(x-1)(x+2)$$

$$(x+4)(x-1)(x+2)\left(\frac{x}{(x+4)(x-1)} + \frac{x}{(x+2)(x+4)}\right) =$$

$$(x+4)(x-1)(x+2)\left(\frac{2x}{(x+2)(x-1)} - \frac{1}{(x+2)(x+4)}\right)$$

$$x(x+2) + x(x-1) = 2x(x+4) - (x-1)$$

$$x^2 + 2x + x^2 - x = 2x^2 + 8x - x + 1$$

$$2x^2 + x = 2x^2 + 7x + 1$$

$$x = 7x + 1$$

$$-6x = 1$$

$$x = -\frac{1}{6}$$

This checks, so the solution is $-\frac{1}{6}$.

50. 3

51. To avoid division by 0, we must have $x + 2 \neq 0$ and $x - 2 \neq 0$, or $x \neq -2$ and $x \neq 2$.

$$\frac{x^2}{x^2 - 4} = \frac{x}{x+2} - \frac{2x}{2-x}$$

$$\frac{x^2}{x^2 - 4} = \frac{x}{x+2} - \frac{2x}{2-x} \cdot \frac{-1}{-1}$$

$$\frac{x^2}{(x+2)(x-2)} = \frac{x}{x+2} - \frac{-2x}{x-2},$$

$$\text{LCD} = (x+2)(x-2)$$

$$(x+2)(x-2) \cdot \frac{x^2}{(x+2)(x-2)} =$$

$$(x+2)(x-2)\left(\frac{x}{x+2} - \frac{-2x}{x-2}\right)$$

$$x^2 = x(x-2) - (-2x)(x+2)$$

$$x^2 = x^2 - 2x + 2x^2 + 4x$$

$$x^2 = 3x^2 + 2x$$

$$0 = 2x^2 + 2x$$

$$0 = 2x(x+1)$$

$$2x = 0 \ \ or \ \ x + 1 = 0$$

$$x = 0 \ \ or \ \ \ \ \ \ x = -1$$

Both of these check, so the solutions are -1 and 0.

52. 4

53. To avoid division by 0, we must have $x - 1 \neq 0$, or $x \neq 1$.

$$\frac{1}{x-1} + x - 5 = \frac{5x-4}{x-1} - 6, \ \text{LCD} = x - 1$$

$$(x-1)\left(\frac{1}{x-1} + x - 5\right) = (x-1)\left(\frac{5x-4}{x-1} - 6\right)$$

$$1 + x(x-1) - 5(x-1) = 5x - 4 - 6(x-1)$$

$$1 + x^2 - x - 5x + 5 = 5x - 4 - 6x + 6$$

$$x^2 - 6x + 6 = -x + 2$$

$$x^2 - 5x + 4 = 0$$

$$(x-1)(x-4) = 0$$

$$x - 1 = 0 \ \ or \ \ x - 4 = 0$$

$$x = 1 \ \ or \ \ \ \ \ \ x = 4$$

Because of the restriction $x \neq 1$, we must reject the number 1 as a solution. The number 4 checks, so it is the solution.

54. -6

55.

56.

Exercise Set 6.7

1. *Familiarize.* Let x = the number.

Translate.

A number	minus	four times its reciprocal	is	3.
↓	↓	↓	↓	↓
x	$-$	$4 \cdot \dfrac{1}{x}$	$=$	3

Carry out.

$$x - \frac{4}{x} = 3$$

$$x\left(x - \frac{4}{x}\right) = x \cdot 3 \ \ \text{Multiplying by the LCD}$$

$$x^2 - 4 = 3x$$

$$x^2 - 3x - 4 = 0$$

$$(x-4)(x+1) = 0$$

$$x - 4 = 0 \ \ or \ \ x + 1 = 0$$

$$x = 4 \ \ or \ \ \ \ \ \ x = -1$$

Check. Four times the reciprocal of 4 is $4 \cdot \frac{1}{4}$, or 1. Since $4 - 1 = 3$, the number 4 is a solution. Four times the reciprocal of -1 is $4(-1)$, or -4. Since $-1 - (-4) = -1 + 4$, or 3, the number -1 is a solution.

State. The solutions are 4 and -1.

2. -1, 5

3. *Familiarize.* Let x = the number.

Translate. We reword the problem.

A number	plus	its reciprocal	is	2.
↓	↓	↓	↓	↓
x	$+$	$\dfrac{1}{x}$	$=$	2

Carry out. We solve the equation.

$$x + \frac{1}{x} = 2$$

$$x\left(x + \frac{1}{x}\right) = x \cdot 2 \ \ \text{Multiplying by the LCD}$$

$$x^2 + 1 = 2x$$

$$x^2 - 2x + 1 = 0$$

$$(x-1)(x-1) = 0$$

$$x - 1 = 0 \ \ or \ \ x - 1 = 0$$

$$x = 1 \ \ or \ \ \ \ \ \ x = 1$$

Check. The reciprocal of 1 is 1. Since $1 + 1 = 2$, the number 1 is a solution.

State. The number is 1.

4. 1, 5

5. *Familiarize.* The job takes Fontella 4 hours working alone and Omar 5 hours working alone. Then in 1 hour Fontella does $\frac{1}{4}$ of the job and Omar does $\frac{1}{5}$ of the job. Working together, they can do $\frac{1}{4} + \frac{1}{5}$, or $\frac{9}{20}$ of the job in 1 hour. In two hours, Fontella does $2\left(\frac{1}{4}\right)$ of the job and Omar does $2\left(\frac{1}{5}\right)$ of the job. Working together they can do $2\left(\frac{1}{4}\right) + 2\left(\frac{1}{5}\right)$, or $\frac{9}{10}$ of the job in 2 hours. In 3 hours they can do $3\left(\frac{1}{4}\right) + 3\left(\frac{1}{5}\right)$, or $1\frac{7}{20}$ of the job which is more of the job then needs to be done. The answer is somewhere between 2 hr and 3 hr.

Translate. If they work together t hours, then Fontella does $t\left(\frac{1}{4}\right)$ of the job and Omar does $t\left(\frac{1}{5}\right)$ of the job. We want some number t such that

$$\left(\frac{1}{4} + \frac{1}{5}\right)t = 1, \ \text{or} \ \frac{9}{20} \cdot t = 1.$$

Carry out. We solve the equation.

$$\frac{9}{20} \cdot t = 1$$

$$\frac{20}{9} \cdot \frac{9}{20} \cdot t = \frac{20}{9} \cdot 1$$

$$t = \frac{20}{9}, \text{ or } 2\frac{2}{9}$$

Check. The check can be done by repeating the computations. We also have a partial check in that we expected from our familiarization step that the answer would be between 2 hr and 3 hr.

State. Working together, it takes them $2\frac{2}{9}$ hr to complete the job.

6. $6\frac{6}{7}$ hr

7. ***Familiarize***. The job takes Vern 45 min working alone and Nina 60 min working alone. Then in 1 minute Vern does $\frac{1}{45}$ of the job and Nina does $\frac{1}{60}$ of the job. Working together, they can do $\frac{1}{45} + \frac{1}{60}$, or $\frac{7}{180}$ of the job in 1 minute. In 20 minutes, Vern does $\frac{20}{45}$ of the job and Nina does $\frac{20}{60}$ of the job. Working together, they can do $\frac{20}{45} + \frac{20}{60}$, or $\frac{7}{9}$ of the job. In 30 minutes, they can do $\frac{30}{45} + \frac{30}{60}$, or $\frac{7}{6}$ of the job which is more of the job than needs to be done. The answer is somewhere between 20 minutes and 30 minutes.

Translate. If they work together t minutes, then Vern does $t\left(\frac{1}{45}\right)$ of the job and Nina does $t\left(\frac{1}{60}\right)$ of the job. We want some number t such that

$$\left(\frac{1}{45} + \frac{1}{60}\right)t = 1, \text{ or } \frac{7}{180} \cdot t = 1.$$

Carry out. We solve the equation.

$$\frac{7}{180} \cdot t = 1$$

$$\frac{180}{7} \cdot \frac{7}{180} \cdot t = \frac{180}{7} \cdot 1$$

$$t = \frac{180}{7}, \text{ or } 25\frac{5}{7}$$

Check. The check can be done by repeating the computations. We also have a partial check in that we expected from our familiarization step that the answer would be between 20 minutes and 30 minutes.

State. It would take them $25\frac{5}{7}$ minutes to complete the job working together.

8. $1\frac{5}{7}$ hr

9. ***Familiarize***. The job takes Kenny Dewitt 8 hours working alone and Betty Wohat 6 hours working alone. Then in 1 hour Kenny does $\frac{1}{8}$ of the job and Betty does $\frac{1}{6}$ of the job. Working together they can do $\frac{1}{8} + \frac{1}{6}$, or $\frac{7}{24}$ of the job in 1 hour. In two hours, Kenny does $2\left(\frac{1}{8}\right)$ of the job and Betty does $2\left(\frac{1}{6}\right)$ of the job. Working together they can and do $2\left(\frac{1}{8}\right) + 2\left(\frac{1}{6}\right)$, or $\frac{7}{12}$ of the job in two hours. In five hours they can do $5\left(\frac{1}{8}\right) + 5\left(\frac{1}{6}\right)$, or $\frac{35}{24}$, or $1\frac{11}{24}$ of the job which is more of the job than needs to be done. The answer is somewhere between 2 hr and 5 hr.

Translate. If they work together t hours, Kenny does $t\left(\frac{1}{8}\right)$ of the job and Betty does $t\left(\frac{1}{6}\right)$ of the job. We want some number t such that

$$\left(\frac{1}{8} + \frac{1}{6}\right)t = 1, \text{ or } \frac{7}{24} \cdot t = 1.$$

Carry out. We solve the equation.

$$\frac{7}{24} \cdot t = 1$$

$$\frac{24}{7} \cdot \frac{7}{24} \cdot t = \frac{24}{7} \cdot 1$$

$$t = \frac{24}{7}, \text{ or } 3\frac{3}{7}$$

Check. The check can be done by repeating the computations. We also have a partial check in that we expected from our familiarization step that the answer would be between 2 hr and 5 hr.

State. Working together, it takes them $3\frac{3}{7}$ hr to complete the job.

10. $20\frac{4}{7}$ hr

11. ***Familiarize***. Let t = the number of minutes it takes Nicole and Glen to weed the garden, working together.

Translate. We use the work principle.

$$\left(\frac{1}{50} + \frac{1}{40}\right)t = 1, \text{ or } \frac{9}{200} \cdot t = 1$$

Carry out. We solve the equation.

$$\frac{9}{200} \cdot t = 1$$

$$\frac{200}{9} \cdot \frac{9}{200} \cdot t = \frac{200}{9} \cdot 1$$

$$t = \frac{200}{9}, \text{ or } 22\frac{2}{9}$$

Check. In $\frac{200}{9}$ min, the portion of the job done is $\frac{1}{50} \cdot \frac{200}{9} + \frac{1}{40} \cdot \frac{200}{9} = \frac{4}{9} + \frac{5}{9} = 1$. The answer checks.

State. It would take $22\frac{2}{9}$ min to weed the garden if Nicole and Glen worked together.

12. $11\frac{1}{9}$ min

13. Familiarize. Let $t =$ the number of minutes it would take the two machines to copy the dissertation, working together.

Translate. We use the work principle.

$$\left(\frac{1}{12} + \frac{1}{20}\right)t = 1, \text{ or } \frac{8}{60} \cdot t = 1, \text{ or } \frac{2}{15} \cdot t = 1$$

Carry out. We solve the equation.

$$\frac{2}{15} \cdot 1 = 1$$

$$\frac{15}{2} \cdot \frac{2}{15} \cdot t = \frac{15}{2} \cdot 1$$

$$t = \frac{15}{2}, \text{ or } 7.5$$

Check. In $\frac{15}{2}$ min, the portion of the job done is $\frac{1}{12} \cdot \frac{15}{2} + \frac{1}{20} \cdot \frac{15}{2} = \frac{5}{8} + \frac{3}{8} = 1$. The answer checks.

State. It would take the two machines 7.5 min to copy the dissertation, working together.

14. $4\frac{4}{9}$ min

15. Familiarize. We complete the table shown in the text.

$$d = r \cdot t$$

	Distance	Speed	Time
Truck	350	r	$\dfrac{350}{r}$
Train	150	$r - 40$	$\dfrac{150}{r-40}$

Translate. Since the times must be the same for both vehicles, we have the equation

$$\frac{350}{r} = \frac{150}{r-40}.$$

Carry out. We first multiply by the LCD, $r(r-40)$.

$$r(r-40) \cdot \frac{350}{r} = r(r-40) \cdot \frac{150}{r-40}$$

$$350(r-40) = 150r$$

$$350r - 14{,}000 = 150r$$

$$-14{,}000 = -200r$$

$$70 = r$$

If the truck's speed is 70 mph, then the speed of the train is $70 - 40$, or 30 mph.

Check. First note that the speed of the truck, 70 mph, is 40 mph faster than the speed of the train, 30 mph. If the truck travels 350 mi at 70 mph, it travels for 350/70, or 5 hr. If the train travels 150 mi at 30 mph, it travels for 150/30, or 5 hr. Since the times are the same, the speeds check.

State. The speed of the truck is 70 mph, and the speed of the train is 30 mph.

16. AMTRAK: 80 km/h; B & M: 66 km/h

17. Familiarize. Let $r =$ Kelly's speed, in km/h, and $t =$ the time the bicyclists travel, in hours. Organize the information in a table.

	Distance	Speed	Time
Hank	42	$r - 5$	t
Kelly	57	r	t

Translate. We can replace the t's in the table above using the formula $r = d/t$.

	Distance	Speed	Time
Hank	42	$r - 5$	$\dfrac{42}{r-5}$
Kelly	57	r	$\dfrac{57}{r}$

Since the times are the same for both bicyclists, we have the equation

$$\frac{42}{r-5} = \frac{57}{r}.$$

Carry out. We first multiply by the LCD, $r(r-5)$.

$$r(r-5) \cdot \frac{42}{r-5} = r(r-5) \cdot \frac{57}{r}$$

$$42r = 57(r-5)$$

$$42r = 57r - 285$$

$$-15r = -285$$

$$r = 19$$

If $r = 19$, then $r - 5 = 14$.

Check. If Hank's speed is 14 km/h and Kelly's speed is 19 km/h, then Hank bicycles 5 km/h slower than

Kelly. Hank's time is 42/14, or 3 hr. Kelly's time is 57/19, or 3 hr. Since the times are the same, the answer checks.

State. Hank travels at 14 km/h, and Kelly travels at 19 km/h.

18. Bill: 50 mph; Hillary: 80 mph

19. *Familiarize*. Let r = Ralph's speed, in km/h. Then Bonnie's speed is $r + 3$. Also set t = the time, in hours, that Ralph and Bonnie walk. We organize the information in a table.

	Distance	Speed	Time
Ralph	7.5	r	t
Bonnie	12	$r+3$	t

Translate. We can replace the t's in the table shown above using the formula $r = d/t$.

	Distance	Speed	Time
Ralph	7.5	r	$\dfrac{7.5}{r}$
Bonnie	12	$r+3$	$\dfrac{12}{r+3}$

Since the times are the same for both walkers, we have the equation

$$\frac{7.5}{r} = \frac{12}{r+3}.$$

Carry out. We first multiply by the LCD, $r(r+3)$.

$$r(r+3) \cdot \frac{7.5}{r} = r(r+3) \cdot \frac{12}{r+3}$$
$$7.5(r+3) = 12r$$
$$7.5r + 22.5 = 12r$$
$$22.5 = 4.5r$$
$$5 = r$$

If $r = 5$, then $r + 3 = 8$.

Check. If Ralph's speed is 5 km/h and Bonnie's speed is 8 km/h, then Bonnie walks 3 km/h faster than Ralph. Ralph's time is 7.5/5, or 1.5 hr. Bonnie's time is 12/8, or 1.5 hr. Since the times are the same, the answer checks.

State. Ralph's speed is 5 km/h, and Bonnie's speed is 8 km/h.

20. Sally: 12 km/h; Gerard: 16 km/h

21. *Familiarize*. Let t = the time it takes Caledonia to drive to town and organize the given information in a table.

	Distance	Speed	Time
Caledonia	15	r	t
Manley	20	r	$t+1$

Translate. We can replace the r's in the table above using the formula $r = d/t$.

	Distance	Speed	Time
Caledonia	15	$\dfrac{15}{t}$	t
Manley	20	$\dfrac{20}{t+1}$	$t+1$

Since the speeds are the same for both riders, we have the equation

$$\frac{15}{t} = \frac{20}{t+1}.$$

Carry out. We multiply by the LCD, $t(t+1)$.

$$t(t+1) \cdot \frac{15}{t} = t(t+1) \cdot \frac{20}{t+1}$$
$$15(t+1) = 20t$$
$$15t + 15 = 20t$$
$$15 = 5t$$
$$3 = t$$

If $t = 3$, then $t + 1 = 3 + 1$, or 4.

Check. If Caledonia's time is 3 hr and Manley's time is 4 hr, then Manley's time is 1 hr more than Caledonia's. Caledonia's speed is 15/3, or 5 mph. Manley's speed is 20/4, or 5 mph. Since the speeds are the same, the answer checks.

State. It takes Caledonia 3 hr to drive to town.

22. $1\frac{1}{3}$ hr

23. We write a proportion and then solve it.

$$\frac{b}{6} = \frac{7}{4}$$
$$b = \frac{7}{4} \cdot 6$$
$$b = \frac{42}{4}, \text{ or } 10.5$$

$\left(\text{Note that the proportions } \dfrac{6}{b} = \dfrac{4}{7}, \dfrac{b}{7} = \dfrac{6}{4}, \text{ or } \dfrac{7}{b} = \dfrac{4}{6} \right.$
could also be used. $\Big)$

24. 6.75

25. We write a proportion and then solve it.

$$\frac{4}{f} = \frac{6}{4}$$

$$4f \cdot \frac{4}{f} = 4f \cdot \frac{6}{4}$$

$$16 = 6f$$

$$\frac{8}{3} = f \qquad \text{Simplifying}$$

$\left(\text{One of the following proportions could also be used:}\right.$
$\left.\frac{f}{4} = \frac{4}{6}, \; \frac{4}{f} = \frac{9}{6}, \; \frac{f}{4} = \frac{6}{9}, \; \frac{4}{9} = \frac{f}{6}, \; \frac{9}{4} = \frac{6}{f}\right)$

26. 7.5

27. We write a proportion and then solve it.

$$\frac{4}{10} = \frac{6}{l}$$

$$10l \cdot \frac{4}{10} = 10l \cdot \frac{6}{l}$$

$$4l = 60$$

$$l = 15 \text{ ft}$$

$\left(\text{One of the following proportions could also be used:}\right.$
$\left.\frac{4}{6} = \frac{10}{l}, \; \frac{10}{4} = \frac{l}{6}, \text{ or } \frac{6}{4} = \frac{l}{10}\right)$

28. 4.5 ft

29.
$$\frac{a}{b} = \frac{c}{d}$$

$$\frac{8}{5} = \frac{6}{d}$$

$$5d \cdot \frac{8}{5} = 5d \cdot \frac{6}{d}$$

$$8d = 30$$

$$d = \frac{30}{8} = \frac{15}{4} \text{ cm, or } 3.75 \text{ cm}$$

30. $\frac{90}{7}$ cm

31. Let $c = b + 2$ and $d = b - 2$.

$$\frac{a}{b} = \frac{c}{d}$$

$$\frac{15}{b} = \frac{b+2}{b-2}$$

$$b(b-2) \cdot \frac{15}{b} = b(b-2) \cdot \frac{b+2}{b-2}$$

$$15(b-2) = b(b+2)$$

$$15b - 30 = b^2 + 2b$$

$$0 = b^2 - 13b + 30$$

$$0 = (b-3)(b-10)$$

$$b - 3 = 0 \; \text{ or } \; b - 10 = 0$$

$$b = 3 \; \text{ or } \qquad b = 10$$

If $b = 3$ m, then $c = 3 + 2$, or 5 m and $d = 3 - 2$, or 1 m.

If $b = 10$ m, then $c = 10 + 2$, or 12 m and $d = 10 - 2$, or 8 m.

32. $b = 3$ m, $c = 6$ m, $d = 1$ m; $b = 12$ m, $c = 15$, m, $d = 10$ m

33. *Familiarize*. The coffee beans from 14 trees are required to produce 7.7 kilograms of coffee, and we wish to find how many trees are required to produce 308 kilograms of coffee. We can set up ratios:

$$\frac{T}{308} \qquad \frac{14}{7.7}$$

Translate. Assuming the two ratios are the same, we can translate to a proportion.

$$\begin{array}{c} \text{Trees} \to \\ \text{Kilograms} \to \end{array} \frac{T}{308} = \frac{14}{7.7} \begin{array}{c} \leftarrow \text{Trees} \\ \leftarrow \text{Kilograms} \end{array}$$

Carry out. We solve the proportion.

$$308 \cdot \frac{T}{308} = 308 \cdot \frac{14}{7.7}$$

$$T = \frac{4312}{7.7}$$

$$T = 560$$

Check. $\frac{560}{308} = 1.8\overline{1} \qquad \frac{14}{7.7} = 1.8\overline{1}$

The ratios are the same.

State. 560 trees are required to produce 308 kg of coffee.

34. 702 km

35. *Familiarize*. 10 cm^3 of human blood contains 1.2 grams of hemoglobin, and we wish to find how many grams of hemoglobin are contained in 16 cm^3 of the same blood. We can set up ratios:

$$\frac{H}{16} \qquad \frac{1.2}{10}$$

Translate. Assuming the two ratios are the same, we can translate to a proportion.

$$\begin{array}{c} \text{Grams} \to \\ \text{cm}^3 \to \end{array} \frac{H}{16} = \frac{1.2}{10} \begin{array}{c} \leftarrow \text{Grams} \\ \leftarrow \text{cm}^3 \end{array}$$

Carry out. We solve the proportion.

We multiply by 16 to get H alone.

$$16 \cdot \frac{H}{16} = 16 \cdot \frac{1.2}{10}$$

$$H = \frac{19.2}{10}$$

$$H = 1.92$$

Check.
$$\frac{1.92}{16} = 0.12 \qquad \frac{1.2}{10} = 0.12$$
The ratios are the same.

State. 16 cm^3 of the same blood would contain 1.92 grams of hemoglobin.

36. $21\frac{2}{3}$ cups

37. Familiarize. U.S. women earn 77 cents for each dollar earned by a man. This gives us one ratio, expressed in dollars: $\frac{0.77}{1}$. If a male sales manager earns \$42,000, we want to find how much a female would earn for comparable work. This gives us a second ratio, also expressed in dollars: $\frac{F}{42,000}$.

Translate. Assuming the two ratios are the same, we can translate to a proportion.

$$\begin{array}{c} \text{Female's} \\ \text{earnings} \\ \text{Male's earnings} \end{array} \begin{array}{c} \rightarrow \\ \rightarrow \end{array} \frac{0.77}{1} = \frac{F}{42,000} \begin{array}{c} \leftarrow \\ \leftarrow \end{array} \begin{array}{c} \text{Female's} \\ \text{earnings} \\ \text{Male's earnings} \end{array}$$

Carry out. We solve the proportion.
$$42,000 \cdot \frac{0.77}{1} = 42,000 \cdot \frac{F}{42,000}$$
$$32,340 = F$$

Check.
$$\frac{0.77}{1} = 0.77 \qquad \frac{32,340}{42,000} = 0.77$$
The ratios are the same.

State. If a male sales manager earns \$42,000, a female would earn \$32,340 for comparable work.

38. $1\frac{11}{39}$ kg

39. Familiarize. The ratio of deer tagged to the total number of deer in the preserve, D, is $\frac{318}{D}$. Of the 168 deer caught later, 56 are tagged. The ratio of tagged deer to deer caught is $\frac{56}{168}$.

Translate. We translate to a proportion.

$$\begin{array}{c} \text{Deer originally} \\ \text{tagged} \\ \text{Deer} \\ \text{in preserve} \end{array} \begin{array}{c} \rightarrow \\ \rightarrow \end{array} \frac{318}{D} = \frac{56}{168} \begin{array}{c} \leftarrow \\ \leftarrow \end{array} \begin{array}{c} \text{Tagged deer} \\ \text{caught later} \\ \text{Deer} \\ \text{caught later} \end{array}$$

Carry out. We solve the proportion. We multiply by the LCD, 168D.

$$168D \cdot \frac{318}{D} = 168D \cdot \frac{56}{168}$$
$$168 \cdot 318 = D \cdot 56$$
$$\frac{166 \cdot 318}{56} = D$$
$$954 = D$$

Check.
$$\frac{318}{954} = 0.\overline{3} \qquad \frac{56}{168} = 0.\overline{3}$$
The ratios are the same.

State. We estimate that there are 954 deer in the preserve.

40. 184

41. Familiarize. Let D = the number of defective bulbs you would expect in a sample of 1288 bulbs. We set up two ratios:
$$\frac{6}{184} \qquad \frac{D}{1288}$$
Translate. Assuming the ratios are the same, we can translate to a proportion.

$$\begin{array}{c} \text{Defective} \rightarrow \\ \text{Sample} \rightarrow \end{array} \frac{6}{184} = \frac{D}{1288} \begin{array}{c} \leftarrow \text{Defective} \\ \leftarrow \text{Sample} \end{array}$$

Carry out. We solve the proportion.
$$184(1288) \cdot \frac{6}{184} = 184(1288) \cdot \frac{D}{1288}$$
$$1288 \cdot 6 = 184 \cdot D$$
$$\frac{1288 \cdot 6}{184} = D$$
$$42 = D$$

Check.
$$\frac{6}{184} \approx 0.0326 \qquad \frac{42}{1288} \approx 0.0326$$
The ratios are the same.

State. You would expect 42 defective bulbs in a sample of 1288 bulbs.

42. 287

43. Familiarize. Let M = the number of miles Emmanuel will drive in 4 years if he continues to drive at the current rate. We set up two ratios:
$$\frac{16,000}{1\frac{1}{2}} \qquad \frac{M}{4}$$

Translate. We write a proportion.

$$\begin{array}{c} \text{Miles} \rightarrow \\ \text{Years} \rightarrow \end{array} \frac{16,000}{1.5} = \frac{M}{4} \begin{array}{c} \leftarrow \text{Miles} \\ \leftarrow \text{Years} \end{array}$$

Carry out. We solve the proportion.

$$1.5(4) \cdot \frac{16,000}{1.5} = 1.5(4) \cdot \frac{M}{4}$$
$$64,000 = 1.5M$$
$$42,666.\overline{6} = M$$

If this possible answer is correct, Emmanuel will not exceed the 45,000 miles allowed for four years.

Check.

$$\frac{16,000}{1.5} = 10,666.\overline{6} \qquad \frac{42,666.\overline{6}}{4} = 10,666.\overline{6}$$

The ratios are the same.

State. At this rate, Emmanuel will not exceed the mileage allowed for four years.

44. 20

45. ***Familiarize***. The ratio of foxes tagged to the total number of foxes in the county, F, is $\frac{25}{F}$. Of the 36 foxes caught later, 4 had tags. The ratio of tagged foxes to foxes caught is $\frac{4}{36}$.

Translate. Assuming the two ratios are the same, we can translate to a proportion.

$$\begin{array}{ccc}
\text{Foxes tagged} & & \text{Tagged foxes} \\
\text{originally} \rightarrow & \dfrac{25}{F} = \dfrac{4}{36} & \leftarrow \text{caught later} \\
\text{Foxes} \rightarrow & & \leftarrow \text{Foxes} \\
\text{in county} & & \text{caught later}
\end{array}$$

Carry out. We solve the proportion.

$$36F \cdot \frac{25}{F} = 36F \cdot \frac{4}{36}$$
$$900 = 4F$$
$$225 = F$$

Check.

$$\frac{25}{225} = \frac{1}{9} \qquad \frac{4}{36} = \frac{1}{9}$$

The ratios are the same.

State. We estimate that there are 225 foxes in the county.

46. a) 1.92 tons

b) 28.8 lb

47. ***Familiarize***. The ratio of the weight of an object on Mars to the weight of an object on the earth is 0.4 to 1.

a) We wish to find how much a 12-ton rocket would weigh on Mars.

b) We wish to find how much a 120-lb astronaut would weigh on Mars.

We can set up ratios.

$$\frac{0.4}{1} \qquad \frac{T}{12} \qquad \frac{P}{120}$$

Translate. Assuming the ratios are the same, we can translate to proportions.

a)
$$\begin{array}{cc}
\text{Weight} & \text{Weight} \\
\text{on Mars} \rightarrow & \dfrac{0.4}{1} = \dfrac{T}{12} \leftarrow \text{on Mars} \\
\text{Weight} \rightarrow & \qquad \leftarrow \text{Weight} \\
\text{on earth} & \text{on earth}
\end{array}$$

b)
$$\begin{array}{cc}
\text{Weight} & \text{Weight} \\
\text{on Mars} \rightarrow & \dfrac{0.4}{1} = \dfrac{P}{120} \leftarrow \text{on Mars} \\
\text{Weight} \rightarrow & \qquad \leftarrow \text{Weight} \\
\text{on earth} & \text{on earth}
\end{array}$$

Carry out. We solve each proportion.

a) $\quad \dfrac{0.4}{1} = \dfrac{T}{12}$ b) $\quad \dfrac{0.4}{1} = \dfrac{P}{120}$

$\quad 12(0.4) = T \qquad\qquad 120(0.4) = P$

$\qquad 4.8 = T \qquad\qquad\qquad 48 = P$

Check. $\dfrac{0.4}{1} = 0.4$, $\dfrac{4.8}{12} = 0.4$, and $\dfrac{48}{120} = 0.4$.

The ratios are the same.

State.

a) A 12-ton rocket would weigh 4.8 tons on Mars.

b) A 120-lb astronaut would weigh 48 lb on Mars.

48. $\dfrac{36}{68}$

49. Writing exercise

50. Writing exercise

51. Graph: $y = 2x - 6$.

We select some x-values and compute y-values.

If $x = 1$, then $y = 2 \cdot 1 - 6 = -4$.

If $x = 3$, then $y = 2 \cdot 3 - 6 = 0$.

If $x = 5$, then $y = 2 \cdot 5 - 6 = 4$.

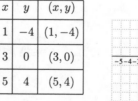

x	y	(x,y)
1	-4	$(1,-4)$
3	0	$(3,0)$
5	4	$(5,4)$

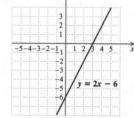

52.

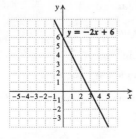

53. Graph: $3x + 2y = 12$.

We can replace either variable with a number and then calculate the other coordinate. We will find the intercepts and one other point.

If $y = 0$, we have:
$$3x + 2 \cdot 0 = 12$$
$$3x = 12$$
$$x = 4$$

The x-intercept is $(4, 0)$.

If $x = 0$, we have:
$$3 \cdot 0 + 2y = 12$$
$$2y = 12$$
$$y = 6$$

The y-intercept is $(0, 6)$.

If $y = -3$, we have:
$$3x + 2(-3) = 12$$
$$3x - 6 = 12$$
$$3x = 18$$
$$x = 6$$

The point $(6, -3)$ is on the graph.

We plot these points and draw a line through them.

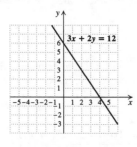

54.

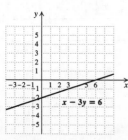

55. Graph: $y = -\dfrac{3}{4}x + 2$

We select some x-values and compute y-values. We use multiples of 4 to avoid fractions.

If $x = -4$, then $y = -\dfrac{3}{4}(-4) + 2 = 5$.

If $x = 0$, then $y = -\dfrac{3}{4} \cdot 0 + 2 = 2$.

If $x = 4$, then $y = -\dfrac{3}{4} \cdot 4 + 2 = -1$.

x	y	(x, y)
-4	5	$(-4, 5)$
0	2	$(0, 2)$
4	-1	$(4, -1)$

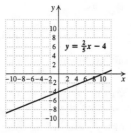

56.

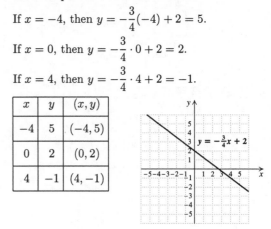

57. Writing exercise

58. Writing exercise

59. *Familiarize.* Let $t =$ the time, in hours, it takes Michelle to wax the car alone. Then $\dfrac{t}{2} =$ Sal's time alone, and $t - 2 =$ Kristen's time alone. In 1 hr they do $\dfrac{1}{t} + \dfrac{1}{\frac{t}{2}} + \dfrac{1}{t-2}$, or $\dfrac{1}{t} + \dfrac{2}{t} + \dfrac{1}{t-2}$ of the job working together. The entire job takes 1 hr and 20 min, or $\dfrac{4}{3}$ hr.

Translate. To get an entire job, we multiply the amount of work done in 1 hr by the number of hours required to complete the job.

$$\frac{4}{3}\left(\frac{1}{t} + \frac{2}{t} + \frac{1}{t-2}\right) = 1$$

Carry out. We solve the equation.

$$3 \cdot \frac{4}{3}\left(\frac{1}{t} + \frac{2}{t} + \frac{1}{t-2}\right) = 3 \cdot 1$$

$$4\left(\frac{1}{t} + \frac{2}{t} + \frac{1}{t-2}\right) = 3$$

$$4\left(\frac{3}{t} + \frac{1}{t-2}\right) = 3 \quad \text{Adding: } \frac{1}{t} + \frac{2}{t} = \frac{3}{t}$$

$$\frac{12}{t} + \frac{4}{t-2} = 3$$

$$t(t-2)\left(\frac{12}{t} + \frac{4}{t-2}\right) = t(t-2)(3)$$

$$12(t-2) + t \cdot 4 = 3t(t-2)$$

$$12t - 24 + 4t = 3t^2 - 6t$$

$$16t - 24 = 3t^2 - 6t$$

$$0 = 3t^2 - 22t + 24$$

$$0 = (3t-4)(t-6)$$

$$3t - 4 = 0 \quad or \quad t - 6 = 0$$
$$3t = 4 \quad or \qquad t = 6$$
$$t = \frac{4}{3} \quad or \qquad t = 6$$

Check. If $t = \frac{4}{3}$, then $t - 2 = -\frac{2}{3}$. Since time cannot be negative in this problem, $\frac{4}{3}$ cannot be a solution. If $t = 6$, then $t/2 = 3$ and $t - 2 = 4$. If Michelle, Sal, and Kristen can do the job in 6 hr, 3 hr, and 4 hr, respectively, then in one hour they do $\frac{1}{6} + \frac{1}{3} + \frac{1}{4}$, or $\frac{3}{4}$ of the job working together and in $\frac{4}{3}$ hr they do $\frac{4}{3} \cdot \frac{3}{4}$, or 1 entire job. The answer checks.

State. Working alone, the job would take Michelle 6 hr, Sal 3 hr, and Kristen 4 hr.

60. $9\frac{3}{13}$ days

61. Familiarize. Let $t =$ the number of hours it takes to wire one house, working together. We want to find the number of hours it takes to wire two houses, working together.

Translate. We write an equation.

$$\frac{t}{28} + \frac{t}{34} = 2$$

Carry out. We solve the equation.

$$\frac{t}{28} + \frac{t}{34} = 2, \text{ LCD} = 476$$

$$476\left(\frac{t}{28} + \frac{t}{34}\right) = 476 \cdot 2$$

$$17t + 14t = 952$$

$$31t = 952$$

$$t = \frac{952}{31}, \text{ or } 32\frac{22}{31}$$

Check. If $30\frac{22}{31}$ hr, Janet does $\frac{952}{31} \cdot \frac{1}{28} = \frac{34}{31}$ of one complete job and Linus does $\frac{952}{31} \cdot \frac{1}{34}$, or $\frac{28}{31}$ of one complete job. Together they do $\frac{34}{31} + \frac{28}{31}$, or $\frac{62}{31}$, or 2 complete jobs. The answer checks.

State. It will take Janet and Linus $30\frac{22}{31}$ hr to wire two houses, working together.

62. Ann: 6 hr; Betty: 12 hr

63. Familiarize. We will begin by finding how long it will take Alma and Kevin to grade a batch of exams, working together. Then we will find what percentage of the job was done by Alma.

Translate. We use the work principle to find how long it will take Alma and Kevin to do the job, working together.

$$\left(\frac{1}{3} + \frac{1}{4}\right)t = 1, \text{ or } \frac{7}{12} \cdot t = 1$$

Carry out. We solve the equation.

$$\frac{7}{12} \cdot t = 1$$

$$\frac{12}{7} \cdot \frac{7}{12} \cdot t = \frac{12}{7} \cdot 1$$

$$t = \frac{12}{7}$$

Now, since Alma can do the job alone in 3 hr, she does $\frac{1}{3}$ of the job in 1 hr and in $\frac{12}{7}$ hr she does $\frac{12}{7} \cdot \frac{1}{3} \approx 0.57 \approx 57\%$ of the job.

Check. We can repeat the calculations. The answer checks.

State. About 57% of the exams will have been graded by Alma.

64. 12 hr

65. Familiarize. We organize the information in a table. Let $r =$ the speed of the current and $t =$ the time it takes to travel upstream.

Translate.

	Distance	Speed	Time
Upstream	24	$10-r$	t
Downstream	24	$10+r$	$5-t$

From the rows of the table we get two equations:

$$24 = (10 - r)t$$
$$24 = (10 + r)(5 - t)$$

We solve each equation for t and set the results equal:

Solving $24 = (10 - r)t$ for t: $t = \frac{24}{10 - r}$

Solving $24 = (10+r)(5-t)$ for t: $t = 5 - \dfrac{24}{10+r}$

Then $\dfrac{24}{10-r} = 5 - \dfrac{24}{10+r}$.

Carry out. We first multiply on both sides of the equation by the LCD, $(10-r)(10+r)$:

$$(10-r)(10+r) \cdot \frac{24}{10-r} = (10-r)(10+r)\left(5 - \frac{24}{10+r}\right)$$
$$24(10+r) = 5(10-r)(10+r) - 24(10-r)$$
$$240 + 24r = 500 - 5r^2 - 240 + 24r$$
$$240 + 24r = 260 - 5r^2 + 24r$$
$$5r^2 - 20 = 0$$
$$5(r^2 - 4) = 0$$
$$5(r+2)(r-2) = 0$$
$$r+2 = 0 \quad or \quad r-2 = 0$$
$$r = -2 \quad or \qquad r = 2$$

Check. We only check 2 since the speed of the current cannot be negative. If $r = 2$, then the speed upstream is $10 - 2$, or 8 mph and the time is $\dfrac{24}{8}$, or 3 hours. If $r = 2$, then the speed downstream is $10 + 2$, or 12 mph and the time is $\dfrac{24}{12}$, or 2 hours. The sum of 3 hr and 2 hr is 5 hr. This checks.

State. The speed of the current is 2 mph.

66. 270

67. Familiarize. We organize the information in a table. Let $r = $ the speed on the first part of the trip and $t = $ the time driven at that speed.

	Distance	Speed	Time
First part	30	r	t
Second part	30	$r+15$	$1-t$

Translate. From the rows of the table we obtain two equations:

$$30 = rt$$
$$30 = (r+15)(1-t)$$

We solve each equation for t and set the results equal:

Solving $30 = rt$ for t: $t = \dfrac{30}{r}$

Solving $20 = (r+15)(1-t)$ for t: $t = 1 - \dfrac{20}{r+15}$

Then $\dfrac{30}{r} = 1 - \dfrac{20}{r+15}$.

Carry out. We first multiply the equation by the LCD, $r(r+15)$.

$$r(r+15) \cdot \frac{30}{r} = r(r+15)\left(1 - \frac{20}{r+15}\right)$$
$$30(r+15) = r(r+15) - 20r$$
$$30r + 450 = r^2 + 15r - 20r$$
$$0 = r^2 - 35r - 450$$
$$0 = (r-45)(r+10)$$
$$r - 45 = 0 \quad or \quad r + 10 = 0$$
$$r = 45 \quad or \qquad r = -10$$

Check. Since the speed cannot be negative, we only check 45. If $r = 45$, then the time for the first part is $\dfrac{30}{45}$, or $\dfrac{2}{3}$ hr. If $r = 45$, then $r + 15 = 60$ and the time for the second part is $\dfrac{20}{60}$, or $\dfrac{1}{3}$ hr. The total time is $\dfrac{2}{3} + \dfrac{1}{3}$, or 1 hour. The value checks.

State. The speed for the first 30 miles was 45 mph.

68. $66\dfrac{2}{3}$ ft

69. Familiarize. Let $t = $ the number of minutes after 5:00 at which the hands will first be together. When the minute hand moves through t minutes, the hour hand moves through $t/12$ minutes. At 5:00 the hour hand is on the 25-minute mark, so at t minutes after 5:00 it is at $25 + t/12$.

Translate. We equate the positions of the minute hand and the hour hand.

$$t = 25 + \frac{t}{12}$$

Carry out. We solve the equation.

$$t = 25 + \frac{t}{12}$$
$$12 \cdot t = 12\left(25 + \frac{t}{12}\right)$$
$$12t = 300 + t$$
$$11t = 300$$
$$t = \frac{300}{11}, \text{ or } 27\frac{3}{11}$$

Check. When the minute hand is at $27\dfrac{3}{11}$ minutes after 5:00, the hour hand is at $25 + \dfrac{\frac{300}{11}}{12} =$

$25 + \dfrac{300}{11} \cdot \dfrac{1}{12} = 25 + \dfrac{25}{11} = 25 + 2\dfrac{3}{11} = 27\dfrac{3}{11}$ minutes after 5:00 also. The answer checks.

State. After 5:00 the hands on a clock will first be together in $27\dfrac{3}{11}$ minutes or at $27\dfrac{3}{11}$ minutes after 5:00.

70. $\dfrac{D}{B} = \dfrac{C}{A}$; $\dfrac{A}{C} = \dfrac{B}{D}$; $\dfrac{D}{C} = \dfrac{B}{A}$

71. Writing exercise

72. Writing exercise

Exercise Set 6.8

1. $3x + 5 = 7x + 1$

$4 = 4x$ Adding $-3x - 1$

$1 = x$ Dividing by 4

The solution is 1.

2. $-\dfrac{13}{4}$

3. $x^2 = 5x - 6$

$x^2 - 5x + 6 = 0$ Adding $-5x + 6$

$(x - 2)(x - 3) = 0$

$x - 2 = 0 \ \ or \ \ x - 3 = 0$

$x = 2 \ \ or \ \ \ \ \ \ x = 3$

The solutions are 2 and 3.

4. $-1, \ 4$

5. $2(x - 1) = 2x - 5$

$2x - 2 = 2x - 5$

$-2 = -5$ Subtracting $2x$

There are no replacements for x that make $-2 = -5$ true. The equation has no solution. It is a contradiction.

6.   All real numbers; identity

7. $\dfrac{5}{x + 1} = \dfrac{5}{8}$

Since the numerators are equal, the denominators must also be equal. Then we have:

$x + 1 = 8$

$x = 7$

Check:

$$\dfrac{5}{x+1} = \dfrac{5}{8}$$

$$\dfrac{5}{7+1} \ \overset{?}{\vert} \ \dfrac{5}{8}$$

$$\dfrac{5}{8} \ \bigg\vert \ \dfrac{5}{8} \quad \text{TRUE}$$

The solution is 7.

8. -7

9. $4(2x - 1) = 3x - 5$

$8x - 4 = 3x - 5$

$5x = -1$ Adding $-3x + 4$

$x = -\dfrac{1}{5}$

The solution is $-\dfrac{1}{5}$.

10. $-\dfrac{1}{6}$

11. $\dfrac{3}{10x} - \dfrac{4}{5x} = 6$ Note that $x \neq 0$;

                       LCD $= 10x$.

$10x\left(\dfrac{3}{10x} - \dfrac{4}{5x}\right) = 10x \cdot 6$

$10x \cdot \dfrac{3}{10x} - 10x \cdot \dfrac{4}{5x} = 60x$

$3 - 8 = 60x$

$-5 = 60x$

$-\dfrac{1}{12} = x$

This number checks. The solution is $-\dfrac{1}{12}$.

12. $\dfrac{4}{33}$

13. $2t^2 - 7t + 3 = 0$

$(2t - 1)(t - 3) = 0$

$2t - 1 = 0 \ \ or \ \ t - 3 = 0$

$2t = 1 \ \ or \ \ \ \ \ \ \ t = 3$

$t = \dfrac{1}{2} \ \ or \ \ \ \ \ \ \ t = 3$

The solutions are $\dfrac{1}{2}$ and 3.

14. $\dfrac{1}{3}, \ 2$

15. $\dfrac{2}{x} - 1 = \dfrac{1 - x}{x}$, LCD $= x$

$x\left(\dfrac{2}{x} - 1\right) = x \cdot \dfrac{1 - x}{x}$ Multiplying by the LCD

$x \cdot \dfrac{2}{x} - x \cdot 1 = 1 - x$

$2 - x = 1 - x$

$2 = 1$ Adding x

There are no replacements for x that make $2 = 1$ true. The equation has no solution. It is a contradiction.

16. $\emptyset$; contradiction

17.
$$1 - \frac{3}{7n} = \frac{5}{14}, \text{ LCD } = 14n$$
$$14n\left(1 - \frac{3}{7n}\right) = 14n \cdot \frac{5}{14}$$
$$14n \cdot 1 - 14n \cdot \frac{3}{7n} = 5n$$
$$14n - 6 = 5n$$
$$-6 = -9n$$
$$\frac{2}{3} = n$$

The number $\frac{2}{3}$ checks and is the solution.

18. $\frac{10}{47}$

19. $\frac{x}{x-2} = \frac{2}{x-2}$

Since the denominator are equal, the numerators must be equal. Then we have $x = 2$. However, note that the denominators are 0 when $x = 2$. Thus, the equation has no solution. It is a contradiction.

20. 5

21. $2\{(x+2) - 3\} = 3(x+2) - (x+8)$
$$2\{x - 1\} = 3x + 6 - x - 8$$
$$2x - 2 = 2x - 2$$
$$-2 = -2 \quad \text{Subtracting } 2x$$

$-2 = -2$ is true for all replacements for x. All real numbers are solutions. The equation is an identity.

22. All real numbers; identity

23.
$$\frac{x}{x+1} = \frac{x}{x+2},$$
$$\text{LCD } = (x+1)(x+2)$$
$$(x+1)(x+2) \cdot \frac{x}{x+1} = (x+1)(x+2) \cdot \frac{x}{x+2}$$
$$x(x+2) = x(x+1)$$
$$x^2 + 2x = x^2 + x$$
$$2x = x \quad \text{Subtracting } x^2$$
$$x = 0 \quad \text{Subtracting } x$$

The number 0 checks and is the solution.

24. 0

25.
$$\frac{x+3}{3x+4} = \frac{2}{x+2},$$
$$\text{LCD } = (3x+4)(x+2)$$
$$(3x+4)(x+2) \cdot \frac{x+3}{3x+4} = (3x+4)(x+2) \cdot \frac{2}{x+2}$$
$$(x+2)(x+3) = 2(3x+4)$$
$$x^2 + 5x + 6 = 6x + 8$$
$$x^2 - x - 2 = 0$$
$$(x-2)(x+1) = 0$$
$$x - 2 = 0 \quad or \quad x + 1 = 0$$
$$x = 2 \quad or \qquad x = -1$$

Both numbers check. The solutions are 2 and -1.

26. -5, -1

27.
$$s = \frac{1}{2}gt^2$$
$$2s = gt^2 \quad \text{Multiplying by 2}$$
$$\frac{2s}{t^2} = g \quad \text{Dividing by } t^2$$

28. $h = \frac{2A}{b}$

29.
$$S = 2\pi rh$$
$$\frac{S}{2\pi r} = h \quad \text{Multiplying by } \frac{1}{2\pi r}$$

30. $t = \frac{A-P}{Pr}$, or $\frac{A}{Pr} - \frac{1}{r}$

31.
$$A = P + Prt$$
$$A = P(1 + rt) \quad \text{Factoring}$$
$$\frac{A}{1+rt} = P \quad \text{Dividing by } 1 + rt$$

32. $r = \frac{L}{l-S}$

33.
$$A = \frac{1}{2}h(b_1 + b_2)$$
$$2A = h(b_1 + b_2) \quad \text{Multiplying by 2}$$
$$\frac{2A}{b_1 + b_2} = h \quad \text{Dividing by } b_1 + b_2$$

34. $m = \frac{T}{g+f}$

35. $\dfrac{1}{180} = \dfrac{n-2}{s}$

$\dfrac{s}{180} = n - 2$ Multiplying by s

$\dfrac{s}{180} + 2 = n$ Adding 2

36. $a = \dfrac{2S}{n} - l$

37. $\dfrac{m}{n} = p - q$

$m = n(p - q)$ Multiplying by n

$\dfrac{m}{p - q} = n$ Dividing by $p - q$

38. $t = \dfrac{M - g}{r + s}$

39. $\dfrac{1}{R} = \dfrac{1}{r_1} + \dfrac{1}{r_2}$

$Rr_1r_2 \cdot \dfrac{1}{R} = Rr_1r_2\left(\dfrac{1}{r_1} + \dfrac{1}{r_2}\right)$ Multiplying by
the LCD, Rr_1r_2, to clear fractions

$r_1r_2 = Rr_1r_2 \cdot \dfrac{1}{r_1} + Rr_1r_2 \cdot \dfrac{1}{r_2}$

$r_1r_2 = Rr_2 + Rr_1$

$r_1r_2 = R(r_2 + r_1)$

$\dfrac{r_1r_2}{r_2 + r_1} = R$

40. $f = \dfrac{pq}{q + p}$

41. $S = 2\pi r(r + h)$

$\dfrac{S}{2\pi r} = r + h$ Dividing by $2\pi r$

$\dfrac{S}{2\pi r} - r = h$, or

$\dfrac{S - 2\pi r^2}{2\pi r} = h$

42. $a = \dfrac{d}{b - c}$

43. $\dfrac{m}{n} = r$

$m = nr$ Multiplying by n

$\dfrac{m}{r} = n$ Dividing by r

44. $v = \dfrac{r}{s + t}$

45. Writing exercise

46. Writing exercise

47. Graph: $y = \dfrac{4}{5}x + 1$

x	y
-5	-3
0	1
5	5

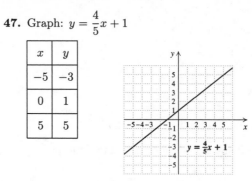

48.

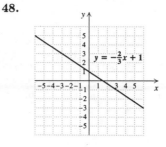

49. $-\dfrac{3}{5}x = \dfrac{9}{10}$

$-\dfrac{5}{3}\left(-\dfrac{3}{5}x\right) = -\dfrac{5}{3} \cdot \dfrac{9}{10}$

$x = -\dfrac{45}{30}$, or $-\dfrac{3}{2}$

The solution is $-\dfrac{3}{2}$.

50.

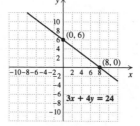

51. $x^2 - 13x + 30$

We look for two factors of 30 whose sum is -13. The numbers we need are -10 and -3.

$x^2 - 13x - 30 = (x - 10)(x - 3)$

52. $-3x^3 - 5x^2 + 5$

53. Writing exercise

54. Writing exercise

55. Substitute 8 for c and 224 for d and solve for a.

$$c = \frac{a}{a+12} \cdot d$$

$$8 = \frac{a}{a+12} \cdot 24$$

$$\frac{1}{3} = \frac{a}{a+12} \qquad \text{Dividing by 24}$$

$$3(a+12) \cdot \frac{1}{3} = 3(a+12) \cdot \frac{a}{a+12}$$

$$a + 12 = 3a$$

$$12 = 2a$$

$$6 = a$$

The child is 6 years old.

56. $x = -p$ or $x = -q$

57.
$$\frac{a}{b} = \frac{b}{a}, \ \ \text{LCD} = ab$$

$$ab \cdot \frac{a}{b} = ab \cdot \frac{b}{a}$$

$$a^2 = b^2$$

$$a^2 - b^2 = 0$$

$$(a+b)(a-b) = 0$$

$$a + b = 0 \quad or \quad a - b = 0$$

$$a = -b \ \ or \qquad a = b$$

58. $l = 0$ or $l = w$

59.
$$(s+t)^2 = 4st$$

$$s^2 + 2st + t^2 = 4st$$

$$s^2 - 2st + t^2 = 0$$

$$(s-t)^2 = 0$$

$$s - t = 0 \ \ or \ \ s - t = 0$$

$$s = t \ \ or \qquad s = t$$

We have $s = t$.

60. $n_2 = \dfrac{n_1 p_2 R + p_1 p_2 n_1}{p_1 p_2 - p_1 R}$

61.
$$u = -F\left(E - \frac{P}{T}\right)$$

$$u = -EF + \frac{FP}{T}$$

$$T \cdot u = T\left(-EF + \frac{FP}{T}\right)$$

$$Tu = -EFT + FP$$

$$Tu + EFT = FP$$

$$T(u + EF) = FP$$

$$T = \frac{FP}{u + EF}$$

62. $v = \dfrac{Nbf_2 - bf_1 - df_1}{Nf_2 - 1}$

63. When $C = F$, we have

$$C = \frac{5}{9}(C - 32)$$

$$9C = 5(C - 32)$$

$$9C = 5C - 160$$

$$4C = -160$$

$$C = -40$$

At $-40°$ the Fahrenheit and Celsius readings are the same.

Chapter 7

Functions and Graphs

Exercise Set 7.1

1. The correspondence is not a function, because a member of the domain (3) corresponds to more than one member of the range.

2. Yes

3. The correspondence is a function, because each member of the domain corresponds to just one member of the range.

4. Yes

5. The correspondence is a function, because each member of the domain corresponds to just one member of the range.

6. No

7. This correspondence is a function, because each Christmas tree has only one price.

8. Function

9. The correspondence is not a function, since it is reasonable to assume that at least one member of a rock band plays more than one instrument.

   The correspondence is a relation, since it is reasonable to assume that each member of a rock band plays at least one instrument.

10. Function

11. This correspondence is a function, because each number in the domain, when squared and then increased by 4, corresponds to only one number in the range.

12. Function

13. a) Locate 1 on the horizontal axis and then find the point on the graph for which 1 is the first coordinate. From that point, look to the vertical axis to find the corresponding y-coordinate, -2. Thus, $f(1) = -2$.

    b) To determine which member(s) of the domain are paired with 2, locate 2 on the vertical axis. From there look left and right to the graph to find any points for which 2 is the second coordinate. One such point exists. Its first coordinate is 4. Thus, the x-value for which $f(x) = 2$ is 4.

14. a) -1; b) -3

15. a) Locate 1 on the horizontal axis and then find the point on the graph for which 1 is the first coordinate. From that point, look to the vertical axis to find the corresponding y-coordinate, 3. Thus, $f(1) = 3$.

    b) To determine which member(s) of the domain are paired with 2, locate 2 on the vertical axis. From there look left and right to the graph to find any points for which 2 is the second coordinate. One such point exists. Its first coordinate is 3. Thus, the x-value for which $f(x) = 2$ is 3.

16. a) 1; b) -1

17. a) Locate 1 on the horizontal axis and the find the point on the graph for which 1 is the first coordinate. From that point, look to the vertical axis to find the corresponding y-coordinate. It appears to be -2. Thus, $f(1) = -2$.

    b) To determine which member(s) of the domain are paired with 2, locate 2 on the vertical axis. From there look left and right to the graph to find any points for which 2 is the second coordinate. One such point exists. Its first coordinate is -2, so the x-value for which $f(x) = 2$ is -2.

18. a) 3; b) 0

19. a) Locate 1 on the horizontal axis and then find the point on the graph for which 1 is the first coordinate. From that point, look to the vertical axis to find the corresponding y-coordinate, 3. Thus, $f(1) = 3$.

    b) To determine which member(s) of the domain are paired with 2, locate 2 on the vertical axis. From there look left and right to the graph to find any points for which 2 is the second coordinate. One such point exists. Its first coordinate is -3. Thus, the x-value for which $f(x) = 2$ is -3.

20. a) 4; b) -1

21. a) Locate 1 on the horizontal axis and then find the point on the graph for which 1 is the first coordinate. From that point, look to the vertical axis to find the corresponding y-coordinate, 1. Thus, $f(1) = 1$.

b) To determine which member(s) of the domain are paired with 2, locate 2 on the vertical axis. From there look left and right to the graph to find any points for which 2 is the second coordinate. One such point exists. Its first coordinate is 3. Thus, the x-value for which $f(x) = 2$ is 3.

22. a) 3; b) $-2, 0$

23. a) Locate 1 on the horizontal axis and then find the point on the graph for which 1 is the first coordinate. From that point, look to the vertical axis to find the corresponding y-coordinate, 4. Thus, $f(1) = 4$.

b) To determine which member(s) of the domain are paired with 2, locate 2 on the vertical axis. From there look left and right to the graph to find any points for which 2 is the second coordinate. There are two such points, $(-1, 2)$ and $(3, 2)$. Thus, the x-values for which $f(x) = 2$ are -1 and 3.

24. a) 2; b) $-5, 1$

25. a) Locate 1 on the horizontal axis and then find the point on the graph for which 1 is the first coordinate. From that point, look to the vertical axis to find the corresponding y-coordinate, 1. Thus, $f(1) = 1$.

b) To determine which member(s) of the domain are paired with 2, locate 2 on the vertical axis. From there look left and right to the graph to find any points for which 2 is the second coordinate. All points in the set $\{x | 2 < x \le 5\}$ satisfy this condition. These are the x-values for which $f(x) = 2$.

26. a) 2; b) $\{x | 0 < x \le 2\}$

27. $g(x) = x + 3$

a) $g(0) = 0 + 3 = 3$

b) $g(-4) = -4 + 3 = -1$

c) $g(-7) = -7 + 3 = -4$

d) $g(8) = 8 + 3 = 11$

e) $g(a + 2) = a + 2 + 3 = a + 5$

28. a) 2; b) 6; c) -5; d) -6; e) $a - 3$

29. $f(n) = 5n^2 + 4n$

a) $f(0) = 5 \cdot 0^2 + 4 \cdot 0 = 0 + 0 = 0$

b) $f(-1) = 5(-1)^2 + 4(-1) = 5 - 4 = 1$

c) $f(3) = 5 \cdot 3^2 + 4 \cdot 3 = 45 + 12 = 57$

d) $f(t) = 5t^2 + 4t$

e) $f(2a) = 5(2a)^2 + 4 \cdot 2a = 5 \cdot 4a^2 + 8a = 20a^2 + 8a$

30. a) 0; b) 5; c) 21; d) $3t^2 - 2t$; e) $12a^2 - 4a$

31. $f(x) = \dfrac{x - 3}{2x - 5}$

a) $f(0) = \dfrac{0 - 3}{2 \cdot 0 - 5} = \dfrac{-3}{0 - 5} = \dfrac{-3}{-5} = \dfrac{3}{5}$

b) $f(4) = \dfrac{4 - 3}{2 \cdot 4 - 5} = \dfrac{1}{8 - 5} = \dfrac{1}{3}$

c) $f(-1) = \dfrac{-1 - 3}{2(-1) - 5} = \dfrac{-4}{-2 - 5} = \dfrac{-4}{-7} = \dfrac{4}{7}$

d) $f(3) = \dfrac{3 - 3}{2 \cdot 3 - 5} = \dfrac{0}{6 - 5} = \dfrac{0}{1} = 0$

e) $f(x + 2) = \dfrac{x + 2 - 3}{2(x + 2) - 5} = \dfrac{x - 1}{2x + 4 - 5} = \dfrac{x - 1}{2x - 1}$

32. a) $\dfrac{26}{25}$; b) $\dfrac{2}{9}$; c) $-\dfrac{5}{12}$; d) $-\dfrac{7}{3}$; e) $\dfrac{3x + 5}{2x + 11}$

33. $A(s) = s^2 \dfrac{\sqrt{3}}{4}$

$A(4) = 4^2 \dfrac{\sqrt{3}}{4} = 4\sqrt{3} \approx 6.93$

The area is $4\sqrt{3}$ cm$^2 \approx 6.93$ cm^2.

34. $9\sqrt{3}$ in$^2 \approx 15.59$ in^2

35. $V(r) = 4\pi r^2$

$V(3) = 4\pi(3)^2 = 36\pi$

The area is 36π in$^2 \approx 113.10$ in^2.

36. 100π cm$^2 \approx 314.16$ cm^2

37. $F(C) = \dfrac{9}{5}C + 32$

$F(-10) = \dfrac{9}{5}(-10) + 32 = -18 + 32 = 14$

The equivalent temperature is 14°F.

38. 41°F

39. $H(x) = 2.75x + 71.48$

$H(32) = 2.75(32) + 71.48 = 159.48$

The predicted height is 159.48 cm.

40. 167.73 cm

41. Locate the point that is directly above 225. Then estimate its second coordinate by moving horizontally from the point to the vertical axis. The rate is about 75 per 10,000 men.

42. 125 per 10,000 men

43. Locate the point on the graph that is directly above '60. Then estimate its second coordinate by moving horizontally from the point to the vertical axis. In 1960, about 56% of Americans were willing to vote for a woman for president. That is, $P(1960) \approx 56\%$.

44. 92%

45. Plot and connect the points, using body weight as the first coordinate and the corresponding number of drinks as the second coordinate.

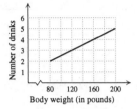

To estimate the number of drinks that a 140-lb-person would have to drink to be considered intoxicated, first locate the point that is directly above 140. Then estimate its second coordinate by moving horizontally from the point to the vertical axis. Read the approximate function value there. The estimated number of drinks is 3.5.

46. 3 drinks

47. Plot and connect the points, using the year as the first coordinate and the corresponding number of reported cases of AIDS as the second coordinate.

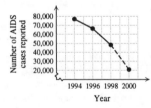

To estimate the number of cases of AIDS reported in 2001, extend the graph and extrapolate. It appears that about 21,000 cases of AIDS were reported in 2000.

48. About 57,000 cases

49. Plot and connect the points, using the year as the first coordinate and the population as the second.

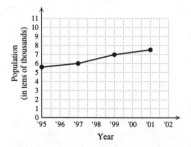

To estimate what the population was in 1998, first locate the point that is directly above 1998. Then

estimate its second coordinate by moving horizontally from the point to the vertical axis. Read the approximate function value there. The population was about 65,000.

50. About 80,000

51. Plot and connect the points, using the year as the first coordinate and the sales total as the second coordinate.

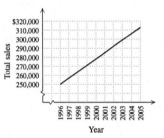

To predict the total sales for 2005, first locate the point directly above 2005. Then estimate its second coordinate by moving horizontally to the vertical axis. Read the approximate function value there. The predicted 2005 sales total is about $313,000.

52. About $271,000

53. Writing exercise

54. Writing exercise

55. $\dfrac{10 - 3^2}{9 - 2 \cdot 3} = \dfrac{10 - 9}{9 - 6} = \dfrac{1}{3}$

56. -1

57.
$$S = 2lh + 2lw + 2wh$$
$$S - 2wh = 2lh + 2lw$$
$$S - 2wh = l(2h + 2w)$$
$$\frac{S - 2wh}{2h + 2w} = l$$

58. $w = \dfrac{S - 2lh}{2l + 2h}$

59.
$$2x + 3y = 6$$
$$3y = 6 - 2x$$
$$y = \frac{6 - 2x}{3}, \text{ or } 2 - \frac{2}{3}x$$

60. $y = \dfrac{5}{4}x - 2$

61. Writing exercise

62. Writing exercise

63. To find $f(g(-4))$, we first find $g(-4)$:

$g(-4) = 2(-4) + 5 = -8 + 5 = -3$.

Then $f(g(-4)) = f(-3) = 3(-3)^2 - 1 = 3 \cdot 9 - 1 = 27 - 1 = 26$.

To find $g(f(-4))$, we first find $f(-4)$:

$f(-4) = 3(-4)^2 - 1 = 3 \cdot 16 - 1 = 48 - 1 = 47$.

Then $g(f(-4)) = g(47) = 2 \cdot 47 + 5 = 94 + 5 = 99$.

64. 26; 9

65. Locate the highest point on the graph. Then move horizontally to the vertical axis and read the corresponding pressure. It is about 22 mm.

66. About 2 minutes, 50 seconds.

67. Writing exercise

68. 1 every 3 minutes

69.

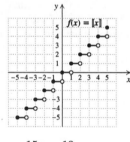

70. $g(x) = \dfrac{15}{4}x - \dfrac{13}{4}$

71. Graph the energy expenditures for walking and for bicycling on the same axes. Using the information given we plot and connect the points $\left(2\frac{1}{2}, 210\right)$ and $\left(3\frac{3}{4}, 300\right)$ for walking. We use the points $\left(5\frac{1}{2}, 210\right)$ and $(13, 660)$ for bicycling.

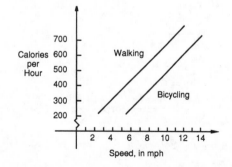

From the graph we see that walking $4\frac{1}{2}$ mph burns about 350 calories per hour and bicycling 14 mph burns about 725 calories per hour. Walking for two

hours at $4\frac{1}{2}$ mph, then, would burn about $2 \cdot 350$, or 700 calories. Thus, bicycling 14 mph for one hour burns more calories than walking $4\frac{1}{2}$ mph for two hours.

Exercise Set 7.2

1. The domain is the set of all first coordinates, $\{2, 9, -2, -4\}$.

The range is the set of all second coordinates, $\{8, 3, 10, 4\}$.

2. Domain: $\{1, 2, 3, 4\}$; range: $\{2, 3, 4, 5\}$

3. The domain is the set of all first coordinates, $\{0, 4, -5, -1\}$.

The range is the set of all second coordinates, $\{0, -2\}$.

4. Domain: $\{3, 2, 1, 0\}$; range: $\{7\}$

5. The function f can be written as $\{(-4, -2), (-2, -1), (0, 0), (2, 1), (4, 2)\}$.
The domain is the set of all first coordinates, $\{-4, -2, 0, 2, 4\}$ and the range is the set of all second coordinates, $\{-2, -1, 0, 1, 2\}$.

6. Domain: $\{-4, -2, 0, 3, 5\}$; range: $\{4, 1, 3, -2, 0\}$

7. The function f can be written as $\{(-5, -1), (-3, -1), (-1, -1), (0, 1), (2, 1), (4, 1)\}$.
The domain is the set of all first coordinates, $\{-5, -3, -1, 0, 2, 4\}$ and the range is the set of all second coordinates, $\{-1, 1\}$.

8. Domain: $\{-3, -2, -1, 0, 1, 2, 3\}$; range: $\{-2, -1, 1, 2\}$

9. The domain of the function is the set of all x-values that are in the graph, $\{x| -4 \le x \le 3\}$.

The range is the set of all y-values that are in the graph, $\{y| -3 \le y \le 4\}$.

10. Domain: $\{x| -4 \le x \le 3\}$; range: $\{y|0 \le y \le 2\}$

11. The domain of the function is the set of all x-values that are in the graph, $\{x| -4 \le x \le 5\}$.

The range is the set of all y-values that are in the graph, $\{y| -2 \le y \le 4\}$.

12. Domain: $\{x| -2 \le x \le 5\}$; range: $\{y| -2 \le y \le 4\}$

13. The domain of the function is the set of all x-values that are in the graph, $\{x| -4 \le x \le 4\}$.

The range is the set of all y-values that are in the graph, $\{-3, -1, 1\}$.

14. Domain: $\{x|-3 \le x \le 5\}$; range: $\{-2, 1, 4\}$

15. For any x-value and for any y-value there is a point on the graph. thus,

Domain of $f = \{x|x$ is a real number$\}$ and

Range of $f = \{y|y$ is a real number$\}$.

16. Domain: $\{x|x$ is a real number$\}$;
range: $\{y|y$ is a real number$\}$

17. For any x-value there is a point on the graph. Thus,

Domain of $f = \{x|x$ is a real number$\}$.

The only y-value on the graph is 4. Thus,

Range of $f = \{4\}$.

18. Domain: $\{x|x$ is a real number$\}$; range: $\{-2\}$

19. For an x-value there is a point on the graph. Thus,

Domain of $f = \{x|x$ is a real number$\}$.

The function has no y-values less than 1 and every y-value greater than or equal to 1 corresponds to a member of the domain. Thus,

Range of $f = \{y|y \ge 1\}$.

20. Domain: $\{x|x$ is a real number$\}$; range: $\{y|y \le 4\}$

21. The hole in the graph at $(-2, -4)$ indicates that the function is not defined for $x = -2$. For any other x-value there is a point on the graph. Thus,

Domain of $f = \{x|x$ is a real number and $x \ne -2\}$.

There is no function value at $(-2, -4)$, so -4 is not in the range of the function. For any other y-value there is a point on the graph. Thus,

Range of $f = \{y|y$ is a real number and $y \ne -4\}$.

22. Domain: $\{x|x$ is a real number and $x \ne 5\}$;

range: $\{y|y$ is a real number and $y \ne 2\}$

23. The function has no x-values less than 0 and every x-value greater than or equal to 0 corresponds to a member of the domain. Thus,

Domain of $f = \{x|x \ge 0\}$.

The function has no y-values less than 0 and every y-value greater than or equal to 0 corresponds to a member of the range. Thus,

Range of $f = \{y|y \ge 0\}$.

24. Domain: $\{x|x \le 3\}$; range: $\{y|y \ge 0\}$

25. $f(x) = \dfrac{5}{x-3}$

Since $\dfrac{5}{x-3}$ cannot be computed when the denominator is 0, we find the x-value that causes $x - 3$ to be 0:

$x - 3 = 0$

$x = 3$ Adding 3 to both sides

Thus, 3 is not in the domain of f, while all other real numbers are. The domain of f is $\{x|x$ is a real number and $x \ne 3\}$.

26. $\{x|x$ is a real number and $x \ne 6\}$

27. $f(x) = \dfrac{3}{2x-1}$

Since $\dfrac{3}{2x-1}$ cannot be computed when the denominator is 0, we find the x-value that causes $2x - 1$ to be 0:

$2x - 1 = 0$

$2x = 1$

$x = \dfrac{1}{2}$

Thus, $\dfrac{1}{2}$ is not in the domain of f, while all other real numbers are. The domain of f is $\left\{x|x \text{ is a real number } and \text{ } x \ne \dfrac{1}{2}\right\}$.

28. $\left\{x\middle|x \text{ is a real number } and \text{ } x \ne -\dfrac{3}{4}\right\}$

29. $f(x) = 2x + 1$

Since we can compute $2x + 1$ for any real number x, the domain is the set of all real numbers.

30. All real numbers

31. $g(x) = |5 - x|$

Since we can compute $|5 - x|$ for any real number x, the domain is the set of all real numbers.

32. All real numbers

33. $f(x) = \dfrac{5}{x^2 - 9}$

The expression $\dfrac{5}{x^2 - 9}$ is undefined when $x^2 - 9 = 0$.

$x^2 - 9 = 0$

$(x + 3)(x - 3) = 0$

$x + 3 = 0$ or $x - 3 = 0$

$x = -3$ or $x = 3$

Thus, Domain of $f = \{x|x$ is a real number and $x \ne -3$ and $x \ne 3\}$.

34. $\{x | x$ is a real number and $x \neq 1\}$

35. $f(x) = x^2 - 9$

Since we can compute $x^2 - 9$ for any real number x, the domain is the set of all real numbers.

36. All real numbers

37. $f(x) = \dfrac{2x - 7}{x^2 + 8x + 7}$

The expression $\dfrac{2x - 7}{x^2 + 8x + 7}$ is undefined when $x^2 + 8x + 7 = 0$.

$$x^2 + 8x + 7 = 0$$
$$(x + 1)(x + 7) = 0$$
$$x + 1 = 0 \quad or \quad x + 7 = 0$$
$$x = -1 \quad or \quad x = -7$$

Thus, Domain of $f = \{x | x$ is a real number and $x \neq -1$ and $x \neq -7\}$.

38. $\left\{x \middle| x \text{ is a real number } and \text{ } x \neq \dfrac{3}{2} \text{ } and \text{ } x \neq -1\right\}$

39. $R(t) = 46.8 - 0.075t$

If we assume the function is not valid for years before 1930, we must have $t \geq 0$. In addition, $R(t)$ must be positive, so we have:

$$46.8 - 0.075t > 0$$
$$-0.075t > -46.8$$
$$t < 624$$

Then the domain of the function is $\{t | 0 \leq t < 624\}$.

40. $\left\{t \middle| 0 \leq t < 513\dfrac{1}{3}\right\}$

41. $A(p) = -2.5p + 26.5$

The price must be positive, so we have $p > \$0$. In addition $A(p)$ must be nonnegative, so we have:

$$-2.5p + 26.5 \geq 0$$
$$26.5 \geq 2.5p$$
$$10.6 \geq p$$

Then the domain of the function is $\{p | \$0 < p \leq \$10.60\}$.

42. $\{p | p \geq \$5.50\}$

43. $P(d) = 0.03d + 1$

The depth must be nonnegative, so we have $d \geq 0$. In addition, $P(d)$ must be nonnegative, so we have:

$$0.03d + 1 \geq 0$$
$$0.03d \geq -1$$
$$d \geq -33.\overline{3}$$

Then we have $d \geq 0$ and $d \geq -33.\overline{3}$, so the domain of the function is $\{d | d \geq 0\}$.

44. $\{s | s > 0\}$

45. $h(t) = -16t^2 + 64t + 80$

The time cannot be negative, so we have $t \geq 0$. The height cannot be negative either, so an upper limit for t will be the positive value of t for which $h(t) = 0$.

$$-16t^2 + 64t + 80 = 0$$
$$-16(t^2 - 4t - 5) = 0$$
$$-16(t - 5)(t + 1) = 0$$
$$t - 5 = 0 \quad or \quad t + 1 = 0$$
$$t = 5 \quad or \quad \quad t = -1$$

We know that -1 is not in the domain of the function. We also see that 5 is an upper limit for t. Then the domain of the function is $\{t | 0 \leq t \leq 5\}$.

46. $\{t | 0 \leq t \leq 7\}$

47. $f(x) = \begin{cases} x, & \text{if } x < 0 \\ 2x + 1, & \text{if } x \geq 0 \end{cases}$

a) Since $-5 < 0$, we use the equation $f(x) = x$.

Thus, $f(-5) = -5$.

b) Since $0 \geq 0$, we use the equation $f(x) = 2x + 1$.

$f(0) = 2 \cdot 0 + 1 = 0 + 1 = 1$

c) Since $10 \geq 0$, we use the equation $f(x) = 2x + 1$.

$f(10) = 2 \cdot 10 + 1 = 20 + 1 = 21$

48. a) -5; b) 0; c) 18

49. $G(x) = \begin{cases} x - 5, & \text{if } x < -1 \\ x, & \text{if } -1 \leq x \leq 2 \\ x + 2, & \text{if } x > 2 \end{cases}$

a) Since $-1 \leq 0 \leq 2$, we use the equation $G(x) = x$.

$G(0) = 0$

b) Since $-1 \leq 2 \leq 2$, we use the equation $G(x) = x$.

$G(2) = 2$

c) Since $5 > 2$, we use the equation $G(x) = x + 2$.

$G(5) = 5 + 2 = 7$

50. a) -2; b) 3; c) -50

51. $f(x) = \begin{cases} x^2 - 10, & \text{if } x < -10 \\ x^2, & \text{if } -10 \le x \le 10 \\ x^2 + 10, & \text{if } x > 10 \end{cases}$

a) Since $-10 \le -10 \le 10$, we use the equation $f(x) = x^2$.

$f(-10) = (-10)^2 = 100$

b) Since $-10 \le 10 \le 10$, we use the equation $f(x) = x^2$.

$f(10) = 10^2 = 100$

c) Since $11 > 10$, we use the equation $f(x) = x^2 + 10$.

$f(11) = 11^2 + 10 = 121 + 10 = 131$

52. a) -3; b) 9; c) 23

53. Writing exercise

54. Writing exercise

55. $y = 2x - 3$

First we plot the y-intercept, $(0, -3)$. We can think of the slope as $\frac{2}{1}$. Starting at $(0, -3)$, find a second point by moving up 2 units and to the right 1 unit to the point $(1, -1)$. In a similar manner we can move from $(1, -1)$ to $(2, 1)$. Then we connect the points.

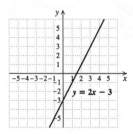

56.

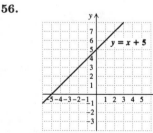

57. $\frac{2}{3}x - 4$

The slope is $\frac{2}{3}$, and the y-intercept is $(0, -4)$.

58. Slope: $-\frac{1}{4}$; y-intercept: $(0, 6)$

59. $y = \frac{4}{3}x$, or $y = \frac{4}{3}x + 0$

The slope is $\frac{4}{3}$ and the y-intercept is $(0, 0)$.

60. Slope: -5; y-intercept: $(0, 0)$

61. Writing exercise

62. Writing exercise

63.

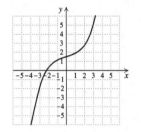

64.

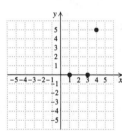

65.

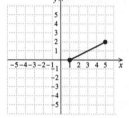

66.

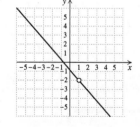

67. The graph indicates that the function is not defined for $x = 0$. For any other x-value there is a point on the graph. Thus,

Domain of $f = \{x | x \text{ is a real number } and \ x \ne 0\}$.

The graph also indicates that the function is not defined for $y = 0$. For any other y-value there is a point on the graph. Thus,

Domain of $f = \{y | y$ is a real number $and\ y \neq 0\}$.

68. Domain: $\{x | x \leq -2\ or\ x \geq 2\}$; range: $\{y | y \geq 0\}$

69. The function has no x-values for $-2 \leq x \leq 0$. For any other x-value there is a point on the graph. Thus, the domain of the function is $\{x | x < -2\ or\ x > 0\}$.

The function has no y-values for $-2 \leq y \leq 3$. Every other y-value corresponds to a member of the range. Then the range is $\{y | y < -2\ or\ y > 3\}$.

70. Domain: $\{x | x$ is a real number$\}$;

range: $\{x | x$ is a real number$\}$

71.

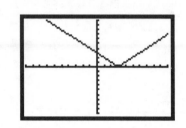

From the graph we see that the domain of f is $\{x | x$ is a real number$\}$ and the range is $\{y | y \geq 0\}$.

72. Domain: $\{x | x$ is a real number$\}$; range: $\{y | y \geq -3\}$

73.

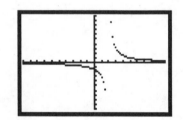

From the graph we see that the domain of f is $\{x | x$ is a real number $and\ y \neq 2\}$ and the range is $\{y | y$ is a real number $and\ y \neq 0\}$.

74. Domain: $\{x | x$ is a real number $and\ x \neq -3\}$;

range: $\{y | y$ is a real number $and\ y \neq 0\}$

75. We graph the function $h(t) = -16t^2 + 64t + 80$ in the window $[0, 5, -5, 150]$ with Xscl = 1 and Yscl = 15.

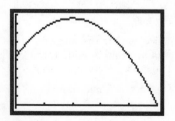

From the graph we estimate that the range of the function is $\{h | 0 \leq h \leq 144\}$.

76. $\{h | 0 \leq h \leq 324\}$

77.

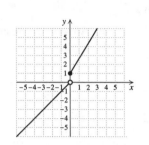

78.

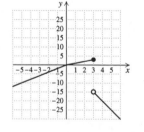

79.

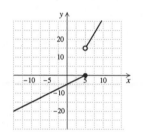

80.

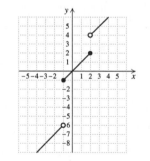

81.

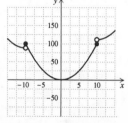

82.

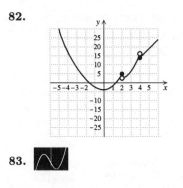

83.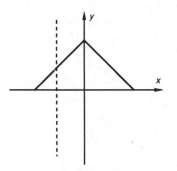

Exercise Set 7.3

1. We can use the vertical line test:

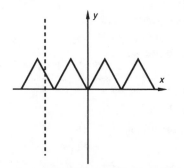

Visualize moving this vertical line across the graph. No vertical line will intersect the graph more than once. Thus, the graph is a graph of a function.

2. No

3. We can use the vertical line test:

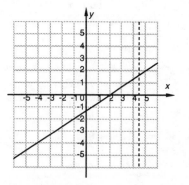

Visualize moving this vertical line across the graph. No vertical line will intersect the graph more than once. Thus, the graph is a graph of a function.

4. No

5. We can use the vertical line test.

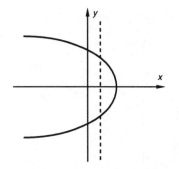

It is possible for a vertical line to intersect the graph more than once. Thus this is not the graph of a function.

6. Yes

7. We can use the vertical line test.

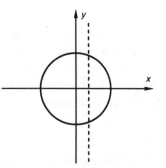

It is possible for a vertical line to intersect the graph more than once. Thus this is not a graph of a function.

8. Yes

9. We can use the vertical line test.

Visualize moving this vertical line across the graph. No vertical line will intersect the graph more than once. Thus, the graph is a graph of a function.

10. Yes

11. We can use the vertical line test.

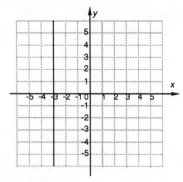

The vertical line $x = -3$ intersects the graph more than once. Thus this is not a graph of a function.

12. Yes

13. $F(t) = -5000t + 90,000$

a) -5000 signifies that the truck's value depreciates $5000 per year; 90,000 signifies that the original value of the truck was $90,000.

b) We find the value of t for which $F(t) = 0$.
$$0 = -5000t + 90,000$$
$$5000t = 90,000$$
$$t = 18$$
It will take 18 yr for the truck to depreciate completely.

c) The truck's value goes from $90,000 when $t = 0$ to $0 when $t = 18$, so the domain of F is $\{x | 0 \le t \le 18\}$.

14. a) -2000 signifies that the color separator's value depreciates $2000 per year; 15,000 signifies that the original value of the separator was $15,000.

b) 7.5 yr; c) $\{t | 0 \le t \le 7.5\}$

15. $v(n) = -150n + 900$

a) -150 signifies that the snowblower's value depreciates $150 per winter of use; 900 signifies that the original value of the snowblower was $900.

b) We find the value of n for which $v(n) = 300$.
$$300 = -150n + 900$$
$$-600 = -150n$$
$$4 = n$$
The snowblower's trade-in value will be $300 after 4 winters of use.

c) First we find the value of n for which $v(n) = 0$.
$$0 = -150n + 900$$
$$-900 = -150n$$
$$6 = n$$
The value of the snowblower goes from $900 when $n = 0$ to $0 when $n = 6$. Since the snowblower is used only in the winter we express the domain of v as $\{0, 1, 2, 3, 4, 5, 6\}$.

16. a) -300 signifies that the mower's value depreciates $300 per summer of use; 2400 signifies that the original value of the mower was $2400.

b) after 4 summers of use;

c) $\{0, 1, 2, 3, 4, 5, 6, 7, 8\}$

17. *Familiarize*. A monthly fee is charged after the purchase of the phone. After one month of service, the total cost will be $50 + $25 = $75. After two months, the total cost will be $50 + 2 \cdot $25 = $100. We can generalize this with a model, letting $C(t)$ represent the total cost, in dollars, for t months of service.

Translate. We reword the problem and translate.

Total cost	is	cost of phone	plus	$25 per month.
$C(t)$	$=$	50	$+$	$25 \cdot t$

Carry out. The model can be written $C(t) = 25t + 50$. To find the time required for the total cost to reach $150, substitute 150 for $C(t)$ and solve for t.
$$C(t) = 25t + 50$$
$$150 = 25t + 50$$
$$100 = 25t$$
$$4 = t$$

Check. We evaluate.
$$C(4) = 25 \cdot 4 + 50$$
$$= 100 + 50$$
$$= 150$$
The answer checks.

State. It takes 4 months for the total cost to reach $150.

18. $C(t) = 40t + 60$; 5 months

19. *Familiarize*. Tina's hair is initially 1 in. long, and it grows at a rate of $\dfrac{1}{2}$ in. per month. After 1 month the length of her hair will be $1 + \dfrac{1}{2}$, or $1\dfrac{1}{2}$ in. After 2 months the length will be $1 + 2 \cdot \dfrac{1}{2}$, or 2 in. This can be generalized with a model, letting $L(t)$

represent the length of Tina's hair, in inches, after t months.

Translate.

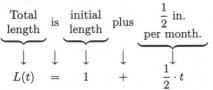

$$L(t) = 1 + \frac{1}{2} \cdot t$$

Carry out. The model can be written $L(t) = \frac{1}{2}t + 1$. To find the time required for the hair to be 3 in. long, substitute 3 for $L(t)$ and solve for t.

$$L(t) = \frac{1}{2}t + 1$$

$$3 = \frac{1}{2}t + 1$$

$$2 = \frac{1}{2}t$$

$$4 = t \qquad \text{Multiplying by 2}$$

Check. We evaluate.

$$L(4) = \frac{1}{2} \cdot 4 + 1$$

$$= 2 + 1$$

$$= 3$$

The answer checks.

State. Tina's hair will be 3 in. long after 4 months.

20. $L(t) = \frac{1}{8}t + 2$; 12 days after the lawn is cut

21. Familiarize. A charge for each mile is applied after an initial charge. For a 1 mile taxi ride, the total cost is $\$2 + \$0.75 = \$2.75$. For a two mile ride the cost is $\$2 + 2(\$0.75) = \$3.50$. We can generalize this with a model, letting $C(d)$ represent the total cost, in dollars, for a taxi ride of d miles.

Translate.

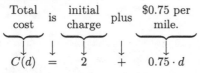

$$C(d) = 2 + 0.75 \cdot d$$

Carry out. The model can be written $C(d) = 0.75d + 2$. To find the length of a taxi ride when the cost is $\$4.25$, substitute 4.25 for $C(d)$ and solve for d.

$$C(d) = 0.75d + 2$$

$$4.25 = 0.75d + 2$$

$$2.25 = 0.75d$$

$$3 = d$$

Check. We evaluate.

$$C(3) = 0.75(3) + 2$$

$$= 2.25 + 2$$

$$= 4.25$$

State. The length of a taxi ride is 3 mi when the cost is $\$4.25$.

22. $D(t) = \frac{1}{5}t + 20$; 40 yr after 1960, or in 2000

23. a) We form pairs of the type (t, R) where t is the number of years since 1930 and R is the record. We have two pairs, $(0, 46.8)$ and $(40, 43.8)$. These are two points on the graph of the linear function we are seeking. We use the point-slope form to write an equation relating R and t:

$$m = \frac{43.8 - 46.8}{40 - 0} = \frac{-3}{40} = -0.075$$

$$R - 46.8 = -0.075(t - 0)$$

$$R - 46.8 = -0.075t$$

$$R = -0.075t + 46.8$$

$$R(t) = -0.075t + 46.8 \quad \text{Using function notation}$$

b) 2003 is 73 years since 1930, so to predict the record in 2003, we find $R(73)$:

$$R(73) = -0.075(73) + 46.8$$

$$= 41.325$$

The predicted record is 41.325 seconds in 2003.

2006 is 76 years since 1930, so to predict the record in 2006, we find $R(76)$:

$$R(76) = -0.075(76) + 46.8$$

$$= 41.1$$

The predicted record is 41.1 seconds in 2006.

c) Substitute 40 for $R(t)$ and solve for t:

$$40 = -0.075t + 46.8$$

$$-6.8 = -0.075t$$

$$91 \approx t$$

The record will be 40 seconds about 91 years after 1930, or in 2021.

24. a) $R(t) = -0.0075t + 3.85$; b) 3.31 min; 3.28 min; c) 2030

25. a) We form the pairs $(0, 178.6)$ and $(8, 243.1)$.

Use the point-slope form to write an equation relating A and t:
$$m = \frac{243.1 - 178.6}{8 - 0} = \frac{64.5}{8} = 8.0625$$
$$A - 178.6 = 8.0625(x - 0)$$
$$A - 178.6 = 8.0625x$$
$$A = 8.0625t + 178.6$$
$$A(t) = 8.0625t + 178.6 \quad \text{Using}$$
$$\text{function notation}$$

b) 2008 is 16 years since 1992, so we find $A(16)$:
$$A(16) = 8.0625(16) + 178.6$$
$$= 307.6$$
We predict that the amount of PAC contributions in 2008 will be \$307.6 million.

26. a) $N(t) = 4.25t + 43.8$; b) 94.8 million tons

27. a) We form the pairs $(0, 74.9)$ and $(3, 77.5)$.

Use the point-slope form to write an equation relating A and t:
$$m = \frac{77.5 - 74.9}{3 - 0} = \frac{2.6}{3} = \frac{26}{30} = \frac{13}{15}$$
$$A - 74.9 = \frac{13}{15}(t - 0)$$
$$A - 74.9 = \frac{13}{15}t$$
$$A = \frac{13}{15}t + 74.9$$
$$A(t) = \frac{13}{15}t + 74.9 \quad \text{Using function}$$
$$\text{notation}$$

b) 2006 is 12 years since 1994, so we find $A(12)$:
$$A(12) = \frac{13}{15}(12) + 74.9$$
$$= 85.3$$
We predict that there will be 85.3 million acres of land in the national park system in 2006.

28. a) $E(t) = \frac{2}{35}t + 78.8$; b) 79.8 years

29. $f(x) = \frac{1}{3}x - 7$

The function is in the form $f(x) = mx + b$, so it is a linear function. We can compute $\frac{1}{3}x - 7$ for any value of x, so the domain is the set of all real numbers.

30. Rational function;
$\{x | x$ is a real number $and \ x \neq -1\}$

31. $p(x) = x^2 + x + 1$

The function is in the form $(x) = ax^2 + bx + c$, $a \neq 0$, so it is a quadratic function. We can compute $x^2 + x + 1$ for any value of x, so the domain is the set of all real numbers.

32. Absolute-value function; all real numbers

33. $f(t) = \frac{12}{3t + 4}$

The function is described by a rational equation, so it is a rational function. The expression $\frac{12}{3t + 4}$ is undefined when $t = -\frac{4}{3}$, so the domain is $\left\{t \middle| t \text{ is a real number } and \ t \neq -\frac{4}{3}\right\}$.

34. Linear function; all real numbers

35. $f(x) = 0.02x^4 - 0.1x + 1.7$

The function is described by a polynomial equation that is neither linear nor quadratic, so it is a polynomial function. We can compute $0.02x^4 - 0.1x + 1.7$ for any value of x, so the domain is the set of all real numbers.

36. Absolute-value function; all real numbers

37. $f(x) = \frac{x}{2x - 5}$

The function is described by a rational equation, so it is a rational function. The expression $\frac{x}{2x - 5}$ is undefined when $x = \frac{5}{2}$, so the domain is $\left\{x \middle| x \text{ is a real number } and \ x \neq \frac{5}{2}\right\}$.

38. Rational function;
$\left\{x \middle| x \text{ is a real number } and \ x \neq \frac{4}{3}\right\}$

39. $f(n) = \frac{4n - 7}{n^2 + 3n + 2}$

The function is described by a rational equation, so it is a rational function. The expression $\frac{4n - 7}{n^2 + 3n + 2}$ is undefined for values of n that make the denominator 0. We find those values:
$$n^2 + 3n + 2 = 0$$
$$(n + 1)(n + 2) = 0$$
$$n + 1 = 0 \quad or \quad n + 2 = 0$$
$$n = -1 \quad or \quad n = -2$$
Then the domain is $\{n | n$ is a real number and $n \neq -1 \ and \ n \neq -2\}$.

40. Rational function;
$\{x|x \text{ is a real number } and \ x \neq -1$
$and \ x \neq 1\}$

41. $f(n) = 200 - 0.1n$

The function can be written in the form $f(n) = mn + b$, so it is a linear function. We can compute $200 - 0.1n$ for any value of n, so the domain is the set of all real numbers.

42. Quadratic function; all real numbers

43. The function has no y-values less than 0 and every y-value greater than or equal to 0 corresponds to a member of the domain. Thus, the range is $\{y|y \geq 0\}$.

44. $\{y|y \geq 0\}$

45. Every y-value corresponds to a member of the domain, so the range is the set of all real numbers.

46. $\{y|y \geq 3\}$

47. There is no y-value greater than 0 and every y-value less than or equal to 0 corresponds to a member of the domain. Thus, the range is $\{y|y \leq 0\}$.

48. All real numbers

49.

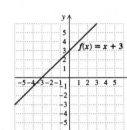

For any x-value and for any y-value there is a point on the graph. Thus,

Domain of $f = \{x|x \text{ is a real number}\}$ and

Range of $f = \{y|y \text{ is a real number}\}$.

50.

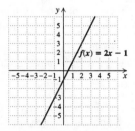

Domain: $\{x|x \text{ is a real number}\}$;

range: $\{y|y \text{ is a real number}\}$

51.

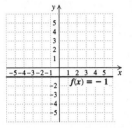

For any x-value there is a point on the graph, so

Domain of $f = \{x|x \text{ is a real number}\}$.

The only y-value on the graph is -1, so

Range of $f = \{-1\}$.

52.

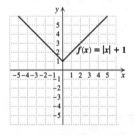

Domain: $\{x|x \text{ is a real number}\}$;

range: $\{2\}$

53.

For any x-value there is a point on the graph, so

Domain of $f = \{x|x \text{ is a real number}\}$.

There is no y-value less than 1 and every y-value greater than or equal to 1 corresponds to a member of the domain. Thus,

Range of $f = \{y|y \geq 1\}$.

54.

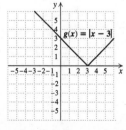

Domain: $\{x|x \text{ is a real number}\}$;

range: $\{y|y \geq 0\}$

55.

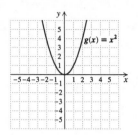

For any x-value there is a point on the graph, so

Domain of $g = \{x|x$ is a real number$\}$.

There is no y-value less than 0 and every y-value greater than or equal to 0 corresponds to a member of the domain. Thus,

Range of $g = \{y|y \geq 0\}$.

56.

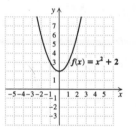

Domain: $\{x|x$ is a real number$\}$;

range: $\{y|y \geq 2\}$

57. Writing exercise

58. Writing exercise

59. $(x^2 + 2x + 7) + (3x^2 - 8)$
$= (x^2 + 3x^2) + 2x + (7 - 8)$
$= 4x^2 + 2x - 1$

60. $2x^3 - x^2 - x + 7$

61. $(2x + 1)(x - 7) = 2x^2 - 14x + x - 7$
$= 2x^2 - 13x - 7$

62. $x^2 + x - 12$

63. $(x^3 + x^2 - 4x + 7) - (3x^2 - x + 2)$
$= x^3 + x^2 - 4x + 7 - 3x^2 + x - 2$
$= x^3 + (x^2 - 3x^2) + (-4x + x) + (7 - 2)$
$= x^3 - 2x^2 - 3x + 5$

64. $x^3 + 2x^2 + x + 4$

65. Writing exercise

66. Writing exercise

67. Writing exercise

68. \$1350

69. *Familiarize*. The Celsius temperature C corresponding to a given Fahrenheit temperature F can be modeled by a line that contains the points $(32, 0)$ and $(212, 100)$.

***Translate*.** We find an equation relating C and F.

$$m = \frac{100 - 0}{212 - 32} = \frac{100}{180} = \frac{5}{9}$$

$$C - 0 = \frac{5}{9}(F - 32)$$

$$C = \frac{5}{9}(F - 32)$$

***Carry out*.** Using function notation we have $C(F) = \frac{5}{9}(F - 32)$. To find the Celsius temperature that corresponds to a Fahrenheit temperature of $70°$, we find $C(70)$:

$$C(70) = \frac{5}{9}(70 - 32) = \frac{5}{9}(38) \approx 21.1.$$

***Check*.** We can repeat our calculations. We could also graph the function and determine that $(70, 21.1)$ is on the graph.

***State*.** A temperature of about $21.1°C$ corresponds to a temperature of $70°F$.

70. \$11,000

71. *Familiarize*. The percentage of premiums paid out in benefits in 1993 was $103.6/124.7 \approx 83.1\%$. In 1996 the percentage was $113.8/137.1 \approx 83.0\%$. The percentage P of premiums, in dollars, paid out in benefits t years after 1993 can be modeled by a line that contains the points $(0, 83.1)$ and $(3, 83.0)$.

***Translate*.** We find an equation relating P and t.

$$m = \frac{83.0 - 83.1}{3 - 0} = \frac{-0.1}{3} = -\frac{1}{30}$$

We know the slope and the y-intercept, so we use the slope-intercept equation.

$$P = -\frac{1}{30}t + 83.1$$

***Carry out*.** Using function notation we have $P(t) = -\frac{1}{30}t + 83.1$. Since 2005 is 12 years after 1993, we find $P(12)$ to predict the percentage of premiums that will be paid out in benefits in 2005:

$$P(12) = -\frac{1}{30}(12) + 83.1 = 82.7$$

***Check*.** We can repeat the calculations. We could also graph the function and determine that $(12, 82.7)$ is on the graph.

***State*.** We predict that 82.7% will be paid out in benefits in 2005. (Answers may vary slightly depending on when and how rounding occurred.)

72.

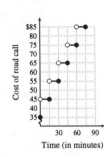

73. We graph $C(t) = 0.50t + 3$, where t represents the number of 15-min units of time, as a series of steps. The cost is constant within each 15-min unit of time. Thus,

for $0 < t \le 1$, $C(t) = 0.5(1) + 3 = \$3.50$;

for $1 < t \le 2$, $C(t) = 0.5(2) + 3 = \$4.00$;

for $2 < t \le 3$, $C(t) = 0.5(3) + 3 = \$4.50$;

and so on. We draw the graph. An open circle at a point indicates that the point is not on the graph.

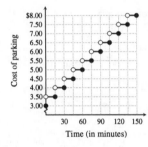

74. False

75. Let $c = 2$ and $d = 3$. Then $f(cd) = f(2 \cdot 3) = f(6) = m \cdot 6 + b = 6m + b$, but $f(c)f(d) = f(2)f(3) = (m \cdot 2 + b)(m \cdot 3 + b) = 6m^2 + 5mb + b^2$. Thus, the given statement is false.

76. False

77. Let $c = 5$ and $d = 2$. Then $f(c - d) = f(5 - 2) = f(3) = m \cdot 3 + b = 3m + b$, but $f(c) - f(d) = f(5) - f(2) = (m \cdot 5 + b) - (m \cdot 2 + b) = 5m + b - 2m - b = 3m$. Thus, the given statement is false.

78. a) $g(x) = x - 8$; b) -10; c) 83

79. Writing exercise

80.

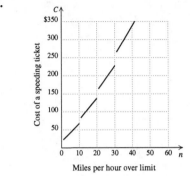

Exercise Set 7.4

1. Since $f(2) = -3 \cdot 2 + 1 = -5$, and $g(2) = 2^2 + 2 = 6$, we have $f(2) + g(2) = -5 + 6 = 1$.

2. 7

3. Since $f(5) = -3 \cdot 5 + 1 = -14$ and $g(5) = 5^2 + 2 = 27$, we have $f(5) - g(5) = -14 - 27 = -41$.

4. -29

5. Since $f(-1) = -3(-1) + 1 = 4$ and $g(-1) = (-1)^2 + 2 = 3$, we have $f(-1) \cdot g(-1) = 4 \cdot 3 = 12$.

6. 42

7. Since $f(-4) = -3(-4) + 1 = 13$ and $g(-4) = (-4)^2 + 2 = 18$, we have $f(-4)/g(-4) = 13/18$.

8. $-\dfrac{8}{11}$

9. Since $g(1) = 1^2 + 2 = 3$ and $f(1) = -3 \cdot 1 + 1 = -2$, we have $g(1) - f(1) = 3 - (-2) = 3 + 2 = 5$.

10. $-\dfrac{6}{5}$

11. $(f + g)(x) = f(x) + g(x) = (-3x + 1) + (x^2 + 2) = x^2 - 3x + 3$

12. $x^2 + 3x + 1$

13. $(F + G)(x) = F(x) + G(x)$
$= x^2 - 2 + 5 - x$
$= x^2 - x + 3$

14. $a^2 - a + 3$

15. Using our work in Exercise 13, we have
$$(F + G)(-4) = (-4)^2 - (-4) + 3$$
$$= 16 + 4 + 3$$
$$= 23.$$

16. 33

17.
$$(F - G)(x) = F(x) - G(x)$$
$$= x^2 - 2 - (5 - x)$$
$$= x^2 - 2 - 5 + x$$
$$= x^2 + x - 7$$
Then we have
$$(F - G)(3) = 3^2 + 3 - 7$$
$$= 9 + 3 - 7$$
$$= 5.$$

18. -1

19.
$$(F \cdot G)(x) = F(x) \cdot G(x)$$
$$= (x^2 - 2)(5 - x)$$
$$= 5x^2 - x^3 - 10 + 2x$$
Then we have
$$(F \cdot G)(-3) = 5(-3)^2 - (-3)^3 - 10 + 2(-3)$$
$$= 5 \cdot 9 - (-27) - 10 - 6$$
$$= 45 + 27 - 10 - 6$$
$$= 56.$$

20. 126

21.
$$(F/G)(x) = F(x)/G(x)$$
$$= \frac{x^2 - 2}{5 - x}, \ x \neq 5$$

22. $-x^2 - x + 7$

23. Using our work in Exercise 21, we have
$$(F/G)(-2) = \frac{(-2)^2 - 2}{5 - (-2)} = \frac{4 - 2}{5 + 2} = \frac{2}{7}.$$

24. $-\dfrac{1}{6}$

25.
$$N(1980) = (R + W)(1980)$$
$$= R(1980) + W(1980)$$
$$\approx 0.75 + 2.5$$
$$\approx 3.25$$
We estimate that 3.25 million U.S. women had children in 1980.

26. $1.3 + 2.6 \approx 3.9$

27. The number of women under 30 who gave birth dropped from 1990 to 1998.

28. Women 30 and over

29. $(n + l)(98) = n(98) + l(98)$

From the middle line of the graph, we can see that $n(98) + l(98) \approx 50$ million.

This represents the total number of passengers serviced by Newark and LaGuardia airports in 1998.

30. About 41 million; the total number of passengers serviced by Kennedy and LaGuardia airports in 1998

31.
$$(k - l)(94) = k(94) - l(94)$$
$$\approx 29 - 21$$
$$\approx 8 \text{ million}$$

This represents how many more passengers used Kennedy airport than LaGuardia airport in 1994.

32. About 1 million; how many more passengers used Kennedy airport than Newark airport in 1994

33. $(n + l + k)(99) = n(99) + l(99) + k(99)$

From the top line of the graph, we can see that $n(99) + l(99) + k(99) \approx 89$ million.

This represents the number of passengers serviced by Newark, LaGuardia, and Kennedy airports in 1999.

34. About 69 million; the number of passengers serviced by Newark, LaGuardia, and Kennedy airports in 1998

35. The domain of f and of g is all real numbers. Thus, Domain of $f + g =$ Domain of $f - g =$ Domain of $f \cdot g = \{x | x$ is a real number$\}$.

36. $\{x | x$ is a real number$\}$

37. Because division by 0 is undefined, we have

Domain of $f = \{x | x$ is a real number *and* $x \neq 3\}$,

and

Domain of $g = \{x | x$ is a real number$\}$.

Thus, Domain of $f + g =$ Domain of $f - g =$ Domain of $f \cdot g = \{x | x$ is a real number *and* $x \neq 3\}$.

38. $\{x | x$ is a real number *and* $x \neq 9\}$

39. Because division by 0 is undefined, we have

Domain of $f = \{x | x$ is a real number *and* $x \neq 0\}$,

and

Domain of $g = \{x | x$ is a real number$\}$.

Thus, Domain of $f + g =$ Domain of $f - g =$ Domain of $f \cdot g = \{x | x$ is a real number *and* $x \neq 0\}$.

40. $\{x | x$ is a real number *and* $x \neq 0\}$

41. Because division by 0 is undefined, we have

Domain of $f = \{x | x$ is a real number *and* $x \neq 1\}$,

and

Domain of $g = \{x | x$ is a real number$\}$.

Thus, Domain of $f + g$ = Domain of $f - g$ = Domain of $f \cdot g = \{x | x$ is a real number *and* $x \neq 1\}$.

42. $\{x | x$ is a real number *and* $x \neq 6\}$

43. Because division by 0 is undefined, we have

Domain of $f = \{x | x$ is a real number *and* $x \neq 2\}$,

and

Domain of $g = \{x | x$ is a real number *and* $x \neq 4\}$.

Thus, Domain of $f + g$ = Domain of $f - g$ = Domain of $f \cdot g = \{x | x$ is a real number *and* $x \neq 2$ *and* $x \neq 4\}$.

44. $\{x | x$ is a real number *and* $x \neq 3$ *and* $x \neq 2\}$

45. Domain of f = Domain of g =

$\{x | x$ is a real number$\}$.

Since $g(x) = 0$ when $x - 3 = 0$, we have $g(x) = 0$ when $x = 3$. We conclude that Domain of f/g = $\{x | x$ is a real number *and* $x \neq 3\}$.

46. $\{x | x$ is a real number *and* $x \neq 5\}$

47. Domain of f = Domain of g =

$\{x | x$ is a real number$\}$.

Since $g(x) = 0$ when $2x - 8 = 0$, we have $g(x) = 0$ when $x = 4$. We conclude that Domain of f/g = $\{x | x$ is a real number *and* $x \neq 4\}$.

48. $\{x | x$ is a real number *and* $x \neq 3\}$

49. Domain of $f = \{x | x$ is a real number *and* $x \neq 4\}$.

Domain of $g = \{x | x$ is a real number$\}$.

Since $g(x) = 0$ when $5 - x = 0$, we have $g(x) = 0$ when $x = 5$. We conclude that Domain of f/g = $\{x | x$ is a real number *and* $x \neq 4$ *and* $x \neq 5\}$.

50. $\{x | x$ is a real number *and* $x \neq 2$ *and* $x \neq 7\}$

51. Domain of $f = \{x | x$ is a real number *and* $x \neq -1\}$.

Domain of $g = \{x | x$ is a real number$\}$.

Since $g(x) = 0$ when $2x + 5 = 0$, we have $g(x) = 0$ when $x = -\dfrac{5}{2}$. We conclude that Domain of f/g = $\left\{x \middle| x$ is a real number *and* $x \neq -1$ *and* $x \neq -\dfrac{5}{2}\right\}$.

52. $\left\{x \middle| x$ is a real number *and* $x \neq 2$ *and* $x \neq -\dfrac{7}{3}\right\}$

53. $(F + G)(5) = F(5) + G(5) = 1 + 3 = 4$

$(F + G)(7) = F(7) + G(7) = -1 + 4 = 3$

54. 0; 2

55. $(G - F)(7) = G(7) - F(7) = 4 - (-1) = 4 + 1 = 5$

$(G - F)(3) = G(3) - F(3) = 1 - 2 = -1$

56. 2; $-\dfrac{1}{4}$

57. From the graph we see that Domain of

$F = \{x | 0 \leq x \leq 9\}$ and Domain of

$G = \{x | 3 \leq x \leq 10\}$. Then Domain of

$F + G = \{x | 3 \leq x \leq 9\}$. Since $G(x)$ is never 0, Domain of $F/G = \{x | 3 \leq x \leq 9\}$.

58. $\{x | 3 \leq x \leq 9\}$; $\{x | 3 \leq x \leq 9\}$;

$\{x | 3 \leq x \leq 9$ *and* $x \neq 6$ *and* $x \neq 8\}$

59. We use $(F + G)(x) = F(x) + G(x)$.

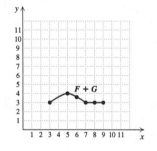

60.

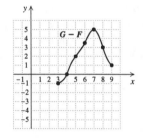

61. Writing exercise

62. Writing exercise

63. $4x - 7y = 8$

$\quad 4x = 7y + 8 \qquad$ Adding $7y$

$\quad \dfrac{1}{4} \cdot 4x = \dfrac{1}{4}(7y + 8) \quad$ Multiplying by $\dfrac{1}{4}$

$\quad\quad x = \dfrac{7}{4}y + 2$

64. $y = \dfrac{3}{8}x - \dfrac{5}{8}$

65. $5x + 2y = -3$

$\quad\quad 2y = -5x - 3$ $\quad$ Subtracting $5x$

$\quad\quad \dfrac{1}{2} \cdot 2y = \dfrac{1}{2}(-5x - 3)$ $\quad$ Multiplying by $\dfrac{1}{2}$

$\quad\quad\quad y = -\dfrac{5}{2}x - \dfrac{3}{2}$

66. $y = -\dfrac{5}{6}y - \dfrac{1}{3}$

67. Let n represent the number; $2n + 5 = 49$.

68. Let n represent the number; $\dfrac{1}{2}n - 3 = 57$.

69. Let n represent the first integer; $x + (x + 1) = 145$.

70. Let x represent the number; $x - (-x) = 20$

71. Writing exercise

72. Writing exercise

73. Domain of $f = \left\{ x \middle| x \text{ is a real number } and \right.$

$\left. x \neq -\dfrac{5}{2} \right\}$; domain of $g = \{ x | x \text{ is a real number } and$

$x \neq -3 \}$; $g(x) = 0$ when $x^4 - 1 = 0$, or when $x = 1$ or $x = -1$.

Then domain of $f/g = \left\{ x \middle| x \text{ is a real number } and \right.$

$\left. x \neq -\dfrac{5}{2} \text{ and } x \neq -3 \text{ and } x \neq 1 \text{ and } x \neq -1 \right\}$.

74. $\{ x | x \text{ is a real number } and \ x \neq 4 \text{ and } x \neq 3 \text{ and } x \neq 2 \text{ and } x \neq -2 \}$

75. Answers may vary.

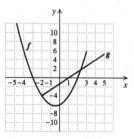

76. $\left\{ x \middle| x \text{ is a real number } and \ -1 < x < 5 \text{ and } x \neq \dfrac{3}{2} \right\}$

77. The domain of each function is the set of first coordinates for that function.

Domain of $f = \{ -2, -1, 0, 1, 2 \}$ and

Domain of $g = \{ -4, -3, -2, -1, 0, 1 \}$.

Domain of $f + g = $ Domain of $f - g = $

Domain of $f \cdot g = \{ -2, -1, 0, 1 \}$.

Since $g(-1) = 0$, we conclude that Domain of $f/g = \{ -2, 0, 1 \}$.

78. 5; 15; 2/3

79. Answers may vary. $f(x) = \dfrac{1}{x + 2}$, $g(x) = \dfrac{1}{x - 5}$

80. Dot mode

81.

82.

Exercise Set 7.5

1. $\quad y = kx$

$\quad 28 = k \cdot 4$ $\quad$ Substituting

$\quad\quad 7 = k$

The variation constant is 7.
The equation of variation is $y = 7x$.

2. $k = \dfrac{5}{12}$; $y = \dfrac{5}{12}x$

3. $\quad y = kx$

$\quad 3.4 = k \cdot 2$ $\quad$ Substituting

$\quad 1.7 = k$

The variation constant is 1.7.
The equation of variation is $y = 1.7x$.

4. $k = \dfrac{2}{5}$; $y = \dfrac{2}{5}x$

5. $\quad y = kx$

$\quad 2 = k \cdot \dfrac{1}{3}$ $\quad$ Substituting

$\quad 6 = k$ $\quad\quad$ Multiplying by 3

The variation constant is 6.

The equation of variation is $y = 6x$.

6. $k = 1.8$; $y = 1.8x$

7. *Familiarize*. Because of the phrase "$d \dots$ varies directly as $\dots m$," we express the distance as a function of the mass. Thus we have $d(m) = km$. We know that $d(3) = 20$.

Translate. We find the variation constant and then find the equation of variation.

$$d(m) = km$$

$$d(3) = k \cdot 3 \quad \text{Replacing } m \text{ with } 3$$

$$20 = k \cdot 3 \quad \text{Replacing } d(3) \text{ with } 20$$

$$\frac{20}{3} = k \quad \text{Variation constant}$$

The equation of variation is $d(m) = \frac{20}{3}m$.

Carry out. We compute $d(5)$.

$$d(m) = \frac{20}{3}m$$

$$d(5) = \frac{20}{3} \cdot 5 \quad \text{Replacing } m \text{ with } 5$$

$$d(5) = \frac{100}{3}, \text{ or } 33\frac{1}{3}$$

Check. Reexamine the calculations. Note that the answer seems reasonable since $\frac{3}{20}$ and $\frac{5}{100/3}$ are equal.

State. The spring is stretched $33\frac{1}{3}$ cm by a hanging object with mass 5 kg.

8. 6 amperes

9. Familiarize. Because N varies directly as the number of people P using the cans, we write N as a function of P: $N(P) = kP$. We know that $N(250) = 60,000$.

Translate.

$$N(P) = kP$$

$$N(250) = k \cdot 250 \quad \text{Replacing } P \text{ with } 250$$

$$60,000 = k \cdot 250 \quad \text{Replacing } N(250) \text{ with } 60,000$$

$$\frac{60,000}{250} = k$$

$$240 = k \quad \text{Variation constant}$$

$$N(P) = 240P \quad \text{Equation of variation}$$

Carry out. Find $N(1,008,000)$.

$$N(P) = 240P$$

$$N(1,008,000) = 240 \cdot 1,008,000$$

$$= 241,920,000$$

Check. Reexamine the calculation.

State. 241,920,000 aluminum cans are used each year in Dallas.

10. $4.29

11. Since we have direct variation and $48 = \frac{1}{2} \cdot 96$, then the result is $\frac{1}{2} \cdot 64$ kg, or 32 kg. We could also do this problem as follows.

Familiarize. Because W varies directly as the total mass, we write $W(m) = km$. We know that $W(96) = 64$.

Translate.

$$W(m) = km$$

$$W(96) = k \cdot 96 \quad \text{Replacing } m \text{ with } 96$$

$$64 = k \cdot 96 \quad \text{Replacing } W(96) \text{ with } 64$$

$$\frac{2}{3} = k \quad \text{Variation constant}$$

$$W(m) = \frac{2}{3}m \quad \text{Equation of variation}$$

Carry out. Find $W(48)$.

$$W(m) = \frac{2}{3}m$$

$$W(48) = \frac{2}{3} \cdot 48$$

$$= 32$$

Check. Reexamine the calculations.

State. There are 32 kg of water in a 64 kg person.

12. 40 lb

13. Familiarize. Because the f-stop varies directly as F, we write $f(F) = kF$. We know that $F(150) = 6.3$.

Translate.

$$f(F) = kF$$

$$f(150) = k \cdot 150 \quad \text{Replacing } F \text{ with } 150$$

$$6.3 = k \cdot 150 \quad \text{Replacing } f(150) \text{ with } 6.3$$

$$0.042 = k \quad \text{Variation constant}$$

$$f(F) = 0.042F \quad \text{Equation of variation}$$

Carry out. Find $f(80)$.

$$f(F) = 0.042F$$

$$f(80) = 0.042(80)$$

$$= 3.36$$

Check. Reexamine the calculations.

State. An 80 mm focal length has an f-stop of 3.36.

14. 7,700,000 tons

15.

$$y = \frac{k}{x}$$

$$3 = \frac{k}{20} \quad \text{Substituting}$$

$$60 = k$$

The variation constant is 60.

The equation of variation is $y = \frac{60}{x}$.

16. $k = 64;\ y = \dfrac{64}{x}$

17. $\quad y = \dfrac{k}{x}$

$\quad 28 = \dfrac{k}{4}\quad$ Substituting

$\quad 112 = k$

The variation constant is 112.

The equation of variation is $y = \dfrac{112}{x}$.

18. $k = 45;\ y = \dfrac{45}{x}$

19. $\quad y = \dfrac{k}{x}$

$\quad 27 = \dfrac{k}{\dfrac{1}{3}}\quad$ Substituting

$\quad 9 = k$

The variation constant is 9.

The equation of variation is $y = \dfrac{9}{x}$.

20. $k = 9;\ y = \dfrac{9}{x}$

21. *Familiarize*. Because of the phrase "t varies inversely as $\ldots u$," we write $t(u) = k/u$. We know that $t(4) = 70$.

***Translate*.** We find the variation constant and then we find the equation of variation.

$\quad t(u) = \dfrac{k}{u}$

$\quad t(4) = \dfrac{k}{4}\qquad$ Replacing u with 4

$\quad 70 = \dfrac{k}{4}\qquad$ Replacing $t(4)$ with 70

$\quad 280 = k\qquad$ Variation constant

$\quad t(u) = \dfrac{280}{u}\qquad$ Equation of variation

***Carry out*.** We find $t(14)$.

$\quad t(14) = \dfrac{280}{14} = 20$

***Check*.** Reexamine the calculations. Note that, as expected, as the UV rating increases, the time it takes to burn goes down.

***State*.** It will take 20 min to burn when the UV rating is 14.

22. $\dfrac{2}{9}$ ampere

23. *Familiarize*. Because V varies inversely as P, we write $V(P) = k/P$. We know that $V(32) = 200$.

***Translate*.**

$\quad V(P) = \dfrac{k}{P}$

$\quad V(32) = \dfrac{k}{32}\qquad$ Replacing P with 32

$\quad 200 = \dfrac{k}{32}\qquad$ Replacing $V(32)$ with 200

$\quad 6400 = k\qquad$ Variation constant

$\quad V(P) = \dfrac{6400}{P}\qquad$ Equation of variation

***Carry out*.** Find $V(40)$.

$\quad V(40) = \dfrac{6400}{40}$

$\qquad\quad = 160$

***Check*.** Reexamine the calculations.

***State*.** The volume will be 160 cm^3.

24. 27 min

25. *Familiarize*. Because T varies inversely as P, we write $T(p) = k/p$. We know that $T(7) = 5$.

***Translate*.** We find the variation constant and the equation of variation.

$\quad T(P) = \dfrac{k}{p}$

$\quad T(7) = \dfrac{k}{7}\qquad$ Replacing P with 7

$\quad 5 = \dfrac{k}{7}\qquad$ Replacing $T(P)$ with 5

$\quad 35 = k\qquad$ Variation constant

$\quad T(P) = \dfrac{35}{P}\qquad$ Equation of variation

***Carry out*.** We find $T(10)$.

$\quad T(10) = \dfrac{35}{10}$

$\qquad\quad = 3.5$

***Check*.** Reexamine the calculations.

***State*.** It would take 3.5 hr for 10 volunteers to complete the job.

26. 450 meters

27. $y = kx^2$

$6 = k \cdot 3^2$ Substituting

$6 = 9k$

$\dfrac{6}{9} = k$

$\dfrac{2}{3} = k$ Variation constant

The equation of variation is $y = \dfrac{2}{3}x^2$.

28. $y = 15x^2$

29. $y = \dfrac{k}{x^2}$

$6 = \dfrac{k}{3^2}$ Substituting

$6 = \dfrac{k}{9}$

$6 \cdot 9 = k$

$54 = k$ Variation constant

The equation of variation is $y = \dfrac{54}{x^2}$.

30. $y = \dfrac{0.0015}{x^2}$

31. $y = kxz^2$

$105 = k \cdot 14 \cdot 5^2$ Substituting 105 for y, 14 for x, and 5 for z

$105 = 350k$

$\dfrac{105}{350} = k$

$0.3 = k$

The equation of variation is $y = 0.3xz^2$.

32. $y = \dfrac{xz}{w}$

33. $y = k \cdot \dfrac{wx^2}{z}$

$49 = k \cdot \dfrac{3 \cdot 7^2}{12}$ Substituting

$4 = k$ Variation constant

The equation of variation is $y = \dfrac{4wx^2}{z}$.

34. $y = \dfrac{6x}{wz^2}$

35. *Familiarize.* I varies inversely as d^2, so we write $I = k/d^2$. We know that $I = 90$ when $d = 5$.

Translate. Find k.

$I = \dfrac{k}{d^2}$

$90 = \dfrac{k}{5^2}$

$2250 = k$

$I = \dfrac{2250}{d^2}$ Equation of variation

Carry out. Substitute 7.5 for d and find for I.

$I = \dfrac{2250}{(7.5)^2} = \dfrac{2250}{56.25} = 40$

Check. Reexamine the calculations.

State. The intensity is 40 W/m^2 at a distance of 7.5 m from the bulb.

36. 72 ft

37. *Familiarize.* Because V varies directly as T and inversely as P, we write $V = kT/P$. We know that $V = 231$ when $T = 42$ and $P = 20$.

Translate. Find k and the equation of variation.

$V = \dfrac{kT}{P}$

$231 = \dfrac{k \cdot 42}{20}$

$\dfrac{20}{42} \cdot 231 = k$

$110 = k$

$V = \dfrac{110T}{P}$ Equation of variation

Carry out. Substitute 30 for T and 15 for P and find V.

$V = \dfrac{110 \cdot 30}{15} = 220$

Check. Reexamine the calculations.

State. The volume is 220 cm^3 when $T = 30°$and $P = 15$ kg/cm^2.

38. 2.56 W/m^2

39. *Familiarize.* The drag W varies jointly as the surface area A and velocity v, so we write $W = kAv$. We know that $W = 222$ when $A = 37.8$ and $v = 40$.

Translate. Find k.

$W = kAv$

$222 = k(37.8)(40)$

$\dfrac{222}{37.8(40)} = k$

$\dfrac{37}{252} = k$

$W = \dfrac{37}{252}Av$ Equation of variation

Carry out. Substitute 51 for A and 430 for W and solve for v.

$$430 = \frac{37}{252} \cdot 51 \cdot v$$
$$57.42 \text{ mph} \approx v$$

(If we had used the rounded value 0.1468 for k, the resulting speed would have been approximately 57.43 mph.)

Check. Reexamine the calculations.

State. The car must travel about 57.42 mph.

40. About 28.5 ft^2; answers may vary slightly due to rounding differences.

41. Writing exercise

42. Writing exercise

43. Writing exercise

44. Writing exercise

45. $2x - 5 = 8$
$$2x = 13 \quad \text{Adding 5}$$
$$x = \frac{13}{2} \quad \text{Dividing by 2}$$

The solution is $\frac{13}{2}$.

46. $\frac{5}{3}$

47.
$$\frac{1}{x+1} = \frac{3}{x}, \text{LCD} = x(x+1)$$
$$x(x+1) \cdot \frac{1}{x+1} = x(x+1) \cdot \frac{3}{x} \quad \text{Multiplying by}$$
$$\text{the LCD}$$
$$x = 3(x+1)$$
$$x = 3x + 3$$
$$-2x = 3$$
$$x = -\frac{3}{2}$$

The number $-\frac{3}{2}$ checks. It is the solution.

48. $-\frac{5}{3}, \frac{7}{2}$

49. $3a + 1 = 3(a+1)$
$$3a + 1 = 3a + 3$$
$$1 = 3 \quad \text{Subtracting } 3a$$

There is no value of a for which $1 = 3$. The equation has no solution.

50. All real numbers

51. Writing exercise

52. Writing exercise

53. $S = kv^6$

54. $P^2 = kt$

55. $I = \dfrac{k}{d^2}$

56. $B = kN$

57. $P = kv^3$

58. $P = 8S$

59. $C = kr$

From the formula for the circumference of a circle we know that $k = 2\pi$, and we have $C = 2\pi r$.

60. $A = \pi r^2$

61. $V = kr^3$

From the formula for the volume of a sphere we know that $k = \frac{4}{3}\pi$, and we have $V = \frac{4}{3}\pi r^3$.

62. Q varies directly as the square of p and inversely as the cube of q.

63. $W = \dfrac{km_1 M_1}{d^2}$

W varies jointly as m_1 and M_1 and inversely as the square of d.

64. About 1.697 m

65. Let V represent the volume and p represent the price of a jar of peanut butter.
$$V = kp$$
$$\pi\left(\frac{3}{2}\right)^2 (4) = k(1.2)$$
$$7.5\pi = k$$
$$V = 7.5\pi p$$
$$\pi(3)^2(6) = 7.5\pi p$$
$$7.2 = p$$

The bigger jar should cost \$7.20.

66. $d(s) = \dfrac{28}{s}$; 70 yd

Chapter 8

Systems of Equations and Problem Solving

1. We use alphabetical order for the variables. We replace x by 1 and y by 2.

$$\begin{array}{c|c}
\underline{4x - y = 2} & \underline{10x - 3y = 4} \\
4 \cdot 1 - 2 \ ? \ 2 & 10 \cdot 1 - 3 \cdot 2 \ ? \ 4 \\
4 - 2 & 10 - 6 \\
\hline
2 \ \big| \ 2 \ \text{ TRUE} & 4 \ \big| \ 4 \ \text{ TRUE}
\end{array}$$

The pair $(1, 2)$ makes both equations true, so it is a solution of the system.

2. Yes

3. We use alphabetical order for the variables. We replace x by 2 and y by 5.

$$\begin{array}{c|c}
\underline{y = 3x - 1} & \underline{2x + y = 4} \\
5 \ ? \ 3 \cdot 2 - 1 & 2 \cdot 2 + 5 \ ? \ 4 \\
6 - 1 & 4 + 5 \\
\hline
5 \ \big| \ 5 \quad \text{ TRUE} & 9 \ \big| \ 4 \ \text{ FALSE}
\end{array}$$

The pair $(2, 5)$ is not a solution of $2x + y = 4$. Therefore, it is not a solution of the system of equations.

4. No

5. We replace x by 1 and y by 5.

$$\begin{array}{c|c}
\underline{x + y = 6} & \underline{y = 2x + 3} \\
1 + 5 \ ? \ 6 & 5 \ ? \ 2 \cdot 1 + 3 \\
6 \ \big| \ 6 \ \text{ TRUE} & 2 + 3 \\
& \hline
& 5 \ \big| \ 5 \quad \text{ TRUE}
\end{array}$$

The pair $(1, 5)$ makes both equations true, so it is a solution of the system.

6. Yes

7. Observe that if we multiply both sides of the first equation by 2, we get the second equation. Thus, if we find that the given point makes the one equation true, we will also know that it makes the other equation true. We replace x by 3 and y by 1 in the first equation.

$$\begin{array}{c}
\underline{3x + 4y = 13} \\
3 \cdot 3 + 4 \cdot 1 \ ? \ 13 \\
9 + 4 \\
\hline
13 \ \big| \ 13 \quad \text{TRUE}
\end{array}$$

The pair $(3, 1)$ makes both equations true, so it is a solution of the system.

8. Yes

9. Graph both equations.

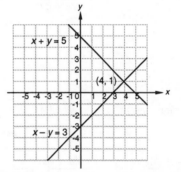

The solution (point of intersection) is apparently $(4, 1)$.

Check:

$$\begin{array}{c|c}
\underline{x - y = 3} & \underline{x + y = 5} \\
4 - 1 \ ? \ 3 & 4 + 1 \ ? \ 5 \\
\hline
3 \ \big| \ 3 \ \text{ TRUE} & 5 \ \big| \ 5 \ \text{ TRUE}
\end{array}$$

The solution is $(4, 1)$.

10. $(3,1)$

11. Graph the equations.

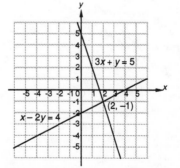

The solution (point of intersection) is apparently $(2, -1)$.

Check:

$3x + y = 5$			$x - 2y = 4$	
$3 \cdot 2 + (-1)$? 5			$2 - 2(-1)$? 4	
$6 - 1$			$2 + 2$	
	5	5 TRUE		4	4 TRUE

The solution is $(2, -1)$.

12. $(3, 2)$

13. Graph both equations.

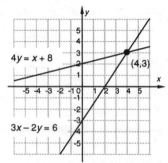

The solution (point of intersection) is apparently $(4, 3)$.

Check:

$4y = x + 8$		$3x - 2y = 6$	
$4 \cdot 3$? $4 + 8$		$3 \cdot 4 - 2 \cdot 3$? 6	
12	12 TRUE	$12 - 6$	
		6	6 TRUE

The solution is $(4, 3)$.

14. $(1, -5)$

15. Graph both equations.

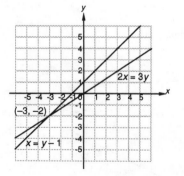

The solution (point of intersection) is apparently $(-3, -2)$.

Check:

$x = y - 1$			$2x = 3y$	
-3 ? $-2 - 1$			$2(-3)$? $3(-2)$	
-3	-3	TRUE	-6	-6 TRUE

The solution is $(-3, -2)$.

16. $(2, 1)$

17. Graph both equations.

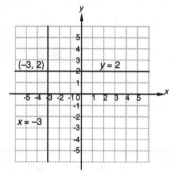

The ordered pair $(-3, 2)$ checks in both equations. It is the solution.

18. $(4, -5)$

19. Graph both equations.

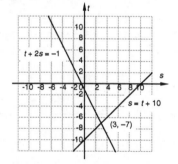

The solution (point of intersection) is apparently $(3, -7)$.

Check:

$t + 2s = -1$		$s = t + 10$	
$-7 + 2 \cdot 3$? -1		3 ? $-7 + 10$	
$-7 + 6$		3	3 TRUE
-1	-1 TRUE		

The solution is $(3, -7)$.

20. $(5, -8)$

21. Graph both equations.

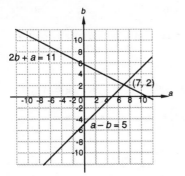

The solution (point of intersection) is apparently $(7, 2)$.

Check:

$2b + a = 11$		$a - b = 5$	
$2 \cdot 2 + 7$? 11		$7 - 2$? 5	
$4 + 7$			5 | TRUE
11 | 11 TRUE			

The solution is $(7,2)$.

22. $(3, -2)$

23. Graph both equations.

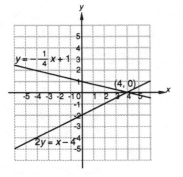

The solution (point of intersection) is apparently $(4, 0)$.

Check:

$y = -\frac{1}{4}x + 1$		$2y = x - 4$	
0 ? $-\frac{1}{4} \cdot 4 + 1$		$2 \cdot 0$? $4 - 4$	
$-1 + 1$		0 | 0 TRUE	
0 | 0 TRUE			

The solution is $(4,0)$.

24. No solution

25. Graph both equations.

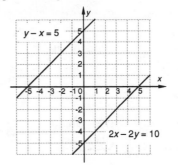

The lines are parallel. The system has no solution.

26. $(3, -4)$

27. Graph both equations.

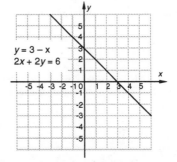

The graphs are the same. Any solution of one equation is a solution of the other. Each equation has infinitely many solutions. The solution set is the set of all pairs (x, y) for which $y = 3 - x$, or $\{(x, y)|y = 3 - x\}$. (In place of $y = 3 - x$ we could have used $2x + 2y = 6$ since the two equations are equivalent.)

28. $\{(x, y)|2x - 3y = 6\}$

29. A system of equations is consistent if it has at least one solution. Of the systems under consideration, only the one in Exercise 25 has no solution. Therefore, all except the system in Exercise 25 are consistent.

30. All except 24

31. A system of two equations in two variables is dependent if it has infinitely many solutions. Only the system in Exercise 27 is dependent.

32. 28

33. *Familiarize*. Let $x =$ the larger number and $y =$ the smaller number.

Translate.

The difference between two numbers is 11.

Rewording:

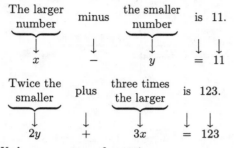

We have a system of equations:

$$x - y = 11,$$
$$3x + 2y = 123$$

34. Let x = the first number and y = the second number.

$$x + y = -42,$$
$$x - y = 52$$

35. *Familiarize*. Let x = the number of less expensive brushes sold and y = the number of more expensive brushes sold.

Translate. We organize the information in a table.

Kind of brush	Less expensive	More expensive	Total
Number sold	x	y	45
Price	$8.50	$9.75	
Amount taken in	$8.50x$	$9.75y$	398.75

The "Number sold" row of the table gives us one equation:

$$x + y = 45$$

The "Amount taken in" row gives us a second equation:

$$8.50x + 9.75y = 398.75$$

We have a system of equations:

$$x + y = 45,$$
$$8.50x + 9.75y = 398.75$$

We can multiply both sides of the second equation by 100 to clear the decimals:

$$x + y = 45,$$
$$850x + 975y = 39,875$$

36. Let x = the number of polarfleece neckwarmers sold and y = the number of wool neckwarmers sold.

$$x + y = 40,$$
$$9.9x + 12.75y = 421.65$$

37. *Familiarize*. Let x = the measure of one angle and y = the measure of the other angle.

Translate.

Two angles are supplementary.

Rewording:

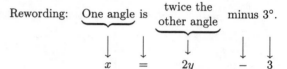

One angle is 3° less than twice the other.

Rewording:

We have a system of equations:

$$x + y = 180,$$
$$x = 2y - 3$$

38. Let x = the measure of the first angle and y = the measure of the second angle.

$$x + y = 90,$$
$$x + \frac{1}{2}y = 64$$

39. *Familiarize*. Let g = the number of field goals and t = the number of free throws made.

Translate. We organize the information in a table.

Kind of shot	Field goal	Free throw	Total
Number scored	g	t	64
Points per score	2	1	
Points scored	$2g$	t	100

From the "Number scored" row of the table we get one equation:

$$g + t = 64$$

The "Points scored" row gives us another equation:

$$2g + t = 100$$

We have a system of equations:

$$g + t = 64,$$
$$2g + t = 100$$

40. Let x = the number of children's plates and y = the number of adult's plates served.

$$x + y = 250,$$
$$3.5x + 7y = 1347.5$$

41. *Familiarize*. Let h = the number of vials of Humulin Insulin sold and n = the number of vials of Novolin Insulin sold.

***Translate*.** We organize the information in a table.

Brand	Humulin	Novolin	Total
Number sold	h	n	50
Price	\$21.95	\$20.95	
Amount taken in	21.95h	20.95n	1077.50

The "Number sold" row of the table gives us one equation:

$$h + n = 50$$

The "Amount taken in" row gives us a second equation:

$$21.95h + 20.95n = 1077.50$$

We have a system of equations:

$$h + n = 50$$
$$21.95h + 20.95n = 1077.50$$

We can multiply both sides of the second equation by 100 to clear the decimals:

$$h + n = 50$$
$$2195h + 2095n = 107,750$$

42. Let l = the length, in feet, and w = the width, in feet.

$$2l + 2w = 288,$$
$$l = w + 44$$

43. *Familiarize*. The tennis court is a rectangle with perimeter 228 ft. Let l = the length, in feet, and w = width, in feet. Recall that for a rectangle with length l and width w, the perimeter P is given by $P = 2l + 2w$.

***Translate*.** The formula for perimeter gives us one equation:

$$2l + 2w = 228$$

The statement relating width and length gives us another equation:

The width is 42 ft less than the length.

$$w = l - 42$$

We have a system of equations:

$$2l + 2w = 228,$$
$$w = l - 42$$

44. Let x = the number of 2-pointers scored and y = the number of 3-pointers scored.

$$x + y = 40,$$
$$2x + 3y = 89$$

45. *Familiarize*. Let w = the number of wins and t = the number of ties. Then the total number of points received from wins was $2w$ and the total number of points received from ties was t.

***Translate*.**

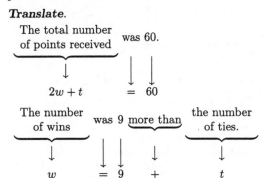

We have a system of equations:

$$2w + t = 60,$$
$$w = 9 + t$$

46. Let x = the number of 30-sec commercials and y = the number of 60-sec commercials.

$$x + y = 12,$$
$$30x + 60y = 600$$

47. *Familiarize*. Let x = the number of ounces of lemon juice and y = the number of ounces of linseed oil to be used.

***Translate*.**

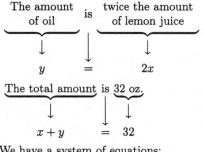

We have a system of equations:

$$y = 2x,$$
$$x + y = 32$$

48. Let l = the number of pallets of lumber produced and p = the number of pallets of plywood produced.

$$l + p = 42,$$
$$25l + 40p = 1245$$

49. *Familiarize*. Let x = the number of general-interest films rented and y = the number of children's films rented. Then $3x$ is taken in from the general-interest rentals and $1.5y$ is taken in from the children's rentals.

Translate.

The number of videos rented is 77.

$$x + y \qquad = 77$$

The amount taken in is \$213.

$$3x + 1.5y \qquad = 213$$

We have a system of equations:

$$x + y = 77,$$
$$3x + 1.5y = 213$$

Clearing decimals we have

$$x + y = 77,$$
$$30x + 15y = 2130$$

50. Let $c =$ the number of coach-class seats and $f =$ the number of first-class seats.

$$c + f = 152,$$
$$c = 5 + 6f$$

51. Writing exercise

52. Writing exercise

53. $\quad 2(4x - 3) - 7x = 9$

$\qquad 8x - 6 - 7x = 9 \qquad$ Removing parentheses

$\qquad\qquad x - 6 = 9 \qquad$ Collecting like terms

$\qquad\qquad\qquad x = 15 \qquad$ Adding 6 to both sides

The solution is 15.

54. $\dfrac{19}{12}$

55. $\quad 4x - 5x = 8x - 9 + 11x$

$\qquad\qquad -x = 19x - 9 \qquad$ Collecting like terms

$\qquad\quad -20x = -9 \qquad$ Adding $-19x$ to both sides

$\qquad\qquad x = \dfrac{9}{20} \qquad$ Multiplying both sides by $-\dfrac{1}{20}$

The solution is $\dfrac{9}{20}$.

56. $\dfrac{13}{3}$

57. $\quad 3x + 4y = 7$

$\qquad\quad 4y = -3y + 7 \qquad$ Adding $-3x$ to both sides

$\qquad\quad y = \dfrac{1}{4}(-3x + 7) \qquad$ Multiplying both sides by $\dfrac{1}{4}$

$\qquad\quad y = -\dfrac{3}{4}x + \dfrac{7}{4}$

58. $y = \dfrac{2}{5}x - \dfrac{9}{5}$

59. Writing exercise

60. Writing exercise

61. The line representing the number of schools with CD-ROMs first lies above the line representing the number of schools with modems in 1994, so this is the year during which the number of schools with CD-ROMs first exceeded the number of schools with modems.

62. 15,000 schools per year

63. a) There are many correct answers. One can be found by expressing the sum and difference of the two numbers:

$$x + y = 6,$$
$$x - y = 4$$

b) There are many correct answers. For example, write an equation in two variables. Then write a second equation by multiplying the left side of the first equation by one nonzero constant and multiplying the right side by another nonzero constant.

$$x + y = 1,$$
$$2x + 2y = 3$$

c) There are many correct answers. One can be found by writing an equation in two variables and then writing a nonzero constant multiple of that equation:

$$x + y = 1,$$
$$2x + 2y = 2$$

64. a) Answers may vary; $(4, -5)$

b) Infinitely many

65. Substitute 4 for x and -5 for y in the first equation:

$$A(4) - 6(-5) = 13$$
$$4A + 30 = 13$$
$$4A = -17$$
$$A = -\frac{17}{4}$$

Substitute 4 for x and -5 for y in the second equation:

$$4 - B(-5) = -8$$
$$4 + 5B = -8$$
$$5B = -12$$
$$B = -\frac{12}{5}$$

We have $A = -\dfrac{17}{4}$, $B = -\dfrac{12}{5}$.

66. Let $x =$ Burl's age now and $y =$ his son's age now.

$$x = 2y,$$
$$x - 10 = 3(y - 10)$$

67. *Familiarize*. Let $x =$ the number of years Lou has taught and $y =$ the number of years Juanita has taught. Two years ago, Lou and Juanita had taught $x - 2$ and $y - 2$ years, respectively.

Translate.

$$\underbrace{\text{Together, the number}}_{\substack{\downarrow \\ x + y}} \underbrace{\text{is}}_{} \underbrace{46.}_{\substack{\downarrow \downarrow \\ = 46}}$$

Two years ago
$$\underbrace{\text{Lou had taught 2.5 times as many years as Juanita.}}_{x - 2 = 2.5(y - 2)}$$

We have a system of equations:

$$x + y = 46,$$
$$x - 2 = 2.5(y - 2)$$

68. Let $l =$ the original length, in inches, and $w =$ the original width, in inches.

$$2l + 2w = 156,$$
$$l = 4(w - 6)$$

69. *Familiarize*. Let $b =$ the number of ounces of baking soda and $v =$ the number of ounces of vinegar to be used. The amount of baking soda in the mixture will be four times the amount of vinegar.

Translate.

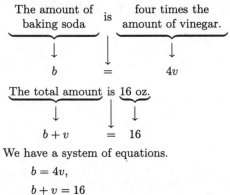

$$\underbrace{\text{The amount of}}_{\substack{\downarrow \\ b}} \underbrace{\text{is}}_{\substack{\downarrow \\ =}} \underbrace{\text{four times the}}_{\substack{\downarrow \\ 4v}}$$
baking soda   amount of vinegar.

$$\underbrace{\text{The total amount}}_{\substack{\downarrow \\ b + v}} \underbrace{\text{is}}_{\substack{\downarrow \\ =}} \underbrace{16 \text{ oz.}}_{\substack{\downarrow \\ 16}}$$

We have a system of equations.

$$b = 4v,$$
$$b + v = 16$$

70. $(-5, 5)$, $(3, 3)$

71. Graph both equations.

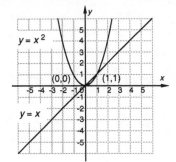

The solutions are apparently $(0, 0)$ and $(1, 1)$. Both pairs check.

72. $(-0.39, -1.10)$

73. $(0.07, -7.95)$

74. $(-0.13, 0.67)$

75. $(0.02, 1.25)$

Exercise Set 8.2

1. $\quad y = 5 - 4x, \quad (1)$
$\quad 2x - 3y = 13 \quad (2)$

We substitute $5 - 4x$ for y in the second equation and solve for x.

$$2x - 3y = 13 \quad (2)$$
$$2x - 3(5 - 4x) = 13 \quad \text{Substituting}$$
$$2x - 15 + 12x = 13$$
$$14x - 15 = 13$$
$$14x = 28$$
$$x = 2$$

Next we substitute 2 for x in either equation of the original system and solve for y.

$$y = 5 - 4x \quad (1)$$
$$y = 5 - 4 \cdot 2 \quad \text{Substituting}$$
$$y = 5 - 8$$
$$y = -3$$

We check the ordered pair $(2, -3)$.

$$\begin{array}{c|c} y = 5 - 4x \\ \hline -3 \; ? \; 5 - 4 \cdot 2 \\ \quad\;\; 5 - 8 \\ \hline -3 \;\big|\; -3 \qquad \text{TRUE} \end{array}$$

$$2x - 3y = 13$$

$$\begin{array}{c|c} 2 \cdot 2 - 3(-3) \ ? \ 13 & \\ 4 + 9 & \\ 13 & 13 \quad \text{TRUE} \end{array}$$

Since $(2, -3)$ checks, it is the solution.

2. $(-4, 3)$

3. $2y + x = 9,$ (1)
  $x = 3y - 3$ (2)

We substitute $3y - 3$ for x in the first equation and solve for y.

$$2y + x = 9 \qquad (1)$$
$$2y + (3y - 3) = 9 \qquad \text{Substituting}$$
$$5y - 3 = 9$$
$$5y = 12$$
$$y = \frac{12}{5}$$

Next we substitute $\dfrac{12}{5}$ for y in either equation of the original system and solve for x.

$$x = 3y - 3 \qquad\qquad (2)$$
$$x = 3 \cdot \frac{12}{5} - 3 = \frac{36}{5} - \frac{15}{5} = \frac{21}{5}$$

We check the ordered pair $\left(\dfrac{21}{5}, \dfrac{12}{5} \right)$.

$$2y + x = 9$$

$$\begin{array}{c|c} 2 \cdot \dfrac{12}{5} + \dfrac{21}{15} \ ? \ 9 & \\[2mm] \dfrac{24}{5} + \dfrac{21}{5} & \\[2mm] \dfrac{45}{5} & \\[2mm] 9 & 9 \quad \text{TRUE} \end{array}$$

$$x = 3y - 3$$

$$\begin{array}{c|c} \dfrac{21}{5} \ ? \ 3 \cdot \dfrac{12}{5} - 3 & \\[2mm] & \dfrac{36}{5} - \dfrac{15}{5} \\[2mm] \dfrac{21}{5} & \dfrac{21}{5} \qquad \text{TRUE} \end{array}$$

Since $\left(\dfrac{21}{5}, \dfrac{12}{5} \right)$ checks, it is the solution.

4. $(-3, -15)$

5. $3s - 4t = 14,$ (1)
  $5s + t = 8$ (2)

We solve the second equation for t.

$$5s + t = 8 \qquad\qquad (2)$$
$$t = 8 - 5s \qquad\qquad (3)$$

We substitute $8 - 5s$ for t in the first equation and solve for s.

$$3s - 4t = 14 \qquad (1)$$
$$3s - 4(8 - 5s) = 14 \qquad \text{Substituting}$$
$$3s - 32 + 20s = 14$$
$$23s - 32 = 14$$
$$23s = 46$$
$$s = 2$$

Next we substitute 2 for s in Equation (1), (2), or (3). It is easiest to use Equation (3) since it is already solved for t.

$$t = 8 - 5 \cdot 2 = 8 - 10 = -2$$

We check the ordered pair $(2, -2)$.

$$3s - 4t = 14$$

$$\begin{array}{c|c} 3 \cdot 2 - 4(-2) \ ? \ 14 & \\ 6 + 8 & \\ 14 & 14 \quad \text{TRUE} \end{array}$$

$$5s + t = 8$$

$$\begin{array}{c|c} 5 \cdot 2 + (-2) \ ? \ 8 & \\ 10 - 2 & \\ 8 & 8 \quad \text{TRUE} \end{array}$$

Since $(2, -2)$ checks, it is the solution.

6. $(2, -7)$

7. $4x - 2y = 6,$ (1)
  $2x - 3 = y$ (2)

We substitute $2x - 3$ for y in the first equation and solve for x.

$$4x - 2y = 6 \qquad (1)$$
$$4x - 2(2x - 3) = 6$$
$$4x - 4x + 6 = 6$$
$$6 = 6$$

We have an identity, or an equation that is always true. The equations are dependent and the solution set is infinite: $\{(x, y) | 2x - 3 = y\}$.

8. No solution

9. $-5s + t = 11$,   (1)

$4s + 12t = 4$ (2)

We solve the first equation for t.

$-5s + t = 11$ (1)

$t = 5s + 11$ (3)

We substitute $5s + 11$ for t in the second equation and solve for s.

$4s + 12t = 4$ (2)

$4s + 12(5s + 11) = 4$

$4s + 60s + 132 = 4$

$64s + 132 = 4$

$64s = -128$

$s = -2$

Next we substitute -2 for s in Equation (3).

$t = 5s + 11 = 5(-2) + 11 = -10 + 11 = 1$

We check the ordered pair $(-2, 1)$.

$$\frac{-5s + t = 11}{}$$

$-5(-2) + 1$? 11

$10 + 1$

11 | 11 TRUE

$$\frac{4s + 12t = 4}{}$$

$4(-2) + 12 \cdot 1$? 4

$-8 + 12$

4 | 4 TRUE

Since $(-2, 1)$ checks, it is the solution.

10. $(4, -1)$

11. $2x + 2y = 2$,   (1)

$3x - y = 1$ (2)

We solve the second equation for y.

$3x - y = 1$ (2)

$-y = -3x + 1$

$y = 3x - 1$ (3)

We substitute $3x - 1$ for y in the first equation and solve for x.

$2x + 2y = 2$ (1)

$2x + 2(3x - 1) = 2$

$2x + 6x - 2 = 2$

$8x - 2 = 2$

$8x = 4$

$x = \frac{1}{2}$

Next we substitute $\frac{1}{2}$ for x in Equation (3).

$y = 3x - 1 = 3 \cdot \frac{1}{2} - 1 = \frac{3}{2} - 1 = \frac{1}{2}$

The ordered pair $\left(\frac{1}{2}, \frac{1}{2}\right)$ checks in both equations. It is the solution.

12. $(3, -2)$

13. $3a - b = 7$,   (1)

$2a + 2b = 5$ (2)

We solve the first equation for b.

$3a - b = 7$ (1)

$-b = -3a + 7$

$b = 3a - 7$ (3)

We substitute $3a - 7$ for b in the second equation and solve for a.

$2a + 2b = 5$ (2)

$2a + 2(3a - 7) = 5$

$2a + 6a - 14 = 5$

$8a - 14 = 5$

$8a = 19$

$a = \frac{19}{8}$

We substitute $\frac{19}{8}$ for a in Equation (3).

$b = 3a - 7 = 3 \cdot \frac{19}{8} - 7 = \frac{57}{8} - \frac{56}{8} = \frac{1}{8}$

The ordered pair $\left(\frac{19}{8}, \frac{1}{8}\right)$ checks in both equations. It is the solution.

14. $\left(\frac{25}{23}, -\frac{11}{23}\right)$

15. $2x - 3 = y$ (1)

$y - 2x = 1$,   (2)

We substitute $2x - 3$ for y in the second equation and solve for x.

$y - 2x = 1$ (2)

$2x - 3 - 2x = 1$ Substituting

$-3 = 1$ Collecting like terms

We have a contradiction, or an equation that is always false. Therefore, there is no solution.

16. $\{(a, b) | a - 2b = 3\}$

17. $x + 3y = 7$ (1)

$\underline{-x + 4y = 7}$ (2)

$0 + 7y = 14$ Adding

$7y = 14$

$y = 2$

Substitute 2 for y in one of the original equations and solve for x.

$$x + 3y = 7 \quad (1)$$
$$x + 3 \cdot 2 = 7 \quad \text{Substituting}$$
$$x + 6 = 7$$
$$x = 1$$

Check:

$x + 3y = 7$		$-x + 4y = 7$	
$1 + 3 \cdot 2 \; ? \; 7$		$-1 + 4 \cdot 2 \; ? \; 7$	
$1 + 6$		$-1 + 8$	
	$7 \mid 7$ TRUE		$7 \mid 7$ TRUE

Since $(1, 2)$ checks, it is the solution.

18. $(2, 7)$

19.
$$2x + y = 6 \quad (1)$$
$$\underline{x - y = 3} \quad (2)$$
$$3x + 0 = 9 \quad \text{Adding}$$
$$3x = 9$$
$$x = 3$$

Substitute 3 for x in one of the original equations and solve for y.

$$2x + y = 6 \quad (1)$$
$$2 \cdot 3 + y = 6 \quad \text{Substituting}$$
$$6 + y = 6$$
$$y = 0$$

We obtain $(3, 0)$. This checks, so it is the solution.

20. $(10, 2)$

21.
$$9x + 3y = -3 \quad (1)$$
$$\underline{2x - 3y = -8} \quad (2)$$
$$11x + 0 = -11 \quad \text{Adding}$$
$$11x = -11$$
$$x = -1$$

Substitute -1 for x in Equation (1) and solve for y.

$$9x + 3y = -3$$
$$9(-1) + 3y = -3 \quad \text{Substituting}$$
$$-9 + 3y = -3$$
$$3y = 6$$
$$y = 2$$

We obtain $(-1, 2)$. This checks, so it is the solution.

22. $\left(\dfrac{1}{2}, -5 \right)$

23.
$$5x + 3y = 19, \quad (1)$$
$$2x - 5y = 11 \quad (2)$$

We multiply twice to make two terms become opposites.

From (1): $25x + 15y = 95$ Multiplying by 5
From (2): $\underline{6x - 15y = 33}$ Multiplying by 3
$$31x + 0 = 128 \quad \text{Adding}$$
$$x = \frac{128}{31}$$

Substitute $\dfrac{128}{31}$ for x in Equation (1) and solve for y.

$$5x + 3y = 19$$
$$5 \cdot \frac{128}{31} + 3y = 19 \quad \text{Substituting}$$
$$\frac{640}{31} + 3y = \frac{589}{31}$$
$$3y = -\frac{51}{31}$$
$$\frac{1}{3} \cdot 3y = \frac{1}{3} \cdot \left(-\frac{51}{31} \right)$$
$$y = -\frac{17}{31}$$

We obtain $\left(\dfrac{128}{31}, -\dfrac{17}{31} \right)$. This checks, so it is the solution.

24. $\left(\dfrac{10}{21}, \dfrac{11}{14} \right)$

25.
$$5r - 3s = 24, \quad (1)$$
$$3r + 5s = 28 \quad (2)$$

We multiply twice to make two terms become additive inverses.

From (1): $25r - 15s = 120$ Multiplying by 5
From (2): $\underline{9r + 15s = 84}$ Multiplying by 3
$$34r + 0 = 204 \quad \text{Adding}$$
$$r = 6$$

Substitute 6 for r in Equation (2) and solve for s.

$$3r + 5s = 28$$
$$3 \cdot 6 + 5s = 28 \quad \text{Substituting}$$
$$18 + 5s = 28$$
$$5s = 10$$
$$s = 2$$

We obtain $(6, 2)$. This checks, so it is the solution.

26. $(1, 3)$

27. $6s + 9t = 12$, (1)

$4s + 6t = 5$ (2)

We multiply twice to make two terms become opposites.

From (1): $12s + 18t = 24$ Multiplying by 2

From (2): $-12s - 18t = -15$ Multiplying by -3

$0 = 9$

We get a contradiction, or an equation that is always false. The system has no solution.

28. No solution

29. $\dfrac{1}{2}x - \dfrac{1}{6}y = 3$ (1)

$\dfrac{2}{5}x + \dfrac{1}{2}y = 2$, (2)

We first multiply each equation by the LCM of the denominators to clear fractions.

$3x - y = 18$ (3) Multiplying (1) by 6

$4x + 5y = 20$ (4) Multiplying (2) by 10

We multiply by 5 on both sides of Equation (3) and then add.

$15x - 5y = 90$ Multiplying (3) by 5

$4x + 5y = 20$ (4)

$19x + 0 = 110$ Adding

$x = \dfrac{110}{19}$

Substitute $\dfrac{110}{19}$ for x in one of the equations in which the fractions were cleared and solve for y.

$3x - y = 18$ (3)

$3\left(\dfrac{110}{19}\right) - y = 18$ Substituting

$\dfrac{330}{19} - y = \dfrac{342}{19}$

$-y = \dfrac{12}{19}$

$y = -\dfrac{12}{19}$

We obtain $\left(\dfrac{110}{19}, -\dfrac{12}{19}\right)$. This checks, so it is the solution.

30. $(12, 15)$

31. $\dfrac{x}{2} + \dfrac{y}{3} = \dfrac{7}{6}$, (1)

$\dfrac{2x}{3} + \dfrac{3y}{4} = \dfrac{5}{4}$ (2)

We first multiply each equation by the LCM of the denominators to clear fractions.

$3x + 2y = 7$ (3) Multiplying (1) by 6

$8x + 9y = 15$ (4) Multiplying (2) by 12

We multiply twice to make two terms become opposites.

From (1): $104a - 56b = 72$ Multiplying by 8

From (2): $-14a + 56b = -42$ Multiplying by -7

$90a = 30$ Adding

$a = \dfrac{1}{3}$

Substitute 3 for x in one of the equations in which the fractions were cleared and solve for y.

$3x + 2y = 7$ (3)

$3 \cdot 3 + 2y = 7$ Substituting

$9 + 2y = 7$

$2y = -2$

$y = -1$

We obtain $(3, -1)$. This checks, so it is the solution.

32. $(-2, 3)$

33. $12x - 6y = -15$, (1)

$-4x + 2y = 5$ (2)

Observe that, if we multiply Equation (1) by $-\dfrac{1}{3}$, we obtain Equation (2). Thus, any pair that is a solution of Equation (1) is also a solution of Equation (2). The equations are dependent and the solution set is infinite: $\{(x, y) | -4x + 2y = 5\}$.

34. $\{(s, t) | 6s + 9t = 12\}$

35. $0.2a + 0.3b = 1$,

$0.3a - 0.2b = 4$,

We first multiply each equation by 10 to clear decimals.

$2a + 3b = 10$ (1)

$3a - 2b = 40$ (2)

We multiply so that the b-terms can be eliminated.

From (1): $4a + 6b = 20$ Multiplying by 2

From (2): $9a - 6b = 120$ Multiplying by 3

$13a + 0 = 140$ Adding

$a = \dfrac{140}{13}$

Substitute $\dfrac{140}{13}$ for a in Equation (1) and solve for b.

$$2a + 3b = 10$$

$$2 \cdot \frac{140}{13} + 3b = 10 \qquad \text{Substituting}$$

$$\frac{280}{13} + 3b = \frac{130}{13}$$

$$3b = -\frac{150}{13}$$

$$b = -\frac{50}{13}$$

We obtain $\left(\frac{140}{13}, -\frac{50}{13}\right)$. This checks, so it is the solution.

36. $(2, 3)$

37. $a - 2b = 16, \quad (1)$
$b + 3 = 3a \quad (2)$

We will use the substitution method. First solve Equation (1) for a.

$$a - 2b = 16$$

$$a = 2b + 16 \quad (3)$$

Now substitute $2b + 16$ for a in Equation (2) and solve for b.

$$b + 3 = 3a \qquad (2)$$

$$b + 3 = 3(2b + 16) \quad \text{Substituting}$$

$$b + 3 = 6b + 48$$

$$-45 = 5b$$

$$-9 = b$$

Substitute -9 for b in Equation (3).

$$a = 2(-9) + 16 = -2$$

We obtain $(-2, -9)$. This checks, so it is the solution.

38. $\left(\frac{1}{2}, -\frac{1}{2}\right)$

39. $10x + y = 306, \quad (1)$
$10y + x = 90 \quad (2)$

We will use the substitution method. First solve Equation (1) for y.

$$10x + y = 306$$

$$y = -10x + 306 \quad (3)$$

Now substitute $-10x + 306$ for y in Equation (2) and solve for y.

$$10y + x = 90 \qquad (2)$$

$$10(-10x + 306) + x = 90 \qquad \text{Substituting}$$

$$-100x + 3060 + x = 90$$

$$-99x + 3060 = 90$$

$$-99x = -2970$$

$$x = 30$$

Substitute 30 for x in Equation (3).

$$y = -10 \cdot 30 + 306 = 6$$

We obtain $(30, 6)$. This checks, so it is the solution.

40. $\left(-\frac{4}{3}, -\frac{19}{3}\right)$

41. $3y = x - 2, \quad (1)$
$x = 2 + 3y \quad (2)$

We will use the substitution method. Substitute $2 + 3y$ for x in the first equation and solve for y.

$$3y = x - 2 \qquad (1)$$

$$3y = 2 + 3y - 2 \quad \text{Substituting}$$

$$3y = 3y \qquad \text{Collecting like terms}$$

We get an identity. The system is dependent and the solution set is infinite: $\{(x, y) | x = 2 + 3y\}$.

42. No solution

43. $3s - 7t = 5,$
$7t - 3s = 8$

First we rewrite the second equation with the variables in a different order. Then we use the elimination method.

$$3s - 7t = 5, \quad (1)$$

$$\underline{-3s + 7t = 8} \quad (2)$$

$$0 = 13$$

We get a contradiction, so the system has no solution.

44. $\{(s, t) | 2s - 13t = 120\}$

45. $0.05x + 0.25y = 22, \quad (1)$
$0.15x + 0.05y = 24 \quad (2)$

We first multiply each equation by 100 to clear decimals.

$$5x + 25y = 2200$$

$$15x + 5y = 2400$$

We multiply by -5 on both sides of the second equation and add.

$$5x + 25y = 2200$$

$$\underline{-75x - 25y = -12{,}000} \quad \text{Multiplying (2) by } -5$$

$$-70x = -9800 \quad \text{Adding}$$

$$x = \frac{-9800}{-70}$$

$$x = 140$$

Substitute 140 for x in one of the equations in which the decimals were cleared and solve for y.

$$5x + 25y = 2200 \quad (1)$$
$$5 \cdot 140 + 25y = 2200 \quad \text{Substituting}$$
$$700 + 25y = 2200$$
$$25y = 1500$$
$$y = 60$$

We obtain $(140, 60)$. This checks, so it is the solution.

46. No solution

47. $13a - 7b = 9, \quad (1)$
$2a - 8b = 6 \quad (2)$

We will use the elimination method. First we multiply so that the b-terms can be eliminated.

From (1): $104a - 56b = 72$ Multiplying by 8
From (2): $\underline{-14a + 56b = -42}$ Multiplying by -7
$90a = 30$ Adding
$$a = \frac{1}{3}$$

Substitute $\frac{1}{3}$ for a in one of the equations and solve for b.

$$2a - 8b = 6 \quad (2)$$
$$2 \cdot \frac{1}{3} - 8b = 6$$
$$\frac{2}{3} - 8b = 6$$
$$-8b = \frac{16}{3}$$
$$b = -\frac{2}{3}$$

We obtain $\left(\frac{1}{3}, -\frac{2}{3}\right)$. This checks, so it is the solution.

48. $\left(-\dfrac{13}{45}, -\dfrac{37}{45}\right)$

49. Writing exercise

50. Writing exercise

51. *Familiarize.* Let m = the number of $\frac{1}{4}$-mi units traveled after the first $\frac{1}{2}$ mi. The total distance traveled will be $\frac{1}{2}$ mi $+ m \cdot \frac{1}{4}$ mi.

Translate.

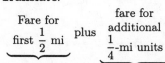

Carry out. We solve the equation.

$$1 + 0.3m = 5.20$$
$$0.3m = 4.20$$
$$m = 14$$

If the taxi travels the first $\frac{1}{2}$ mi plus 14 additional $\frac{1}{4}$-mi units, then it travels a total of $\frac{1}{2} + 14 \cdot \frac{1}{4}$, or $\frac{1}{2} + \frac{7}{2}$, or 4 mi.

Check. We have 4 mi $= \frac{1}{2}$ mi $+ \frac{7}{2}$ mi $= \frac{1}{2}$ mi $+ 14 \cdot \frac{1}{4}$ mi. The fare for traveling this distance is $\$1.00 + \$0.30(14) = \$1.00 + \$4.20 = \$5.20$. The answer checks.

State. It is 4 mi from Johnson Street to Elm Street.

52. 86

53. *Familiarize.* Let a = the amount spent to remodel bathrooms, in billions of dollars. Then $2a$ = the amount spent to remodel kitchens. The sum of these two amounts is $\$35$ billion.

Translate.

Amount spent on bathrooms	plus	amount spent on kitchens	is	$35 billion.
a	$+$	$2a$	$=$	35

Carry out. We solve the equation.

$$a + 2a = 35$$
$$3a = 35 \quad \text{Combining like terms}$$
$$a = \frac{35}{3}, \text{ or } 11\frac{2}{3}$$

If $a = \frac{35}{3}$, then $2a = 2 \cdot \frac{35}{3} = \frac{70}{3} = 23\frac{1}{3}$.

Check. $\frac{70}{3}$ is twice $\frac{35}{3}$, and $\frac{35}{3} + \frac{70}{3} = \frac{105}{3} = 35$. The answer checks.

State. $\$11\frac{2}{3}$ billion was spent to remodel bathrooms, and $\$23\frac{1}{3}$ billion was spent to remodel kitchens.

54. 30 m, 90 m, 360 m

55. *Familiarize.* The total cost is the daily charge plus the mileage charge. The mileage charge is the cost per mile times the number of miles driven. Let m = the number of miles that can be driven for $\$80$.

Translate. We reword the problem.

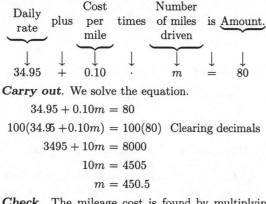

$$34.95 + 0.10 \cdot m = 80$$

Carry out. We solve the equation.

$$34.95 + 0.10m = 80$$
$$100(34.95 + 0.10m) = 100(80) \quad \text{Clearing decimals}$$
$$3495 + 10m = 8000$$
$$10m = 4505$$
$$m = 450.5$$

Check. The mileage cost is found by multiplying 450.5 by $0.10 obtaining $45.05. Then we add $45.05 to $34.95, the daily rate, and get $80.

State. The businessperson can drive 450.5 mi on the car-rental allotment.

56. 460.5 mi

57. Writing exercise

58. Writing exercise

59. First write $f(x) = mx + b$ as $y = mx + b$. Then substitute 1 for x and 2 for y to get one equation and also substitute -3 for x and 4 for y to get a second equation:

$$2 = m \cdot 1 + b$$
$$4 = m(-3) + b$$

Solve the resulting system of equations.

$$2 = m + b$$
$$4 = -3m + b$$

Multiply the second equation by -1 and add.

$$\begin{array}{r} 2 = m + b \\ -4 = 3m - b \\ \hline -2 = 4m \\ -\dfrac{1}{2} = m \end{array}$$

Substitute $-\dfrac{1}{2}$ for m in the first equation and solve for b.

$$2 = -\frac{1}{2} + b$$
$$\frac{5}{2} = b$$

Thus, $m = -\dfrac{1}{2}$ and $b = \dfrac{5}{2}$.

60. $p = 2,\ q = -\dfrac{1}{3}$

61. Substitute -4 for x and -3 for y in both equations and solve for a and b.

$$-4a - 3b = -26, \quad (1)$$
$$-4b + 3a = 7 \quad\quad (2)$$

$$\begin{array}{rl} -12a - 9b = -78 & \text{Multiplying (1) by 3} \\ 12a - 16b = 28 & \text{Multiplying (2) by 4} \\ \hline -25b = -50 \\ b = 2 \end{array}$$

Substitute 2 for b in Equation (2).

$$-4 \cdot 2 + 3a = 7$$
$$3a = 15$$
$$a = 5$$

Thus, $a = 5$ and $b = 2$.

62. $\left(\dfrac{a + 2b}{7}, \dfrac{a - 5b}{7} \right)$

63.
$$\frac{x + y}{2} - \frac{x - y}{5} = 1,$$
$$\frac{x - y}{2} + \frac{x + y}{6} = -2$$

After clearing fractions we have:

$$3x + 7y = 10, \quad (1)$$
$$4x - 2y = -12 \quad (2)$$

$$\begin{array}{rl} 6x + 14y = 20 & \text{Multiplying (1) by 2} \\ 28x - 14y = -84 & \text{Multiplying (2) by 7} \\ \hline 34x = -64 \\ x = -\dfrac{32}{17} \end{array}$$

Substitute $-\dfrac{32}{17}$ for x in Equation (1).

$$3\left(-\frac{32}{17} \right) + 7y = 10$$
$$7y = \frac{266}{17}$$
$$y = \frac{38}{17}$$

The solution is $\left(-\dfrac{32}{17}, \dfrac{38}{17} \right)$.

64. $(23.118879, -12.039964)$

65.
$$\frac{2}{x} + \frac{1}{y} = 0, \qquad\qquad 2 \cdot \frac{1}{x} + \frac{1}{y} = 0,$$
$$\text{or}$$
$$\frac{5}{x} + \frac{2}{y} = -5 \qquad 5 \cdot \frac{1}{x} + 2 \cdot \frac{1}{y} = -5$$

Substitute u for $\frac{1}{x}$ and v for $\frac{1}{y}$.

$$2u + v = 0, \quad (1)$$
$$5u + 2v = -5 \quad (2)$$

$$-4u - 2v = 0 \quad \text{Multiplying (1) by } -2$$
$$\underline{5u + 2v = -5 \quad (2)}$$
$$u = -5$$

Substitute -5 for u in Equation (1).

$$2(-5) + v = 0$$
$$-10 + v = 0$$
$$v = 10$$

If $u = -5$, then $\frac{1}{x} = -5$. Thus $x = -\frac{1}{5}$.

If $v = 10$, then $\frac{1}{y} = 10$. Thus $y = \frac{1}{10}$.

The solution is $\left(-\frac{1}{5}, \frac{1}{10}\right)$.

66. $\left(-\frac{1}{4}, -\frac{1}{2}\right)$

67. Writing exercise

Exercise Set 8.3

1. The Familiarize and Translate steps were done in Exercise 33 of Exercise Set 8.1

Carry out. We solve the system of equations

$$x - y = 11, \quad (1)$$
$$3x + 2y = 123 \quad (2)$$

where $x =$ the larger number and $y =$ the smaller number. We use elimination.

$$2x - 2y = 22 \quad \text{Multiplying (1) by 2}$$
$$\underline{3x + 2y = 123}$$
$$5x = 145$$
$$x = 29$$

Substitute 29 for x in (1) and solve for y.

$$29 - y = 11$$
$$-y = -18$$
$$y = 18$$

Check. The difference between the numbers is $29 - 18$, or 11. Also $2 \cdot 18 + 3 \cdot 29 = 36 + 87 = 123$. The numbers check.

State. The larger number is 29, and the smaller is 18.

2. $5, -47$

3. The Familiarize and Translate steps were done in Exercise 35 of Exercise Set 8.1

Carry out. We solve the system of equations

$$x + y = 45, \quad (1)$$
$$850x + 975y = 39,875 \quad (2)$$

where $x =$ the number of less expensive brushes sold and $y =$ the number of more expensive brushes sold. We use elimination. Begin by multiplying Equation (1) by -850.

$$-850x - 850y = -38,250 \quad \text{Multiplying (1)}$$
$$\underline{850x + 975y = 39,875}$$
$$125y = 1625$$
$$y = 13$$

Substitute 13 for y in (1) and solve for x.

$$x + 13 = 45$$
$$x = 32$$

Check. The number of brushes sold is $32 + 13$, or 45. The amount taken in was $\$8.50(32) + \$9.75(13) = \$272 + \$126.75 = \$398.75$. The answer checks.

State. 32 of the less expensive brushes were sold, and 13 of the more expensive brushes were sold.

4. 31 polarfleece; 9 wool

5. The Familiarize and Translate steps were done in Exercise 37 of Exercise Set 8.1

Carry out. We solve the system of equations

$$x + y = 180, \quad (1)$$
$$x = 2y - 3 \quad (2)$$

where $x =$ the measure of one angle and $y =$ the measure of the other angle. We use substitution.

Substitute $2y - 3$ for x in (1) and solve for y.

$$2y - 3 + y = 180$$
$$3y - 3 = 180$$
$$3y = 183$$
$$y = 61$$

Now substitute 61 for y in (2).

$$x = 2 \cdot 61 - 3 = 122 - 3 = 119$$

Check. The sum of the angle measures is $119° + 61°$, or $180°$, so the angles are supplementary. Also $2 \cdot 61° - 3° = 122° - 3° = 119°$. The answer checks.

State. The measures of the angles are $119°$ and $61°$.

6. $38°, 52°$

7. The Familiarize and Translate steps were done in Exercise 39 of Exercise Set 8.1

Carry out. We solve the system of equations

$$g + t = 64, \quad (1)$$
$$2g + t = 100 \quad (2)$$

where g = the number of field goals and t = the number of free throws Chamberlain made. We use elimination.

$$-g - t = -64 \quad \text{Multiplying (1) by } -1$$
$$\underline{2g + t = \ 100}$$
$$g \quad = \ \ 36$$

Substitute 36 for g in (1) and solve for t.

$$36 + t = 64$$
$$t = 28$$

Check. The total number of scores was $36 + 28$, or 64. The total number of points was $2 \cdot 36 + 28 = 72 + 28 = 100$. The answer checks.

State. Chamberlain made 36 field goals and 28 free throws.

8. 115 children's plates; 135 adult's plates

9. The Familiarize and Translate steps were done in Exercise 41 of Exercise Set 8.1

Carry out. We solve the system of equations

$$h + n = 50, \quad (1)$$
$$2195h + 2095n = 107,750 \quad (2)$$

where h = the number of vials of Humulin Insulin sold and n = the number of vials of Novolin Insulin sold. We use elimination.

$$-2095h - 2095n = -104,750 \quad \text{Multiplying (1)}$$
$$\underline{2195h + 2095n = \ \ 107,750} \quad \text{by } -2095$$
$$100h \qquad = \qquad 3000$$
$$h = \qquad 30$$

Substitute 30 for h in (1) and solve for n.

$$30 + n = 50$$
$$n = 20$$

Check. A total of $30 + 20$, or 50 vials, was sold. The amount collected was $21.95(30) + 20.95(20) = 658.50 + 419 = 1077.50$. The answer checks.

State. 30 vials of Humulin Insulin and 20 vials of Novolin Insulin were sold.

10. Length: 94 ft, width: 50 ft

11. The Familiarize and Translate steps were done in Exercise 43 of Exercise Set 8.1

Carry out. We solve the system of equations

$$2l + 2w = 228, \quad (1)$$
$$w = l - 42 \quad (2)$$

where l = the length, in feet, and w = the width, in feet, of the tennis court. We use substitution.

Substitute $l - 42$ for w in (1) and solve for l.

$$2l + 2(l - 42)w = 228$$
$$2l + 2l - 84 = 228$$
$$4l - 84 = 228$$
$$4l = 312$$
$$l = 78$$

Now substitute 78 for l in (2).

$$w = 78 - 42 = 36$$

Check.The perimeter is $2 \cdot 78 \text{ ft} + 2 \cdot 36 \text{ ft} = 156 \text{ ft} + 72 \text{ ft} = 228 \text{ ft}$. The width, 36 ft, is 42 ft less than the length, 78 ft. The answer checks.

State. The length of the tennis court is 78 ft, and the width is 36 ft.

12. 31 two-point field goals; 9 three-point field goals

13. The Familiarize and Translate steps were done in Exercise 45 of Exercise Set 8.1.

Carry out. We solve the system of equations

$$2w + t = 60, \quad (1)$$
$$w = 9 + t \quad (2)$$

where w = the number of wins and t = the number of ties. We use substitution.

Substitute $9 + t$ for w in (1) and solve for t.

$$2(9 + t) + t = 60$$
$$18 + 2t + t = 60$$
$$18 + 3t = 60$$
$$3t = 42$$
$$t = 14$$

Now substitute 14 for t in (2).

$$w = 9 + 14 = 23$$

Check. The total number of points is $2 \cdot 23 + 14 = 46 + 14 = 60$. The number of wins, 23, is nine more than the number of ties, 14. The answer checks.

State. The Wildcats had 23 wins and 14 ties.

14. 4 30-sec commercials; 8 60-sec commercials

15. The Familiarize and Translate steps were done in Exercise 47 of Exercise Set 8.1.

Carry out. We solve the system of equations

$$y = 2x, \quad (1)$$
$$x + y = 32 \quad (2)$$

where x = the number of ounces of lemon juice and y = the number of ounces of linseed oil to be used. We use substitution.

Substitute $2x$ for y in (2) and solve for x.

$$x + 2x = 32$$
$$3x = 32$$
$$x = \frac{32}{3}, \text{or} 10\frac{2}{3}$$

Now substitute $\frac{32}{3}$ for x in (1).

$$y = 2 \cdot \frac{32}{3} = \frac{64}{3}, \text{ or } 21\frac{1}{3}$$

Check. The amount of oil, $\frac{64}{3}$ oz, is twice the amount
of lemon juice, $\frac{32}{3}$ oz. The mixture contains
$\frac{32}{3}$ oz $+ \frac{64}{3}$ oz $= \frac{96}{3}$ oz $= 32$ oz. The answer checks.

State. $10\frac{2}{3}$ oz of lemon juice and $21\frac{1}{3}$ oz of linseed
oil are needed.

16. 29 pallets of lumbe; 13 pallets of plywood

17. The Familiarize and Translate steps were done in
Exercise 49 of Exercise Set 8.1.

Carry out. We solve the system of equations

$$x + y = 77, \qquad (1)$$
$$30x + 15y = 2130 \quad (2)$$

where x = the number of general-interest films
rented and y = the number of children's films rented.
We use elimination.

$$-15x - 15y = -1155 \quad \text{Multiplying (1) by } -15$$
$$\underline{30x + 15y = 2130}$$
$$15x = 975$$
$$x = 65$$

Substitute 65 for x in (1) and solve for y.

$$65 + y = 77$$
$$y = 12$$

Check. The total number of films rented is $65 +$
12, or 77. The total amount taken in was $\$3(65) +$
$\$1.50(12) = \$195 + \$18 = \$213$. The answer checks.

State. 65 general-interest videos and 12 children's
videos were rented.

18. 131 coach-class seats; 21 first-class seats

19. **Familiarize**. Let f = the number of boxes of Flair
pens sold and u = the number of four-packs of Uniball pens sold.

Translate. We organize the information in a table.

	Flair boxes	Uniball four-packs	Total
Number sold	f	u	40
Price	$12	$8	
Total cost	$12f$	$8u$	372

We get one equation from the "Number sold" row of
the table:

$$f + u = 40$$

The "Total cost" row yields a second equation:

$$12f + 8u = 372$$

We have translated to a system of equations:

$$f + u = 40, \quad (1)$$
$$12f + 8u = 372 \quad (2)$$

Carry out. We solve the system of equations using
the elimination method.

$$-8f - 8u = -320 \quad \text{Multiplying (1) by } -8$$
$$\underline{12f + 8u = 372}$$
$$4f = 52$$
$$f = 13$$

Now substitute 13 for f in (1) and solve for u.

$$13 + u = 40$$
$$u = 27$$

Check. The total number of boxes and four-packs
sold is $13 + 27$, or 40. The total cost of these purchases is $\$12 \cdot 13 + \$8 \cdot 27 = \$156 + \$216 = \$372$. The
answer checks.

State. 13 boxes of Flair pens and 27 four-packs of
Uniball pens were sold.

20. 18 graph-paper notebooks; 32 college-ruled notebooks

21. **Familiarize**. Let k = the number of pounds of
Kenyan French Roast coffee and s = the number of
pounds of Sumatran coffee to be used in the mixture.
The value of the mixture will be $\$8.40(20)$, or $\$168$.

Translate. We organize the information in a table.

	Kenyan	Sumatran	Mixture
Number of pounds	k	s	20
Price per pound	$9	$8	$8.40
Value of coffee	$9k$	$8s$	168

The "Number of pounds" row of the table gives us
one equation:

$$k + s = 20$$

The "Value of coffee" row yields a second equation:

$$9k + 8s = 168$$

We have translated to a system of equations:

$$k + s = 20, \quad (1)$$
$$9k + 8s = 168 \quad (2)$$

Carry out. We use the elimination method to solve the system of equations.

$$-8k - 8s = -160 \quad \text{Multiplying (1) by } -8$$
$$\underline{9k + 8s = 168}$$
$$k = 8$$

Substitute 8 for k in (1) and solve for s.

$$8 + s = 20$$
$$s = 12$$

Check. The total mixture contains 8 lb + 12 lb, or 20 lb. Its value is $\$9 \cdot 8 + \$8 \cdot 12 = \$72 + \$96 = \$168$. The answer checks.

State. 8 lb of Kenyan French Roast coffee and 12 lb of Sumatran coffee should be used.

22. 20 lb of cashews; 30 lb of Brazil nuts

23. Observe that the average of 40% and 10% is 25%: $\frac{40\% + 10\%}{2} = \frac{50\%}{2} = 25\%$. Thus, the caterer should use equal parts of the 40% and 10% mixtures. Since a 10-lb mixture is desired, the caterer should use 5 lb each of the 40% and the 10% mixture.

24. 150 lb of soybean meal; 220 lb of corn meal

25. ***Familiarize***. Let $x =$ the number of liters of 25% solution and $y =$ the number of liters of 50% solution to be used. The mixture contains 40%(10 L), or $0.4(10 \text{ L}) = 4$ L of acid.

Translate. We organize the information in a table.

	25% solution	50% solution	Mixture
Number of liters	x	y	10
Percent of acid	25%	50%	40%
Amount of acid	$0.25x$	$0.5y$	4 L

We get one equation from the "Number of liters" row of the table.

$$x + y = 10$$

The last row of the table yields a second equation.

$$0.25x + 0.5y = 4$$

After clearing decimals, we have the problem translated to a system of equations:

$$x + y = 10, \quad (1)$$
$$25x + 50y = 400 \quad (2)$$

Carry out. We use the elimination method to solve the system of equations.

$$-25x - 25y = -250 \quad \text{Multiplying (1) by } -25$$
$$\underline{25x + 50y = 400}$$
$$25y = 150$$
$$y = 6$$

Substitute 6 for y in (1) and solve for x.

$$x + 6 = 10$$
$$x = 4$$

Check. The total amount of the mixture is 4 lb + 6 lb, or 10 lb. The amount of acid in the mixture is $0.25(4 \text{ L}) + 0.5(6 \text{ L}) = 1 \text{ L} + 3 \text{ L} = 4$ L. The answer checks.

State. 4 L of the 25% solution and 6 L of the 50% solution should be mixed.

26. 12 lb of Deep Thought Granola; 8 lb of Oat Dream Granola

27. ***Familiarize***. Let $x =$ the amount of the 6% loan and $y =$ the amount of the 9% loan. Recall that the formula for simple interest is

$$\text{Interest} = \text{Principal} \cdot \text{Rate} \cdot \text{Time}.$$

Translate. We organize the information in a table.

	6% loan	9% loan	Total
Principal	x	y	$12,000
Interest Rate	6%	9%	
Time	1 yr	1 yr	
Interest	$0.06x$	$0.09y$	$855

The "Principal" row of the table gives us one equation:

$$x + y = 12,000$$

The last row of the table yields another equation:

$$0.06x + 0.09y = 855$$

After clearing decimals, we have the problem translated to a system of equations:

$$x + y = 12,000 \quad (1)$$
$$6x + 9y = 85,500 \quad (2)$$

Carry out. We use the elimination method to solve the system of equations.

$$-6x - 6y = -72,000 \quad \text{Multiplying (1) by } -6$$
$$\underline{6x + 9y = 85,500}$$
$$3y = 13,500$$
$$y = 4500$$

Substitute 4500 for y in (1) and solve for x.

$$x + 4500 = 12,000$$
$$x = 7500$$

Check. The loans total $7500 + $4500, or $12,000. The total interest is 0.06($7500) + 0.09($4500) = $450 + $405 = $855. The answer checks.

State. The 6% loan was for $7500, and the 9% loan was for $4500.

28. $6800 at 9%; $8200 at 10%

29. ***Familiarize***. Let x = the number of liters of Arctic Antifreeze and y = the number of liters of Frost-No-More in the mixture. The amount of alcohol in the mixture is 0.15(20 L) = 3 L.

Translate. We organize the information in a table.

	18% solution	10% solution	Mixture
Number of liters	x	y	20
Percent of alcohol	18%	10%	15%
Amount of alcohol	$0.18x$	$0.1y$	3

We get one equation from the "Number of liters" row of the table:

$$x + y = 20$$

The last row of the table yields a second equation:

$$0.18x + 0.1y = 3$$

After clearing decimals we have the problem translated to a system of equations:

$$x + y = 20, \quad (1)$$
$$18x + 10y = 300 \quad (2)$$

Carry out. We use the elimination method to solve the system of equations.

$$-10x - 10y = -200 \quad \text{Multiplying (1) by } -10$$
$$\underline{18x + 10y = 300}$$
$$8x = 100$$
$$x = 12.5$$

Substitute 12.5 for x in (1) and solve for y.

$$12.5 + y = 20$$
$$y = 7.5$$

Check. The total amount of the mixture is 12.5 L + 7.5 L or 20 L. The amount of alcohol in the mixture is 0.18(12.5 L) + 0.1(7.5 L) = 2.25 L + 0.75 L = 3 L. The answer checks.

State. 12.5 L of Arctic Antifreeze and 7.5 L of Frost-No-More should be used.

30. $169\frac{3}{13}$ lb of whole milk; $30\frac{10}{13}$ lb of cream

31. ***Familiarize***. Let l = the length, in meters, and w = the width, in meters. Recall that the formula for the perimeter P of a rectangle with length l and width w is $P = 2l + 2w$.

Translate.

$$\underbrace{\text{The perimeter}}_{\downarrow} \ \underbrace{\text{is}}_{\downarrow} \ \underbrace{\text{190 m.}}_{\downarrow}$$
$$2l + 2w \quad = \quad 190$$

$$\underbrace{\text{The width}}_{\downarrow} \ \underbrace{\text{is}}_{\downarrow} \ \underbrace{\text{one-fourth}}_{\downarrow} \ \underbrace{\text{of}}_{\downarrow} \ \underbrace{\text{the length.}}_{\downarrow}$$
$$w \quad = \quad \frac{1}{4} \quad \cdot \quad l$$

We have translated to a system of equations:

$$2l + 2w = 190, \quad (1)$$
$$w = \frac{1}{4}l$$

Carry out. We use the substitution method to solve the system of equations.

Substitute $\frac{1}{4}l$ for w in (1) and solve for l.

$$2l + 2\left(\frac{1}{4}l\right) = 190$$
$$2l + \frac{1}{2}l = 190$$
$$\frac{5}{2}l = 190$$
$$l = \frac{2}{5} \cdot 190 = 76$$

Now substitute 76 for l in (2).

$$l = \frac{1}{4} \cdot 76 = 19$$

Check. The perimeter is 2·76 m + 2·19 m = 152 m + 38 m = 190 m. The width, 19 m, is one-fourth the length, 76 m. The answer checks.

State. The length is 76 m, and the width is 19 m.

32. Length: 265 ft, width: 165 ft

33. *Familiarize*. The change from the $9.25 purchase is $20 − $9.25, or $10.75. Let x = the number of quarters and y = the number of fifty-cent pieces. The total value of the quarters, in dollars, is $0.25x$ and the total value of the fifty-cent pieces, in dollars, is $0.50y$.

Translate.

The total number of coins is 30.

$$x + y \quad = \quad 30$$

The total value of the coins is $10.75.

$$0.25x + 0.50y \quad = \quad 10.75$$

After clearing decimals we have the following system of equations:

$$x + y = 30, \quad (1)$$
$$25x + 50y = 1075 \quad (2)$$

Carry out. We use the elimination method to solve the system of equations.

$$\begin{array}{r} -25x - 25y = -750 \quad \text{Multiplying (1) by } -25 \\ \underline{25x + 50y = 1075} \\ 25y = 325 \\ y = 13 \end{array}$$

Substitute 13 for y in (1) and solve for x.

$$x + 13 = 30$$
$$x = 17$$

Check. The total number of coins is $17 + 13$, or 30. The total value of the coins is $\$0.25(17) + \$0.50(13) = \$4.25 + \$6.50 = \$10.75$. The answer checks.

State. There were 17 quarters and 13 fifty-cent pieces.

34. $5 bills; 15 $1 bills

35. *Familiarize*. We first make a drawing.

Slow train
d kilometers   75 km/h $(t + 2)$ hr

Fast train
d kilometers   125 km/h t hr

From the drawing we see that the distances are the same. Now complete the chart.

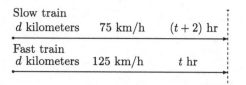

$$d = r \cdot t$$

	Distance	Rate	Time
Slow train	d	75	$t + 2$
Fast train	d	125	t

$\rightarrow d = 75(t+2)$

$\rightarrow d = 125t$

Translate. Using $d = rt$ in each row of the table, we get a system of equations:

$$d = 75(t + 2),$$
$$d = 125t$$

Carry out. We solve the system of equations.

$$\begin{array}{rl} 125t = 75(t + 2) & \text{Using substitution} \\ 125t = 75t + 150 & \\ 50t = 150 & \\ t = 3 & \end{array}$$

Then $d = 125t = 125 \cdot 3 = 375$

Check. At 125 km/h, in 3 hr the fast train will travel $125 \cdot 3 = 375$ km. At 75 km/h, in $3 + 2$, or 5 hr the slow train will travel $75 \cdot 5 = 375$ km. The numbers check.

State. The trains will meet 375 km from the station.

36. 3 hr

37. *Familiarize*. We first make a drawing. Let d = the distance and r = the speed of the canoe in still water. Then when the canoe travels downstream its speed is $r + 6$, and its speed upstream is $r − 6$. From the drawing we see that the distances are the same.

Downstream, 6 mph current

d mi, $r + 6$, 4 hr

Upstream, 6 mph current

d mi, $r − 6$, 10 hr

Organize the information in a table.

	Distance	Rate	Time
With current	d	$r + 6$	4
Against current	d	$r − 6$	10

Translate. Using $d = rt$ in each row of the table, we get a system of equations:

$$d = 4(r + 6), \qquad d = 4r + 24,$$
$$\text{or}$$
$$d = 10(r − 6) \qquad d = 10r − 60$$

Carry out. Solve the system of equations.

$$\begin{array}{rl} 4r + 24 = 10r − 60 & \text{Using substitution} \\ 24 = 6r − 60 & \\ 84 = 6r & \\ 14 = r & \end{array}$$

Check. When $r = 14$, then $r + 6 = 14 + 6 = 20$, and the distance traveled in 4 hr is $4 \cdot 20 = 80$ km. Also, $r − 6 = 14 − 6 = 8$, and the distance traveled in 10 hr is $8 \cdot 10 = 80$ km. The answer checks.

State. The speed of the canoe in still water is 14 km/h.

38. 24 mph

39. *Familiarize*. We make a drawing. Note that the plane's speed traveling toward London is $360 + 50$, or 410 mph, and the speed traveling toward New York City is $360 - 50$, or 310 mph. Also, when the plane is d mi from New York City, it is $3458 - d$ mi from London.

New York City   London
310 mph t hours t hours   410 mph

$\longleftarrow$————— 3458 mi—————$\longrightarrow$

$\longleftarrow$—— d ——$\longrightarrow$——— 3458 mi $-d$———$\longrightarrow$

Organize the information in a table.

	Distance	Rate	Time
Toward NYC	d	310	t
Toward London	$3458 - d$	410	t

Translate. Using $d = rt$ in each row of the table, we get a system of equations:

$$d = 310t, \quad (1)$$
$$3458 - d = 410t \quad (2)$$

Carry out. We solve the system of equations.

$$3458 - 310t = 410t \quad \text{Using substitution}$$
$$3458 = 720t$$
$$4.8028 \approx t$$

Substitute 4.8028 for t in (1).

$$d \approx 310(4.8028) \approx 1489$$

Check. If the plane is 1489 mi from New York City, it can return to New York City, flying at 310 mph, in $1489/310 \approx 4.8$ hr. If the plane is $3458 - 1489$, or 1969 mi from London, it can fly to London, traveling at 410 mph, in $1969/410 \approx 4.8$ hr. Since the times are the same, the answer checks.

State. The point of no return is about 1489 mi from New York City.

40. About 1524 mi

41. Writing exercise

42. Writing exercise

43.
$$2x - 3y + 12 = 2 \cdot 5 - 3 \cdot 2 + 12$$
$$= 10 - 6 + 12$$
$$= 4 + 12$$
$$= 16$$

44. 11

45.
$$5a - 7b + 3c = 5(-2) - 7(3) + 3 \cdot 1$$
$$= -10 - 21 + 3$$
$$= -31 + 3$$
$$= -28$$

46. -10

47.
$$4 - 2y + 3z = 4 - 2 \cdot \frac{1}{3} + 3 \cdot \frac{1}{4}$$
$$= 4 - \frac{2}{3} + \frac{3}{4}$$
$$= \frac{48}{12} - \frac{8}{12} + \frac{9}{12}$$
$$= \frac{40}{12} + \frac{9}{12}$$
$$= \frac{49}{12}$$

48. $\dfrac{13}{10}$

49. Writing exercise

50. Writing exercise

51. The Familiarize and Translate steps were done in Exercise 66 of Exercise Set 8.1.

Carry out. We solve the system of equations

$$x = 2y, \quad (1)$$
$$x + 20 = 3y \quad (2)$$

where x = Burl's age now and y = his son's age now.

$$2y + 20 = 3y \quad \text{Substituting } 2y \text{ for } x \text{ in (2)}$$
$$20 = y$$

$$x = 2 \cdot 20 \quad \text{Substituting 20 for } y \text{ in (1)}$$
$$x = 40$$

Check. Burl's age now, 40, is twice his son's age now, 20. Ten years ago Burl was 30 and his son was 10, and $30 = 3 \cdot 10$. The numbers check.

State. Now Burl is 40 and his son is 20.

52. Lou: 32 years, Juanita: 14 years

53. The Familiarize and Translate steps were done in Exercise 68 of Exercise Set 8.1.

Carry out. We solve the system of equations

$$2l + 2w = 156, \quad (1)$$
$$l = 4(w - 6) \quad (2)$$

where $l =$ length, in inches, and $w =$ width, in inches.

$2 \cdot 4(w - 6) + 2w = 156$ Substituting $4(w - 6)$
 for l in (1)

$$8w - 48 + 2w = 156$$

$$10w - 48 = 156$$

$$10w = 204$$

$$w = \frac{204}{10}, \text{ or } \frac{102}{5}$$

$l = 4\left(\frac{102}{5} - 6\right)$ Substituting $\frac{102}{5}$ for w
$\phantom{l = 4\left(\frac{102}{5} - 6\right)}$ in (2)

$$l = 4\left(\frac{102}{5} - \frac{30}{5}\right)$$

$$l = 4\left(\frac{72}{5}\right)$$

$$l = \frac{288}{5}$$

Check. The perimeter of a rectangle with

width $\dfrac{102}{5}$ in. and length $\dfrac{288}{5}$ in. is

$2\left(\dfrac{288}{5}\right) + 2\left(\dfrac{102}{5}\right) = \dfrac{576}{5} + \dfrac{204}{5} = \dfrac{780}{5} = 156$ in.

If 6 in. is cut off the width, the new width is

$\dfrac{102}{5} - 6 = \dfrac{102}{5} - \dfrac{30}{5} = \dfrac{72}{5}$. The length, $\dfrac{288}{5}$, is

$4\left(\dfrac{72}{5}\right)$. The numbers check.

State. The original piece of posterboard had width

$\dfrac{102}{5}$ in. and length $\dfrac{288}{5}$ in.

54. $\dfrac{64}{5}$ oz of baking soda; $\dfrac{16}{5}$ oz of vinegar

55. Familiarize. Let $k =$ the number of pounds of Kona coffee that must be added to the Mexican coffee, and $m =$ the number of pounds of coffee in the mixture.

Translate. We organize the information in a table.

	Mexican	Kona	Mixture
Number of pounds	40	k	m
Percent of Kona	0%	100%	30%
Amount of Kona	0	k	$0.3m$

We get one equation from the "Number of pounds" row of the table:

$$40 + k = m$$

The last row of the table gives us a second equation:

$$k = 0.3m$$

After clearing the decimal we have the problem translated to a system of equations:

$$40 + k = m, \quad (1)$$

$$10k = 3m \quad (2)$$

Carry out. We use substitution to solve the system of equations. First we substitute $40 + k$ for m in (2).

$$10k = 3m \qquad (2)$$

$10k = 3(40 + k)$ Substituting

$$10k = 120 + 3k$$

$$7k = 120$$

$$k = \frac{120}{7}$$

Although the problem asks only for k, the amount of Kona coffee that should be used, we will also find m in order to check the answer.

$$40 + k = m \quad (1)$$

$40 + \dfrac{120}{7} = m$ Substituting $\dfrac{120}{7}$ for k

$$\frac{280}{7} + \frac{120}{7} = m$$

$$\frac{400}{7} = m$$

Check. If $\dfrac{400}{7}$ lb of coffee contain $\dfrac{120}{7}$ lb of Kona coffee, then the percent of Kona beans in the mixture is $\dfrac{120/7}{400/7} = \dfrac{120}{7} \cdot \dfrac{7}{400} = \dfrac{3}{10}$, or 30%. The answer checks.

State. $\dfrac{120}{7}$ lb of Kona coffee should be added to the Mexican coffee.

56. 1.8 L

57. Familiarize. Let $d =$ the distance, in km, that Natalie jogs in a trip to school, and let $t =$ the time, in hr, that she jogs. We organize the information in a table.

	Distance	Rate	Time
Jogging	d	8	t
Walking	$6 - d$	4	$1 - t$

Translate. Using $d = rt$ in each row of the table we get a system of equations:

$$d = 8t, \qquad (1)$$

$$6 - d = 4(1 - t) \quad (2)$$

Carry out. We use substitution to solve the system of equations.

$6 - 8t = 4(1 - t)$ Substituting $8t$ for d in (2)

$6 - 8t = 4 - 4t$

$2 - 8t = -4t$

$2 = 4t$

$\frac{1}{2} = t$

Substitute $\frac{1}{2}$ for t in (1).

$d = 8 \cdot \frac{1}{2} = 4$

Check. If Natalie jogs 4 km in $\frac{1}{2}$ hr, then she walks $6 - 4$ or 2 km, in $1 - \frac{1}{2}$, or $\frac{1}{2}$ hr. At a rate of 8 km/h, in $\frac{1}{2}$ hr she can jog $8 \cdot \frac{1}{2}$, or 4 km. At a rate of 4 km/h, in $\frac{1}{2}$ hr she can walk $4 \cdot \frac{1}{2}$, or 2 km. Then the total time is $\frac{1}{2}$ hr $+ \frac{1}{2}$ hr, or 1 hr, and the total distance is 4 km$+$2 km, or 6 km. The answer checks.

State. Natalie jogs 4 km in a trip to school.

58. 180

59. *Familiarize*. Let $x =$ the ten's digit and $y =$ the unit's digit. Then the number is $10x + y$. If the digits are interchanged, the new number is $10y + x$.

Translate.

Ten's digit is 2 more than 3 times unit's digit.

$x = 2 + 3 \cdot y$

If the digits are interchanged,

new number is half of given number minus 13.

$10y + x = \frac{1}{2} \cdot (10x + y) - 13$

The system of equations is

$x = 2 + 3y$,   (1)

$10y + x = \frac{1}{2}(10x + y) - 13$ (2)

Carry out. We use the substitution method. Substitute $2 + 3y$ for x in (2).

$10y + (2 + 3y) = \frac{1}{2}[10(2 + 3y) + y] - 13$

$13y + 2 = \frac{1}{2}[20 + 30y + y] - 13$

$13y + 2 = \frac{1}{2}[20 + 31y] - 13$

$13y + 2 = 10 + \frac{31}{2}y - 13$

$13y + 2 = \frac{31}{2}y - 3$

$5 = \frac{5}{2}y$

$2 = y$

$x = 2 + 3 \cdot 2$ Substituting 2 for y in (1)

$x = 2 + 6$

$x = 8$

Check. If $x = 8$ and $y = 2$, the given number is 82 and the new number is 28. In the given number the ten's digit, 8, is two more than three times the unit's digit, 2. The new number is 13 less than one-half the given number: $28 = \frac{1}{2}(82) - 13$. The values check.

State. The given integer is 82.

60. First train: 36 km/h: second train: 54 km/h

61. *Familiarize*. Let $x =$ the number of gallons of pure brown and $y =$ the number of gallons of neutral stain that should be added to the original 0.5 gal. Note that a total of 1 gal of stain needs to be added to bring the amount of stain up to 1.5 gal. The original 0.5 gal of stain contains 20%(0.5 gal), or 0.2(0.5 gal) = 0.1 gal of brown stain. The final solution contains 60%(1.5 gal), or 0.6(1.5 gal) = 0.9 gal of brown stain. This is composed of the original 0.1 gal and the x gal that are added.

Translate.

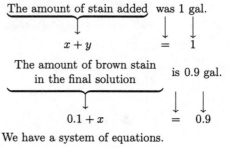

The amount of stain added was 1 gal.

$x + y = 1$

The amount of brown stain in the final solution is 0.9 gal.

$0.1 + x = 0.9$

We have a system of equations.

$x + y = 1$,   (1)

$0.1 + x = 0.9$ (2)

Carry out. First we solve (2) for x.

$$0.1 + x = 0.9$$
$$x = 0.8$$

Then substitute 0.8 for x in (1) and solve for y.

$$0.8 + y = 1$$
$$y = 0.2$$

Check. Total amount of stain: $0.5 + 0.8 + 0.2 = 1.5$ gal

Total amount of brown stain: $0.1 + 0.8 = 0.9$ gal

Total amount of neutral stain: $0.8(0.5) + 0.2 = 0.4 + 0.2 = 0.6$ gal $= 0.4(1.5$ gal$)$

The answer checks.

State. 0.8 gal of pure brown and 0.2 gal of neutral stain should be added.

62. City: 261 miles; highway: 204 miles

63. Observe that if 100% acetone is added to water to create a 10% acetone solution, then the ratio of acetone to water is 10% to 90%, or 10 to 90, or 1 to 9. Thus, for each liter of acetone, 9 liters of water are required. If 5 extra liters of acetone are added to the vat, then $9 \cdot 5$, or 45 L of additional water must be added to bring the concentration down to 10%.

64. 3 girls, 4 boys

65. The 1.5 gal mixture contains $0.1 + x$ gal of pure brown stain. (See Exercise 61.). Thus, the function $P(x) = \dfrac{0.1 + x}{1.5}$ gives the percentage of brown in the mixture as a decimal quantity. Using the Intersect feature, we confirm that when $x = 0.8$, then $P(x) = 0.6$ or 60%.

Exercise Set 8.4

1. Substitute $(2, -1, -2)$ into the three equations, using alphabetical order.

$$\frac{x + y - 2z = 5}{2 + (-1) - 2(-2)\ ?\ 5}$$
$$\begin{array}{c|c} 2 - 1 + 4 & \\ 5 & 5 \quad \text{TRUE} \end{array}$$

$$\frac{2x - y - z = 7}{2 \cdot 2 - (-1) - (-2)\ ?\ 7}$$
$$\begin{array}{c|c} 4 + 1 + 2 & \\ 7 & 7 \quad \text{TRUE} \end{array}$$

$$\frac{-x - 2y + 3z = 6}{-2 - 2(-1) + 3(-2)\ ?\ 6}$$
$$\begin{array}{c|c} -2 + 2 - 6 & \\ -6 & 6 \quad \text{FALSE} \end{array}$$

The triple $(2, -1, -2)$ does not make the third equation true, so it is not a solution of the system.

2. Yes

3.
$$x + y + z = 6, \quad (1)$$
$$2x - y + 3z = 9, \quad (2)$$
$$-x + 2y + 2z = 9 \quad (3)$$

1., 2.   The equations are already in standard form with no fractions or decimals.

3. Add Equations (1) and (2) to eliminate y:
$$\begin{array}{ll} x + y + z = 6 & (1) \\ \underline{2x - y + 3z = 9} & (2) \\ 3x \quad\;\; + 4z = 15 & (4) \quad \text{Adding} \end{array}$$

4. Use a different pair of equations and eliminate y:
$$\begin{array}{ll} 4x - 2y + 6z = 18 & \text{Multiplying (2) by 2} \\ \underline{-x + 2y + 2z = 9} & (3) \\ 3x \quad\;\; + 8z = 27 & (5) \end{array}$$

5. Now solve the system of Equations (4) and (5).
$$3x + 4z = 15 \quad (4)$$
$$3x + 8z = 27 \quad (5)$$

$$\begin{array}{ll} -3x - 4z = -15 & \text{Multiplying (4) by } -1 \\ \underline{3x + 8z = 27} & \\ 4z = 12 & \\ z = 3 & \end{array}$$

$$3x + 4 \cdot 3 = 15 \quad \text{Substituting 3 for } z \text{ in (4)}$$
$$3x + 12 = 15$$
$$3x = 3$$
$$x = 1$$

6. Substitute in one of the original equations to find y.
$$1 + y + 3 = 6 \quad \text{Substituting 1 for } x \text{ and 3 for } z \text{ in (1)}$$
$$y + 4 = 6$$
$$y = 2$$

We obtain $(1, 2, 3)$. This checks, so it is the solution.

4. $(4, 0, 2)$

5. $2x - y - 3z = -1,$ (1)

     $2x - y + z = -9,$ (2)

     $x + 2y - 4z = 17$ (3)

1., 2. The equations are already in standard form with no fractions or decimals.

3., 4. We eliminate z from two different pairs of equations.

$$2x - y - 3z = -1 \quad (1)$$
$$\underline{6x - 3y + 3z = -27} \quad \text{Multiplying (2) by 3}$$
$$8x - 4y = -28 \quad (4) \quad \text{Adding}$$

$$8x - 4y + 4z = -36 \quad \text{Multiplying (2) by 4}$$
$$\underline{x + 2y - 4z = 17} \quad (3)$$
$$9x - 2y = -19 \quad (5) \quad \text{Adding}$$

5. Now solve the system of Equations (4) and (5).

$$8x - 4y = -28 \quad (4)$$
$$9x - 2y = -19 \quad (5)$$

$$8x - 4y = -28 \quad (4)$$
$$\underline{-18x + 4y = 38} \quad \text{Multiplying (5) by } -2$$
$$-10x = 10 \quad \text{Adding}$$
$$x = -1$$

$$8(-1) - 4y = -28 \quad \text{Substituting } -1 \text{ for } x \text{ in (4)}$$
$$-8 - 4y = -28$$
$$-4y = -20$$
$$y = 5$$

6. Substitute in one of the original equations to find z.

$$2(-1) - 5 + z = -9 \quad \text{Substituting } -1 \text{ for } x \text{ and 5 for } y \text{ in (2)}$$
$$-2 - 5 + z = -9$$
$$-7 + z = -9$$
$$z = -2$$

We obtain $(-1, 5, -2)$. This checks, so it is the solution.

6. $(2, -2, 2)$

7. $2x - 3y + z = 5,$ (1)

     $x + 3y + 8z = 22,$ (2)

     $3x - y + 2z = 12$ (3)

1., 2. The equations are already in standard form with no fractions or decimals.

3., 4. We eliminate y from two different pairs of equations.

$$2x - 3y + z = 5 \quad (1)$$
$$\underline{x + 3y + 8z = 22} \quad (2)$$
$$3x + 9z = 27 \quad (4) \quad \text{Adding}$$

$$x + 3y + 8z = 22 \quad (2)$$
$$\underline{9x - 3y + 6z = 36} \quad \text{Multiplying (3) by 3}$$
$$10x + 14z = 58 \quad (5) \quad \text{Adding}$$

5. Solve the system of Equations (4) and (5).

$$3x + 9z = 27 \quad (4)$$
$$10x + 14z = 58 \quad (5)$$

$$30x + 90z = 270 \quad \text{Multiplying (4) by 10}$$
$$\underline{-30x - 42z = -174} \quad \text{Multiplying (5) by } -3$$
$$48z = 96 \quad \text{Adding}$$
$$z = 2$$

$$3x + 9 \cdot 2 = 27 \quad \text{Substituting 2 for } z \text{ in (4)}$$
$$3x + 18 = 27$$
$$3x = 9$$
$$x = 3$$

6. Substitute in one of the original equations to find y.

$$2 \cdot 3 - 3y + 2 = 5 \quad \text{Substituting 3 for } x \text{ and 2 for } z \text{ in (1)}$$
$$-3y + 8 = 5$$
$$-3y = -3$$
$$y = 1$$

We obtain $(3, 1, 2)$. This checks, so it is the solution.

8. $(3, -2, 1)$

9. $3a - 2b + 7c = 13,$ (1)

     $a + 8b - 6c = -47,$ (2)

     $7a - 9b - 9c = -3$ (3)

1., 2. The equations are already in standard form with no fractions or decimals.

3., 4. We eliminate a from two different pairs of equations.

$$3a - 2b + 7c = 13 \quad (1)$$
$$\underline{-3a - 24b + 18c = 141} \quad \text{Multiplying (2) by } -3$$
$$-26b + 25c = 154 \quad (4) \quad \text{Adding}$$

$$-7a - 56b + 42c = 329 \quad \text{Multiplying (2) by } -7$$
$$\underline{7a - 9b - 9c = -3} \quad (3)$$
$$-65b + 33c = 326 \quad (5) \quad \text{Adding}$$

5. Now solve the system of Equations (4) and (5).

$$-26b + 25c = 154 \quad (4)$$
$$-65b + 33c = 326 \quad (5)$$

$$-130b + 125c = 770 \quad \text{Multiplying (4) by 5}$$
$$\underline{130b - 66c = -652} \quad \text{Multiplying (5) by } -2$$
$$59c = 118$$
$$c = 2$$

$$-26b + 25 \cdot 2 = 154 \quad \text{Substituting 2 for } c$$
$$\text{in (4)}$$
$$-26b + 50 = 154$$
$$-26b = 104$$
$$b = -4$$

6. Substitute in one of the original equations to find a.
$$a + 8(-4) - 6(2) = -47 \quad \text{Substituting } -4$$
$$\text{for } b \text{ and 2 for } c$$
$$\text{in (2)}$$
$$a - 32 - 12 = -47$$
$$a - 44 = -47$$
$$a = -3$$

We obtain $(-3, -4, 2)$. This checks, so it is the solution.

10. $(7, -3, -4)$

11. $2x + 3y + z = 17, \quad (1)$
$x - 3y + 2z = -8, \quad (2)$
$5x - 2y + 3z = 5 \quad (3)$

1., 2. The equations are already in standard form with no fractions or decimals.

3., 4. We eliminate y from two different pairs of equations.
$$2x + 3y + z = 17 \quad (1)$$
$$\underline{x - 3y + 2z = -8} \quad (2)$$
$$3x + 3z = 9 \quad (4) \quad \text{Adding}$$

$$4x + 6y + 2z = 34 \quad \text{Multiplying (1) by 2}$$
$$\underline{15x - 6y + 9z = 15} \quad \text{Multiplying (3) by 3}$$
$$19x + 11z = 49 \quad (5) \quad \text{Adding}$$

5. Now solve the system of Equations (4) and (5).
$$3x + 3z = 9 \quad (4)$$
$$19x + 11z = 49 \quad (5)$$

$$33x + 33z = 99 \quad \text{Multiplying (4) by 11}$$
$$\underline{-57x - 33z = -147} \quad \text{Multiplying (5) by } -3$$
$$-24x = -48$$
$$x = 2$$

$$3 \cdot 2 + 3z = 9 \quad \text{Substituting 2 for } x \text{ in (4)}$$
$$6 + 3z = 9$$
$$3z = 3$$
$$z = 1$$

6. Substitute in one of the original equations to find y.
$$2 \cdot 2 + 3y + 1 = 17 \quad \text{Substituting 2 for } x \text{ and}$$
$$1 \text{ for } z \text{ in (1)}$$
$$3y + 5 = 17$$
$$3y = 12$$
$$y = 4$$

We obtain $(2, 4, 1)$. This checks, so it is the solution.

12. $(2, 1, 3)$

13. $2x + y + z = -2, \quad (1)$
$2x - y + 3z = 6, \quad (2)$
$3x - 5y + 4z = 7 \quad (3)$

1., 2. The equations are already in standard form with no fractions or decimals.

3., 4. We eliminate y from two different pairs of equations.
$$2x + y + z = -2 \quad (1)$$
$$\underline{2x - y + 3z = 6} \quad (2)$$
$$4x + 4z = 4 \quad (4) \quad \text{Adding}$$

$$10x + 5y + 5z = -10 \quad \text{Multiplying (1) by 5}$$
$$\underline{3x - 5y + 4z = 7} \quad (3)$$
$$13x + 9z = -3 \quad (5) \quad \text{Adding}$$

5. Now solve the system of Equations (4) and (5).
$$4x + 4z = 4 \quad (4)$$
$$13x + 9z = -3 \quad (5)$$

$$36x + 36z = 36 \quad \text{Multiplying (4) by 9}$$
$$\underline{-52x - 36z = 12} \quad \text{Multiplying (5) by } -4$$
$$-16x = 48 \quad \text{Adding}$$
$$x = -3$$

$$4(-3) + 4z = 4 \quad \text{Substituting } -3 \text{ for } x \text{ in (4)}$$
$$-12 + 4z = 4$$
$$4z = 16$$
$$z = 4$$

6. Substitute in one of the original equations to find y.
$$2(-3) + y + 4 = -2 \quad \text{Substituting } -3 \text{ for}$$
$$x \text{ and 4 for } z \text{ in (1)}$$
$$y - 2 = -2$$
$$y = 0$$

We obtain $(-3, 0, 4)$. This checks, so it is the solution.

14. $(2, -5, 6)$

15.

$$x - y + z = 4, \quad (1)$$
$$5x + 2y - 3z = 2, \quad (2)$$
$$4x + 3y - 4z = -2 \quad (3)$$

1., 2. The equations are already in standard form with no fractions or decimals.

3., 4. We eliminate z from two different pairs of equations.

$$\begin{array}{ll} 3x - 3y + 3z = 12 & \text{Multiplying (1) by 3} \\ \underline{5x + 2y - 3z = \ 2} & (2) \\ 8x - y \qquad = 14 & (4) \quad \text{Adding} \end{array}$$

$$\begin{array}{ll} 4x - 4y + 4z = 16 & \text{Multiplying (1) by 4} \\ \underline{4x + 3y - 4z = -2} & (3) \\ 8x - y \qquad = 14 & (5) \quad \text{Adding} \end{array}$$

5. Now solve the system of Equations (4) and (5).

$$8x - y = 14 \quad (4)$$
$$8x - y = 14 \quad (5)$$

$$\begin{array}{ll} 8x - y = \ \ 14 & (4) \\ \underline{-8x + y = -14} & \text{Multiplying (5) by } -1 \\ 0 = \ \ \ 0 & (6) \end{array}$$

Equation (6) indicates Equations (1), (2), and (3) are dependent. (Note that if Equation (1) is subtracted from Equation (2), the result is Equation (3).) We could also have concluded that the equations are dependent by observing that Equations (4) and (5) are identical.

16. The equations are dependent.

17.

$$a + 2b + c = 1, \quad (1)$$
$$7a + 3b - c = -2, \quad (2)$$
$$a + 5b + 3c = 2 \quad (3)$$

1., 2. The equations are already in standard form with no fractions or decimals.

3., 4. We eliminate c from two different pairs of equations.

$$\begin{array}{ll} a + 2b + c = \ \ 1 & (1) \\ \underline{7a + 3b - c = -2} & (2) \\ 8a + 5b \qquad = -1 & (4) \end{array}$$

$$\begin{array}{ll} 21a + 9b - 3c = -6 & \text{Multiplying (2) by 3} \\ \underline{a + 5b + 3c = \ \ 2} & \\ 22a + 14b \qquad = -4 & (5) \end{array}$$

5. Now solve the system of Equations (4) and (5).

$$8a + 5b = -1 \quad (4)$$
$$22a + 14b = -4 \quad (5)$$

$$\begin{array}{ll} 112a + 70b = -14 & \text{Multiplying (4) by 14} \\ \underline{-110a - 70b = \ \ 20} & \text{Multiplying (5) by } -5 \\ 2a \qquad = \ \ 6 & \\ a = \ \ 3 & \end{array}$$

$$\begin{array}{ll} 8 \cdot 3 + 5b = -1 & \text{Substituting in (4)} \\ 24 + 5b = -1 & \\ 5b = -25 & \\ b = -5 & \end{array}$$

6. Substitute in one of the original equations to find c.

$$\begin{array}{ll} 3 + 2(-5) + c = 1 & \text{Substituting in (1)} \\ -7 + c = 1 & \\ c = 8 & \end{array}$$

We obtain $(3, -5, 8)$. This checks, so it is the solution.

18. $\left(\dfrac{1}{2}, 4, -6 \right)$

19.

$$5x + 3y + \frac{1}{2}z = \frac{7}{2},$$
$$0.5x - 0.9y - 0.2z = 0.3,$$
$$3x - 2.4y + 0.4z = -1$$

1. All equations are already in standard form.

2. Multiply the first equation by 2 to clear the fractions. Also, multiply the second and third equations by 10 to clear the decimals.

$$10x + 6y + z = 7, \quad (1)$$
$$5x - 9y - 2z = 3, \quad (2)$$
$$30x - 24y + 4z = -10 \quad (3)$$

3., 4. We eliminate z from two different pairs of equations.

$$\begin{array}{ll} 20x + 12y + 2z = 14 & \text{Multiplying (1) by 2} \\ \underline{5x - 9y - 2z = \ 3} & (2) \\ 25x + 3y \qquad = 17 & (4) \end{array}$$

$$\begin{array}{ll} 10x - 18y - 4z = \ \ 6 & \text{Multiplying(2) by 2} \\ \underline{30x - 24y + 4z = -10} & (3) \\ 40x - 42y \qquad = -4 & (5) \end{array}$$

5. Now solve the system of Equations (4) and (5).

$$25x + 3y = 17 \quad (4)$$
$$40x - 42y = -4 \quad (5)$$

$$\begin{array}{ll} 350x + 42y = 238 & \text{Multiplying (4) by 14} \\ \underline{40x - 42y = -4} & (5) \\ 390x \qquad = 234 & \\ x = \dfrac{3}{5} & \end{array}$$

$$25\left(\frac{3}{5}\right) + 3y = 17 \quad \text{Substituting in (4)}$$
$$15 + 3y = 17$$
$$3y = 2$$
$$y = \frac{2}{3}$$

6. Substitute in one of the original equations to find z.

$$10\left(\frac{3}{5}\right) + 6\left(\frac{2}{3}\right) + z = 7 \quad \text{Substituting in (1)}$$
$$6 + 4 + z = 7$$
$$10 + z = 7$$
$$z = -3$$

We obtain $\left(\frac{3}{5}, \frac{2}{3}, -3\right)$. This checks, so it is the solution.

20. $\left(\frac{1}{2}, \frac{1}{3}, \frac{1}{6}\right)$

21. $\begin{aligned} 3p \qquad + 2r &= 11, \quad (1) \\ q - 7r &= 4, \quad (2) \\ p - 6q \qquad &= 1 \quad (3) \end{aligned}$

1., 2. The equations are already in standard form with no fractions or decimals.

3., 4. Note that there is no q in Equation (1). We will use Equations (2) and (3) to obtain another equation with no q-term.

$$\begin{array}{ll} 6q - 42r = 24 & \text{Multiplying (2) by 6} \\ \underline{p - 6q \qquad\quad = 1} & (3) \\ p \qquad - 42r = 25 & (4) \end{array}$$

5. Solve the system of Equations (1) and (4).

$$\begin{aligned} 3p + 2r &= 11 \quad (1) \\ p - 42r &= 25 \quad (4) \end{aligned}$$

$$\begin{array}{ll} 3p + \quad 2r = \quad 11 & (1) \\ \underline{-3p + 126r = -75} & \text{Multiplying (4) by } -3 \\ 128r = -64 \end{array}$$
$$r = -\frac{1}{2}$$

$$3p + 2\left(-\frac{1}{2}\right) = 11 \quad \text{Substituting in (1)}$$
$$3p - 1 = 11$$
$$3p = 12$$
$$p = 4$$

6. Substitute in Equation (2) or (3) to find q.

$$q - 7\left(-\frac{1}{2}\right) = 4 \quad \text{Substituting in (2)}$$
$$q + \frac{7}{2} = 4$$
$$q = \frac{1}{2}$$

We obtain $\left(4, \frac{1}{2}, -\frac{1}{2}\right)$. This checks, so it is the solution.

22. $\left(\frac{1}{2}, \frac{2}{3}, -\frac{5}{6}\right)$

23. $\begin{aligned} x + \quad y + z &= 105, \quad (1) \\ 10y - z &= 11, \quad (2) \\ 2x - \quad 3y \qquad &= 7 \quad (3) \end{aligned}$

1., 2. The equations are already in standard form with no fractions or decimals.

3., 4. Note that there is no z in Equation (3). We will use Equations (1) and (2) to obtain another equation with no z-term.

$$\begin{array}{ll} x + \quad y + z = 105 & (1) \\ \underline{10y - z = \quad 11} & (2) \\ x + 11y \qquad = 116 & (4) \end{array}$$

5. Now solve the system of Equations (3) and (4).

$$\begin{aligned} 2x - 3y &= 7 \quad (3) \\ x + 11y &= 116 \quad (4) \end{aligned}$$

$$\begin{array}{ll} 2x - \quad 3y = \quad 7 & (3) \\ \underline{-2x - 22y = -232} & \text{Multiplying (4) by } -2 \\ - 25y = -225 \\ y = \quad 9 \end{array}$$

$$x + 11 \cdot 9 = 116 \quad \text{Substituting in (4)}$$
$$x + 99 = 116$$
$$x = 17$$

6. Substitute in Equation (1) or (2) to find z.

$$17 + 9 + z = 105 \quad \text{Substituting in (1)}$$
$$26 + z = 105$$
$$z = 79$$

We obtain $(17, 9, 79)$. This checks, so it is the solution.

24. $(15, 33, 9)$

25. $\begin{aligned} 2a - \quad 3b \qquad &= 2, \quad (1) \\ 7a \qquad + 4c &= \frac{3}{4}, \quad (2) \\ -3b + 2c &= 1 \quad (3) \end{aligned}$

1. The equations are already in standard form.

2. Multiply Equation (2) by 4 to clear the fraction. The resulting system is

$$2a - 3b \qquad = 2, \quad (1)$$
$$28a \qquad + 16c = 3, \quad (4)$$
$$-3b + 2c = 1 \quad (3)$$

3. Note that there is no b in Equation (2). We will use Equations (1) and (3) to obtain another equation with no b-term.

$$2a - 3b \qquad = \quad 2 \quad (1)$$
$$\underline{\qquad 3b - 2c = -1} \quad \text{Multiplying (3) by } -1$$
$$2a \qquad - 2c = \quad 1 \quad (5)$$

5. Now solve the system of Equations (4) and (5).

$$28a + 16c = 3 \quad (4)$$
$$2a - 2c = 1 \quad (5)$$

$$28a + 16c = \quad 3 \quad (4)$$
$$\underline{16a - 16c = \quad 8} \quad \text{Multiplying (5) by 8}$$
$$44a \qquad = 11$$
$$a = \frac{1}{4}$$

$$2 \cdot \frac{1}{4} - 2c = 1 \quad \text{Substituting } \frac{1}{4} \text{ for } a \text{ in (5)}$$
$$\frac{1}{2} - 2c = 1$$
$$-2c = \frac{1}{2}$$
$$c = -\frac{1}{4}$$

6. Substitute in Equation (1) or (2) to find b.

$$2\left(\frac{1}{4}\right) - 3b = 2 \quad \text{Substituting } \frac{1}{4} \text{ for } a \text{ in (1)}$$
$$\frac{1}{2} - 3b = 2$$
$$-3b = \frac{3}{2}$$
$$b = -\frac{1}{2}$$

We obtain $\left(\frac{1}{4}, -\frac{1}{2}, -\frac{1}{4}\right)$. This checks, so it is the solution.

26. $(3, 4, -1)$

27. $x + y + z = 182, \quad (1)$
$y = 2 + 3x, \qquad (2)$
$z = 80 + x \qquad (3)$

Observe, from Equations (2) and (3), that we can substitute $2+3x$ for y and $80+x$ for z in Equation (1) and solve for x.

$$x + y + x = 182$$
$$x + (2 + 3x) + (80 + x) = 182$$
$$5x + 82 = 182$$
$$5x = 100$$
$$x = 20$$

Now substitute 20 for x in Equation (2).

$$y = 2 + 3x = 2 + 3 \cdot 20 = 2 + 60 = 62$$

Finally, substitute 20 for x in Equation (3).

$$z = 80 + x = 80 + 20 = 100.$$

We obtain $(20, 62, 100)$. This checks, so it is the solution.

28. $(2, 5, -3)$

29. $x + y \qquad = 0, \quad (1)$
$x \qquad + z = 1, \quad (2)$
$2x + y + z = 2 \quad (3)$

1., 2. The equations are already in standard form with no fractions or decimals.

3., 4. Note that there is no z in Equation (1). We will use Equations (2) and (3) to obtain another equation with no z-term.

$$-x \qquad - z = -1 \quad \text{Multiplying (2) by } -1$$
$$\underline{2x + y + z = \quad 2} \quad (3)$$
$$x + y \qquad = \quad 1 \quad (4)$$

5. Now solve the system of Equations (1) and (4).

$$x + y = 0 \quad (1)$$
$$x + y = 1 \quad (4)$$

$$x + y = \quad 0 \quad (1)$$
$$\underline{-x - y = -1} \quad \text{Multiplying (4) by } -1$$
$$0 = -1 \quad \text{Adding}$$

We get a false equation, or contradiction. There is no solution.

30. No solution

31. $y + z = 1, \quad (1)$
$x + y + z = 1, \quad (2)$
$x + 2y + 2z = 2 \quad (3)$

1., 2. The equations are already in standard form with no fractions or decimals.

3., 4. Note that there is no x in Equation (1). We will use Equations (2) and (3) to obtain another equation with no x-term.

$$\begin{array}{ll} -x - y - z = -1 & \text{Multiplying (2)} \\ & \text{by } -1 \\ \underline{x + 2y + 2z = 2} & (3) \\ y + z = 1 & (4) \end{array}$$

Equations (1) and (4) are identical. This means that Equations (1), (2), and (3) are dependent. (We have seen that if Equation (2) is multiplied by -1 and added to Equation (3), the result is Equation (1).)

32. The equations are dependent.

33. Writing exercise

34. Writing exercise

35. Let x represent the larger number and y represent the smaller number. Then we have $x = 2y$.

36. Let x represent the first number and y represent the second number; $x + y = 3x$

37. Let x, $x + 1$, and $x + 2$ represent the numbers. Then we have $x + (x + 1) + (x + 2) = 45$.

38. Let x and y represent the numbers; $x + 2y = 17$

39. Let x and y represent the first two numbers and let z represent the third number. Then we have $x + y = 5z$.

40. Let x and y represent the numbers; $xy = 2(x + y)$

41. Writing exercise

42. Writing exercise

43. $\dfrac{x + 2}{3} - \dfrac{y + 4}{2} + \dfrac{z + 1}{6} = 0,$

$\dfrac{x - 4}{3} + \dfrac{y + 1}{4} - \dfrac{z - 2}{2} = -1,$

$\dfrac{x + 1}{2} + \dfrac{y}{2} + \dfrac{z - 1}{4} = \dfrac{3}{4}$

1., 2. We clear fractions and write each equation in standard form.

To clear fractions, we multiply both sides of each equation by the LCM of its denominators. The LCM's are 6, 12, and 4, respectively.

$$6\left(\frac{x + 2}{3} - \frac{y + 4}{2} + \frac{z + 1}{6}\right) = 6 \cdot 0$$
$$2(x + 2) - 3(y + 4) + (z + 1) = 0$$
$$2x + 4 - 3y - 12 + z + 1 = 0$$
$$2x - 3y + z = 7$$

$$12\left(\frac{x - 4}{3} + \frac{y + 1}{4} - \frac{z - 2}{2}\right) = 12 \cdot (-1)$$
$$4(x - 4) + 3(y + 1) - 6(z - 2) = -12$$
$$4x - 16 + 3y + 3 - 6z + 12 = -12$$
$$4x + 3y - 6z = -11$$

$$4\left(\frac{x + 1}{2} + \frac{y}{2} + \frac{z - 1}{4}\right) = 4 \cdot \frac{3}{4}$$
$$2(x + 1) + 2(y) + (z - 1) = 3$$
$$2x + 2 + 2y + z - 1 = 3$$
$$2x + 2y + z = 2$$

The resulting system is

$$\begin{array}{rcrl} 2x - 3y + z &=& 7, & (1) \\ 4x + 3y - 6z &=& -11, & (2) \\ 2x + 2y + z &=& 2 & (3) \end{array}$$

3., 4. We eliminate z from two different pairs of equations.

$$\begin{array}{ll} 12x - 18y + 6z = 42 & \text{Multiplying (1) by 6} \\ \underline{4x + 3y - 6z = -11} & (2) \\ 16x - 15y = 31 & (4) \quad \text{Adding} \end{array}$$

$$\begin{array}{ll} 2x - 3y + z = 7 & (1) \\ \underline{-2x - 2y - z = -2} & \text{Multiplying (3) by } -1 \\ - 5y = 5 & (5) \quad \text{Adding} \end{array}$$

5. Solve (5) for y: $\quad -5y = 5$

$$\phantom{\text{Solve (5) for } y: \quad} y = -1$$

Substitute -1 for y in (4):

$$16x - 15(-1) = 31$$
$$16x + 15 = 31$$
$$16x = 16$$
$$x = 1$$

6. Substitute 1 for x and -1 for y in (1):

$$2 \cdot 1 - 3(-1) + z = 7$$
$$5 + z = 7$$
$$z = 2$$

We obtain $(1, -1, 2)$. This checks, so it is the solution.

44. $(1, -2, 4, -1)$

45.
$$\begin{array}{rcrl} w + x - y + z &=& 0, & (1) \\ w - 2x - 2y - z &=& -5, & (2) \\ w - 3x - y + z &=& 4, & (3) \\ 2w - x - y + 3z &=& 7 & (4) \end{array}$$

The equations are already in standard form with no fractions or decimals.

Start by eliminating z from three different pairs of equations.

$$\begin{array}{ll} w + x - y + z = 0 & (1) \\ \underline{w - 2x - 2y - z = -5} & (2) \\ 2w - x - 3y = -5 & (5) \text{ Adding} \end{array}$$

$$\begin{array}{ll} w - 2x - 2y - z = -5 & (2) \\ \underline{w - 3x - y + z = 4} & (3) \\ 2w - 5x - 3y = -1 & (6) \text{ Adding} \end{array}$$

$$\begin{array}{ll} 3w - 6x - 6y - 3z = -15 & \text{Multiplying (2) by 3} \\ \underline{2w - x - y + 3z = 7} & (4) \\ 5w - 7x - 7y = -8 & (7) \text{ Adding} \end{array}$$

Now solve the system of equations (5), (6), and (7).

$$2w - x - 3y = -5, \quad (5)$$
$$2w - 5x - 3y = -1, \quad (6)$$
$$5w - 7x - 7y = -8. \quad (7)$$

$$\begin{array}{ll} 2w - x - 3y = -5 & (5) \\ \underline{-2w + 5x + 3y = 1} & \text{Multiplying (6) by } -1 \\ 4x = -4 \\ x = -1 \end{array}$$

Substituting -1 for x in (5) and (7) and simplifying, we have

$$2w - 3y = -6, \quad (8)$$
$$5w - 7y = -15. \quad (9)$$

Now solve the system of Equations (8) and (9).

$$\begin{array}{ll} 10w - 15y = -30 & \text{Multiplying (8) by 5} \\ \underline{-10w + 14y = 30} & \text{Multiplying (9) by } -2 \\ -y = 0 \\ y = 0 \end{array}$$

Substitute 0 for y in Equation (8) or (9) and solve for w.

$$2w - 3 \cdot 0 = -6 \quad \text{Substituting in (8)}$$
$$2w = -6$$
$$w = -3$$

Substitute in one of the original equations to find z.

$$-3 - 1 - 0 + z = 0 \quad \text{Substituting in (1)}$$
$$-4 + z = 0$$
$$z = 4$$

We obtain $(-3, -1, 0, 4)$. This checks, so it is the solution.

46. $\left(-1, \dfrac{1}{5}, -\dfrac{1}{2}\right)$

47. $\dfrac{2}{x} + \dfrac{2}{y} - \dfrac{3}{z} = 3,$

$\dfrac{1}{x} - \dfrac{2}{y} - \dfrac{3}{z} = 9,$

$\dfrac{7}{x} - \dfrac{2}{y} + \dfrac{9}{z} = -39$

Let u represent $\dfrac{1}{x}$, v represent $\dfrac{1}{y}$, and w represent $\dfrac{1}{z}$. Substituting, we have

$$2u + 2v - 3w = 3, \quad (1)$$
$$u - 2v - 3w = 9, \quad (2)$$
$$7u - 2v + 9w = -39 \quad (3)$$

1., 2. The equations in u, v, and w are in standard form with no fractions or decimals.

3., 4. We eliminate v from two different pairs of equations.

$$\begin{array}{ll} 2u + 2v - 3w = 3 & (1) \\ \underline{u - 2v - 3w = 9} & (2) \\ 3u - 6w = 12 & (4) \text{ Adding} \end{array}$$

$$\begin{array}{ll} 2u + 2v - 3w = 3 & (1) \\ \underline{7u - 2v + 9w = -39} & (3) \\ 9u + 6w = -36 & (5) \text{ Adding} \end{array}$$

5. Now solve the system of Equations (4) and (5).

$$\begin{array}{ll} 3u - 6w = 12, & (4) \\ \underline{9u + 6w = -36} & (5) \\ 12u = -24 \\ u = -2 \end{array}$$

$$3(-2) - 6w = 12 \quad \text{Substituting in (4)}$$
$$-6 - 6w = 12$$
$$-6w = 18$$
$$w = -3$$

6. Substitute in Equation (1), (2), or (3) to find v.

$$2(-2) + 2v - 3(-3) = 3 \quad \text{Substituting in (1)}$$
$$2v + 5 = 3$$
$$2v = -2$$
$$v = -1$$

Solve for x, y, and z. We substitute -2 for u, -1 for v, and -3 for w.

$$u = \dfrac{1}{x} \qquad v = \dfrac{1}{y} \qquad w = \dfrac{1}{z}$$
$$-2 = \dfrac{1}{x} \qquad -1 = \dfrac{1}{y} \qquad -3 = \dfrac{1}{z}$$
$$x = \dfrac{1}{2} \qquad y = -1 \qquad z = -\dfrac{1}{3}$$

We obtain $\left(-\dfrac{1}{2}, -1, -\dfrac{1}{3}\right)$. This checks, so it is the solution.

48. 12

49.
$$5x - 6y + kz = -5, \quad (1)$$
$$x + 3y - 2z = 2, \quad (2)$$
$$2x - y + 4z = -1 \quad (3)$$

Eliminate y from two different pairs of equations.

$$\begin{array}{ll} 5x - 6y + kz = -5 & (1) \\ \underline{2x + 6y - 4z = 4} & \text{Multiplying (2) by 2} \\ 7x + (k-4)z = -1 & (4) \end{array}$$

$$\begin{array}{ll} x + 3y - 2z = 2 & (2) \\ \underline{6x - 3y + 12z = -3} & \text{Multiplying (3) by 3} \\ 7x + 10z = -1 & (5) \end{array}$$

Solve the system of Equations (4) and (5).

$$7x + (k-4)z = -1 \quad (4)$$
$$7x + 10z = -1 \quad (5)$$

$$\begin{array}{ll} -7x - (k-4)z = 1 & \text{Multiplying (4) by } -1 \\ \underline{7x + 10z = -1} & (5) \\ (-k + 14)z = 0 & (6) \end{array}$$

The system is dependent for the value of k that makes Equation (6) true. This occurs when $-k + 14$ is 0. We solve for k:

$$-k + 14 = 0$$
$$14 = k$$

50. $3x + 4y + 2z = 12$

51. $z = b - mx - ny$

Three solutions are $(1, 1, 2)$, $(3, 2, -6)$, and $\left(\dfrac{3}{2}, 1, 1\right)$. We substitute for x, y, and z and then solve for b, m, and n.

$$2 = b - m - n,$$
$$-6 = b - 3m - 2n,$$
$$1 = b - \frac{3}{2}m - n$$

1., 2. Write the equations in standard form. Also, clear the fraction in the last equation.

$$b - m - n = 2, \quad (1)$$
$$b - 3m - 2n = -6, \quad (2)$$
$$2b - 3m - 2n = 2 \quad (3)$$

3., 4. Eliminate b from two different pairs of equations.

$$\begin{array}{ll} b - m - n = 2 & (1) \\ \underline{-b + 3m + 2n = 6} & \text{Multiplying (2) by } -1 \\ 2m + n = 8 & (4) \quad \text{Adding} \end{array}$$

$$\begin{array}{ll} -2b + 2m + 2n = -4 & \text{Multiplying (1) by } -2 \\ \underline{2b - 3m - 2n = 2} & (3) \\ -m = -2 & (5) \quad \text{Adding} \end{array}$$

5. We solve Equation (5) for m:

$$-m = -2$$
$$m = 2$$

Substitute in Equation (4) and solve for n.

$$2 \cdot 2 + n = 8$$
$$4 + n = 8$$
$$n = 4$$

6. Substitute in one of the original equations to find b.

$$b - 2 - 4 = 2 \quad \text{Substituting 2 for } m$$
$$ \text{and 4 for } n \text{ in (1)}$$
$$b - 6 = 2$$
$$b = 8$$

The solution is $(8, 2, 4)$, so the equation is $z = 8 - 2x - 4y$.

52. Answers may vary.
$$x + y + z = 1,$$
$$2x + 2y + 2z = 2,$$
$$x + y + z = 3$$

Exercise Set 8.5

1. **Familiarize.** Let $x =$ the first number, $y =$ the second number, and $z =$ the third number.

Translate.

The sum of three numbers is 57.
$$x + y + z = 57$$

The second is 3 more than the first.
$$y = 3 + x$$

The third is 6 more than the first.
$$z = 6 + x$$

We now have a system of equations.

$$\begin{array}{lll} x + y + z = 57, & \text{or} & x + y + z = 57, \\ y = 3 + x & & -x + y = 3, \\ z = 6 + x & & -x + z = 6 \end{array}$$

Carry out. Solving the system we get $(16, 19, 22)$.

Check. The sum of the three numbers is $16 + 19 + 22$, or 57. The second number, 19, is three more than the first number, 16. The third number, 22, is 6 more than the first number, 16. The numbers check.

State. The numbers are 16, 19, and 22.

2. 4, 2, −1

3. Familiarize. Let x = the first number, y = the second number, and z = the third number.

Translate.

The sum of three numbers is 26.

$$x + y + z = 26$$

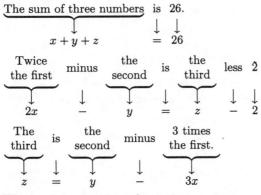

We now have a system of equations.

$$x + y + z = 26, \quad \text{or} \quad x + y + z = 26,$$
$$2x - y = z - 2, \qquad 2x - y - z = -2,$$
$$z = y - 3x \qquad\qquad 3x - y + z = 0$$

Carry out. Solving the system we get $(8, 21, -3)$.

Check. The sum of the numbers is $8 + 21 - 3$, or 26. Twice the first minus the second is $2 \cdot 8 - 21$, or −5, which is 2 less than the third. The second minus three times the first is $21 - 3 \cdot 8$, or −3, which is the third. The numbers check.

State. The numbers are 8, 21, and −3.

4. 17, 9, 79

5. Familiarize. We first make a drawing.

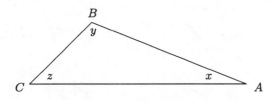

We let x, y, and z represent the measures of angles A, B, and C, respectively. The measures of the angles of a triangle add up to 180°.

Translate.

The sum of the measures is 180°.

$$x + y + z = 180$$

The measure of angle B is three times the measure of angle A.

$$y = 3x$$

The measure of angle C is 20° more than the measure of angle A.

$$z = x + 20$$

We now have a system of equations.

$$x + y + z = 180,$$
$$y = 3x,$$
$$z = x + 20$$

Carry out. Solving the system we get $(32, 96, 52)$.

Check. The sum of the measures is $32° + 96° + 52°$, or 180°. Three times the measure of angle A is $3 \cdot 32°$, or 96°, the measure of angle B. 20° more than the measure of angle A is $32° + 20°$, or 52°, the measure of angle C. The numbers check.

State. The measures of angles A, B, and C are 32°, 96°, and 52°, respectively.

6. 25°, 50°, 105°

7. Familiarize. Let x = the cost of automatic transmission, y = the cost of power door locks, and z = the cost of air conditioning. The prices of the options are added to the basic price of \$12,685.

Translate.

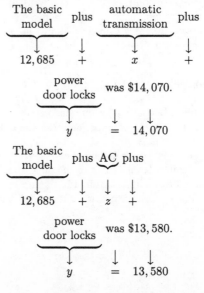

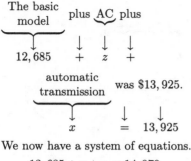

The basic model plus AC plus

$$12,685 + z +$$

automatic transmission was $13,925.

$$x = 13,925$$

We now have a system of equations.

$$12,685 + x + y = 14,070,$$
$$12,685 + z + y = 13,580,$$
$$12,685 + z + x = 13,925$$

Carry out. Solving the system we get $(865, 520, 375)$.

Check. The basic model with automatic transmission and power door locks costs $12,685 + $865 + 520, or $14,070. The basic model with AC and power door locks costs $12,685 + $375 + 520, or $13,580. The basic model with AC and automatic transmission costs $12,685 + $375 + 865, or $13,925. The numbers check.

State. Automatic transmission costs $865, power door locks cost $520, and AC costs $375.

8. A: 1500; B: 1900; C: 2300

9. We know that Elrod, Dot, and Wendy can weld 74 linear feet per hour when working together. We also know that Elrod and Dot together can weld 44 linear feet per hour, which leads to the conclusion that Wendy can weld $74 - 44$, or 30 linear feet per hour alone. We also know that Elrod and Wendy together can weld 50 linear feet per hour. This, along with the earlier conclusion that Wendy can weld 30 linear feet per hour alone, leads to two conclusions: Elrod can weld $50 - 30$, or 20 linear feet per hour alone and Dot can weld $74 - 50$, or 24 linear feet per hour alone.

10. Sven: 220; Tillie: 250; Isaiah: 270

11. Familiarize. Let $x =$ the number of 10-oz cups, $y =$ the number of 14-oz cups, and $z =$ the number of 20-oz cups that Kyle filled. Note that five 96-oz pots contain $5 \cdot 96$ oz, or 480 oz of coffee. Also, x 10-oz cups contain a total of $10x$ oz of coffee and bring in $1.05x$, y 14-oz cups contain $14y$ oz and bring in $1.35y$, and z 20-oz cups contain $20z$ oz and bring in $1.65z$.

Translate.

The total number of coffees served was 34.

$$x + y + z = 34$$

The total amount of coffee served was 480 oz.

$$10x + 14y + 20z = 480$$

The total amount collected was $45.

$$1.05x + 1.35y + 1.65z = 45$$

Now we have a system of equations.

$$x + y + z = 34,$$
$$10x + 14y + 20z = 480,$$
$$1.05x + 1.35y + 1.65z = 45$$

Carry out. Solving the system we get $(11, 15, 8)$.

Check. The total number of coffees served was $11 + 15 + 8$, or 34, The total amount of coffee served was $10 \cdot 11 + 14 \cdot 15 + 20 \cdot 8 = 110 + 210 + 160 = 480$ oz. The total amount collected was $1.05(11) + 1.35(15) + 1.65(8) = 11.55 + 20.25 + 13.20 = 45$. The numbers check.

State. Kyle filled 11 10-oz cups, 15 14-oz cups, and 8 20-oz cups.

12. Newspaper: $41.1 billion; television: $36 billion; radio: $7.7 billion

13. Familiarize. Let $x =$ the amount invested in the first fund, $y =$ the amount invested in the second fund, and $z =$ the amount invested in the third fund. Then the earnings from the investments were $0.1x$, $0.06y$, and $0.15z$.

Translate.

The total amount invested was $80,000.

$$x + y + z = 80,000$$

The total earnings were $8850.

$$0.1x + 0.06y + 0.15z = 8850$$

The earnings from the first fund were $750 more than the earnings from the third fund.

$$0.1x = 750 + 0.15z$$

Now we have a system of equations.

$$x + y + z = 80,000$$
$$0.1x + 0.06y + 0.15z = 8850,$$
$$0.1x = 750 + 0.15z$$

Carry out. Solving the system we get $(45,000, 10,000, 25,000)$.

Check. The total investment was $45,000+ $10,000 + $25,000, or $80,000. The total earnings were $0.1(\$45,000) + 0.06(10,000) + 0.15(25,000) = \$4500+\$600+\$3750 = \$8850$. The earnings from the first fund, $4500, were $750 more than the earnings from the second fund, $3750.

State. $45,000 was invested in the first fund, $10,000 in the second fund, and $25,000 in the third fund.

14. 10 small drinks, 25 medium drinks, 5 large drinks

15. Familiarize. Let $r =$ the number of servings of roast beef, $p =$ the number of baked potatoes, and $b =$ the number of servings of broccoli. Then r servings of roast beef contain $300r$ Calories, $20r$ g of protein, and no vitamin C. In p baked potatoes there are $100p$ Calories, $5p$ g of protein, and $20p$ mg of vitamin C. And b servings of broccoli contain $50b$ Calories, $5b$ g of protein, and $100b$ mg of vitamin C. The patient requires 800 Calories, 55 g of protein, and 220 mg of vitamin C.

Translate. Write equations for the total number of calories, the total amount of protein, and the total amount of vitamin C.

$$300r + 100p + 50b = 800 \quad \text{(Calories)}$$
$$20r + 5p + 5b = 55 \quad \text{(protein)}$$
$$20p + 100b = 220 \quad \text{(vitamin C)}$$

We now have a system of equations.

Carry out. Solving the system we get $(2, 1, 2)$.

Check. Two servings of roast beef provide 600 Calories, 40 g of protein, and no vitamin C. One baked potato provides 100 Calories, 5 g of protein, and 20 mg of vitamin C. And 2 servings of broccoli provide 100 Calories, 10 g of protein, and 200 mg of vitamin C. Together, then, they provide 800 Calories, 55 g of protein, and 220 mg of vitamin C. The values check.

State. The dietician should prepare 2 servings of roast beef, 1 baked potato, and 2 servings of broccoli.

16. $1\frac{1}{8}$ servings of roast beef; $2\frac{3}{4}$ baked potatoes; $3\frac{3}{4}$ servings of asparagus

17. Let x, y, and z represent the average number of times a man, a woman, and a one-year-old child cry each month, respectively.

Translate.

The sum of the averages is 71.7.

$$x + y + z = 71.7$$

The number of times a one-year-old cries is 46.4 times more than the number of times a man cries.

$$z = 46.4 + x$$

The number of times a one-year-old cries is 28.3 times more than the number of times a man and a woman cry.

$$z = 28.3 + x + y$$

Now we have a system of equations.

$$x + y + z = 71.7,$$
$$z = 46.4 + x,$$
$$z = 28.3 + x + y$$

Carry out. Solving the system, we get $(3.6, 18.1, 50)$.

Check. The sum of the average number times a man, a woman, and a one-year-old child cry each month is $3.6+18.1+50 = 71.7$. The number of times a one-year-old child cries, 50, is 46.4 more than 3.6, the average number of times a man cries each month and is 28.3 more than $3.6+18.1$, or 21.7, the average number of times a man and a woman cry. These numbers check.

State. In a month, a man cries an average of 3.6 times, a woman cries 18.1 times, and a one-year old child cries 50 times.

18. Asian-American: 385; African-American: 200; Caucasian: 154

19. Familiarize. Let x, y, and z represent the number of 2-point field goals, 3-point field goals, and 1-point foul shots made, respectively. The total number of points scored from each of these types of goals is $2x$, $3y$, and z.

Translate.

The total number of points was 92.

$$2x + 3y + z = 92$$

The total number of baskets was 50.

$$x + y + z = 50$$

The number of $\underbrace{\text{2-pointers}}$ $\underset{\downarrow}{\text{was}}$ $\underset{\downarrow}{19}$ $\underbrace{\text{more}}_{\text{than}}$ $\underbrace{\text{the number of}}_{\text{foul shots.}}$

$$x \qquad = \quad 19 \quad + \qquad z$$

Now we have a system of equations.

$$2x + 3y + z = 92,$$
$$x + y + z = 50,$$
$$x = 19 + z$$

Carry out. Solving the system we get $(32, 5, 13)$.

Check. The total number of points was $2 \cdot 32 + 3 \cdot 5 + 13 = 64 + 15 + 13 = 92$. The number of baskets was $32 + 5 + 13$, or 50. The number of 2-pointers, 32, was 19 more than the number of foul shots, 13. The numbers check.

State. The Knicks made 32 two-point field goals, 5 three-point field goals, and 13 foul shots.

20. 1869

21. Writing exercise

22. Writing exercise

23. $5(-3) + 7 = -15 + 7 = -8$

24. 33

25. $-6(8) + (-7) = -48 + (-7) = -55$

26. -71

27. $-7(2x - 3y + 5z) = -7 \cdot 2x - 7(-3y) - 7(5z)$
$$= -14x + 21y - 35z$$

28. $-24a - 42b + 54c$

29. $\quad -4(2a + 5b) + 3a + 20b$
$$= -8a - 20b + 3a + 20b$$
$$= -8a + 3a - 20b + 20b$$
$$= -5a$$

30. $11x$

31. Writing exercise

32. Writing exercise

33. **Familiarize**. Let $x =$ the one's digit, $y =$ the ten's digit, and $z =$ the hundred's digit. Then the number is represented by $100z + 10y + x$. When the digits are reversed, the resulting number is represented by $100x + 10y + z$.

Translate.

The sum of the digits $\underset{\downarrow}{\text{is}}$ $\underset{\downarrow}{14}$.

$$x + y + z \qquad = \quad 14$$

$\underbrace{\text{The ten's digit}}$ $\underset{\downarrow}{\text{is}}$ $\underset{\downarrow}{2}$ $\underbrace{\text{more}}_{\text{than}}$ $\underbrace{\text{the one's}}_{\text{digit.}}$

$$y \qquad = \quad 2 \quad + \qquad x$$

$\underbrace{\text{The number}}$ $\underbrace{\text{is the}}_{\text{same as}}$ $\underbrace{\text{the number with the}}_{\text{digits reversed.}}$

$$100z + 10y + x \qquad = \qquad 100x + 10y + z$$

Now we have a system of equations.

$$x + y + z = 14,$$
$$y = 2 + x,$$
$$100z + 10y + x = 100x + 10y + z$$

Carry out. Solving the system we get $(4, 6, 4)$.

Check. If the number is 464, then the sum of the digits is $4 + 6 + 4$, or 14. The ten's digit, 6, is 2 more than the one's digit, 4. If the digits are reversed the number is unchanged The result checks.

State. The number is 464.

34. 20

35. **Familiarize**. Let $x =$ the number of adults, $y =$ the number of students, and $z =$ the number of children in attendance.

Translate. The given information gives rise to two equations.

$\underbrace{\begin{array}{c}\text{The total number}\\\text{in attendance}\end{array}}$ $\underset{\downarrow}{\text{was}}$ $\underset{\downarrow}{100}$.

$$x + y + z \qquad = \quad 100$$

$\underbrace{\begin{array}{c}\text{The total amount}\\\text{taken in}\end{array}}$ $\underset{\downarrow}{\text{was}}$ $\underset{\downarrow}{\$100}$.

$$10x + 3y + 0.5z \qquad = \quad 100$$

Now we have a system of equations.

$$x + y + z = 100,$$
$$10x + 3y + 0.5z = 100$$

Multiply the second equation by 2 to clear the decimal:

$$x + y + z = 100, \quad (1)$$
$$20x + 6y + z = 200. \quad (2)$$

Carry out. We use the elimination method.

$-x - y - z = -100$ Multiplying (1) by -1

$\underline{20x + 6y + z = 200}$ (2)

$19x + 5y \quad\quad = 100$ (3)

In (3), note that 5 is a factor of both $5y$ and 100. Therefore, 5 must also be a factor of $19x$, and hence of x, since 5 is not a factor of 19. Then for some positive integer n, $x = 5n$. (We require $n > 0$, since the number of adults clearly cannot be negative and must also be nonzero since the exercise states that the audience consists of *adults*, students, and children.) We have

$$19 \cdot 5n + 5y = 100, \text{ or}$$

$$19n + y = 20.\quad \text{Dividing by 5 on}$$
$$\text{both sides}$$

Since n and y must both be positive, $n = 1$. Otherwise, $19n + y$ would be greater than 20. Then $x = 5 \cdot 1$, or 5.

$$19 \cdot 5 + 5y = 100 \quad \text{Substituting in (3)}$$

$$95 + 5y = 100$$

$$5y = 5$$

$$y = 1$$

$$5 + 1 + z = 100 \quad \text{Substituting in (1)}$$

$$6 + z = 100$$

$$z = 94$$

Check. The number of people in attendance was $5 + 1 + 94$, or 100. The amount of money taken in was $\$10 \cdot 5 + \$3 \cdot 1 + \$0.50(94) = \$50 + \$3 + \$47 = \$100$. The numbers check.

State. There were 5 adults, 1 student, and 94 children.

36. 35

37. ***Familiarize.*** We first make a drawing with additional labels.

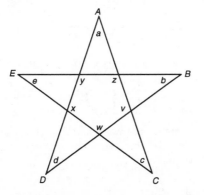

We let a, b, c, d, and e represent the angle measures at the tips of the star. We also label the interior angles of the pentagon v, w, x, y, and z. We recall the following geometric fact:

The sum of the measures of the interior angles of a polygon of n sides is given by $(n-2)180°$.

Using this fact we know:

1. The sum of the angle measures of a triangle is $(3-2)180°$, or $180°$.

2. The sum of the angle measures of a pentagon is $(5-2)180°$, or $3(180°)$.

Translate. Using fact (1) listed above we obtain a system of 5 equations.

$$a + v + d = 180$$

$$b + w + e = 180$$

$$c + x + a = 180$$

$$d + y + b = 180$$

$$e + z + c = 180$$

Carry out. Adding we obtain

$$2a + 2b + 2c + 2d + 2e + v + w + x + y + z = 5(180)$$

$$2(a + b + c + d + e) + (v + w + x + y + z) = 5(180)$$

Using fact (2) listed above we substitute $3(180)$ for $(v + w + x + y + z)$ and solve for $(a + b + c + d + e)$.

$$2(a + b + c + d + e) + 3(180) = 5(180)$$

$$2(a + b + c + d + e) = 2(180)$$

$$a + b + c + d + e = 180$$

Check. We should repeat the above calculations.

State. The sum of the angle measures at the tips of the star is $180°$.

Exercise Set 8.6

1. $9x - 2y = 5,$

$3x - 3y = 11$

Write a matrix using only the constants.

$$\begin{bmatrix} 9 & -2 & \vdots & 5 \\ 3 & -3 & \vdots & 11 \end{bmatrix}$$

Multiply row 2 by 3 to make the first number in row 2 a multiple of 9.

$$\begin{bmatrix} 9 & -2 & \vdots & 5 \\ 9 & -9 & \vdots & 33 \end{bmatrix} \quad \text{New Row 2} = 3(\text{Row 2})$$

Multiply row 1 by -1 and add it to row 2.

$$\begin{bmatrix} 9 & -2 & \vdots & 5 \\ 0 & -7 & \vdots & 28 \end{bmatrix}$$ New Row 2 $= -1$(Row 1) + Row 2

Reinserting the variables, we have

$$9x - 2y = 5, \quad (1)$$
$$-7y = 28. \quad (2)$$

Solve Equation (2) for y.

$$-7y = 28$$
$$y = -4$$

Substitute -4 for y in Equation (1) and solve for x.

$$9x - 2y = 5$$
$$9x - 2(-4) = 5$$
$$9x + 8 = 5$$
$$9x = -3$$
$$x = -\frac{1}{3}$$

The solution is $\left(-\dfrac{1}{3}, -4\right)$.

2. $(2, -1)$

3. $x + 4y = 8,$
$\quad 3x + 5y = 3$

We first write a matrix using only the constants.

$$\begin{bmatrix} 1 & 4 & \vdots & 8 \\ 3 & 5 & \vdots & 3 \end{bmatrix}$$

Multiply the first row by -3 and add it to the second row.

$$\begin{bmatrix} 1 & 4 & \vdots & 8 \\ 0 & -7 & \vdots & -21 \end{bmatrix}$$ New Row 2 $= -3$(Row 1) + Row 2

Reinserting the variables, we have

$$x + 4y = 8, \quad (1)$$
$$-7y = -21. \quad (2)$$

Solve Equation (2) for y.

$$-7y = -21$$
$$y = 3$$

Substitute 3 for y in Equation (1) and solve for x.

$$x + 4 \cdot 3 = 8$$
$$x + 12 = 8$$
$$x = -4$$

The solution is $(-4, 3)$.

4. $(-3, 2)$

5. $6x - 2y = 4,$
$\quad 7x + y = 13$

Write a matrix using only the constants.

$$\begin{bmatrix} 6 & -2 & \vdots & 4 \\ 7 & 1 & \vdots & 13 \end{bmatrix}$$

Multiply the second row by 6 to make the first number in row 2 a multiple of 6.

$$\begin{bmatrix} 6 & -2 & \vdots & 4 \\ 42 & 6 & \vdots & 78 \end{bmatrix}$$ New Row 2 $= 6$(Row 2)

Now multiply the first row by -7 and add it to the second row.

$$\begin{bmatrix} 6 & -2 & \vdots & 4 \\ 0 & 20 & \vdots & 50 \end{bmatrix}$$ New Row 2 $= -7$(Row 1) + Row 2

Reinserting the variables, we have

$$6x - 2y = 4, \quad (1)$$
$$20y = 50. \quad (2)$$

Solve Equation (2) for y.

$$20y = 50$$
$$y = \frac{5}{2}$$

Substitute $\dfrac{5}{2}$ for y in Equation (1) and solve for x.

$$6x - 2y = 4$$
$$6x - 2\left(\frac{5}{2}\right) = 4$$
$$6x - 5 = 4$$
$$6x = 9$$
$$x = \frac{3}{2}$$

The solution is $\left(\dfrac{3}{2}, \dfrac{5}{2}\right)$.

6. $\left(-1, \dfrac{5}{2}\right)$

7. $3x + 2y + 2z = 3,$
$\quad x + 2y - z = 5,$
$\quad 2x - 4y + z = 0$

We first write a matrix using only the constants.

$$\begin{bmatrix} 3 & 2 & 2 & \vdots & 3 \\ 1 & 2 & -1 & \vdots & 5 \\ 2 & -4 & 1 & \vdots & 0 \end{bmatrix}$$

First interchange rows 1 and 2 so that each number below the first number in the first row is a multiple of that number.

$$\begin{bmatrix} 1 & 2 & -1 & \vdots & 5 \\ 3 & 2 & 2 & \vdots & 3 \\ 2 & -4 & 1 & \vdots & 0 \end{bmatrix}$$

Multiply row 1 by -3 and add it to row 2.

Multiply row 1 by -2 and add it to row 3.

$$\begin{bmatrix} 1 & 2 & -1 & | & 5 \\ 0 & -4 & 5 & | & -12 \\ 0 & -8 & 3 & | & -10 \end{bmatrix}$$

Multiply row 2 by -2 and add it to row 3.

$$\begin{bmatrix} 1 & 2 & -1 & | & 5 \\ 0 & -4 & 5 & | & -12 \\ 0 & 0 & -7 & | & 14 \end{bmatrix}$$

Reinserting the variables, we have

$$x + 2y - z = 5, \quad (1)$$
$$-4y + 5z = -12, \quad (2)$$
$$-7z = 14. \quad (3)$$

Solve (3) for z.

$$-7z = 14$$
$$z = -2$$

Substitute -2 for z in (2) and solve for y.

$$-4y + 5(-2) = -12$$
$$-4y - 10 = -12$$
$$-4y = -2$$
$$y = \frac{1}{2}$$

Substitute $\frac{1}{2}$ for y and -2 for z in (1) and solve for x.

$$x + 2 \cdot \frac{1}{2} - (-2) = 5$$
$$x + 1 + 2 = 5$$
$$x + 3 = 5$$
$$x = 2$$

The solution is $\left(2, \frac{1}{2}, -2\right)$.

8. $\left(\frac{3}{2}, -4, -3\right)$

9. $p - 2q - 3r = 3,$
$\quad 2p - q - 2r = 4,$
$\quad 4p + 5q + 6r = 4$

We first write a matrix using only the constants.

$$\begin{bmatrix} 1 & -2 & -3 & | & 3 \\ 2 & -1 & -2 & | & 4 \\ 4 & 5 & 6 & | & 4 \end{bmatrix}$$

$$\begin{bmatrix} 1 & -2 & -3 & | & 3 \\ 0 & 3 & 4 & | & -2 \\ 0 & 13 & 18 & | & -8 \end{bmatrix} \begin{array}{l} \\ \text{New Row 2 =} \\ -2(\text{Row 1}) + \text{Row 2} \\ \text{New Row 3 =} \\ -4(\text{Row 1}) + \text{Row 3} \end{array}$$

$$\begin{bmatrix} 1 & -2 & -3 & | & 3 \\ 0 & 3 & 4 & | & -2 \\ 0 & 39 & 54 & | & -24 \end{bmatrix} \text{New Row 3 = 3(Row 3)}$$

$$\begin{bmatrix} 1 & -2 & -3 & | & 3 \\ 0 & 3 & 4 & | & -2 \\ 0 & 0 & 2 & | & 2 \end{bmatrix} \begin{array}{l} \text{New Row 3 =} \\ -13(\text{Row 2}) + \text{Row 3} \end{array}$$

Reinserting the variables, we have

$$p - 2q - 3r = 3, \quad (1)$$
$$3q + 4r = -2, \quad (2)$$
$$2r = 2 \quad (3)$$

Solve (3) for r.

$$2r = 2$$
$$r = 1$$

Substitute 1 for r in (2) and solve for q.

$$3q + 4 \cdot 1 = -2$$
$$3q + 4 = -2$$
$$3q = -6$$
$$q = -2$$

Substitute -2 for q and 1 for r in (1) and solve for p.

$$p - 2(-2) - 3 \cdot 1 = 3$$
$$p + 4 - 3 = 3$$
$$p + 1 = 3$$
$$p = 2$$

The solution is $(2, -2, 1)$.

10. $(-1, 2, -2)$

11. $3p \quad + 2r = 11,$
$\quad q - 7r = 4,$
$\quad p - 6q \quad = 1$

We first write a matrix using only the constants.

$$\begin{bmatrix} 3 & 0 & 2 & | & 11 \\ 0 & 1 & -7 & | & 4 \\ 1 & -6 & 0 & | & 1 \end{bmatrix}$$

$$\begin{bmatrix} 1 & -6 & 0 & | & 1 \\ 0 & 1 & -7 & | & 4 \\ 3 & 0 & 2 & | & 11 \end{bmatrix} \begin{array}{l} \text{Interchange} \\ \text{Row 1 and Row 3} \end{array}$$

$$\begin{bmatrix} 1 & -6 & 0 & | & 1 \\ 0 & 1 & -7 & | & 4 \\ 0 & 18 & 2 & | & 8 \end{bmatrix} \begin{array}{l} \text{New Row 3 = } -3(\text{Row 1}) + \\ \text{Row 3} \end{array}$$

$$\begin{bmatrix} 1 & -6 & 0 & | & 1 \\ 0 & 1 & -7 & | & 4 \\ 0 & 0 & 128 & | & -64 \end{bmatrix}$$ New Row 3 =
$$-18(\text{Row } 2) + \text{Row } 3$$

Reinserting the variables, we have

$$p - 6q \qquad = 1, \qquad (1)$$
$$q - \quad 7r = 4, \qquad (2)$$
$$128r = -64. \quad (3)$$

Solve (3) for r.

$$128r = -64$$
$$r = -\frac{1}{2}$$

Substitute $-\frac{1}{2}$ for r in (2) and solve for q.

$$q - 7r = 4$$
$$q - 7\left(-\frac{1}{2}\right) = 4$$
$$q + \frac{7}{2} = 4$$
$$q = \frac{1}{2}$$

Substitute $\frac{1}{2}$ for q in (1) and solve for p.

$$p - 6 \cdot \frac{1}{2} = 1$$
$$p - 3 = 1$$
$$p = 4$$

The solution is $\left(4, \frac{1}{2}, -\frac{1}{2}\right)$.

12. $\left(\frac{1}{2}, \frac{2}{3}, -\frac{5}{6}\right)$

13. We will rewrite the equations with the variables in alphabetical order:

$$-2w + 2x + 2y - 2z = -10,$$
$$w + x + y + z = -5,$$
$$3w + x - y + 4z = -2,$$
$$w + 3x - 2y + 2z = -6$$

Write a matrix using only the constants.

$$\begin{bmatrix} -2 & 2 & 2 & -2 & | & -10 \\ 1 & 1 & 1 & 1 & | & -5 \\ 3 & 1 & -1 & 4 & | & -2 \\ 1 & 3 & -2 & 2 & | & -6 \end{bmatrix}$$

$$\begin{bmatrix} -1 & 1 & 1 & -1 & | & -5 \\ 1 & 1 & 1 & 1 & | & -5 \\ 3 & 1 & -1 & 4 & | & -2 \\ 1 & 3 & -2 & 2 & | & -6 \end{bmatrix}$$ New Row 1 =
$$\frac{1}{2}(\text{Row } 1)$$

$$\begin{bmatrix} -1 & 1 & 1 & -1 & | & -5 \\ 0 & 2 & 2 & 0 & | & -10 \\ 0 & 4 & 2 & 1 & | & -17 \\ 0 & 4 & -1 & 1 & | & -11 \end{bmatrix}$$ New Row 2 =
Row 1 + Row 2
New Row 3 =
3(Row 1) +Row 3
New Row 4 =
Row 1 + Row 4

$$\begin{bmatrix} -1 & 1 & 1 & -1 & | & -5 \\ 0 & 2 & 2 & 0 & | & -10 \\ 0 & 0 & -2 & 1 & | & 3 \\ 0 & 0 & -5 & 1 & | & 9 \end{bmatrix}$$ New Row 3 =
$$-2(\text{Row } 2) + \text{Row } 3$$
New Row 4 =
$$-2(\text{Row } 2) + \text{Row } 4$$

$$\begin{bmatrix} -1 & 1 & 1 & -1 & | & -5 \\ 0 & 2 & 2 & 0 & | & -10 \\ 0 & 0 & -2 & 1 & | & 3 \\ 0 & 0 & -10 & 2 & | & 18 \end{bmatrix}$$ New Row 4 =
2(Row 4)

$$\begin{bmatrix} -1 & 1 & 1 & -1 & | & -5 \\ 0 & 2 & 2 & 0 & | & -10 \\ 0 & 0 & -2 & 1 & | & 3 \\ 0 & 0 & 0 & -3 & | & 3 \end{bmatrix}$$ New Row 4 =
$$-5(\text{Row } 3) + \text{Row } 4$$

Reinserting the variables, we have

$$-w + x + y - z = -5, \qquad (1)$$
$$2x + 2y \qquad = -10, \qquad (2)$$
$$-2y + z = 3, \qquad (3)$$
$$-3z = 3. \qquad (4)$$

Solve (4) for z.

$$-3z = 3$$
$$z = -1$$

Substitute -1 for z in (3) and solve for y.

$$-2y + (-1) = 3$$
$$-2y = 4$$
$$y = -2$$

Substitute -2 for y in (2) and solve for x.

$$2x + 2(-2) = -10$$
$$2x - 4 = -10$$
$$2x = -6$$
$$x = -3$$

Substitute -3 for x, -2 for y, and -1 for z in (1) and solve for w.

$$-w + (-3) + (-2) - (-1) = -5$$
$$-w - 3 - 2 + 1 = -5$$
$$-w - 4 = -5$$
$$-w = -1$$
$$w = 1$$

The solution is $(1, -3, -2, -1)$.

14. $(7,4,5,6)$

15. *Familiarize.* Let $d = $ the number of dimes and $n = $ the number of nickels. The value of d dimes is $\$0.10d$, and the value of n nickels is $\$0.05n$.

Translate.

Total number of coins is 34.

$$d + n \quad = \quad 34$$

Total value of coins is $\$1.90$.

$$0.10d + 0.05n \quad = \quad 1.90$$

After clearing decimals, we have this system.

$$d + \quad n = \quad 34,$$
$$10d + 5n = 190$$

Carry out. Solve using matrices.

$$\begin{bmatrix} 1 & 1 & | & 34 \\ 10 & 5 & | & 190 \end{bmatrix}$$

$$\begin{bmatrix} 1 & 1 & | & 34 \\ 0 & -5 & | & -150 \end{bmatrix} \text{New Row 2} = -10(\text{Row 1}) + \text{Row 2}$$

Reinserting the variables, we have

$$d + n = \quad 34, \quad (1)$$
$$-5n = -150 \quad (2)$$

Solve (2) for n.

$$-5n = -150$$
$$n = 30$$

$$d + 30 = 34 \quad \text{Back-substituting}$$
$$d = 4$$

Check. The sum of the two numbers is 34. The total value is $\$0.10(4) + \$0.50(30) = \$0.40 + \$1.50 = \$1.90$. The numbers check.

State. There are 4 dimes and 30 nickels.

16. 21 dimes; 22 quarters

17. *Familiarize.* We let x represent the number of pounds of the $\$4.05$ kind and y represent the number of pounds of the $\$2.70$ kind of granola. We organize the information in a table.

Granola	Number of pounds	Price per pound	Value
$\$4.05$ kind	x	$\$4.05$	$\$4.05x$
$\$2.70$ kind	y	$\$2.70$	$\$2.70y$
Mixture	15	$\$3.15$	$\$3.15 \times 15$ or $\$47.25$

Translate.

Total number of pounds is 15.

$$x + y \quad = \quad 15$$

Total value of mixture is $\$47.25$.

$$4.05x + 2.70y \quad = \quad 47.25$$

After clearing decimals, we have this system:

$$x + \quad y = \quad 15,$$
$$405x + 270y = 4725$$

Carry out. Solve using matrices.

$$\begin{bmatrix} 1 & 1 & | & 15 \\ 405 & 270 & | & 4725 \end{bmatrix}$$

$$\begin{bmatrix} 1 & 1 & | & 15 \\ 0 & -135 & | & -1350 \end{bmatrix} \text{New Row 2} = -405(\text{Row 1}) + \text{Row 2}$$

Reinserting the variables, we have

$$x + y = 15, \quad (1)$$
$$-135y = -1350 \quad (2)$$

Solve (2) for y.

$$-135y = -1350$$
$$y = 10$$

Back-substitute 10 for y in (1) and solve for x.

$$x + 10 = 15$$
$$x = 5$$

Check. The sum of the numbers is 15. The total value is $\$4.05(5) + \$2.70(10)$, or $\$20.25 + \$27.00$, or $\$47.25$. The numbers check.

State. 5 pounds of the $\$4.05$ per lb granola and 10 pounds of the $\$2.70$ per lb granola should be used.

18. 14 pounds of nuts; 6 pounds of oats

19. *Familiarize*.   We let x, y, and z represent the amounts invested at 7%, 8%, and 9%, respectively. Recall the formula for simple interest:

Interest = Principal $\times$ Rate $\times$ Time

Translate. We organize the information in a table.

	First Investment	Second Investment	Third Investment	Total
P	x	y	z	$2500
R	7%	8%	9%	
T	1 yr	1 yr	1 yr	
I	$0.07x$	$0.08y$	$0.09z$	$212

The first row gives us one equation:

$$x + y + z = 2500$$

The last row gives a second equation:

$$0.07x + 0.08y + 0.09z = 212$$

Amount invested at 9%	is	$1100	more than	amount invested at 8%.
$\downarrow$	$\downarrow$	$\downarrow$	$\downarrow$	$\downarrow$
z	=	\$1100	+	y

After clearing decimals, we have this system:

$$x + y + z = 2500,$$
$$7x + 8y + 9z = 21,200,$$
$$-y + z = 1100$$

Carry out. Solve using matrices.

$$\begin{bmatrix} 1 & 1 & 1 & \vdots & 2500 \\ 7 & 8 & 9 & \vdots & 21,200 \\ 0 & -1 & 1 & \vdots & 1100 \end{bmatrix}$$

$$\begin{bmatrix} 1 & 1 & 1 & \vdots & 2500 \\ 0 & 1 & 2 & \vdots & 3700 \\ 0 & -1 & 1 & \vdots & 1100 \end{bmatrix}$$ New Row 2 = -7(Row 1) + Row 2

$$\begin{bmatrix} 1 & 1 & 1 & \vdots & 2500 \\ 0 & 1 & 2 & \vdots & 3700 \\ 0 & 0 & 3 & \vdots & 4800 \end{bmatrix}$$ New Row 3 = Row 2 + Row 3

Reinserting the variables, we have

$$x + y + z = 2500, \quad (1)$$
$$y + 2z = 3700, \quad (2)$$
$$3z = 4800 \quad (3)$$

Solve (3) for z.

$$3z = 4800$$
$$z = 1600$$

Back-substitute 1600 for z in (2) and solve for y.

$$y + 2 \cdot 1600 = 3700$$
$$y + 3200 = 3700$$
$$y = 500$$

Back-substitute 500 for y and 1600 for z in (1) and solve for x.

$$x + 500 + 1600 = 2500$$
$$x + 2100 = 2500$$
$$x = 400$$

Check.   The total investment is $400 + $500 + $1600, or $2500. The total interest is 0.07($400) + 0.08($500) + 0.09($1600) = $28 + $40 + $144 = $212. The amount invested at 9%, $1600, is $1100 more than the amount invested at 8%, $500. The numbers check.

State. $400 is invested at 7%, $500 is invested at 8%, and $1600 is invested at 9%.

20. $500 at 8%; $400 at 9%; $2300 at 10%

21. Writing exercise

22. Writing exercise

23. $5(-3) - (-7)4 = -15 - (-28) = -15 + 28 = 13$

24. -22

25. $-2(5 \cdot 3 - 4 \cdot 6) - 3(2 \cdot 7 - 15) + 4(3 \cdot 8 - 5 \cdot 4)$
$= -2(15 - 24) - 3(14 - 15) + 4(24 - 20)$
$= -2(-9) - 3(-1) + 4(4)$
$= 18 + 3 + 16$
$= 21 + 16$
$= 37$

26. 422

27. Writing exercise

28. Writing exercise

29. *Familiarize*. Let w, x, y, and z represent the thousand's, hundred's, ten's, and one's digits, respectively.

Translate.

The sum of the digits	is	10.
$\downarrow$	$\downarrow$	$\downarrow$
$w + x + y + z$	=	10

Twice the sum of the thousand's and ten's digits	is	the sum of the hundred's and one's digits	less one.
$\downarrow$	$\downarrow$	$\downarrow$	$\downarrow$
$2(w + y)$	=	$x + z$	$- 1$

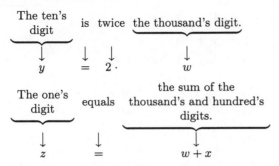

We have a system of equations which can be written as

$$w + x + y + z = 10,$$
$$2w - x + 2y - z = -1,$$
$$-2w \quad\ + y \quad\quad = 0,$$
$$w + x \quad\quad - z = 0.$$

Carry out. We can use matrices to solve the system. We get $(1, 3, 2, 4)$.

Check. The sum of the digits is 10. Twice the sum of 1 and 2 is 6. This is one less than the sum of 3 and 4. The ten's digit, 2, is twice the thousand's digit, 1. The one's digit, 4, equals $1 + 3$. The numbers check.

State. The number is 1324.

30. $x = \dfrac{ce - bf}{ae - bd}, \ y = \dfrac{af - cd}{ae - bd}$

Exercise Set 8.7

1. $\begin{vmatrix} 5 & 1 \\ 2 & 4 \end{vmatrix} = 5 \cdot 4 - 2 \cdot 1 = 20 - 2 = 18$

2. -13

3. $\begin{vmatrix} 6 & -9 \\ 2 & 3 \end{vmatrix} = 6 \cdot 3 - 2(-9) = 18 + 18 = 36$

4. 29

5. $\begin{vmatrix} 1 & 4 & 0 \\ 0 & -1 & 2 \\ 3 & -2 & 1 \end{vmatrix}$

$= 1 \begin{vmatrix} -1 & 2 \\ -2 & 1 \end{vmatrix} - 0 \begin{vmatrix} 4 & 0 \\ -2 & 1 \end{vmatrix} + 3 \begin{vmatrix} 4 & 0 \\ -1 & 2 \end{vmatrix}$

$= 1[-1 \cdot 1 - (-2) \cdot 2] - 0 + 3[4 \cdot 2 - (-1) \cdot 0]$

$= 1 \cdot 3 - 0 + 3 \cdot 8$

$= 3 - 0 + 24$

$= 27$

6. 1

7. $\begin{vmatrix} -1 & -2 & -3 \\ 3 & 4 & 2 \\ 0 & 1 & 2 \end{vmatrix}$

$= -1 \begin{vmatrix} 4 & 2 \\ 1 & 2 \end{vmatrix} - 3 \begin{vmatrix} -2 & -3 \\ 1 & 2 \end{vmatrix} + 0 \begin{vmatrix} -2 & -3 \\ 4 & 2 \end{vmatrix}$

$= -1[4 \cdot 2 - 1 \cdot 2] - 3[-2 \cdot 2 - 1(-3)] + 0$

$= -1 \cdot 6 - 3 \cdot (-1) + 0$

$= -6 + 3 + 0$

$= -3$

8. 3

9. $\begin{vmatrix} -4 & -2 & 3 \\ -3 & 1 & 2 \\ 3 & 4 & -2 \end{vmatrix}$

$= -4 \begin{vmatrix} 1 & 2 \\ 4 & -2 \end{vmatrix} - (-3) \begin{vmatrix} -2 & 3 \\ 4 & -2 \end{vmatrix} + 3 \begin{vmatrix} -2 & 3 \\ 1 & 2 \end{vmatrix}$

$= -4[1(-2) - 4 \cdot 2] + 3[-2(-2) - 4 \cdot 3] +$
$\qquad 3(-2 \cdot 2 - 1 \cdot 3)$

$= -4(-10) + 3(-8) + 3(-7)$

$= 40 - 24 - 21 = -5$

10. -6

11. $5x + 8y = 1,$
$\quad\ 3x + 7y = 5$

We compute D, D_x, and D_y.

$D = \begin{vmatrix} 5 & 8 \\ 3 & 7 \end{vmatrix} = 35 - 24 = 11$

$D_x = \begin{vmatrix} 1 & 8 \\ 5 & 7 \end{vmatrix} = 7 - 40 = -33$

$D_y = \begin{vmatrix} 5 & 1 \\ 3 & 5 \end{vmatrix} = 25 - 3 = 22$

Then,

$$x = \frac{D_x}{D} = \frac{-33}{11} = -3$$

and

$$y = \frac{D_y}{D} = \frac{22}{11} = 2.$$

The solution is $(-3, 2)$.

12. $(2, 0)$

13. $5x - 4y = -3,$
$\qquad 7x + 2y = 6$

We compute D, D_x, and D_y.

$$D = \begin{vmatrix} 5 & -4 \\ 7 & 2 \end{vmatrix} = 10 - (-28) = 38$$

$$D_x = \begin{vmatrix} -3 & -4 \\ 6 & 2 \end{vmatrix} = -6 - (-24) = 18$$

$$D_y = \begin{vmatrix} 5 & -3 \\ 7 & 6 \end{vmatrix} = 30 - (-21) = 51$$

Then,

$$x = \frac{D_x}{D} = \frac{18}{38} = \frac{9}{19}$$

and

$$y = \frac{D_y}{D} = \frac{51}{38}.$$

The solution is $\left(\dfrac{9}{19}, \dfrac{51}{38} \right)$.

14. $\left(-\dfrac{25}{2}, -\dfrac{11}{2} \right)$

15. $\quad 3x - y + 2z = 1,$
$\qquad x - y + 2z = 3,$
$\quad -2x + 3y + z = 1$

We compute D, D_x, and D_y.

$$D = \begin{vmatrix} 3 & -1 & 2 \\ 1 & -1 & 2 \\ -2 & 3 & 1 \end{vmatrix}$$

$$= 3 \begin{vmatrix} -1 & 2 \\ 3 & 1 \end{vmatrix} - 1 \begin{vmatrix} -1 & 2 \\ 3 & 1 \end{vmatrix} - 2 \begin{vmatrix} -1 & 2 \\ -1 & 2 \end{vmatrix}$$

$$= 3(-7) - 1(-7) - 2(0)$$
$$= -21 + 7 - 0$$
$$= -14$$

$$D_x = \begin{vmatrix} 1 & -1 & 2 \\ 3 & -1 & 2 \\ 1 & 3 & 1 \end{vmatrix}$$

$$= 1 \begin{vmatrix} -1 & 2 \\ 3 & 1 \end{vmatrix} - 3 \begin{vmatrix} -1 & 2 \\ 3 & 1 \end{vmatrix} + 1 \begin{vmatrix} -1 & 2 \\ -1 & 2 \end{vmatrix}$$

$$= 1(-7) - 3(-7) + 1(0)$$
$$= -7 + 21 + 0$$
$$= 14$$

$$D_y = \begin{vmatrix} 3 & 1 & 2 \\ 1 & 3 & 2 \\ -2 & 1 & 1 \end{vmatrix}$$

$$= 3 \begin{vmatrix} 3 & 2 \\ 1 & 1 \end{vmatrix} - 1 \begin{vmatrix} 1 & 2 \\ 1 & 1 \end{vmatrix} - 2 \begin{vmatrix} 1 & 2 \\ 3 & 2 \end{vmatrix}$$

$$= 3 \cdot 1 - 1(-1) - 2(-4)$$
$$= 3 + 1 + 8$$
$$= 12$$

Then,

$$x = \frac{D_x}{D} = \frac{14}{-14} = -1$$

and

$$y = \frac{D_y}{D} = \frac{12}{-14} = -\frac{6}{7}.$$

Substitute in the third equation to find z.

$$-2(-1) + 3\left(-\frac{6}{7} \right) + z = 1$$

$$2 - \frac{18}{7} + z = 1$$

$$-\frac{4}{7} + z = 1$$

$$z = \frac{11}{7}$$

The solution is $\left(-1, -\dfrac{6}{7}, \dfrac{11}{7} \right)$.

16. $\left(\dfrac{3}{2}, \dfrac{13}{14}, \dfrac{33}{14} \right)$

17. $\quad 2x - 3y + 5z = 27,$
$\qquad x + 2y - z = -4,$
$\quad 5x - y + 4z = 27$

We compute D, D_x, and D_y.

$$D = \begin{vmatrix} 2 & -3 & 5 \\ 1 & 2 & -1 \\ 5 & -1 & 4 \end{vmatrix}$$

$$= 2 \begin{vmatrix} 2 & -1 \\ -1 & 4 \end{vmatrix} - 1 \begin{vmatrix} -3 & 5 \\ -1 & 4 \end{vmatrix} + 5 \begin{vmatrix} -3 & 5 \\ 2 & -1 \end{vmatrix}$$

$$= 2(7) - 1(-7) + 5(-7)$$
$$= 14 + 7 - 35$$
$$= -14$$

$$D_x = \begin{vmatrix} 27 & -3 & 5 \\ -4 & 2 & -1 \\ 27 & -1 & 4 \end{vmatrix}$$

$$= 27 \begin{vmatrix} 2 & -1 \\ -1 & 4 \end{vmatrix} - (-4) \begin{vmatrix} -3 & 5 \\ -1 & 4 \end{vmatrix} + 27 \begin{vmatrix} -3 & 5 \\ 2 & -1 \end{vmatrix}$$

$$= 27(7) + 4(-7) + 27(-7)$$

$$= 189 - 28 - 189$$

$$= -28$$

$$D_y = \begin{vmatrix} 2 & 27 & 5 \\ 1 & -4 & -1 \\ 5 & 27 & 4 \end{vmatrix}$$

$$= 2 \begin{vmatrix} -4 & -1 \\ 27 & 4 \end{vmatrix} - 1 \begin{vmatrix} 27 & 5 \\ 27 & 4 \end{vmatrix} + 5 \begin{vmatrix} 27 & 5 \\ -4 & -1 \end{vmatrix}$$

$$= 2(11) - 1(-27) + 5(-7)$$

$$= 22 + 27 - 35$$

$$= 14$$

Then,

$$x = \frac{D_x}{D} = \frac{-28}{-14} = 2,$$

and

$$y = \frac{D_y}{D} = \frac{14}{-14} = -1.$$

We substitute in the second equation to find z.

$$2 + 2(-1) - z = -4$$
$$2 - 2 - z = -4$$
$$-z = -4$$
$$z = 4$$

The solution is $(2, -1, 4)$.

18. $(-3, 2, 1)$

19. $r - 2s + 3t = 6,$
$2r - s - t = -3,$
$r + s + t = 6$

We compute D, D_r, and D_s.

$$D = \begin{vmatrix} 1 & -2 & 3 \\ 2 & -1 & -1 \\ 1 & 1 & 1 \end{vmatrix}$$

$$= 1 \begin{vmatrix} -1 & -1 \\ 1 & 1 \end{vmatrix} - 2 \begin{vmatrix} -2 & 3 \\ 1 & 1 \end{vmatrix} + 1 \begin{vmatrix} -2 & 3 \\ -1 & -1 \end{vmatrix}$$

$$= 1(0) - 2(-5) + 1(5)$$

$$= 0 + 10 + 5$$

$$= 15$$

$$D_r = \begin{vmatrix} 6 & -2 & 3 \\ -3 & -1 & -1 \\ 6 & 1 & 1 \end{vmatrix}$$

$$= 6 \begin{vmatrix} -1 & -1 \\ 1 & 1 \end{vmatrix} - (-3) \begin{vmatrix} -2 & 3 \\ 1 & 1 \end{vmatrix} + 6 \begin{vmatrix} -2 & 3 \\ -1 & -1 \end{vmatrix}$$

$$= 6(0) + 3(-5) + 6(5)$$

$$= 0 - 15 + 30$$

$$= 15$$

$$D_s = \begin{vmatrix} 1 & 6 & 3 \\ 2 & -3 & -1 \\ 1 & 6 & 1 \end{vmatrix}$$

$$= 1 \begin{vmatrix} -3 & -1 \\ 6 & 1 \end{vmatrix} - 2 \begin{vmatrix} 6 & 3 \\ 6 & 1 \end{vmatrix} + 1 \begin{vmatrix} 6 & 3 \\ -3 & -1 \end{vmatrix}$$

$$= 1(3) - 2(-12) + 1(3)$$

$$= 3 + 24 + 3$$

$$= 30$$

Then,

$$r = \frac{D_r}{D} = \frac{15}{15} = 1,$$

and

$$s = \frac{D_s}{D} = \frac{30}{15} = 2.$$

Substitute in the third equation to find t.

$$1 + 2 + t = 6$$
$$3 + t = 6$$
$$t = 3$$

The solution is $(1, 2, 3)$.

20. $(3, 4, -1)$

21. Writing exercise

22. Writing exercise

23. $0.5x - 2.34 + 2.4x = 7.8x - 9$

$$2.9x - 2.34 = 7.8x - 9$$

$$6.66 = 4.9x$$

$$\frac{6.66}{4.9} = x$$

$$\frac{666}{490} = x$$

$$\frac{333}{245} = x$$

The solution is $\dfrac{333}{245}$.

24. -12

25. *Familiarize.* We first make a drawing.

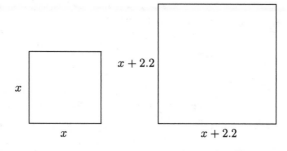

Let x represent the length of a side of the smaller square and $x + 2.2$ the length of a side of the larger square. The perimeter of the smaller square is $4x$. The perimeter of the larger square is $4(x + 2.2)$.

Translate.

$$\underbrace{\text{The sum of the perimeters}}_{4x + 4(x + 2.2)} \text{ is } \underset{=}{\downarrow} \ \underset{32.8}{\downarrow}$$

Carry out. We solve the equation.

$$4x + 4x + 8.8 = 32.8$$

$$8x = 24$$

$$x = 3$$

Check. If $x = 3$ ft, then $x + 2.2 = 5.2$ ft. The perimeters are $4 \cdot 3$, or 12 ft, and $4(5.2)$, or 20.8 ft. The sum of the two perimeters is $12 + 20.8$, or 32.8 ft. The values check.

State. The wire should be cut into two pieces, one measuring 12 ft and the other 20.8 ft.

26. 18 scientific calculators; 27 graphing calculators

27. *Familiarize.* Let x represent the number of rolls of insulation required for the Mazzas' attic and let y represent the number of rolls required for the Kranepools' attic.

Translate.

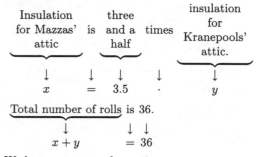

We have a system of equations:

$$x = 3.5y, \quad (1)$$

$$x + y = 36 \quad (2)$$

Carry out. We use the substitution method to solve the system of equations. First we substitute $3.5y$ for x in Equation (2).

$$x + y = 36 \quad (2)$$

$$3.5y + y = 36 \quad \text{Substituting}$$

$$4.5y = 36$$

$$y = 8$$

Now substitute 8 for y in Equation (1).

$$x = 3.5(8) = 28$$

Check. The number 28 is three and a half times 8. Also, the total number of rolls is $28 + 8$, or 36. The answer checks.

State. The Mazzas' attic requires 28 rolls of insulation, and the Kranepools' attic requires 8 rolls.

28. 17 buckets; 11 dinners

29. Writing exercise

30. Writing exercise

31. $\begin{vmatrix} y & -2 \\ 4 & 3 \end{vmatrix} = 44$

$$y \cdot 3 - 4(-2) = 44 \quad \text{Evaluating the determinant}$$

$$3y + 8 = 44$$

$$3y = 36$$

$$y = 12$$

32. 3

33. $\begin{vmatrix} m+1 & -2 \\ m-2 & 1 \end{vmatrix} = 27$

$\quad (m+1)(1) - (m-2)(-2) = 27 \quad$ Evaluating

$\qquad\qquad\qquad\qquad\qquad\qquad$ the determinant

$\qquad\qquad m + 1 + 2m - 4 = 27$

$\qquad\qquad\qquad\qquad\qquad 3m = 30$

$\qquad\qquad\qquad\qquad\qquad m = 10$

34. $\begin{vmatrix} x & y & 1 \\ x_1 & y_1 & 1 \\ x_2 & y_2 & 1 \end{vmatrix} = 0$

is equivalent to

$x \begin{vmatrix} y_1 & 1 \\ y_2 & 1 \end{vmatrix} - x_1 \begin{vmatrix} y & 1 \\ y_2 & 1 \end{vmatrix} + x_2 \begin{vmatrix} y & 1 \\ y_1 & 1 \end{vmatrix} = 0$

or

$x(y_1 - y_2) - x_1(y - y_2) + x_2(y - y_1) = 0$

or

$xy_1 - xy_2 - x_1 y + x_1 y_2 + x_2 y - x_2 y_1 = 0. \quad (1)$

Since the slope of the line through (x_1, y_1) and (x_2, y_2) is $\dfrac{y_2 - y_1}{x_2 - x_1}$, an equation of the line through (x_1, y_1) and (x_2, y_2) is

$y - y_1 = \dfrac{y_2 - y_1}{x_2 - x_1}(x - x_1)$

which is equivalent to

$(x_2 - x_1)(y - y_1) = (y_2 - y_1)(x - x_1)$

or

$x_2 y - x_2 y_1 - x_1 y + x_1 y_1 = y_2 x - y_2 x_1 - y_1 x + y_1 x_1$

or

$x_2 y - x_2 y_1 - x_1 y - x y_2 + x_1 y_2 + x y_1 = 0. \quad (2)$

Equations (1) and (2) are equivalent.

Exercise Set 8.8

1. $C(x) = 45x + 300,000 \qquad R(x) = 65x$

a) $P(x) = R(x) - C(x)$

$\qquad\quad = 65x - (45x + 300,000)$

$\qquad\quad = 65x - 45x - 300,000$

$\qquad\quad = 20x - 300,000$

b) To find the break-even point we solve the system

$\qquad R(x) = 65x,$

$\qquad C(x) = 45x + 300,000.$

Since $R(x) = C(x)$ at the break-even point, we can rewrite the system:

$\qquad R(x) = 65x, \qquad\qquad (1)$

$\qquad R(x) = 45x + 300,000 \quad (2)$

We solve using substitution.

$\qquad 65x = 45x + 300,000 \quad$ Substituting $65x$ for

$\qquad\qquad\qquad\qquad\qquad\qquad R(x)$ in (2)

$\qquad 20x = 300,000$

$\qquad\quad x = 15,000$

Thus, 15,000 units must be produced and sold in order to break even.

2. a) $P(x) = 45x - 270,000;$ b) 6000 units

3. $C(x) = 10x + 120,000 \qquad R(x) = 60x$

a) $P(x) = R(x) - C(x)$

$\qquad\quad = 60x - (10x + 120,000)$

$\qquad\quad = 60x - 10x - 120,000$

$\qquad\quad = 50x - 120,000$

b) Solve the system

$\qquad R(x) = 60x,$

$\qquad C(x) = 10x + 120,000.$

Since both $R(x)$ and $C(x)$ are in dollars and they are equal at the break-even point, we can rewrite the system:

$\qquad d = 60x, \qquad\qquad (1)$

$\qquad d = 10x + 120,000 \quad (2)$

We solve using substitution.

$\qquad 60x = 10x + 120,000 \quad$ Substituting $60x$

$\qquad\qquad\qquad\qquad\qquad\qquad$ for d in (2)

$\qquad 50x = 120,000$

$\qquad\quad x = 2400$

Thus, 2400 units must be produced and sold in order to break even.

4. a) $P(x) = 55x - 49,500;$ b) 900 units

5. $C(x) = 40x + 22,500 \qquad R(x) = 85x$

a) $P(x) = R(x) - C(x)$

$\qquad\quad = 85x - (40x + 22,500)$

$\qquad\quad = 85x - 40x - 22,500$

$\qquad\quad = 45x - 22,500$

b) Solve the system

$\qquad R(x) = 85x,$

$\qquad C(x) = 40x + 22,500.$

Since both $R(x)$ and $C(x)$ are in dollars and they are equal at the break-even point, we can rewrite the system:

$$d = 85x, \qquad (1)$$
$$d = 40x + 22,500 \quad (2)$$

We solve using substitution.

$$85x = 40x + 22,500 \quad \text{Substituting } 85x \text{ for}$$
$$\qquad\qquad\qquad\qquad\qquad d \text{ in (2)}$$
$$45x = 22,500$$
$$x = 500$$

Thus, 500 units must be produced and sold in order to break even.

6. a) $P(x) = 80x - 10,000$; b) 125 units

7. $C(x) = 22x + 16,000 \qquad R(x) = 40x$

a) $P(x) = R(x) - C(x)$
$$= 40x - (22x + 16,000)$$
$$= 40x - 22x - 16,000$$
$$= 18x - 16,000$$

b) Solve the system

$$R(x) = 40x,$$
$$C(x) = 22x + 16,000.$$

Since both $R(x)$ and $C(x)$ are in dollars and they are equal at the break-even point, we can rewrite the system:

$$d = 40x, \qquad (1)$$
$$d = 22x + 16,000 \quad (2)$$

We solve using substitution.

$$40x = 22x + 16,000 \quad \text{Substituting } 40x \text{ for}$$
$$\qquad\qquad\qquad\qquad\qquad d \text{ in (2)}$$
$$18x = 16,000$$
$$x \approx 889 \text{ units}$$

Thus, 889 units must be produced and sold in order to break even.

8. a) $P(x) = 40x - 75,000$; b) 1875 units

9. $C(x) = 75x + 100,000 \qquad R(x) = 125x$

a) $P(x) = R(x) - C(x)$
$$= 125x - (75x + 100,000)$$
$$= 125x - 75x - 100,000$$
$$= 50x - 100,000$$

b) Solve the system

$$R(x) = 125x,$$
$$C(x) = 75x + 100,000.$$

Since $R(x) = C(x)$ at the break-even point, we can rewrite the system:

$$R(x) = 125x, \qquad (1)$$
$$R(x) = 75x + 100,000 \quad (2)$$

We solve using substitution.

$$125x = 75x + 100,000 \quad \text{Substituting } 125x$$
$$\qquad\qquad\qquad\qquad\qquad \text{for } R(x) \text{ in (2)}$$
$$50x = 100,000$$
$$x = 2000$$

To break even 2000 units must be produced and sold.

10. a) $P(x) = 30x - 120,000$; b) 4000 units

11. $D(p) = 1000 - 10p,$
$\quad S(p) = 230 + p$

Since both demand and supply are quantities, the system can be rewritten:

$$q = 1000 - 10p, \quad (1)$$
$$q = 230 + p \qquad (2)$$

Substitute $1000 - 10p$ for q in (2) and solve.

$$1000 - 10p = 230 + p$$
$$770 = 11p$$
$$70 = p$$

The equilibrium price is \$70 per unit. To find the equilibrium quantity we substitute \$70 into either $D(p)$ or $S(p)$.

$$D(70) = 1000 - 10 \cdot 70 = 1000 - 700 = 300$$

The equilibrium quantity is 300 units.

The equilibrium point is (\$70, 300).

12. (\$10, 1400)

13. $D(p) = 760 - 13p,$
$\quad S(p) = 430 + 2p$

Rewrite the system:

$$q = 760 - 13p, \quad (1)$$
$$q = 430 + 2p \qquad (2)$$

Substitute $760 - 13p$ for q in (2) and solve.

$$760 - 13p = 430 + 2p$$
$$330 = 15p$$
$$22 = p$$

The equilibrium price is \$22 per unit.

To find the equilibrium quantity we substitute \$22 into either $D(p)$ or $S(p)$.

$$S(22) = 430 + 2(22) = 430 + 44 = 474$$

The equilibrium quantity is 474 units.

The equilibrium point is (\$22, 474).

14. (\$10, 370)

15. $D(p) = 7500 - 25p$,

$S(p) = 6000 + 5p$

Rewrite the system:

$q = 7500 - 25p$,   (1)

$q = 6000 + 5p$ (2)

Substitute $7500 - 25p$ for q in (2) and solve.

$7500 - 25p = 6000 + 5p$

$1500 = 30p$

$50 = p$

The equilibrium price is $50 per unit.

To find the equilibrium quantity we substitute $50 into either $D(p)$ or $S(p)$.

$D(50) = 7500 - 25(50) = 7500 - 1250 = 6250$

The equilibrium quantity is 6250 units.

The equilibrium point is ($50, 6250$).

16. ($40, 7600$)

17. $D(p) = 1600 - 53p$,

$S(p) = 320 + 75p$

Rewrite the system:

$q = 1600 - 53p$,   (1)

$q = 320 + 75p$ (2)

Substitute $1600 - 53p$ for q in (2) and solve.

$1600 - 53p = 320 + 75p$

$1280 = 128p$

$10 = p$

The equilibrium price is $10 per unit.

To find the equilibrium quantity we substitute $10 into either $D(p)$ or $S(p)$.

$S(10) = 320 + 75(10) = 320 + 750 = 1070$

The equilibrium quantity is 1070 units.

The equilibrium point is ($10, 1070$).

18. ($36, 4060$)

19. a) $C(x) =$ Fixed costs + Variable costs

$C(x) = 125,100 + 450x$,

where x is the number of computers produced.

b) Each computer sells for $800. The total revenue is 800 times the number of computers sold. We assume that all computers produced are sold.

$R(x) = 800x$

c) $P(x) = R(x) - C(x)$

$P(x) = 800x - (125,100 + 450x)$

$= 800x - 125,100 - 450x$

$= 350x - 125,100$

d) $P(x) = 350x - 125,100$

$P(100) = 350(100) - 125,100$

$= 35,000 - 125,100$

$= -90,000$

The company will realize a $90,000 loss when 100 computers are produced and sold.

$P(400) = 350(400) - 125,100$

$= 140,000 - 125,100$

$= 14,900$

The company will realize a profit of $14,900 from the production and sale of 400 computers.

e) Solve the system

$R(x) = 800x$,

$C(x) = 125,100 + 450x$.

Since both $R(x)$ and $C(x)$ are in dollars and they are equal at the break-even point, we can rewrite the system:

$d = 800x$,   (1)

$d = 125,100 + 450x$ (2)

We solve using substitution.

$800x = 125,100 + 450x$ Substituting $800x$ for d in (2)

$350x = 125,100$

$x \approx 358$ Rounding up

The firm will break even if it produces and sells 358 computers and takes in a total of $R(358) = 800 \cdot 358 = \$286,400$ in revenue. Thus, the break-even point is (358 computers, $286,400).

20. a) $C(x) = 22,500 + 40x$, where x is the number of lamps produced;

b) $R(x) = 85x$;

c) $P(x) = 45x - 22,500$;

d) $112,500 profit, $4500 loss;

e) (500 lamps, $42,500)

21. a) $C(x) =$ Fixed costs + Variable costs

$C(x) = 16,404 + 6x$,

where x is the number of caps produced, in dozens.

b) Each dozen caps sell for $18. The total revenue is 18 times the number of caps sold, in dozens. We assume that all caps produced are sold.

$R(x) = 18x$

c) $P(x) = R(x) - C(x)$

$P(x) = 18x - (16,404 + 6x)$

$= 18x - 16,404 - 6x$

$= 12x - 16,404$

d) $P(3000) = 12(3000) - 16,404$

$\qquad = 36,000 - 16,404$

$\qquad = 19,596$

The company will realize a profit of $19,596 when 3000 dozen caps are produced and sold.

$P(1000) = 12(1000) - 16,404$

$\qquad = 12,000 - 16,404$

$\qquad = -4404$

The company will realize a $4404 loss when 1000 dozen caps are produced and sold.

e) Solve the system

$\qquad R(x) = 18x,$

$\qquad C(x) = 16,404 + 6x.$

Since both $R(x)$ and $C(x)$ are in dollars and they are equal at the break-even point, we can rewrite the system:

$\qquad d = 18x, \qquad\qquad (1)$

$\qquad d = 16,404 + 6x \quad (2)$

We solve using substitution.

$18x = 16,404 + 6x$ Substituting $18x$ for d
$\qquad\qquad\qquad\qquad$ in (2)

$12x = 16,404$

$\quad x = 1367$

The firm will break even if it produces and sells 1367 dozen caps and takes in a total of $R(1367) = 18 \cdot 1367 = \$24,606$ in revenue. Thus, the break-even point is (1367 dozen caps, $24,606).

22. a) $C(x) = 10,000 + 30x$, where x is the number of sport coats produced;

b) $R(x) = 80x$;

c) $P(x) = 50x - 10,000$;

d) $90,000 profit; $7500 loss;

e) (200 sport coats, $16,000)

23. Writing exercise

24. Writing exercise

25. $3x - 9 = 27$

$\quad 3x = 36$ Adding 9 to both sides

$\quad\ x = 12$ Dividing both sides by 3

The solution is 12.

26. 15

27. $4x - 5 = 7x - 13$

$\quad -5 = 3x - 13$ Subtracting $4x$ from both sides

$\quad\ 8 = 3x$ Adding 13 to both sides

$\quad \dfrac{8}{3} = x$ Dividing both sides by 3

The solution is $\dfrac{8}{3}$.

28. 4

29. $7 - 2(x - 8) = 14$

$\quad 7 - 2x + 16 = 14$ Removing parentheses

$\quad -2x + 23 = 14$ Collecting like terms

$\quad -2x = -9$ Subtracting 23 from both sides

$\quad x = \dfrac{9}{2}$ Dividing both sides by -2

The solution is $\dfrac{9}{2}$.

30. $\dfrac{1}{3}$

31. Writing exercise

32. Writing exercise

33. The supply function contains the points ($2, 100) and ($8, 500). We find its equation:

$$m = \frac{500 - 100}{8 - 2} = \frac{400}{6} = \frac{200}{3}$$

$y - y_1 = m(x - x_1)$ Point-slope form

$y - 100 = \dfrac{200}{3}(x - 2)$

$y - 100 = \dfrac{200}{3}x - \dfrac{400}{3}$

$\qquad y = \dfrac{200}{3}x - \dfrac{100}{3}$

We can equivalently express supply S as a function of price p:

$$S(p) = \frac{200}{3}p - \frac{100}{3}$$

The demand function contains the points ($1, 500) and ($9, 100). We find its equation:

$$m = \frac{100 - 500}{9 - 1} = \frac{-400}{8} = -50$$

$y - y_1 = m(x - x_1)$

$y - 500 = -50(x - 1)$

$y - 500 = -50x + 50$

$\qquad y = -50x + 550$

We can equivalently express demand D as a function of price p:

$$D(p) = -50p + 550$$

We have a system of equations

$$S(p) = \frac{200}{3}p - \frac{100}{3},$$
$$D(p) = -50p + 550.$$

Rewrite the system:

$$q = \frac{200}{3}p - \frac{100}{3}, \quad (1)$$
$$q = -50p + 550 \quad (2)$$

Substitute $\frac{200}{3}p - \frac{100}{3}$ for q in (2) and solve.

$$\frac{200}{3}p - \frac{100}{3} = -50p + 550$$

$200p - 100 = -150p + 1650$ Multiplying by 3
to clear fractions

$$350p - 100 = 1650$$
$$350p = 1750$$
$$p = 5$$

The equilibrium price is $5 per unit.

To find the equilibrium quantity, we substitute $5 into either $S(p)$ or $D(p)$.

$$D(5) = -50(5) + 550 = -250 + 550 = 300$$

The equilibrium quantity is 300 units.

The equilibrium point is ($5, 300$).

34. 308 pairs

35. a) Use a grapher to find the first coordinate of the point of intersection of $y_1 = -14.97x + 987.35$ and $y_2 = 98.55x - 5.13$, to the nearest hundredth. It is 8.74, so the price per unit that should be charged is $8.74.

    b) Use a grapher to find the first coordinate of the point of intersection of $y_1 = 87,985 + 5.15x$ and $y_2 = 8.74x$. It is about 24,508.4, so 24,509 units must be sold in order to break even.

36. a) (4526 units,$4,390,220); b) $870

Chapter 9

Inequalities and Problem Solving

Exercise Set 9.1

---

1. $y < 6$

Graph: The solutions consist of all real numbers less than 6, so we shade all numbers to the left of 5 and use an open circle at 6 to indicate that it is not a solution.

Set builder notation: $\{y|y < 6\}$

Interval notation: $(-\infty, 6)$

2.

$\{x|x > 4\}$; $(4, \infty)$

3. $x \geq -4$

Graph: We shade all numbers to the right of -4 and use a solid endpoint at -4 to indicate that it is also a solution.

Set builder notation: $\{x|x \geq -4\}$

Interval notation: $[-4, \infty)$

4.

$\{t|t \leq 6\}$; $(-\infty, 6]$

5. $t > -3$

Graph: We shade all numbers to the right of -3 and use an open circle at -3 to indicate that it is not a solution.

Set builder notation: $\{t|t > -3\}$

Interval notation: $(-3, \infty)$

6.

$\{y|y < -3\}$; $(-\infty, -3)$

7. $x \leq -7$

Graph: We shade all numbers to the left of -7 and use a solid endpoint at -7 to indicate that it is also a solution.

Set builder notation: $\{x|x \leq -7\}$

Interval notation: $(-\infty, -7]$

8.

$\{x|x \geq -6\}$; $[-6, \infty)$

9.
$$y - 9 > -18$$
$$y - 9 + 9 > -18 + 9 \quad \text{Adding 9}$$
$$y > -9$$

The solution set is $\{y|y > -9\}$, or $(-9, \infty)$.

10. $\{y|y > -6\}$, or $(-6, \infty)$

11.
$$y - 20 \leq -6$$
$$y - 20 + 20 \leq -6 + 20 \quad \text{Adding 20}$$
$$y \leq 14$$

The solution set is $\{y|y \leq 14\}$, or $(-\infty, 14]$.

12. $\{x|x \leq 9\}$, or $(-\infty, 9]$

13.
$$9t < -81$$
$$\frac{1}{9} \cdot 9t < \frac{1}{9}(-81) \quad \text{Multiplying by } \frac{1}{9}$$
$$t < -9$$

The solution set is $\{t|t < -9\}$, or $(-\infty, -9)$.

14. $\{x|x \geq 3\}$, or $[3, \infty)$

15.
$$-8y \leq 3.2$$
$$-\frac{1}{8}(-8y) \geq -\frac{1}{8}(3.2) \quad \text{Multiplying by } -\frac{1}{8} \text{ and reversing the inequality symbol}$$
$$y \geq -0.4$$

The solution set is $\{y|y \geq -0.4\}$, or $[-0.4, \infty)$.

16. $\{x | x \le 0.9\}$, or $(-\infty, 0.9]$

17.
$$-\frac{5}{6}y \le -\frac{3}{4}$$

$$-\frac{6}{5}\left(-\frac{5}{6}y\right) \ge -\frac{6}{5}\left(-\frac{3}{4}\right) \quad \text{Multiplying by } -\frac{6}{5}$$
$$\text{and reversing the}$$
$$\text{inequality symbol}$$

$$y \ge \frac{9}{10}$$

The solution set is $\left\{y \middle| y \ge \frac{9}{10}\right\}$, or $\left[\frac{9}{10}, \infty\right)$.

18. $\left\{x \middle| x \le \frac{5}{6}\right\}$, or $\left(-\infty, \frac{5}{6}\right]$

19.
$$5y + 13 > 28$$
$$5y + 13 + (-13) > 28 + (-13) \quad \text{Adding } -13$$
$$5y > 15$$
$$\frac{1}{5} \cdot 5y > \frac{1}{5} \cdot 15 \quad \text{Multiplying by } \frac{1}{5}$$
$$y > 3$$

The solution set is $\{y | y > 3\}$, or $(3, \infty)$.

20. $\{x | x < 6\}$, or $(-\infty, 6)$

21.
$$-9x + 3x \ge -24$$
$$-6x \ge -24 \quad \text{Combining like terms}$$
$$-\frac{1}{6}(-6x) \le -\frac{1}{6}(-24) \quad \text{Multiplying by } -\frac{1}{6}$$
$$\text{and reversing the}$$
$$\text{inequality symbol}$$
$$x \le 4$$

The solution set is $\{x | x \le 4\}$, or $(-\infty, 4]$.

22. $\{y | y \le -3\}$, or $(-\infty, -3]$

23. $f(x) = 8x - 9$, $g(x) = 3x - 11$
$$f(x) < g(x)$$
$$8x - 9 < 3x - 11$$
$$5x - 9 < -11 \quad \text{Adding } -3x$$
$$5x < -2 \quad \text{Adding } 9$$
$$x < -\frac{2}{5} \quad \text{Multiplying by } \frac{1}{5}$$

The solution set is $\left\{x \middle| x < -\frac{2}{5}\right\}$, or $\left(-\infty, -\frac{2}{5}\right)$.

24. $\left\{x \middle| x > \frac{2}{3}\right\}$, or $\left(\frac{2}{3}, \infty\right)$

25. $f(x) = 0.4x + 5$, $g(x) = 1.2x - 4$
$$g(x) \ge f(x)$$
$$1.2x - 4 \ge 0.4x + 5$$
$$0.8x - 4 \ge 5 \quad \text{Adding } -0.4x$$
$$0.8x \ge 9 \quad \text{Adding } 4$$
$$x \ge 11.25 \quad \text{Multiplying by } \frac{1}{0.8}$$

The solution set is $\{x | x \ge 11.25\}$, or $[11.25, \infty)$.

26. $\left\{x \middle| x \ge \frac{1}{2}\right\}$, or $\left[\frac{1}{2}, \infty\right)$

27.
$$4(3y - 2) \ge 9(2y + 5)$$
$$12y - 8 \ge 18y + 45$$
$$-6y - 8 \ge 45$$
$$-6y \ge 53$$
$$y \le -\frac{53}{6}$$

The solution set is $\left\{y \middle| y \le -\frac{53}{6}\right\}$, or $\left(-\infty, -\frac{53}{6}\right]$.

28. $\left\{m \middle| m \le \frac{49}{10}\right\}$, or $\left(-\infty, \frac{49}{10}\right]$

29. $5(t - 3) + 4t < 2(7 + 2t)$

$5t - 15 + 4t < 14 + 4t$

$9t - 15 < 14 + 4t$

$5t - 15 < 14$

$5t < 29$

$t < \dfrac{29}{5}$

The solution set is $\left\{ t \middle| t < \dfrac{29}{5} \right\}$, or $\left(-\infty, \dfrac{29}{5} \right)$.

30. $\left\{ x \middle| x > -\dfrac{2}{17} \right\}$, or $\left(-\dfrac{2}{17}, \infty \right)$

31. $5[3m - (m + 4)] > -2(m - 4)$

$5(3m - m - 4) > -2(m - 4)$

$5(2m - 4) > -2(m - 4)$

$10m - 20 > -2m + 8$

$12m - 20 > 8$

$12m > 28$

$m > \dfrac{28}{12}$

$m > \dfrac{7}{3}$

The solution set is $\left\{ m \middle| m > \dfrac{7}{3} \right\}$, or $\left(\dfrac{7}{3}, \infty \right)$.

32. $\left\{ x \middle| x \le -\dfrac{23}{2} \right\}$, or $\left(-\infty, -\dfrac{23}{2} \right]$

33. $19 - (2x + 3) \le 2(x + 3) + x$

$19 - 2x - 3 \le 2x + 6 + x$

$16 - 2x \le 3x + 6$

$16 - 5x \le 6$

$-5x \le -10$

$x \ge 2$

The solution set is $\{ x | x \ge 2 \}$, or $[2, \infty)$.

34. $\{ c | c \le 1 \}$, or $(-\infty, 1]$

35. $\dfrac{1}{4}(8y + 4) - 17 < -\dfrac{1}{2}(4y - 8)$

$2y + 1 - 17 < -2y + 4$

$2y - 16 < -2y + 4$

$4y - 16 < 4$

$4y < 20$

$y < 5$

The solution set is $\{ y | y < 5 \}$, or $(-\infty, 5)$.

36. $\{ x | x > 6 \}$, or $(6, \infty)$

37. $2[8 - 4(3 - x)] - 2 \ge 8[2(4x - 3) + 7] - 50$

$2[8 - 12 + 4x] - 2 \ge 8[8x - 6 + 7] - 50$

$2[-4 + 4x] - 2 \ge 8[8x + 1] - 50$

$-8 + 8x - 2 \ge 64x + 8 - 50$

$8x - 10 \ge 64x - 42$

$-56x - 10 \ge -42$

$-56x \ge -32$

$x \le \dfrac{32}{56}$

$x \le \dfrac{4}{7}$

The solution set is $\left\{ x \middle| x \le \dfrac{4}{7} \right\}$, or $\left(-\infty, \dfrac{4}{7} \right]$.

38. $\left\{ t \middle| t \ge -\dfrac{27}{19} \right\}$, or $\left[-\dfrac{27}{19}, \infty \right)$

39. *Familiarize.* Let m = the number of peak local minutes used. Then the charge for the minutes used is \$0.022m$ and the total monthly charge is \$13.55 + \$0.022m$.

Translate. We write an inequality stating that the monthly charge is at least \$39.40.

$13.55 + 0.022m \ge 39.40$

Carry out.

$13.55 + 0.022m \ge 39.40$

$0.022m \ge 25.85$

$m \ge 1175$

Check. We can do a partial check by substituting a value for m less than 1175. When $m = 1174$, the monthly charge is \$13.55 + \$0.022(1174) \approx \$39.38$. This is less than the maximum charge of \$39.40. We cannot check all possible values for m, so we stop here.

State. A customer must speak on the phone for 1175 local peak minutes or more if the maximum charge is to apply.

40. For 5170 local off-peak minutes or more

41. *Familiarize.* Let c = the number of checks per month. Then the Anywhere plan will cost \$0.20c$ per month and the Acu-checking plan will cost \$2 + \$0.12c$ per month.

Translate. We write an inequality stating that the Acu-checking plan costs less than the Anywhere plan.

$2 + 0.12c < 0.20c$

Carry out.

$2 + 0.12c < 0.20c$

$2 < 0.08c$

$25 < c$

Check. We can do a partial check by substituting a value for c less than 25 and a value for c greater than 25. When $c = 24$, the Acu-checking plan costs $\$2+\$0.12(24)$, or $\$4.88$, and the Anywhere plan costs $\$0.20(24)$, or $\$4.80$, so the Anywhere plan is less expensive. When $c = 26$, the Acu-checking plan costs $\$2+\$0.12(26)$, or $\$5.12$, and the Anywhere plan costs $\$0.20(26)$, or $\$5.20$, so Acu-checking is less expensive. We cannot check all possible values for c, so we stop here.

State. The Acu-checking plan costs less for more than 25 checks per month.

42. Times greater than 4.25 hr

43. *Familiarize.* We list the given information in a table.

Plan A: Monthly Income	Plan B: Monthly Income
$400 salary	$610
8% of sales	5% of sales
Total: 400 + 8% of sales	Total: 610 + 5% of sales

Suppose Toni had gross sales of $5000 one month. Then under plan A she would earn

$\$400 + 0.08(\$5000)$, or $\$800$.

Under plan B she would earn

$\$610 + 0.05(\$5000)$, or $\$860$.

This shows that, for gross sales of $5000, plan B is better.

If Toni had gross sales of $10,000 one month, then under plan A she would earn

$\$400 + 0.08(\$10,000)$, or $\$1200$.

Under plan B she would earn

$\$610 + 0.05(\$10,000)$, or $\$1110$.

This shows that, for gross sales of $10,000, plan A is better. To determine all values for which plan A is better we solve an inequality.

Translate.

Income from plan A	is greater than	income from plan B.
$400 + 0.08s$	$>$	$610 + 0.05s$

Carry out.

$$400 + 0.08s > 610 + 0.05s$$
$$400 + 0.03s > 610$$
$$0.03s > 210$$
$$s > 7000$$

Check. For $s = \$7000$, the income from plan A is

$\$400 + 0.08(\$7000)$, or $\$960$

and the income from plan B is

$\$610 + 0.05(\$7000)$, or $\$960$.

This shows that for sales of $7000 Toni's income is the same from each plan. In the Familiarize step we shows that, for a value less than $7000, plan B is better and, for a value greater than $7000, plan A is better. Since we cannot check all possible values, we stop here.

State. Toni should select plan A for gross sales greater than $7000.

44. Values of n greater than $85\frac{5}{7}$

45. *Familiarize.* Let $m =$ the amount of the medical bills. Then under plan A Giselle would pay $\$50 + 0.2(m - \$50)$. Under plan B she would pay $\$250 + 0.1(m - \$250)$.

Translate. We write an inequality stating than the cost of plan B is less than the cost of plan A.

$$250 + 0.1(m - 250) < 50 + 0.2(m - 50)$$

Carry out.

$$250 + 0.1(m - 250) < 50 + 0.2(m - 50)$$
$$250 + 0.1m - 25 < 50 + 0.2m - 10$$
$$225 + 0.1m < 40 + 0.2m$$
$$185 + 0.1m < 0.2m$$
$$185 < 0.1m$$
$$1850 < m$$

Check. We can do a partial check by substituting a value for m less than $1850 and a value for m greater than $1850. When $m = \$1840$, plan A costs $\$50 + 0.2(\$1840 - \$50)$, or $\$408$, and plan B costs $\$250+0.1(\$1840-\$250)$, or $\$409$. When $m = \$1860$, plan A costs $\$50+0.2(\$1860-\$50)$, or $\$412$, and plan B costs $\$250 + 0.1(\$1860 - \$250)$, or $\$411$, so plan B will save Giselle money. We cannot check all possible values for m, so we stop here.

State. Plan B will save Giselle money for medical bills greater than $1850.

46. Parties of more than 80

47. *Familiarize.* Let $n =$ the number of people who attend. Then the total receipts are $\$6 \cdot n$, and the amount of receipts over $750 is $\$6 \cdot n - \$750$. The band will receive $750 plus 15% of $\$6 \cdot n - \$750$, or $\$750 + 0.15(\$6 \cdot n - \$750)$.

Translate. We write an inequality stating that the amount the band receives is at least $1200.

$$750 + 0.15(6n - 750) \geq 1200$$

Carry out.

$$750 + 0.15(6n - 750) \geq 1200$$
$$750 + 0.9n - 112.5 \geq 1200$$
$$0.9n + 637.5 \geq 1200$$
$$0.9n \geq 562.5$$
$$n \geq 625$$

Check. When $n = 625$, the band receives $\$75+$ $0.15(\$6 \cdot 625 - \$750)$, or $\$750 + 0.15(\$3000)$, or $\$750 + \$450$, or $\$1200$. When $n = 626$, the band receives $\$750 + 0.15(\$6 \cdot 626 - \$750)$, or $\$750 + 0.15(\$3006)$, or $\$750 + \$450.90$, or $\$1200.90$. Since the band receives exactly $\$1200$ when 625 people attend and more than $\$1200$ when 626 people attend, we have performed a partial check. We cannot check all possible solutions, so we stop here.

State. At least 625 people must attend in order for the band to receive at least $\$1200$.

48. a) Fahrenheit temperatures less than $1945.4°$

b) Fahrenheit temperatures less than $1761.44°$

49. a) ***Familiarize***. Find the values of x for which $R(x) < C(x)$.

Translate.

$$26x < 90,000 + 15x$$

Carry out.

$$11x < 90,000$$
$$x < 8181\frac{9}{11}$$

Check. $R\left(8181\frac{9}{11}\right) = \$212,727.27 = C\left(8181\frac{9}{11}\right)$.

Calculate $R(x)$ and $C(x)$ for some x greater than $8181\frac{9}{11}$ and for some x less than $8181\frac{9}{11}$.

Suppose $x = 8200$:

$$R(x) = 26(8200) = 213,200 \quad \text{and}$$
$$C(x) = 90,000 + 15(8200) = 213,000.$$

In this case $R(x) > C(x)$.

Suppose $x = 8000$:

$$R(x) = 26(8000) = 208,000 \quad \text{and}$$
$$C(x) = 90,000 + 15(8000) = 210,000.$$

In this case $R(x) < C(x)$.

Then for $x < 8181\frac{9}{11}$, $R(x) < C(x)$.

State. We will state the result in terms of integers, since the company cannot sell a fraction of a radio. For 8181 or fewer radios the company loses money.

b) Our check in part a) shows that for $x > 8181\frac{9}{11}$, $R(x) > C(x)$ and the company makes a profit. Again, we will state the result in terms of an integer. For more than 8181 radios the company makes money.

50. a) $\{p | p < 10\}$; b) $\{p | p > 10\}$

51. Writing exercise

52. Writing exercise

53. $f(x) = \dfrac{3}{x - 2}$

Since $\dfrac{3}{x - 2}$ cannot be computed when $x - 2$ is 0, we solve an equation:

$$x - 2 = 0$$
$$x = 2$$

The domain is $\{x | x \text{ is a real number } and\ x \neq 2\}$.

54. $\{x | x \text{ is a real number } and\ x \neq -3\}$

55. $f(x) = \dfrac{5x}{7 - 2x}$

Since $\dfrac{5x}{7 - 2x}$ cannot be computed when $7 - 2x$ is 0, we solve an equation:

$$7 - 2x = 0$$
$$7 = 2x$$
$$\frac{7}{2} = x$$

The domain is $\left\{x \middle| x \text{ is a real number } and\ x \neq \frac{7}{2}\right\}$.

56. $\left\{x \middle| x \text{ is a real number } and\ x \neq \frac{9}{4}\right\}$

57. $9x - 2(x - 5) = 9x - 2x + 10 = 7x + 10$

58. $22x - 7$

59. Writing exercise

60. Writing exercise

61. $\quad 3ax + 2x \geq 5ax - 4$

$$2x - 2ax \geq -4$$
$$2x(1 - a) \geq -4$$
$$x(1 - a) \geq -2$$
$$x \leq -\frac{2}{1 - a}, \text{ or } \frac{2}{a - 1}$$

We reversed the inequality symbol when we divided because when $a > 1$, then $1 - a < 0$.

The solution set is $\left\{x \middle| x \leq \dfrac{2}{a - 1}\right\}$.

62. $\left\{y \middle| y \geq -\dfrac{10}{b+4}\right\}$

63. $a(by - 2) \geq b(2y + 5)$

$aby - 2a \geq 2by + 5b$

$aby - 2by \geq 2a + 5b$

$y(ab - 2b) \geq 2a + 5b$

$y \geq \dfrac{2a + 5b}{ab - 2b}, \text{ or } \dfrac{2a + 5b}{b(a - 2)}$

The inequality symbol remained unchanged when we divided because when $a > 2$ and $b > 0$, then $ab - 2b > 0$.

The solution set is $\left\{y \middle| y \geq \dfrac{2a + 5b}{b(a - 2)}\right\}$.

64. $\left\{x \middle| x < \dfrac{4c + 3d}{6c - 2d}\right\}$

65. $c(2 - 5x) + dx > m(4 + 2x)$

$2c - 5cx + dx > 4m + 2mx$

$-5cx + dx - 2mx > 4m - 2c$

$x(-5c + d - 2m) > 4m - 2c$

$x[d - (5c + 2m)] > 4m - 2c$

$x > \dfrac{4m - 2c}{d - (5c + 2m)}$

The inequality symbol remained unchanged when we divided because when $5c + 2m < d$, then $d - (5c + 2m) > 0$.

The solution set is $\left\{x \middle| x > \dfrac{4m - 2c}{d - (5c + 2m)}\right\}$.

66. $\left\{x \middle| x < \dfrac{-3a + 2d}{c - (4a + 5d)}\right\}$

67. False. If $a = 2$, $b = 3$, $c = 4$, and $d = 5$, then $2 < 3$ and $4 < 5$ but $2 - 4 = 3 - 5$.

68. False, because $-3 < -2$, but $9 > 4$.

69. Writing exercise

70. Writing exercise

71. $x + 5 \leq 5 + x$

$5 \leq 5$ Subtracting x

We get an inequality that is true for all real numbers x. Thus the solution set is all real numbers.

72. $\emptyset$

73. $0^2 = 0$, $x^2 > 0$ for $x \neq 0$

The solution is $\{x | x \text{ is a real number and } x \neq 0\}$.

74. a) $\{x | x < 4\}$, or $(-\infty, 4)$

b) $\{x | x \geq 2\}$, or $[2, \infty)$

c) $\{x | x \geq 3.2\}$, or $[3.2, \infty)$

75.

Exercise Set 9.2

1. $\{7, 9, 11\} \cap \{9, 11, 13\}$

The numbers 9 and 11 are common to both sets, so the intersection is $\{9, 11\}$.

2. $\{2, 4, 8, 9, 10\}$

3. $\{1, 5, 10, 15\} \cup \{5, 15, 20\}$

The numbers in either or both sets are 1, 5, 10, 15, and 20, so the union is $\{1, 5, 10, 15, 20\}$.

4. $\{5\}$

5. $\{a, b, c, d, e, f\} \cap \{b, d, f\}$

The letters b, d, and f are common to both sets, so the intersection is $\{b, d, f\}$.

6. $\{a, b, c\}$

7. $\{r, s, t\} \cup \{r, u, t, s, v\}$

The letters in either or both sets are r, s, t, u, and v, so the union is $\{r, s, t, u, v\}$.

8. $\{m, o, p\}$

9. $\{3, 6, 9, 12\} \cap \{5, 10, 15\}$

There are no numbers common to both sets, so the solution set has no members. It is $\emptyset$.

10. $\{1, 4, 5, 6, 8, 9\}$

11. $\{3, 5, 7\} \cup \emptyset$

The numbers in either or both sets are 3, 5, and 7, so the union is $\{3, 5, 7\}$.

12. $\emptyset$

13. $3 < x < 8$

This inequality is an abbreviation for the conjunction $3 < x$ *and* $x < 8$. The graph is the intersection of two separate solution sets: $\{x | 3 < x\} \cap \{x | x < 8\} = \{x | 3 < x < 8\}$.

Interval notation: $(3, 8)$

14.

$[0, 4]$

15. $-6 \leq y \leq -2$

This inequality is an abbreviation for the conjunction $-6 \leq y$ *and* $y \leq -2$.

Interval notation: $[-6, -2]$

16.

$[-9, -5)$

17. $x < -2$ *or* $x > 3$

The graph of this disjunction is the union of the graphs of the individual solution sets $\{x | x < -2\}$ and $\{x | x > 3\}$.

Interval notation: $(-\infty, -2) \cup (3, \infty)$

18.

$(-\infty, -5) \cup (1, \infty)$

19. $x \leq -1$ *or* $x > 5$

Interval notation: $(-\infty, -1] \cup (5, \infty)$

20.

$(-\infty, -5] \cup (2, \infty)$

21. $-4 \leq -x < 2$

$\quad 4 \geq x > -2$ Multiplying by -1 and reversing the inequality symbols

$-2 < x \leq 4$ Rewriting

Interval notation: $(-2, 4]$

22.

$(-7, -2)$

23. $x > -2$ *and* $x < 4$

This conjunction can be abbreviated as $-2 < x < 4$.

Interval notation: $(-2, 4)$

24.

$(-3, 1]$

25. $5 > a$ *or* $a > 7$

Interval notation: $(-\infty, 5) \cup (7, \infty)$

26.

$(-\infty, -3) \cup [2, \infty)$

27. $x \geq 5$ *or* $-x \geq 4$

Multiplying the second inequality by -1 and reversing the inequality symbols, we get $x \geq 5$ *or* $x \leq -4$.

Interval notation: $(-\infty - 4] \cup [5, \infty)$

28.

$(-\infty, -6) \cup (-3, \infty)$

29. $4 > y$ *and* $y \geq -6$

This conjunction can be abbreviated as $-6 \leq x < 4$.

Interval notation: $[-6, 4)$

30.

$(-6, 0]$

31. $x < 7$ *and* $x \geq 3$

This conjunction can be abbreviated as $3 \leq x < 7$.

Interval notation: $[3, 7)$

32.

$[-3, 3)$

33. $t < 2 \ or \ t < 5$

Observe that every number that is less than 2 is also less than 5. Then $t < 2 \ or \ t < 5$ is equivalent to $t < 5$ and the graph of this disjunction is the set $\{t | t < 5\}$.

Interval notation: $(-\infty, 5)$

34.

$(-1, \infty)$

35. $x > -1 \ or \ x \le 3$

The graph of this disjunction is the union of the graphs of the individual solution sets:

$\{x | x > -1\} \cup \{x | x \le 3\} = $ the set of all real numbers.

Interval notation: $(-\infty, \infty)$

36.

$(-\infty, \infty)$

37. $x \ge 5 \ and \ x > 7$

The graph of this conjunction is the intersection of two separate solution sets: $\{x | x \ge 5\} \cap \{x | x > 7\} = \{x | x > 7\}$.

Interval notation: $(7, \infty)$

38.

$(-\infty, -4]$

39. $-1 < t + 2 < 7$

$-1 - 2 < t < 7 - 2$

$-3 < t < 5$

The solution set is $\{t | -3 < t < 5\}$, or $(-3, 5)$.

40. $\{t | -4 < t \le 4\}$, or $(-4, 4]$

41. $2 < x + 3 \ and \ x + 1 \le 5$

$-1 < x \qquad and \qquad x \le 4$

We can abbreviate the answer as $-1 < x \le 4$. The solution set is $\{x | -1 < x \le 4\}$, or $(-1, 4]$.

42. $\{x | -3 < x < 7\}$, or $(-3, 7)$

43. $-7 \le 2a - 3 \ and \ 3a + 1 < 7$

$-4 \le 2a \qquad and \qquad 3a < 6$

$-2 \le a \qquad and \qquad a < 2$

We can abbreviate the answer as $-2 \le a < 2$. The solution set is $\{a | -2 \le a < 2\}$, or $[-2, 2)$.

44. $\{n | -2 \le n \le 4\}$, or $[-2, 4]$

45. $x + 7 \le -2 \ or \ x + 7 \ge -3$

Observe that any real number is either less than or equal to -2 or greater than or equal to -3. Then the solution set is $\{x | x \text{ is a real number}\}$, or $(-\infty, \infty)$.

46. $\{x | x < -8 \ or \ x \ge -1\}$, or $(-\infty, -8) \cup [-1, \infty)$

47. $2 \le 3x - 1 \le 8$

$3 \le 3x \le 9$

$1 \le x \le 3$

The solution set is $\{x | 1 \le x \le 3\}$, or $[1, 3]$.

48. $\{x | 1 \le x \le 4\}$, or $[1, 4]$

49. $-21 \le -2x - 7 < 0$

$-14 \le -2x < 7$

$7 \ge x > -\dfrac{7}{2}$, or

$-\dfrac{7}{2} < x \le 7$

The solution set is $\left\{x \ \middle| \ -\dfrac{7}{2} < x \le 7\right\}$, or $\left(-\dfrac{7}{2}, 7\right]$.

50. $\left\{t\middle| -4 < t \le -\dfrac{10}{3}\right\}$, or $\left(-4, -\dfrac{10}{3}\right]$

51. $3x - 1 \le 2$ or $3x - 1 \ge 8$

$\qquad 3x \le 3$ or $\qquad 3x \ge 9$

$\qquad x \le 1$ or $\qquad x \ge 3$

The solution set is $\{x|x \le 1 \ or \ x \ge 3\}$, or $(-\infty, 1] \cup [3, \infty)$.

52. $\{x|x \le 1 \ or \ x \ge 5\}$, or $(-\infty, 1] \cup [5, \infty)$

53. $2x - 7 < -1$ or $2x - 7 > 1$

$\qquad 2x < 6$ or $\qquad 2x > 8$

$\qquad x < 3$ or $\qquad x > 4$

The solution set is $\{x|x < 3 \ or \ x > 4\}$, or $(-\infty, 3) \cup (4, \infty)$.

54. $\left\{x\middle| x < -4 \ or \ x > \dfrac{2}{3}\right\}$, or $(-\infty, -4) \cup \left(\dfrac{2}{3}, \infty\right)$

55. $6 > 2a - 1$ or $-4 \le -3a + 2$

$\quad 7 > 2a \qquad$ or $\ -6 \le -3a$

$\quad \dfrac{7}{2} > a \qquad$ or $\qquad 2 \ge a$

The solution set is $\left\{a\middle| \dfrac{7}{2} > a\right\} \cup \{a|2 \ge a\} =$

$\left\{a\middle| \dfrac{7}{2} > a\right\}$, or $\left\{a\middle| a < \dfrac{7}{2}\right\}$, or $\left(-\infty, \dfrac{7}{2}\right)$.

56. The set of all real numbers, or $(-\infty, \infty)$

57. $a + 4 < -1$ and $3a - 5 < 7$

$\quad a < -5$ and $\qquad 3a < 12$

$\quad a < -5$ and $\qquad a < 4$

The solution set is $\{a|a < -5\} \cap \{a|a < 4\} = \{a|a < -5\}$, or $(-\infty, -5)$.

58. $\{a|a > 4\}$, or $(4, \infty)$

59. $3x + 2 < 2$ or $4 - 2x < 14$

$\quad 3x < 0$ or $\quad -2x < 10$

$\quad x < 0$ or $\qquad x > -5$

The solution set is $\{x|x < 0\} \cup \{x|x > -5\} =$ the set of all real numbers, or $(-\infty, \infty)$.

60. $\{x|x \le -2 \ or \ x > 3\}$, or $(-\infty, -2] \cup (3, \infty)$

61. $2t - 7 \le 5$ or $5 - 2t > 3$

$\quad 2t \le 12$ or $\quad -2t > -2$

$\quad t \le 6$ or $\qquad t < 1$

The solution set is $\{t|t \le 6\} \cup \{t|t < 1\} = \{t|t \le 6\}$, or $(-\infty, 6]$.

62. $\{a|a \ge -1\}$, or $[-1, \infty)$

63. $f(x) = \dfrac{9}{x + 7}$

$f(x)$ cannot be computed when the denominator is 0. Since $x + 7 = 0$ is equivalent to $x = -7$, we have Domain of $f = \{x|x$ is a real number $and \ x \ne -7\} = (-\infty, -7) \cup (-7, \infty)$.

64. $(-\infty, -3) \cup (-3, \infty)$

65. $f(x) = \sqrt{x - 6}$

The expression $\sqrt{x - 6}$ is not a real number when $x - 6$ is negative. Thus, the domain of f is the set of all x-values for which $x - 6 \ge 0$. Since $x - 6 \ge 0$ is equivalent to $x \ge 6$, we have Domain of $f = [6, \infty)$.

66. $[2, \infty)$

67. $f(x) = \dfrac{x + 3}{2x - 5}$

$f(x)$ cannot be computed when the denominator is 0.

Since $2x - 5 = 0$ is equivalent to $x = \dfrac{5}{2}$, we have

Domain of $f = \left\{x\middle| x$ is a real number and $x \ne \dfrac{5}{2}\right\}$, or

$\left(-\infty, \dfrac{5}{2}\right) \cup \left(\dfrac{5}{2}, \infty\right)$.

68. $\left(-\infty, -\dfrac{4}{3}\right) \cup \left(-\dfrac{4}{3}, \infty\right)$

69. $f(x) = \sqrt{2x + 8}$

The expression $\sqrt{2x + 8}$ is not a real number when $2x + 8$ is negative. Thus, the domain of f is the set of all x-values for which $2x + 8 \geq 0$. Since $2x + 8 \geq 0$ is equivalent to $x \geq -4$, we have Domain of $f = [-4, \infty)$.

70. $(-\infty, 2]$

71. $f(x) = \sqrt{8 - 2x}$

The expression $\sqrt{8 - 2x}$ is not a real number when $8 - 2x$ is negative. Thus, the domain of f is the set of all x-values for which $8 - 2x \geq 0$. Since $8 - 2x \geq 0$ is equivalent to $x \leq 4$, we have Domain of $f = (-\infty, 4]$.

72. $(-\infty, 5]$

73. Writing exercise

74. Writing exercise

75. Graph: $y = 5$

The graph of any constant function $y = c$ is a horizontal line that crosses the vertical axis at $(0, c)$. Thus, the graph of $y = 5$ is a horizontal line that crosses the vertical axis at $(0, 5)$.

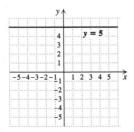

76.

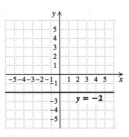

77. Graph $f(x) = |x|$

We make a table of values, plot points, and draw the graph.

x	$f(x)$
-5	5
-2	2
0	0
1	1
4	4

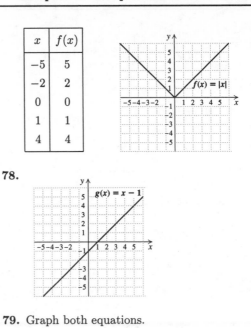

78.

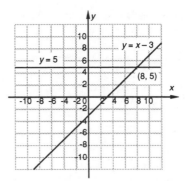

79. Graph both equations.

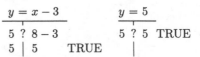

The solution (point of intersection) is apparently $(8, 5)$.

$$\begin{array}{c|c}
y = x - 3 & y = 5 \\
\hline
5 ? 8 - 3 & 5 ? 5 \quad \text{TRUE} \\
5 \mid 5 \quad \text{TRUE} &
\end{array}$$

The solution is $(8, 5)$.

80. $(-5, -3)$

81. Writing exercise

82. Writing exercise

83. From the graph we observe that the values of x for which $2x - 5 > -7$ *and* $2x - 5 < 7$ are $\{x \mid -1 < x < 6\}$, or $(-1, 6)$.

84. $\{x \mid x < -3 \text{ or } x > 6\}$, or $(-\infty, -3) \cup (6, \infty)$.

85. *Familiarize*. Let $c = $ the number of crossings per year. Then at the \$3 per crossing rate, the total cost of c crossings is \$3c. Two six-month passes cost

$2 \cdot \$15$, or \$30. The additional \$0.50 per crossing toll brings the total cost of c crossings to $\$30 + \$0.50c$. A one-year pass costs \$150 regardless of the number of crossings.

Translate. We write an inequality that states that the cost of c crossings per year using the six-month passes is less than the cost using the \$3 per crossing toll and is less than the cost using the one-year pass.

$$30 + 0.50c < 3c \ and \ 30 + 0.50c < 150$$

Carry out. We solve the inequality.

$$
\begin{array}{lll}
30 + 0.50c < 3c & and & 30 + 0.50c < 150 \\
30 < 2.5c & and & 0.50c < 120 \\
12 < c & and & c < 240
\end{array}
$$

This result can be written as $12 < c < 240$.

Check. When we substitute values of c less than 12, between 12 and 240, and greater than 240, we find that the result checks. Since we cannot check every possible value of c, we stop here.

State. For more than 12 crossings but less than 240 crossings per year the six-month passes are the most economical choice.

86. $0 \text{ ft} \leq d \leq 198 \text{ ft}$

87. Solve $32 < f(x) < 46$, or $32 < 2(x + 10) < 46$.

$$
\begin{aligned}
32 &< 2(x + 10) < 46 \\
32 &< 2x + 20 < 46 \\
12 &< 2x < 26 \\
6 &< x < 13
\end{aligned}
$$

For U.S. dress sizes between 6 and 13, dress sizes in Italy will be between 32 and 46.

88. From 2011 through 2036

89. a) Substitute $\frac{5}{9}(F - 32)$ for C in the given inequality.

$$1063 \leq \frac{5}{9}(F - 32) < 2660$$

$$9 \cdot 1063 \leq 9 \cdot \frac{5}{9}(F - 32) < 9 \cdot 2660$$

$$9567 \leq 5(F - 32) < 23{,}940$$

$$9567 \leq 5F - 160 < 23{,}940$$

$$9727 \leq 5F < 24{,}100$$

$$1945.4 \leq F < 4820$$

The inequality for Fahrenheit temperatures is $1945.4° \leq F < 4820°$.

b) Substitute $\frac{5}{9}(F - 32)$ for C in the given inequality.

$$960.8 \leq \frac{5}{9}(F - 32) < 2180$$

$$9(960.8) \leq 9 \cdot \frac{5}{9}(F - 32) < 9 \cdot 2180$$

$$8647.2 \leq 5(F - 32) < 19{,}620$$

$$8647.2 \leq 5F - 160 < 19{,}620$$

$$8807.2 \leq 5F < 19{,}780$$

$$1761.44 \leq F < 3956$$

The inequality for Fahrenheit temperatures is $1761.44° \leq F < 3956°$.

90. $1965 \leq y \leq 1981$

91. $4a - 2 \leq a + 1 \leq 3a + 4$

$$4a - 2 \leq a + 1 \ and \ a + 1 \leq 3a + 4$$

$$
\begin{array}{lll}
3a \leq 3 & and & -3 \leq 2a \\
a \leq 1 & and & -\dfrac{3}{2} \leq a
\end{array}
$$

The solution set is $\left\{ a \,\middle|\, -\dfrac{3}{2} \leq a \leq 1 \right\}$, or $\left[-\dfrac{3}{2}, 1 \right]$.

92. $\left\{ m \,\middle|\, m < \dfrac{6}{5} \right\}$, or $\left(-\infty, \dfrac{6}{5} \right)$

93. $x - 10 < 5x + 6 \leq x + 10$

$$
\begin{aligned}
-10 &< 4x + 6 \leq 10 \\
-16 &< 4x \leq 4 \\
-4 &< x \leq 1
\end{aligned}
$$

The solution set is $\{x \,|\, -4 < x \leq 1\}$, or $(-4, 1]$.

94. $\left\{ x \,\middle|\, -\dfrac{1}{8} < x < \dfrac{1}{2} \right\}$, or $\left(-\dfrac{1}{8}, \dfrac{1}{2} \right)$

95. If $-b < -a$, then $-1(-b) > -1(-a)$, or $b > a$, or $a < b$. The statement is true.

96. False

97. Let $a = 5$, $c = 12$, and $b = 2$. Then $a < c$ and $b < c$, but $a \not< b$. The given statement is false.

98. True

99. $f(x) = \dfrac{\sqrt{5+2x}}{x-1}$

The expression $\sqrt{5+2x}$ is not a real number when $5 + 2x$ is negative. Then for $5 + 2x \geq 0$, or for $x \geq -\dfrac{5}{2}$, the numerator of $f(x)$ is a real number. In addition, $f(x)$ cannot be computed when the denominator is 0. Since $x - 1 = 0$ is equivalent to $x = 1$, we have Domain of $f = \left\{ x \middle| x \geq -\dfrac{5}{2} \ and \ x \neq 1 \right\}$, or $\left[-\dfrac{5}{2}, 1 \right) \cup (1, \infty)$.

100. $(-\infty, -7) \cup \left(-7, \dfrac{3}{4} \right]$

101. Observe that the graph of y_2 lies on or above the graph of y_1 and below the graph of y_3 for x in the interval $[-3, 4)$.

102.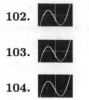

103.

104.

Exercise Set 9.3

1. $|x| = 4$

   $x = -4 \ or \ x = 4$ Using the absolute-value principle

   The solution set is $\{-4, 4\}$.

2. $\{-9, 9\}$

3. $|x| = -5$

   The absolute value of a number is always nonnegative. Therefore, the solution set is $\emptyset$.

4. $\emptyset$

5. $|y| = 7.3$

   $y = -7.3 \ or \ y = 7.3$ Using the absolute-value principle

   The solution set is $\{-7.3, 7.3\}$.

6. $\{0\}$

7. $|m| = 0$

   $m = 0$

   $\{0\}$

   The only number whose absolute value is 0 is 0. The solution set is $\{0\}$.

8. $\{-5.5, 5.5\}$

9. $|5x + 2| = 7$

   $5x + 2 = -7 \ or \ 5x + 2 = 7$ Absolute-value principle

   $5x = -9 \ or \qquad 5x = 5$

   $x = -\dfrac{9}{5} \ or \qquad x = 1$

   The solution set is $\left\{ -\dfrac{9}{5}, 1 \right\}$.

10. $\left\{ -\dfrac{1}{2}, \dfrac{7}{2} \right\}$

11. $|7x - 2| = -9$

   Absolute value is always nonnegative, so the equation has no solution. The solution set is $\emptyset$.

12. $\emptyset$

13. $|x - 3| = 8$

   $x - 3 = -8 \ or \ x - 3 = 8$ Absolute value principle

   $x = -5 \ or \qquad x = 11$

   The solution set is $\{-5, 11\}$.

14. $\{-4, 8\}$

15. $|x - 6| = 1$

   $x - 6 = -1 \ or \ x - 6 = 1$

   $x = 5 \qquad or \qquad x = 7$

   The solution set is $\{5, 7\}$.

16. $\{2, 8\}$

17. $|x - 4| = 5$

   $x - 4 = -5 \ or \ x - 4 = 5$

   $x = -1 \ or \qquad x = 9$

   The solution set is $\{-1, 9\}$.

18. $\{-2, 16\}$

19. $|2y| - 5 = 13$

   $|2y| = 18$ Adding 5

   $2y = -18 \ or \ 2y = 18$

   $y = -9 \ or \ y = 9$

   The solution set is $\{-9, 9\}$.

20. $\{-8, 8\}$

21. $7|z| + 2 = 16$ Adding -2

$\qquad 7|z| = 14$ Multiplying by $\frac{1}{7}$

$\qquad |z| = 2$

$z = -2 \ or \ z = 2$

The solution set is $\{-2, 2\}$.

22. $\left\{ -\frac{11}{5}, \frac{11}{5} \right\}$

23. $\left| \frac{4 - 5x}{6} \right| = 3$

$\qquad \frac{4 - 5x}{6} = -3 \quad or \quad \frac{4 - 5x}{6} = 3$

$\qquad 4 - 5x = -18 \quad or \quad 4 - 5x = 18$

$\qquad -5x = -22 \quad or \quad -5x = 14$

$\qquad x = \frac{22}{5} \quad or \quad x = -\frac{14}{5}$

The solution set is $\left\{ -\frac{14}{5}, \frac{22}{5} \right\}$.

24. $\{-7, 8\}$

25. $|t - 7| + 1 = 4$ Adding -1

$\qquad |t - 7| = 3$

$t - 7 = -3 \ or \ t - 7 = 3$

$\qquad t = 4 \quad or \quad t = 10$

The solution set is $\{4, 10\}$.

26. $\{-12, 2\}$

27. $3|2x - 5| - 7 = -1$

$\qquad 3|2x - 5| = 6$

$\qquad |2x - 5| = 2$

$2x - 5 = -2 \ or \ 2x - 5 = 2$

$\qquad 2x = 3 \quad or \quad 2x = 7$

$\qquad x = \frac{3}{2} \quad or \quad x = \frac{7}{2}$

The solution set is $\left\{ \frac{3}{2}, \frac{7}{2} \right\}$.

28. $\left\{ -\frac{1}{3}, 3 \right\}$

29. $|3x - 4| = 8$

$3x - 4 = -8 \ or \ 3x - 4 = 8$

$\qquad 3x = -4 \quad or \quad 3x = 12$

$\qquad x = -\frac{4}{3} \quad or \quad x = 4$

The solution set is $\left\{ -\frac{4}{3}, 4 \right\}$.

30. $\left\{ -\frac{3}{2}, \frac{17}{2} \right\}$

31. $|x| - 2 = 6.3$

$\qquad |x| = 8.3$

$x = -8.3 \ or \ x = 8.3$

The solution set is $\{-8.3, 8.3\}$.

32. $\{-11, 11\}$

33. $\left| \frac{3x - 2}{5} \right| = 2$

$\qquad \frac{3x - 2}{5} = -2 \quad or \quad \frac{3x - 2}{5} = 2$

$\qquad 3x - 2 = -10 \quad or \quad 3x - 2 = 10$

$\qquad 3x = -8 \quad or \quad 3x = 12$

$\qquad x = -\frac{8}{3} \quad or \quad x = 4$

The solution set is $\left\{ -\frac{8}{3}, 4 \right\}$.

34. $\{-1, 2\}$

35. $|x + 4| = |2x - 7|$

$\qquad x + 4 = 2x - 7 \quad or \quad x + 4 = -(2x - 7)$

$\qquad 4 = x - 7 \quad or \quad x + 4 = -2x + 7$

$\qquad 11 = x \quad or \quad 3x + 4 = 7$

$\qquad\qquad\qquad\qquad\quad 3x = 3$

$\qquad\qquad\qquad\qquad\quad x = 1$

The solution set is $\{1, 11\}$.

36. $\left\{ -\frac{11}{2}, \frac{1}{4} \right\}$

37. $|x - 9| = |x + 6|$

$\quad x - 9 = x + 6 \quad or \quad x - 9 = -(x + 6)$

$\quad -9 = 6 \qquad or \quad x - 9 = -x - 6$

False — $\qquad\qquad 2x - 9 = -6$

yields no $\qquad\qquad\quad 2x = 3$

solution $\qquad\qquad\qquad x = \frac{3}{2}$

The solution set is $\left\{ \frac{3}{2} \right\}$.

38. $\left\{ -\frac{1}{2} \right\}$

39. $|5t + 7| = |4t + 3|$

$\quad 5t + 7 = 4t + 3 \quad or \quad 5t + 7 = -(4t + 3)$

$\quad t + 7 = 3 \qquad or \quad 5t + 7 = -4t - 3$

$\quad t = -4 \qquad or \quad 9t + 7 = -3$

$\qquad\qquad\qquad\qquad\quad 9t = -10$

$\qquad\qquad\qquad\qquad\quad t = -\frac{10}{9}$

The solution set is $\left\{ -4, -\frac{10}{9} \right\}$.

40. $\left\{-\dfrac{3}{5}, 5\right\}$

41. $|n - 3| = |3 - n|$

$n - 3 = 3 - n$ or $n - 3 = -(3 - n)$

$2n - 3 = 3$ or $n - 3 = -3 + n$

$2n = 6$ or $-3 = -3$

$n = 3$ True for all real values of n

The solution set is the set of all real numbers.

42. The set of all real numbers

43. $|7 - a| = |a + 5|$

$7 - a = a + 5$ or $7 - a = -(a + 5)$

$7 = 2a + 5$ or $7 - a = -a - 5$

$2 = 2a$ or $7 = -5$

$1 = a$ False

The solution set is $\{1\}$.

44. $\left\{-\dfrac{1}{2}\right\}$

45. $\left|\dfrac{1}{2}x - 5\right| = \left|\dfrac{1}{4}x + 3\right|$

$\dfrac{1}{2}x - 5 = \dfrac{1}{4}x + 3$ or $\dfrac{1}{2}x - 5 = -\left(\dfrac{1}{4}x + 3\right)$

$\dfrac{1}{4}x - 5 = 3$ or $\dfrac{1}{2}x - 5 = -\dfrac{1}{4}x - 3$

$\dfrac{1}{4}x = 8$ or $\dfrac{3}{4}x - 5 = -3$

$x = 32$ or $\dfrac{3}{4}x = 2$

$x = \dfrac{8}{3}$

The solution set is $\left\{32, \dfrac{8}{3}\right\}$.

46. $\left\{-\dfrac{48}{37}, -\dfrac{144}{5}\right\}$

47. $|a| \le 7$

$-7 \le a \le 7$ Part (b)

The solution set is $\{a| -7 \le a \le 7\}$, or $[-7, 7]$.

48. $\{x| -2 < x < 2\}$, or $(-2, 2)$

49. $|x| > 8$

$x < -8$ or $8 < x$ Part (c)

The solution set is $\{x|x < -8$ or $x > 8\}$, or $(-\infty, -8) \cup (8, \infty)$.

50. $\{a|a \le -3$ or $a \ge 3\}$, or $(-\infty, -3] \cup [3, \infty)$

51. $|t| > 0$

$t < 0$ or $0 < t$ Part (c)

The solution set is $\{t|t < 0$ or $t > 0\}$, or $\{t|t \ne 0\}$, or $(-\infty, 0) \cup (0, \infty)$.

52. $\{t|t \le -1.7$ or $t \ge 1.7\}$, or $(-\infty, -1.7] \cup [1.7, \infty)$

53. $|x - 3| < 5$

$-5 < x - 3 < 5$ Part (b)

$-2 < x < 8$

The solution set is $\{x| -2 < x < 8\}$, or $(-2, 8)$.

54. $\{x| -2 < x < 4\}$, or $(-2, 4)$

55. $|x + 2| \le 6$

$-6 \le x + 2 \le 6$ Part (b)

$-8 \le x \le 4$ Adding -2

The solution set is $\{x| -8 \le x \le 4\}$, or $[-8, 4]$.

56. $\{x| -5 \le x \le -3\}$, or $[-5, -3]$

57. $|x - 3| + 2 > 7$

$|x - 3| > 5$ Adding -2

$x - 3 < -5$ or $5 < x - 3$ Part (c)

$x < -2$ or $8 < x$

The solution set is $\{x|x < -2$ or $x > 8\}$, or $(-\infty, -2) \cup (8, \infty)$.

58. The set of all real numbers, or $(-\infty, \infty)$

59. $|2y - 7| > -5$

Since absolute value is never negative, any value of $2y - 7$, and hence any value of y, will satisfy the inequality. The solution set is the set of all real numbers, or $(-\infty, \infty)$.

60. $\left\{ y \middle| y < -\dfrac{4}{3} \text{ or } y > 4 \right\}$, or $\left(-\infty, -\dfrac{4}{3} \right) \cup (4, \infty)$

61. $|3a - 4| + 2 \geq 8$

$|3a - 4| \geq 6$ Adding -2

$3a - 4 \leq -6 \quad \text{or} \quad 6 \leq 3a - 4$ Part (c)

$3a \leq -2 \quad \text{or} \quad 10 \leq 3a$

$a \leq -\dfrac{2}{3} \quad \text{or} \quad \dfrac{10}{3} \leq a$

The solution set is $\left\{ a \middle| a \leq -\dfrac{2}{3} \text{ or } a \geq \dfrac{10}{3} \right\}$, or

$\left(-\infty, -\dfrac{2}{3} \right] \cup \left[\dfrac{10}{3}, \infty \right)$.

62. $\left\{ a \middle| a \leq -\dfrac{3}{2} \text{ or } a \geq \dfrac{13}{2} \right\}$, or $\left(-\infty, -\dfrac{3}{2} \right] \cup \left[\dfrac{13}{2}, \infty \right)$

63. $|y - 3| < 12$

$-12 < y - 3 < 12$ Part (b)

$-9 < y < 15$ Adding 3

The solution set is $\{ y | -9 < y < 15 \}$, or $(-9, 15)$.

64. $\{ p | -1 < p < 5 \}$ or $(-1, 5)$

65. $9 - |x + 4| \leq 5$

$-|x + 4| \leq -4$

$|x + 4| \geq 4$ Multiplying by -1

$x + 4 \leq -4 \text{ or } 4 \leq x + 4$ Part (c)

$x \leq -8 \text{ or } 0 \leq x$

The solution set is $\{ x | x \leq -8 \text{ or } x \geq 0 \}$, or $(-\infty, -8] \cup [0, \infty)$.

66. $\{ x | x \leq 2 \text{ or } x \geq 8 \}$, or $(-\infty, 2] \cup [8, \infty)$

67. $|4 - 3y| > 8$

$4 - 3y < -8 \quad \text{or} \quad 8 < 4 - 3y$ Part (c)

$-3y < -12 \quad \text{or} \quad 4 < -3y$ Adding -4

$y > 4 \quad \text{or} \quad -\dfrac{4}{3} > y$ Multiplying by $-\dfrac{1}{3}$

The solution set is $\left\{ y \middle| y < -\dfrac{4}{3} \text{ or } y > 4 \right\}$, or

$\left(-\infty, -\dfrac{4}{3} \right) \cup (4, \infty)$.

68. $\emptyset$

69. $|3 - 4x| < -5$

Absolute value is always nonnegative, so the inequality has no solution. The solution set is $\emptyset$.

70. $\left\{ a \middle| -\dfrac{7}{2} \leq a \leq 6 \right\}$, or $\left[-\dfrac{7}{2}, 6 \right]$

71. $\left| \dfrac{2 - 5x}{4} \right| \geq \dfrac{2}{3}$

$\dfrac{2 - 5x}{4} \leq -\dfrac{2}{3} \quad \text{or} \quad \dfrac{2}{3} \leq \dfrac{2 - 5x}{4}$ Part (c)

$2 - 5x \leq -\dfrac{8}{3} \quad \text{or} \quad \dfrac{8}{3} \leq 2 - 5x$ Multiplying by 4

$-5x \leq -\dfrac{14}{3} \quad \text{or} \quad \dfrac{2}{3} \leq -5x$ Adding -2

$x \geq \dfrac{14}{15} \quad \text{or} \quad -\dfrac{2}{15} \geq x$ Multiplying by $-\dfrac{1}{5}$

The solution set is $\left\{ x \middle| x \leq -\dfrac{2}{15} \text{ or } x \geq \dfrac{14}{15} \right\}$, or

$\left(-\infty, -\dfrac{2}{15} \right] \cup \left[\dfrac{14}{15}, \infty \right)$.

72. $\left\{ x \middle| x < -\dfrac{43}{24} \text{ or } x > \dfrac{9}{8} \right\}$, or $\left(-\infty, -\dfrac{43}{24} \right) \cup \left(\dfrac{9}{8}, \infty \right)$.

73. $|m + 5| + 9 \le 16$

$|m + 5| \le 7$ Adding -9

$-7 \le m + 5 \le 7$

$-12 \le m \le 2$

The solution set is $\{m| -12 \le m \le 2\}$, or $[-12, 2]$.

74. $\{t| t \le 6 \ or \ t \ge 8\}$, or $(-\infty, 6] \cup [8, \infty)$

75. $25 - 2|a + 3| > 19$

$-2|a + 3| > -6$

$|a + 3| < 3$ Multiplying by $-\dfrac{1}{2}$

$-3 < a + 3 < 3$ Part (b)

$-6 < a < 0$

The solution set is $\{a| -6 < a < 0\}$, or $(-6, 0)$.

76. $\left\{a \middle| -\dfrac{13}{2} < a < \dfrac{5}{2}\right\}$, or $\left(-\dfrac{13}{2}, \dfrac{5}{2}\right)$.

77. $|2x - 3| \le 4$

$-4 \le 2x - 3 \le 4$ Part (b)

$-1 \le 2x \le 7$ Adding 3

$-\dfrac{1}{2} \le x \le \dfrac{7}{2}$ Multiplying by $\dfrac{1}{2}$

The solution set is $\left\{x \middle| -\dfrac{1}{2} \le x \le \dfrac{7}{2}\right\}$, or $\left[-\dfrac{1}{2}, \dfrac{7}{2}\right]$.

78. $\left\{x \middle| -1 \le x \le \dfrac{1}{5}\right\}$, or $\left[-1, \dfrac{1}{5}\right]$

79. $2 + |3x - 4| \ge 13$

$|3x - 4| \ge 11$

$3x - 4 \le -11$ or $11 \le 3x - 4$ Part (c)

$3x \le -7$ or $15 \le 3x$

$x \le -\dfrac{7}{3}$ or $5 \le x$

The solution set is $\left\{x \middle| x \le -\dfrac{7}{3} \ or \ x \ge 5\right\}$, or

$\left(-\infty, -\dfrac{7}{3}\right] \cup [5, \infty)$.

80. $\left\{x \middle| x \le -\dfrac{23}{9} \ or \ x \ge 3\right\}$, or $\left(-\infty, -\dfrac{23}{9}\right] \cup [3, \infty)$

81. $7 + |2x - 1| < 16$

$|2x - 1| < 9$

$-9 < 2x - 1 < 9$ Part (b)

$-8 < 2x < 10$

$-4 < x < 5$

The solution set is $\{x| -4 < x < 5\}$, or $(-4, 5)$.

82. $\left\{x \middle| -\dfrac{16}{3} < x < 4\right\}$, or $\left(-\dfrac{16}{3}, 4\right)$

83. Writing exercise

84. Writing exercise

85. $2x - 3y = 7$,   (1)

$3x + 2y = -10$ (2)

We will use the elimination method. First, multiply equation (1) by 2 and equation (2) by 3 and add to eliminate a variable.

$4x - 6y = 14$

$\underline{9x + 6y = -30}$

$13x = -16$

$x = -\dfrac{16}{13}$

Now substitute $-\dfrac{16}{13}$ for x in either of the original equations and solve for y.

$$3x + 2y = -10 \qquad (2)$$

$$3\left(-\frac{16}{13}\right) + 2y = -10$$

$$-\frac{48}{13} + 2y = -10$$

$$2y = -\frac{130}{13} + \frac{48}{13}$$

$$2y = -\frac{82}{13}$$

$$y = -\frac{41}{13}$$

The solution is $\left(-\dfrac{16}{13}, -\dfrac{41}{13}\right)$.

86. $(-2, -3)$

87. $x = -2 + 3y$, (1)

$x - 2y = 2$ (2)

We will use the substitution method. We substitute $-2 + 3y$ for x in equation (2).

$$x - 2y = 2 \quad (2)$$

$$(-2 + 3y) - 2y = 2 \quad \text{Substituting}$$

$$-2 + y = 2$$

$$y = 4$$

Now substitute 4 for y in equation (1) and find x.

$$x = -2 + 3 \cdot 4 = -2 + 12 = 10$$

The solution is $(10, 4)$.

88. $(-1, 7)$

89. Graph both equations.

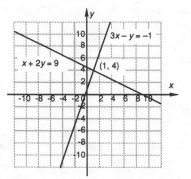

The solution (point of intersection) is apparently $(1, 4)$.

Check:

$x + 2y = 9$	$3x - y = -1$
$1 + 2 \cdot 4 \; ? \; 9$	$3 \cdot 1 - 4 \; ? \; -1$
$1 + 8$	$3 - 4$
$9 \;\big|\; 9$ TRUE	$-1 \;\big|\; -1$ TRUE

The solution is $(1, 4)$.

90. $(24, -41)$

91. Writing exercise

92. Writing exercise

93. From the definition of absolute value, $|3t - 5| = 3t - 5$ only when $3t - 5 \geq 0$. Solve $3t - 5 \geq 0$.

$$3t - 5 \geq 0$$

$$3t \geq 5$$

$$t \geq \frac{5}{3}$$

The solution set is $\left\{t \middle| t \geq \dfrac{5}{3}\right\}$, or $\left[\dfrac{5}{3}, \infty\right)$.

94. $\{x | x \text{ is a real number}\}$, or $(-\infty, \infty)$

95. $2 \leq |x - 1| \leq 5$

$2 \leq |x - 1|$ *and* $|x - 1| \leq 5$.

For $2 \leq |x - 1|$:

$x - 1 \leq -2$ *or* $2 \leq x - 1$

$x \leq -1$ *or* $3 \leq x$

The solution set of $2 \leq |x-1|$ is $\{x | x \leq -1 \text{ or } x \geq 3\}$.

For $|x - 1| \leq 5$:

$-5 \leq x - 1 \leq 5$

$-4 \leq x \leq 6$

The solution set of $|x - 1| \leq 5$ is $\{x | -4 \leq x \leq 6\}$.

The solution set of $2 \leq |x - 1| \leq 5$ is

$\{x | x \leq -1 \text{ or } x \geq 3\} \cap \{x | -4 \leq x \leq 6\}$

$= \{x | -4 \leq x \leq -1 \text{ or } 3 \leq x \leq 6\}$, *or*

$[-4, -1] \cup [3, 6]$.

96. $\left\{-\dfrac{1}{7}, \dfrac{7}{3}\right\}$

97. $t - 2 \leq |t - 3|$

$t - 3 \leq -(t - 2)$ *or* $t - 2 \leq t - 3$

$t - 3 \leq -t + 2$ *or* $-2 \leq -3$

$2t - 3 \leq 2$ $\qquad\qquad$ False

$2t \leq 5$

$t \leq \dfrac{5}{2}$

The solution set is $\left\{t \middle| t \leq \dfrac{5}{2}\right\}$, or $\left(-\infty, \dfrac{5}{2}\right]$.

98. $|x| < 3$

99. Using part (b), we find that $-5 \leq y \leq 5$ is equivalent to $|y| \leq 5$.

100. $|x| \geq 6$

101. $x < -4$ *or* $4 < x$

$|x| > 4$ Using part (c)

102. $|x + 3| > 5$

103. $-5 < x < 1$

$-3 < x + 2 < 3$ Adding 2

$|x + 2| < 3$ Using part (b)

104. $|x - 7| < 2$, or $|7 - x| < 2$

105. The distance from x to 5 is $|x - 5|$ or $|5 - x|$, so we have $|x - 5| < 1$, or $|5 - x| < 1$.

106. $|x - 3| \leq 4$

107. The length of the segment from -4 to 8 is $|-4-8| = |-12| = 12$ units. The midpoint of the segment is $\frac{-4+8}{2} = \frac{4}{2} = 2$. Thus, the interval extends 12/2, or 6, units on each side of 2. An inequality for which the open interval is the solution set is $|x - 2| < 6$.

108. $|x + 4| < 3$

109. The length of the segment from 2 to 12 is $|2 - 12| = |-10| = 10$ units. The midpoint of the segment is $\frac{2+12}{2} = \frac{14}{2} = 7$. Thus, the interval extends 10/2, or 5, units on each side of 7. An inequality for which the closed interval is the solution set is $|x - 7| \leq 5$.

110. $\left\{ d \Big| 5\frac{1}{2} \text{ ft} \leq d \leq 6\frac{1}{2} \text{ ft} \right\}$, or $\left[5\frac{1}{2} \text{ ft}, 6\frac{1}{2} \text{ ft} \right]$

111. Graph $g(x) = 4$ or the same axes as $f(x) = |2x - 6|$.

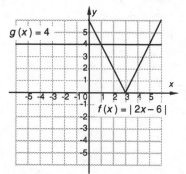

The solution set consists of the x-values for which $(x, f(x))$ is on or below the horizontal line $g(x) = 4$. These x-values comprise the interval $[1, 5]$.

112. Writing exercise

113.

114.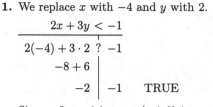

115. Writing exercise

Exercise Set 9.4

1. We replace x with -4 and y with 2.

$$
\begin{array}{c|c}
\multicolumn{2}{c}{2x + 3y < -1} \\
\hline
2(-4) + 3 \cdot 2 \ ? \ -1 & \\
-8 + 6 & \\
-2 & -1 \quad \text{TRUE}
\end{array}
$$

Since $-2 < -1$ is true, $(-4, 2)$ is a solution.

2. No

3. We replace x with 8 and y with 14.

$$
\begin{array}{c|c}
\multicolumn{2}{c}{2y - 3x \geq 9} \\
\hline
2 \cdot 14 - 3 \cdot 8 \ ? \ 9 & \\
28 - 24 & \\
4 & 9 \quad \text{FALSE}
\end{array}
$$

Since $4 > 9$ is false, $(8, 14)$ is not a solution.

4. Yes

5. Graph: $y > \frac{1}{2}x$

We first graph the line $y = \frac{1}{2}x$. We draw the line dashed since the inequality symbol is $>$. To determine which half-plane to shade, test a point not on the line. We try $(0, 1)$:

$$
\begin{array}{c|c}
\multicolumn{2}{c}{y > \frac{1}{2}x} \\
\hline
1 \ ? \ \frac{1}{2} \cdot 0 & \\
1 & 0 \quad \text{TRUE}
\end{array}
$$

Since $1 > 0$ is true, (0.1) is a solution as are all of the points in the half-plane containing $(0, 1)$. We shade that half-plane and obtain the graph.

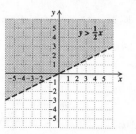

6.

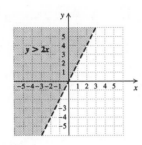

7. Graph: $y \geq x - 3$

First graph the line $y = x - 3$. Draw it solid since the inequality symbol is $\geq$. Test the point $(0,0)$ to determine if it is a solution.

$$\frac{y \geq x - 3}{0 \ ? \ 0 - 3}$$
$$0 \ \Big| \ -3 \qquad \text{TRUE}$$

Since $0 \geq -3$ is true, we shade the half-plane that contains $(0,0)$ and obtain the graph.

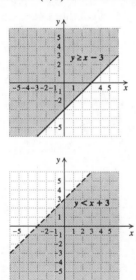

8.

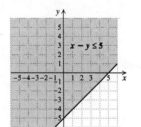

9. Graph: $y \leq x + 4$

First graph the line $y = x + 4$. Draw it solid since the inequality symbol is $\leq$. Test the point $(0,0)$ to determine if it is a solution.

$$\frac{y \leq x + 4}{0 \ ? \ 0 + 4}$$
$$0 \ \Big| \ 4 \qquad \qquad \text{TRUE}$$

Since $0 \leq 4$ is true, we shade the half-plane that contains $(0,0)$ and obtain the graph.

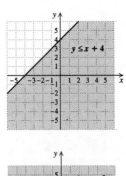

10.

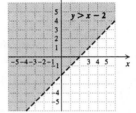

11. Graph: $x - y \leq 5$

First graph the line $x - y = 5$. Draw a solid line since the inequality symbol is $\leq$. Test the point $(0,0)$ to determine if it is a solution.

$$\frac{x - y \leq 5}{0 - 0 \ ? \ 5}$$
$$0 \ \Big| \ 5 \qquad \text{TRUE}$$

Since $0 \leq 5$ is true, we shade the half-plane that contains $(0,0)$ and obtain the graph.

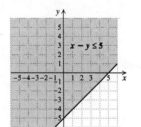

12.

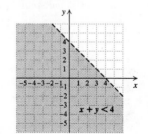

13. Graph: $2x + 3y < 6$

First graph $2x + 3y = 6$. Draw the line dashed since the inequality symbol is $<$. Test the point $(0,0)$ to determine if it is a solution.

$$\begin{array}{c} 2x + 3y < 6 \\ \hline 2 \cdot 0 + 3 \cdot 0 \ ? \ 6 \\ 0 \ \bigm| \ 6 \qquad \text{TRUE} \end{array}$$

Since $0 < 6$ is true, we shade the half-plane containing $(0,0)$ and obtain the graph.

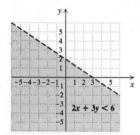

14.

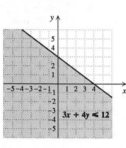

15. Graph: $2x - y \le 4$

We first graph $2x - y = 4$. Draw the line solid since the inequality symbol is $\le$. Test the point $(0,0)$ to determine if it is a solution.

$$\begin{array}{c} 2x - y \le 4 \\ \hline 2 \cdot 0 - 0 \ ? \ 4 \\ 0 \ \bigm| \ 4 \qquad \text{TRUE} \end{array}$$

Since $0 \le 4$ is true, we shade the half-plane containing $(0,0)$ and obtain the graph.

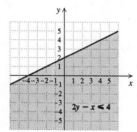

16.

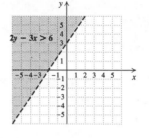

17.   Graph: $2x - 2y \ge 8 + 2y$

$$2x - 4y \ge 8$$

First graph $2x - 4y = 8$. Draw the line solid since the inequality symbol is $\ge$. Test the point $(0,0)$ to determine if it is a solution.

$$\begin{array}{c} 2x - 4y \ge 8 \\ \hline 2 \cdot 0 - 4 \cdot 0 \ ? \ 8 \\ 0 \ \bigm| \ 8 \qquad \text{FALSE} \end{array}$$

Since $0 \ge 8$ is false, we shade the half-plane that does not contain $(0,0)$ and obtain the graph.

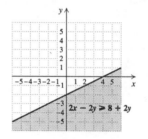

18.

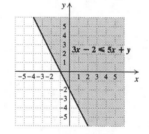

19. Graph: $y \ge 2$

We first graph $y = 2$. Draw the line solid since the inequality symbol is $\ge$. Test the point $(0,0)$ to determine if it is a solution.

$$\begin{array}{c} y \ge 2 \\ \hline 0 \ ? \ 2 \qquad \text{FALSE} \end{array}$$

Since $0 \ge 2$ is false, we shade the half-plane that does not contain $(0,0)$ and obtain the graph.

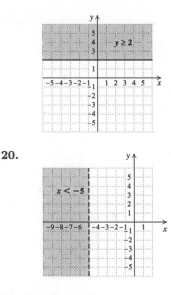

20.

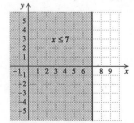

21. Graph: $x \leq 7$

We first graph $x = 7$. We draw the line solid since the inequality symbol is $\leq$. Test the point $(0,0)$ to determine if it is a solution.

$$\frac{x \leq 7}{0 \; ? \; 7} \qquad \text{TRUE}$$

Since $0 \leq 7$ is true, we shade the half-plane containing $(0,0)$ and obtain the graph.

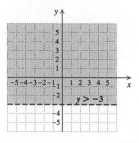

22.

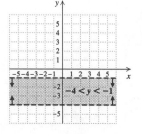

23. Graph: $-2 < y < 6$

This is a system of inequalities:

$$-2 < y,$$
$$y < 6$$

We graph the equation $-2 = y$ and see that the graph of $-2 < y$ is the half-plane above the line $-2 = y$. We also graph $y = 6$ and see that the graph of $y < 6$ is the half-plane below the line $y = 6$.

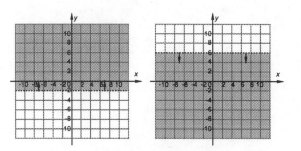

Finally, we shade the intersection of these graphs.

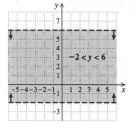

24.

25. Graph: $-4 \leq x \leq 5$

This is a system of inequalities:

$$-4 \leq x,$$
$$x \leq 5$$

Graph $-4 \leq x$ and $x \leq 5$.

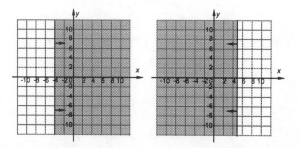

Then we shade the intersection of these graphs.

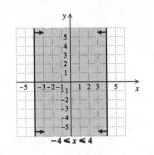

$-4 \leq x \leq 4$

28.

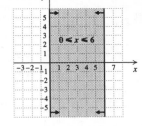

$0 \leq x \leq 6$

26.

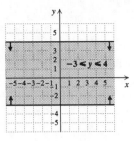

$-3 < y \leq 4$

29. Graph: $y < x$,

$y > -x + 2$

We graph the lines $y = x$ and $y = -x + 2$, using dashed lines. We indicate the region for each inequality by the arrows at the ends of the lines. Note where the regions overlap and shade the region of solutions.

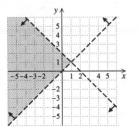

27. Graph: $0 \leq y \leq 3$

This is a system of inequalities:

$0 \leq y,$

$y \leq 3$

Graph $0 \leq y$ and $y \leq 3$.

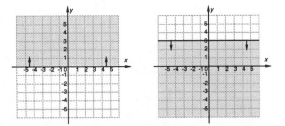

Then we shade the intersection of these graphs.

30.

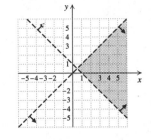

31. Graph: $y \geq x$,

$y \leq 2x - 4$

Graph $y = x$ and $y = 2x - 4$, using solid lines. Indicate the region for each inequality by arrows, and shade the region where they overlap.

$0 \leq y \leq 3$

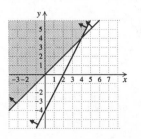

32.

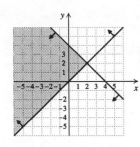

33. Graph: $y \leq -3$,

    $x \geq -1$

Graph $y = -3$ and $x = -1$ using solid lines. Indicate the region for each inequality by arrows, and shade the region where they overlap.

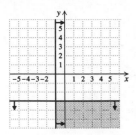

34.

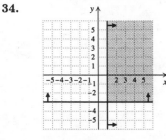

35.   Graph: $x > -4$,

    $y < -2x + 3$

Graph the lines $x = -4$ and $y = -2x + 3$, using dashed lines. Indicate the region for each inequality by arrows, and shade the region where they overlap.

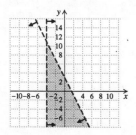

36.

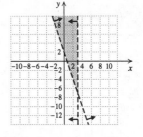

37. Graph: $y \leq 3$,

    $y \geq -x + 2$

Graph the lines $y = 3$ and $y = -x + 2$, using solid lines. Indicate the region for each inequality by arrows, and shade the region where they overlap.

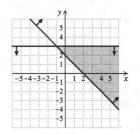

38.

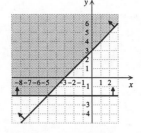

39. Graph: $x + y \leq 6$,

    $x - y \leq 4$

Graph the lines $x + y = 6$ and $x - y = 4$, using solid lines. Indicate the region for each inequality by arrows, and shade the region where they overlap.

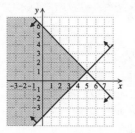

40.

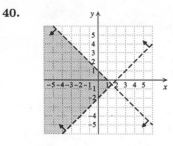

41. Graph: $y + 3x > 0,$
$$y + 3x < 2$$

Graph the lines $y + 3x = 0$ and $y + 3x = 2$, using dashed lines. Indicate the region for each inequality by arrows, and shade the region where they overlap.

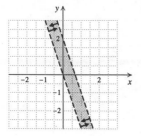

42.

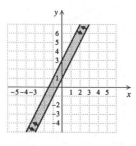

43. Graph: $y \le 2x - 1,$ (1)
$$y \ge -2x + 1, \quad (2)$$
$$x \le 3 \quad (3)$$

Graph the lines $y = 2x - 1$, $y = -2x + 1$, and $x = 3$ using solid lines. Indicate the region for each inequality by arrows, and shade the region where they overlap.

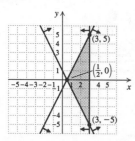

To find the vertex we solve three different systems of related equations.

From (1) and (2) we have $y = 2x - 1,$
$$y = -2x + 1.$$

Solving, we obtain the vertex $\left(\frac{1}{2}, 0\right)$.

From (1) and (3) we have $y = 2x - 1,$
$$x = 3.$$

Solving, we obtain the vertex $(3, 5)$.

From (2) and (3) we have $y = -2x + 1,$
$$x = 3.$$

Solving, we obtain the vertex $(3, -5)$.

44.

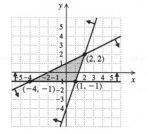

45. Graph: $x + 2y \le 12,$ (1)
$$2x + y \le 12 \quad (2)$$
$$x \ge 0, \quad (3)$$
$$y \ge 0 \quad (4)$$

Graph the lines $x + 2y = 12$, $2x + y = 12$, $x = 0$, and $y = 0$ using solid lines. Indicate the region for each inequality by arrows, and shade the region where they overlap.

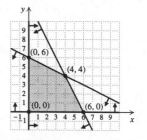

To find the vertices we solve four different systems of equations.

From (1) and (2) we have $x + 2y = 12,$
$$2x + y = 12.$$

Solving, we obtain the vertex $(4, 4)$.

From (1) and (3) we have $x + 2y = 12,$
$$x = 0.$$

Solving, we obtain the vertex $(0, 6)$.

From (2) and (4) we have $2x + y = 12,$
$$y = 0.$$
Solving, we obtain the vertex $(6, 0)$.

From (3) and (4) we have $x = 0,$
$$y = 0.$$
Solving, we obtain the vertex $(0, 0)$.

46.

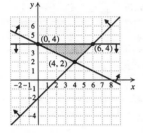

47.   Graph: $8x + 5y \leq 40,$ (1)
$$x + 2y \leq 8 \qquad (2)$$
$$x \geq 0, \qquad (3)$$
$$y \geq 0 \qquad (4)$$

Graph the lines $8x + 5y = 40$, $x + 2y = 8$, $x = 0$, and $y = 0$ using solid lines. Indicate the region for each inequality by arrows, and shade the region where they overlap.

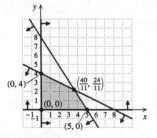

To find the vertices we solve four different systems of equations.

From (1) and (2) we have $8x + 5y = 40,$
$$x + 2y = 8.$$
Solving, we obtain the vertex $\left(\dfrac{40}{11}, \dfrac{24}{11} \right)$.

From (1) and (4) we have $8x + 5y = 40,$
$$y = 0.$$
Solving, we obtain the vertex $(5, 0)$.

From (2) and (3) we have $x + 2y = 8,$
$$x = 0.$$
Solving, we obtain the vertex $(0, 4)$.

From (3) and (4) we have $x = 0,$
$$y = 0.$$
Solving, we obtain the vertex $(0, 0)$.

48.

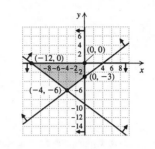

49.   Graph: $y - x \geq 1,$ (1)
$$y - x \leq 3, \qquad (2)$$
$$2 \leq x \leq 5 \qquad (3)$$

Think of (3) as two inequalities:
$$2 \leq x, \qquad (4)$$
$$x \leq 5 \qquad (5)$$

Graph the lines $y - x = 1$, $y - x = 3$, $x = 2$, and $x = 5$, using solid lines. Indicate the region for each inequality by arrows, and shade the region where they overlap.

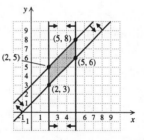

To find the vertices we solve four different systems of equations.

From (1) and (4) we have $y - x = 1,$
$$x = 2.$$
Solving, we obtain the vertex $(2, 3)$.

From (1) and (5) we have $y - x = 1,$
$$x = 5.$$
Solving, we obtain the vertex $(5, 6)$.

From (2) and (4) we have $y - x = 3,$
$$x = 2.$$
Solving, we obtain the vertex $(2, 5)$.

From (2) and (5) we have $y - x = 3,$
$$x = 5.$$
Solving, we obtain the vertex $(5, 8)$.

50.

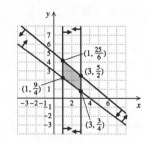

51. Writing exercise

52. Writing exercise

53. *Familiarize*. We let x and y represent the number of pounds of peanuts and fancy nuts in the mixture, respectively. We organize the given information in a table.

Type of nuts	Peanuts	Fancy	Mixture
Amount	x	y	10
Price per pound	$2.50	$7	
Value	$2.5x$	$7y$	40

Translate. We get a system of equations from the first and third rows of the table.
$$x + y = 10,$$
$$2.5x + 7y = 40$$
Clearing decimals we have
$$x + y = 10, \quad (1)$$
$$25x + 70y = 400. \quad (2)$$

Carry out. We use the elimination method. Multiply Equation (1) by -25 and add.
$$-25x - 25y = -250$$
$$\underline{25x + 70y = 400}$$
$$45y = 150$$
$$y = \frac{10}{3}, \quad \text{or } 3\frac{1}{3}$$

Substitute $\frac{10}{3}$ for y in Equation (1) and solve for x.
$$x + y = 10$$
$$x + \frac{10}{3} = 10$$
$$x = \frac{20}{3}, \quad \text{or } 6\frac{2}{3}$$

Check. The sum of $6\frac{2}{3}$ and $3\frac{1}{3}$ is 10. The value of the mixture is $2.5\left(\frac{20}{3}\right) + 7\left(\frac{10}{3}\right)$, or $\frac{50}{3} + \frac{70}{3}$, or $40. These numbers check.

State. $6\frac{2}{3}$ lb of peanuts and $3\frac{1}{3}$ lb of fancy nuts should be used.

54. Hendersons: 10 bags; Savickis: 4 bags

55. *Familiarize*. Let x = the number of cardholders tickets that were sold and y = the number of non-cardholders tickets. We arrange the information in a table.

	Card-holders	Non-card-holders	Total
Price	$1.25	$2	
Number sold	x	y	203
Money taken in	$1.25x$	$2y$	$310

Translate. The last two rows of the table give us two equations. The total number of tickets sold was 203, so we have
$$x + y = 203.$$
The total amount of money collected was $310, so we have
$$1.25x + 2y = 310.$$
We can multiply the second equation on both sides by 100 to clear decimals. The resulting system is
$$x + y = 203, \quad (1)$$
$$125x + 200y = 31,000. \quad (2)$$

Carry out. We use the elimination method. We multiply on both sides of Equation (1) by -125 and then add.
$$-125x - 125y = -25,375 \quad \text{Multiplying by } -125$$
$$\underline{125x + 200y = 31,000}$$
$$75y = 5625$$
$$y = 75$$
We go back to Equation (1) and substitute 75 for y.
$$x + y = 203$$
$$x + 75 = 203$$
$$x = 128$$

Check. The number of tickets sold was $128 + 75$, or 203. The money collected was $1.25(128) + $2(75)$, or $160 + 150, or $310. These numbers check.

State. 128 cardholders tickets and 75 non-cardholders tickets were sold.

56. 70 student tickets; 130 adult tickets

57. *Familiarize*. The formula for the area of a triangle with base b and height h is $A = \frac{1}{2}bh$.

Translate. Substitute 200 for A and 16 for b in the formula.

$$A = \frac{1}{2}bh$$

$$200 = \frac{1}{2} \cdot 16 \cdot h$$

Carry out. We solve the equation.

$$200 = \frac{1}{2} \cdot 16 \cdot h$$

$$200 = 8h \qquad \text{Multiplying}$$

$$25 = h \qquad \text{Dividing by 8 on both sides}$$

Check. The area of a triangle with base 16 ft and height 25 ft is $\frac{1}{2} \cdot 16 \cdot 25$, or 200 ft^2. The answer checks.

State. The seed can fill a triangle that is 25 ft tall.

58. 11%

59. Writing exercise

60. Writing exercise

61. Graph: $x + y > 8$,
$\quad\quad\quad x + y \leq -2$

Graph the line $x + y = 8$ using a dashed line and graph $x + y = -2$, using a solid line. Indicate the region for each inequality by arrows. The regions do not overlap (the solution set is $\emptyset$), so we do not shade any portion of the graph.

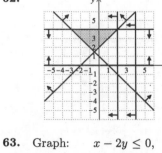

62.

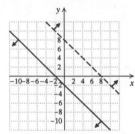

63. Graph: $x - 2y \leq 0$,
$\quad\quad\quad\quad -2x + y \leq 2$,
$\quad\quad\quad\quad\quad\quad x \leq 2$,
$\quad\quad\quad\quad\quad\quad y \leq 2$,
$\quad\quad\quad\quad x + y \leq 4$

Graph the five inequalities above, and shade the region where they overlap.

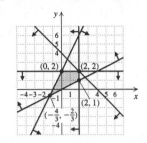

64. $x \geq -2$,
$\quad y \leq 2$,
$\quad x \leq 0$,
$\quad y \geq 0$;

$\quad x \geq 0$,
$\quad y \leq 2$,
$\quad x \leq 2$,
$\quad y \geq 0$;

$\quad x \geq 0$,
$\quad y \leq 0$,
$\quad x \leq 2$,
$\quad y \geq -2$;

$\quad x \geq -2$,
$\quad y \leq 0$,
$\quad x \leq 0$,
$\quad y \geq -2$

65. Both the width and the height must be positive, but they must be less than 62 in. in order to be checked as luggage, so we have:

$$0 < w \leq 62,$$
$$0 < h \leq 62$$

The girth is represented by $2w + 2h$ and the length is 62 in. In order to meet postal regulations the sum of the girth and the length cannot exceed 108 in., so we have:

$$62 + 2w + 2h \leq 108, \text{ or}$$
$$2w + 2h \leq 46, \text{ or}$$
$$w + h \leq 23$$

Thus, have a system of inequalities:

$$0 < w \leq 62,$$
$$0 < h \leq 62,$$
$$w + h \leq 23$$

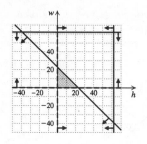

66. $2w + t \geq 60,$

      $w \geq 0,$

      $t \geq 0$

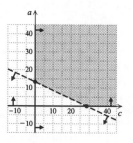

67.   Graph: $35c + 75a > 1000,$

      $c \geq 0,$

      $a \geq 0$

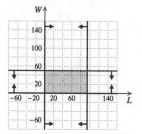

68. $0 < L \leq 94,$

      $0 < W \leq 50$

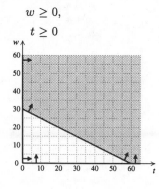

69. a) $3x + 6y > 2$

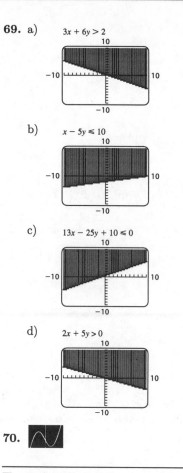

b) $x - 5y \leq 10$

c) $13x - 25y + 10 \leq 0$

d) $2x + 5y > 0$

70.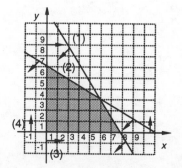

Exercise Set 9.5

1. Find the maximum and minimum values of
$F = 2x + 14y,$

subject to

     $5x + 3y \leq 34,$ (1)

     $3x + 5y \leq 30,$ (2)

        $x \geq 0,$ (3)

        $y \geq 0.$ (4)

Graph the system of inequalities and find the coordinates of the vertices.

To find one vertex we solve the system

$$x = 0,$$
$$y = 0.$$

This vertex is $(0, 0)$.

To find a second vertex we solve the system

$$5x + 3y = 34,$$
$$y = 0.$$

This vertex is $\left(\frac{34}{5}, 0\right)$.

To find a third vertex we solve the system

$$5x + 3y = 34,$$
$$3x + 5y = 30.$$

This vertex is $(5, 3)$.

To find the fourth vertex we solve the system

$$3x + 5y = 30,$$
$$x = 0.$$

This vertex is $(0, 6)$.

Now find the value of F at each of these points.

Vertex (x, y)	$F = 2x + 14y$
$(0, 0)$	$2 \cdot 0 + 14 \cdot 0 = 0 + 0 = 0$ ← Minimum
$\left(\frac{34}{5}, 0\right)$	$2 \cdot \frac{34}{5} + 14 \cdot 0 = \frac{68}{5} + 0 = 13\frac{3}{5}$
$(5, 3)$	$2 \cdot 5 + 14 \cdot 3 = 10 + 42 = 52$
$(0, 6)$	$2 \cdot 0 + 14 \cdot 6 = 0 + 84 = 84$ ← Maximum

The maximum value of F is 84 when $x = 0$ and $y = 6$. The minimum value of F is 0 when $x = 0$ and $y = 0$.

2. The maximum is 38 when $x = 2$ and $y = 3$; the minimum is 0 when $x = 0$ and $y = 0$.

3. Find the maximum and minimum values of
$$P = 8x - y + 20,$$

subject to

$$6x + 8y \le 48, \quad (1)$$
$$0 \le y \le 4, \qquad (2)$$
$$0 \le x \le 7. \qquad (3)$$

Think of (2) as $0 \le y$, (4)

$y \le 4$. (5)

Think of (3) as $0 \le x$, (6)

$x \le 7$. (7)

Graph the system of inequalities.

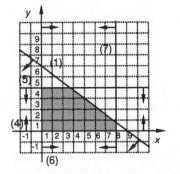

To determine the coordinates of the vertices, we solve the following systems:

$$x = 0, \qquad\qquad x = 7, \qquad\qquad 6x + 8y = 48,$$
$$y = 0; \qquad\qquad y = 0; \qquad\qquad x = 7;$$

$$6x + 8y = 48, \qquad\qquad x = 0,$$
$$y = 4; \qquad\qquad\qquad y = 4$$

The vertices are $(0, 0)$, $(7, 0)$, $\left(7, \frac{3}{4}\right)$, $\left(\frac{8}{3}, 4\right)$, and $(0, 4)$, respectively. Compute the value of P at each of these points.

Vertex (x, y)	$P = 8x - y + 20$
$(0, 0)$	$8 \cdot 0 - 0 + 20 =$ $0 - 0 + 20 = 20$
$(7, 0)$	$8 \cdot 7 - 0 + 20 =$ $56 - 0 + 20 =$ 76 ← Maximum
$\left(7, \frac{3}{4}\right)$	$8 \cdot 7 - \frac{3}{4} + 20 =$ $56 - \frac{3}{4} + 20 = 75\frac{1}{4}$
$\left(\frac{8}{3}, 4\right)$	$8 \cdot \frac{8}{3} - 4 + 20 =$ $\frac{64}{3} - 4 + 20 = 37\frac{1}{3}$
$(0, 4)$	$8 \cdot 0 - 4 + 20 =$ $0 - 4 + 20 =$ 16 ← Minimum

The maximum is 76 when $x = 7$ and $y = 0$. The minimum is 16 when $x = 0$ and $y = 4$.

4. The maximum is 124 when $x = 3$ and $y = 0$; the minimum is 40 when $x = 0$ and $y = 4$.

5. Find the maximum and minimum values of
$$F = 2y - 3x,$$

subject to

$$y \leq 2x + 1, \quad (1)$$

$$y \geq -2x + 3, \quad (2)$$

$$x \leq 3 \quad\quad\quad (3)$$

Graph the system of inequalities and find the coordinates of the vertices.

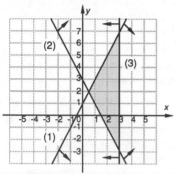

To determine the coordinates of the vertices, we solve the following systems:

$$y = 2x + 1, \quad\quad y = 2x + 1, \quad\quad y = -2x + 3,$$

$$y = -2x + 3; \quad\quad x = 3; \quad\quad\quad x = 3$$

The solutions of the systems are $\left(\frac{1}{2}, 2\right)$, $(3, 7)$, and $(3, -3)$, respectively. Now find the value of F at each of these points.

Vertex (x, y)	$F = 2y - 3x$	
$\left(\frac{1}{2}, 2\right)$	$2 \cdot 2 - 3 \cdot \dfrac{1}{2} = \dfrac{5}{2}$	
$(3, 7)$	$2 \cdot 7 - 3 \cdot 3 = 5$	←Maximum
$(3, -3)$	$2(-3) - 3 \cdot 3 = -15$	←Minimum

The maximum value is 5 when $x = 3$ and $y = 7$. The minimum value is -15 when $x = 3$ and $y = -3$.

6. The maximum is 51 when $x = 5$ and $y = 11$; the minimum is 12 when $x = \dfrac{2}{3}$ and $y = \dfrac{7}{3}$.

7. *Familiarize*. Let $x =$ the number of orders of chili and $y =$ the number of burritos sold each day.

Translate. The profit P is given by

$$P = \$1.65x + \$1.05y.$$

We wish to maximize P subject to these facts (constraints) about x and y:

$$10 \leq x \leq 40,$$

$$30 \leq y \leq 70,$$

$$x + y \leq 90.$$

Carry out. We graph the system of inequalities, determine the vertices, and evaluate P at each vertex.

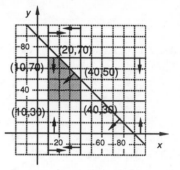

Vertex	$P = \$1.65x + \$1.05y$
$(10, 30)$	$\$1.65(10) + \$1.05(30) = \$48$
$(40, 30)$	$\$1.65(40) + \$1.05(30) = \$97.50$
$(40, 50)$	$\$1.65(40) + \$1.05(50) = \$118.50$
$(20, 70)$	$\$1.65(20) + \$1.05(70) = \$106.50$
$(10, 70)$	$\$1.65(10) + \$1.05(70) = \$90$

The greatest profit in the table is $118.50, obtained when 40 orders of chili and 50 burritos are sold.

Check. Go over the algebra and arithmetic.

State. The maximum profit occurs when 40 orders of chili and 50 burritos are sold.

8. 100 units of lumber; 300 units of plywood

9. *Familiarize*. Let $x =$ the number of motorcycles manufactured and $y =$ the number of bicycles manufactured each month.

Translate. The profit P is given by

$$P = \$1340x + \$200y.$$

We wish to maximize P subject to these facts (constraints) about x and y.

$y \leq 3x$	The number of bicycles cannot exceed three times the number of motorcycles.
$0 \leq x \leq 60$	No more than 60 motorcycles will be produced.
$0 \leq y \leq 120$	No more than 120 bicycles can be produced.
$x + y \leq 160$	Total production cannot exceed 160.

Carry out. We graph the system of inequalities, determine the vertices, and evaluate P at each vertex.

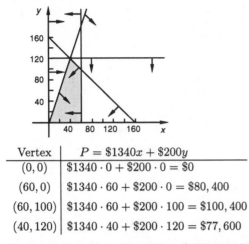

Vertex	$P = \$1340x + \$200y$
$(0,0)$	$\$1340 \cdot 0 + \$200 \cdot 0 = \$0$
$(60,0)$	$\$1340 \cdot 60 + \$200 \cdot 0 = \$80,400$
$(60,100)$	$\$1340 \cdot 60 + \$200 \cdot 100 = \$100,400$
$(40,120)$	$\$1340 \cdot 40 + \$200 \cdot 120 = \$77,600$

The greatest profit in the table is $100,400, obtained when 60 motorcycles and 100 bicycles are manufactured.

Check. Go over the algebra and arithmetic.

State. The maximum profit occurs when 60 motorcycles and 100 bicycles are manufactured.

10. Car: 9 gal; moped: 3 gal

11. *Familiarize*. We organize the information in a table. Let $x = $ the number of matching questions and $y = $ the number of essay questions you answer.

Type	Number of points for each	Number answered	Total points
Matching	10	$3 \leq x \leq 12$	$10x$
Essay	25	$4 \leq y \leq 15$	$25y$
Total		$x + y \leq 20$	$10x + 25y$

Since Phil can answer no more than a total of 20 questions, we have the inequality $x + y \leq 20$ in the "Number answered" column. The expression $10x + 25y$ in the "Total points" column gives the total score on the test.

Translate. The score S is given by

$$S = 10x + 25y.$$

We wish to maximize S subject to these facts (constraints) about x and y.

$$3 \leq x \leq 12,$$
$$4 \leq y \leq 15,$$
$$x + y \leq 20.$$

Carry out. We graph the system of inequalities, determine the vertices, and evaluate S at each vertex.

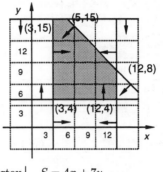

Vertex	$S = 4x + 7y$
$(3,4)$	$10 \cdot 3 + 25 \cdot 4 = 130$
$(3,15)$	$10 \cdot 3 + 25 \cdot 15 = 405$
$(5,15)$	$10 \cdot 5 + 25 \cdot 15 = 425$
$(12,8)$	$10 \cdot 12 + 25 \cdot 8 = 320$
$(12,4)$	$10 \cdot 12 + 25 \cdot 4 = 220$

The greatest score in the table is 425, obtained when 5 matching questions and 15 essay questions are answered correctly.

Check. Go over the algebra and arithmetic.

State. The maximum score is 425 points when 5 matching questions and 15 essay questions are answered correctly.

12. The maximum score is 102 points when 8 short-answer questions and 10 word problems are answered correctly.

13. In order to earn the most interest Rosa should invest the entire $40,000. She should also invest as much as possible in the type of investment that has the higher interest rate. Thus, she should invest $22,000 in corporate bonds and the remaining $18,000 in municipal bonds. The maximum income is $0.08(\$22,000) + 0.075(\$18,000) = \$3110$.

We can also solve this problem as follows.

Let $x = $ the amount invested in corporate bonds and $y = $ the amount invested in municipal bonds. Find the maximum value of

$$I = 0.08x + 0.075y$$

subject to

$$x + y \leq \$40,000,$$
$$\$6000 \leq x \leq \$22,000$$
$$0 \leq y \leq \$30,000.$$

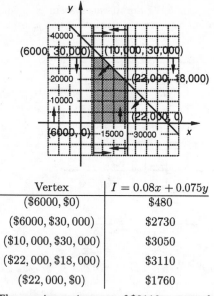

Vertex	$I = 0.08x + 0.075y$
($6000, $0)	$480
($6000, $30,000)	$2730
($10,000, $30,000)	$3050
($22,000, $18,000)	$3110
($22,000, $0)	$1760

The maximum income of $3110 occurs when $22,000 is invested in corporate bonds and $18,000 is invested in municipal bonds.

14. 80 acres of Merlot grapes; 160 acres of Cabernet grapes

15. *Familiarize*. Let $x =$ the number of batches of Hawaiian Blend and $y =$ the number of batches of Classic Blend that are made. We organize the information in a table.

	Blend		Number of lb
	Hawaiian	Classic	available
Number of batches	x	y	
Sumatran coffee	12	16	1440
Kona coffee	8	4	700
Profit per batch	$90	$55	

***Translate*.** The profit P is given by

$P = 90x + 55y.$

We wish to maximum P subject to these facts (constraints) about x and y.

$$12x + 16y \leq 1440,$$
$$8x + 4y \leq 700,$$
$$x \geq 0,$$
$$y \geq 0.$$

***Carry out*.** We graph the system of inequalities, determine the vertices, and evaluate P at each vertex.

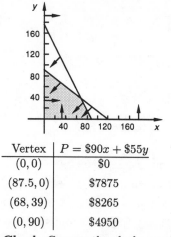

Vertex	$P = \$90x + \$55y$
$(0, 0)$	$0
$(87.5, 0)$	$7875
$(68, 39)$	$8265
$(0, 90)$	$4950

***Check*.** Go over the algebra and arithmetic.

***State*.** The maximum profit of $8265 occurs when 68 batches of Hawaiian Blend and 39 batches of Classic Blend are made.

16. The maximum interest income is $1395 when $7000 is invested in City Bank and $15,000 is invested in the Southwick Credit Union.

17. *Familiarize*. Let $x =$ the number of knit suits and $y =$ the number of worsted suits made per day.

***Translate*.** The profit P is given by

$P = \$68x + \$62y.$

We wish to maximize P subject to these facts (constraints) about x and y.

$$2x + 4y \leq 20,$$
$$4x + 2y \leq 16,$$
$$x \geq 0,$$
$$y \geq 0.$$

***Carry out*.** Graph the system of inequalities, determine the vertices, the evaluate P at each vertex.

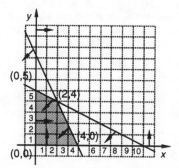

Vertex	$P = \$68x + \$62y$
$(0,0)$	$\$68 \cdot 0 + \$62 \cdot 0 = \$0$
$(0,5)$	$\$68 \cdot 0 + \$62 \cdot 5 = \$310$
$(2,4)$	$\$68 \cdot 2 + \$62 \cdot 4 = \$384$
$(4,0)$	$\$68 \cdot 4 + \$62 \cdot 0 = \$272$

Check. Go over the algebra and arithmetic.

State. The maximum profit per day is $384 when 2 knit suits and 4 worsted suits are made.

18. The company will have a maximum income of $200 when 200 Biscuit Jumbos and 0 Mitimites are made.

19. Writing exercise

20. Writing exercise

21.
$$5x^3 - 4x^2 - 7x + 2$$
$$= 5(-2)^3 - 4(-2)^2 - 7(-2) + 2$$
$$= 5(-8) - 4(4) - 7(-2) + 2$$
$$= -40 - 16 + 14 + 2$$
$$= -40$$

22. 46

23. $3(2x - 5) + 4(x + 5) = 6x - 15 + 4x + 20 = 10x + 5$

24. $26t + 20$

25. $6x - 3(x + 2) = 6x - 3x - 6 = 3x - 6$

26. $2t + 2$

27. Writing exercise

28. Writing exercise

29. *Familiarize*. Let x represent the number of T3 planes and y represent the number of S5 planes. Organize the information in a table.

Plane	Number of planes	Passengers		
		First	Tourist	Economy
T3	x	$40x$	$40x$	$120x$
S5	y	$80y$	$30y$	$40y$

Plane	Cost per mile
T3	$30x$
S5	$25y$

Translate. Suppose C is the total cost per mile. Then $C = 30x + 25y$. We wish to minimize C subject to these facts (constraints) about x and y.

$$40x + 80y \geq 2000,$$
$$40x + 30y \geq 1500,$$
$$120x + 40y \geq 2400,$$
$$x \geq 0,$$
$$y \geq 0$$

Carry out. Graph the system of inequalities, determine the vertices, and evaluate C at each vertex.

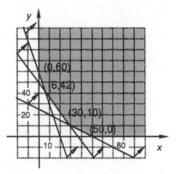

Vertex	$C = 30x + 25y$
$(0,60)$	$30(0) + 25(60) = 1500$
$(6,42)$	$30(6) + 25(42) = 1230$
$(30,10)$	$30(30) + 25(10) = 1150$
$(50,0)$	$30(50) + 25(0) = 1500$

Check. Go over the algebra and arithmetic.

State. In order to minimize the operating cost, 30 T3 planes and 10 S5 planes should be used.

30. 30 S5's; 15 T4's

31. *Familiarize*. Let $x =$ the number of chairs and $y =$ the number of sofas produced.

Translate. Find the maximum value of
$$I = \$80x + \$1200y$$
subject to
$$20x + 100y \leq 1900,$$
$$x + 50y \leq 500,$$
$$2x + 20y \leq 240,$$
$$x \geq 0,$$
$$y \geq 0.$$

Carry out. Graph the system of inequalities, determine the vertices, and evaluate I at each vertex.

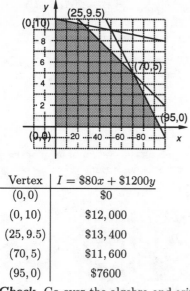

Vertex	$I = \$80x + \$1200y$
$(0,0)$	$\$0$
$(0,10)$	$\$12,000$
$(25,9.5)$	$\$13,400$
$(70,5)$	$\$11,600$
$(95,0)$	$\$7600$

Check. Go over the algebra and arithmetic.

State.   The maximum income of $\$13,400$ occurs when 25 chairs and 9.5 sofas are made. A more practical answer is that the maximum income of $\$12,800$ is achieved when 25 chairs and 9 sofas are made.

Chapter 10

Exponents and Radicals

1. The square roots of 16 are 4 and -4, because $4^2 = 16$ and $(-4)^2 = 16$.

2. $7, -7$

3. The square roots of 144 are 12 and -12, because $12^2 = 144$ and $(-12)^2 = 144$.

4. $3, -3$

5. The square roots of 81 are 9 and -9, because $9^2 = 81$ and $(-9)^2 = 81$.

6. $20, -20$

7. The square roots of 900 are 30 and -30, because $30^2 = 900$ and $(-30)^2 = 900$.

8. $15, -15$

9. $-\sqrt{\dfrac{49}{36}} = -\dfrac{7}{6}$ Since $\sqrt{\dfrac{49}{36}} = \dfrac{7}{6}$, $-\sqrt{\dfrac{49}{36}} = -\dfrac{7}{6}$.

10. $-\dfrac{19}{3}$

11. $\sqrt{441} = 21$ Remember, $\sqrt{}$ indicates the principle square root.

12. 14

13. $-\sqrt{\dfrac{16}{81}} = -\dfrac{4}{9}$ Since $\sqrt{\dfrac{16}{81}} = \dfrac{4}{9}$, $-\sqrt{\dfrac{16}{81}} = -\dfrac{4}{9}$.

14. $-\dfrac{3}{4}$

15. $\sqrt{0.09} = 0.3$

16. 0.6

17. $-\sqrt{0.0049} = -0.07$

18. 0.12

19. $5\sqrt{p^2 + 4}$

The radicand is the expression written under the radical sign, $p^2 + 4$.

Since the index is not written, we know it is 2.

20. $y^2 - 8$; 2

21. $x^2 y^2 \sqrt{\dfrac{x}{y + 4}}$

The radicand is the expression written under the radical sign, $\dfrac{x}{y + 4}$.

The index is 3.

22. $\dfrac{a}{a^2 - b}$; 3

23. $\begin{aligned} f(t) &= \sqrt{5t - 10} \\ f(6) &= \sqrt{5 \cdot 6 - 10} = \sqrt{20} \\ f(2) &= \sqrt{5 \cdot 2 - 10} = \sqrt{0} = 0 \\ f(1) &= \sqrt{5 \cdot 1 - 10} = \sqrt{-5} \end{aligned}$

Since negative numbers do not have real-number square roots, $f(1)$ does not exist.

$f(-1) = \sqrt{5(-1) - 10} = \sqrt{-15}$

Since negative numbers do not have real-number square roots, $f(-1)$ does not exist.

24. $\sqrt{11}$; does not exist; $\sqrt{11}$; 12

25. $\begin{aligned} t(x) &= -\sqrt{2x + 1} \\ t(4) &= -\sqrt{2 \cdot 4 + 1} = -\sqrt{9} = -3 \\ t(-1) &= -\sqrt{2(-1) + 1} = -\sqrt{-1}; \\ & \quad t(-1) \text{ does not exist.} \end{aligned}$

$t\left(-\dfrac{1}{2}\right) = -\sqrt{2\left(-\dfrac{1}{2}\right) + 1} = -\sqrt{0} = 0$

26. $\sqrt{12}$; does not exist; $\sqrt{30}$; does not exist

27. $\begin{aligned} f(t) &= \sqrt{t^2 + 1} \\ f(0) &= \sqrt{0^2 + 1} = \sqrt{1} = 1 \\ f(-1) &= \sqrt{(-1)^2 + 1} = \sqrt{2} \\ f(-10) &= \sqrt{(-10)^2 + 1} = \sqrt{101} \end{aligned}$

28. $-2; -5; -4$

29. $g(x) = \sqrt{x^3 + 9}$

$g(-2) = \sqrt{(-2)^3 + 9} = \sqrt{1} = 1$

$g(-3) = \sqrt{(-3)^3 + 9} = \sqrt{-18};$

$g(-3)$ does not exist.

$g(3) = \sqrt{3^3 + 9} = \sqrt{36} = 6$

30. Does not exist; $\sqrt{17}$; $\sqrt{54}$

31. $\sqrt{36x^2} = \sqrt{(6x)^2} = |6x| = 6|x|$

Since x might be negative, absolute-value notation is necessary.

32. $5|t|$

33. $\sqrt{(-6b)^2} = |-6b| = |-6| \cdot |b| = 6|b|$

Since b might be negative, absolute-value notation is necessary.

34. $7|c|$

35. $\sqrt{(7 - t)^2} = |7 - t|$

Since $7-t$ might be negative, absolute-value notation is necessary.

36. $|a + 1|$

37. $\sqrt{y^2 + 16y + 64} = \sqrt{(y + 8)^2} = |y + 8|$

Since $y + 8$ might be negative, absolute-value notation is necessary.

38. $|x - 2|$

39. $\sqrt{9x^2 - 30x + 25} = \sqrt{(3x - 5)^2} = |3x - 5|$

Since $3x - 5$ might be negative, absolute-value notation is necessary.

40. $|2x + 7|$

41. $-\sqrt[4]{625} = -5$ Since $5^4 = 625$

42. 4

43. $-\sqrt[5]{3^5} = -3$

44. -1

45. $\sqrt[5]{-\dfrac{1}{32}} = -\dfrac{1}{2}$ Since $\left(-\dfrac{1}{2}\right)^5 = -\dfrac{1}{32}$

46. $-\dfrac{2}{3}$

47. $\sqrt[8]{y^8} = |y|$

The index is even. Use absolute-value notation since y could have a negative value.

48. $|x|$

49. $\sqrt[4]{(7b)^4} = |7b| = 7|b|$

The index is even. Use absolute-value notation since b could have a negative value.

50. $5|a|$

51. $\sqrt[12]{(-10)^{12}} = |-10| = 10$

52. 6

53. $\sqrt[1976]{(2a + b)^{1976}} = |2a + b|$

The index is even. Use absolute-value notation since $2a + b$ could have a negative value.

54. $|a + b|$

55. $\sqrt{x^{10}} = |x^5|$ Note that $(x^5)^2 = x^{10}$; x^5 could have a negative value.

56. $|a^{11}|$

57. $\sqrt{a^{14}} = |a^7|$ Note that $(a^7)^2 = a^{14}$; a^7 could have a negative value.

58. x^8

59. $\sqrt{25t^2} = \sqrt{(5t)^2} = 5t$ Assuming t is nonnegative

60. $4x$

61. $\sqrt{(7c)^2} = 7c$ Assuming c is nonnegative

62. $6b$

63. $\sqrt{(5 + b)^2} = 5 + b$ Assuming $5 + b$ is nonnegative

64. $a + 1$

65. $\sqrt{9x^2 + 36x + 36} = \sqrt{9(x^2 + 4x + 4)} = \sqrt{[3(x + 2)]^2} = 3(x + 2)$, or $3x + 6$

66. $2(x + 1)$, or $2x + 2$

67. $\sqrt{25t^2 - 20t + 4} = \sqrt{(5t - 2)^2} = 5t - 2$

68. $3t - 2$

69. $-\sqrt[3]{64} = -4$ $(4^3 = 64)$

70. 3

71. $\sqrt[4]{81x^4} = \sqrt[4]{(3x)^4} = 3x$

72. $2x$

73. $-\sqrt[5]{-100,000} = -(-10) = 10$ $[(-10)^5 = -100,000]$

74. -6

75. $-\sqrt[3]{-64x^3} = -(-4x)$ $[(-4x)^3 = -64x^3]$
$$= 4x$$

76. $5y$

77. $\sqrt{a^{14}} = \sqrt{(a^7)^2} = a^7$

78. a^{11}

79. $\sqrt{(x+3)^{10}} = \sqrt{[(x+3)^5]^2} = (x+3)^5$

80. $(x-2)^4$

81. $\qquad f(x) = \sqrt[3]{x+1}$
$$f(7) = \sqrt[3]{7+1} = \sqrt[3]{8} = 2$$
$$f(26) = \sqrt[3]{26+1} = \sqrt[3]{27} = 3$$
$$f(-9) = \sqrt[3]{-9+1} = \sqrt[3]{-8} = -2$$
$$f(-65) = \sqrt[3]{-65+1} = \sqrt[3]{-64} = -4$$

82. $1;\ 5;\ 3;\ -5$

83. $\qquad g(t) = \sqrt[4]{t-3}$
$$g(19) = \sqrt[4]{19-3} = \sqrt[4]{16} = 2$$
$$g(-13) = \sqrt[4]{-13-3} = \sqrt[4]{-16};$$
$$\qquad g(-13) \text{ does not exist.}$$
$$g(1) = \sqrt[4]{1-3} = \sqrt[4]{-2};$$
$$\qquad g(1) \text{ does not exist.}$$
$$g(84) = \sqrt[4]{84-3} = \sqrt[4]{81} = 3$$

84. $1;\ 2;\ \text{does not exist};\ 3$

85. $f(x) = \sqrt{x-5}$

Since the index is even, the radicand, $x-5$, must be nonnegative. We solve the inequality:
$$x - 5 \geq 0$$
$$x \geq 5$$
Domain of $f = \{x | x \geq 5\}$, or $[5, \infty)$

86. $\{x | x \geq -8\}$, or $[-8, \infty)$

87. $g(t) = \sqrt[4]{t+3}$

Since the index is even, the radicand, $t+3$, must be nonnegative. We solve the inequality:
$$t + 3 \geq 0$$
$$t \geq -3$$
Domain of $g = \{t | t \geq -3\}$, or $[-3, \infty)$

88. $\{x | x \geq 7\}$, or $[7, \infty)$

89. $g(x) = \sqrt[4]{5-x}$

Since the index is even, the radicand, $5-x$, must be nonnegative. We solve the inequality:
$$5 - x \geq 0$$
$$5 \geq x$$
Domain of $g = \{x | x \leq 5\}$, or $(-\infty, 5]$

90. $\{t | t \text{ is a real number}\}$, or $(-\infty, \infty)$

91. $f(t) = \sqrt[5]{2t+9}$

Since the index is odd, the radicand can be any real number.

Domain of $f = \{t | t \text{ is a real number}\}$, or $(-\infty, \infty)$

92. $\left\{ t | t \geq -\dfrac{5}{2} \right\}$, or $\left[-\dfrac{5}{2}, \infty \right)$

93. $h(z) = -\sqrt[6]{5z+3}$

Since the index is even, the radicand, $5z+3$, must be nonnegative. We solve the inequality:
$$5z + 3 \geq 0$$
$$5z \geq -3$$
$$z \geq -\frac{3}{5}$$
Domain of $h = \left\{ z | z \geq -\dfrac{3}{5} \right\}$, or $\left[-\dfrac{3}{5}, \infty \right)$

94. $\left\{ x | x \geq \dfrac{5}{7} \right\}$, or $\left[\dfrac{5}{7}, \infty \right)$

95. $f(t) = 7 + \sqrt[8]{t^8}$

Since we can compute $7 + \sqrt[8]{t^8}$ for any real number t, the domain is the set of real numbers, or $\{x | x \text{ is a real number}\}$, or $(-\infty, \infty)$.

96. $\{x | x \text{ is a real number}\}$, or $(-\infty, \infty)$

97. Writing exercise

98. Writing exercise

99. $(a^3 b^2 c^5)^3 = a^{3\cdot 3} b^{2\cdot 3} c^{5\cdot 3} = a^9 b^6 c^{15}$

100. $10a^{10}b^9$

101. $(2a^{-2} b^3 c^{-4})^{-3} = 2^{-3} a^{-2(-3)} b^{3(-3)} c^{-4(-3)} =$
$$\frac{1}{2^3} a^6 b^{-9} c^{12} = \frac{a^6 c^{12}}{8 b^9}$$

102. $\dfrac{x^6 y^2}{25 z^4}$

103. $\dfrac{8x^{-2} y^5}{4x^{-6} z^{-2}} = \dfrac{8}{4} x^{-2-(-6)} y^5 z^2 = 2x^4 y^5 z^2$

104. $\dfrac{5c^3}{a^4b^7}$

105. Writing exercise

106. Writing exercise

107. $N = 2.5\sqrt{A}$

    a) $N = 2.5\sqrt{25} = 2.5(5) = 12.5 \approx 13$

    b) $N = 2.5\sqrt{36} = 2.5(6) = 15$

    c) $N = 2.5\sqrt{49} = 2.5(7) = 17.5 \approx 18$

    d) $N = 2.5\sqrt{64} = 2.5(8) = 20$

108. $\{x | x \geq -5\}$, or $[-5, \infty)$

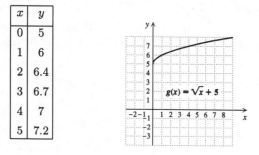

109. $g(x) = \sqrt{x} + 5$

Since the index is even, th radicand, x, must be non-negative, so we have $x \geq 0$.

Domain of $g = \{x | x \geq 0\}$, or $[0, \infty)$

Make a table of values, keeping in mind that x must be nonnegative. Plot these points and draw the graph.

x	y
0	5
1	6
2	6.4
3	6.7
4	7
5	7.2

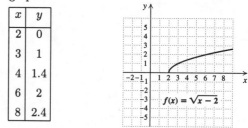

110. $\{x | x \geq 0\}$, or $[0, \infty)$

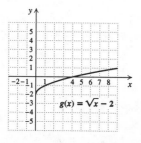

111. $f(x) = \sqrt{x - 2}$

Since the index is even, the radicand, $x - 2$, must be nonnegative. We solve the inequality.

$$x - 2 \geq 0$$
$$x \geq 2$$

Domain of $f = \{x | x \geq 2\}$, or $[2, \infty)$

Make a table of values, keeping in mind that x must be 2 or greater. Plot these points and draw the graph.

x	y
2	0
3	1
4	1.4
6	2
8	2.4

112. $\{x | -3 \leq x < 2\}$, or $[-3, 2)$

113. $g(x) = \dfrac{\sqrt[4]{5 - x}}{\sqrt[6]{x + 4}}$

The radical expression in the numerator has an even index, so the radicand, $5 - x$, must be nonnegative. We solve the inequality:

$$5 - x \geq 0$$
$$5 \geq x$$

The radical expression in the denominator also has an even index, so the radicand, $x + 4$, must be nonnegative in order for $\sqrt[6]{x + 4}$ to exist. In addition, the denominator cannot be zero, so the radicand must be positive. We solve the inequality:

$$x + 4 > 0$$
$$x > -4$$

We have $x \leq 5$ *and* $x > -4$ so

Domain of $g = \{x | -4 < x \leq 5\}$, or $(-4, 5]$.

114.

115.

Exercise Set 10.2

1. $x^{1/4} = \sqrt[4]{x}$

2. $\sqrt[5]{y}$

3. $(16)^{1/2} = \sqrt{16} = 4$

4. 2

5. $81^{1/4} = \sqrt[4]{81} = 3$

6. 2

7. $9^{1/2} = \sqrt{9} = 3$

8. 5

9. $(xyz)^{1/3} = \sqrt[3]{xyz}$

10. $\sqrt[4]{ab}$

11. $(a^2b^2)^{1/5} = \sqrt[5]{a^2b^2}$

12. $\sqrt[4]{x^3y^3}$

13. $a^{2/3} = \sqrt[3]{a^2}$

14. $\sqrt{b^3}$

15. $16^{3/4} = \sqrt[4]{16^3} = (\sqrt[4]{16})^3 = 2^3 = 8$

16. 128

17. $49^{3/2} = \sqrt{49^3} = (\sqrt{49})^3 = 7^3 = 343$

18. 81

19. $9^{5/2} = \sqrt{9^5} = (\sqrt{9})^5 = 3^5 = 243$

20. 729

21. $(81x)^{3/4} = \sqrt[4]{(81x)^3} = \sqrt[4]{81^3x^3}$, or $\sqrt[4]{81^3} \cdot \sqrt[4]{x^3} = (\sqrt[4]{81})^3 \cdot \sqrt[4]{x^3} = 3^3\sqrt[4]{x^3} = 27\sqrt[4]{x^3}$

22. $25\sqrt[3]{a^2}$

23. $(25x^4)^{3/2} = \sqrt{(25x^4)^3} = \sqrt{25^3 \cdot x^{12}} = \sqrt{25^3} \cdot \sqrt{x^{12}} = (\sqrt{25})^3x^6 = 5^3x^6 = 125x^6$

24. $27y^9$

25. $\sqrt[3]{20} = 20^{1/3}$

26. $19^{1/3}$

27. $\sqrt{17} = 17^{1/2}$

28. $6^{1/2}$

29. $\sqrt{x^3} = x^{3/2}$

30. $a^{5/2}$

31. $\sqrt[5]{m^2} = m^{2/5}$

32. $n^{4/5}$

33. $\sqrt[4]{cd} = (cd)^{1/4}$ Parentheses are required.

34. $(xy)^{1/5}$

35. $\sqrt[5]{xy^2z} = (xy^2z)^{1/5}$

36. $(x^3y^2z^2)^{1/7}$

37. $(\sqrt{3mn})^3 = (3mn)^{3/2}$

38. $(7xy)^{4/3}$

39. $(\sqrt[7]{8x^2y})^5 = (8x^2y)^{5/7}$

40. $(2a^5b)^{7/6}$

41. $\dfrac{2x}{\sqrt[3]{z^2}} = \dfrac{2x}{z^{2/3}}$

42. $\dfrac{3a}{c^{2/5}}$

43. $x^{-1/3} = \dfrac{1}{x^{1/3}}$

44. $\dfrac{1}{y^{1/4}}$

45. $(2rs)^{-3/4} = \dfrac{1}{(2rs)^{3/4}}$

46. $\dfrac{1}{(5xy)^{5/6}}$

47. $\left(\dfrac{1}{8}\right)^{-2/3} = \left(\dfrac{8}{1}\right)^{2/3} = (2^3)^{2/3} = 2^{\frac{3}{1} \cdot \frac{2}{3}} = 2^2 = 4$

48. 8

49. $\dfrac{1}{a^{-5/7}} = a^{5/7}$

50. $a^{3/5}$

51. $2a^{3/4}b^{-1/2}c^{2/3} = 2 \cdot a^{3/4} \cdot \dfrac{1}{b^{1/2}} \cdot c^{2/3} = \dfrac{2a^{3/4}c^{2/3}}{b^{1/2}}$

52. $\dfrac{5y^{4/5}z}{x^{2/3}}$

53. $2^{-1/3}x^4y^{-2/7} = \dfrac{1}{2^{1/3}} \cdot x^4 \cdot \dfrac{1}{y^{2/7}} = \dfrac{x^4}{2^{1/3}y^{2/7}}$

54. $\dfrac{a^3}{3^{5/2}b^{7/3}}$

55. $\left(\dfrac{7x}{8yx}\right)^{-3/5} = \left(\dfrac{8yz}{7x}\right)^{3/5}$ Finding the reciprocal of the base and changing the sign of the exponent

56. $\left(\dfrac{3c}{2ab}\right)^{5/6}$

57. $\dfrac{7x}{\sqrt[3]{z}} = \dfrac{7x}{z^{1/3}}$

58. $\dfrac{6a}{b^{1/4}}$

59. $\dfrac{5a}{3c^{-1/2}} = \dfrac{5a}{3} \cdot c^{1/2} = \dfrac{5ac^{1/2}}{3}$

60. $\dfrac{2x^{1/3}z}{5}$

61. $5^{3/4} \cdot 5^{1/8} = 5^{3/4+1/8} = 5^{6/8+1/8} = 5^{7/8}$

We added exponents after finding a common denominator.

62. $11^{7/6}$

63. $\dfrac{3^{5/8}}{3^{-1/8}} = 3^{5/8-(-1/8)} = 3^{5/8+1/8} = 3^{6/8} = 3^{3/4}$

We subtracted exponents and simplified.

64. $8^{9/11}$

65. $\dfrac{4.1^{-1/6}}{4.1^{-2/3}} = 4.1^{-1/6-(-2/3)} = 4.1^{-1/6+2/3} =$

$4.1^{-1/6+4/6} = 4.1^{3/6} = 4.1^{1/2}$

We subtracted exponents after finding a common denominator. Then we simplified.

66. $\dfrac{1}{2.3^{1/10}}$

67. $(10^{3/5})^{2/5} = 10^{3/5 \cdot 2/5} = 10^{6/25}$

We multiplied exponents.

68. $5^{15/28}$

69. $a^{2/3} \cdot a^{5/4} = a^{2/3+5/4} = a^{8/12+15/12} = a^{23/12}$

We added exponents after finding a common denominator.

70. $x^{17/12}$

71. $(64^{3/4})^{4/3} = 64^{\frac{3}{4} \cdot \frac{4}{3}} = 64^1 = 64$

72. $\dfrac{1}{27}$

73. $(m^{2/3}n^{-1/4})^{1/2} = m^{2/3 \cdot 1/2}n^{-1/4 \cdot 1/2} = m^{1/3}n^{-1/8} =$

$m^{1/3} \cdot \dfrac{1}{n^{1/8}} = \dfrac{m^{1/3}}{n^{1/8}}$

74. $\dfrac{y^{1/10}}{x^{1/12}}$

75. $\sqrt[6]{a^2} = a^{2/6}$ Converting to exponential notation

     $= a^{1/3}$ Simplifying the exponent

     $= \sqrt[3]{a}$ Returning to radical notation

76. $\sqrt[3]{t^2}$

77. $\sqrt[3]{x^{15}} = x^{15/3}$ Converting to exponential notation

     $= x^5$ Simplifying

78. a^3

79. $\sqrt[6]{x^{18}} = x^{18/6}$ Converting to exponential notation

     $= x^3$ Simplifying

80. a^2

81. $(\sqrt[3]{ab})^{15} = (ab)^{15/3}$ Converting to exponential notation

     $= (ab)^5$ Simplifying the exponent

     $= a^5b^5$ Using the law of exponents

82. x^2y^2

83. $\sqrt[8]{(3x)^2} = (3x)^{2/8}$ Converting to exponential notation

     $= (3x)^{1/4}$ Simplifying the exponent

     $= \sqrt[4]{3x}$ Returning to radical notation

84. $\sqrt{7a}$

85. $(\sqrt[10]{3a})^5 = (3a)^{5/10}$ Converting to exponential notation

     $= (3a)^{1/2}$ Simplifying the exponent

     $= \sqrt{3a}$ Returning to radical notation

86. $\sqrt[4]{8x^3}$

87. $\sqrt[4]{\sqrt{x}} = \sqrt[4]{x^{1/2}}$ Converting to

     $= (x^{1/2})^{1/4}$ exponential notation

     $= x^{1/8}$ Using a law of exponents

     $= \sqrt[8]{x}$ Returning to radical notation

88. $\sqrt[18]{m}$

89. $\sqrt{(ab)^6} = (ab)^{6/2}$ Converting to exponential notation

     $= (ab)^3$ Using the laws

     $= a^3b^3$ of exponents

90. x^3y^3

91. $(\sqrt[3]{x^2y^5})^{12} = (x^2y^5)^{12/3}$ Converting to exponential notation

     $= (x^2y^5)^4$ Simplifying the exponent

     $= x^8y^{20}$ Using the laws of exponents

92. $a^6 b^{12}$

93. $\sqrt[3]{\sqrt[4]{xy}} = \sqrt[3]{(xy)^{1/4}}$ Converting to

$\quad = [(xy)^{1/4}]^{1/3}$ exponential notation

$\quad = (xy)^{1/12}$ Using a law of exponents

$\quad = \sqrt[12]{xy}$ Returning to radical notation

94. $\sqrt[10]{2a}$

95. Writing exercise

96. Writing exercise

97. $\quad 3x(x^3 - 2x^2) + 4x^2(2x^2 + 5x)$

$\quad = 3x^4 - 6x^3 + 8x^4 + 20x^3$

$\quad = 11x^4 + 14x^3$

98. $-3t^6 + 28t^5 - 20t^4$

99. $\quad (3a - 4b)(5a + 3b)$

$\quad = 3a \cdot 5a + 3a \cdot 3b - 4b \cdot 5a - 4b \cdot 3b$

$\quad = 15a^2 + 9ab - 20ab - 12b^2$

$\quad = 15a^2 - 11ab - 12b^2$

100. $49x^2 - 14xy + y^2$

101. *Familiarize*. Let $p =$ the selling price of the home.

Translate.

$\underbrace{0.5\% \text{ of the selling price}}$ is \$467.50

$\quad\quad\quad \downarrow \quad\quad\quad\quad\quad \downarrow \quad \downarrow$

$\quad\quad\quad 0.005p \quad\quad\quad = \quad 467.50$

Carry out. We solve the equation.

$\quad 0.005p = 467.50$

$\quad\quad p = 93,500$ Dividing by 0.005

Check.   0.5% of \$93,500 is 0.005(\$93,500), or \$467.50. The answer checks.

State. The selling price of the home was \$93,500.

102. 0, 1

103. Writing exercise

104. Writing exercise

105. $\sqrt[5]{x^2 y \sqrt{xy}} = \sqrt[5]{x^2 y (xy)^{1/2}} = \sqrt[5]{x^2 y x^{1/2} y^{1/2}} =$

$\sqrt[5]{x^{5/2} y^{3/2}} = (x^{5/2} y^{3/2})^{1/5} = x^{5/10} y^{3/10} =$

$(x^5 y^3)^{1/10} = \sqrt[10]{x^5 y^3}$

106. $\sqrt[6]{x^5}$

107. $\sqrt[4]{\sqrt[3]{8x^3 y^6}} = \sqrt[4]{(2^3 x^3 y^6)^{1/3}} = \sqrt[4]{2^{3/3} x^{3/3} y^{6/3}} =$

$\sqrt[4]{2xy^2}$

108. $\sqrt[6]{p+q}$

109. $f(x) = 262 \cdot 2^{x/12}$

$\quad f(12) = 262 \cdot 2^{12/12}$

$\quad\quad\quad = 262 \cdot 2^1$

$\quad\quad\quad = 262 \cdot 2$

$\quad\quad\quad = 524$ cycles per second

110. 1760 cycles per second

111. $2^{7/12} \approx 1.498 \approx 1.5$ so the G that is 7 half steps above middle C has a frequency that is about 1.5 times that of middle C.

112. $2^{4/12} \approx 1.2599 \approx 1.25$ so the C sharp that is 4 half steps above concert A has a frequency that is 125% of, or 25% greater than, that of concert A.

113. a) $L = \dfrac{(0.000169)60^{2.27}}{1} \approx 1.8$ m

b) $L = \dfrac{(0.000169)75^{2.27}}{0.9906} \approx 3.1$ m

c) $L = \dfrac{(0.000169)80^{2.27}}{2.4} \approx 1.5$ m

d) $L = \dfrac{(0.000169)100^{2.27}}{1.1} \approx 5.3$ m

114. About 7.937×10^{-13} to 1

115. $m = m_0 (1 - v^2 c^{-2})^{1/2}$

$m = 8 \left[1 - \left(\dfrac{9}{5} \times 10^8 \right)^2 (3 \times 10^8)^{-2} \right]^{1/2}$

$\quad = 8 \left[1 - \dfrac{\left(\dfrac{9}{5} \times 10^8 \right)^2}{(3 \times 10^8)^2} \right]^{1/2}$

$\quad = 8 \left[1 - \dfrac{\dfrac{81}{25} \times 10^{16}}{9 \times 10^6} \right]^{1/2}$

$\quad = 8 \left[1 - \dfrac{81}{25} \cdot \dfrac{1}{9} \right]^{1/2}$

$\quad = 8 \left[1 - \dfrac{9}{25} \right]^{1/2}$

$\quad = 8 \left(\dfrac{16}{25} \right)^{1/2}$

$\quad = 8 \cdot \dfrac{4}{5} = \dfrac{32}{5}$

$\quad = 6.4$

The particle's new mass is 6.4 mg.

116. $y_1 = x^{1/2}$, $y_2 = 3x^{2/5}$,
$y_3 = x^{4/7}$, $y_4 = \frac{1}{5}x^{3/4}$

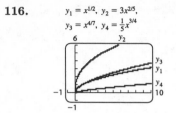

Exercise Set 10.3

1. $\sqrt{10}\sqrt{7} = \sqrt{10\cdot 7} = \sqrt{70}$

2. $\sqrt{35}$

3. $\sqrt[3]{2}\sqrt[3]{5} = \sqrt[3]{2\cdot 5} = \sqrt[3]{10}$

4. $\sqrt[3]{14}$

5. $\sqrt[4]{8}\sqrt[4]{9} = \sqrt[4]{8\cdot 9} = \sqrt[4]{72}$

6. $\sqrt[4]{18}$

7. $\sqrt{5a}\sqrt{6b} = \sqrt{5a\cdot 6b} = \sqrt{30ab}$

8. $\sqrt{26xy}$

9. $\sqrt[5]{9t^2}\sqrt[5]{2t} = \sqrt[5]{9t^2\cdot 2t} = \sqrt[5]{18t^3}$

10. $\sqrt[5]{80y^4}$

11. $\sqrt{x-a}\sqrt{x+a} = \sqrt{(x-a)(x+a)} = \sqrt{x^2 - a^2}$

12. $\sqrt{y^2 - b^2}$

13. $\sqrt[3]{0.5x}\sqrt[3]{0.2x} = \sqrt[3]{0.5x\cdot 0.2x} = \sqrt[3]{0.1x^2}$

14. $\sqrt[3]{0.21y^2}$

15. $\sqrt[4]{x-1}\sqrt[4]{x^2+x+1} = \sqrt[4]{(x-1)(x^2+x+1)} = \sqrt[4]{x^3 - 1}$

16. $\sqrt[5]{(x-2)^3}$

17. $\sqrt{\dfrac{x}{6}}\sqrt{\dfrac{7}{y}} = \sqrt{\dfrac{x}{6}\cdot\dfrac{7}{y}} = \sqrt{\dfrac{7x}{6y}}$

18. $\sqrt{\dfrac{7s}{11t}}$

19. $\sqrt[7]{\dfrac{x-3}{4}}\sqrt[7]{\dfrac{5}{x+2}} = \sqrt[7]{\dfrac{x-3}{4}\cdot\dfrac{5}{x+2}} = \sqrt[7]{\dfrac{5x-15}{4x+8}}$

20. $\sqrt[6]{\dfrac{3a}{b^2 - 4}}$

21. $\sqrt{50}$

$= \sqrt{25\cdot 2}$ 25 is the largest perfect square factor of 50.

$= \sqrt{25}\cdot\sqrt{2}$

$= 5\sqrt{2}$

22. $3\sqrt{3}$

23. $\sqrt{28}$

$= \sqrt{4\cdot 7}$ 4 is the largest perfect square factor of 28.

$= \sqrt{4}\cdot\sqrt{7}$

$= 2\sqrt{7}$

24. $3\sqrt{5}$

25. $\sqrt{8} = \sqrt{4\cdot 2} = \sqrt{4}\cdot\sqrt{2} = 2\sqrt{2}$

26. $3\sqrt{2}$

27. $\sqrt{198} = \sqrt{9\cdot 22} = \sqrt{9}\cdot\sqrt{22} = 3\sqrt{22}$

28. $5\sqrt{13}$

29. $\sqrt{36a^4 b}$

$= \sqrt{36a^4\cdot b}$ $36a^4$ is a perfect square.

$= \sqrt{36a^4}\cdot\sqrt{b}$ Factoring into two radicals

$= 6a^2\sqrt{b}$ Taking the square root of $36a^4$

30. $5y^4\sqrt{7}$

31. $\sqrt[3]{8x^3y^2}$

$= \sqrt[3]{8x^3\cdot y^2}$ $8x^3$ is a perfect cube.

$= \sqrt[3]{8x^3}\cdot\sqrt[3]{y^2}$ Factoring into two radicals

$= 2x\sqrt[3]{y^2}$ Taking the cube root of $8x^3$

32. $3b^2\sqrt[3]{a}$

33. $\sqrt[3]{-16x^6}$

$= \sqrt[3]{-8x^6\cdot 2}$ $-8x^6$ is a perfect cube.

$= \sqrt[3]{-8x^6}\cdot\sqrt[3]{2}$

$= -2x^2\sqrt[3]{2}$ Taking the cube root of $-8x^6$

34. $-2a^2\sqrt[3]{4}$

35. $f(x) = \sqrt[3]{125x^5}$

$= \sqrt[3]{125x^3\cdot x^2}$

$= \sqrt[3]{125x^3}\cdot\sqrt[3]{x^2}$

$= 5x\sqrt[3]{x^2}$

36. $2x^2\sqrt[3]{2}$

37. $f(x) = \sqrt{49(x+5)^2}$ $\quad 49(x+5)^2$ is a perfect square.

$\quad\quad = |7(x+5)|$, or $7|x+5|$

38. $9|x-1|$

39. $f(x) = \sqrt{5x^2 - 10x + 5}$

$\quad\quad = \sqrt{5(x^2 - 2x + 1)}$

$\quad\quad = \sqrt{5(x-1)^2}$

$\quad\quad = \sqrt{(x-1)^2} \cdot \sqrt{5}$

$\quad\quad = |x-1|\sqrt{5}$

40. $|x+2|\sqrt{2}$

41. $\quad \sqrt{a^3 b^4}$

$\quad = \sqrt{a^2 \cdot a \cdot b^4}\quad$ Identifying the largest even powers of a and b

$\quad = \sqrt{a^2}\sqrt{b^4}\sqrt{a}\quad$ Factoring into several radicals

$\quad = ab^2\sqrt{a}$

42. $x^3 y^4\sqrt{y}$

43. $\quad \sqrt[3]{x^5 y^6 z^{10}}$

$\quad = \sqrt[3]{x^3 \cdot x^2 \cdot y^6 \cdot z^9 \cdot z}\quad$ Identifying the largest perfect-cube powers of x, y, and z

$\quad = \sqrt[3]{x^3} \cdot \sqrt[3]{y^6} \cdot \sqrt[3]{z^9} \cdot \sqrt[3]{x^2 z}\quad$ Factoring into several radicals

$\quad = xy^2 z^3 \sqrt[3]{x^2 z}$

44. $a^2 b^2 c^4 \sqrt[3]{bc}$

45. $\sqrt[5]{-32a^7 b^{11}} = \sqrt[5]{-32 \cdot a^5 \cdot a^2 \cdot b^{10} \cdot b} = \sqrt[5]{-32}\sqrt[5]{a^5}\sqrt[5]{b^{10}}\sqrt[5]{a^2 b} = -2ab^2\sqrt[5]{a^2 b}$

46. $2xy^2\sqrt[4]{xy^3}$

47. $\sqrt[5]{a^6 b^8 c^9} = \sqrt[5]{a^5 \cdot a \cdot b^5 \cdot b^3 \cdot c^5 \cdot c^4} = \sqrt[5]{a^5}\sqrt[5]{b^5}\sqrt[5]{c^5}\sqrt[5]{ab^3 c^4} = abc\sqrt[5]{ab^3 c^4}$

48. $x^2 yz^3 \sqrt[5]{x^3 y^3 z^2}$

49. $\sqrt[4]{810x^9} = \sqrt[4]{81 \cdot 10 \cdot x^8 \cdot x} = \sqrt[4]{81} \cdot \sqrt[4]{x^8} \cdot \sqrt[4]{10x} = 3x^2\sqrt[4]{10x}$

50. $-2a^4\sqrt[3]{10a^2}$

51. $\sqrt{15}\sqrt{5} = \sqrt{15 \cdot 5} = \sqrt{75} = \sqrt{25 \cdot 3} = 5\sqrt{3}$

52. $3\sqrt{2}$

53. $\sqrt{10}\sqrt{14} = \sqrt{10 \cdot 14} = \sqrt{140} = \sqrt{4 \cdot 35} = 2\sqrt{35}$

54. $3\sqrt{35}$

55. $\sqrt[3]{2}\sqrt[3]{4} = \sqrt[3]{2 \cdot 4} = \sqrt[3]{8} = 2$

56. 3

57. $\sqrt{18a^3}\sqrt{18a^3} = \sqrt{(18a^3)^2} = 18a^3$

58. $75x^7$

59. $\sqrt[3]{5a^2}\sqrt[3]{2a} = \sqrt[3]{5a^2 \cdot 2a} = \sqrt[3]{10a^3} = \sqrt[3]{a^3 \cdot 10} = a\sqrt[3]{10}$

60. $x\sqrt[3]{21}$

61. $\sqrt{3x^5}\sqrt{15x^2} = \sqrt{45x^7} = \sqrt{9x^6 \cdot 5x} = 3x^3\sqrt{5x}$

62. $5a^5\sqrt{3}$

63. $\sqrt[3]{s^2 t^4}\sqrt[3]{s^4 t^6} = \sqrt[3]{s^6 t^{10}} = \sqrt[3]{s^6 t^9 \cdot t} = s^2 t^3 \sqrt[3]{t}$

64. $xy^3\sqrt[3]{xy}$

65. $\sqrt[3]{(x+5)^2}\sqrt[3]{(x+5)^4} = \sqrt[3]{(x+5)^6} = (x+5)^2$

66. $(a-b)^4$

67. $\sqrt[4]{12a^3 b^7}\sqrt[4]{4a^2 b^5} = \sqrt[4]{48a^5 b^{12}} = \sqrt[4]{16a^4 b^{12} \cdot 3a} = 2ab^3\sqrt[4]{3a}$

68. $3x^2 y^2 \sqrt[4]{xy^3}$

69. $\sqrt[5]{x^3(y+z)^4}\sqrt[5]{x^3(y+z)^6} = \sqrt[5]{x^6(y+z)^{10}} = \sqrt[5]{x^5(y+z)^{10} \cdot x} = x(y+z)^2\sqrt[5]{x}$

70. $a^2(b-c)\sqrt[5]{(b-c)^3}$

71. Writing exercise

72. Writing exercise

73. $\quad \dfrac{3x}{16y} + \dfrac{5y}{64x}$, LCD is $64xy$

$\quad = \dfrac{3x}{16y} \cdot \dfrac{4x}{4x} + \dfrac{5y}{64x} \cdot \dfrac{y}{y}$

$\quad = \dfrac{12x^2}{64xy} + \dfrac{5y^2}{64xy}$

$\quad = \dfrac{12x^2 + 5y^2}{64xy}$

74. $\dfrac{2a + 6b^3}{a^4 b^4}$

75.
$$\frac{4}{x^2 - 9} - \frac{7}{2x - 6}$$

$$= \frac{4}{(x+3)(x-3)} - \frac{7}{2(x-3)},$$
$$\text{LCD is } 2(x+3)(x-3)$$

$$= \frac{4}{(x+3)(x-3)} \cdot \frac{2}{2} - \frac{7}{2(x-3)} \cdot \frac{x+3}{x+3}$$

$$= \frac{8}{2(x+3)(x-3)} - \frac{7(x+3)}{2(x+3)(x-3)}$$

$$= \frac{8 - 7(x+3)}{2(x+3)(x-3)}$$

$$= \frac{8 - 7x - 21}{2(x+3)(x-3)}$$

$$= \frac{-7x - 13}{2(x+3)(x-3)}$$

76. $\dfrac{-3x + 1}{2(x+5)(x-5)}$

77. $\dfrac{9a^4 b^7}{3a^2 b^5} = \dfrac{9}{3} a^{4-2} b^{7-5} = 3a^2 b^2$

78. $3ab^5$

79. Writing exercise

80. Writing exercise

81. $r(L) = 2\sqrt{5L}$

a) $r(L) = 2\sqrt{5 \cdot 20}$
$$= 2\sqrt{100}$$
$$= 2 \cdot 10 = 20 \text{ mph}$$

b) $r(L) = 2\sqrt{5 \cdot 70}$
$$= 2\sqrt{350}$$
$$\approx 37.4 \text{ mph}\quad \text{Multiplying and rounding}$$

c) $r(L) = 2\sqrt{5 \cdot 90}$
$$= 2\sqrt{450}$$
$$\approx 42.4 \text{ mph}\quad \text{Multiplying and rounding}$$

82. a) $-3.3°$C; b) $-16.6°$C; c) $-25.5°$C; d) $-54.0°$C

83. $(\sqrt{r^3 t})^7 = \sqrt{(r^3 t)^7} = \sqrt{r^{21} t^7} =$
$$\sqrt{r^{20} \cdot r \cdot t^6 \cdot t} = \sqrt{r^{20}} \sqrt{t^6} \sqrt{rt} = r^{10} t^3 \sqrt{rt}$$

84. $25x^5 \sqrt[3]{25x}$

85. $(\sqrt[3]{a^2 b^4})^5 = \sqrt[3]{(a^2 b^4)^5} = \sqrt[3]{a^{10} b^{20}} =$
$$\sqrt[3]{a^9 \cdot a \cdot b^{18} \cdot b^2} = \sqrt[3]{a^9} \sqrt[3]{b^{18}} \sqrt[3]{ab^2} = a^3 b^6 \sqrt[3]{ab^2}$$

86. $a^{10} b^{17} \sqrt{ab}$

87.

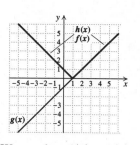

We see that $f(x) = h(x)$ and $f(x) \neq g(x)$.

88.

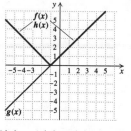

$$f(x) = h(x); \; f(x) \neq g(x)$$

89. $f(t) = \sqrt{t^2 - 3t - 4}$

We must have $t^2 - 3t - 4 \geq 0$, or $(t-4)(t+1) \geq 0$.
We graph $y = t^2 - 3t - 4$.

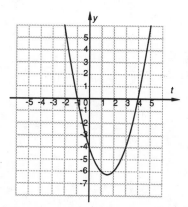

From the graph we see that $y \geq 0$ for $t \leq -1$ or $t \geq 4$, so the domain of f is $\{t | t \leq -1 \text{ or } t \geq 4\}$, or $(-\infty, -1] \cup [4, \infty)$.

90. $\{x | x \leq 2 \text{ or } x \geq 4\}$, or $(-\infty, 2] \cup [4, \infty)$

91. $\sqrt[3]{5x^{k+1}}\sqrt[3]{25x^k} = 5x^7$

$\sqrt[3]{5x^{k+1} \cdot 25x^k} = 5x^7$

$\sqrt[3]{125x^{2k+1}} = 5x^7$

$\sqrt[3]{125}\sqrt[3]{x^{2k+1}} = 5x^7$

$5\sqrt[3]{x^{2k+1}} = 5x^7$

$\sqrt[3]{x^{2k+1}} = x^7$

$(x^{2k+1})^{1/3} = x^7$

$x^{\frac{2k+1}{3}} = x^7$

Since the base is the same, the exponents must be equal. We have:

$\dfrac{2k+1}{3} = 7$

$2k + 1 = 21$

$2k = 20$

$k = 10$

92. 6

93.

94. Writing exercise

Exercise Set 10.4

1. $\sqrt{\dfrac{25}{36}} = \dfrac{\sqrt{25}}{\sqrt{36}} = \dfrac{5}{6}$

2. $\dfrac{10}{9}$

3. $\sqrt[3]{\dfrac{64}{27}} = \dfrac{\sqrt[3]{64}}{\sqrt[3]{27}} = \dfrac{4}{3}$

4. $\dfrac{7}{10}$

5. $\sqrt{\dfrac{49}{y^2}} = \dfrac{\sqrt{49}}{\sqrt{y^2}} = \dfrac{7}{y}$

6. $\dfrac{11}{x}$

7. $\sqrt{\dfrac{25y^3}{x^4}} = \dfrac{\sqrt{25y^3}}{\sqrt{x^4}} = \dfrac{\sqrt{25y^2 \cdot y}}{\sqrt{x^4}} = \dfrac{\sqrt{25y^2}\sqrt{y}}{\sqrt{x^4}} =$

$\dfrac{5y\sqrt{y}}{x^2}$

8. $\dfrac{6a^2\sqrt{a}}{b^3}$

9. $\sqrt[3]{\dfrac{27a^4}{8b^3}} = \dfrac{\sqrt[3]{27a^4}}{\sqrt[3]{8b^3}} = \dfrac{\sqrt[3]{27a^3 \cdot a}}{\sqrt[3]{8b^3}} = \dfrac{\sqrt[3]{27a^3}\sqrt[3]{a}}{\sqrt[3]{8b^3}} =$

$\dfrac{3a\sqrt[3]{a}}{2b}$

10. $\dfrac{2x^2\sqrt[3]{x}}{3y^2}$

11. $\sqrt[4]{\dfrac{16a^4}{b^4c^8}} = \dfrac{\sqrt[4]{16a^4}}{\sqrt[4]{b^4c^8}} = \dfrac{2a}{bc^2}$

12. $\dfrac{3x}{y^2z}$

13. $\sqrt[4]{\dfrac{a^5b^8}{c^{10}}} = \dfrac{\sqrt[4]{a^5b^8}}{\sqrt[4]{c^{10}}} = \dfrac{\sqrt[4]{a^4b^8 \cdot a}}{\sqrt[4]{c^8 \cdot c^2}} = \dfrac{\sqrt[4]{a^4b^8}\sqrt[4]{a}}{\sqrt[4]{c^8}\sqrt[4]{c^2}} =$

$\dfrac{ab^2\sqrt[4]{a}}{c^2\sqrt[4]{c^2}}$, or $\dfrac{ab^2}{c^2}\sqrt[4]{\dfrac{a}{c^2}}$

14. $\dfrac{x^2y^3}{z}\sqrt[4]{\dfrac{x}{z^2}}$

15. $\sqrt[5]{\dfrac{32x^6}{y^{11}}} = \dfrac{\sqrt[5]{32x^6}}{\sqrt[5]{y^{11}}} = \dfrac{\sqrt[5]{32x^5 \cdot x}}{\sqrt[5]{y^{10} \cdot y}} = \dfrac{\sqrt[5]{32x^5}\sqrt[5]{x}}{\sqrt[5]{y^{10}}\sqrt[5]{y}} =$

$\dfrac{2x\sqrt[5]{x}}{y^2\sqrt[5]{y}}$, or $\dfrac{2x}{y^2}\sqrt[5]{\dfrac{x}{y}}$

16. $\dfrac{3a}{b^2}\sqrt[5]{\dfrac{a^4}{b^3}}$

17. $\sqrt[6]{\dfrac{x^6y^8}{z^{15}}} = \dfrac{\sqrt[6]{x^6y^8}}{\sqrt[6]{z^{15}}} = \dfrac{\sqrt[6]{x^6y^6 \cdot y^2}}{\sqrt[6]{z^{12} \cdot z^3}} = \dfrac{\sqrt[6]{x^6y^6}\sqrt[6]{y^2}}{\sqrt[6]{z^{12}}\sqrt[6]{z^3}} =$

$\dfrac{xy\sqrt[6]{y^2}}{z^2\sqrt[6]{z^3}}$, or $\dfrac{xy}{z^2}\sqrt[6]{\dfrac{y^2}{z^3}}$

18. $\dfrac{ab^2}{c^2}\sqrt[6]{\dfrac{a^3}{c}}$

19. $\dfrac{\sqrt{35x}}{\sqrt{7x}} = \sqrt{\dfrac{35x}{7x}} = \sqrt{5}$

20. $\sqrt{7}$

21. $\dfrac{\sqrt[3]{270}}{\sqrt[3]{10}} = \sqrt[3]{\dfrac{270}{10}} = \sqrt[3]{27} = 3$

22. 2

23. $\dfrac{\sqrt{40xy^3}}{\sqrt{8x}} = \sqrt{\dfrac{40xy^3}{8x}} = \sqrt{5y^3} = \sqrt{y^2 \cdot 5y} =$

$\sqrt{y^2}\sqrt{5y} = y\sqrt{5y}$

24. $2b\sqrt{2b}$

25. $\dfrac{\sqrt[3]{96a^4b^2}}{\sqrt[3]{12a^2b}} = \sqrt[3]{\dfrac{96a^4b^2}{12a^2b}} = \sqrt[3]{8a^2b} = \sqrt[3]{8}\,\sqrt[3]{a^2b} =$
 $2\sqrt[3]{a^2b}$

26. $3xy\sqrt[3]{y^2}$

27. $\dfrac{\sqrt{100ab}}{5\sqrt{2}} = \dfrac{1}{5}\dfrac{\sqrt{100ab}}{\sqrt{2}} = \dfrac{1}{5}\sqrt{\dfrac{100ab}{2}} = \dfrac{1}{5}\sqrt{50ab} =$
 $\dfrac{1}{5}\sqrt{25\cdot 2ab} = \dfrac{1}{5}\cdot 5\sqrt{2ab} = \sqrt{2ab}$

28. $\dfrac{5}{3}\sqrt{ab}$

29. $\dfrac{\sqrt[4]{48x^9y^{13}}}{\sqrt[4]{3xy^{-2}}} = \sqrt[4]{\dfrac{48x^9y^{13}}{3xy^{-2}}} = \sqrt[4]{16x^8y^{15}} =$
 $\sqrt[4]{16x^8y^{12}}\,\sqrt[4]{y^3} = 2x^2y^3\sqrt[4]{y^3}$

30. $2a^2b^6$

31. $\dfrac{\sqrt[3]{x^3-y^3}}{\sqrt[3]{x-y}} = \sqrt[3]{\dfrac{x^3-y^3}{x-y}} =$
 $\sqrt[3]{\dfrac{(x-y)(x^2+xy+y^2)}{x-y}} =$
 $\sqrt[3]{\dfrac{(x\!-\!y)(x^2+xy+y^2)}{x\!-\!y}} = \sqrt[3]{x^2+xy+y^2}$

32. $\sqrt[3]{r^2-rs+s^2}$

33. $\sqrt{\dfrac{5}{7}} = \sqrt{\dfrac{5}{7}\cdot\dfrac{7}{7}} = \sqrt{\dfrac{35}{49}} = \dfrac{\sqrt{35}}{\sqrt{49}} = \dfrac{\sqrt{35}}{7}$

34. $\dfrac{\sqrt{66}}{6}$

35. $\dfrac{6\sqrt{5}}{5\sqrt{3}} = \dfrac{6\sqrt{5}}{5\sqrt{3}}\cdot\dfrac{\sqrt{3}}{\sqrt{3}} = \dfrac{6\sqrt{15}}{5\cdot 3} = \dfrac{2\sqrt{15}}{5}$

36. $\dfrac{2\sqrt{10}}{3}$

37. $\sqrt[3]{\dfrac{16}{9}} = \sqrt[3]{\dfrac{16}{9}\cdot\dfrac{3}{3}} = \sqrt[3]{\dfrac{48}{27}} = \dfrac{\sqrt[3]{8\cdot 6}}{\sqrt[3]{27}} = \dfrac{2\sqrt[3]{6}}{3}$

38. $\dfrac{\sqrt[3]{6}}{3}$

39. $\dfrac{\sqrt[3]{3a}}{\sqrt[3]{5c}} = \dfrac{\sqrt[3]{3a}}{\sqrt[3]{5c}}\cdot\dfrac{\sqrt[3]{5^2c^2}}{\sqrt[3]{5^2c^2}} = \dfrac{\sqrt[3]{75ac^2}}{\sqrt[3]{5^3c^3}} = \dfrac{\sqrt[3]{75ac^2}}{5c}$

40. $\dfrac{\sqrt[3]{63xy^2}}{3y}$

41. $\dfrac{\sqrt[3]{5y^4}}{\sqrt[3]{6x^4}} = \dfrac{\sqrt[3]{5y^4}}{\sqrt[3]{6x^4}}\cdot\dfrac{\sqrt[3]{36x^2}}{\sqrt[3]{36x^2}} = \dfrac{\sqrt[3]{y^3\cdot 180x^2y}}{\sqrt[3]{216x^6}} =$
 $\dfrac{y\sqrt[3]{180x^2y}}{6x^2}$

42. $\dfrac{a\sqrt[3]{147ab}}{7b}$

43. $\sqrt[3]{\dfrac{2}{x^2y}} = \sqrt[3]{\dfrac{2}{x^2y}\cdot\dfrac{xy^2}{xy^2}} = \sqrt[3]{\dfrac{2xy^2}{x^3y^3}} = \dfrac{\sqrt[3]{2xy^2}}{\sqrt[3]{x^3y^3}} =$
 $\dfrac{\sqrt[3]{2xy^2}}{xy}$

44. $\dfrac{\sqrt[3]{5a^2b}}{ab}$

45. $\sqrt{\dfrac{7a}{18}} = \sqrt{\dfrac{7a}{18}\cdot\dfrac{2}{2}} = \sqrt{\dfrac{14a}{36}} = \dfrac{\sqrt{14a}}{\sqrt{36}} = \dfrac{\sqrt{14a}}{6}$

46. $\dfrac{\sqrt{30x}}{10}$

47. $\sqrt{\dfrac{9}{20x^2y}} = \sqrt{\dfrac{9}{20x^2y}\cdot\dfrac{5y}{5y}} = \sqrt{\dfrac{9\cdot 5y}{100x^2y^2}} =$
 $\dfrac{\sqrt{9\cdot 5y}}{\sqrt{100x^2y^2}} = \dfrac{3\sqrt{5y}}{10xy}$

48. $\dfrac{\sqrt{14b}}{8ab}$

49. $\sqrt{\dfrac{10ab^2}{72a^3b}} = \sqrt{\dfrac{5b}{36a^2}} = \dfrac{\sqrt{5b}}{6a}$

50. $\dfrac{\sqrt{7x}}{5y^2}$

51. $\dfrac{\sqrt{5}}{\sqrt{7x}} = \dfrac{\sqrt{5}}{\sqrt{7x}}\cdot\dfrac{\sqrt{5}}{\sqrt{5}} = \dfrac{\sqrt{25}}{\sqrt{35x}} = \dfrac{5}{\sqrt{35x}}$

52. $\dfrac{10}{\sqrt{30x}}$

53. $\sqrt{\dfrac{14}{21}} = \sqrt{\dfrac{2}{3}} = \sqrt{\dfrac{2}{3}\cdot\dfrac{2}{2}} = \sqrt{\dfrac{4}{6}} = \dfrac{\sqrt{4}}{\sqrt{6}} = \dfrac{2}{\sqrt{6}}$

54. $\dfrac{2}{\sqrt{5}}$

55. $\dfrac{4\sqrt{13}}{3\sqrt{7}} = \dfrac{4\sqrt{13}}{3\sqrt{7}}\cdot\dfrac{\sqrt{13}}{\sqrt{13}} = \dfrac{4\sqrt{169}}{3\sqrt{91}} = \dfrac{4\cdot 13}{3\sqrt{91}} = \dfrac{52}{3\sqrt{91}}$

56. $\dfrac{105}{2\sqrt{105}}$

57. $\dfrac{\sqrt[3]{7}}{\sqrt[3]{2}} = \dfrac{\sqrt[3]{7}}{\sqrt[3]{2}}\cdot\dfrac{\sqrt[3]{7^2}}{\sqrt[3]{7^2}} = \dfrac{\sqrt[3]{7^3}}{\sqrt[3]{98}} = \dfrac{7}{\sqrt[3]{98}}$

58. $\dfrac{5}{\sqrt[3]{100}}$

59. $\sqrt{\dfrac{7x}{3y}} = \sqrt{\dfrac{7x}{3y} \cdot \dfrac{7x}{7x}} = \dfrac{\sqrt{(7x)^2}}{\sqrt{21xy}} = \dfrac{7x}{\sqrt{21xy}}$

60. $\dfrac{6a}{\sqrt{30ab}}$

61. $\sqrt[3]{\dfrac{2a^5}{5b}} = \sqrt[3]{\dfrac{2a^5}{5b} \cdot \dfrac{4a}{4a}} = \sqrt[3]{\dfrac{8a^6}{20ab}} = \dfrac{2a^2}{\sqrt[3]{20ab}}$

62. $\dfrac{2a^2}{\sqrt[3]{28a^2b}}$

63. $\sqrt{\dfrac{x^3y}{2}} = \sqrt{\dfrac{x^3y}{2} \cdot \dfrac{xy}{xy}} = \sqrt{\dfrac{x^4y^2}{2xy}} = \dfrac{\sqrt{x^4y^2}}{\sqrt{2xy}} = \dfrac{x^2y}{\sqrt{2xy}}$

64. $\dfrac{ab^3}{\sqrt{3ab}}$

65. Writing exercise

66. Writing exercise

67. $\dfrac{3}{x-5} \cdot \dfrac{x-1}{x+5} = \dfrac{3(x-1)}{(x-5)(x+5)}$

68. $\dfrac{7(x-2)}{(x+4)(x-4)}$

69. $\dfrac{a^2 - 8a + 7}{a^2 - 49} = \dfrac{(a-1)(a-7)}{(a+7)(a-7)}$
$\phantom{\dfrac{a^2 - 8a + 7}{a^2 - 49}} = \dfrac{(a-1)(a\!\!\!\!\diagup 7)}{(a+7)(a\!\!\!\!\diagup 7)}$
$\phantom{\dfrac{a^2 - 8a + 7}{a^2 - 49}} = \dfrac{a-1}{a+7}$

70. $\dfrac{t+11}{t+2}$

71. $(5a^3b^4)^3 = 5^3(a^3)^3(b^4)^3 = 125a^{3\cdot3}b^{4\cdot3} = 125a^9b^{12}$

72. $225x^{10}y^6$

73. Writing exercise

74. Writing exercise

75. a) $T = 2\pi\sqrt{\dfrac{65}{980}} \approx 1.62$ sec

b) $T = 2\pi\sqrt{\dfrac{98}{980}} \approx 1.99$ sec

c) $T = 2\pi\sqrt{\dfrac{120}{980}} \approx 2.20$ sec

76. a^3bxy^2

77. $\dfrac{(\sqrt[3]{81mn^2})^2}{(\sqrt[3]{mn})^2} = \dfrac{\sqrt[3]{(81mn^2)^2}}{\sqrt[3]{(mn)^2}}$
$\phantom{\dfrac{(\sqrt[3]{81mn^2})^2}{(\sqrt[3]{mn})^2}} = \dfrac{\sqrt[3]{6561m^2n^4}}{\sqrt[3]{m^2n^2}}$
$\phantom{\dfrac{(\sqrt[3]{81mn^2})^2}{(\sqrt[3]{mn})^2}} = \sqrt[3]{\dfrac{6561m^2n^4}{m^2n^2}}$
$\phantom{\dfrac{(\sqrt[3]{81mn^2})^2}{(\sqrt[3]{mn})^2}} = \sqrt[3]{6561n^2}$
$\phantom{\dfrac{(\sqrt[3]{81mn^2})^2}{(\sqrt[3]{mn})^2}} = \sqrt[3]{729 \cdot 9n^2}$
$\phantom{\dfrac{(\sqrt[3]{81mn^2})^2}{(\sqrt[3]{mn})^2}} = \sqrt[3]{729}\,\sqrt[3]{9n^2}$
$\phantom{\dfrac{(\sqrt[3]{81mn^2})^2}{(\sqrt[3]{mn})^2}} = 9\sqrt[3]{9n^2}$

78. $2yz\sqrt{2z}$

79. $\sqrt{a^2 - 3} - \dfrac{a^2}{\sqrt{a^2 - 3}}$
$= \sqrt{a^2 - 3} - \dfrac{a^2}{\sqrt{a^2 - 3}} \cdot \dfrac{\sqrt{a^2 - 3}}{\sqrt{a^2 - 3}}$
$= \sqrt{a^2 - 3} - \dfrac{a^2\sqrt{a^2 - 3}}{a^2 - 3}$
$= \sqrt{a^2 - 3} \cdot \dfrac{a^2 - 3}{a^2 - 3} - \dfrac{a^2\sqrt{a^2 - 3}}{a^2 - 3}$
$= \dfrac{a^2\sqrt{a^2 - 3} - 3\sqrt{a^2 - 3} - a^2\sqrt{a^2 - 3}}{a^2 - 3}$
$= \dfrac{-3\sqrt{a^2 - 3}}{a^2 - 3}$, or $\dfrac{-3}{\sqrt{a^2 - 3}}$

80. $\dfrac{(5x + 4y - 3)\sqrt{xy}}{xy}$

81. Step 1: $\sqrt[n]{x} = x^{1/n}$, by definition;

Step 2: $\left(\dfrac{x}{y}\right)^n = \dfrac{x^n}{y^n}$, raising a quotient to a power;

Step 3: $x^{1/n} = \sqrt[n]{x}$, by definition

82. A number c is the nth root of a/b if $c^n = a/b$. Let $c = \sqrt[n]{a}/\sqrt[n]{b}$.
$$c^n = \left(\dfrac{\sqrt[n]{a}}{\sqrt[n]{b}}\right)^n = \left(\dfrac{a^{1/n}}{b^{1/n}}\right)^n = \dfrac{(a^{1/n})^n}{(b^{1/n})^n} = \dfrac{a}{b}$$

83. $f(x) = \sqrt{18x^3}$, $g(x) = \sqrt{2x}$

$(f/g)(x) = \dfrac{f(x)}{g(x)} = \dfrac{\sqrt{18x^3}}{\sqrt{2x}} = \sqrt{\dfrac{18x^3}{2x}} = \sqrt{9x^2} = 3x$

$\sqrt{2x}$ is defined for $2x \geq 0$, or $x \geq 0$. To avoid division by 0, we must exclude 0 from the domain. Thus, the domain of $f/g = \{x | x$ is a real number and $x > 0\}$, or $(0, \infty)$.

84. $(f/g)(t) = \dfrac{1}{5t}$, where t is a real number and $t > 0$

85. $f(x) = \sqrt{x^2 - 9}$, $g(x) = \sqrt{x - 3}$

$(f/g)(x) = \dfrac{f(x)}{g(x)} = \dfrac{\sqrt{x^2 - 9}}{\sqrt{x - 3}} = \sqrt{\dfrac{x^2 - 9}{x - 3}} =$

$\sqrt{\dfrac{(x+3)(x-3)}{x - 3}} = \sqrt{x + 3}$

$\sqrt{x - 3}$ is defined for $x - 3 \geq 0$, or $x \geq 3$. To avoid division by 0 we must exclude 3 from the domain. Thus, the domain of $f/g = \{x | x$ is a real number and $x > 3\}$, or $(3, \infty)$.

Exercise Set 10.5

1. $3\sqrt{7} + 2\sqrt{7} = (3 + 2)\sqrt{7} = 5\sqrt{7}$

2. $17\sqrt{5}$

3. $9\sqrt[3]{5} - 6\sqrt[3]{5} = (9 - 6)\sqrt[3]{5} = 3\sqrt[3]{5}$

4. $8\sqrt[5]{2}$

5. $4\sqrt[3]{y} + 9\sqrt[3]{y} = (4 + 9)\sqrt[3]{y} = 13\sqrt[3]{y}$

6. $6\sqrt[4]{t}$

7. $8\sqrt{2} - 6\sqrt{2} + 5\sqrt{2} = (8 - 6 + 5)\sqrt{2} = 7\sqrt{2}$

8. $7\sqrt{6}$

9. $9\sqrt[3]{7} - \sqrt{3} + 4\sqrt[3]{7} + 2\sqrt{3} =$
$(9 + 4)\sqrt[3]{7} + (-1 + 2)\sqrt{3} = 13\sqrt[3]{7} + \sqrt{3}$

10. $6\sqrt{7} + \sqrt[4]{11}$

11. $\quad 8\sqrt{27} - 3\sqrt{3}$
$= 8\sqrt{9 \cdot 3} - 3\sqrt{3}$ ⠀Factoring the
$= 8\sqrt{9} \cdot \sqrt{3} - 3\sqrt{3}$ ⠀first radical
$= 8 \cdot 3\sqrt{3} - 3\sqrt{3}$ ⠀Taking the square root of 9
$= 24\sqrt{3} - 3\sqrt{3}$
$= 21\sqrt{3}$ ⠀Combining like radicals

12. $41\sqrt{2}$

13. $\quad 3\sqrt{45} + 7\sqrt{20}$
$= 3\sqrt{9 \cdot 5} + 7\sqrt{4 \cdot 5}$ ⠀Factoring the
$= 3\sqrt{9} \cdot \sqrt{5} + 7\sqrt{4} \cdot \sqrt{5}$ ⠀radicals
$= 3 \cdot 3\sqrt{5} + 7 \cdot 2\sqrt{5}$ ⠀Taking the square roots
$= 9\sqrt{5} + 14\sqrt{5}$
$= 23\sqrt{5}$ ⠀Combining like radicals

14. $58\sqrt{3}$

15. $3\sqrt[3]{16} + \sqrt[3]{54} = 3\sqrt[3]{8 \cdot 2} + \sqrt[3]{27 \cdot 2} =$
$3\sqrt[3]{8} \cdot \sqrt[3]{2} + \sqrt[3]{27} \cdot \sqrt[3]{2} = 3 \cdot 2\sqrt[3]{2} + 3\sqrt[3]{2} =$
$6\sqrt[3]{2} + 3\sqrt[3]{2} = 9\sqrt[3]{2}$

16. -7

17. $\sqrt{5a} + 2\sqrt{45a^3} = \sqrt{5a} + 2\sqrt{9a^2 \cdot 5a} =$
$\sqrt{5a} + 2\sqrt{9a^2} \cdot \sqrt{5a} = \sqrt{5a} + 2 \cdot 3a\sqrt{5a} =$
$\sqrt{5a} + 6a\sqrt{5a} = (1 + 6a)\sqrt{5a}$

18. $(4x - 2)\sqrt{3x}$

19. $\sqrt[3]{6x^4} + \sqrt[3]{48x} = \sqrt[3]{x^3 \cdot 6x} + \sqrt[3]{8 \cdot 6x} =$
$\sqrt[3]{x^3} \cdot \sqrt[3]{6x} + \sqrt[3]{8} \cdot \sqrt[3]{6x} = x\sqrt[3]{6x} + 2\sqrt[3]{6x} =$
$(x + 2)\sqrt[3]{6x}$

20. $(3 - x)\sqrt[3]{2x}$

21. $\sqrt{4a - 4} + \sqrt{a - 1} = \sqrt{4(a - 4)} + \sqrt{a - 1} =$
$\sqrt{4}\sqrt{a - 1} + \sqrt{a - 1} = 2\sqrt{a - 1} + \sqrt{a - 1} = 3\sqrt{a - 1}$

22. $4\sqrt{y + 3}$

23. $\sqrt{x^3 - x^2} + \sqrt{9x - 9} = \sqrt{x^2(x - 1)} + \sqrt{9(x - 1)} =$
$\sqrt{x^2} \cdot \sqrt{x - 1} + \sqrt{9} \cdot \sqrt{x - 1} =$
$x\sqrt{x - 1} + 3\sqrt{x - 1} = (x + 3)\sqrt{x - 1}$

24. $(2 - x)\sqrt{x - 1}$

25. $\sqrt{7}(3 - \sqrt{7}) = \sqrt{7} \cdot 3 - \sqrt{7} \cdot \sqrt{7} = 3\sqrt{7} - 7$

26. $4\sqrt{3} + 3$

27. $4\sqrt{2}(\sqrt{3} - \sqrt{5}) = 4\sqrt{2} \cdot \sqrt{3} - 4\sqrt{2} \cdot \sqrt{5} = 4\sqrt{6} - 4\sqrt{10}$

28. $15 - 3\sqrt{10}$

29. $\sqrt{3}(2\sqrt{5} - 3\sqrt{4}) = \sqrt{3}(2\sqrt{5} - 3 \cdot 2) =$
$\sqrt{3} \cdot 2\sqrt{5} - \sqrt{3} \cdot 6 = 2\sqrt{15} - 6\sqrt{3}$

30. $6\sqrt{5} - 4$

31. $\sqrt[3]{2}(\sqrt[3]{4} - 2\sqrt[3]{32}) = \sqrt[3]{2} \cdot \sqrt[3]{4} - \sqrt[3]{2} \cdot 2\sqrt[3]{32} =$
$\sqrt[3]{8} - 2\sqrt[3]{64} = 2 - 2 \cdot 4 = 2 - 8 = -6$

32. $3 - 4\sqrt[3]{63}$

33. $\sqrt[3]{a}(\sqrt[3]{a^2} + \sqrt[3]{24a^2}) = \sqrt[3]{a} \cdot \sqrt[3]{a^2} + \sqrt[3]{a}\sqrt[3]{24a^2} =$
$\sqrt[3]{a^3} + \sqrt[3]{24a^3} = \sqrt[3]{a^3} + \sqrt[3]{8a^3 \cdot 3} =$
$a + 2a\sqrt[3]{3}$

34. $-2x\sqrt[3]{3}$

35. $(5+\sqrt{6})(5-\sqrt{6}) = 5^2 - (\sqrt{6})^2 = 25 - 6 = 19$

36. -1

37. $(3-2\sqrt{7})(3+2\sqrt{7}) = 3^2 - (2\sqrt{7})^2 = 9 - 4 \cdot 7 =$
$9 - 28 = -19$

38. -2

39. $(3+\sqrt{5})^2 = 3^2 + 2 \cdot 3 \cdot \sqrt{5} + (\sqrt{5})^2 = 9 + 6\sqrt{5} + 5 =$
$14 + 6\sqrt{5}$

40. $52 + 14\sqrt{3}$

41. $(2\sqrt{7} - 4\sqrt{2})(3\sqrt{7} + 6\sqrt{2}) =$
$2\sqrt{7} \cdot 3\sqrt{7} + 2\sqrt{7} \cdot 6\sqrt{2} - 4\sqrt{2} \cdot 3\sqrt{7} - 4\sqrt{2} \cdot 6\sqrt{2} =$
$6 \cdot 7 + 12\sqrt{14} - 12\sqrt{14} - 24 \cdot 2 =$
$42 + 12\sqrt{14} - 12\sqrt{14} - 48 = -6$

42. $24 - 7\sqrt{15}$

43. $(2\sqrt[3]{3} - \sqrt[3]{2})(\sqrt[3]{3} + 2\sqrt[3]{2}) =$
$2\sqrt[3]{3} \cdot \sqrt[3]{3} + 2\sqrt[3]{3} \cdot 2\sqrt[3]{2} - \sqrt[3]{2} \cdot \sqrt[3]{3} - \sqrt[3]{2} \cdot 2\sqrt[3]{2} =$
$2\sqrt[3]{9} + 4\sqrt[3]{6} - \sqrt[3]{6} - 2\sqrt[3]{4} = 2\sqrt[3]{9} + 3\sqrt[3]{6} - 2\sqrt[3]{4}$

44. $6\sqrt[4]{63} - 9\sqrt[4]{42} + 2\sqrt[4]{54} - 3\sqrt[4]{36}$

45. $(\sqrt{3x} + \sqrt{y})^2$
$= (\sqrt{3x})^2 + 2 \cdot \sqrt{3x} \cdot \sqrt{y} + (\sqrt{y})^2$ Squaring a binomial
$= 3x + 2\sqrt{3xy} + y$

46. $t - 2\sqrt{2rt} + 2r$

47. $\dfrac{2}{3+\sqrt{5}} = \dfrac{2}{3+\sqrt{5}} \cdot \dfrac{3-\sqrt{5}}{3-\sqrt{5}} =$
$\dfrac{2(3-\sqrt{5})}{(3+\sqrt{5})(3-\sqrt{5})} = \dfrac{6-2\sqrt{5}}{3^2 - (\sqrt{5})^2} =$
$\dfrac{6-2\sqrt{5}}{9-5} = \dfrac{6-2\sqrt{5}}{4} = \dfrac{2(3-\sqrt{5})}{2 \cdot 2} =$
$\dfrac{3-\sqrt{5}}{2}$

48. $\dfrac{4+\sqrt{7}}{3}$

49. $\dfrac{2+\sqrt{5}}{6-\sqrt{3}} = \dfrac{2+\sqrt{5}}{6-\sqrt{3}} \cdot \dfrac{6+\sqrt{3}}{6+\sqrt{3}} =$
$\dfrac{(2+\sqrt{5})(6+\sqrt{3})}{(6-\sqrt{3})(6+\sqrt{3})} = \dfrac{12+2\sqrt{3}+6\sqrt{5}+\sqrt{15}}{36-3} =$
$\dfrac{12+2\sqrt{3}+6\sqrt{5}+\sqrt{15}}{33}$

50. $\dfrac{3-\sqrt{5}+3\sqrt{2}-\sqrt{10}}{4}$

51. $\dfrac{\sqrt{a}}{\sqrt{a}+\sqrt{b}} = \dfrac{\sqrt{a}}{\sqrt{a}+\sqrt{b}} \cdot \dfrac{\sqrt{a}-\sqrt{b}}{\sqrt{a}-\sqrt{b}} =$
$\dfrac{\sqrt{a}(\sqrt{a}-\sqrt{b})}{(\sqrt{a}+\sqrt{b})(\sqrt{a}-\sqrt{b})} = \dfrac{a-\sqrt{ab}}{a-b}$

52. $\dfrac{\sqrt{xz}+z}{x-z}$

53. $\dfrac{\sqrt{7}-\sqrt{3}}{\sqrt{3}-\sqrt{7}} = \dfrac{-1(\sqrt{3}-\sqrt{7})}{\sqrt{3}-\sqrt{7}} = -1 \cdot \dfrac{\sqrt{3}-\sqrt{7}}{\sqrt{3}-\sqrt{7}} =$
$-1 \cdot 1 = -1$

54. $\dfrac{\sqrt{35}-\sqrt{14}+5-\sqrt{10}}{3}$

55. $\dfrac{3\sqrt{2}-\sqrt{7}}{4\sqrt{2}+\sqrt{5}} = \dfrac{3\sqrt{2}-\sqrt{7}}{4\sqrt{2}+\sqrt{5}} \cdot \dfrac{4\sqrt{2}-\sqrt{5}}{4\sqrt{2}-\sqrt{5}} =$
$\dfrac{(3\sqrt{2}-\sqrt{7})(4\sqrt{2}-\sqrt{5})}{(4\sqrt{2}+\sqrt{5})(4\sqrt{2}-\sqrt{5})} =$
$\dfrac{12 \cdot 2 - 3\sqrt{10} - 4\sqrt{14} + \sqrt{35}}{16 \cdot 2 - 5} =$
$\dfrac{24 - 3\sqrt{10} - 4\sqrt{14} + \sqrt{35}}{32 - 5} =$
$\dfrac{24 - 3\sqrt{10} - 4\sqrt{14} + \sqrt{35}}{27}$

56. $\dfrac{30 + 25\sqrt{6} - 2\sqrt{33} - 5\sqrt{22}}{-38}$, or
$\dfrac{-30 - 25\sqrt{6} + 2\sqrt{33} + 5\sqrt{22}}{38}$

57. $\dfrac{5\sqrt{3}-3\sqrt{2}}{3\sqrt{2}-2\sqrt{3}} = \dfrac{5\sqrt{3}-3\sqrt{2}}{3\sqrt{2}-2\sqrt{3}} \cdot \dfrac{3\sqrt{2}+2\sqrt{3}}{3\sqrt{2}+2\sqrt{3}} =$
$\dfrac{15\sqrt{6} + 10 \cdot 3 - 9 \cdot 2 - 6\sqrt{6}}{9 \cdot 2 - 4 \cdot 3} =$
$\dfrac{15\sqrt{6} + 30 - 18 - 6\sqrt{6}}{18 - 12} = \dfrac{9\sqrt{6} + 12}{6} =$
$\dfrac{3(3\sqrt{6} + 4)}{3 \cdot 2} = \dfrac{3\sqrt{6} + 4}{2}$

58. $\dfrac{4\sqrt{6}+9}{3}$

59. $\dfrac{\sqrt{7}+2}{5} = \dfrac{\sqrt{7}+2}{5} \cdot \dfrac{\sqrt{7}-2}{\sqrt{7}-2} =$
$\dfrac{(\sqrt{7}+2)(\sqrt{7}-2)}{5(\sqrt{7}-2)} = \dfrac{(\sqrt{7})^2 - 2^2}{5\sqrt{7} - 10} =$
$\dfrac{7-4}{5\sqrt{7}-10} = \dfrac{3}{5\sqrt{7}-10}$

60. $\dfrac{1}{2\sqrt{3}-2}$

61. $\dfrac{\sqrt{6}-2}{\sqrt{3}+7}=\dfrac{\sqrt{6}-2}{\sqrt{3}+7}\cdot\dfrac{\sqrt{6}+2}{\sqrt{6}+2}=$

$\dfrac{(\sqrt{6}-2)(\sqrt{6}+2)}{(\sqrt{3}+7)(\sqrt{6}+2)}=\dfrac{6-4}{\sqrt{18}+2\sqrt{3}+7\sqrt{6}+14}=$

$\dfrac{2}{3\sqrt{2}+2\sqrt{3}+7\sqrt{6}+14}$

62. $\dfrac{-6}{2\sqrt{5}-4\sqrt{2}-3\sqrt{10}+12}$, or $\dfrac{6}{-2\sqrt{5}+4\sqrt{2}+3\sqrt{10}-12}$

63. $\dfrac{\sqrt{x}-\sqrt{y}}{\sqrt{x}+\sqrt{y}}=\dfrac{\sqrt{x}-\sqrt{y}}{\sqrt{x}+\sqrt{y}}\cdot\dfrac{\sqrt{x}+\sqrt{y}}{\sqrt{x}+\sqrt{y}}=$

$\dfrac{(\sqrt{x}-\sqrt{y})(\sqrt{x}+\sqrt{y})}{(\sqrt{x}+\sqrt{y})(\sqrt{x}+\sqrt{y})}=\dfrac{x-y}{x+2\sqrt{xy}+y}$

64. $\dfrac{a-b}{a-2\sqrt{ab}+b}$

65. $\sqrt{a}\,\sqrt[4]{a^3}$

$= a^{1/2}\cdot a^{3/4}$ Converting to exponential notation

$= a^{5/4}$ Adding exponents

$= a^{1+1/4}$ Writing 5/4 as a mixed number

$= a\cdot a^{1/4}$ Factoring

$= a\sqrt[4]{a}$ Returning to radical notation

66. $x\sqrt{x}$

67. $\sqrt[5]{b^2}\sqrt{b^3}$

$= b^{2/5}\cdot b^{3/2}$ Converting to exponential notation

$= b^{19/10}$ Adding exponents

$= b^{1+9/10}$ Writing 19/10 as a mixed number

$= b\cdot b^{9/10}$ Factoring

$= b\sqrt[10]{b^9}$ Returning to radical notation

68. $a\sqrt[12]{a^5}$

69. $\sqrt{xy^3}\sqrt[3]{x^2y} = (xy^3)^{1/2}(x^2y)^{1/3}$

$= (xy^3)^{3/6}(x^2y)^{2/6}$

$= [(xy^3)^3(x^2y)^2]^{1/6}$

$= \sqrt[6]{x^3y^9\cdot x^4y^2}$

$= \sqrt[6]{x^7y^{11}}$

$= \sqrt[6]{x^6y^6\cdot xy^5}$

$= xy\sqrt[6]{xy^5}$

70. $a\sqrt[10]{ab^7}$

71. $\sqrt[4]{9ab^3}\sqrt{3a^4b} = (9ab^3)^{1/4}(3a^4b)^{1/2}$

$= (9ab^3)^{1/4}(3a^4b)^{2/4}$

$= [(9ab^3)(3a^4b)^2]^{1/4}$

$= \sqrt[4]{9ab^3\cdot 9a^8b^2}$

$= \sqrt[4]{81a^9b^5}$

$= \sqrt[4]{81a^8b^4\cdot ab}$

$= 3a^2b\sqrt[4]{ab}$

72. $2xy^2\sqrt[6]{2x^5y}$

73. $\sqrt[3]{xy^2z}\sqrt{x^3yz^2} = (xy^2z)^{1/3}(x^3yz^2)^{1/2}$

$= (xy^2z)^{2/6}(x^3yz^2)^{3/6}$

$= [(xy^2z)^2(x^3yz^2)^3]^{1/6}$

$= \sqrt[6]{x^2y^4z^2\cdot x^9y^3z^6}$

$= \sqrt[6]{x^{11}y^7z^8}$

$= \sqrt[6]{x^6y^6z^6\cdot x^5yz^2}$

$= xyz\sqrt[6]{x^5yz^2}$

74. $a^2b^2c^2\sqrt[6]{a^2bc^2}$

75. $\dfrac{\sqrt[3]{x^2}}{\sqrt[5]{x}}$

$= \dfrac{x^{2/3}}{x^{1/5}}$ Converting to exponential notation

$= x^{2/3-1/5}$ Subtracting exponents

$= x^{7/15}$ Converting back

$= \sqrt[15]{x^7}$ to radical notation

76. $\sqrt[12]{a^5}$

77. $\dfrac{\sqrt[5]{a^4b}}{\sqrt[3]{ab}}$

$= \dfrac{(a^4b)^{1/5}}{(ab)^{1/3}}$ Converting to exponential notation

$= \dfrac{a^{4/5}b^{1/5}}{a^{1/3}b^{1/3}}$ Using the product and power rules

$= a^{4/5-1/3}b^{1/5-1/3}$ Subtracting exponents

$= a^{7/15}b^{-2/15}$

$= (a^7b^{-2})^{1/15}$ Converting back

$= \sqrt[15]{a^7b^{-2}}$, or   to radical notation

$\sqrt[15]{\dfrac{a^7}{b^2}}$

78. $\sqrt[12]{x^2 y^5}$

79. $\dfrac{\sqrt[5]{x^3 y^4}}{\sqrt{xy}}$

$= \dfrac{(x^3 y^4)^{1/5}}{(xy)^{1/2}}$ Converting to exponential notation

$= \dfrac{x^{3/5} y^{4/5}}{x^{1/2} y^{1/2}}$

$= x^{3/5 - 1/2} y^{4/5 - 1/2}$ Subtracting exponents

$= x^{1/10} y^{3/10}$

$= (xy^3)^{1/10}$ Converting back to

$= \sqrt[10]{xy^3}$ radical notation

80. $\sqrt[10]{ab^9}$

81. $\dfrac{\sqrt[3]{(2+5x)^2}}{\sqrt[4]{2+5x}}$

$= \dfrac{(2+5x)^{2/3}}{(2+5x)^{1/4}}$ Converting to exponential notation

$= (2+5x)^{2/3 - 1/4}$ Subtracting exponents

$= (2+5x)^{5/12}$ Converting back to

$= \sqrt[12]{(2+5x)^5}$ radical notation

82. $\sqrt[20]{(3x-1)^3}$

83. $\dfrac{\sqrt[4]{(5+3x)^3}}{\sqrt[3]{(5+3x)^2}}$

$= \dfrac{(5+3x)^{3/4}}{(5+3x)^{2/3}}$ Converting to exponential notation

$= (5+3x)^{3/4 - 2/3}$ Subtracting exponents

$= (5+3x)^{1/12}$ Converting back

$= \sqrt[12]{5+3x}$ to radical notation

84. $\sqrt[15]{(2x+1)^4}$

85. $\sqrt[3]{x^2 y}\left(\sqrt{xy} - \sqrt[5]{xy^3}\right)$

$= (x^2 y)^{1/3} \left[(xy)^{1/2} - (xy^3)^{1/5}\right]$

$= x^{2/3} y^{1/3} \left(x^{1/2} y^{1/2} - x^{1/5} y^{3/5}\right)$

$= x^{2/3} y^{1/3} x^{1/2} y^{1/2} - x^{2/3} y^{1/3} x^{1/5} y^{3/5}$

$= x^{2/3 + 1/2} y^{1/3 + 1/2} - x^{2/3 + 1/5} y^{1/3 + 3/5}$

$= x^{7/6} y^{5/6} - x^{13/15} y^{14/15}$

$= x^{1\frac{1}{6}} y^{\frac{5}{6}} - x^{13/15} y^{14/15}$

                 Writing a mixed numeral

$= x \cdot x^{1/6} y^{5/6} - x^{13/15} y^{14/15}$

$= x(xy^5)^{1/6} - (x^{13} y^{14})^{1/15}$

$= x \sqrt[6]{xy^5} - \sqrt[15]{x^{13} y^{14}}$

86. $a\sqrt[12]{a^2 b^7} - \sqrt[20]{a^{18} b^{13}}$

87. $(m + \sqrt[3]{n^2})(2m + \sqrt[4]{n})$

$= (m + n^{2/3})(2m + n^{1/4})$ Converting to exponential notation

$= 2m^2 + mn^{1/4} + 2mn^{2/3} + n^{2/3} n^{1/4}$ Using FOIL

$= 2m^2 + mn^{1/4} + 2mn^{2/3} + n^{2/3 + 1/4}$ Adding exponents

$= 2m^2 + mn^{1/4} + 2mn^{2/3} + n^{11/12}$

$= 2m^2 + m\sqrt[4]{n} + 2m\sqrt[3]{n^2} + \sqrt[12]{n^{11}}$ Converting back to radical notation

88. $3r^2 - r\sqrt[5]{s} - 3r\sqrt[4]{s^3} + \sqrt[20]{s^{19}}$

89. $f(x) = \sqrt[4]{x}, \ g(x) = \sqrt[4]{2x} - \sqrt[4]{x^{11}}$

$(f \cdot g)(x) = \sqrt[4]{x}\left(\sqrt[4]{2x} - \sqrt[4]{x^{11}}\right)$

$= \sqrt[4]{2x^2} - \sqrt[4]{x^{12}}$

$= \sqrt[4]{2x^2} - x^3$

90. $x^2 + \sqrt[4]{3x^3}$

91. $f(x) = x + \sqrt{7}, \ g(x) = x - \sqrt{7}$

$(f \cdot g)(x) = (x + \sqrt{7})(x - \sqrt{7})$

$= x^2 - (\sqrt{7})^2$

$= x^2 - 7$

92. $x^2 + x\sqrt{6} - x\sqrt{2} - 2\sqrt{3}$

93. $f(x) = x^2$

$f(5 - \sqrt{2}) = (5 - \sqrt{2})^2 = 25 - 10\sqrt{2} + (\sqrt{2})^2 =$
$25 - 10\sqrt{2} + 2 = 27 - 10\sqrt{2}$

94. $52 + 14\sqrt{3}$

95. $f(x) = x^2$

$f(\sqrt{3} + \sqrt{5}) = (\sqrt{3} + \sqrt{5})^2 =$
$(\sqrt{3})^2 + 2 \cdot \sqrt{3} \cdot \sqrt{5} + (\sqrt{5})^2 =$
$3 + 2\sqrt{15} + 5 = 8 + 2\sqrt{15}$

96. $9 - 6\sqrt{2}$

97. Writing exercise

98. Writing exercise

99.

$$\frac{12x}{x-4} - \frac{3x^2}{x+4} = \frac{384}{x^2 - 16}$$

$$\frac{12x}{x-4} - \frac{3x^2}{x+4} = \frac{384}{(x+4)(x-4)},$$

$$\text{LCM is } (x+4)(x-4).$$

Note that $x \neq -4$ and $x \neq 4$.

$$(x+4)(x-4)\left[\frac{12x}{x-4} - \frac{3x^2}{x+4}\right] =$$

$$(x+4)(x-4) \cdot \frac{384}{(x+4)(x-4)}$$

$$12x(x+4) - 3x^2(x-4) = 384$$

$$12x^2 + 48x - 3x^3 + 12x^2 = 384$$

$$-3x^3 + 24x^2 + 48x - 384 = 0$$

$$-3(x^3 - 8x^2 - 16x + 128) = 0$$

$$-3[x^2(x-8) - 16(x-8)] = 0$$

$$-3(x-8)(x^2 - 16) = 0$$

$$-3(x-8)(x+4)(x-4) = 0$$

$$x - 8 = 0 \text{ or } x + 4 = 0 \text{ or } x - 4 = 0$$

$$x = 8 \text{ or } \quad x = -4 \text{ or } \quad x = 4$$

Check: For 8:

$$\frac{12x}{x-4} - \frac{3x^2}{x+4} = \frac{384}{x^2-16}$$

$$\begin{array}{c|c} \dfrac{12 \cdot 8}{8-4} - \dfrac{3 \cdot 8^2}{8+4} & \dfrac{384}{8^2 - 16} \\[2mm] \dfrac{96}{4} - \dfrac{192}{12} & \dfrac{384}{48} \\[2mm] 24 - 16 & 8 \\[2mm] 8 & \text{TRUE} \end{array}$$

8 is a solution.

For -4:

$$\frac{12x}{x-4} - \frac{3x^2}{x+4} = \frac{384}{x^2-16}$$

$$\begin{array}{c|c} \dfrac{12(-4)}{-4-4} - \dfrac{3(-4)^2}{-4+4} & \dfrac{384}{(-4)^2 - 16} \\[2mm] \dfrac{-48}{-8} - \dfrac{48}{0} & \dfrac{384}{16-16} \quad \text{UNDEFINED} \end{array}$$

-4 is not a solution.

For 4:

$$\frac{12x}{x-4} - \frac{3x^2}{x+4} = \frac{384}{x^2-16}$$

$$\begin{array}{c|c} \dfrac{12 \cdot 4}{4-4} - \dfrac{3 \cdot 4^2}{4+4} & \dfrac{384}{4^2 - 16} \\[2mm] \dfrac{48}{0} - \dfrac{48}{8} & \dfrac{384}{16-16} \quad \text{UNDEFINED} \end{array}$$

4 is not a solution.

The checks confirm that -4 and 4 are not solutions. The solution is 8.

100. $\dfrac{15}{2}$

101. *Familiarize*. Let x and y represent the width and length of the rectangle, respectively.

Translate. We write two equations.

$$\underbrace{\text{The width}}_{} \ \underbrace{\text{is}}_{} \ \underbrace{\text{one-fourth}}_{} \ \underbrace{\text{the length.}}_{}$$
$$\downarrow \qquad \downarrow \qquad \downarrow \qquad \downarrow$$
$$x \qquad = \qquad \frac{1}{4} \cdot \qquad y$$

$$\underbrace{\text{The area}}_{} \ \underbrace{\text{is}}_{} \ \underbrace{\text{twice}}_{} \ \underbrace{\text{the perimeter.}}_{}$$
$$\downarrow \qquad \downarrow \qquad \downarrow \qquad \downarrow$$
$$xy \qquad = \qquad 2 \cdot \qquad (2x + 2y)$$

Carry out. Solving the system of equations we get (5,20).

Check. The width, 5, is one-fourth the length, 20. The area is $5 \cdot 20$, or 100. The perimeter is $2 \cdot 5 + 2 \cdot 20$, or 50. Since $100 = 2 \cdot 50$, the area is twice the perimeter. The values check.

State. The width is 5, and the length is 20.

102. $-5, 4$

103.
$$5x^2 - 6x + 1 = 0$$

$$(5x - 1)(x - 1) = 0$$

$$5x - 1 = 0 \text{ or } x - 1 = 0$$

$$5x = 1 \text{ or } \qquad x = 1$$

$$x = \frac{1}{5} \text{ or } \qquad x = 1$$

The solutions are $\dfrac{1}{5}$ and 1.

104. $\dfrac{1}{7}, 1$

105. Writing exercise

106. Writing exercise

107. To add radical expressions, the <u>indices</u> and the <u>radicands</u> must be the same.

108. indices

109. To add rational expressions, the <u>denominators</u> must be the same.

110. bases

111. $f(x) = \sqrt{20x^2 + 4x^3} - 3x\sqrt{45 + 9x} + \sqrt{5x^2 + x^3}$
$= \sqrt{4x^2(5+x)} - 3x\sqrt{9(5+x)} + \sqrt{x^2(5+x)}$
$= \sqrt{4x^2}\sqrt{5+x} - 3x\sqrt{9}\sqrt{5+x} + \sqrt{x^2}\sqrt{5+x}$
$= 2x\sqrt{5+x} - 3x \cdot 3\sqrt{5+x} + x\sqrt{5+x}$
$= 2x\sqrt{5+x} - 9x\sqrt{5+x} + x\sqrt{5+x}$
$= -6x\sqrt{5+x}$

112. $2x\sqrt{x-1}$

113. $f(x) = \sqrt[4]{x^5 - x^4} + 3\sqrt[4]{x^9 - x^8}$
$= \sqrt[4]{x^4(x-1)} + 3\sqrt[4]{x^8(x-1)}$
$= \sqrt[4]{x^4} \cdot \sqrt[4]{x-1} + 3\sqrt[4]{x^8}\sqrt[4]{x-1}$
$= x\sqrt[4]{x-1} + 3x^2\sqrt[4]{x-1}$
$= (x + 3x^2)\sqrt[4]{x-1}$

114. $(2x - 2x^2)\sqrt[4]{1+x}$

115. $\frac{1}{2}\sqrt{36a^5bc^4} - \frac{1}{2}\sqrt[3]{64a^4bc^6} + \frac{1}{6}\sqrt{144a^3bc^6} =$

$\frac{1}{2}\sqrt{36a^4c^4 \cdot ab} - \frac{1}{2}\sqrt[3]{64a^3c^6 \cdot ab} + \frac{1}{6}\sqrt{144a^2c^6 \cdot ab} =$

$\frac{1}{2}(6a^2c^2)\sqrt{ab} - \frac{1}{2}(4ac^2)\sqrt[3]{ab} + \frac{1}{6}(12ac^3)\sqrt{ab} =$

$3a^2c^2\sqrt{ab} - 2ac^2\sqrt[3]{ab} + 2ac^3\sqrt{ab}$

$(3a^2c^2 + 2ac^3)\sqrt{ab} - 2ac^2\sqrt[3]{ab}$, or

$ac^2[(3a + 2c)\sqrt{ab} - 2\sqrt[3]{ab}]$

116. $(7x^2 - 2y^2)\sqrt{x+y}$

117. $\quad \sqrt{27a^5(b+1)}\sqrt[3]{81a(b+1)^4}$
$= [27a^5(b+1)]^{1/2}[81a(b+1)^4]^{1/3}$
$= [27a^5(b+1)]^{3/6}[81a(b+1)^4]^{2/6}$
$= \{[3^3a^5(b+1)]^3[3^4a(b+1)^4]^2\}^{1/6}$
$= \sqrt[6]{3^9a^{15}(b+1)^3 \cdot 3^8a^2(b+1)^8}$
$= \sqrt[6]{3^{17}a^{17}(b+1)^{11}}$
$= \sqrt[6]{3^{12}a^{12}(b+1)^6 \cdot 3^5a^5(b+1)^5}$
$= 3^2a^2(b+1)\sqrt[6]{3^5a^5(b+1)^5}$, or
$\quad 9a^2(b+1)\sqrt[6]{243a^5(b+1)^5}$

118. $4x(y+z)^3\sqrt[6]{2x(y+z)}$

119. $\dfrac{\dfrac{1}{\sqrt{w}} - \sqrt{w}}{\dfrac{\sqrt{w}+1}{\sqrt{w}}} = \dfrac{\dfrac{1}{\sqrt{w}} - \sqrt{w}}{\dfrac{\sqrt{w}+1}{\sqrt{w}}} \cdot \dfrac{\sqrt{w}}{\sqrt{w}} = \dfrac{1-w}{\sqrt{w}+1} =$

$\dfrac{1-w}{\sqrt{w}+1} \cdot \dfrac{\sqrt{w}-1}{\sqrt{w}-1} = \dfrac{\sqrt{w}-1-w\sqrt{w}+w}{w-1} =$

$\dfrac{(w-1) - \sqrt{w}(w-1)}{w-1} = \dfrac{(w-1)(1-\sqrt{w})}{w-1} =$

$1 - \sqrt{w}$

120. $\dfrac{7\sqrt{3}}{39}$

121. $x - 5 = (\sqrt{x})^2 - (\sqrt{5})^2 = (\sqrt{x} + \sqrt{5})(\sqrt{x} - \sqrt{5})$

122. $(\sqrt{y} + \sqrt{7})(\sqrt{y} - \sqrt{7})$

123. $x - a = (\sqrt{x})^2 - (\sqrt{a})^2 = (\sqrt{x} + \sqrt{a})(\sqrt{x} - \sqrt{a})$

124. 6

125. $(\sqrt{x+2} - \sqrt{x-2})^2 =$
$x + 2 - 2\sqrt{(x+2)(x-2)} + x - 2 =$
$x + 2 - 2\sqrt{x^2 - 4} + x - 2 = 2x - 2\sqrt{x^2 - 4}$

126. $\dfrac{ab + (a-b)\sqrt{a+b} - a - b}{a + b - b^2}$

127. $\dfrac{b + \sqrt{b}}{1 + b + \sqrt{b}} = \dfrac{b + \sqrt{b}}{(1+b) + \sqrt{b}} \cdot \dfrac{(1+b) - \sqrt{b}}{(1+b) - \sqrt{b}}$

$\qquad = \dfrac{(b + \sqrt{b})(1 + b - \sqrt{b})}{(1+b)^2 - (\sqrt{b})^2}$

$\qquad = \dfrac{b + b^2 - b\sqrt{b} + \sqrt{b} + b\sqrt{b} - b}{1 + 2b + b^2 - b}$

$\qquad = \dfrac{b^2 + \sqrt{b}}{1 + b + b^2}$

128. $\dfrac{1}{\sqrt{y+18} + \sqrt{y}}$

129. $\dfrac{\sqrt{x+6} - 5}{\sqrt{x+6} + 5} = \dfrac{\sqrt{x+6} - 5}{\sqrt{x+6} + 5} \cdot \dfrac{\sqrt{x+6} + 5}{\sqrt{x+6} + 5}$

$\qquad = \dfrac{(x+6) - 25}{(x+6) + 10\sqrt{x+6} + 25}$

$\qquad = \dfrac{x - 19}{x + 10\sqrt{x+6} + 31}$

130.

Exercise Set 10.6

1. $\sqrt{x+3} = 5$

   $(\sqrt{x+3})^2 = 5^2$ Principle of powers (squaring)

   $x + 3 = 25$

   $x = 22$

   Check: $\dfrac{\sqrt{x+3} = 5}{\sqrt{22+3} \;?\; 5}$
   $\phantom{\sqrt{22+3} \;?\;}\sqrt{25}$
   $\phantom{\sqrt{22+3} \;?\;}5 \;\Big|\; 5 \qquad$ TRUE

   The solution is 22.

2. $\dfrac{63}{5}$

3. $\sqrt{2x} - 1 = 2$

   $\sqrt{2x} = 3$ Adding to isolate the radical

   $(\sqrt{2x})^2 = 3^2$ Principle of powers (squaring)

   $2x = 9$

   $x = \dfrac{9}{2}$

   Check: $\dfrac{\sqrt{2x} - 1 = 2}{\;}$
   $\sqrt{2 \cdot \dfrac{9}{2} - 1} \;?\; 2$
   $\phantom{\sqrt{2 \cdot }}\sqrt{9} - 1$
   $\phantom{\sqrt{2 \cdot }}3 - 1$
   $\phantom{\sqrt{2 \cdot }}2 \;\Big|\; 2 \qquad$ TRUE

   The solution is $\dfrac{9}{2}$.

4. $\dfrac{25}{3}$

5. $\sqrt{x-2} - 7 = -4$

   $\sqrt{x-2} = 3$ Adding to isolate the radical

   $(\sqrt{x-2})^2 = 3^2$ Principle of powers (squaring)

   $x - 2 = 9$

   $x = 11$

   Check: $\dfrac{\sqrt{x-2} - 7 = -4}{\sqrt{11-2} - 7 \;?\; -4}$
   $\phantom{\sqrt{11-2} - }\sqrt{9} - 7$
   $\phantom{\sqrt{11-2} - }3 - 7$
   $\phantom{\sqrt{11-2} - }-4 \;\Big|\; -4 \qquad$ TRUE

   The solution is 11.

6. 168

7. $\sqrt{y+4} + 6 = 7$

   $\sqrt{y+4} = 1$ Adding to isolate the radical

   $(\sqrt{y+4})^2 = 1^2$ Principle of powers (squaring)

   $y + 4 = 1$

   $y = -3$

   Check: $\dfrac{\sqrt{y+4} + 6 = 7}{\sqrt{-3+4} + 6 \;?\; 7}$
   $\phantom{\sqrt{-3+4} + }\sqrt{1} + 6$
   $\phantom{\sqrt{-3+4} + }1 + 6$
   $\phantom{\sqrt{-3+4} + }7 \;\Big|\; 7 \qquad$ TRUE

   The solution is -3.

8. 56

9. $\sqrt[3]{x-2} = 3$

   $(\sqrt[3]{x-2})^3 = 3^3$

   $x - 2 = 27$

   $x = 29$

   Check: $\dfrac{\sqrt[3]{x-2} = 3}{\sqrt[3]{29-2} \;?\; 3}$
   $\phantom{\sqrt[3]{29-2} \;?\;}\sqrt[3]{27}$
   $\phantom{\sqrt[3]{29-2} \;?\;}3 \;\Big|\; 3 \qquad$ TRUE

   The solution is 29.

10. 3

11. $\sqrt[4]{x+3} = 2$

   $(\sqrt[4]{x+3})^4 = 2^4$

   $x + 3 = 16$

   $x = 13$

   Check: $\dfrac{\sqrt[4]{x+3} = 2}{\sqrt[4]{13+3} \;?\; 2}$
   $\phantom{\sqrt[4]{13+3} \;?\;}\sqrt[4]{16}$
   $\phantom{\sqrt[4]{13+3} \;?\;}2 \;\Big|\; 2 \qquad$ TRUE

   The solution is 13.

12. 82

13. $8\sqrt{y} = y$

   $(8\sqrt{y})^2 = y^2$

   $64y = y^2$

   $0 = y^2 - 64y$

   $0 = y(y - 64)$

   $y = 0 \;$ or $\; y - 64 = 0$

   $y = 0 \;$ or $\; y = 64$

Check:

For 0: $8\sqrt{y} = y$

$$\frac{}{8\sqrt{0} \ ? \ 0}$$
$$8 \cdot 0$$
$$0 \ | \ 0 \qquad \text{TRUE}$$

For 64: $8\sqrt{y} = y$

$$\frac{}{8\sqrt{64} \ ? \ 64}$$
$$8 \cdot 8$$
$$64 \ | \ 64 \qquad \text{TRUE}$$

The solutions are 0 and 64.

14. 0, 9

15. $3x^{1/2} + 12 = 9$

$$3\sqrt{x} + 12 = 9$$
$$3\sqrt{x} = -3$$
$$\sqrt{x} = -1$$

Since the principal square root is never negative, this equation has no solution.

16. 64

17. $\sqrt[3]{y} = -4$

$$(\sqrt[3]{y})^3 = (-4)^3$$
$$y = -64$$

Check: $\sqrt[3]{y} = -4$

$$\frac{}{\sqrt[3]{-64} \ ? \ -4}$$
$$-4 \ | \ -4 \qquad \text{TRUE}$$

The solution is -64.

18. -27

19. $x^{1/4} - 2 = 1$

$$x^{1/4} = 3$$
$$(x^{1/4})^4 = 3^4$$
$$x = 81$$

Check: $x^{1/4} - 2 = 1$

$$\frac{}{81^{1/4} - 2 \ ? \ 1}$$
$$3 - 2 \ | \ \cdot$$
$$1 \ | \ 1 \qquad \text{TRUE}$$

The solution is 81.

20. 125

21. $(y - 3)^{1/2} = -2$

$$\sqrt{y - 3} = -2$$

This equation has no solution, since the principal square root is never negative.

22. No solution

23. $\sqrt[4]{3x + 1} - 4 = -1$

$$\sqrt[4]{3x + 1} = 3$$
$$(\sqrt[4]{3x + 1})^4 = 3^4$$
$$3x + 1 = 81$$
$$3x = 80$$
$$x = \frac{80}{3}$$

Check: $\sqrt[4]{3x + 1} - 4 = -1$

$$\frac{}{\sqrt[4]{3 \cdot \frac{80}{3}} - 4 \ ? \ -1}$$
$$\sqrt[4]{81} - 4$$
$$3 - 4$$
$$-1 \ | \ -1 \qquad \text{TRUE}$$

The solution is $\frac{80}{3}$.

24. 39

25. $(x + 7)^{1/3} = 4$

$$[(x + 7)^{1/3}]^3 = 4^3$$
$$x + 7 = 64$$
$$x = 57$$

Check: $(x + 7)^{1/3} = 4$

$$\frac{}{(57 + 7)^{1/3} \ ? \ 4}$$
$$64^{1/3}$$
$$4 \ | \ 4 \qquad \text{TRUE}$$

The solution is 57.

26. 88

27. $\sqrt[3]{3y + 6} + 2 = 3$

$$\sqrt[3]{3y + 6} = 1$$
$$(\sqrt[3]{3y + 6})^3 = 1^3$$
$$3y + 6 = 1$$
$$3y = -5$$
$$y = -\frac{5}{3}$$

Check: $\sqrt[3]{3y + 6} + 2 = 3$

$$\frac{}{\sqrt[3]{3\left(-\frac{5}{3}\right) + 6} + 2 \ ? \ 3}$$
$$\sqrt[3]{1} + 2$$
$$1 + 2$$
$$3 \ | \ 3 \qquad \text{TRUE}$$

The solution is $-\frac{5}{3}$.

28. -6

29.
$$\sqrt{3t+4} = \sqrt{4t+3}$$
$$(\sqrt{3t+4})^2 = (\sqrt{4t+3})^2$$
$$3t+4 = 4t+3$$
$$4 = t+3$$
$$1 = t$$

Check:
$$\sqrt{3t+4} = \sqrt{4t+3}$$
$$\overline{\sqrt{3\cdot 1+4} \ ? \ \sqrt{4\cdot 1+3}}$$
$$\sqrt{7} \ | \ \sqrt{7} \qquad \text{TRUE}$$

The solution is 1.

30. 5

31.
$$3(4-t)^{1/4} = 6^{1/4}$$
$$[3(4-t)^{1/4}]^4 = (6^{1/4})^4$$
$$81(4-t) = 6$$
$$324 - 81t = 6$$
$$-81t = -318$$
$$t = \frac{106}{27}$$

The number $\dfrac{106}{27}$ checks and is the solution.

32. $\dfrac{1}{2}$

33.
$$3 + \sqrt{5-x} = x$$
$$\sqrt{5-x} = x-3$$
$$(\sqrt{5-x})^2 = (x-3)^2$$
$$5-x = x^2 - 6x + 9$$
$$0 = x^2 - 5x + 4$$
$$0 = (x-1)(x-4)$$
$$x-1 = 0 \ or \ x-4 = 0$$
$$x = 1 \ or \qquad x = 4$$

Check:
For 1:
$$3 + \sqrt{5-x} = x$$
$$\overline{3 + \sqrt{5-1} \ ? \ 1}$$
$$3 + \sqrt{4} \ |$$
$$3 + 2 \ |$$
$$5 \ | \ 1 \qquad \text{FALSE}$$

For 4:
$$3 + \sqrt{5-x} = x$$
$$\overline{3 + \sqrt{5-4} \ ? \ 4}$$
$$3 + \sqrt{1} \ |$$
$$3 + 1 \ |$$
$$4 \ | \ 4 \qquad \text{TRUE}$$

Since 4 checks but 1 does not, the solution is 4.

34. 5

35.
$$\sqrt{4x-3} = 2 + \sqrt{2x-5} \qquad \text{One radical is already isolated.}$$
$$(\sqrt{4x-3})^2 = (2 + \sqrt{2x-5})^2 \qquad \text{Squaring both sides}$$
$$4x-3 = 4 + 4\sqrt{2x-5} + 2x - 5$$
$$2x - 2 = 4\sqrt{2x-5}$$
$$x - 1 = 2\sqrt{2x-5}$$
$$x^2 - 2x + 1 = 8x - 20$$
$$x^2 - 10x + 21 = 0$$
$$(x-7)(x-3) = 0$$
$$x - 7 = 0 \ or \ x - 3 = 0$$
$$x = 7 \ or \qquad x = 3$$

Both numbers check. The solutions are 7 and 3.

36. 7

37.
$$\sqrt{20-x} + 8 = \sqrt{9-x} + 11$$
$$\sqrt{20-x} = \sqrt{9-x} + 3 \qquad \text{Isolating one radical}$$
$$(\sqrt{20-x})^2 = (\sqrt{9-x} + 3)^2 \qquad \text{Squaring both sides}$$
$$20 - x = 9 - x + 6\sqrt{9-x} + 9$$
$$2 = 6\sqrt{9-x} \qquad \text{Isolating the remaining radical}$$
$$1 = 3\sqrt{9-x} \qquad \text{Multiplying by } \frac{1}{2}$$
$$1^2 = (3\sqrt{9-x})^2 \qquad \text{Squaring both sides}$$
$$1 = 9(9-x)$$
$$1 = 81 - 9x$$
$$-80 = -9x$$
$$\frac{80}{9} = x$$

The number $\dfrac{80}{9}$ checks and is the solution.

38. $\dfrac{15}{4}$

39. $\sqrt{x+2} + \sqrt{3x+4} = 2$

$\sqrt{x+2} = 2 - \sqrt{3x+4}$ Isolating one radical

$(\sqrt{x+2})^2 = (2 - \sqrt{3x+4})^2$

$x + 2 = 4 - 4\sqrt{3x+4} + 3x + 4$

$-2x - 6 = -4\sqrt{3x+4}$ Isolating the remaining radical

$x + 3 = 2\sqrt{3x+4}$ Multiplying by $-\dfrac{1}{2}$

$(x+3)^2 = (2\sqrt{3x+4})^2$

$x^2 + 6x + 9 = 4(3x+4)$

$x^2 + 6x + 9 = 12x + 16$

$x^2 - 6x - 7 = 0$

$(x-7)(x+1) = 0$

$x - 7 = 0 \ \ or \ \ x + 1 = 0$

$x = 7 \ \ or \ \ x = -1$

Check:

For 7:

$$\sqrt{x+2} + \sqrt{3x+4} = 2$$

$\sqrt{7+2} + \sqrt{3\cdot 7 + 4} \ ? \ 2$

$\sqrt{9} + \sqrt{25} \ \Big| $

$8 \ \Big| \ 2$ FALSE

For -1:

$$\sqrt{x+2} + \sqrt{3x+4} = 2$$

$\sqrt{-1+2} + \sqrt{3\cdot(-1)+4} \ ? \ 2$

$\sqrt{1} + \sqrt{1} \ \Big|$

$2 \ \Big| \ 2$ TRUE

Since -1 checks but 7 does not, the solution is -1.

40. $\dfrac{1}{3}, -1$

41. We must have $f(x) = 2$, or $\sqrt{x} + \sqrt{x-9} = 1$.

$\sqrt{x} + \sqrt{x-9} = 1$

$\sqrt{x-9} = 1 - \sqrt{x}$ Isolating one radical term

$(\sqrt{x-9})^2 = (1 - \sqrt{x})^2$

$x - 9 = 1 - 2\sqrt{x} + x$

$-10 = -2\sqrt{x}$ Isolating the remaining radical term

$5 = \sqrt{x}$

$25 = x$

This value does not check. There is no solution, so there is no value of x for which $f(x) = 1$.

42. 9

43. $\sqrt{a-2} - \sqrt{4a+1} = -3$

$\sqrt{a-2} = \sqrt{4a+1} - 3$

$(\sqrt{a-2})^2 = (\sqrt{4a+1} - 3)^2$

$a - 2 = 4a + 1 - 6\sqrt{4a+1} + 9$

$-3a - 12 = -6\sqrt{4a+1}$

$a + 4 = 2\sqrt{4a+1}$

$(a+4)^2 = (2\sqrt{4a+1})^2$

$a^2 + 8a + 16 = 4(4a+1)$

$a^2 + 8a + 16 = 16a + 4$

$a^2 - 8a + 12 = 0$

$(a-2)(a-6) = 0$

$a - 2 = 0 \ \ or \ \ a - 6 = 0$

$a = 2 \ \ or \ \ \ \ \ \ a = 6$

Both numbers check, so we have $f(a) = -3$ when $a = 2$ and when $a = 6$.

44. 1

45. We must have $\sqrt{2x-3} = \sqrt{x+7} - 2$.

$\sqrt{2x-3} = \sqrt{x+7} - 2$

$(\sqrt{2x-3})^2 = (\sqrt{x+7} - 2)^2$

$2x - 3 = x + 7 - 4\sqrt{x+7} + 4$

$x - 14 = -4\sqrt{x+7}$

$(x-14)^2 = (-4\sqrt{x+7})^2$

$x^2 - 28x + 196 = 16(x+7)$

$x^2 - 28x + 196 = 16x + 112$

$x^2 - 44x + 84 = 0$

$(x-2)(x-42) = 0$

$x = 2 \ \ or \ \ x = 42$

Since 2 checks but 42 does not, we have $f(x) = g(x)$ when $x = 2$.

46. 10

47. We must have $4 - \sqrt{a-3} = (a+5)^{1/2}$.

$4 - \sqrt{a-3} = (a+5)^{1/2}$

$(4 - \sqrt{a-3})^2 = [(a+5)^{1/2}]^2$

$16 - 8\sqrt{a-3} + a - 3 = a + 5$

$-8\sqrt{a-3} = -8$

$\sqrt{a-3} = 1$

$(\sqrt{a-3})^2 = 1^2$

$a - 3 = 1$

$a = 4$

The number 4 checks, so we have $f(a) = g(a)$ when $a = 4$.

48. 15

49. Writing exercise

50. Writing exercise

51. *Familiarize*. Let h = the height of the triangle, in inches. Then $h + 2$ = the base. Recall that the formula for the area of a triangle with base b and height h is $A = \frac{1}{2}bh$.

Translate. Substitute in the formula.
$$31\frac{1}{2} = \frac{1}{2}(h + 2)(h)$$

Carry out. We solve the equation.
$$31\frac{1}{2} = \frac{1}{2}(h + 2)(h)$$
$$\frac{63}{2} = \frac{1}{2}(h + 2)(h)$$
$$63 = (h + 2)(h) \quad \text{Multiplying by 2}$$
$$63 = h^2 + 2h$$
$$0 = h^2 + 2h - 63$$
$$0 = (h + 9)(h - 7)$$
$$h + 9 = 0 \quad or \quad h - 7 = 0$$
$$h = -9 \quad or \qquad h = 7$$

Check. Since the height of the triangle cannot be negative we check only 7. If the height is 7 in., then the base is 7 + 2, or 9 in., and the area is $\frac{1}{2} \cdot 9 \cdot 7 = \frac{63}{2} = 31\frac{1}{2}$ in². The answer checks.

State. The height of the triangle is 7 in., and the base is 9 in.

52. 8

53. *Familiarize*. Let t = the time, in hours, it takes Gonzalo to sew the quilt. Then $t - 6$ = the time it takes Elaine to sew the quilt. In 4 hours Gonzalo does $\frac{4}{t}$ of the job and Elaine does $\frac{4}{t-6}$ of the job.

Translate. Together, in 4 hr one entire job is done.
$$\frac{4}{t} + \frac{4}{t-6} = 1$$

Carry out. We solve the equation. The LCD is $t(t - 6)$.

$$\frac{4}{t} + \frac{4}{t-6} = 1$$
$$t(t-6)\left(\frac{4}{t} + \frac{4}{t-6}\right) = t(t-6) \cdot 1$$
$$4(t - 6) + 4t = t^2 - 6t$$
$$4t - 24 + 4t = t^2 - 6t$$
$$8t - 24 = t^2 - 6t$$
$$0 = t^2 - 14t + 24$$
$$0 = (t - 2)(t - 12)$$
$$t - 2 = 0 \quad or \quad t - 12 = 0$$
$$t = 2 \quad or \qquad t = 12$$

Check. If $t = 2$, then $t - 6 = 2 - 6 = -4$. Since time cannot be negative in this application, 2 is not a solution. If Gonzalo sews the quilt in 12 hr, then Elaine sews it in $12 - 6$, or 6 hr. In 4 hr Gonzalo does 4/12, or 1/3 of the job and Elaine does 4/6, or 2/3 of the job. Together they do $1/3 + 2/3$, or 1 entire job. The answer checks.

State. It would take Elaine 6 hr and Gonzalo 12 hr to sew the quilt working alone.

54. $\frac{165}{52}$ hr, or approximately 3.2 hr

55. Graph $y > 3x + 5$.

First graph the related equation, $y = 3x + 5$. Use a dashed line since the inequality symbol is $>$. Then test a point not on the line to determine if it is a solution of the inequality. We use $(0, 0)$.

$$\frac{y > 3x + 5}{0 \;?\; 3 \cdot 0 + 5}$$
$$0 \;\Big|\; 5 \qquad\qquad \text{FALSE}$$

Since $0 > 5$ is false, we shade the half-plane that does not contain $(0, 0)$.

56.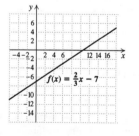

57. Writing exercise

58. Writing exercise

59. $S(t) = 1.087.7 \sqrt{\dfrac{9t + 2617}{2457}}$

Substitute 1502.3 for $S(t)$ and solve for t.

$$1502.3 = 1087.7 \sqrt{\dfrac{9t + 2617}{2457}}$$

$$1.3812 \approx \sqrt{\dfrac{9t + 2617}{2457}} \quad \text{Dividing by 1087.7}$$

$$(1.3812)^2 \approx \left(\sqrt{\dfrac{9t + 2617}{2457}} \right)^2$$

$$1.9077 \approx \dfrac{9t + 2617}{2457}$$

$$4687.2189 \approx 9t + 2617$$

$$2070.2189 \approx 9t$$

$$230.0243 \approx t$$

The temperature is about 230.0°C.

60. About 524.8°C

61.
$$S = 1087.7 \sqrt{\dfrac{9t + 2617}{2457}}$$

$$\dfrac{S}{1087.7} = \sqrt{\dfrac{9t + 2617}{2457}}$$

$$\left(\dfrac{S}{1087.7} \right)^2 = \left(\sqrt{\dfrac{9t + 2617}{2457}} \right)^2$$

$$\dfrac{S^2}{1087.7^2} = \dfrac{9t + 2617}{2457}$$

$$\dfrac{2457 S^2}{1087.7^2} = 9t + 2617$$

$$\dfrac{2457 S^2}{1087.7^2} - 2617 = 9t$$

$$\dfrac{1}{9} \left(\dfrac{2457 S^2}{1087.7^2} - 2617 \right) = t$$

62. About 4166 rpm

63. $d(n) = 0.75 \sqrt{2.8n}$

Substitute 84 for $d(n)$ and solve for n.

$$84 = 0.75 \sqrt{2.8n}$$

$$112 = \sqrt{2.8n}$$

$$(112)^2 = (\sqrt{2.8n})^2$$

$$12,544 = 2.8n$$

$$4480 = n$$

About 4480 rpm will produce peak performance.

64. $h = \dfrac{v^2 r}{2gr - v^2}$

65.
$$v = \sqrt{2gr} \sqrt{\dfrac{h}{r + h}}$$

$$v^2 = 2gr \cdot \dfrac{h}{r + h} \quad \text{Squaring both sides}$$

$$v^2(r + h) = 2grh \quad \text{Multiplying by } r+h$$

$$v^2 r + v^2 h = 2grh$$

$$v^2 h = 2grh - v^2 r$$

$$v^2 h = r(2gh - v^2)$$

$$\dfrac{v^2 h}{2gh - v^2} = r$$

66. 22,500 ft

67.
$$D(h) = 1.2\sqrt{h}$$

$$10.2 = 1.2\sqrt{h}$$

$$8.5 = \sqrt{h}$$

$$(8.5)^2 = (\sqrt{h})^2$$

$$72.25 = h$$

The sailor must climb 72.25 ft above sea level.

68. $-\dfrac{8}{9}$

69.
$$\left(\dfrac{z}{4} - 5 \right)^{2/3} = \dfrac{1}{25}$$

$$\left[\left(\dfrac{z}{4} - 5 \right)^{2/3} \right]^3 = \left(\dfrac{1}{25} \right)^3$$

$$\left(\dfrac{z}{4} - 5 \right)^2 = \dfrac{1}{15,625}$$

$$\dfrac{z^2}{16} - \dfrac{5}{2}z + 25 = \dfrac{1}{15,625}$$

$$15,625z^2 - 625,000z + 6,250,000 = 16$$

$$15,625z^2 - 625,000z + 6,249,984 = 0$$

$$(125z - 2504)(125z - 2496) = 0$$

$$125z - 2504 = 0 \quad \text{or} \quad 125z - 2496 = 0$$

$$125z = 2504 \quad \text{or} \qquad 125z = 2496$$

$$z = \dfrac{2504}{125} \quad \text{or} \qquad z = \dfrac{2496}{125}$$

Both numbers check. The solutions are $\dfrac{2504}{125}$ and $\dfrac{2496}{125}$.

70. $-8, 8$

71.
$$\sqrt{\sqrt{y}+49}=7$$
$$(\sqrt{\sqrt{y}+49})^2=7^2$$
$$\sqrt{y}+49=49$$
$$\sqrt{y}=0$$
$$(\sqrt{y})^2=0^2$$
$$y=0$$

This number 0 checks and is the solution.

72. $-1,6$

73.
$$\sqrt{8-b}=b\sqrt{8-b}$$
$$(\sqrt{8-b})^2=(b\sqrt{8-b})^2$$
$$(8-b)=b^2(8-b)$$
$$0=b^2(8-b)-(8-b)$$
$$0=(8-b)(b^2-1)$$
$$0=(8-b)(b+1)(b-1)$$
$$8-b=0 \ \ \text{or} \ \ b+1=0 \ \text{or} \ \ b-1=0$$
$$8=b \ \ \text{or} \qquad b=-1 \ \text{or} \quad b=1$$

Since the numbers 8 and 1 check but -1 does not, 8 and 1 are the solutions.

74. $(2,0)$

75. We find the values of x for which $g(x)=0$.
$$6x^{1/2}+6x^{-1/2}-37=0$$
$$6\sqrt{x}+\frac{6}{\sqrt{x}}=37$$
$$\left(6\sqrt{x}+\frac{6}{\sqrt{x}}\right)^2=37^2$$
$$36x+72+\frac{36}{x}=1369$$
$$36x^2+72x+36=1369x \quad \text{Multiplying by } x$$
$$36x^2-1297x+36=0$$
$$(36x-1)(x-36)=0$$
$$36x-1=0 \quad or \ \ x-36=0$$
$$36x=1 \quad or \qquad x=36$$
$$x=\frac{1}{36} \quad or \qquad x=36$$

Both numbers check. The x-intercepts are $\left(\frac{1}{36},0\right)$ and $(36,0)$.

76. $(0,0), \left(\dfrac{125}{4},0\right)$

77.

78. Writing exercise

79.

Exercise Set 10.7

1. $a=5, \quad b=3$

Find c.
$$c^2=a^2+b^2 \quad \text{Pythagorean equation}$$
$$c^2=5^2+3^2 \quad \text{Substituting}$$
$$c^2=25+9$$
$$c^2=34$$
$$c=\sqrt{34} \qquad \text{Exact answer}$$
$$c\approx 5.831 \qquad \text{Approximation}$$

2. $\sqrt{164}$; 12.806

3. $a=9, \quad b=9$

Observe that the legs have the same length, so this is an isosceles right triangle. Then we know that the length of the hypotenuse is the length of a leg times $\sqrt{2}$, or $9\sqrt{2}$, or approximately 12.728.

4. $10\sqrt{2}$; 14.142

5. $b=12, \quad c=13$

Find a.
$$a^2+b^2=c^2 \quad \text{Pythagorean equation}$$
$$a^2+12^2=13^2 \quad \text{Substituting}$$
$$a^2+144=169$$
$$a^2=25$$
$$a=5$$

6. $\sqrt{119}$; 10.909

7. $c=6, \quad a=\sqrt{5}$

Find b.
$$c^2=a^2+b^2$$
$$(\sqrt{5})^2+b^2=6^2$$
$$5+b^2=36$$
$$b^2=31$$
$$b=\sqrt{31} \qquad \text{Exact answer}$$
$$b\approx 5.568 \qquad \text{Approximation}$$

8. 4

9. $b = 2$, $c = \sqrt{15}$

Find a.

$a^2 + b^2 = c^2$ Pythagorean equation

$a^2 + 2^2 = (\sqrt{15})^2$ Substituting

$a^2 + 4 = 15$

$a^2 = 11$

$a = \sqrt{11}$ Exact answer

$a \approx 3.317$ Approximation

10. $\sqrt{19}$; 4.359

11. $a = 1$, $c = \sqrt{2}$

Observe that the length of the hypotenuse, $\sqrt{2}$, is $\sqrt{2}$ times the length of the given leg, 1. Thus, we have an isosceles right triangle and the length of the other leg is also 1.

12. $\sqrt{3}$; 1.732

13. We make a drawing and let $d =$ the length of the guy wire.

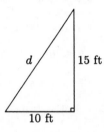

We use the Pythagorean equation to find d.

$d^2 = 10^2 + 15^2$

$d^2 = 100 + 225$

$d^2 = 325$

$d = \sqrt{325}$

$d \approx 18.028$

The wire is $\sqrt{325}$ ft, or about 18.028 ft long.

14. $\sqrt{8450}$ ft; 91.924 ft

15. We first make a drawing and let $d =$ the distance, in feet, to second base. A right triangle is formed in which the length of the leg from second base to third base is 90 ft. The length of the leg from third base to where the catcher fields the ball is $90 - 10$, or 80 ft.

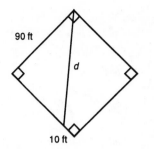

We substitute these values into the Pythagorean equation to find d.

$d^2 = 90^2 + 80^2$

$d^2 = 8100 + 6400$

$d^2 = 14,500$

$d = \sqrt{14,500}$

Exact answer: $d = \sqrt{14,500}$ ft

Approximation: $d \approx 120.416$ ft

16. 12 in.

17. We make a drawing.

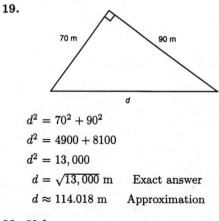

We use the Pythagorean equation to find w.

$w^2 + 15^2 = 25^2$

$w^2 + 225 = 625$

$w^2 = 400$

$w = 20$

The width is 20 in.

18. $\sqrt{340} + 8$ ft; 26.439 ft

19.

$d^2 = 70^2 + 90^2$

$d^2 = 4900 + 8100$

$d^2 = 13,000$

$d = \sqrt{13,000}$ m   Exact answer

$d \approx 114.018$ m   Approximation

20. 50 ft

21. Since one acute angle is 45°, this is an isosceles right triangle with $b = 5$. Then $a = 5$ also. We substitute to find c.

$$c = a\sqrt{2}$$
$$c = 5\sqrt{2}$$

Exact answer: $a = 5$, $c = 5\sqrt{2}$

Approximation: $c \approx 7.071$

22. $a = 14$, $c = 14\sqrt{2} \approx 19.799$

23. This is a 30-60-90 right triangle with $c = 14$. We substitute to find a and b.

$$c = 2a$$
$$14 = 2a$$
$$7 = a$$

$$b = a\sqrt{3}$$
$$b = 7\sqrt{3}$$

Exact answer: $a = 7$, $b = 7\sqrt{3}$

Approximation: $b \approx 12.124$

24. $a = 9$, $b = 9\sqrt{3} \approx 15.588$

25. This is a 30-60-90 right triangle with $b = 15$. We substitute to find a and c.

$$b = a\sqrt{3}$$
$$15 = a\sqrt{3}$$
$$\frac{15}{\sqrt{3}} = a$$
$$\frac{15\sqrt{3}}{3} = a \qquad \text{Rationalizing the denominator}$$
$$5\sqrt{3} = a \qquad \text{Simplifying}$$
$$c = 2a$$
$$c = 2 \cdot 5\sqrt{3}$$
$$c = 10\sqrt{3}$$

Exact answer: $a = 5\sqrt{3}$, $c = 10\sqrt{3}$

Approximations: $a \approx 8.660$, $c \approx 17.321$

26. $a = 4\sqrt{2} \approx 5.657$, $b = 4\sqrt{2} \approx 5.657$

27. This is an isosceles right triangle with $c = 13$. We substitute to find a.

$$a = \frac{c\sqrt{2}}{2}$$
$$a = \frac{13\sqrt{2}}{2}$$

Since $a = b$, we have $b = \dfrac{13\sqrt{2}}{2}$ also.

Exact answer: $a = \dfrac{13\sqrt{2}}{2}$, $b = \dfrac{13\sqrt{2}}{2}$

Approximations: $a \approx 9.192$, $b \approx 9.192$

28. $a = \dfrac{7\sqrt{3}}{3} \approx 4.041$, $c = \dfrac{14\sqrt{3}}{3} \approx 8.083$

29. This is a 30-60-90 triangle with $a = 14$. We substitute to find b and c.

$$b = a\sqrt{3} \qquad\qquad c = 2a$$
$$b = 14\sqrt{3} \qquad\qquad c = 2 \cdot 14$$
$$c = 28$$

Exact answer: $b = 14\sqrt{3}$, $c = 28$

Approximation: $b \approx 24.249$

30. $b = 9\sqrt{3} \approx 15.588$; $c = 18$

31.

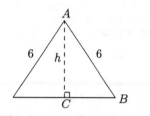

This is an equilateral triangle, so all the angles are 60°. The altitude bisects one angle and one side. Then triangle ABC is a 30-60-90 right triangle with the shorter leg of length $6/2$, or 3, and hypotenuse of length 6. We substitute to find the length of the other leg.

$$b = a\sqrt{3}$$
$$h = 3\sqrt{3} \qquad \text{Substituting } h \text{ for } b \text{ and 3 for } a$$

Exact answer: $h = 3\sqrt{3}$

Approximation: $h \approx 5.196$

32. $5\sqrt{3} \approx 8.660$

33.

Triangle ABC is an isosceles right triangle with $a = 13$. We substitute to find c.

$$c = a\sqrt{2}$$
$$c = 13\sqrt{2}$$

Exact answer: $c = 13\sqrt{2}$

Approximation: $c \approx 18.385$

34. $7\sqrt{2} \approx 9.899$

35.

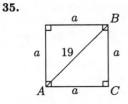

Triangle ABC is an isosceles right triangle with $c = 19$. We substitute to find a.

$$a = \frac{c\sqrt{2}}{2}$$
$$a = \frac{19\sqrt{2}}{2}$$

Exact answer: $a = \dfrac{19\sqrt{2}}{2}$

Approximation: $a \approx 13.435$

36. $\dfrac{15\sqrt{2}}{2} \approx 10.607$

37. We will express all distances in feet. Recall that 1 mi = 5280 ft.

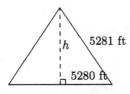

We use the Pythagorean equation to find h.

$$h^2 + (5280)^2 = (5281)^2$$
$$h^2 + 27{,}878{,}400 = 27{,}888{,}961$$
$$h^2 = 10{,}561$$
$$h = \sqrt{10{,}561}$$
$$h \approx 102.767$$

The height of the bulge is $\sqrt{10{,}561}$ ft, or about 102.767 ft.

38. Neither; they have the same area, 300 ft^2.

39.

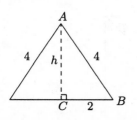

The entrance is an equilateral triangle, so all the angles are 60°. The altitude bisects one angle and one

side. Then triangle ABC is a 30-60-90 right triangle with the shorter leg of length 4/2, or 2, and hypotenuse of length 4. We substitute to find h, the height of the tent.

$$b = a\sqrt{3}$$
$$h = 2\sqrt{3} \qquad \text{Substituting } h \text{ for } b \text{ and } 2 \text{ for } a$$

Exact answer: $h = 2\sqrt{3}$ ft

Approximation: $h \approx 3.464$ ft

40. $d = s + s\sqrt{2}$

41.

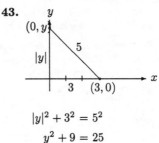

Triangle ABC is an isosceles right triangle with $c = 8\sqrt{2}$. We substitute to find a.

$$a = \frac{c\sqrt{2}}{2} = \frac{8\sqrt{2} \cdot \sqrt{2}}{2} = \frac{8 \cdot 2}{2} = 8$$

The length of a side of the square is 8 ft.

42. $\sqrt{181}$ cm ≈ 13.454 cm

43.

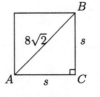

$$|y|^2 + 3^2 = 5^2$$
$$y^2 + 9 = 25$$
$$y^2 = 16$$
$$y = \pm 4$$

The points are $(0, -4)$ and $(0, 4)$.

44. $(3, 0)$, $(-3, 0)$

45. Writing exercise

46. Writing exercise

47. $47(-1)^{19} = 47(-1) = -47$

48. 5

49. $x^3 - 9x = x \cdot x^2 - 9 \cdot x = x(x^2 - 9)$

50. $7a(a + 2)(a - 2)$

51. $|3x - 5| = 7$

$3x - 5 = 7 \quad or \quad 3x - 5 = -7$

$3x = 12 \quad or \qquad 3x = -2$

$x = 4 \quad or \qquad x = -\dfrac{2}{3}$

The solution set is $\left\{4, -\dfrac{2}{3}\right\}$.

52. $\left\{10, -\dfrac{4}{3}\right\}$

53. Writing exercise

54. Writing exercise

55.

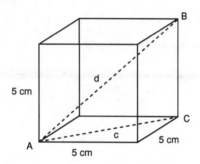

First find the length of a diagonal of the base of the cube. It is the hypotenuse of an isosceles right triangle with $a = 5$. Then $c = a\sqrt{2} = 5\sqrt{2}$ cm.

Triangle ABC is a right triangle with legs of $5\sqrt{2}$ cm and 5 cm and hypotenuse d. Use the Pythagorean equation to find d, the length of the diagonal that connects two opposite corners of the cube.

$d^2 = (5\sqrt{2})^2 + 5^2$

$d^2 = 25 \cdot 2 + 25$

$d^2 = 50 + 25$

$d^2 = 75$

$d = \sqrt{75}$

Exact answer: $d = \sqrt{75}$ cm

56. 9

57.

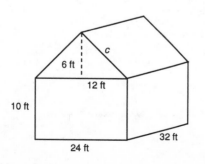

The area to be painted consists of two 10 ft by 24 ft rectangles, two 10 ft by 32 ft rectangles, and two triangles with height 6 ft and base 24 ft. The area of the two 10 ft by 24 ft rectangles is $2 \cdot 10$ ft $\cdot 24$ ft $= 480$ ft^2. The area of the two 10 ft by 32 ft rectangles is $2 \cdot 10$ ft $\cdot 32$ ft $= 640$ ft^2. The area of the two triangles is $2 \cdot \dfrac{1}{2} \cdot 24$ ft $\cdot 6$ ft $= 144$ ft^2. Thus, the total area to be painted is 480 ft$^2 + 640$ ft$^2 + 144$ ft$^2 = 1264$ ft^2.

One gallon of paint covers 275 ft^2, so we divide to determine how many gallons of paint are required: $\dfrac{1264}{275} \approx 4.6$. Thus, 4 gallons of paint should be bought to paint the house. This answer assumes that the total area of the doors and windows is 164 ft^2 or more. ($4 \cdot 275 = 1100$ and $1264 = 1100 + 164$)

58. 49.5 ft by 49.5 ft

59. First we find the radius of a circle with an area of 6160 ft^2.

$A = \pi r^2$

$6160 = \pi r^2$

$\dfrac{6160}{\pi} = r^2$

$\sqrt{\dfrac{6160}{\pi}} = r$

$44.28 \approx r$

Now we make a drawing. Let $s =$ the length of a side of the room.

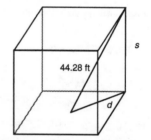

We make a drawing of the floor of the room to help us find d.

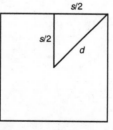

We have an isosceles right triangle, so $d = \dfrac{s}{2} \cdot \sqrt{2}$, or $\dfrac{s\sqrt{2}}{2}$.

Now we use the Pythagorean theorem to find s.

$$d^2 + s^2 = (44.28)^2$$

$$\left(\frac{s\sqrt{2}}{2}\right)^2 + s^2 = (44.28)^2 \quad \text{Substituting } \frac{s\sqrt{2}}{2}$$
$$\text{for } d.$$

$$\frac{s^2}{2} + s^2 = (44.28)^2$$

$$\frac{3s^2}{2} = (44.28)^2$$

$$s^2 = \frac{2}{3}(44.28)^2$$

$$s = 44.28\sqrt{\frac{2}{3}}$$

$$s \approx 36.15$$

The dimensions of the room are 36.15 ft by 36.15 ft by 36.15 ft.

Exercise Set 10.8

1. $\sqrt{-25} = \sqrt{-1 \cdot 25} = \sqrt{-1} \cdot \sqrt{25} = i \cdot 5 = 5i$

2. $6i$

3. $\sqrt{-13} = \sqrt{-1 \cdot 13} = \sqrt{-1} \cdot \sqrt{13} = i\sqrt{13}$, or $\sqrt{13}i$

4. $i\sqrt{19}$, or $\sqrt{19}i$

5. $\sqrt{-18} = \sqrt{-1} \cdot \sqrt{9} \cdot \sqrt{2} = i \cdot 3 \cdot \sqrt{2} = 3i\sqrt{2}$, or $3\sqrt{2}i$

6. $7i\sqrt{2}$, or $7\sqrt{2}i$

7. $\sqrt{-3} = \sqrt{-1 \cdot 3} = \sqrt{-1} \cdot \sqrt{3} = i\sqrt{3}$, or $\sqrt{3}i$

8. $2i$

9. $\sqrt{-81} = \sqrt{-1 \cdot 81} = \sqrt{-1} \cdot \sqrt{81} = i \cdot 9 = 9i$

10. $3i\sqrt{3}$, or $3\sqrt{3}i$

11. $\sqrt{-300} = \sqrt{-1} \cdot \sqrt{100} \cdot \sqrt{3} = i \cdot 10 \cdot \sqrt{3} = 10i\sqrt{3}$, or $10\sqrt{3}i$

12. $-5i\sqrt{3}$, or $-5\sqrt{3}i$

13. $-\sqrt{-49} = -\sqrt{-1 \cdot 49} = -\sqrt{-1} \cdot \sqrt{49} = -i \cdot 7 = -7i$

14. $-5i\sqrt{5}$, or $-5\sqrt{5}i$

15. $4 - \sqrt{-60} = 4 - \sqrt{-1 \cdot 60} = 4 - \sqrt{-1} \cdot \sqrt{60} = 4 - i \cdot 2\sqrt{15} = 4 - 2\sqrt{15}i$, or $4 - 2i\sqrt{15}$

16. $6 - 2i\sqrt{21}$, or $6 - 2\sqrt{21}i$

17. $\sqrt{-4} + \sqrt{-12} = \sqrt{-1 \cdot 4} + \sqrt{-1 \cdot 12} = \sqrt{-1} \cdot \sqrt{4} + \sqrt{-1} \cdot \sqrt{12} = i \cdot 2 + i \cdot 2\sqrt{3} = (2 + 2\sqrt{3})i$

18. $(-2\sqrt{19} + 5\sqrt{5})i$

19. $\sqrt{-72} - \sqrt{-25} = \sqrt{-1 \cdot 36 \cdot 2} - \sqrt{-1 \cdot 25} = \sqrt{-1}\sqrt{36 \cdot 2} - \sqrt{-1}\sqrt{25} = i \cdot 6\sqrt{2} - i \cdot 5 = (6\sqrt{2} - 5)i$

20. $(3\sqrt{2} - 10)i$

21. $(7 + 8i) + (5 + 3i)$
$= (7 + 5) + (8 + 3)i$ Combining the real and the imaginary parts
$= 12 + 11i$

22. $7 + 4i$

23. $(9 + 8i) - (5 + 3i) = (9 - 5) + (8 - 3)i$
$= 4 + 5i$

24. $7 + 3i$

25. $(5 - 3i) - (9 + 2i) = (5 - 9) + (-3 - 2)i$
$= -4 - 5i$

26. $2 - i$

27. $(-2 + 6i) - (-7 + i) = -2 - (-7) + (6 - 1)i$
$= 5 + 5i$

28. $-12 - 5i$

29. $6i \cdot 9i = 54 \cdot i^2$
$= 54 \cdot (-1) \quad\quad i^2 = -1$
$= -54$

30. -42

31. $7i \cdot (-8i) = -56 \cdot i^2$
$= -56 \cdot (-1) \quad\quad i^2 = -1$
$= 56$

32. -24

33. $\sqrt{-49}\sqrt{-25} = \sqrt{-1} \cdot \sqrt{49} \cdot \sqrt{-1} \cdot \sqrt{25}$
$= i \cdot 7 \cdot i \cdot 5$
$= i^2 \cdot 35$
$= -1 \cdot 35$
$= -35$

34. -18

35. $\sqrt{-6}\sqrt{-7} = \sqrt{-1} \cdot \sqrt{6} \cdot \sqrt{-1} \cdot \sqrt{7}$
$$= i \cdot \sqrt{6} \cdot i \cdot \sqrt{7}$$
$$= i^2 \cdot \sqrt{42}$$
$$= -1 \cdot \sqrt{42}$$
$$= -\sqrt{42}$$

36. $-\sqrt{10}$

37. $\sqrt{-15}\sqrt{-10} = \sqrt{-1} \cdot \sqrt{15} \cdot \sqrt{-1} \cdot \sqrt{10}$
$$= i \cdot \sqrt{15} \cdot i \cdot \sqrt{10}$$
$$= i^2 \cdot \sqrt{150}$$
$$= -\sqrt{25 \cdot 6}$$
$$= -5\sqrt{6}$$

38. $-3\sqrt{14}$

39. $\quad 2i(7 + 3i)$
$$= 2i \cdot 7 + 2i \cdot 3i \quad \text{Using the distributive law}$$
$$= 14i + 6i^2$$
$$= 14i - 6 \qquad i^2 = -1$$
$$= -6 + 14i$$

40. $-30 + 10i$

41. $-4i(6 - 5i) = -4i \cdot 6 - 4i(-5i)$
$$= -24i + 20i^2$$
$$= -24i - 20$$
$$= -20 - 24i$$

42. $-28 - 21i$

43. $\quad (1 + 5i)(4 + 3i)$
$$= 4 + 3i + 20i + 15i^2 \quad \text{Using FOIL}$$
$$= 4 + 3i + 20i - 15 \quad i^2 = -1$$
$$= -11 + 23i$$

44. $1 + 5i$

45. $(5 - 6i)(2 + 5i) = 10 + 25i - 12i - 30i^2$
$$= 10 + 25i - 12i + 30$$
$$= 40 + 13i$$

46. $38 + 9i$

47. $(-4 + 5i)(3 - 4i) = -12 + 16i + 15i - 20i^2$
$$= -12 + 16i + 15i + 20$$
$$= 8 + 31i$$

48. $2 - 46i$

49. $(7 - 3i)(4 - 7i) = 28 - 49i - 12i + 21i^2 =$
$28 - 49i - 12i - 21 = 7 - 61i$

50. $5 - 37i$

51. $(-3 + 6i)(-3 + 4i) = 9 - 12i - 18i + 24i^2 =$
$9 - 12i - 18i - 24 = -15 - 30i$

52. $-11 - 16i$

53. $(2 + 9i)(-3 - 5i) = -6 - 10i - 27i - 45i^2 =$
$-6 - 10i - 27i + 45 = 39 - 37i$

54. $13 - 47i$

55. $\quad (1 - 2i)^2$
$$= 1^2 - 2 \cdot 1 \cdot 2i + (2i)^2 \quad \text{Squaring a binomial}$$
$$= 1 - 4i + 4i^2$$
$$= 1 - 4i - 4 \qquad i^2 = -1$$
$$= -3 - 4i$$

56. $12 - 16i$

57. $\quad (3 + 2i)^2$
$$= 3^2 + 2 \cdot 3 \cdot 2i + (2i)^2 \quad \text{Squaring a binomial}$$
$$= 9 + 12i + 4i^2$$
$$= 9 + 12i - 4 \qquad i^2 = -1$$
$$= 5 + 12i$$

58. $-5 + 12i$

59. $(-5 - 2i)^2 = 25 + 20i + 4i^2 = 25 + 20i - 4 =$
$21 + 20i$

60. $-5 - 12i$

61. $\quad \dfrac{3}{2 - i}$
$$= \frac{3}{2 - i} \cdot \frac{2 + i}{2 + i} \quad \text{Multiplying by 1, using the conjugate}$$
$$= \frac{6 + 3i}{4 - i^2} \quad \text{Multiplying}$$
$$= \frac{6 + 3i}{4 - (-1)} \qquad i^2 = -1$$
$$= \frac{6 + 3i}{5}$$
$$= \frac{6}{5} + \frac{3}{5}i$$

62. $\dfrac{6}{5} - \dfrac{2}{5}i$

63. $\dfrac{3i}{5+2i}$

$= \dfrac{3i}{5+2i} \cdot \dfrac{5-2i}{5-2i}$ Multiplying by 1, using the conjugate

$= \dfrac{15i - 6i^2}{25 - 4i^2}$ Multiplying

$= \dfrac{15i + 6}{25 + 4}$

$= \dfrac{15i + 6}{29}$

$= \dfrac{6}{29} + \dfrac{15}{29}i$

64. $-\dfrac{6}{17} + \dfrac{10}{17}i$

65. $\dfrac{7}{9i} = \dfrac{7}{9i} \cdot \dfrac{i}{i} = \dfrac{7i}{9i^2} = \dfrac{7i}{-9} = -\dfrac{7}{9}i$

66. $-\dfrac{5}{8}i$

67. $\dfrac{5-3i}{4i} = \dfrac{5-3i}{4i} \cdot \dfrac{i}{i} = \dfrac{5i - 3i^2}{4i^2} = \dfrac{5i + 3}{-4} =$

$-\dfrac{3}{4} - \dfrac{5}{4}i$

68. $\dfrac{7}{5} - \dfrac{2}{5}i$

69. $\dfrac{7i + 14}{7i} = \dfrac{7i}{7i} + \dfrac{14}{7i} = 1 + \dfrac{2}{i} = 1 + \dfrac{2}{i} \cdot \dfrac{i}{i} =$

$1 + \dfrac{2i}{i^2} = 1 + \dfrac{2i}{-1} = 1 - 2i$

70. $2 - i$

71. $\dfrac{4+5i}{3-7i} = \dfrac{4+5i}{3-7i} \cdot \dfrac{3+7i}{3+7i} = \dfrac{12 + 28i + 15i + 35i^2}{9 - 49i^2} =$

$\dfrac{12 + 28i + 15i - 35}{9 + 49} = \dfrac{-23 + 43i}{58} = -\dfrac{23}{58} + \dfrac{43}{58}i$

72. $\dfrac{23}{65} + \dfrac{41}{65}i$

73. $\dfrac{3-2i}{4+3i} = \dfrac{3-2i}{4+3i} \cdot \dfrac{4-3i}{4-3i} = \dfrac{12 - 9i - 8i + 6i^2}{16 - 9i^2} =$

$\dfrac{12 - 9i - 8i - 6}{16 + 9} = \dfrac{6 - 17i}{25} = \dfrac{6}{25} - \dfrac{17}{25}i$

74. $\dfrac{1}{15} - \dfrac{4}{5}i$

75. $i^7 = i^6 \cdot i = (i^2)^3 \cdot i = (-1)^3 \cdot i = -1 \cdot i = -i$

76. $-i$

77. $i^{24} = (i^2)^{12} = (-1)^{12} = 1$

78. $-i$

79. $i^{42} = (i^2)^{21} = (-1)^{21} = -1$

80. 1

81. $i^9 = (i^2)^4 \cdot i = (-1)^4 \cdot i = 1 \cdot i = i$

82. i

83. $i^6 = (i^2)^3 = (-1)^3 = -1$

84. 1

85. $(5i)^3 = 5^3 \cdot i^3 = 125 \cdot i^2 \cdot i = 125(-1)(i) = -125i$

86. $-243i$

87. $i^2 + i^4 = -1 + (i^2)^2 = -1 + (-1)^2 = -1 + 1 = 0$

88. i

89. Writing exercise

90. Writing exercise

91. $f(x) = x^2 - 3x$, $g(x) = 2x - 5$

$(f + g)(-2) = f(-2) + g(-2)$

$\quad\quad = (-2)^2 - 3(-2) + 2(-2) - 5$

$\quad\quad = 4 + 6 - 4 - 5$

$\quad\quad = 1$

92. 1

93. $(f \cdot g)(5) = f(5)g(5)$

$\quad\quad = (5^2 - 3 \cdot 5)(2 \cdot 5 - 5)$

$\quad\quad = (25 - 15)(10 - 5)$

$\quad\quad = 10 \cdot 5$

$\quad\quad = 50$

94. 0

95. $28 = 3x^2 - 17x$

$0 = 3x^2 - 17x - 28$

$0 = (3x + 4)(x - 7)$

$3x + 4 = 0 \quad or \quad x - 7 = 0$

$3x = -4 \quad or \quad\quad x = 7$

$x = -\dfrac{4}{3} \quad or \quad\quad x = 7$

Both values check. The solutions are $-\dfrac{4}{3}$ and 7.

96. $\left\{x \,\middle|\, -\dfrac{29}{3} < x < 5\right\}$, or $\left(-\dfrac{29}{3}, 5\right)$

97. Writing exercise

98. Writing exercise

99. $g(3i) = \dfrac{(3i)^4 - (3i)^2}{3i - 1} = \dfrac{81i^4 - 9i^2}{-1 + 3i} = \dfrac{81 + 9}{-1 + 3i} =$

$\dfrac{90}{-1 + 3i} = \dfrac{90}{-1 + 3i} \cdot \dfrac{-1 - 3i}{-1 - 3i} = \dfrac{90(-1 - 3i)}{1 - 9i^2} =$

$\dfrac{90(-1 - 3i)}{1 + 9} = \dfrac{90(-1 - 3i)}{10} = \dfrac{9 \cdot \cancel{10}(-1 - 3i)}{\cancel{10}} =$

$9(-1 - 3i) = -9 - 27i$

100. $-2 + 4i$

101. First we simplify $g(z)$.

$g(z) = \dfrac{z^4 - z^2}{z - 1} = \dfrac{z^2(z^2 - 1)}{z - 1} = \dfrac{z^2(z + 1)(z - 1)}{z - 1} =$

$\dfrac{z^2(z + 1)\cancel{(z - 1)}}{\cancel{z - 1}} = z^2(z + 1)$

Now we substitute.

$g(5i - 1) = (5i - 1)^2(5i - 1 + 1) =$

$(25i^2 - 10i + 1)(5i) =$

$(-25 - 10i + 1)(5i) = (-24 - 10i)(5i) =$

$-120i - 50i^2 = 50 - 120i$

102. $-51 - 21i$

103. $\dfrac{1}{\dfrac{1 - i}{10} - \left(\dfrac{1 - i}{10}\right)^2} = \dfrac{1}{\dfrac{1 - i}{10} - \left(\dfrac{-2i}{100}\right)} =$

$\dfrac{1}{\dfrac{1 - i}{10} + \dfrac{i}{50}} = \dfrac{1}{\dfrac{1 - i}{10} + \dfrac{i}{50}} \cdot \dfrac{50}{50} = \dfrac{50}{5 - 5i + i} =$

$\dfrac{50}{5 - 4i} = \dfrac{50}{5 - 4i} \cdot \dfrac{5 + 4i}{5 + 4i} = \dfrac{250 + 200i}{41} = \dfrac{250}{41} + \dfrac{200}{41}i$

104. 0

105. $(1 - i)^3(1 + i)^3 =$

$(1 - i)(1 + i) \cdot (1 - i)(1 + i) \cdot (1 - i)(1 + i) =$

$(1 - i^2)(1 - i^2)(1 - i^2) = (1 + 1)(1 + 1)(1 + 1) =$

$2 \cdot 2 \cdot 2 = 8$

106. $-1 - \sqrt{5}i$

107. $\dfrac{6}{1 + \dfrac{3}{i}} = \dfrac{6}{\dfrac{i + 3}{i}} = \dfrac{6i}{i + 3} = \dfrac{6i}{i + 3} \cdot \dfrac{-i + 3}{-i + 3} =$

$\dfrac{-6i^2 + 18i}{-i^2 + 9} = \dfrac{6 + 18i}{10} = \dfrac{6}{10} + \dfrac{18}{10}i = \dfrac{3}{5} + \dfrac{9}{5}i$

108. $-\dfrac{2}{3}i$

109. $\dfrac{i - i^{38}}{1 + i} = \dfrac{i - (i^2)^{19}}{1 + i} = \dfrac{i - (-1)^{19}}{1 + i} = \dfrac{i - (-1)}{1 + i} =$

$\dfrac{i + 1}{1 + i} = 1$

Chapter 11

Quadratic Functions and Equations

Exercise Set 11.1

1.
$$7x^2 = 21$$
$$x^2 = 3 \qquad \text{Multiplying by } \frac{1}{7}$$
$$x = \sqrt{3} \text{ or } x = -\sqrt{3} \quad \text{Using the principle} \\ \text{of square roots}$$
The solutions are $\sqrt{3}$ and $-\sqrt{3}$, or $\pm\sqrt{3}$.

2. $\pm\sqrt{5}$

3.
$$25x^2 + 4 = 0$$
$$x^2 = -\frac{4}{25} \qquad \text{Isolating } x^2$$
$$x = \sqrt{-\frac{4}{25}} \text{ or } x = -\sqrt{-\frac{4}{25}} \quad \text{Principle of} \\ \text{square roots}$$
$$x = \sqrt{\frac{4}{25}}\sqrt{-1} \text{ or } x = -\sqrt{\frac{4}{25}}\sqrt{-1}$$
$$x = \frac{2}{5}i \text{ or } x = -\frac{2}{5}i$$
The solutions are $\frac{2}{5}i$ and $-\frac{2}{5}i$, or $\pm\frac{2}{5}i$.

4. $\pm\frac{4}{3}i$

5. $3t^2 - 2 = 0$
$$3t^2 = 2$$
$$t^2 = \frac{2}{3}$$
$$t = \sqrt{\frac{2}{3}} \quad \text{ or } \quad t = -\sqrt{\frac{2}{3}} \quad \text{Principle of} \\ \text{square roots}$$
$$t = \sqrt{\frac{2}{3}\cdot\frac{3}{3}} \quad \text{ or } \quad t = -\sqrt{\frac{2}{3}\cdot\frac{3}{3}} \quad \text{Rationalizing} \\ \text{denominators}$$
$$t = \frac{\sqrt{6}}{3} \quad \text{ or } \quad t = \frac{-\sqrt{6}}{3}$$
The solutions are $\sqrt{\frac{2}{3}}$ and $-\sqrt{\frac{2}{3}}$. This can also be written as $\pm\sqrt{\frac{2}{3}}$ or, if we rationalize the denominator, $\pm\frac{\sqrt{6}}{3}$.

6. $\pm\sqrt{\frac{7}{5}}$, or $\pm\frac{\sqrt{35}}{5}$

7. $(x+2)^2 = 25$
$$x + 2 = 5 \text{ or } x + 2 = -5 \quad \text{Principle of square} \\ \text{roots}$$
$$x = 3 \text{ or } x = -7$$
The solutions are 3 and -7.

8. $-6, 8$

9.
$$(a+5)^2 = 8$$
$$a + 5 = \sqrt{8} \text{ or } a + 5 = -\sqrt{8} \quad \text{Principle of} \\ \text{square roots}$$
$$a + 5 = 2\sqrt{2} \text{ or } a + 5 = -2\sqrt{2} \quad (\sqrt{8} = \sqrt{4\cdot2} = \\ 2\sqrt{2})$$
$$a = -5 + 2\sqrt{2} \text{ or } a = -5 - 2\sqrt{2}$$
The solutions are $-5 + 2\sqrt{2}$ and $-5 - 2\sqrt{2}$, or $-5 \pm 2\sqrt{2}$.

10. $13 \pm 3\sqrt{2}$

11. $(x-1)^2 = -49$
$$x - 1 = \sqrt{-49} \quad \text{ or } \quad x - 1 = -\sqrt{-49}$$
$$x - 1 = 7i \qquad \text{ or } \quad x - 1 = -7i$$
$$x = 1 + 7i \quad \text{ or } \qquad x = 1 - 7i$$
The solutions are $1 + 7i$ and $1 - 7i$, or $1 \pm 7i$.

12. $-1 \pm 3i$

13. $\left(t + \frac{3}{2}\right)^2 = \frac{7}{2}$
$$t + \frac{3}{2} = \sqrt{\frac{7}{2}} \text{ or } t + \frac{3}{2} = -\sqrt{\frac{7}{2}}$$
$$t + \frac{3}{2} = \sqrt{\frac{7}{2}\cdot\frac{2}{2}} \text{ or } t + \frac{3}{2} = -\sqrt{\frac{7}{2}\cdot\frac{2}{2}}$$
$$t + \frac{3}{2} = \frac{\sqrt{14}}{2} \text{ or } t + \frac{3}{2} = -\frac{\sqrt{14}}{2}$$
$$t = -\frac{3}{2} + \frac{\sqrt{14}}{2} \text{ or } t = -\frac{3}{2} - \frac{\sqrt{14}}{2}$$
$$t = \frac{-3 + \sqrt{14}}{2} \text{ or } t = \frac{-3 - \sqrt{14}}{2}$$
The solutions are $\frac{-3 + \sqrt{14}}{2}$ and $\frac{-3 - \sqrt{14}}{2}$, or $\frac{-3 \pm \sqrt{14}}{2}$.

14. $\dfrac{-3 \pm \sqrt{17}}{4}$

15.
$$x^2 - 6x + 9 = 100$$
$$(x - 3)^2 = 100$$
$$x - 3 = 10 \ or \ x - 3 = -10$$
$$x = 13 \ or \ x = -7$$
The solutions are 13 and -7.

16. $-3, 13$

17.
$$f(x) = 16$$
$$(x - 5)^2 = 16 \qquad \text{Substituting}$$
$$x - 5 = 4 \ or \ x - 5 = -4$$
$$x = 9 \ or \ x = 1$$
The solutions are 9 and 1.

18. $-3, 7$

19.
$$F(t) = 13$$
$$(t + 4)^2 = 13 \quad \text{Substituting}$$
$$t + 4 = \sqrt{13} \qquad or \ t + 4 = -\sqrt{13}$$
$$t = -4 + \sqrt{13} \ or \qquad t = -4 - \sqrt{13}$$
The solutions are $-4 + \sqrt{13}$ and $-4 - \sqrt{13}$, or $-4 \pm \sqrt{13}$.

20. $-6 \pm \sqrt{15}$

21. $g(x) = x^2 + 14x + 49$

Observe first that $g(0) = 49$. Also observe that when $x = -14$, then $x^2 + 14x = (-14)^2 - (14)(14) = (14)^2 - (14)^2 = 0$, so $g(-14) = 49$ as well. Thus, we have $x = 0$ or $x = 14$.

We can also do this problem as follows.
$$g(x) = 49$$
$$x^2 + 14x + 49 = 49 \quad \text{Substituting}$$
$$(x + 7)^2 = 49$$
$$x + 7 = 7 \ or \ x + 7 = -7$$
$$x = 0 \ or \qquad x = -14$$
The solutions are -1 and -14.

22. $-7, -1$

23. $x^2 + 8x$

We take half the coefficient of x and square it:
Half of 8 is 4, and $4^2 = 16$. We add 16.
$x^2 + 8x + 16, \ (x + 4)^2$

24. $x^2 + 16x + 64, \ (x + 8)^2$

25. $x^2 - 6x$

We take half the coefficient of x and square it:
Half of -6 is -3, and $(-3)^2 = 9$. We add 9.
$x^2 - 6x + 9, \ (x - 3)^2$

26. $x^2 - 10x + 25, \ (x - 5)^2$

27. $x^2 - 24x$

We take half the coefficient of x and square it:
$\dfrac{1}{2}(-24) = -12$ and $(-12)^2 = 144$. We add 144.
$x^2 - 24x + 144, \ (x - 12)^2$

28. $x^2 - 18x + 81, \ (x - 9)^2$

29. $t^2 + 9t$

$\dfrac{1}{2} \cdot 9 = \dfrac{9}{2}$, and $\left(\dfrac{9}{2}\right)^2 = \dfrac{81}{4}$. We add $\dfrac{81}{4}$.
$t^2 + 9t + \dfrac{81}{4}, \ \left(t + \dfrac{9}{2}\right)^2$

30. $t^2 + 3t + \dfrac{9}{4}, \ \left(t + \dfrac{3}{2}\right)^2$

31. $x^2 - 3x$

We take half the coefficient of x and square it:
$\dfrac{1}{2}(-3) = -\dfrac{3}{2}$ and $\left(-\dfrac{3}{2}\right)^2 = \dfrac{9}{4}$. We add $\dfrac{9}{4}$.
$x^2 - 3x + \dfrac{9}{4}, \ \left(x - \dfrac{3}{2}\right)^2$

32. $x^2 - 7x + \dfrac{49}{4}, \ \left(x - \dfrac{7}{2}\right)^2$

33. $x^2 + \dfrac{2}{3}x$

$\dfrac{1}{2} \cdot \dfrac{2}{3} = \dfrac{1}{3}$, and $\left(\dfrac{1}{3}\right)^2 = \dfrac{1}{9}$. We add $\dfrac{1}{9}$.
$x^2 + \dfrac{2}{3}x + \dfrac{1}{9}, \ \left(x + \dfrac{1}{3}\right)^2$

34. $x^2 + \dfrac{2}{5}x + \dfrac{1}{25}, \ \left(x + \dfrac{1}{5}\right)^2$

35. $t^2 - \dfrac{5}{3}t$

$\dfrac{1}{2}\left(-\dfrac{5}{3}\right) = -\dfrac{5}{6}$, and $\left(-\dfrac{5}{6}\right)^2 = \dfrac{25}{36}$. We add $\dfrac{25}{36}$.
$t^2 - \dfrac{5}{3}t + \dfrac{25}{36}, \ \left(t - \dfrac{5}{6}\right)^2$

36. $t^2 - \dfrac{5}{6}t + \dfrac{25}{144}, \left(t - \dfrac{5}{12}\right)^2$

37. $x^2 + \dfrac{9}{5}x$

$\dfrac{1}{2} \cdot \dfrac{9}{5} = \dfrac{9}{10},$ and $\left(\dfrac{9}{10}\right)^2 = \dfrac{81}{100}.$ We add $\dfrac{81}{100}.$

$x^2 + \dfrac{9}{5}x + \dfrac{81}{100}, \left(x + \dfrac{9}{10}\right)^2$

38. $x^2 + \dfrac{9}{4}x + \dfrac{81}{64}, \left(x + \dfrac{9}{8}\right)^2$

39. $\qquad x^2 + 6x = 7$

$\qquad x^2 + 6x + 9 = 7 + 9 \qquad$ Adding 9 to both sides
$\qquad\qquad\qquad\qquad\qquad$ to complete the square

$\qquad (x + 3)^2 = 16 \qquad$ Factoring

$\qquad x + 3 = \pm 4 \qquad$ Principle of square roots

$\qquad\qquad x = -3 \pm 4$

$x = -3 + 4 \;\; or \;\; x = -3 - 4$

$x = 1 \qquad or \;\; x = -7$

The solutions are 1 and -7.

40. $-9, 1$

41. $\qquad x^2 - 10x = 22$

$\qquad x^2 - 10x + 25 = 22 + 25 \qquad$ Adding 25 to both
$\qquad\qquad\qquad\qquad\qquad\quad$ sides to complete the square

$\qquad (x - 5)^2 = 47$

$\qquad x - 5 = \pm\sqrt{47} \qquad$ Principle of square
$\qquad\qquad\qquad\qquad\qquad$ roots

$\qquad\qquad x = 5 \pm \sqrt{47}$

The solutions are $5 \pm \sqrt{47}$.

42. $2 \pm i\sqrt{5}$

43. $\qquad x^2 + 8x + 7 = 0$

$\qquad x^2 + 8x = -7 \qquad$ Adding -7 to both sides

$\qquad x^2 + 8x + 16 = -7 + 16 \qquad$ Completing the square

$\qquad (x + 4)^2 = 9$

$\qquad x + 4 = \pm 3$

$\qquad\qquad x = -4 \pm 3$

$x = -4 - 3 \;\; or \;\; x = -4 + 3$

$x = -7 \qquad or \;\; x = -1$

The solutions are -7 and -1.

44. $-9, -1$

45. $\qquad x^2 - 10x + 21 = 0$

$\qquad x^2 - 10x = -21$

$\qquad x^2 - 10x + 25 = -21 + 25$

$\qquad (x - 5)^2 = 4$

$\qquad x - 5 = \pm 2$

$\qquad\qquad x = 5 \pm 2$

$x = 5 - 2 \;\; or \;\; x = 5 + 2$

$x = 3 \qquad or \;\; x = 7$

The solutions are 3 and 7.

46. $4, 6$

47. $\qquad t^2 + 5t + 3 = 0$

$\qquad t^2 + 5t = -3$

$\qquad t^2 + 5t + \dfrac{25}{4} = -3 + \dfrac{25}{4}$

$\qquad \left(t + \dfrac{5}{2}\right)^2 = \dfrac{13}{4}$

$\qquad t + \dfrac{5}{2} = \pm\dfrac{\sqrt{13}}{2}$

$\qquad t = -\dfrac{5}{2} \pm \dfrac{\sqrt{13}}{2}$

$\qquad t = \dfrac{-5 \pm \sqrt{13}}{2}$

The solutions are $\dfrac{-5 \pm \sqrt{13}}{2}$.

48. $-3 \pm \sqrt{2}$

49. $\qquad x^2 + 10 = 6x$

$\qquad x^2 - 6x = -10$

$\qquad x^2 - 6x + 9 = -10 + 9$

$\qquad (x - 3)^2 = -1$

$\qquad x - 3 = \pm\sqrt{-1}$

$\qquad x - 3 = \pm i$

$\qquad\qquad x = 3 \pm i$

The solutions are $3 \pm i$.

50. $5 \pm \sqrt{2}$

51. $\qquad s^2 + 4s + 13 = 0$

$\qquad s^2 + 4s = -13$

$\qquad s^2 + 4s + 4 = -13 + 4$

$\qquad (s + 2)^2 = -9$

$\qquad s + 2 = \pm\sqrt{-9}$

$\qquad s + 2 = \pm 3i$

$\qquad\qquad s = -2 \pm 3i$

The solutions are $-2 \pm 3i$.

52. $-6 \pm \sqrt{11}$

53. $2x^2 - 5x - 3 = 0$

$$2x^2 - 5x = 3$$

$$x^2 - \frac{5}{2}x = \frac{3}{2} \quad \text{Dividing both sides by 2}$$

$$x^2 - \frac{5}{2}x + \frac{25}{16} = \frac{3}{2} + \frac{25}{16}$$

$$\left(x - \frac{5}{4}\right)^2 = \frac{49}{16}$$

$$x - \frac{5}{4} = \pm\frac{7}{4}$$

$$x = \frac{5}{4} \pm \frac{7}{4}$$

$$x = \frac{5}{4} - \frac{7}{4} \quad or \quad x = \frac{5}{4} + \frac{7}{4}$$

$$x = -\frac{1}{2} \quad or \quad x = 3$$

The solutions are $-\frac{1}{2}$ and 3.

54. $-2, \frac{1}{3}$

55. $4x^2 + 8x + 3 = 0$

$$4x^2 + 8x = -3$$

$$x^2 + 2x = -\frac{3}{4}$$

$$x^2 + 2x + 1 = -\frac{3}{4} + 1$$

$$(x + 1)^2 = \frac{1}{4}$$

$$x + 1 = \pm\frac{1}{2}$$

$$x = -1 \pm \frac{1}{2}$$

$$x = -1 - \frac{1}{2} \quad or \quad x = -1 + \frac{1}{2}$$

$$x = -\frac{3}{2} \quad or \quad x = -\frac{1}{2}$$

The solutions are $-\frac{3}{2}$ and $-\frac{1}{2}$.

56. $-\frac{4}{3}, -\frac{2}{3}$

57. $6x^2 - x = 15$

$$x^2 - \frac{1}{6}x = \frac{5}{2}$$

$$x^2 - \frac{1}{6}x + \frac{1}{144} = \frac{5}{2} + \frac{1}{144}$$

$$\left(x - \frac{1}{12}\right)^2 = \frac{361}{144}$$

$$x - \frac{1}{12} = \pm\frac{19}{12}$$

$$x = \frac{1}{12} \pm \frac{19}{12}$$

$$x = \frac{1}{12} + \frac{19}{12} \quad or \quad x = \frac{1}{12} - \frac{19}{12}$$

$$x = \frac{20}{12} \quad or \quad x = -\frac{18}{12}$$

$$x = \frac{5}{3} \quad or \quad x = -\frac{3}{2}$$

The solutions are $\frac{5}{3}$ and $-\frac{3}{2}$.

58. $-\frac{1}{2}, \frac{2}{3}$

59. $2x^2 + 4x + 1 = 0$

$$2x^2 + 4x = -1$$

$$x^2 + 2x = -\frac{1}{2}$$

$$x^2 + 2x + 1 = -\frac{1}{2} + 1$$

$$(x + 1)^2 = \frac{1}{2}$$

$$x + 1 = \pm\sqrt{\frac{1}{2}}$$

$$x + 1 = \pm\frac{\sqrt{2}}{2} \quad \text{Rationalizing the denominator}$$

$$x = -1 \pm \frac{\sqrt{2}}{2}$$

The solutions are $-1 \pm \frac{\sqrt{2}}{2}$, or $\frac{-2 \pm \sqrt{2}}{2}$.

60. $-2, -\frac{1}{2}$

61. $3x^2 - 5x - 3 = 0$

$$3x^2 - 5x = 3$$

$$x^2 - \frac{5}{3}x = 1$$

$$x^2 - \frac{5}{3}x + \frac{25}{36} = 1 + \frac{25}{36}$$

$$\left(x - \frac{5}{6}\right)^2 = \frac{61}{36}$$

$$x - \frac{5}{6} = \pm\frac{\sqrt{61}}{6}$$

$$x = \frac{5 \pm \sqrt{61}}{6}$$

The solutions are $\dfrac{5 \pm \sqrt{61}}{6}$.

62. $\dfrac{3 \pm \sqrt{13}}{4}$

63. *Familiarize*. We are already familiar with the compound-interest formula.

***Translate*.** We substitute into the formula.

$$A = P(1 + r)^t$$

$$2420 = 2000(1 + r)^2$$

***Carry out*.** We solve for r.

$$2420 = 2000(1 + r)^2$$

$$\frac{2420}{2000} = (1 + r)^2$$

$$\frac{121}{100} = (1 + r)^2$$

$$\pm\sqrt{\frac{121}{100}} = 1 + r$$

$$\pm\frac{11}{10} = 1 + r$$

$$-\frac{10}{10} + \frac{11}{10} = r$$

$$\frac{1}{10} = r \ or \ -\frac{21}{10} = r$$

***Check*.** Since the interest rate cannot be negative, we need only check $\dfrac{1}{10}$, or 10%. If $2000 were invested at 10% interest, compounded annually, then in 2 years it would grow to $2000(1.1)^2$, or $2420. The number 10% checks.

***State*.** The interest rate is 10%.

64. 6.25%

65. *Familiarize*. We are already familiar with the compound-interest formula.

***Translate*.** We substitute into the formula.

$$A = P(1 + r)^t$$

$$1805 = 1280(1 + r)^2$$

***Carry out*.** We solve for r.

$$1805 = 1280(1 + r)^2$$

$$\frac{1805}{1280} = (1 + r)^2$$

$$\frac{361}{256} = (1 + r)^2$$

$$\pm\frac{19}{16} = 1 + r$$

$$-\frac{16}{16} \pm \frac{19}{16} = r$$

$$\frac{3}{16} = r \ or \ -\frac{35}{16} = r$$

***Check*.** Since the interest rate cannot be negative, we need only check $\dfrac{3}{16}$ or 18.75%. If $1280 were invested at 18.75% interest, compounded annually, then in 2 years it would grow to $1280(1.1875)^2$, or $1805. The number 18.75% checks.

***State*.** The interest rate is 18.75%.

66. 20%

67. *Familiarize*. We are already familiar with the compound-interest formula.

***Translate*.** We substitute into the formula.

$$A = P(1 + r)^t$$

$$6760 = 6250(1 + r)^2$$

***Carry out*.** We solve for r.

$$\frac{6760}{6250} = (1 + r)^2$$

$$\frac{676}{625} = (1 + r)^2$$

$$\pm\frac{26}{25} = 1 + r$$

$$-\frac{25}{25} \pm \frac{26}{25} = r$$

$$\frac{1}{25} = r \ or \ -\frac{51}{25} = r$$

***Check*.** Since the interest rate cannot be negative, we need only check $\dfrac{1}{25}$, or 4%. If $6250 were invested at 4% interest, compounded annually, then in 2 years it would grow to $6250(1.04)^2$, or $6760. The number 4% checks.

***State*.** The interest rate is 4%.

68. 8%

69. *Familiarize*. We will use the formula $s = 16t^2$.

***Translate*.** We substitute into the formula.

$$s = 16t^2$$

$$1815 = 16t^2$$

Carry out. We solve for t.

$$1815 = 16t^2$$

$$\frac{1815}{16} = t^2$$

$$\sqrt{\frac{1815}{16}} = t \quad \text{Principle of square roots;}$$
$$\text{rejecting the negative}$$
$$\text{square root}$$

$$10.7 \approx t$$

Check. Since $16(10.7)^2 = 1831.84 \approx 1815$, our answer checks.

State. It would take an object about 10.7 sec to fall freely from the top of the CN Tower.

70. About 6.8 sec

71. *Familiarize.* We will use the formula $s = 16t^2$.

Translate. We substitute into the formula.

$$s = 16t^2$$
$$640 = 16t^2$$

Carry out. We solve for t.

$$640 = 16t^2$$
$$40 = t^2$$
$$\sqrt{40} = t \quad \text{Principle of square roots;}$$
$$\text{rejecting the negative square}$$
$$\text{root}$$
$$6.3 \approx t$$

Check. Since $16(6.3)^2 = 635.04 \approx 640$, our answer checks.

State. It would take an object about 6.3 sec to fall freely from the top of the Gateway Arch.

72. About 9.5 sec

73. Writing exercise

74. Writing exercise

75.
$$at^2 - bt = 3 \cdot 4^2 - 5 \cdot 4$$
$$= 3 \cdot 16 - 5 \cdot 4$$
$$= 48 - 20$$
$$= 28$$

76. -92

77. $\sqrt[3]{270} = \sqrt[3]{27 \cdot 10} = \sqrt[3]{27}\sqrt[3]{10} = 3\sqrt[3]{10}$

78. $4\sqrt{5}$

79. $f(x) = \sqrt{3x - 5}$

$f(10) = \sqrt{3 \cdot 10 - 5} = \sqrt{30 - 5} = \sqrt{25} = 5$

80. 7

81. Writing exercise

82. Writing exercise

83. In order for $x^2 + bx + 81$ to be a square, the following must be true:

$$\left(\frac{b}{2}\right)^2 = 81$$

$$\frac{b^2}{4} = 81$$

$$b^2 = 324$$

$$b = 18 \text{ or } b = -18$$

84. ± 14

85. We see that x is a factor of each term, so x is also a factor of $f(x)$. We have $f(x) = x(2x^4 - 9x^3 - 66x^2 + 45x + 280)$. Since $x^2 - 5$ is a factor of $f(x)$ it is also a factor of $2x^4 - 9x^3 - 66x^2 + 45x + 280$. We divide to find another factor.

$$
\begin{array}{r}
2x^2 - 9x - 56 \\
x^2 - 5 \overline{\smash{\big)}\, 2x^4 - 9x^3 - 66x^2 + 45x + 280} \\
\underline{2x^4 - 10x^2} \\
-9x^3 - 56x^2 + 45x \\
\underline{-9x^3 + 45x } \\
-56x^2 + 280 \\
\underline{-56x^2 + 280} \\
0
\end{array}
$$

Then we have $f(x) = x(x^2 - 5)(2x^2 - 9x - 56)$, or $f(x) = x(x^2 - 5)(2x + 7)(x - 8)$. Now we find the values of a for which $f(a) = 0$.

$$f(a) = 0$$
$$a(a^2 - 5)(2a + 7)(a - 8) = 0$$

$a=0$ *or* $a^2-5=0$ *or* $2a+7=0$ *or* $a-8=0$

$a=0$ *or* $a^2=5$ *or* $2a=-7$ *or* $a=8$

$a=0$ *or* $a=\pm\sqrt{5}$ *or* $a=-\dfrac{7}{2}$ *or* $a=8$

The solutions are 0, $\sqrt{5}$, $-\sqrt{5}$, $-\dfrac{7}{2}$, and 8.

86. $\dfrac{1}{3}, \pm\dfrac{2\sqrt{6}}{3}i$

87. *Familiarize.* It is helpful to list information in a chart and make a drawing. Let r represent the speed of the fishing boat. Then $r - 7$ represents the speed of the barge.

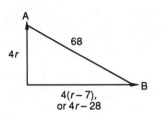

Right triangle with vertex A at top left, hypotenuse labeled 68 going to B at lower right, left side labeled $4r$, bottom labeled $4(r-7)$, or $4r-28$.

Boat	r	t	d
Fishing	r	4	$4r$
Barge	$r-7$	4	$4(r-7)$

Translate. We use the Pythagorean equation:

$$a^2 + b^2 = c^2$$
$$(4r-28)^2 + (4r)^2 = 68^2$$

Carry out.

$$(4r-28)^2 + (4r)^2 = 68^2$$
$$16r^2 - 224r + 784 + 16r^2 = 4624$$
$$32r^2 - 224r - 3840 = 0$$
$$r^2 - 7r - 120 = 0$$
$$(r+8)(r-15) = 0$$
$$r + 8 = 0 \quad or \quad r - 15 = 0$$
$$r = -8 \quad or \quad r = 15$$

Check. We check only 15 since the speeds of the boats cannot be negative. If the speed of the fishing boat is 15 km/h, then the speed of the barge is $15-7$, or 8 km/h, and the distances they travel are $4 \cdot 15$ (or 60) and $4 \cdot 8$ (or 32).

$$60^2 + 32^2 = 3600 + 1024 = 4624 = 68^2$$

The values check.

State. The speed of the fishing boat is 15 km/h, and the speed of the barge is 8 km/h.

88. 5, 6, 7

89.

90.

91. Writing exercise

Exercise Set 11.2

1. $x^2 - 7x - 3 = 0$

$a = 1, b = -7, c = -3$

$$x = \frac{-b \pm \sqrt{b^2 - 4ac}}{2a}$$
$$x = \frac{-(-7) \pm \sqrt{(-7)^2 - 4 \cdot 1 \cdot (-3)}}{2 \cdot 1} = \frac{7 \pm \sqrt{49 + 12}}{2}$$
$$x = \frac{7 \pm \sqrt{61}}{2}$$

The solutions are $\dfrac{7 + \sqrt{61}}{2}$ and $\dfrac{7 - \sqrt{61}}{2}$.

2. $\dfrac{-7 \pm \sqrt{33}}{2}$

3.
$$3p^2 = 18p - 6$$
$$3p^2 - 18p + 6 = 0$$
$$p^2 - 6p + 2 = 0 \quad \text{Dividing by 3}$$
$$a = 1, b = -6, c = 2$$
$$p = \frac{-b \pm \sqrt{b^2 - 4ac}}{2a}$$
$$p = \frac{-(-6) \pm \sqrt{(-6)^2 - 4 \cdot 1 \cdot 2}}{2 \cdot 1} = \frac{6 \pm \sqrt{36 - 8}}{2}$$
$$p = \frac{6 \pm \sqrt{28}}{2} = \frac{6 \pm 2\sqrt{7}}{2}$$
$$p = \frac{2(3 \pm \sqrt{7})}{2} = 3 \pm \sqrt{7}$$

The solutions are $3 + \sqrt{7}$ and $3 - \sqrt{7}$.

4. $1, \dfrac{5}{3}$

5. $x^2 - x + 2 = 0$

$a = 1, b = -1, c = 2$

$$x = \frac{-b \pm \sqrt{b^2 - 4ac}}{2a}$$
$$x = \frac{-(-1) \pm \sqrt{(-1)^2 - 4 \cdot 1 \cdot 2}}{2 \cdot 1} = \frac{1 \pm \sqrt{1 - 8}}{2}$$
$$x = \frac{1 \pm \sqrt{-7}}{2} = \frac{1 \pm i\sqrt{7}}{2}$$

The solutions are $\dfrac{1 + i\sqrt{7}}{2}$ and $\dfrac{1 - i\sqrt{7}}{2}$, or $\dfrac{1}{2} + \dfrac{\sqrt{7}}{2}i$ and $\dfrac{1}{2} - \dfrac{\sqrt{7}}{2}i$.

6. $-\dfrac{1}{2} \pm \dfrac{\sqrt{3}}{2}i$

7.
$$x^2 + 13 = 4x$$
$$x^2 - 4x + 13 = 0$$
$$a = 1, b = -4, c = 13$$

$$x = \frac{-b \pm \sqrt{b^2 - 4ac}}{2a}$$

$$x = \frac{-(-4) \pm \sqrt{(-4)^2 - 4 \cdot 1 \cdot 13}}{2 \cdot 1} = \frac{4 \pm \sqrt{16 - 52}}{2}$$

$$x = \frac{4 \pm \sqrt{-36}}{2} = \frac{4 \pm 6i}{2}$$

$$x = \frac{2(2 \pm 3i)}{2} = 2 \pm 3i$$

The solutions are $2 + 3i$ and $2 - 3i$.

8. $3 \pm 2i$

9.
$$h^2 + 4 = 6h$$
$$h^2 - 6h + 4 = 0$$
$$a = 1, \ b = -6, \ c = 4$$
$$x = \frac{-(-6) \pm \sqrt{(-6)^2 - 4 \cdot 1 \cdot 4}}{2 \cdot 1} = \frac{6 \pm \sqrt{36 - 16}}{2}$$
$$x = \frac{6 \pm \sqrt{20}}{2} = \frac{6 \pm \sqrt{4 \cdot 5}}{2} = \frac{6 \pm 2\sqrt{5}}{2}$$
$$x = 3 \pm \sqrt{5}$$

The solutions are $3 + \sqrt{5}$ and $3 - \sqrt{5}$.

10. $\dfrac{-3 \pm \sqrt{41}}{2}$

11.
$$3 + \frac{8}{x} = \frac{1}{x^2}, \ \text{LCD is } x^2$$
$$x^2\left(3 + \frac{8}{x}\right) = x^2 \cdot \frac{1}{x^2}$$
$$3x^2 + 8x = 1$$
$$3x^2 + 8x - 1 = 0$$
$$a = 3, \ b = 8, \ c = -1$$
$$x = \frac{-8 \pm \sqrt{8^2 - 4 \cdot 3 \cdot (-1)}}{2 \cdot 3} = \frac{-8 \pm \sqrt{64 + 12}}{6}$$
$$x = \frac{-8 \pm \sqrt{76}}{6} = \frac{-8 \pm \sqrt{4 \cdot 19}}{6} = \frac{-8 \pm 2\sqrt{19}}{6}$$
$$x = \frac{-4 \pm \sqrt{19}}{3}$$

The solutions are $\dfrac{-4 + \sqrt{19}}{3}$ and $\dfrac{-4 - \sqrt{19}}{3}$.

12. $\dfrac{9 \pm \sqrt{41}}{4}$

13.
$$3x + x(x - 2) = 4$$
$$3x + x^2 - 2x = 4$$
$$x^2 + x = 4$$
$$x^2 + x - 4 = 0$$
$$a = 1, \ b = 1, \ c = -4$$

$$x = \frac{-1 \pm \sqrt{1^2 - 4 \cdot 1 \cdot (-4)}}{2 \cdot 1} = \frac{-1 \pm \sqrt{1 + 16}}{2}$$

$$x = \frac{-1 \pm \sqrt{17}}{2}$$

The solutions are $\dfrac{-1 + \sqrt{17}}{2}$ and $\dfrac{-1 - \sqrt{17}}{2}$.

14. $\dfrac{-1 \pm \sqrt{21}}{2}$

15.
$$12x^2 + 9t = 1$$
$$12t^2 + 9t - 1 = 0$$
$$a = 12, \ b = 9, \ c = -1$$
$$t = \frac{-9 \pm \sqrt{9^2 - 4 \cdot 12 \cdot (-1)}}{2 \cdot 12} = \frac{-9 \pm \sqrt{81 + 48}}{24}$$
$$t = \frac{-9 \pm \sqrt{129}}{24}$$

The solutions are $\dfrac{-9 + \sqrt{129}}{24}$ and $\dfrac{-9 - \sqrt{129}}{24}$.

16. $-\dfrac{2}{3}, \dfrac{1}{5}$

17.
$$25x^2 - 20x + 4 = 0$$
$$(5x - 2)(5x - 2) = 0$$
$$5x - 2 = 0 \quad or \quad 5x - 2 = 0$$
$$5x = 2 \quad or \quad 5x = 2$$
$$x = \frac{2}{5} \quad or \quad x = \frac{2}{5}$$

The solution is $\dfrac{2}{5}$.

18. $-\dfrac{7}{6}$

19.
$$7x(x + 2) + 5 = 3x(x + 1)$$
$$7x^2 + 14x + 5 = 3x^2 + 3x$$
$$4x^2 + 11x + 5 = 0$$
$$a = 4, \ b = 11, \ c = 5$$
$$x = \frac{-11 \pm \sqrt{11^2 - 4 \cdot 4 \cdot 5}}{2 \cdot 4} = \frac{-11 \pm \sqrt{121 - 80}}{8}$$
$$x = \frac{-11 \pm \sqrt{41}}{8}$$

The solutions are $\dfrac{-11 + \sqrt{41}}{8}$ and $\dfrac{-11 - \sqrt{41}}{8}$.

20. $\dfrac{-3 \pm \sqrt{37}}{2}$

21.
$$14(x - 4) - (x + 2) = (x + 2)(x - 4)$$
$$14x - 56 - x - 2 = x^2 - 2x - 8 \quad \text{Removing}$$
$$\text{parentheses}$$
$$13x - 58 = x^2 - 2x - 8$$
$$0 = x^2 - 15x + 50$$
$$0 = (x - 10)(x - 5)$$

$x - 10 = 0 \quad or \quad x - 5 = 0$

$x = 10 \quad or \qquad x = 5$

The solutions are 10 and 5.

22. $1, 15$

23.
$$5x^2 = 13x + 17$$
$$5x^2 - 13x - 17 = 0$$
$$a = 5, \, b = -13, \, c = -17$$
$$x = \frac{-(-13) \pm \sqrt{(-13)^2 - 4(5)(-17)}}{2 \cdot 5}$$
$$x = \frac{13 \pm \sqrt{169 + 340}}{10} = \frac{13 \pm \sqrt{509}}{10}$$

The solutions are $\dfrac{13 + \sqrt{509}}{10}$ and $\dfrac{13 - \sqrt{509}}{10}$.

24. $\dfrac{4}{3}, 7$

25.
$$x^2 + 9 = 4x$$
$$x^2 - 4x + 9 = 0$$
$$a = 1, \, b = -4, \, c = 9$$
$$x = \frac{-(-4) \pm \sqrt{(-4)^2 - 4 \cdot 1 \cdot 9}}{2 \cdot 1} = \frac{4 \pm \sqrt{16 - 36}}{2}$$
$$x = \frac{4 \pm \sqrt{-20}}{2} = \frac{4 \pm \sqrt{-4 \cdot 5}}{2}$$
$$x = \frac{4 \pm 2i\sqrt{5}}{2} = 2 \pm i\sqrt{5}$$

The solutions are $2 + i\sqrt{5}$ and $2 - i\sqrt{5}$.

26. $\dfrac{3}{2} \pm \dfrac{\sqrt{19}}{2}i$

27.
$$x^3 - 8 = 0$$
$$x^3 - 2^3 = 0$$
$$(x - 2)(x^2 + 2x + 4) = 0$$
$$x - 2 = 0 \quad or \quad x^2 + 2x + 4 = 0$$
$$x = 2 \quad or \quad x = \frac{-2 \pm \sqrt{2^2 - 4 \cdot 1 \cdot 4}}{2 \cdot 1}$$
$$x = 2 \quad or \quad x = \frac{-2 \pm \sqrt{-12}}{2} = \frac{-2 \pm 2i\sqrt{3}}{2}$$
$$x = 2 \quad or \quad x = -1 \pm i\sqrt{3}$$

The solutions are 2, $-1 + i\sqrt{3}$, and $-1 - i\sqrt{3}$.

28. $-1, \dfrac{1}{2} \pm \dfrac{\sqrt{3}}{2}i$

29.
$$f(x) = 0$$
$$3x^2 - 5x - 1 = 0 \qquad \text{Substituting}$$
$$a = 3, \, b = -5, \, c = -1$$

$$x = \frac{-(-5) \pm \sqrt{(-5)^2 - 4 \cdot 3 \cdot (-1)}}{2 \cdot 3}$$
$$x = \frac{5 \pm \sqrt{25 + 12}}{6} = \frac{5 \pm \sqrt{37}}{6}$$

The solutions are $\dfrac{5 + \sqrt{37}}{6}$ and $\dfrac{5 - \sqrt{37}}{6}$.

30. $\dfrac{1 \pm \sqrt{13}}{4}$

31.
$$f(x) = 1$$
$$\frac{7}{x} + \frac{7}{x + 4} = 1 \qquad \text{Substituting}$$
$$x(x + 4)\left(\frac{7}{x} + \frac{7}{x + 4}\right) = x(x + 4) \cdot 1$$
$$\qquad\qquad \text{Multiplying by the LCD}$$
$$7(x + 4) + 7x = x^2 + 4x$$
$$7x + 28 + 7x = x^2 + 4x$$
$$14x + 28 = x^2 + 4x$$
$$0 = x^2 - 10x - 28$$
$$a = 1, \, b = -10, \, c = -28$$
$$x = \frac{-(-10) \pm \sqrt{(-10)^2 - 4 \cdot 1 \cdot (-28)}}{2 \cdot 1}$$
$$x = \frac{10 \pm \sqrt{100 + 112}}{2} = \frac{10 \pm \sqrt{212}}{2}$$
$$x = \frac{10 \pm \sqrt{4 \cdot 53}}{2} = \frac{10 \pm 2\sqrt{53}}{2}$$
$$x = 5 \pm \sqrt{53}$$

The solutions are $5 + \sqrt{53}$ and $5 - \sqrt{53}$.

32. $-2, 3$

33.
$$F(x) = G(x)$$
$$\frac{x + 3}{x} = \frac{x - 4}{3} \qquad \text{Substituting}$$
$$3x\left(\frac{x + 3}{x}\right) = 3x\left(\frac{x - 4}{3}\right) \qquad \begin{array}{l}\text{Multiplying}\\\text{by the LCD}\end{array}$$
$$3x + 9 = x^2 - 4x$$
$$0 = x^2 - 7x - 9$$
$$a = 1, \, b = -7, \, c = -9$$
$$x = \frac{-(-7) \pm \sqrt{(-7)^2 - 4 \cdot 1 \cdot (-9)}}{2 \cdot 1}$$
$$x = \frac{7 \pm \sqrt{49 + 36}}{2} = \frac{7 \pm \sqrt{85}}{2}$$

The solutions are $\dfrac{7 + \sqrt{85}}{2}$ and $\dfrac{7 - \sqrt{85}}{2}$.

34. $\dfrac{3 \pm \sqrt{5}}{2}$

35.
$$f(x) = g(x)$$
$$\frac{15 - 2x}{6} = \frac{3}{x}, \text{ LCD is } 6x$$
$$6x \cdot \frac{15 - 2x}{6} = 6x \cdot \frac{3}{x}$$
$$x(15 - 2x) = 6 \cdot 3$$
$$15x - 2x^2 = 18$$
$$0 = 2x^2 - 15x + 18$$
$$0 = (2x - 3)(x - 6)$$
$$2x - 3 = 0 \quad or \quad x - 6 = 0$$
$$2x = 3 \quad or \quad x = 6$$
$$x = \frac{3}{2} \quad or \quad x = 6$$

The solutions are $\frac{3}{2}$ and 6.

36. $\pm 2\sqrt{7}$

37. $x^2 + 4x - 7 = 0$
$$a = 1, b = 4, c = -7$$
$$x = \frac{-4 \pm \sqrt{4^2 - 4 \cdot 1 \cdot (-7)}}{2 \cdot 1} = \frac{-4 \pm \sqrt{16 + 28}}{2}$$
$$x = \frac{-4 \pm \sqrt{44}}{2}$$

Using a calculator we find that $\frac{-4 + \sqrt{44}}{2} \approx$ 1.31662479 and $\frac{-4 - \sqrt{44}}{2} \approx -5.31662479$.

The solutions are approximately 1.31662479 and -5.31662479.

38. $-0.7639320225, -5.236067978$

39. $x^2 - 6x + 4 = 0$
$$a = 1, b = -6, c = 4$$
$$x = \frac{-(-6) \pm \sqrt{(-6)^2 - 4 \cdot 1 \cdot 4}}{2 \cdot 1} = \frac{6 \pm \sqrt{36 - 16}}{2}$$
$$x = \frac{6 \pm \sqrt{20}}{2}$$

Using a calculator we find that $\frac{6 + \sqrt{20}}{2} \approx$ 5.236067978 and $\frac{6 - \sqrt{20}}{2} \approx 0.7639320225$.

The solutions are approximately 5.236067978 and 0.7639320225.

40. $3.732050808, 0.2679491924$

41. $2x^2 - 3x - 7 = 0$
$$a = 2, b = -3, c = -7$$

$$x = \frac{-(-3) \pm \sqrt{(-3)^2 - 4 \cdot 2 \cdot (-7)}}{2 \cdot 2}$$
$$x = \frac{3 \pm \sqrt{9 + 56}}{4} = \frac{3 \pm \sqrt{65}}{4}$$

Using a calculator we find that $\frac{3 + \sqrt{65}}{4} \approx$ 2.765564437 and $\frac{3 - \sqrt{65}}{4} \approx -1.265564437$.

The solutions are approximately 2.765564437 and -1.265564437.

42. $1.457427108, -0.4574271078$

43. Writing exercise

44. Writing exercise

45. *Familiarize*. Let $x =$ the number of pounds of Kenyan coffee and $y =$ the number of pounds of Peruvian coffee in the mixture. We organize the information in a table.

Type of Coffee	Kenyan	Peruvian	Mixture
Price per pound	$6.75	$11.25	$8.55
Number of pounds	x	y	50
Total cost	$6.75x	$11.25y	8.55×50, or $427.50

Translate. From the last two rows of the table we get a system of equations.
$$x + y = 50,$$
$$6.75x + 11.25y = 427.50$$

Solve. Solving the system of equations, we get $(30, 20)$.

Check. The total number of pounds in the mixture is $30 + 20$, or 50. The total cost of the mixture is $6.75(30) + 11.25(20) = 427.50$. The values check.

State. The mixture should consist of 30 lb of Kenyan coffee and 20 lb of Peruvian coffee.

46. 46 cream-filled, 44 glazed

47. $\sqrt{27a^2b^5} \cdot \sqrt{6a^3b} = \sqrt{27a^2b^5 \cdot 6a^3b} = $ $\sqrt{162a^5b^6} = \sqrt{81a^4b^6 \cdot 2a} = \sqrt{81a^4b^6}\sqrt{2a} = $ $9a^2b^3\sqrt{2a}$

48. $4a^2b^3\sqrt{6}$

49.
$$\dfrac{\dfrac{3}{x-1}}{\dfrac{1}{x+1}+\dfrac{2}{x-1}}$$

$$=\dfrac{\dfrac{3}{x-1}}{\dfrac{1}{x+1}+\dfrac{2}{x-1}}\cdot\dfrac{(x-1)(x+1)}{(x-1)(x+1)}$$

$$=\dfrac{3(x+1)}{x-1+2(x+1)}$$

$$=\dfrac{3x+3}{x-1+2x+2}$$

$$=\dfrac{3x+3}{3x+1},\text{ or }\dfrac{3(x+1)}{3x+1}$$

50. $\dfrac{4b}{3ab^2-4a^2}$

51. Writing exercise

52. Writing exercise

53. $f(x)=\dfrac{x^2}{x-2}+1$

To find the x-coordinates of the x-intercepts of the graph of f, we solve $f(x)=0$.

$$\dfrac{x^2}{x-2}+1=0$$
$$x^2+x-2=0 \quad \text{Multiplying by } x-2$$
$$(x+2)(x-1)=0$$
$$x=-2 \quad or \quad x=1$$

The x-intercepts are $(-2,0)$ and $(1,0)$.

54. $(-5-\sqrt{37},0),\ (-5+\sqrt{37},0)$

55.
$$f(x)=g(x)$$
$$\dfrac{x^2}{x-2}+1=\dfrac{4x-2}{x-2}+\dfrac{x+4}{2}$$

Substituting

$$2(x-2)\left(\dfrac{x^2}{x-2}+1\right)=2(x-2)\left(\dfrac{4x-2}{x-2}+\dfrac{x+4}{2}\right)$$

Multiplying by the LCD

$$2x^2+2(x-2)=2(4x-2)+(x-2)(x+4)$$
$$2x^2+2x-4=8x-4+x^2+2x-8$$
$$2x^2+2x-4=x^2+10x-12$$
$$x^2-8x+8=0$$
$$a=1,\ b=-8,\ c=8$$
$$x=\dfrac{-(-8)\pm\sqrt{(-8)^2-4\cdot1\cdot8}}{2\cdot1}=\dfrac{8\pm\sqrt{64-32}}{2}$$
$$x=\dfrac{8\pm\sqrt{32}}{2}=\dfrac{8\pm\sqrt{16\cdot2}}{2}=\dfrac{8\pm4\sqrt{2}}{2}$$
$$x=4\pm2\sqrt{2}$$

The solutions are $4+2\sqrt{2}$ and $4-2\sqrt{2}$.

56. $1.17539053,\ -0.4253905297$

57. $z^2+0.84z-0.4=0$

$a=1,\ b=0.84,\ c=-0.4$

$$z=\dfrac{-0.84\pm\sqrt{(0.84)^2-4\cdot1\cdot(-0.4)}}{2\cdot1}$$
$$z=\dfrac{-0.84\pm\sqrt{2.3056}}{2}$$
$$z=\dfrac{-0.84+\sqrt{2.3056}}{2}\approx0.3392101158$$
$$z=\dfrac{-0.84-\sqrt{2.3056}}{2}\approx-1.179210116$$

The solutions are approximately 0.3392101158 and -1.179210116.

58. $\sqrt{3},\ \dfrac{3-\sqrt{3}}{2}$

59. $\sqrt{2}x^2+5x+\sqrt{2}=0$

$$x=\dfrac{-5\pm\sqrt{5^2-4\cdot\sqrt{2}\cdot\sqrt{2}}}{2\sqrt{2}}=\dfrac{-5\pm\sqrt{17}}{2\sqrt{2}},\text{ or}$$
$$x=\dfrac{-5\pm\sqrt{17}}{2\sqrt{2}}\cdot\dfrac{\sqrt{2}}{\sqrt{2}}=\dfrac{-5\sqrt{2}\pm\sqrt{34}}{4}$$

The solutions are $\dfrac{-5\sqrt{2}\pm\sqrt{34}}{4}$.

60. $-i\pm i\sqrt{1-i}$

61.
$$kx^2+3x-k=0$$
$$k(-2)^2+3(-2)-k=0 \quad \text{Substituting } -2 \text{ for } x$$
$$4k-6-k=0$$
$$3k=6$$
$$k=2$$
$$2x^2+3x-2=0 \quad \text{Substituting } 2 \text{ for } k$$
$$(2x-1)(x+2)=0$$
$$2x-1=0 \quad or \quad x+2=0$$
$$x=\dfrac{1}{2} \quad or \quad x=-2$$

The other solution is $\dfrac{1}{2}$.

62.

63.

64. Writing exercise

Exercise Set 11.3

1. **Familiarize**. We first make a drawing, labeling it with the known and unknown information. We can also organize the information in a table. We let r represent the speed and t the time for the first part of the trip.

r km/h t hr $r - 4$ km/h $8 - t$ hr
        60 km   24 km

Canoe trip	Distance	Speed	Time
1st part	60	r	t
2nd part	24	$r - 4$	$8 - t$

Translate. Using $r = \dfrac{d}{t}$, we get two equations from the table, $r = \dfrac{60}{t}$ and $r - 4 = \dfrac{24}{8 - t}$.

Carry out. We substitute $\dfrac{60}{t}$ for r in the second equation and solve for t.

$$\frac{60}{t} - 4 = \frac{24}{8 - t}, \text{ LCD is } t(8 - t)$$

$$t(8 - t)\left(\frac{60}{t} - 4\right) = t(8 - t) \cdot \frac{24}{8 - t}$$

$$60(8 - t) - 4t(8 - t) = 24t$$

$$480 - 60t - 32t + 4t^2 = 24t$$

$$4t^2 - 116t + 480 = 0 \qquad \text{Standard form}$$

$$t^2 - 29t + 120 = 0 \qquad \text{Multiplying by } \frac{1}{4}$$

$$(t - 24)(t - 5) = 0$$

$$t = 24 \quad or \quad t = 5$$

Check. Since the time cannot be negative (If $t = 24$, $8 - t = -16$.), we check only 5 hr. If $t = 5$, then $8 - t = 3$. The speed of the first part is $\dfrac{60}{5}$, or 12 km/h. The speed of the second part is $\dfrac{24}{3}$, or 8 km/h. The speed of the second part is 4 km/h slower than the first part. The value checks.

State. The speed of the first part was 12 km/h, and the speed of the second part was 8 km/h.

2. First part: 60 mph; second part 50 mph

3. **Familiarize**. We first make a drawing. We also organize the information in a table. We let $r =$ the speed and $t =$ the time of the slower trip.

280 mi r mph t hr
280 mi $r + 5$ mph $t - 1$ hr

Trip	Distance	Speed	Time
Slower	280	r	t
Faster	280	$r + 5$	$t - 1$

Translate. Using $t = \dfrac{d}{r}$, we get two equations from the table, $t = \dfrac{280}{r}$, and $t - 1 = \dfrac{280}{r + 5}$.

Carry out. We substitute $\dfrac{280}{r}$ for t in the second equation and solve for r.

$$\frac{280}{r} - 1 = \frac{280}{r + 5}, \text{ LCD is } r(r + 5)$$

$$r(r + 5)\left(\frac{280}{r} - 1\right) = r(r + 5) \cdot \frac{280}{r + 5}$$

$$280(r + 5) - r(r + 5) = 280r$$

$$280r + 1400 - r^2 - 5r = 280r$$

$$0 = r^2 + 5r - 1400$$

$$0 = (r - 35)(r + 40)$$

$$r = 35 \quad or \quad r = -40$$

Check. Since negative speed has no meaning in this problem, we check only 35. If $r = 35$, then the time for the slow trip is $\dfrac{280}{35}$, or 8 hours. If $r = 35$ then $r + 5 = 40$ and the time for the fast trip is $\dfrac{280}{40}$, or 7 hours. This is 1 hour less time than the slow trip took, so we have an answer to the problem.

State. The speed is 35 mph.

4. 40 mph

5. **Familiarize**. We make a drawing and then organize the information in a table. We let $r =$ the speed and $t =$ the time of the Cessna.

600 mi r mph t hr
1000 mi $r + 50$ mph $t + 1$ hr

Plane	Distance	Speed	Time
Cessna	600	r	t
Beechcraft	1000	$r + 50$	$t + 1$

Translate. Using $t = d/r$, we get two equations from the table:

$$t = \frac{600}{r} \quad \text{and} \quad t + 1 = \frac{1000}{r + 50}$$

Carry out. We substitute $\dfrac{600}{r}$ for t in the second equation and solve for r.

$$\frac{600}{r} + 1 = \frac{1000}{r+50},$$

LCD is $r(r+50)$

$$r(r+50)\left(\frac{600}{r} + 1\right) = r(r+50) \cdot \frac{1000}{r+50}$$

$$600(r+50) + r(r+50) = 1000r$$

$$600r + 30,000 + r^2 + 50r = 1000r$$

$$r^2 - 350r + 30,000 = 0$$

$$(r-150)(r-200) = 0$$

$$r = 150 \ \ or \ \ r = 200$$

Check. If $r = 150$, then the Cessna's time is $\frac{600}{150}$, or 4 hr and the Beechcraft's time is $\frac{1000}{150+50}$, or $\frac{1000}{200}$, or 5 hr. If $r = 200$, then the Cessna's time is $\frac{600}{200}$, or 3 hr and the Beechcraft's time is $\frac{1000}{200+50}$, or $\frac{1000}{250}$, or 4 hr. Since the Beechcraft's time is 1 hr longer in each case, both values check. There are two solutions.

State. The speed of the Cessna is 150 mph and the speed of the Beechcraft is 200 mph; or the speed of the Cessna is 200 mph and the speed of the Beechcraft is 250 mph.

6. Super-prop; 350 mph, turbo-jet: 400 mph

7. Familiarize. We make a drawing and then organize the information in a table. We let r represent the speed and t the time of the trip to Hillsboro.

Hillsboro

40 mi r mph t hr

40 mi $r-6$ mph $14-t$ hr

Trip	Distance	Speed	Time
To Hillsboro	40	r	t
Return	40	$r-6$	$14-t$

Translate. Using $t = \frac{d}{r}$, we get two equations from the table,

$$t = \frac{40}{r} \ \text{and} \ 14-t = \frac{40}{r-6}.$$

Carry out. We substitute $\frac{40}{r}$ for t in the second equation and solve for r.

$$14 - \frac{40}{r} = \frac{40}{r-6},$$

LCD is $r(r-6)$

$$r(r-6)\left(14 - \frac{40}{r}\right) = r(r-6) \cdot \frac{40}{r-6}$$

$$14r(r-6) - 40(r-6) = 40r$$

$$14r^2 - 84r - 40r + 240 = 40r$$

$$14r^2 - 164r + 240 = 0$$

$$7r^2 - 82r + 120 = 0$$

$$(7r-12)(r-10) = 0$$

$$r = \frac{12}{7} \ \ or \ \ r = 10$$

Check. Since negative speed has no meaning in this problem (If $r = \frac{12}{7}$, then $r - 6 = -\frac{30}{7}$.), we check only 10 mph. If $r = 10$, then the time of the trip to Hillsboro is $\frac{40}{10}$, or 4 hr. The speed of the return trip is $10 - 6$, or 4 mph, and the time is $\frac{40}{4}$, or 10 hr. The total time for the round trip is 4 hr + 10 hr, or 14 hr. The value checks.

State. Naoki's speed on the trip to Hillsboro was 10 mph and it was 4 mph on the return trip.

8. Average speed to Richmond: 60 mph, average speed returning: 50 mph

9. Familiarize. We make a drawing and organize the information in a table. Let r represent the speed of the barge in still water, and let t represent the time of the trip upriver.

24 mi $r-4$ mph t hr
                                          Upriver

Downriver   24 mi $r+4$ mph $5-t$ hr

Trip	Distance	Speed	Time
Upriver	24	$r-4$	t
Downriver	24	$r+4$	$5-t$

Translate. Using $t = \frac{d}{r}$, we get two equations from the table,

$$t = \frac{24}{r-4} \ \text{and} \ 5-t = \frac{24}{r+4}.$$

Carry out. We substitute $\frac{24}{r-4}$ for t in the second equation and solve for r.

$$5 - \frac{24}{r-4} = \frac{24}{r+4},$$

LCD is $(r-4)(r+4)$

$$(r-4)(r+4)\left(5 - \frac{24}{r-4}\right) = (r-4)(r+4)\cdot\frac{24}{r+4}$$

$$5(r-4)(r+4) - 24(r+4) = 24(r-4)$$

$$5r^2 - 80 - 24r - 96 = 24r - 96$$

$$5r^2 - 48r - 80 = 0$$

We use the quadratic formula.

$$r = \frac{-(-48) \pm \sqrt{(-48)^2 - 4\cdot 5\cdot(-80)}}{2\cdot 5}$$

$$r = \frac{48 \pm \sqrt{3904}}{10}$$

$$r \approx 11 \ or \ r \approx -1.5$$

Check. Since negative speed has no meaning in this problem, we check only 11 mph. If $r \approx 11$, then the speed upriver is about $11 - 4$, or 7 mph, and the time is about $\frac{24}{7}$, or 3.4 hr. The speed downriver is about $11+4$, or 15 mph, and the time is about $\frac{24}{15}$, or 1.6 hr. The total time of the round trip is $3.4 + 1.6$, or 5 hr. The value checks.

State. The barge must be able to travel about 11 mph in still water.

10. About 14 mph

11. *Familiarize*. Let x represent the time it takes one well to fill the pool. Then $x - 6$ represents the time it takes the other well to fill the pool. It takes them 4 hr to fill the pool when both wells are working together, so they can fill $\frac{1}{4}$ of the pool in 1 hr. The first well will fill $\frac{1}{x}$ of the pool in 1 hr, and the other well will fill $\frac{1}{x-6}$ of the pool in 1 hr.

Translate. We have an equation.

$$\frac{1}{x} + \frac{1}{x-6} = \frac{1}{4}$$

Carry out. We solve the equation.
We multiply by the LCD, $4x(x-6)$.

$$4x(x-6)\left(\frac{1}{x} + \frac{1}{x-6}\right) = 4x(x-6)\cdot\frac{1}{4}$$

$$4(x-6) + 4x = x(x-6)$$

$$4x - 24 + 4x = x^2 - 6x$$

$$0 = x^2 - 14x + 24$$

$$0 = (x-2)(x-12)$$

$$x = 2 \ or \ x = 12$$

Check. Since negative time has no meaning in this problem, 2 is not a solution $(2 - 6 = -4)$. We check only 12 hr. This is the time it would take the first

well working alone. Then the other well would take $12 - 6$, or 6 hr working alone. The second well would fill $4\left(\frac{1}{6}\right)$, or $\frac{2}{3}$, of the pool in 4 hr, and the first well would fill $4\left(\frac{1}{12}\right)$, or $\frac{1}{3}$, of the pool in 4 hr. Thus in 4 hr they would fill $\frac{2}{3} + \frac{1}{3}$ of the pool. This is all of it, so the numbers check.

State. It takes the first well, working alone, 12 hr to fill the pool.

12. 6 hr

13. We make a drawing and then organize the information in a table. We let r represent Ellen's speed in still water. Then $r - 2$ is the speed upstream and $r+2$ is the speed downstream. Using $t = \frac{d}{r}$, we let $\frac{1}{r-2}$ represent the time upstream and $\frac{1}{r+2}$ represent the time downstream.

1 mi $r - 2$ mph
$\longrightarrow$ Upstream

Downstream $\longleftarrow$ 1 mi $r + 2$ mph

Trip	Distance	Speed	Time
Upstream	1	$r-2$	$\dfrac{1}{r-2}$
Downstream	1	$r+2$	$\dfrac{1}{r+2}$

Translate. The time for the round trip is 1 hour. We now have an equation.

$$\frac{1}{r-2} + \frac{1}{r+2} = 1$$

Carry out. We solve the equation. We multiply by the LCD, $(r-2)(r+2)$.

$$(r-2)(r+2)\left(\frac{1}{r-2} + \frac{1}{r+2}\right) = (r-2)(r+2)\cdot 1$$

$$(r+2) + (r-2) = (r-2)(r+2)$$

$$2r = r^2 - 4$$

$$0 = r^2 - 2r - 4$$

$a = 1$, $b = -2$, $c = -4$

$$r = \frac{-(-2) \pm \sqrt{(-2)^2 - 4\cdot 1(-4)}}{2\cdot 1}$$

$$r = \frac{2 \pm \sqrt{4 + 16}}{2} = \frac{2 \pm \sqrt{20}}{2}$$

$$r = \frac{2 \pm 2\sqrt{5}}{2} = 1 \pm \sqrt{5}$$

$$1 + \sqrt{5} \approx 1 + 2.236 \approx 3.24$$

$$1 - \sqrt{5} \approx 1 - 2.236 \approx -1.24$$

Check. Since negative speed has no meaning in this problem, we check only 3.24 mph. If $r \approx 3.24$, then $r - 2 \approx 1.24$ and $r + 2 \approx 5.24$. The time it takes to travel upstream is approximately $\frac{1}{1.24}$, or 0.806 hr, and the time it takes to travel downstream is approximately $\frac{1}{5.24}$, or 0.191 hr. The total time is 0.997 which is approximately 1 hour. The value checks.

State. Ellen's speed in still water is approximately 3.24 mph.

14. About 9.34 km/h

15.
$$A = 4\pi r^2$$
$$\frac{A}{4\pi} = r^2 \qquad \text{Dividing by } 4\pi$$
$$\frac{1}{2}\sqrt{\frac{A}{\pi}} = r \qquad \text{Taking the positive square root}$$

16. $s = \sqrt{\dfrac{A}{6}}$

17. $A = 2\pi r^2 + 2\pi rh$
$$0 = 2\pi r^2 + 2\pi rh - A \qquad \text{Standard form}$$
$$a = 2\pi, \ b = 2\pi h, \ c = -A$$
$$r = \frac{-2\pi h \pm \sqrt{(2\pi h)^2 - 4 \cdot 2\pi \cdot (-A)}}{2 \cdot 2\pi} \qquad \begin{array}{l}\text{Using the}\\ \text{quadratic formula}\end{array}$$
$$r = \frac{-2\pi h \pm \sqrt{4\pi^2 h^2 + 8\pi A}}{4\pi}$$
$$r = \frac{-2\pi h \pm 2\sqrt{\pi^2 h^2 + 2\pi A}}{4\pi}$$
$$r = \frac{-\pi h \pm \sqrt{\pi^2 h^2 + 2\pi A}}{2\pi}$$

Since taking the negative square root would result in a negative answer, we take the positive one.
$$r = \frac{-\pi h + \sqrt{\pi^2 h^2 + 2\pi A}}{2\pi}$$

18. $r = \sqrt{\dfrac{Gm_1 m_2}{F}}$

19.
$$N = \frac{kQ_1 Q_2}{s^2}$$
$$Ns^2 = kQ_1 Q_2 \qquad \text{Multiplying by } s^2$$
$$s^2 = \frac{kQ_1 Q_2}{N} \qquad \text{Dividing by } N$$
$$s = \sqrt{\frac{kQ_1 Q_2}{N}} \qquad \begin{array}{l}\text{Taking the positive square}\\ \text{root}\end{array}$$

20. $r = \sqrt{\dfrac{A}{\pi}}$

21.
$$T = 2\pi\sqrt{\frac{l}{g}}$$
$$\frac{T}{2\pi} = \sqrt{\frac{l}{g}} \qquad \text{Multiplying by } \frac{1}{2\pi}$$
$$\frac{T^2}{4\pi^2} = \frac{l}{g} \qquad \text{Squaring}$$
$$gT^2 = 4\pi^2 l \qquad \text{Multiplying by } 4\pi^2 g$$
$$g = \frac{4\pi^2 l}{T^2} \qquad \text{Multiplying by } \frac{1}{T^2}$$

22. $b = \sqrt{c^2 - a^2}$

23. $a^2 + b^2 + c^2 = d^2$
$$c^2 = d^2 - a^2 - b^2 \qquad \begin{array}{l}\text{Subtracting } a^2\\ \text{and } b^2\end{array}$$
$$c = \sqrt{d^2 - a^2 - b^2} \qquad \begin{array}{l}\text{Taking the}\\ \text{positive square root}\end{array}$$

24. $k = \dfrac{3 + \sqrt{9 + 8N}}{2}$

25. $s = v_0 t + \dfrac{gt^2}{2}$
$$0 = \frac{gt^2}{2} + v_0 t - s \qquad \text{Standard form}$$
$$a = \frac{g}{2}, \ b = v_0, \ c = -s$$
$$t = \frac{-v_0 \pm \sqrt{v_0^2 - 4\left(\frac{g}{2}\right)(-s)}}{2\left(\frac{g}{2}\right)}$$
$$t = \frac{-v_0 \pm \sqrt{v_0^2 + 2gs}}{g}$$

Since taking the negative square root would result in a negative answer, we take the positive one.
$$t = \frac{-v_0 + \sqrt{v_0^2 + 2gs}}{g}$$

26. $r = \dfrac{-\pi s + \sqrt{\pi^2 s^2 + 4\pi A}}{2\pi}$

27. $N = \dfrac{1}{2}(n^2 - n)$
$$N = \frac{1}{2}n^2 - \frac{1}{2}n$$
$$0 = \frac{1}{2}n^2 - \frac{1}{2}n - N$$
$$a = \frac{1}{2}, \ b = -\frac{1}{2}, \ c = -N$$

$$n = \frac{-\left(-\frac{1}{2}\right) \pm \sqrt{\left(-\frac{1}{2}\right)^2 - 4 \cdot \frac{1}{2} \cdot (-N)}}{2\left(\frac{1}{2}\right)}$$

$$n = \frac{1}{2} \pm \sqrt{\frac{1}{4} + 2N}$$

$$n = \frac{1}{2} \pm \sqrt{\frac{1 + 8N}{4}}$$

$$n = \frac{1}{2} \pm \frac{1}{2}\sqrt{1 + 8N}$$

Since taking the negative square root would result in a negative answer, we take the positive one.

$$n = \frac{1}{2} + \frac{1}{2}\sqrt{1 + 8N}, \text{ or } \frac{1 + \sqrt{1 + 8N}}{2}$$

28. $r = 1 - \sqrt{\dfrac{A}{A_0}}$

29. $V = 3.5\sqrt{h}$

$V = 12.25h$ Squaring

$$\frac{V^2}{12.25} = h$$

30. $L = \dfrac{1}{W^2 C}$

31. $at^2 + bt + c = 0$

The quadratic formula gives the result.

$$t = \frac{-b \pm \sqrt{b^2 - 4ac}}{2a}$$

32. $r = -1 + \dfrac{-P_2 + \sqrt{P_2^2 + 4AP_1}}{2P_1}$

33. a) *Familiarize and Translate.* From Example 4, we know

$$t = \frac{-v_0 + \sqrt{v_0{}^2 + 19.6s}}{9.8}.$$

Carry out. Substituting 500 for s and 0 for v_0, we have

$$t = \frac{0 + \sqrt{0^2 + 19.6(500)}}{9.8}$$

$$t \approx 10.1$$

Check. Substitute 10.1 for t and 0 for v_0 in the original formula. (See Example 4.)

$$s = 4.9t^2 + v_0 t = 4.9(10.1)^2 + 0 \cdot (10.1)^2$$

$$\approx 500$$

The answer checks.

State. It takes about 10.1 sec to reach the ground.

b) *Familiarize and Translate.* From Example 4, we know

$$t = \frac{-v_0 + \sqrt{v_0^2 + 19.6s}}{9.8}.$$

Carry out. Substitute 500 for s and 30 for v_0.

$$t = \frac{-30 + \sqrt{30^2 + 19.6(500)}}{9.8}$$

$$t \approx 7.49$$

Check. Substitute 30 for v_0 and 7.49 for t in the original formula. (See Example 4.)

$$s = 4.9t^2 + v_0 t = 4.9(7.49)^2 + (30)(7.49)$$

$$\approx 500$$

The answer checks.

State. It takes about 7.49 sec to reach the ground.

c) *Familiarize and Translate.* We will use the formula in Example 4, $s = 4.9t^2 + v_0 t$.

Carry out. Substitute 5 for t and 30 for v_0.

$$s = 4.9(5)^2 + 30(5) = 272.5$$

Check. We can substitute 30 for v_0 and 272.5 for s in the form of the formula we used in part (b).

$$t = \frac{-v_0 + \sqrt{v_0^2 + 19.6s}}{9.8}$$

$$= \frac{-30 + \sqrt{(30)^2 + 19.6(272.5)}}{9.8} = 5$$

The answer checks.

State. The object will fall 272.5 m.

34. a) 3.9 sec; b) 1.9 sec c) 79.6 m

35. *Familiarize and Translate.* From Example 4, we know

$$t = \frac{-v_0 + \sqrt{v_0^2 + 19.6s}}{9.8}.$$

Carry out. Substituting 40 for s and 0 for v_0 we have

$$t = \frac{0 + \sqrt{0^2 + 19.6(40)}}{9.8}$$

$$t \approx 2.9$$

Check. Substitute 2.9 for t and 0 for v_0 in the original formula. (See Example 4.)

$$s = 4.9t^2 + v_0 t = 4.9(2.9)^2 + 0(2.9)$$

$$\approx 40$$

The answer checks.

State. He will be falling for about 2.9 sec.

36. 30.625 m

37. *Familiarize and Translate*. From Example 3, we know
$$T = \frac{\sqrt{3V}}{12}.$$
***Carry out*.** Substituting 36 for V, we have
$$T = \frac{\sqrt{3 \cdot 36}}{12}$$
$$T \approx 0.87$$
***Check*.** Substitute 0.87 for T in the original formula. (See Example 3.)
$$48T^2 = V$$
$$48(0.87)^2 = V$$
$$36 \approx V$$
The answer checks.

***State*.** Vince Carter's hang time is about 0.87 sec.

38. 12

39. *Familiarize and Translate*. We will use the formula in Example 4, $s = 4.9t^2 + v_0 t$.

***Carry out*.** Solve the formula for v_0.
$$s - 4.9t^2 = v_0 t$$
$$\frac{s - 4.9t^2}{t} = v_0$$
Now substitute 51.6 for s and 3 for t.
$$\frac{51.6 - 4.9(3)^2}{3} = v_0$$
$$2.5 = v_0$$
***Check*.** Substitute 3 for t and 2.5 for v_0 in the original formula.
$$s = 4.9(3)^2 + 2.5(3) = 51.6$$
The solution checks.

***State*.** The initial velocity is 2.5 m/sec.

40. 3.2 m/sec

41. *Familiarize and Translate*. From Exercise 32 we know that
$$r = -1 + \frac{-P_2 + \sqrt{P_2^2 + 4P_1 A}}{2P_1},$$
where A is the total amount in the account after two years, P_1 is the amount of the original deposit, P_2 is deposited at the beginning of the second year, and r is the annual interest rate.

***Carry out*.** Substitute 3000 for P_1, 1700 for P_2, and 5253.70 for A.
$$r = -1 + \frac{-1700 + \sqrt{(1700)^2 + 4(3000)(5253.70)}}{2(3000)}$$
Using a calculator, we have $r = 0.07$.

***Check*.** Substitute in the original formula in Exercise 32.
$$P_1(1 + r)^2 + P_2(1 + r) = A$$
$$3000(1.07)^2 + 1700(1.07) = A$$
$$5253.70 = A$$
The answer checks.

***State*.** The annual interest rate is 0.07, or 7%.

42. 8.5%

43. Writing exercise

44. Writing exercise

45.
$$b^2 - 4ac = 6^2 - 4 \cdot 5 \cdot 7$$
$$= 36 - 4 \cdot 5 \cdot 7$$
$$= 36 - 140$$
$$= -104$$

46. $2i\sqrt{11}$

47.
$$\frac{x^2 + xy}{2x} = \frac{x(x + y)}{2x}$$
$$= \frac{x(x + y)}{2 \cdot x}$$
$$= \frac{\cancel{x}(x + y)}{2 \cdot \cancel{x}}$$
$$= \frac{x + y}{2}$$

48. $\dfrac{a^2 - b^2}{b}$

49.
$$\frac{3 + \sqrt{45}}{6} = \frac{3 + \sqrt{9 \cdot 5}}{6} = \frac{3 + 3\sqrt{5}}{6} = \frac{\cancel{3}(1 + \sqrt{5})}{\cancel{3} \cdot 2} = \frac{1 + \sqrt{5}}{2}$$

50. $\dfrac{1 - \sqrt{7}}{5}$

51. Writing exercise

52. Writing exercise

53.
$$A = 6.5 - \frac{20.4t}{t^2 + 36}$$
$$(t^2 + 36)A = (t^2 + 36)\left(6.5 - \frac{20.4t}{t^2 + 36}\right)$$
$$At^2 + 36A = (t^2 + 36)(6.5) - (t^2 + 36)\left(\frac{20.4t}{t^2 + 36}\right)$$
$$At^2 + 36A = 6.5t^2 + 234 - 20.4t$$

$$At^2 - 6.5t^2 + 20.4 + 36A - 234 = 0$$

$$(A - 6.5)t^2 + 20.4t + (36A - 234) = 0$$

$$a = A - 6.5,\ b = 20.4,\ c = 36A - 234$$

$$t = \frac{-20.4 \pm \sqrt{(20.4)^2 - 4(A - 6.5)(36A - 234)}}{2(A - 6.5)}$$

$$t = \frac{-20.4 \pm \sqrt{416.16 - 144A^2 + 1872A - 6084}}{2(A - 6.5)}$$

$$t = \frac{-20.4 \pm \sqrt{-144A^2 + 1872A - 5667.84}}{2(A - 6.5)}$$

$$t = \frac{-20.4 \pm \sqrt{144(-A^2 + 13A - 39.36)}}{2(A - 6.5)}$$

$$t = \frac{-20.4 \pm 12\sqrt{-A^2 + 13A - 39.36}}{2(A - 6.5)}$$

$$t = \frac{2(-10.2 \pm 6\sqrt{-A^2 + 13A - 39.36})}{2(A - 6.5)}$$

$$t = \frac{-10.2 \pm 6\sqrt{-A^2 + 13A - 39.36}}{A - 6.5}$$

54. $c = \dfrac{mv}{\sqrt{m^2 - m_0^2}}$

55.

$$\frac{w}{l} = \frac{l}{w + l}$$

$$l(w + l) \cdot \frac{w}{l} = l(w + l) \cdot \frac{l}{w + l}$$

$$w(w + l) = l^2$$

$$w^2 + lw = l^2$$

$$0 = l^2 - lw - w^2$$

Use the quadratic formula with $a = 1$, $b = -w$, and $c = -w^2$.

$$l = \frac{-(-w) \pm \sqrt{(-w)^2 - 4 \cdot 1(-w^2)}}{2 \cdot 1}$$

$$l = \frac{w \pm \sqrt{w^2 + 4w^2}}{2} = \frac{w \pm \sqrt{5w^2}}{2}$$

$$l = \frac{w \pm w\sqrt{5}}{2}$$

Since $\dfrac{w - w\sqrt{5}}{2}$ is negative we use the positive square root:

$$l = \frac{w + w\sqrt{5}}{2}$$

56. $L(A) = \sqrt{\dfrac{A}{2}}$

57. **Familiarize.** Let $a =$ the number. Then $a - 1$ is 1 less than a and the reciprocal of that number is $\dfrac{1}{a - 1}$. Also, 1 more than the number is $a + 1$.

Translate.

The reciprocal of 1 less than a number	is	1 more than the number.
$\downarrow$	$\downarrow$	$\downarrow$
$\dfrac{1}{(a - 1)}$	$=$	$a + 1$

Carry out. We solve the equation.

$$\frac{1}{a - 1} = a + 1, \text{ LCD is } a - 1$$

$$(a - 1) \cdot \frac{1}{a - 1} = (a - 1)(a + 1)$$

$$1 = a^2 - 1$$

$$2 = a^2$$

$$\pm\sqrt{2} = a$$

Check. $\dfrac{1}{\sqrt{2} - 1} \approx 2.4142 \approx \sqrt{2} + 1$ and $\dfrac{1}{-\sqrt{2} - 1} \approx$ $-0.4142 \approx -\sqrt{2} + 1$. The answers check.

State. The numbers are $\sqrt{2}$ and $-\sqrt{2}$, or $\pm\sqrt{2}$.

58. \$2.50

59. $mn^4 - r^2pm^3 - r^2n^2 + p = 0$

Let $u = n^2$. Substitute and rearrange.

$$mu^2 - r^2u - r^2pm^3 + p = 0$$

$$a = m,\ b = -r^2,\ c = -r^2pm^3 + p$$

$$u = \frac{-(-r^2) \pm \sqrt{(-r^2)^2 - 4 \cdot m(-r^2pm^3 + p)}}{2 \cdot m}$$

$$u = \frac{r^2 \pm \sqrt{r^4 + 4m^4r^2p - 4mp}}{2m}$$

$$n^2 = \frac{r^2 \pm \sqrt{r^4 + 4m^4r^2p - 4mp}}{2m}$$

$$n = \pm\sqrt{\frac{r^2 \pm \sqrt{r^4 + 4m^4r^2p - 4mp}}{2m}}$$

60. $d = \dfrac{-\pi h + \sqrt{\pi^2 h^2 + 2\pi A}}{\pi}$

61. Let s represent a length of a side of the cube, let S represent the surface area of the cube, and let A represent the surface area of the sphere. Then the diameter of the sphere is s, so the radius r is $s/2$. From Exercise 15, we know, $A = 4\pi r^2$, so when $r = s/2$ we have $A = 4\pi\left(\dfrac{s}{2}\right)^2 = 4\pi \cdot \dfrac{s^2}{4} = \pi s^2$. From the formula for the surface area of a cube (See Exercise 16.) we know that $S = 6s^2$, so $\dfrac{S}{6} = s^2$ and then $A = \pi \cdot \dfrac{S}{6}$, or $A(S) = \dfrac{\pi S}{6}$.

62. Writing exercise

Exercise Set 11.4

1. $x^2 - 5x + 3 = 0$

$a = 1, b = -5, c = 3$

We substitute and compute the discriminant.

$$b^2 - 4ac = (-5)^2 - 4 \cdot 1 \cdot 3$$
$$= 25 - 12$$
$$= 13$$

Since the discriminant is a positive number that is not a perfect square, there are two irrational solutions.

2. Two irrational

3. $x^2 + 5 = 0$

$a = 1, b = 0, c = 5$

We substitute and compute the discriminant.

$$b^2 - 4ac = 0^2 - 4 \cdot 1 \cdot 5$$
$$= -20$$

Since the discriminant is negative, there are two imaginary-number solutions.

4. Two imaginary

5. $x^2 - 3 = 0$

$a = 1, b = 0, c = -3$

We substitute and compute the discriminant.

$$b^2 - 4ac = 0^2 - 4 \cdot 1 \cdot (-3)$$
$$= 12$$

Since the discriminant is a positive number that is not a perfect square, there are two irrational solutions.

6. Two irrational

7. $4x^2 - 12x + 9 = 0$

$a = 4, b = -12, c = 9$

We substitute and compute the discriminant.

$$b^2 - 4ac = (-12)^2 - 4 \cdot 4 \cdot 9$$
$$= 144 - 144$$
$$= 0$$

Since the discriminant is 0, there is just one solution, and it is a rational number.

8. Two rational

9. $x^2 - 2x + 4 = 0$

$a = 1, b = -2, c = 4$

We substitute and compute the discriminant.

$$b^2 - 4ac = (-2)^2 - 4 \cdot 1 \cdot 4$$
$$= 4 - 16$$
$$= -12$$

Since the discriminant is negative, there are two imaginary-number solutions.

10. Two imaginary

11. $6t^2 - 19t - 20 = 0$

$a = 6, b = -19, c = -20$

We substitute and compute the discriminant.

$$b^2 - 4ac = (-19)^2 - 4 \cdot 6 \cdot (-20)$$
$$= 361 + 480$$
$$= 841$$

Since the discriminant is a positive number and a perfect square, there are two rational solutions.

12. One rational

13. $6x^2 + 5x - 4 = 0$

$a = 6, b = 5, c = -4$

We substitute and compute the discriminant.

$$b^2 - 4ac = 5^2 - 4 \cdot 6 \cdot (-4)$$
$$= 25 + 96 = 121$$

Since the discriminant is a positive number and a perfect square, there are two rational solutions.

14. Two rational

15. $9t^2 - 3t = 0$

Observe that we can factor $9t^2 - 3t$. This tells us that there are two rational solutions. We could also do this problem as follows.

$a = 9, b = -3, c = 0$

We substitute and compute the discriminant.

$$b^2 - 4ac = (-3)^2 - 4 \cdot 9 \cdot 0$$
$$= 9 - 0$$
$$= 9$$

Since the discriminant is a positive number and a perfect square, there are two rational solutions.

16. Two rational

17. $x^2 + 4x = 8$

$x^2 + 4x - 8 = 0$ Standard form

$a = 1, b = 4, c = -8$

We substitute and compute the discriminant.
$$b^2 - 4ac = 4^2 - 4 \cdot 1 \cdot (-8)$$
$$= 16 + 32 = 48$$
Since the discriminant is a positive number that is not a perfect square, there are two irrational solutions.

18. Two irrational

19.
$$2a^2 - 3a = -5$$
$$2a^2 - 3a + 5 = 0 \quad \text{Standard form}$$
$$a = 2, \, b = -3, \, c = 5$$
We substitute and compute the discriminant.
$$b^2 - 4ac = (-3)^2 - 4 \cdot 2 \cdot 5$$
$$= 9 - 40$$
$$= -31$$
Since the discriminant is negative, there are two imaginary-number solutions.

20. Two imaginary

21.
$$y^2 + \frac{9}{4} = 4y$$
$$y^2 - 4y + \frac{9}{4} = 0 \quad \text{Standard form}$$
$$a = 1, \, b = -4, \, c = \frac{9}{4}$$
We substitute and compute the discriminant.
$$b^2 - 4ac = (-4)^2 - 4 \cdot 1 \cdot \frac{9}{4}$$
$$= 16 - 9$$
$$= 7$$
The discriminant is a positive number that is not a perfect square. There are two irrational solutions.

22. Two imaginary

23. The solutions are -7 and 3.
$$x = -7 \ \text{ or } \quad x = 3$$
$$x + 7 = 0 \quad \text{ or } \ x - 3 = 0$$
$$(x + 7)(x - 3) = 0 \quad \text{Principle of zero products}$$
$$x^2 + 4x - 21 = 0 \quad \text{FOIL}$$

24. $x^2 + 2x - 24 = 0$

25. The only solution is 3. It must be a repeated solution.
$$x = 3 \ \text{ or } \quad x = 3$$
$$x - 3 = 0 \ \text{ or } \ x - 3 = 0$$
$$(x - 3)(x - 3) = 0 \quad \text{Principle of zero products}$$
$$x^2 - 6x + 9 = 0 \quad \text{FOIL}$$

26. $x^2 + 10x + 25 = 0$

27. The solutions are -2 and -5.
$$x = -2 \ \text{ or } \quad x = -5$$
$$x + 2 = 0 \ \text{ or } \ x + 5 = 0$$
$$(x + 2)(x + 5) = 0$$
$$x^2 + 7x + 10 = 0$$

28. $x^2 + 4x + 3 = 0$

29. The solutions are 4 and $\frac{2}{3}$.
$$x = 4 \ \text{ or } \quad x = \frac{2}{3}$$
$$x - 4 = 0 \ \text{ or } \ x - \frac{2}{3} = 0$$
$$(x - 4)\left(x - \frac{2}{3}\right) = 0$$
$$x^2 - \frac{2}{3}x - 4x + \frac{8}{3} = 0$$
$$x^2 - \frac{14}{3}x + \frac{8}{3} = 0$$
$$3x^2 - 14x + 8 = 0 \quad \text{Multiplying by 3}$$

30. $4x^2 - 23x + 15 = 0$

31. The solutions are $\frac{1}{2}$ and $\frac{1}{3}$.
$$x = \frac{1}{2} \ \text{ or } \quad x = \frac{1}{3}$$
$$x - \frac{1}{2} = 0 \ \text{ or } \ x - \frac{1}{3} = 0$$
$$\left(x - \frac{1}{2}\right)\left(x - \frac{1}{3}\right) = 0$$
$$x^2 - \frac{1}{3}x - \frac{1}{2}x + \frac{1}{6} = 0$$
$$x^2 - \frac{5}{6}x + \frac{1}{6} = 0$$
$$6x^2 - 5x + 1 = 0 \quad \text{Multiplying by 6}$$

32. $8x^2 + 6x + 1 = 0$

33. The solutions are -0.6 and 1.4.
$$x = -0.6 \ \text{ or } \quad x = 1.4$$
$$x + 0.6 = 0 \ \text{ or } \ x - 1.4 = 0$$
$$(x + 0.6)(x - 1.4) = 0$$
$$x^2 - 1.4x + 0.6x - 0.84 = 0$$
$$x^2 - 0.8x - 0.84 = 0$$

34. $x^2 - 2x - 0.96 = 0$

35. The solutions are $-\sqrt{7}$ and $\sqrt{7}$.

$$x = -\sqrt{7} \quad or \qquad x = \sqrt{7}$$
$$x + \sqrt{7} = 0 \quad or \quad x - \sqrt{7} = 0$$
$$(x + \sqrt{7})(x - \sqrt{7}) = 0$$
$$x^2 - 7 = 0$$

36. $x^2 - 3 = 0$

37. The solutions are $3\sqrt{2}$ and $-3\sqrt{2}$.

$$x = 3\sqrt{2} \quad or \qquad x = -3\sqrt{2}$$
$$x - 3\sqrt{2} = 0 \quad or \quad x + 3\sqrt{2} = 0$$
$$(x - 3\sqrt{2})(x + 3\sqrt{2}) = 0$$
$$x^2 - (3\sqrt{2})^2 = 0$$
$$x^2 - 9 \cdot 2 = 0$$
$$x^2 - 18 = 0$$

38. $x^2 - 20 = 0$

39. The solutions are $3i$ and $-3i$.

$$x = 3i \quad or \qquad x = -3i$$
$$x - 3i = 0 \quad or \quad x + 3i = 0$$
$$(x - 3i)(x + 3i) = 0$$
$$x^2 - (3i)^2 = 0$$
$$x^2 + 9 = 0$$

40. $x^2 + 16 = 0$

41. The solutions are $5 - 2i$ and $5 + 2i$.

$$x = 5 - 2i \quad or \qquad x = 5 + 2i$$
$$x - 5 + 2i = 0 \quad or \quad x - 5 - 2i = 0$$
$$[x + (-5 + 2i)][x + (-5 - 2i)] = 0$$
$$x^2 + x(-5 - 2i) + x(-5 + 2i) + (-5 + 2i)(-5 - 2i) = 0$$
$$x^2 - 5x - 2xi - 5x + 2xi + 25 - 4i^2 = 0$$
$$x^2 - 10x + 29 = 0$$
$$(i^2 = -1)$$

42. $x^2 - 4x + 53 = 0$

43. The solutions are $2 - \sqrt{10}$ and $2 + \sqrt{10}$.

$$x = 2 - \sqrt{10} \quad or \qquad x = 2 + \sqrt{10}$$
$$x - (2 - \sqrt{10}) = 0 \quad or \quad x - (2 + \sqrt{10}) = 0$$
$$[x - (2 - \sqrt{10})][x - (2 + \sqrt{10})] = 0$$
$$x^2 - x(2 + \sqrt{10}) - x(2 - \sqrt{10}) + (2 - \sqrt{10})(2 + \sqrt{10}) = 0$$
$$x^2 - 2x - x\sqrt{10} - 2x + x\sqrt{10} + 4 - 10 = 0$$
$$x^2 - 4x - 6 = 0$$

44. $x^2 - 6x - 5 = 0$

45. The solutions are -2, 1, and 5.

$$x = -2 \quad or \qquad x = 1 \quad or \qquad x = 5$$
$$x + 2 = 0 \quad or \quad x - 1 = 0 \quad or \quad x - 5 = 0$$
$$(x + 2)(x - 1)(x - 5) = 0$$
$$(x^2 + x - 2)(x - 5) = 0$$
$$x^3 + x^2 - 2x - 5x^2 - 5x + 10 = 0$$
$$x^3 - 4x^2 - 7x + 10 = 0$$

46. $x^3 + 3x^2 - 10x = 0$

47. The solutions are -1, 0, and 3.

$$x = -1 \quad or \quad x = 0 \quad or \qquad x = 3$$
$$x + 1 = 0 \quad or \quad x = 0 \quad or \quad x - 3 = 0$$
$$(x + 1)(x)(x - 3) = 0$$
$$(x^2 + x)(x - 3) = 0$$
$$x^3 - 3x^2 + x^2 - 3x = 0$$
$$x^3 - 2x^2 - 3x = 0$$

48. $x^3 - 3x^2 - 4x + 1 = 0$

49. Writing exercise

50. Writing exercise

51. $(3a^2)^4 = 3^4(a^2)^4 = 81a^{2 \cdot 4} = 81a^8$

52. $16x^6$

53. $f(x) = x^2 - 7x - 8$

We find the values of x for which $f(x) = 0$.

$$x^2 - 7x - 8 = 0$$
$$(x - 8)(x + 1) = 0$$
$$x - 8 = 0 \quad or \quad x + 1 = 0$$
$$x = 8 \quad or \qquad x = -1$$

The x-intercepts are $(8, 0)$ and $(-1, 0)$.

54. $(2, 0), (4, 0)$

55. *Familiarize*. Let x and y represent the number of 30-sec and 60-sec commercials, respectively. Then the amount of time for the 30-sec commercials was $30x$ sec, or $\frac{30x}{60} = \frac{x}{2}$ min. The amount of time for the 60-sec commercials was $60x$ sec, or $\frac{60x}{60} = x$ min.

Translate. Rewording, we write two equations. We will express time in minutes.

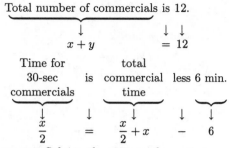

Total number of commercials is 12.

$$x + y \qquad = 12$$

Time for 30-sec commercials is total commercial time less 6 min.

$$\frac{x}{2} = \frac{x}{2} + x - 6$$

Carry out. Solving the system of equations we get $(6, 6)$.

Check. If there are six 30-sec and six 60-sec commercials, the total number of commercials is 12. The amount of time for six 30-sec commercials is 180 sec, or 3 min, and for six 60-sec commercials is 360 sec, or 6 min. The total commercial time is 9 min, and the amount of time for 30-sec commercials is 6 min less than this. The numbers check.

State. There were six 30-sec commercials.

56.

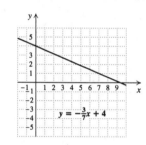

$$y = -\frac{3}{7}x + 4$$

57. Writing exercise

58. Writing exercise

59. The graph includes the points $(-3, 0)$, $(0, -3)$, and $(1, 0)$. Substituting in $y = ax^2 + bx + c$, we have three equations.

$$0 = 9a - 3b + c,$$
$$-3 = \qquad c,$$
$$0 = a + b + c$$

The solution of this system of equations is $a = 1$, $b = 2$, $c = -3$.

60. Consider a quadratic equation in standard form, $ax^2 + bx + c = 0$. The solutions are
$$\frac{-b \pm \sqrt{b^2 - 4ac}}{2a}.$$
The product of the solutions is
$$\left(\frac{-b + \sqrt{b^2 - 4ac}}{2a}\right)\left(\frac{-b - \sqrt{b^2 - 4ac}}{2a}\right) =$$
$$\frac{(-b)^2 - (\sqrt{b^2 - 4ac})^2}{(2a)^2} = \frac{b^2 - (b^2 - 4ac)}{4a^2} = \frac{4ac}{4a^2} =$$
$$\frac{c}{a}.$$

61. a) $kx^2 - 2x + k = 0$; one solution is -3

We first find k by substituting -3 for x.
$$k(-3)^2 - 2(-3) + k = 0$$
$$9k + 6 + k = 0$$
$$10k = -6$$
$$k = -\frac{6}{10}$$
$$k = -\frac{3}{5}$$

b) Now substitute $-\frac{3}{5}$ for k in the original equation.
$$-\frac{3}{5}x^2 - 2x + \left(-\frac{3}{5}\right) = 0$$
$$3x^2 + 10x + 3 = 0 \quad \text{Multiplying by } -5$$
$$(3x + 1)(x + 3) = 0$$
$$x = -\frac{1}{3} \text{ or } x = -3$$

The other solution is $-\frac{1}{3}$.

62. a) 2; b) $1 - i$

63. a) $x^2 - (6 + 3i)x + k = 0$; one solution is 3.

We first find k by substituting 3 for x.
$$3^2 - (6 + 3i)3 + k = 0$$
$$9 - 18 - 9i + k = 0$$
$$-9 - 9i + k = 0$$
$$k = 9 + 9i$$

b) Now we substitute $9 + 9i$ for k in the original equation.
$$x^2 - (6 + 3i)x + (9 + 9i) = 0$$
$$x^2 - (6 + 3i)x + 3(3 + 3i) = 0$$
$$[x - (3 + 3i)][x - 3] = 0$$
$$x = 3 + 3i \text{ or } x = 3$$

The other solution is $3 + 3i$.

64. Consider a quadratic equation in standard form, $ax^2 + bx + c = 0$. The solutions are
$$\frac{-b \pm \sqrt{b^2 - 4ac}}{2a}.$$
The sum of the solutions is
$$\frac{-b + \sqrt{b^2 - 4ac}}{2a} + \frac{-b - \sqrt{b^2 - 4ac}}{2a} = \frac{-2b}{2a} = -\frac{b}{a}.$$

65. The solutions of $ax^2 + bx + c = 0$ are
$$x = \frac{-b \pm \sqrt{b^2 - 4ac}}{2a}.$$ When there is just one solution, $b^2 - 4ac = 0$, so $x = \frac{-b \pm 0}{2a} = -\frac{b}{2a}$.

66. $h = -36$, $k = 15$

67. We substitute $(-3, 0)$, $\left(\dfrac{1}{2}, 0\right)$, and $(0, -12)$ in
$f(x) = ax^2 + bx + c$ and get three equations.
$$0 = 9a - 3b + c,$$
$$0 = \frac{1}{4}a + \frac{1}{2}b + c,$$
$$-12 = c$$
The solution of this system of equations is $a = 8$,
$b = 20$, $c = -12$.

68. $x^4 - 14x^3 + 70x^2 - 126x + 29 = 0$

69. If $1 - \sqrt{5}$ and $3 + 2i$ are two solutions, then $1 + \sqrt{5}$
and $3 - 2i$ are also solutions. The equation of lowest
degree that has these solutions is found as follows.
$$[x - (1-\sqrt{5})][x-(1+\sqrt{5})][x-(3+2i)][x-(3-2i)] = 0$$
$$(x^2 - 2x - 4)(x^2 - 6x + 13) = 0$$
$$x^4 - 8x^3 + 21x^2 - 2x - 52 = 0$$

70. Writing exercise

71. Writing exercise

Exercise Set 11.5

1. $x^4 - 10x^2 + 9 = 0$
Let $u = x^2$ and $u^2 = x^4$.
$u^2 - 10u + 9 = 0$ Substituting
$(u - 1)(u - 9) = 0$
$u - 1 = 0$ or $u - 9 = 0$
$u = 1$ or $u = 9$

Now replace u with x^2 and solve these equations:
$x^2 = 1$ or $x^2 = 9$
$x = \pm 1$ or $x = \pm 3$
The numbers 1, -1, 3, and -3 check. They are the
solutions.

2. $1, -1, 2, -2$

3. $x^4 - 12x^2 + 27 = 0$
Let $u = x^2$ and $u^2 = x^4$..
$u^2 - 12u + 27 = 0$ Substituting u for x^2
$(u - 9)(u - 3) = 0$
$u = 9$ or $u = 3$
Now replace u with x^2 and solve these equations:
$x^2 = 9$ or $x^2 = 3$
$x = \pm 3$ or $x = \pm\sqrt{3}$
The numbers 3, -3, $\sqrt{3}$, and $-\sqrt{3}$ check. They are
the solutions.

4. $\pm 2, \pm\sqrt{5}$

5. $9x^4 - 14x^2 + 5 = 0$
Let $u = x^2$ and $u^2 = x^4$.
$9u^2 - 14u + 5 = 0$ Substituting
$(9u - 5)(u - 1) = 0$
$9u - 5 = 0$ or $u - 1 = 0$
$9u = 5$ or $u = 1$
$u = \dfrac{5}{9}$ or $u = 1$
Now replace u with x^2 and solve these equations:
$x^2 = \dfrac{5}{9}$ or $x^2 = 1$
$x = \pm\dfrac{\sqrt{5}}{3}$ or $x = \pm 1$
The numbers $\dfrac{\sqrt{5}}{3}$, $-\dfrac{\sqrt{5}}{3}$, 1, and -1 check. They are
the solutions.

6. $\dfrac{\sqrt{3}}{2}, -\dfrac{\sqrt{3}}{2}, 2, -2$

7. $x - 4\sqrt{x} - 1 = 0$
Let $u = \sqrt{x}$ and $u^2 = x$.
$u^2 - 4u - 1 = 0$ Substituting
$$u = \frac{-(-4) \pm \sqrt{(-4)^2 - 4 \cdot 1 \cdot (-1)}}{2 \cdot 1}$$
$$u = \frac{4 \pm \sqrt{20}}{2} = \frac{2 \cdot 2 \pm 2\sqrt{5}}{2}$$
$$u = 2 \pm \sqrt{5}$$
$u = 2 + \sqrt{5}$ or $u = 2 - \sqrt{5}$
Replace u with $\sqrt{x}$ and solve these equations.
$\sqrt{x} = 2 + \sqrt{5}$ or $\sqrt{x} = 2 - \sqrt{5}$
$(\sqrt{x})^2 = (2 + \sqrt{5})^2$ No solution:
         $2 - \sqrt{5}$ is negative
$x = 4 + 4\sqrt{5} + 5$
$x = 9 + 4\sqrt{5}$
The number $9 + 4\sqrt{5}$ checks. It is the solution.

8. $8 + 2\sqrt{7}$

9. $(x^2 - 7)^2 - 3(x^2 - 7) + 2 = 0$
Let $u = x^2 - 7$ and $u^2 = (x^2 - 7)^2$.
$u^2 - 3u + 2 = 0$ Substituting
$(u - 1)(u - 2) = 0$

$$u = 1 \quad or \quad u = 2$$
$$x^2 - 7 = 1 \quad or \quad x^2 - 7 = 2 \quad \text{Replacing } u$$
$$\text{with } x^2 - 7$$
$$x^2 = 8 \quad or \quad x^2 = 9$$
$$x = \pm\sqrt{8} \quad or \quad x = \pm 3$$
$$x = \pm 2\sqrt{2} \quad or \quad x = \pm 3$$

The numbers $2\sqrt{2}$, $-2\sqrt{3}$, 3, and -3 check. They are the solutions.

10. $\pm\sqrt{3}, \pm 2$

11. $(1 + \sqrt{x})^2 + 5(1 + \sqrt{x}) + 6 = 0$

Let $u = 1 + \sqrt{x}$ and $u^2 = (1 + \sqrt{x})^2$.

$$u^2 + 5u + 6 = 0 \quad \text{Substituting}$$
$$(u + 3)(u + 2) = 0$$
$$u = -3 \quad or \quad u = -2$$
$$1 + \sqrt{x} = -3 \quad or \quad 1 + \sqrt{x} = -2 \quad \text{Replacing } u$$
$$\text{with } 1 + \sqrt{x}$$
$$\sqrt{x} = -4 \quad or \quad \sqrt{x} = -3$$

Since the principal square root cannot be negative, this equation has no solution.

12. No solution

13. $x^{-2} - x^{-1} - 6 = 0$

Let $u = x^{-1}$ and $u^2 = x^{-2}$.

$$u^2 - u - 6 = 0 \quad \text{Substituting}$$
$$(u - 3)(u + 2) = 0$$
$$u = 3 \quad or \quad u = -2$$

Now we replace u with x^{-1} and solve these equations:
$$x^{-1} = 3 \quad or \quad x^{-1} = -2$$
$$\frac{1}{x} = 3 \quad or \quad \frac{1}{x} = -2$$
$$\frac{1}{3} = x \quad or \quad -\frac{1}{2} = x$$

Both $\frac{1}{3}$ and $-\frac{1}{2}$ check. They are the solutions.

14. $-2, 1$

15. $4x^{-2} + x^{-1} - 5 = 0$

Let $u = x^{-1}$ and $u^2 = x^{-2}$.

$$4u^2 + u - 5 = 0 \quad \text{Substituting}$$
$$(4u + 5)(u - 1) = 0$$
$$u = -\frac{5}{4} \quad or \quad u = 1$$

Now we replace u with x^{-1} and solve these equations:
$$x^{-1} = -\frac{5}{4} \quad or \quad x^{-1} = 1$$
$$\frac{1}{x} = -\frac{5}{4} \quad or \quad \frac{1}{x} = 1$$
$$4 = -5x \quad or \quad 1 = x$$
$$-\frac{4}{5} = x \quad or \quad 1 = x$$

The numbers $-\frac{4}{5}$ and 1 check. They are the solutions.

16. $-\frac{1}{10}, 1$

17. $t^{2/3} + t^{1/3} - 6 = 0$

Let $u = t^{1/3}$ and $u^2 = t^{2/3}$.

$$u^2 + u - 6 = 0 \quad \text{Substituting}$$
$$(u + 3)(u - 2) = 0$$
$$u = -3 \quad or \quad u = 2$$

Now we replace u with $t^{1/3}$ and solve these equations:
$$t^{1/3} = -3 \quad or \quad t^{1/3} = 2$$
$$t = (-3)^3 \quad or \quad t = 2^3 \quad \text{Raising to the}$$
$$\text{third power}$$
$$t = -27 \quad or \quad t = 8$$

Both -27 and 8 check. They are the solutions.

18. $-8, 64$

19. $y^{1/3} - y^{1/6} - 6 = 0$

Let $u = y^{1/6}$ and $u^2 = y^{2/3}$.

$$u^2 - u - 6 = 0 \quad \text{Substituting}$$
$$(u - 3)(u + 2) = 0$$
$$u = 3 \quad or \quad u = -2$$

Now we replace u with $y^{1/6}$ and solve these equations:
$$y^{1/6} = 3 \quad or \quad y^{1/6} = -2$$
$$\sqrt[6]{y} = 3 \quad or \quad \sqrt[6]{y} = -2$$
$$y = 3^6 \qquad \text{This equation has no}$$
$$y = 729 \qquad \text{solution since principal}$$
$$\text{sixth roots are never negative.}$$

The number 729 checks and is the solution.

20. No solution

21.
$$t^{1/3} + 2t^{1/6} = 3$$
$$t^{1/3} + 2t^{1/6} - 3 = 0$$

Let $u = t^{1/6}$ and $u^2 = t^{2/6} = t^{1/3}$.

$$u^2 + 2u - 3 = 0 \quad \text{Substituting}$$
$$(u + 3)(u - 1) = 0$$

$u = -3 \quad or \quad u = 1$

$t^{1/6} = -3 \quad or \quad t^{1/6} = 1 \quad$ Substituting $t^{1/6}$ for u

No solution $\qquad t = 1$

The number 1 checks and is the solution.

22. 16, 81

23. $(3 - \sqrt{x})^2 - 10(3 - \sqrt{x}) + 23 = 0$

Let $u = 3 - \sqrt{x}$ and $u^2 = (3 - \sqrt{x})^2$.

$u^2 - 10u + 23 = 0 \qquad$ Substituting

$$u = \frac{-(-10) \pm \sqrt{(-10)^2 - 4 \cdot 1 \cdot 23}}{2 \cdot 1}$$

$$u = \frac{10 \pm \sqrt{8}}{2} = \frac{2 \cdot 5 \pm 2\sqrt{2}}{2}$$

$$u = 5 \pm \sqrt{2}$$

$u = 5 + \sqrt{2} \quad or \quad u = 5 - \sqrt{2}$

Now we replace u with $3 - \sqrt{x}$ and solve these equations:

$3 - \sqrt{x} = 5 + \sqrt{2} \quad or \quad 3 - \sqrt{x} = 5 - \sqrt{2}$

$-\sqrt{x} = 2 + \sqrt{2} \quad or \quad -\sqrt{x} = 2 - \sqrt{2}$

$\sqrt{x} = -2 - \sqrt{2} \quad or \quad \sqrt{x} = -2 + \sqrt{2}$

Since both $-2 - \sqrt{2}$ and $-2 + \sqrt{2}$ are negative and principal square roots are never negative, the equation has no solution.

24. $4 + 2\sqrt{3}$

25. $16\left(\dfrac{x-1}{x-8}\right)^2 + 8\left(\dfrac{x-1}{x-8}\right) + 1 = 0$

Let $u = \dfrac{x-1}{x-8}$ and $u^2 = \left(\dfrac{x-1}{x-8}\right)^2$.

$16u^2 + 8u + 1 = 0 \quad$ Substituting

$(4u + 1)(4u + 1) = 0$

$u = -\dfrac{1}{4}$

Now we replace u with $\dfrac{x-1}{x-8}$ and solve this equation:

$\dfrac{x-1}{x-8} = -\dfrac{1}{4}$

$4x - 4 = -x + 8 \quad$ Multiplying by $4(x-8)$

$5x = 12$

$x = \dfrac{12}{5}$

The number $\dfrac{12}{5}$ checks and is the solution.

26. $-\dfrac{3}{2}$

27. The x-intercepts occur where $f(x) = 0$. Thus, we must have $5x + 13\sqrt{x} - 6 = 0$.

Let $u = \sqrt{x}$ and $u^2 = x$.

$5u^2 + 13u - 6 = 0 \quad$ Substituting

$(5u - 2)(u + 3) = 0$

$u = \dfrac{2}{5} \ or \ u = -3$

Now replace u with $\sqrt{x}$ and solve these equations:

$\sqrt{x} = \dfrac{2}{5} \quad or \quad \sqrt{x} = -3$

$x = \dfrac{4}{25} \qquad$ No solution

The number $\dfrac{4}{25}$ checks. Thus, the x-intercept is $\left(\dfrac{4}{25}, 0\right)$.

28. $\left(\dfrac{4}{9}, 0\right)$

29. The x-intercepts occur where $f(x) = 0$. Thus, we must have $(x^2 - 3x)^2 - 10(x^2 - 3x) + 24 = 0$.

Let $u = x^2 - 3x$ and $u^2 = (x^2 - 3x)^2$.

$u^2 - 10u + 24 = 0 \quad$ Substituting

$(u - 6)(u - 4) = 0$

$u = 6 \ or \ u = 4$

Now replace u with $x^2 - 3x$ and solve these equations:

$x^2 - 3x = 6 \quad or \qquad x^2 - 3x = 4$

$x^2 - 3x - 6 = 0 \quad or \quad x^2 - 3x - 4 = 0$

$x = \dfrac{-(-3) \pm \sqrt{(-3)^2 - 4(1)(-6)}}{2 \cdot 1} \quad or$

$\qquad\qquad\qquad\qquad (x - 4)(x + 1) = 0$

$x = \dfrac{3 \pm \sqrt{33}}{2} \quad or \quad x = 4 \ or \ x = -1$

All four numbers check. Thus, the x-intercepts are $\left(\dfrac{3 + \sqrt{33}}{2}, 0\right)$, $\left(\dfrac{3 - \sqrt{33}}{2}, 0\right)$, $(4, 0)$, and $(-1, 0)$.

30. $(7, 0)$, $(-1, 0)$, $(5, 0)$, $(1, 0)$

31. The x-intercepts occur where $f(x) = 0$. Thus, we must have $x^{2/5} + x^{1/5} - 6 = 0$.

Let $u = x^{1/5}$ and $u^2 = x^{2/5}$.

$u^2 + u - 6 = 0 \quad$ Substituting

$(u + 3)(u - 2) = 0$

$$u = -3 \quad or \quad u = 2$$
$$x^{1/5} = -3 \quad or \quad x^{1/5} = 2 \quad \text{Replacing } u$$
$$\text{with } x^{1/5}$$
$$x = -243 \ or \quad x = 32 \quad \text{Raising to the fifth}$$
$$\text{power}$$

Both -243 and 32 check. Thus, the x-intercepts are $(-243, 0)$ and $(32, 0)$.

32. $(81, 0)$

33. $f(x) = \left(\dfrac{x^2 + 2}{x}\right)^4 + 7\left(\dfrac{x^2 + 2}{x}\right)^2 + 5$

Observe that, for all real numbers x, each term is positive. Thus, there are no real-number values of x for which $f(x) = 0$ and hence no x-intercepts.

34. No x-intercepts

35. Writing exercise

36. Writing exercise

37. Graph $f(x) = \dfrac{3}{2}x$.

We find some ordered pairs, plot points, and draw the graph.

x	y
-4	-6
-2	-3
0	0
2	3
4	6

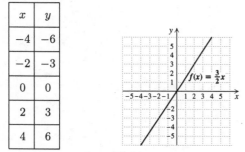

38.

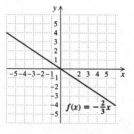

39. Graph $g(x) = \dfrac{2}{x}$.

We find some ordered pairs, plot points, and draw the graph. Note that we cannot use 0 as a first coordinate since division by 0 is undefined.

x	y
-4	$-\dfrac{1}{2}$
-2	-1
$-\dfrac{1}{2}$	-4
$\dfrac{1}{2}$	4
2	1
4	$\dfrac{1}{2}$

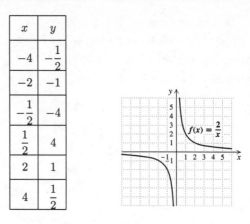

40.

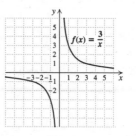

41. *Familiarize.* Let $a =$ the number of liters of solution A in the mixture and $b =$ the number of liters of solution B. We organize the information in a table.

Solution	A	B	Mixture
Number of liters	a	b	12
Percent of alcohol	18%	45%	36%
Amount of alcohol	$0.18a$	$0.45b$	$0.36(12)$, or 4.32 L

From the first row of the table we get one equation:

$$a + b = 12$$

We get a second equation from the last row of the table:

$$0.18a + 0.45b = 4.32$$

After clearing decimals, we have the following system of equations:

$$a + b = 12, \quad (1)$$
$$18a + 45b = 432 \quad (2)$$

Carry out. We use the elimination method. First we multiply equation (1) by -18 and then add.

$$-18a - 18b = -216$$
$$\underline{18a + 45b = 432}$$
$$27b = 216$$
$$b = 8$$

Now we substitute 8 for b in one of the original equations and solve for a.

$$a + b = 12 \quad (1)$$
$$a + 8 = 12$$
$$a = 4$$

Check. If 4 L of solution A and 8 L of solution B are used, the mixture has $4+8$, or 12 L. The amount of alcohol in 4 L of solution A is 0.18(4), or 0.72 L. The amount of alcohol in 8 L of solution B is 0.45(8), or 3.6 L. Then the amount of alcohol in the mixture is $0.72 + 3.6$, or 4.32 L. The answer checks.

State. The mixture should contain 4 L of solution A and 8 L of solution B.

42. $a^2 + a$

43. Writing exercise

44. Writing exercise

45. $5x^4 - 7x^2 + 1 = 0$

Let $u = x^2$ and $u^2 = x^4$.
$$5u^2 - 7u + 1 = 0 \quad \text{Substituting}$$
$$u = \frac{-(-7) \pm \sqrt{(-7)^2 - 4 \cdot 5 \cdot 1}}{2 \cdot 5}$$
$$u = \frac{7 \pm \sqrt{29}}{10}$$
$$x^2 = \frac{7 \pm \sqrt{29}}{10} \quad \text{Replacing } u \text{ with } x^2$$
$$x = \pm \sqrt{\frac{7 \pm \sqrt{29}}{10}}$$

All four numbers check and are the solutions.

46. $\pm \sqrt{\dfrac{-5 \pm \sqrt{37}}{6}}$

47. $(x^2 - 4x - 2)^2 - 13(x^2 - 4x - 2) + 30 = 0$

Let $u = x^2 - 4x - 2$ and $u^2 = (x^2 - 4x - 2)^2$.
$$u^2 - 13u + 30 = 0 \quad \text{Substituting}$$
$$(u - 3)(u - 10) = 0$$
$$u = 3 \quad or \quad u = 10$$
$$x^2 - 4x - 2 = 3 \quad or \quad x^2 - 4x - 2 = 10$$
$$\text{Replacing } u \text{ with } x^2 - 4x - 2$$
$$x^2 - 4x - 5 = 0 \quad or \quad x^2 - 4x - 12 = 0$$
$$(x - 5)(x + 1) = 0 \quad or \quad (x - 6)(x + 2) = 0$$
$$x = 5 \text{ or } x = -1 \text{ or } x = 6 \text{ or } x = -2$$

All four numbers check and are the solutions.

48. $-2, -1, 6, 7$

49. $\dfrac{x}{x-1} - 6\sqrt{\dfrac{x}{x-1}} - 40 = 0$

Let $u = \sqrt{\dfrac{x}{x-1}}$ and $u^2 = \dfrac{x}{x-1}$.
$$u^2 - 6u - 40 = 0 \quad \text{Substituting}$$
$$(u - 10)(u + 4) = 0$$
$$u = 10 \qquad or \qquad u = -4$$
$$\sqrt{\frac{x}{x-1}} = 10 \qquad or \qquad \sqrt{\frac{x}{x-1}} = -4$$
$$\frac{x}{x-1} = 100 \qquad or \qquad \text{No solution}$$
$$x = 100x - 100 \quad \text{Multiplying by } (x-1)$$
$$100 = 99x$$
$$\frac{100}{99} = x$$

The number $\dfrac{100}{99}$ checks. It is the solution.

50. $\dfrac{432}{143}$

51. $a^5(a^2 - 25) + 13a^3(25 - a^2) + 36a(a^2 - 25) = 0$
$$a^5(a^2 - 25) - 13a^3(a^2 - 25) + 36a(a^2 - 25) = 0$$
$$a(a^2 - 25)(a^4 - 13a^2 + 36) = 0$$
$$a(a^2 - 25)(a^2 - 4)(a^2 - 9) = 0$$
$$a = 0 \text{ or } a^2 - 25 = 0 \quad or \ a^2 - 4 = 0 \quad or \ a^2 - 9 = 0$$
$$a = 0 \text{ or } \quad a^2 = 25 \text{ or } \quad a^2 = 4 \quad or \quad a^2 = 9$$
$$a = 0 \text{ or } \quad a = \pm 5 \text{ or } \quad a = \pm 2 \text{ or } \quad a = \pm 3$$

All seven numbers check. The solutions are 0, 5, −5, 2, −2, 3, and −3.

52. 9

53. $x^6 - 28x^3 + 27 = 0$

Let $u = x^3$.
$$u^2 - 28u + 27 = 0$$
$$(u - 27)(u - 1) = 0$$
$$u = 27 \quad or \quad u = 1$$
$$x^3 = 27 \quad or \quad x^3 = 1$$
$$x = 3 \quad or \quad x = 1$$

Both 3 and 1 check. They are the solutions.

54. $-2, 1$

55.

56. $-3, -1, 1, 4$

57. Writing exercise

Exercise Set 11.6

1. $f(x) = x^2$

See Example 1 in the text.

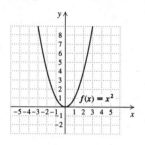

2.

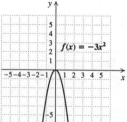

3. $f(x) = -2x^2$

We choose some numbers for x and compute $f(x)$ for each one. Then we plot the ordered pairs $(x, f(x))$ and connect them with a smooth curve.

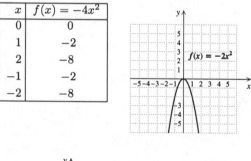

x	$f(x) = -4x^2$
0	0
1	-2
2	-8
-1	-2
-2	-8

4.

5. $g(x) = \dfrac{1}{3}x^2$

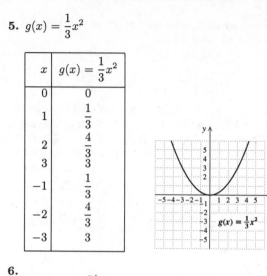

x	$g(x) = \dfrac{1}{3}x^2$
0	0
1	$\dfrac{1}{3}$
2	$\dfrac{4}{3}$
3	3
-1	$\dfrac{1}{3}$
-2	$\dfrac{4}{3}$
-3	3

6.

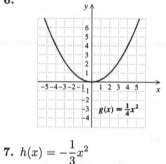

7. $h(x) = -\dfrac{1}{3}x^2$

Observe that the graph of $h(x) = -\dfrac{1}{3}x^2$ is the reflection of the graph of $g(x) = \dfrac{1}{3}x^2$ across the x-axis. We graphed $g(x)$ in Exercise 5, so we can use it to graph $h(x)$. If we did not make this observation we could find some ordered pairs, plot points, and connect them with a smooth curve.

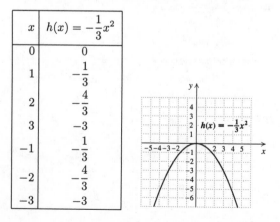

x	$h(x) = -\dfrac{1}{3}x^2$
0	0
1	$-\dfrac{1}{3}$
2	$-\dfrac{4}{3}$
3	-3
-1	$-\dfrac{1}{3}$
-2	$-\dfrac{4}{3}$
-3	-3

8.

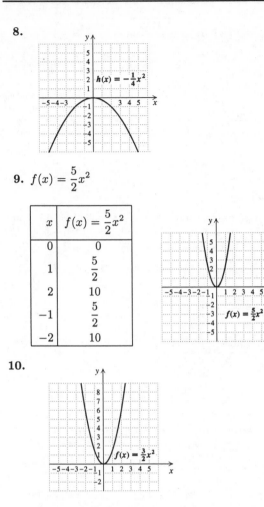

9. $f(x) = \dfrac{5}{2}x^2$

x	$f(x) = \dfrac{5}{2}x^2$
0	0
1	$\dfrac{5}{2}$
2	10
-1	$\dfrac{5}{2}$
-2	10

10.

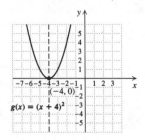

11. $g(x) = (x+4)^2 = [x - (-4)]^2$

We know that the graph of $g(x) = (x+4)^2$ looks like the graph of $f(x) = x^2$ (see Exercise 1) but moved to the left 4 units.

Vertex: $(-4, 0)$, axis of symmetry: $x = -4$

12. Vertex: $(-1, 0)$, axis of symmetry: $x = -1$

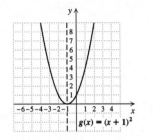

13. $f(x) = (x-1)^2$

The graph of $f(x) = (x-1)^2$ looks like the graph of $f(x) = x^2$ (see Exercise 1) but moved to the right 1 unit.

Vertex: $(1, 0)$, axis of symmetry: $x = 1$

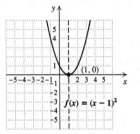

14. Vertex: $(2, 0)$, axis of symmetry: $x = 2$

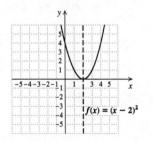

15. $h(x) = (x-3)^2$

The graph of $h(x) = (x-3)^2$ looks like the graph of $f(x) = x^2$ (see Exercise 1) but moved to the right 3 units.

Vertex: $(3, 0)$, axis of symmetry: $x = 3$

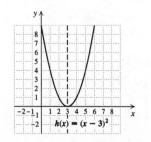

16. Vertex: $(4, 0)$, axis of symmetry: $x = 4$

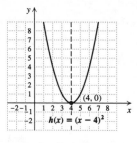

17. $f(x) = -(x + 4)^2 = -[x - (-4)]^2$

The graph of $f(x) = -(x + 4)^2$ looks like the graph of $f(x) = x^2$ (see Exercise 1) but moved to the left 4 units. It will also open downward because of the negative coefficient, -1.

Vertex: $(-4, 0)$, axis of symmetry: $x = -4$

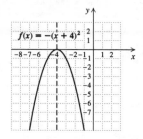

18. Vertex: $(2, 0)$, axis of symmetry: $x = 2$

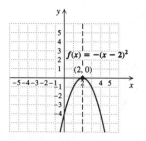

19. $g(x) = -(x - 1)^2$

The graph of $g(x) = -(x - 1)^2$ looks like the graph of $f(x) = x^2$ (see Exercise 1) but moved to the right 1 unit. It will also open downward because of the negative coefficient, -1.

Vertex: $(1, 0)$, axis of symmetry: $x = 1$

20. Vertex: $(-1, 0)$, axis of symmetry: $x = -1$

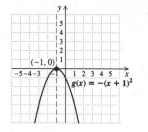

21. $f(x) = 2(x + 1)^2$

The graph of $f(x) = 2(x + 1)^2$ looks like the graph of $h(x) = 2x^2$ (see graph following Example 1) but moved to the left 1 unit.

Vertex: $(-1, 0)$, axis of symmetry: $x = -1$

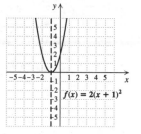

22. Vertex: $(-4, 0)$, axis of symmetry: $x = -4$

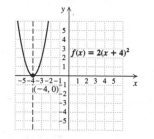

23. $h(x) = -\dfrac{1}{2}(x - 4)^2$

The graph of $h(x) = -\dfrac{1}{2}(x - 4)^2$ looks like the graph of $g(x) = \dfrac{1}{2}x^2$ (see graph following Example 1) but moved to the right 4 units. It will also open downward because of the negative coefficient, $-\dfrac{1}{2}$.

Vertex: $(4, 0)$, axis of symmetry: $x = 4$

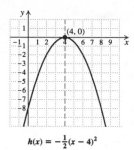

$$h(x) = -\frac{1}{2}(x-4)^2$$

24. Vertex: $(2,0)$, axis of symmetry: $x = 2$

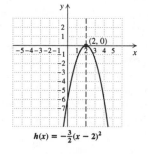

$$h(x) = -\frac{3}{2}(x-2)^2$$

25. $f(x) = \dfrac{1}{2}(x-1)^2$

The graph of $f(x) = \dfrac{1}{2}(x-1)^2$ looks like the graph of $g(x) = \dfrac{1}{2}x^2$ (see graph following Example 1) but moved to the right 1 unit.

Vertex: $(1,0)$, axis of symmetry: $x = 1$

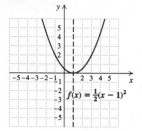

$$f(x) = \tfrac{1}{2}(x-1)^2$$

26. Vertex: $(-2,0)$, axis of symmetry: $x = -2$

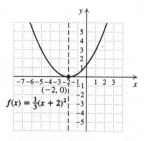

$$f(x) = \tfrac{1}{3}(x+2)^2$$

27. $f(x) = -2(x+5)^2 = -2[x-(-5)]^2$

The graph of $f(x) = -2(x+5)^2$ looks like the graph of $h(x) = 2x^2$ (see graph following Example 1) but

moved to the left 5 units. It will also open downward because of the negative coefficient, -2.

Vertex: $(-5,0)$, axis of symmetry: $x = -5$

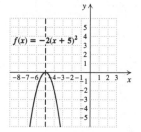

$$f(x) = -2(x+5)^2$$

28. Vertex: $(-7,0)$, axis of symmetry: $x = -7$

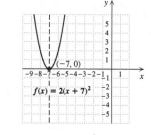

$$f(x) = 2(x+7)^2$$

29. $h(x) = -3\left(x - \dfrac{1}{2}\right)^2$

The graph of $h(x) = -3\left(x - \dfrac{1}{2}\right)^2$ looks like the graph of $f(x) = -3x^2$ (see Exercise 4) but moved to the right $\dfrac{1}{2}$ unit.

Vertex: $\left(\dfrac{1}{2}, 0\right)$, axis of symmetry: $x = \dfrac{1}{2}$

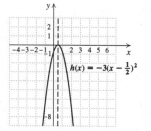

$$h(x) = -3(x - \tfrac{1}{2})^2$$

30. Vertex: $\left(-\dfrac{1}{2}, 0\right)$, axis of symmetry: $x = -\dfrac{1}{2}$

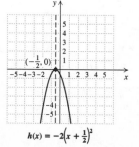

$$h(x) = -2\left(x + \tfrac{1}{2}\right)^2$$

31. $f(x) = (x - 5)^2 + 2$

We know that the graph looks like the graph of $f(x) = x^2$ (see Example 1) but moved to the right 5 units and up 2 units. The vertex is $(5, 2)$, and the axis of symmetry is $x = 5$. Since the coefficient of $(x - 5)^2$ is positive $(1 > 0)$, there is a minimum function value, 2.

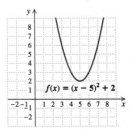

32. Vertex: $(-3, -2)$, axis of symmetry: $x = -3$

Minimum: -2

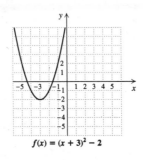

$f(x) = (x + 3)^2 - 2$

33. $f(x) = (x + 1)^2 - 3$

We know that the graph looks like the graph of $f(x) = x^2$ (see Example 1) but moved to the left 1 unit and down 3 units. The vertex is $(-1, -3)$, and the axis of symmetry is $x = -1$. Since the coefficient of $(x + 1)^2$ is positive $(1 > 0)$, there is a minimum function value, -3.

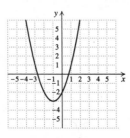

$f(x) = (x + 1)^2 - 3$

34. Vertex: $(1, 2)$, axis of symmetry: $x = 1$

Minimum: 2

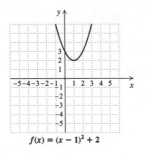

$f(x) = (x - 1)^2 + 2$

35. $g(x) = (x + 4)^2 + 1$

We know that the graph looks like the graph of $f(x) = x^2$ (see Example 1) but moved to the left 4 units and up 1 unit. The vertex is $(-4, 1)$, and the axis of symmetry is $x = -4$. Since the coefficient of $(x + 4)^2$ is positive $(1 > 0)$, there is a minimum function value, 1.

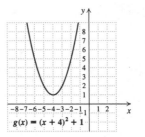

$g(x) = (x + 4)^2 + 1$

36. Vertex: $(2, -4)$, axis of symmetry: $x = 2$

Maximum: -4

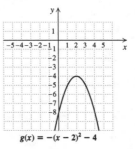

$g(x) = -(x - 2)^2 - 4$

37. $h(x) = -2(x - 1)^2 - 3$

We know that the graph looks like the graph of $h(x) = 2x^2$ (see graph following Example 1) but moved to the right 1 unit and down 3 units and turned upside down. The vertex is $(1, -3)$, and the axis of symmetry is $x = 1$. The maximum function value is -3.

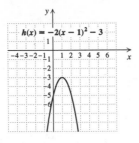

38. Vertex: $(-1, 4)$, axis of symmetry: $x = -1$

Maximum: 4

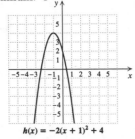

39. $f(x) = 2(x + 4)^2 + 1$

We know that the graph looks like the graph of $f(x) = 2x^2$ (see graph following Example 1) but moved to the left 4 units and up 1 unit. The vertex is $(-4, 1)$, the axis of symmetry is $x = -4$, and the minimum function value is 1.

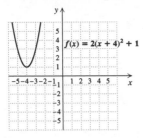

40. Vertex: $(5, -3)$, axis of symmetry: $x = 5$

Minimum: -3

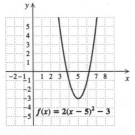

41. $g(x) = -\dfrac{3}{2}(x - 1)^2 + 4$

We know that the graph looks like the graph of $f(x) = \dfrac{3}{2}x^2$ (see Exercise 10) but moved to the right

1 unit and up 4 units and turned upside down. The vertex is $(1, 4)$, the axis of symmetry is $x = 1$, and the maximum function value is 4.

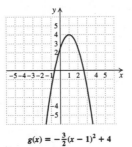

$$g(x) = -\tfrac{3}{2}(x - 1)^2 + 4$$

42. Vertex: $(-2, -3)$, axis of symmetry: $x = -2$

Minimum: -3

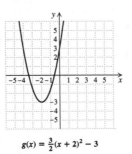

$$g(x) = \tfrac{3}{2}(x + 2)^2 - 3$$

43. $f(x) = 8(x - 9)^2 + 7$

This function is of the form $f(x) = a(x - h)^2 + k$ with $a = 8$, $h = 9$, and $k = 7$. The vertex is (h, k), or $(9, 7)$. The axis of symmetry is $x = h$, or $x = 9$. Since $a > 0$, then k, or 7, is the minimum function value.

44. Vertex: $(-5, -6)$

Axis of symmetry: $x = -5$

Minimum: -6

45. $h(x) = -\dfrac{2}{7}(x + 6)^2 + 11$

This function is of the form $f(x) = a(x - h)^2 + k$ with $a = -\dfrac{2}{7}$, $h = -6$, and $k = 11$. The vertex is (h, k), or $(-6, 11)$. The axis of symmetry is $x = h$, or $x = -6$. Since $a < 0$, then k, or 11, is the maximum function value.

46. Vertex: $(7, -9)$

Axis of symmetry: $x = 7$

Maximum: -9

47. $f(x) = 5\left(x + \dfrac{1}{4}\right)^2 - 13$

This function is of the form $f(x) = a(x - h)^2 + k$ with $a = 5$, $h = -\dfrac{1}{4}$, and $k = -13$. The vertex is (h, k), or $\left(-\dfrac{1}{4}, -13\right)$. The axis of symmetry is $x = h$, or $x = -\dfrac{1}{4}$. Since $a > 0$, then k, or -13, is the minimum function value.

48. Vertex: $\left(\dfrac{1}{4}, 15\right)$

Axis of symmetry: $x = \dfrac{1}{4}$

Minimum: 15

49. $f(x) = \sqrt{2}(x + 4.58)^2 + 65\pi$

This function is of the form $f(x) = a(x - h)^2 + k$ with $a = \sqrt{2}$, $h = -4.58$, and $k = 65\pi$. The vertex is (h, k), or $(-4.58, 65\pi)$. The axis of symmetry is $x = h$, or $x = -4.58$. Since $a > 0$, then k, or 65π, is the minimum function value.

50. Vertex: $(38.2, -\sqrt{34})$

Axis of symmetry: $x = 38.2$

Minimum: $-\sqrt{34}$

51. Writing exercise

52. Writing exercise

53. Graph $2x - 7y = 28$.

Find the x-intercept.

$$2x - 7 \cdot 0 = 28$$
$$2x = 28$$
$$x = 14$$

The x-intercept is $(14, 0)$.

Find the y-intercept.

$$2 \cdot 0 - 7y = 28$$
$$-7y = 28$$
$$y = -4$$

The y-intercept is $(0, -4)$.

Plot the intercepts and draw a line through them. A third point can be plotted as a check.

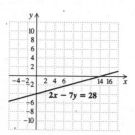

54.

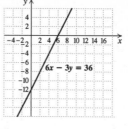

55. $3x + 4y = -19$,   (1)

$7x - 6y = -29$ (2)

Multiply Equation (1) by 3 and multiply Equation (2) by 2. Then add the equations to eliminate the y-term.

$$\begin{aligned} 9x + 12y &= -57 \\ 14x - 12y &= -58 \\ \hline 23x &= -115 \\ x &= -5 \end{aligned}$$

Now substitute -5 for x in one of the original equations and solve for y. We use Equation (1).

$$3(-5) + 4y = -19$$
$$-15 + 4y = -19$$
$$4y = -4$$
$$y = -1$$

The pair $(-5, -1)$ checks and it is the solution.

56. $(-1, 2)$

57. $x^2 + 5x$

We take half the coefficient of x and square it.

$$\dfrac{1}{2} \cdot 5 = \dfrac{5}{2}, \ \left(\dfrac{5}{2}\right)^2 = \dfrac{25}{4}$$

Then we have $x^2 + 5x + \dfrac{25}{4}$.

58. $x^2 - 9x + \dfrac{81}{4}$

59. Writing exercise

60. Writing exercise

61. The equation will be of the form $f(x) = \dfrac{3}{5}(x - h)^2 + k$ with $h = 4$ and $k = 1$:

$$f(x) = \dfrac{3}{5}(x - 4)^2 + 1$$

62. $f(x) = \dfrac{3}{5}(x - 2)^2 + 6$

63. The equation will be of the form $f(x) = \dfrac{3}{5}(x - h)^2 + k$ with $h = 3$ and $k = -1$:

$$f(x) = \frac{3}{5}(x-3)^2 + (-1), \text{ or}$$

$$f(x) = \frac{3}{5}(x-3)^2 - 1$$

64. $f(x) = \frac{3}{5}(x-5)^2 - 6$

65. The equation will be of the form $f(x) = \frac{3}{5}(x-h)^2 + k$ with $h = -2$ and $k = -5$:

$$f(x) = \frac{3}{5}[x - (-2)]^2 + (-5), \text{ or}$$

$$f(x) = \frac{3}{5}(x+2)^2 - 5$$

66. $f(x) = \frac{3}{5}(x+4)^2 - 2$

67. Since there is a maximum at $(5, 0)$, the parabola will have the same shape as $g(x) = -2x^2$. It will be of the form $g(x) = -2(x-h)^2 + k$ with $h = 5$ and $k = 0$: $g(x) = -2(x-5)^2$

68. $f(x) = 2(x-2)^2$

69. Since there is a minimum at $(-4, 0)$, the parabola will have the same shape as $f(x) = 2x^2$. It will be of the form $f(x) = 2(x-h)^2 + k$ with $h = -4$ and $k = 0$: $f(x) = 2[x - (-4)]^2$, or $f(x) = 2(x+4)^2$

70. $g(x) = -2x^2 + 3$

71. Since there is a maximum at $(3, 8)$, the parabola will have the same shape as $g(x) = -2x^2$. It will be of the form $g(x) = -2(x-h)^2 + k$ with $h = 3$ and $k = 8$: $g(x) = -2(x-3)^2 + 8$

72. $f(x) = 2(x+2)^2 + 3$

73. The maximum value of $g(x)$ is 1 and occurs at the point $(5, 1)$, so for $F(x)$ we have $h = 5$ and $k = 1$. $F(x)$ has the same shape as $f(x)$ and has a minimum, so $a = 3$. Thus, $F(x) = 3(x-5)^2 + 1$.

74. $F(x) = -\frac{1}{3}(x+4)^2 - 6$

75. The graph of $y = f(x-1)$ looks like the graph of $y = f(x)$ moved 1 unit to the right.

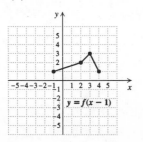

76.

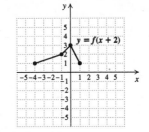

77. The graph of $y = f(x) + 2$ looks like the graph of $y = f(x)$ moved up 2 units.

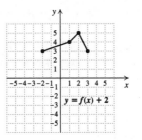

78.

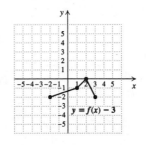

79. The graph of $y = f(x+3) - 2$ looks like the graph of $y = f(x)$ moved 3 units to the left and also moved down 2 units.

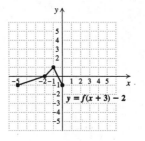

80.

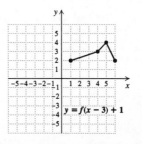

81.

82.

83. Writing exercise

Exercise Set 11.7

1. $f(x) = x^2 + 4x + 5$

$$= (x^2 + 4x + 4 - 4) + 5 \quad \text{Adding } 4 - 4$$
$$= (x^2 + 4x + 4) - 4 + 5 \quad \text{Regrouping}$$
$$= (x + 2)^2 + 1$$

The vertex is $(-2, 1)$, the axis of symmetry is $x = -2$, and the graph opens upward since the coefficient 1 is positive. We plot a few points as a check and draw the curve.

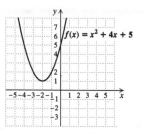

2. Vertex: $(-1, -6)$, axis of symmetry: $x = -1$

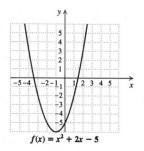

3. $g(x) = x^2 - 6x + 13$

$$= (x^2 - 6x + 9 - 9) + 13 \quad \text{Adding } 9 - 9$$
$$= (x^2 - 6x + 9) - 9 + 13 \quad \text{Regrouping}$$
$$= (x - 3)^2 + 4$$

The vertex is $(3, 4)$, the axis of symmetry is $x = 3$, and the graph opens upward since the coefficient 1 is positive. We plot a few points as a check and draw the curve.

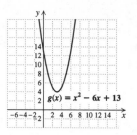

4. Vertex: $(2, 1)$, axis of symmetry: $x = 2$

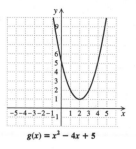

5. $f(x) = x^2 + 8x + 20$

$$= (x^2 + 8x + 16 - 16) + 20 \quad \text{Adding } 16 - 16$$
$$= (x^2 + 8x + 16) - 16 + 20 \quad \text{Regrouping}$$
$$= (x + 4)^2 + 4$$

The vertex is $(-4, 4)$, the axis of symmetry is $x = -4$, and the graph opens upward since the coefficient 1 is positive.

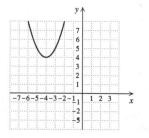

6. Vertex: $(5, -4)$, axis of symmetry: $x = 5$

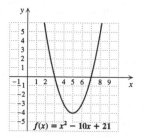

7. $h(x) = 2x^2 - 16x + 25$

$\qquad = 2(x^2 - 8x) + 25$ Factoring 2 from the first two terms

$\qquad = 2(x^2 - 8x + 16 - 16) + 25$ Adding $16-16$ inside the parentheses

$\qquad = 2(x^2 - 8x + 16) + 2(-16) + 25$

$\qquad\qquad\qquad$ Distributing to obtain a trinomial square

$\qquad = 2(x - 4)^2 - 7$

The vertex is $(4, -7)$, the axis of symmetry is $x = 4$, and the graph opens upward since the coefficient 2 is positive.

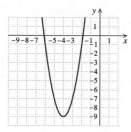

8. Vertex: $(-4, -9)$, axis of symmetry: $x = -4$

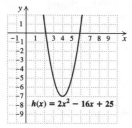

$h(x) = 2x^2 + 16x + 23$

9. $f(x) = -x^2 + 2x + 5$

$\qquad = -(x^2 - 2x) + 5$ Factoring -1 from the first two terms

$\qquad = -(x^2 - 2x + 1 - 1) + 5$

$\qquad\qquad\qquad$ Adding $1 - 1$ inside the parentheses

$\qquad = -(x^2 - 2x + 1) - (-1) + 5$

$\qquad = -(x - 1)^2 + 6$

The vertex is $(1, 6)$, the axis of symmetry is $x = 1$, and the graph opens downward since the coefficient -1 is negative.

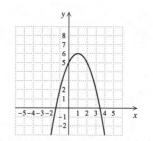

10. Vertex: $(-1, 8)$, axis of symmetry: $x = -1$

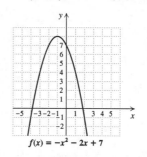

$f(x) = -x^2 - 2x + 7$

11. $g(x) = x^2 + 3x - 10$

$\qquad = \left(x^2 + 3x + \dfrac{9}{4} - \dfrac{9}{4}\right) - 10$

$\qquad = \left(x^2 + 3x + \dfrac{9}{4}\right) - \dfrac{9}{4} - 10$

$\qquad = \left(x + \dfrac{3}{2}\right)^2 - \dfrac{49}{4}$

The vertex is $\left(-\dfrac{3}{2}, -\dfrac{49}{4}\right)$, the axis of symmetry is $x = -\dfrac{3}{2}$, and the graph opens upward since the coefficient 1 is positive.

12. Vertex: $\left(-\dfrac{5}{2}, -\dfrac{9}{4}\right)$, axis of symmetry: $x = -\dfrac{5}{2}$

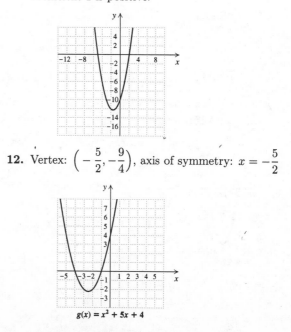

$g(x) = x^2 + 5x + 4$

13. $f(x) = 3x^2 - 24x + 50$

$\qquad = 3(x^2 - 8x) + 50$ Factoring

$\qquad = 3(x^2 - 8x + 16 - 16) + 50$

$\qquad\qquad\qquad\qquad$ Adding $16 - 16$ inside

$\qquad\qquad\qquad\qquad$ the parentheses

$\qquad = 3(x^2 - 8x + 16) - 3 \cdot 16 + 50$

$\qquad = 3(x - 4)^2 + 2$

The vertex is $(4, 2)$, the axis of symmetry is $x = 4$, and the graph opens upward since the coefficient 3 is positive.

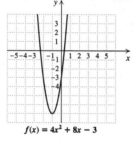

14. Vertex: $(-1, -7)$, axis of symmetry: $x = -1$

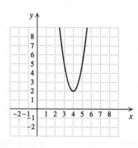

$f(x) = 4x^2 + 8x - 3$

15. $h(x) = x^2 + 7x$

$\qquad = \left(x^2 + 7x + \dfrac{49}{4}\right) - \dfrac{49}{4}$

$\qquad = \left(x + \dfrac{7}{2}\right)^2 - \dfrac{49}{4}$

The vertex is $\left(-\dfrac{7}{2}, -\dfrac{49}{4}\right)$, the axis of symmetry is $x = -\dfrac{7}{2}$, and the graph opens upward since the coefficient 1 is positive.

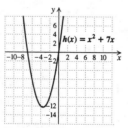

$h(x) = x^2 + 7x$

16. Vertex: $\left(\dfrac{5}{2}, -\dfrac{25}{4}\right)$, axis of symmetry: $x = \dfrac{5}{2}$

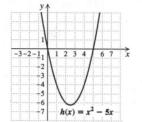

$h(x) = x^2 - 5x$

17. $f(x) = -2x^2 - 4x - 6$

$\qquad = -2(x^2 + 2x) - 6$ Factoring

$\qquad = -2(x^2 + 2x + 1 - 1) - 6$

$\qquad\qquad\qquad\qquad$ Adding $1 - 1$ inside

$\qquad\qquad\qquad\qquad$ the parentheses

$\qquad = -2(x^2 + 2x + 1) - 2(-1) - 6$

$\qquad = -2(x + 1)^2 - 4$

The vertex is $(-1, -4)$, the axis of symmetry is $x = -1$, and the graph opens downward since the coefficient -2 is negative.

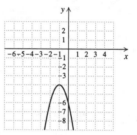

18. Vertex: $(1, 5)$, axis of symmetry: $x = 1$

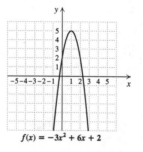

$f(x) = -3x^2 + 6x + 2$

19. $g(x) = 2x^2 - 8x + 3$

$\qquad = 2(x^2 - 4x) + 3$ Factoring

$\qquad = 2(x^2 - 4x + 4 - 4) + 3$

$\qquad\qquad\qquad\qquad$ Adding $4 - 4$ inside

$\qquad\qquad\qquad\qquad$ the parentheses

$\qquad = 2(x^2 - 4x + 4) + 2(-4) + 3$

$\qquad = 2(x - 2)^2 - 5$

The vertex is $(2, -5)$, the axis of symmetry is $x = 2$, and the graph opens upward since the coefficient 2 is positive.

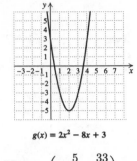

$$g(x) = 2x^2 - 8x + 3$$

20. Vertex: $\left(-\dfrac{5}{4}, -\dfrac{33}{8}\right)$, axis of symmetry: $x = -\dfrac{5}{4}$

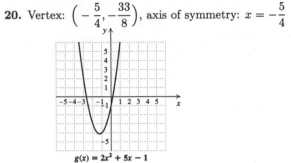

$$g(x) = 2x^2 + 5x - 1$$

21. $f(x) = -3x^2 + 5x - 2$

$$= -3\left(x^2 - \frac{5}{3}x\right) - 2 \qquad \text{Factoring}$$

$$\cdot = -3\left(x^2 - \frac{5}{3}x + \frac{25}{36} - \frac{25}{36}\right) - 2$$

$$\qquad\qquad \text{Adding } \frac{25}{36} - \frac{25}{36} \text{ inside}$$

$$\qquad\qquad \text{the parentheses}$$

$$= -3\left(x^2 - \frac{5}{3}x + \frac{25}{36}\right) - 3\left(-\frac{25}{36}\right) - 2$$

$$= -3\left(x - \frac{5}{6}\right)^2 + \frac{1}{12}$$

The vertex is $\left(\dfrac{5}{6}, \dfrac{1}{12}\right)$, the axis of symmetry is $x = \dfrac{5}{6}$, and the graph opens downward since the coefficient -3 is negative.

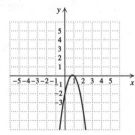

22. Vertex: $\left(-\dfrac{7}{6}, \dfrac{73}{12}\right)$, axis of symmetry: $x = -\dfrac{7}{6}$

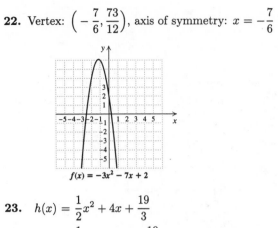

$$f(x) = -3x^2 - 7x + 2$$

23. $h(x) = \dfrac{1}{2}x^2 + 4x + \dfrac{19}{3}$

$$= \frac{1}{2}(x^2 + 8x) + \frac{19}{3} \qquad \text{Factoring}$$

$$= \frac{1}{2}(x^2 + 8x + 16 - 16) + \frac{19}{3}$$

$$\qquad\qquad \text{Adding } 16 - 16 \text{ inside}$$

$$\qquad\qquad \text{the parentheses}$$

$$= \frac{1}{2}(x^2 + 8x + 16) + \frac{1}{2}(-16) + \frac{19}{3}$$

$$= \frac{1}{2}(x + 4)^2 - \frac{5}{3}$$

The vertex is $\left(-4, -\dfrac{5}{3}\right)$, the axis of symmetry is $x = -4$, and the graph opens upward since the coefficient $\dfrac{1}{2}$ is positive.

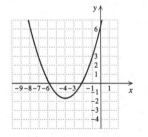

24. Vertex: $\left(3, -\dfrac{5}{2}\right)$, axis of symmetry: $x = 3$

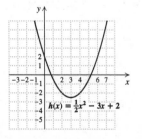

$$h(x) = \tfrac{1}{2}x^2 - 3x + 2$$

25. $f(x) = x^2 - 6x + 3$

To find the x-intercepts, solve the equation $0 = x^2 - 6x + 3$. Use the quadratic formula.

$$x = \frac{-(-6) \pm \sqrt{(-6)^2 - 4 \cdot 1 \cdot 3}}{2 \cdot 1}$$

$$x = \frac{6 \pm \sqrt{24}}{2} = \frac{6 \pm 2\sqrt{6}}{2} = 3 \pm \sqrt{6}$$

The x-intercepts are $(3 - \sqrt{6}, 0)$ and $(3 + \sqrt{6}, 0)$.

The y-intercept is $(0, f(0))$, or $(0, 3)$.

26. $\left(\dfrac{-5 - \sqrt{17}}{2}, 0\right), \left(\dfrac{-5 + \sqrt{17}}{2}, 0\right); (0, 2)$

27. $g(x) = -x^2 + 2x + 3$

To find the x-intercepts, solve the equation $0 = -x^2 + 2x + 3$. We factor.

$$0 = -x^2 + 2x + 3$$
$$0 = x^2 - 2x - 3 \quad \text{Multiplying by } -1$$
$$0 = (x - 3)(x + 1)$$
$$x = 3 \ or \ x = -1$$

The x-intercepts are $(-1, 0)$ and $(3, 0)$.

The y-intercept is $(0, g(0))$, or $(0, 3)$.

28. $(3, 0); (0, 9)$

29. $f(x) = x^2 - 9x$

To find the x-intercepts, solve the equation $0 = x^2 - 9x$. We factor.

$$0 = x^2 - 9x$$
$$0 = x(x - 9)$$
$$x = 0 \ or \ x = 9$$

The x-intercepts are $(0, 0)$ and $(9, 0)$.

Since $(0, 0)$ is an x-intercept, we observe that $(0, 0)$ is also the y-intercept.

30. $(0, 0), (7, 0); (0, 0)$

31. $h(x) = -x^2 + 4x - 4$

To find the x-intercepts, solve the equation $0 = -x^2 + 4x - 4$. We factor.

$$0 = -x^2 + 4x - 4$$
$$0 = x^2 - 4x + 4 \quad \text{Multiplying by } -1$$
$$0 = (x - 2)(x - 2)$$
$$x = 2 \text{ or } x = 2$$

The x-intercept is $(2, 0)$.

The y-intercept is $(0, h(0))$, or $(0, -4)$.

32. $\left(\dfrac{3 - \sqrt{6}}{2}, 0\right), \left(\dfrac{3 + \sqrt{6}}{2}, 0\right); (0, 3)$

33. $f(x) = 2x^2 - 4x + 6$

To find the x-intercepts, solve the equation $0 = 2x^2 - 4x + 6$. We use the quadratic formula.

$$x = \frac{-(-4) \pm \sqrt{(-4)^2 - 4 \cdot 2 \cdot 6}}{2 \cdot 2}$$

$$x = \frac{4 \pm \sqrt{-32}}{4} = \frac{4 \pm 4i\sqrt{2}}{2} = 2 \pm 2i\sqrt{2}$$

There are no real-number solutions, so there is no x-intercept.

The y-intercept is $(0, f(0))$, or $(0, 6)$.

34. No x-intercept; $(0, 2)$

35. Writing exercise

36. Writing exercise

37. $5x - 3y = 16, \quad (1)$
     $4x + 2y = 4 \quad\quad (2)$

Multiply equation (1) by 2 and equation (2) by 3 and add.

$$10x - 6y = 32$$
$$\underline{12x + 6y = 12}$$
$$22x \quad\quad\quad = 44$$
$$x = 2$$

Substitute 2 for x in one of the original equations and solve for y.

$$4x + 2y = 4 \quad (1)$$
$$4 \cdot 2 + 2y = 4$$
$$8 + 2y = 4$$
$$2y = -4$$
$$y = -2$$

The solution is $(2, -2)$.

38. $(7, 1)$

39. $4a - 5b + \ c = 3, \quad (1)$
     $3a - 4b + 2c = 3, \quad (2)$
     $a + \ b - 7c = -2 \quad (3)$

First multiply equation (1) by -2 and add it to equation (2).

$$-8a + 10b - 2c = -6$$
$$\underline{3a - \ 4b + 2c = \ \ 3}$$
$$-5a + \ 6b \quad\quad\ = -3 \quad (4)$$

Next multiply equation (1) by 7 and add it to equation (3).

$$28a - 35b + 7c = 21$$
$$\underline{a + \ \ b - 7c = -2}$$
$$29a - 34b \quad\quad = 19 \quad (5)$$

Now we solve the system of equations (4) and (5). Multiply equation (4) by 29 and equation (5) by 5 and add.

$$-145a + 174b = -87$$
$$\underline{145a - 170b = 95}$$
$$4b = 8$$
$$b = 2$$

Substitute 2 for b in equation (4) and solve for a.

$$-5a + 6 \cdot 2 = -3$$
$$-5a + 12 = -3$$
$$-5a = -15$$
$$a = 3$$

Now substitute 3 for a and 2 for b in equation (1) and solve for c.

$$4 \cdot 3 - 5 \cdot 2 + c = 3$$
$$12 - 10 + c = 3$$
$$2 + c = 3$$
$$c = 1$$

The solution is $(3, 2, 1)$.

40. $(1, -3, 2)$

41.
$$\sqrt{4x - 4} = \sqrt{x + 4} + 1$$
$$4x - 4 = x + 4 + 2\sqrt{x + 4} + 1$$
$$\text{Squaring both sides}$$
$$3x - 9 = 2\sqrt{x + 4}$$
$$9x^2 - 54x + 81 = 4(x + 4) \quad \text{Squaring both sides again}$$
$$9x^2 - 54x + 81 = 4x + 16$$
$$9x^2 - 58x + 65 = 0$$
$$(9x - 13)(x - 5) = 0$$
$$x = \frac{13}{9} \quad \text{or} \quad x = 5$$

Check: For $x = \frac{13}{9}$:

$$\frac{\sqrt{4x - 4} = \sqrt{x + 4} + 1}{\sqrt{4\left(\frac{13}{9}\right) - 4} \ ? \ \sqrt{\frac{13}{9} + 4} + 1}$$

$$\sqrt{\frac{16}{9}} \ \bigg| \ \sqrt{\frac{49}{9}} + 1$$

$$\frac{4}{3} \ \bigg| \ \frac{7}{3} + 1$$

$$\frac{4}{3} \ \bigg| \ \frac{10}{3} \qquad \text{FALSE}$$

For $x = 5$:

$$\frac{\sqrt{4x - 4} = \sqrt{x + 4} + 1}{\sqrt{4 \cdot 5 - 4} \ ? \ \sqrt{5 + 4} + 1}$$

$$\sqrt{16} \ \bigg| \ \sqrt{9} + 1$$

$$4 \ \bigg| \ 3 + 1$$

$$4 \ \bigg| \ 4 \qquad \text{TRUE}$$

5 checks, but $\frac{13}{9}$ does not. The solution is 5.

42. 4

43. Writing exercise

44. Writing exercise

45. a) $f(x) = 2.31x^2 - 3.135x - 5.89$
$$= 2.31(x^2 - 1.357142857x) - 5.89$$
$$= 2.31(x^2 - 1.357142857x +$$
$$0.460459183 - 0.460459183) - 5.89$$
$$= 2.31(x^2 - 1.357142857x + 0.460459183) +$$
$$2.31(-0.460459183) - 5.89$$
$$= 2.31(x - 0.678571428)^2 - 6.953660714$$

Since the coefficient 2.31 is positive, the function has a minimum value. It is -6.953660714.

b) To find the x-intercepts, solve $0 = 2.31x^2 - 3.135x - 5.89$.
$$x = \frac{-(-3.135) \pm \sqrt{(-3.135)^2 - 4(2.31)(-5.89)}}{2(2.31)}$$
$$x \approx \frac{3.135 \pm 8.015723611}{4.62}$$
$$x \approx -1.056433682 \quad \text{or} \quad x \approx 2.413576539$$

The x-intercepts are $(-1.056433682, 0)$ and $(2.413576539, 0)$.

The y-intercept is $(0, f(0))$, or $(0, -5.89)$.

46. a) Maximum: 7.01412766; b) $(-0.400174191, 0)$, $(0.821450787, 0)$; $(0, 6.18)$

47. $f(x) = x^2 - x - 6$

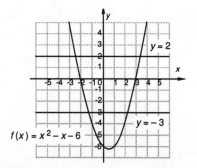

a) The solutions of $x^2 - x - 6 = 2$ are the first coordinates of the points of intersection of the graphs of $f(x) = x^2 - x - 6$ and $y = 2$. From the graph we see that the solutions are approximately -2.4 and 3.4.

b) The solutions of $x^2 - x - 6 = -3$ are the first coordinates of the points of intersection of the graphs of $f(x) = x^2 - x - 6$ and $y = -3$. From the graph we see that the solutions are approximately -1.3 and 2.3.

48. a) $-3, 1$; b) $-3.4, 1.4$; c) $-3.8, 1.8$

49. $f(x) = mx^2 - nx + p$

$= m\left(x^2 - \dfrac{n}{m}x\right) + p$

$= m\left(x^2 - \dfrac{n}{m}x + \dfrac{n^2}{4m^2} - \dfrac{n^2}{4m^2}\right) + p$

$= m\left(x - \dfrac{n}{2m}\right)^2 - \dfrac{n^2}{4m} + p$

$= m\left(x - \dfrac{n}{2m}\right)^2 + \dfrac{-n^2 + 4mp}{4m}$, or

$\quad m\left(x - \dfrac{n}{2m}\right)^2 + \dfrac{4mp - n^2}{4m}$

50. $f(x) = 3\left[x - \left(-\dfrac{m}{6}\right)\right]^2 + \dfrac{11m^2}{12}$

51. The horizontal distance from $(-1, 0)$ to $(3, -5)$ is $|3 - (-1)|$, or 4, so by symmetry the other x-intercept is $(3 + 4, 0)$, or $(7, 0)$. Substituting the three ordered pairs $(-1, 0)$, $(3, -5)$, and $(7, 0)$ in the equation $f(x) = ax^2 + bx + c$ yields a system of equations:

$\quad 0 = a - b + c,$

$\quad -5 = 9a + 3b + c,$

$\quad 0 = 49a + 7b + c$

The solution of this system of equations is $\left(\dfrac{5}{16}, -\dfrac{15}{8}, -\dfrac{35}{16}\right)$, so $f(x) = \dfrac{5}{16}x^2 - \dfrac{15}{8}x - \dfrac{35}{16}$.

52. $f(x) = -0.28x^2 - 0.56x + 6.72$

53. $f(x) = |x^2 - 1|$

We plot some points and draw the curve. Note that it will lie entirely on or above the x-axis since absolute value is never negative.

x	$f(x)$
-3	8
-2	3
-1	0
0	1
1	0
2	3
3	8

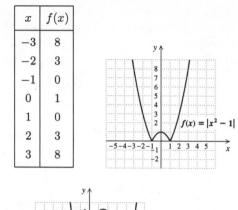

54.

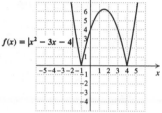

$f(x) = |x^2 - 3x - 4|$

55. $f(x) = |2(x - 3)^2 - 5|$

We plot some points and draw the curve. Note that it will lie entirely on or above the x-axis since absolute value is never negative.

x	$f(x)$
-1	27
0	13
1	3
2	3
3	5
4	3
5	3
6	13

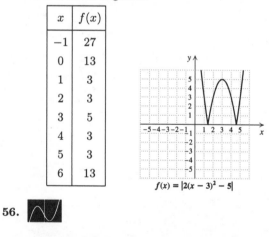

$f(x) = |2(x - 3)^2 - 5|$

56.

Exercise Set 11.8

1. *Familiarize and Translate*. We are given the function $V(x) = x^2 - 6x + 13$.

Carry out. To find the value of x for which $V(x)$ is a minimum, we first find $-\dfrac{b}{2a}$:

$-\dfrac{b}{2a} = -\dfrac{-6}{2 \cdot 1} = 3$

Now we find the minimum value of the function, $V(3)$:

$V(3) = 3^2 - 6 \cdot 3 + 13 = 9 - 18 + 13 = 4$

Check. We can go over the calculations again. We could also solve the problem again by completing the square. The answer checks.

State. The lowest value $V(x)$ will reach is $4. This occurs 3 months after January 2001.

2. $120/bicycle; 350 bicycles

3. Familiarize and Translate. We are given the function $N(x) = -0.4x^2 + 9x + 11$.

Carry out. To find the value of x for which $N(x)$ is a maximum, we first find $-\dfrac{b}{2a}$:

$$-\dfrac{b}{2a} = -\dfrac{9}{2(-0.4)} = 11.25$$

Now we find the maximum value of the function $N(11.25)$:

$$N(11.25) = -0.4(11.25)^2 + 9(11.25) + 11 = 61.625$$

Check. We can go over the calculations again. We could also solve the problem again by completing the square. The answer checks.

State. Daily ticket sales will peak 11 days after the concert was announced. About 62 tickets will be sold that day.

4. $P(x) = -x^2 + 980x - 3000$; $237,000 at $x = 490$

5. Familiarize. We make a drawing and label it.

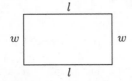

Perimeter: $2l + 2w = 720$ ft

Area: $A = l \cdot w$

Translate. We have a system of equations.

$$2l + 2w = 720,$$
$$A = lw$$

Carry out. Solving the first equation for l, we get $l = 360 - w$. Substituting for l in the second equation we get a quadratic function A:

$$A = (360 - w)w$$
$$A = -w^2 + 360w$$

Completing the square, we get

$$A = -(w - 180)^2 + 32,400.$$

The maximum function value is 32,400. It occurs when w is 180. When $w = 180$, $l = 360 - 180$, or 180.

Check. We check a function value for w less than 180 and for w greater than 180.

$$A(179) = -179^2 + 360(179) = 32,399$$
$$A(181) = -181^2 + 360(181) = 32,399$$

Since 32,400 is greater than these numbers, it looks as though we have a maximum.

State. The maximum area occurs when the dimensions are 180 ft by 180 ft.

6. 21 in. by 21 in.

7. Familiarize. We make a drawing and label it.

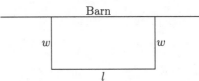

Translate. We have two equations.

$$l + 2w = 40,$$
$$A = lw$$

Carry out. Solve the first equation for l.

$$l = 40 - 2w$$

Substitute for l in the second equation.

$$A = (40 - 2w)w$$
$$A = -2w^2 + 40w$$

Completing the square, we get

$$A = -2(w - 10)^2 + 200.$$

The maximum function value of 200 occurs when $w = 10$. When $w = 10$, $l = 40 - 2 \cdot 10 = 20$.

Check. Check a function value for w less than 10 and for w greater than 10.

$$A(9) = -2 \cdot 9^2 + 40 \cdot 9 = 198$$
$$A(11) = -2 \cdot 11^2 + 40 \cdot 11 = 198$$

Since 200 is greater than these numbers, it looks as though we have a maximum.

State. The maximum area of 200 ft^2 will occur when the dimensions are 10 ft by 20 ft.

8. 450 ft^2; 15 ft by 30 ft

9. Familiarize. Let x represent the height of the file and y represent the width. We make a drawing.

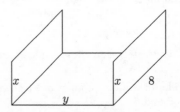

Translate. We have two equations.

$$2x + y = 14$$
$$V = 8xy$$

Carry out. Solve the first equation for y.

$$y = 14 - 2x$$

Substitute for y in the second equation.

$$V = 8x(14 - 2x)$$
$$V = -16x^2 + 112x$$

Completing the square, we get

$$V = -16\left(x - \frac{7}{2}\right)^2 + 196.$$

The maximum function value of 196 occurs when $x = \frac{7}{2}$. When $x = \frac{7}{2}$, $y = 14 - 2 \cdot \frac{7}{2} = 7$.

Check. Check a function value for x less than $\frac{7}{2}$ and for x greater than $\frac{7}{2}$.

$$V(3) = -16 \cdot 3^2 + 112 \cdot 3 = 192$$
$$V(4) = -16 \cdot 4^2 + 112 \cdot 4 = 192$$

Since 196 is greater than these numbers, it looks as though we have a maximum.

State. The file should be $\frac{7}{2}$ in., or 3.5 in., tall.

10. 4 ft by 4 ft

11. *Familiarize*. We let x and y represent the numbers, and we let P represent their product.

Translate. We have two equations.

$$x + y = 18,$$
$$P = xy$$

Carry out. Solving the first equation for y, we get $y = 18 - x$. Substituting for y in the second equation we get a quadratic function P:

$$P = x(18 - x)$$
$$P = -x^2 + 18x$$

Completing the square, we get

$$P = -(x - 9)^2 + 81.$$

The maximum function value is 81. It occurs when $x = 9$. When $x = 9$, $y = 18 - 9$, or 9.

Check. We can check a function value for x less than 9 and for x greater than 9.

$$P(10) = -10^2 + 18 \cdot 10 = 80$$
$$P(8) = -8^2 + 18 \cdot 8 = 80$$

Since 81 is greater than these numbers, it looks as though we have a maximum.

State. The maximum product of 81 occurs for the numbers 9 and 9.

12. 169; 13 and 13

13. *Familiarize*. We let x and y represent the two numbers, and we let P represent their product.

Translate. We have two equations.

$$x - y = 8,$$
$$P = xy$$

Carry out. Solve the first equation for x.

$$x = 8 + y$$

Substitute for x in the second equation.

$$P = (8 + y)y$$
$$P = y^2 + 8y$$

Completing the square, we get

$$P = (y + 4)^2 - 16.$$

The minimum function value is -16. It occurs when $y = -4$. When $y = -4$, $x = 8 + (-4)$, or 4.

Check. Check a function value for y less than -4 and for y greater than -4.

$$P(-5) = (-5)^2 + 8(-5) = -15$$
$$P(-3) = (-3)^2 + 8(-3) = -15$$

Since -16 is less than these numbers, it looks as though we have a minimum.

State. The minimum product of -16 occurs for the numbers 4 and -4.

14. $-\dfrac{49}{4}$; $\dfrac{7}{2}$ and $-\dfrac{7}{2}$

15. From the results of Exercises 11 and 12, we might observe that the numbers are -5 and -5 and that the maximum product is 25. We could also solve this problem as follows.

Familiarize. We let x and y represent the two numbers, and we let P represent their product.

Translate. We have two equations.

$$x + y = -10,$$
$$P = xy$$

Carry out. Solve the first equation for y.

$$y = -10 - x$$

Substitute for y in the second equation.

$$P = x(-10 - x)$$
$$P = -x^2 - 10x$$

Completing the square, we get

$$P = -(x + 5)^2 + 25.$$

The maximum function value is 25. It occurs when $x = -5$. When $x = -5$, $y = -10 - (-5)$, or -5.

Check. Check a function value for x less than -5 and for x greater than -5.

$$P(-6) = -(-6)^2 - 10(-6) = 24$$
$$P(-4) = -(-4)^2 - 10(-4) = 24$$

Since 25 is greater than these numbers, it looks as though we have a maximum.

State. The maximum product of 25 occurs for the numbers -5 and -5.

16. 36; -6 and -6

17. The data points rise, in general. The graph does not appear to represent a quadratic function in which the data points would rise and then fall or vise versa.

18. Not quadratic

19. The data points fall and then rise. The graph appears to represent a quadratic function.

20. Quadratic

21. The data points fall, then rise, and then fall again. The graph does not appear to represent a quadratic function.

22. Not quadratic

23. The data points rise, then fall, then rise again, and finally fall again. The graph does not appear to represent a quadratic function.

24. Quadratic

25. We look for a function of the form $f(x) = ax^2+bx+c$. Substituting the data points, we get

$$4 = a(1)^2 + b(1) + c,$$
$$-2 = a(-1)^2 + b(-1) + c,$$
$$13 = a(2)^2 + b(2) + c,$$

or

$$4 = a + b + c,$$
$$-2 = a - b + c,$$
$$13 = 4a + 2b + c.$$

Solving this system, we get

$$a = 2, b = 3, \text{ and } c = -1.$$

Therefore the function we are looking for is

$$f(x) = 2x^2 + 3x - 1.$$

26. $f(x) = 3x^2 - x + 2$

27. We look for a function of the form $f(x) = ax^2+bx+c$. Substituting the data points, we get

$$0 = a(2)^2 + b(2) + c,$$
$$3 = a(4)^2 + b(4) + c,$$
$$-5 = a(12)^2 + b(12) + c,$$

or

$$0 = 4a + 2b + c,$$
$$3 = 16a + 4b + c,$$
$$-5 = 144a + 12b + c.$$

Solving this system, we get

$$a = -\frac{1}{4}, b = 3, c = -5.$$

Therefore the function we are looking for is

$$f(x) = -\frac{1}{4}x^2 + 3x - 5.$$

28. $f(x) = -\dfrac{1}{3}x^2 + 5x - 12$

29. a) **Familiarize**. We look for a function of the form $A(s) = as^2+bs+c$, where $A(s)$ represents the number of nighttime accidents (for every 200 million km) and s represents the travel speed (in km/h).

Translate. We substitute the given values of s and $A(s)$.

$$400 = a(60)^2 + b(60) + c,$$
$$250 = a(80)^2 + b(80) + c,$$
$$250 = a(100)^2 + b(100) + c,$$

or

$$400 = 3600a + 60b + c,$$
$$250 = 6400a + 80b + c,$$
$$250 = 10,000a + 100b + c.$$

Carry out. Solving the system of equations, we get

$$a = \frac{3}{16}, b = -\frac{135}{4}, c = 1750.$$

Check. Recheck the calculations.

State. The function

$$A(s) = \frac{3}{16}s^2 - \frac{135}{4}s + 1750 \text{ fits the data.}$$

b) Find $A(50)$.

$$A(50) = \frac{3}{16}(50)^2 - \frac{135}{4}(50) + 1750 = 531.25$$

About 531 accidents occur at 50 km/h.

30. a) $A(s) = 0.05x^2 - 5.5x + 250$; b) 100

31. **Familiarize**. Think of a coordinate system placed on the drawing in the text with the origin at the point where the arrow is released. Then three points on the arrow's parabolic path are $(0, 0)$, $(63, 27)$, and $(126, 0)$. We look for a function of the form $h(d) = ad^2+bd+c$, where $h(d)$ represents the arrow's height and d represents the distance the arrow has traveled horizontally.

Translate. We substitute the values given above for d and $h(d)$.

$$0 = a \cdot 0^2 + b \cdot 0 + c,$$
$$27 = a \cdot 63^2 + b \cdot 63 + c,$$
$$0 = a \cdot 126^2 + b \cdot 126 + c$$

or

$$0 = c,$$
$$27 = 3969a + 63b + c,$$
$$0 = 15,876a + 126b + c$$

Carry out. Solving the system of equations, we get $a \approx -0.0068$, $b \approx 0.8571$, and $c = 0$.

Check. Recheck the calculations.

State. The function $h(d) = -0.0068d^2 + 0.8571d$ expresses the arrow's height as a function of the distance it has traveled horizontally.

32. a) $P(d) = \dfrac{1}{64}d^2 + \dfrac{5}{16}d + \dfrac{5}{2}$; b) \$9.94

33. Writing exercise

34. Writing exercise

35.
$$\frac{x}{x^2 + 17x + 72} - \frac{8}{x^2 + 15x + 56}$$
$$= \frac{x}{(x+8)(x+9)} - \frac{8}{(x+8)(x+7)}$$
$$= \frac{x}{(x+8)(x+9)} \cdot \frac{x+7}{x+7} - \frac{8}{(x+8)(x+7)} \cdot \frac{x+9}{x+9}$$
$$= \frac{x(x+7) - 8(x+9)}{(x+8)(x+9)(x+7)}$$
$$= \frac{x^2 + 7x - 8x - 72}{(x+8)(x+9)(x+7)}$$
$$= \frac{x^2 - x - 72}{(x+8)(x+9)(x+7)} = \frac{(x-9)(x+8)}{(x+8)(x+9)(x+7)}$$
$$= \frac{x-9}{(x+9)(x+7)}$$

36. $\dfrac{(x-3)(x+1)}{(x-7)(x+3)}$

37.
$$\frac{t^2 - 4}{t^2 - 7t - 8} \cdot \frac{t^2 - 64}{t^2 - 5t + 6}$$
$$= \frac{(t^2 - 4)(t^2 - 64)}{(t^2 - 7t - 8)(t^2 - 5t + 6)}$$
$$= \frac{(t+2)(t-2)(t+8)(t-8)}{(t+1)(t-8)(t-2)(t-3)}$$
$$= \frac{(t+2)(t-2)(t+8)(t-8)}{(t+1)(t-8)(t-2)(t-3)}$$
$$= \frac{(t+2)(t+8)}{(t+1)(t-3)}$$

38. $\dfrac{2t(t+2)}{(t-3)(t-7)(t+7)}$

39. $5x - 9 < 31$

$$5x < 40$$
$$x < 8$$

The solutions set is $\{x | x < 8\}$, or $(-\infty, 8)$.

40. $\{x | x \geq 10\}$, or $[10, \infty)$

41. Writing exercise

42. Writing exercise

43. ***Familiarize***. We add labels to the drawing in the text.

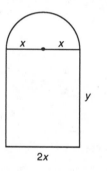

The perimeter of the semicircular portion of the window is $\dfrac{1}{2} \cdot 2\pi x$, or πx. The perimeter of the rectangular portion is $y + 2x + y$, or $2x + 2y$. The area of the semicircular portion of the window is $\dfrac{1}{2} \cdot \pi x^2$, or $\dfrac{\pi}{2}x^2$. The area of the rectangular portion is $2xy$.

Translate. We have two equations, one giving the perimeter of the window and the other giving the area.

$$\pi x + 2x + 2y = 24,$$
$$A = \frac{\pi}{2}x^2 + 2xy$$

Carry out. Solve the first equation for y.

$$\pi x + 2x + 2y = 24$$
$$2y = 24 - \pi x - 2x$$
$$y = 12 - \frac{\pi x}{2} - x$$

Substitute for y in the second equation.

$$A = \frac{\pi}{2}x^2 + 2x\left(12 - \frac{\pi x}{2} - x\right)$$
$$A = \frac{\pi}{2}x^2 + 24x - \pi x^2 - 2x^2$$
$$A = -2x^2 - \frac{\pi}{2}x^2 + 24x$$
$$A = -\left(2x + \frac{\pi}{2}\right)x^2 + 24x$$

Completing the square, we get

$$A = -\left(2 + \frac{\pi}{2}\right)\left(x^2 + \frac{24}{-\left(2 + \frac{\pi}{2}\right)}x\right)$$
$$A = -\left(2 + \frac{\pi}{2}\right)\left(x^2 - \frac{48}{4 + \pi}x\right)$$
$$A = -\left(2 + \frac{\pi}{2}\right)\left(x - \frac{24}{4 + \pi}\right)^2 + \left(\frac{24}{4 + \pi}\right)^2$$

The maximum function value occurs when
$$x = \frac{24}{4 + \pi}. \text{ When } x = \frac{24}{4 + \pi},$$
$$y = 12 - \frac{\pi}{2}\left(\frac{24}{4 + \pi}\right) - \frac{24}{4 + \pi} =$$
$$\frac{48 + 12\pi}{4 + \pi} - \frac{12\pi}{4 + \pi} - \frac{24}{4 + \pi} = \frac{24}{4 + \pi}.$$

Check. Recheck the calculations.

State. The radius of the circular portion of the window and the height of the rectangular portion should each be $\frac{24}{4 + \pi}$ ft.

44. Length of piece used to form circle: $\frac{36\pi}{4 + \pi}$ in.; length of piece used to form square: $\frac{144}{4 + \pi}$ in.

45. **Familiarize**. Let x represent the number of trees added to an acre. Then $20 + x$ represents the total number of trees per acre and $40 - x$ represents the corresponding yield per tree. Let T represent the total yield per acre.

Translate. Since total yield is number of trees times yield per tree we have the following function for total yield per acre.

$$T(x) = (20 + x)(40 - x)$$
$$T(x) = -x^2 + 20x + 800$$

Carry out. Completing the square, we get
$$T(x) = -(x - 10)^2 + 900.$$

The maximum function value of 900 occurs when $x = 10$. When $x = 10$, the number of trees per acre is $20 + 10$, or 30.

Check. We check a function value for x less than 10 and for x greater than 10.

$$T(9) = (20 + 9)(40 - 9) = 899$$
$$T(11) = (20 + 11)(40 - 11) = 899$$

Since 900 is greater than these numbers, it looks as though we have a maximum.

State. The grower should plant 30 trees per acre.

46. $15

47. **Familiarize**. We want to find the maximum value of a function of the form $h(t) = at^2 + bt + c$ that fits the following data.

Time (sec)	Height (ft)
0	0
3	0
3 + 2, or 5	-64

Translate. Substitute the given values for t and $h(t)$.

$$0 = a(0)^2 + b(0) + c,$$
$$0 = a(3)^2 + b(3) + c,$$
$$-64 = a(5)^2 + b(5) + c,$$

or

$$0 = c,$$
$$0 = 9a + 3b + c,$$
$$-64 = 25a + 5b + c.$$

Carry out. Solving the system of equations, we get $a = -6.4$, $b = 19.2$, $c = 0$. The function $h(t) = -6.4t^2 + 19.2t$ fits the data.

Completing the square, we get
$$h(t) = -6.4(t - 1.5)^2 + 14.4.$$

The maximum function value of 14.4 occurs at $t = 1.5$.

Check. Recheck the calculations. Also check a function value for t less than 1.5 and for t greater than 1.5.

$$h(1) = -6.4(1)^2 + 19.2(1) = 12.8$$
$$h(2) = -6.4(2)^2 + 19.2(2) = 12.8$$

Since 14.4 is greater than these numbers, it looks as though we have a maximum.

State. The maximum height above the cliff is 14.4 ft. The maximum height above sea level is $64 + 14.4$, or 78.4 ft.

48. 158 ft

49.

Exercise Set 11.9

1. $(x+4)(x-3) < 0$

The solutions of $(x+4)(x-3) = 0$ are -4 and 3. They are not solutions of the inequality, but they divide the real-number line in a natural way. The product $(x+4)(x-3)$ is positive or negative, for values other than -4 and 3, depending on the signs of the factors $x+4$ and $x-3$.

$x+4 > 0$ when $x > -4$ and $x+4 < 0$ when $x < -4$.

$x-3 > 0$ when $x > 3$ and $x-3 < 0$ when $x < 3$.

We make a diagram.

Sign of $x+4$ $-$ | $+$ | $+$
Sign of $x-3$ $-$ | $-$ | $+$
Sign of product $+$ | $-$ | $+$

$$-4 \qquad 3$$

For the product $(x+4)(x-3)$ to be negative, one factor must be positive and the other negative. We see from the diagram that numbers satisfying $-4 < x < 3$ are solutions. The solution set of the inequality is $(-4, 3)$ or $\{x| -4 < x < 3\}$.

2. $(-\infty, -2) \cup (5, \infty)$, or $\{x | x < -2 \text{ or } x > 5\}$

3. $(x+7)(x-2) \geq 0$

The solutions of $(x+7)(x-2) = 0$ are -7 and 2. They divide the number line into three intervals as shown:

$$A \qquad B \qquad C$$
$$-7 \qquad 2$$

We try test numbers in each interval.

A: Test -8, $f(-8) = (-8+7)(-8-2) = 10$

B: Test 0, $f(0) = (0+7)(0-2) = -14$

C: Test 3, $f(3) = (3+7)(3-2) = 10$

Since $f(-8)$ and $f(3)$ are positive, the function value will be positive for all numbers in the intervals containing -8 and 3. The inequality symbol is $\leq$, so we need to include the intercepts.

The solution set is $(-\infty, -7] \cup [2, \infty)$, or $\{x | x \leq -7 \text{ or } x \geq 2\}$.

4. $[-4, 1]$, or $\{x| -4 \leq x \leq 1\}$

5. $x^2 - x - 2 < 0$

$(x+1)(x-2) < 0$ Factoring

The solutions of $(x+1)(x-2) = 0$ are -1 and 2. They divide the number line into three intervals as shown:

$$A \qquad B \qquad C$$
$$-1 \qquad 2$$

We try test numbers in each interval.

A: Test -2, $f(-2) = (-2+1)(-2-2) = 4$

B: Test 0, $f(0) = (0+1)(0-2) = -2$

C: Test 3, $f(3) = (3+1)(3-2) = 4$

Since $f(0)$ is negative, the function value will be negative for all numbers in the interval containing 0. The solution set is $(-1, 2)$, or $\{x| -1 < x < 2\}$.

6. $(-2, 1)$, or $\{x| -2 < x < 1\}$

7. $\qquad 25 - x^2 \geq 0$

$(5-x)(5+x) \geq 0$

The solutions of $(5-x)(5+x) = 0$ are 5 and -5. They divide the real-number line in a natural way. The product $(5-x)(5+x)$ is positive or negative, for values other than 5 and -5, depending on the signs of the factors $5-x$ and $5+x$.

$5-x > 0$ when $x < 5$ and $5-x < 0$ when $x > 5$.

$5+x > 0$ when $x > -5$ and $5+x < 0$ when $x < -5$.

We make a diagram.

Sign of $5-x$ $+$ | $+$ | $-$
Sign of $5+x$ $-$ | $+$ | $+$
Sign of product $-$ | $+$ | $-$

$$-5 \qquad 5$$

For the product $(5-x)(5+x)$ to be positive, both factors must be positive or both factors must be negative. We see from the diagram that numbers satisfying $-5 < x < 5$ are solutions. The intercepts are also solutions, because the inequality symbol is $\geq$. The solution set of the inequality is $[-5, 5]$, or $\{x| -5 \leq x \leq 5\}$.

8. $[-2, 2]$, or $\{x| -2 \leq x \leq 2\}$

9. $x^2 + 4x + 4 < 0$

$(x+2)^2 < 0$

Observe that $(x+2)^2 \geq 0$ for all values of x. Thus, the solution set is $\emptyset$.

10. $\emptyset$

11.
$$x^2 - 4x < 12$$
$$x^2 - 4x - 12 < 0$$
$$(x - 6)(x + 2) < 0$$

The solutions of $(x - 6)(x + 2) = 0$ are 6 and -2. They are not solutions of the inequality, but they divide the real-number line in a natural way. The product $(x - 6)(x + 2)$ is positive or negative, for values other than 6 and -2, depending on the signs of the factors $x - 6$ and $x + 2$.

$x - 6 > 0$ when $x > 6$ and $x - 6 < 0$ when $x < 6$.

$x + 2 > 0$ when $x > -2$ and $x + 2 < 0$ when $x < -2$.

We make a diagram.

Sign of $x - 6$	$-$	$\mid$	$-$	$\mid$	$+$
Sign of $x + 2$	$-$	$\mid$	$+$	$\mid$	$+$
Sign of product	$+$	$\mid$	$-$	$\mid$	$+$

$$-2 \qquad 6$$

For the product $(x - 6)(x + 2)$ to be negative, one factor must be positive and the other negative. The only situation in the diagram for which this happens is when $-2 < x < 6$. The solution set of the inequality is $(-2, 6)$, or $\{x| -2 < x < 6\}$.

12. $(-\infty, -4) \cup (-2, \infty)$, or $\{x|x < -4 \text{ or } x > -2\}$

13. $3x(x + 2)(x - 2) < 0$

The solutions of $3x(x + 2)(x - 2) = 0$ are 0, -2, and 2. They divide the real-number line into four intervals as shown:

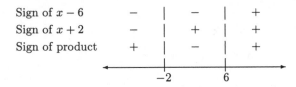

$$-2 \qquad 0 \qquad 2$$

We try test numbers in each interval.

A: Test -3, $f(-3) = 3(-3)(-3 + 2)(-3 - 2) = -45$

B: Test -1, $f(-1) = 3(-1)(-1 + 2)(-1 - 2) = 9$

C: Test 1, $f(1) = 3(1)(1 + 2)(1 - 2) = -9$

D: Test 3, $f(3) = 3(3)(3 + 2)(3 - 2) = 45$

Since $f(-3)$ and $f(1)$ are negative, the function value will be negative for all numbers in the intervals containing -3 and 1. The solution set is $(-\infty, -2) \cup (0, 2)$, or $\{x|x < -2 \text{ or } 0 < x < 2\}$.

14. $(-1, 0) \cup (1, \infty)$, or $\{x| -1 < x < 0 \text{ or } x > 1\}$

15. $(x + 3)(x - 2)(x + 1) > 0$

The solutions of $(x + 3)(x - 2)(x + 1) = 0$ are -3, 2, and -1. They are not solutions of the inequality, but they divide the real-number line in a natural way. The product $(x + 3)(x - 2)(x + 1)$ is positive or negative, for values other than -3, 2, and -1, depending on the signs of the factors $x + 3$, $x - 2$, and $x + 1$.

$x + 3 > 0$ when $x > -3$ and $x + 3 < 0$ when $x < -3$.

$x - 2 > 0$ when $x > 2$ and $x - 2 < 0$ when $x < 2$.

$x + 1 > 0$ when $x > -1$ and $x + 1 < 0$ when $x < -1$.

We make a diagram.

Sign of $x + 3$	$-$	$\mid$	$+$	$\mid$	$+$	$\mid$	$+$
Sign of $x - 2$	$-$	$\mid$	$-$	$\mid$	$-$	$\mid$	$+$
Sign of $x + 1$	$-$	$\mid$	$-$	$\mid$	$+$	$\mid$	$+$
Sign of product	$-$	$\mid$	$+$	$\mid$	$-$	$\mid$	$+$

$$-3 \qquad -1 \qquad 2$$

The product of three numbers is positive when all three are positive or two are negative and one is positive. We see from the diagram that numbers satisfying $-3 < x < -1$ or $x > 2$ are solutions. The solution set of the inequality is $(-3, -1) \cup (2, \infty)$, or $\{x| -3 < x < -1 \text{ or } x > 2\}$.

16. $(-\infty, -2) \cup (1, 4)$, or $\{x|x < -2 \text{ or } 1 < x < 4\}$

17. $(x + 3)(x + 2)(x - 1) < 0$

The solutions of $(x+3)(x+2)(x-1) = 0$ are -3, -2, and 1. They divide the real-number line into four intervals as shown:

$$A \qquad B \qquad C \qquad D$$
$$-3 \quad -2 \qquad \qquad 1$$

We try test numbers in each interval.

A: Test -4, $f(-4) = (-4+3)(-4+2)(-4-1) = -10$

B: Test $-\frac{5}{2}$, $f\left(-\frac{5}{2}\right) = \left(-\frac{5}{2}+3\right)\left(-\frac{5}{2}+2\right)\left(-\frac{5}{2}-1\right) = \frac{7}{8}$

C: Test 0, $f(0) = (0+3)(0+2)(0-1) = -6$

D: Test 2, $f(2) = (2+3)(2+2)(2-1) = 20$

The function value will be negative for all numbers in intervals A and C. The solution set is $(-\infty, -3) \cup (-2, 1)$, or $\{x|x < -3 \text{ or } -2 < x < 1\}$.

18. $(-\infty, -1) \cup (2, 3)$, or $\{x|x < -1 \text{ or } 2 < x < 3\}$

19. $\dfrac{1}{x + 3} < 0$

We write the related equation by changing the $<$ symbol to $=$:

$$\frac{1}{x + 3} = 0$$

We solve the related equation.

$$(x+3) \cdot \frac{1}{x+3} = (x+3) \cdot 0$$
$$1 = 0$$

The related equation has no solution.

Next we find the values that make the denominator 0 by setting the denominate equal to 0 and solving:

$$x + 3 = 0$$
$$x = -3$$

We use -3 to divide the number line into two intervals as shown:

We try test numbers in each interval.

A: Test -4, $\dfrac{1}{-4+3} = \dfrac{1}{-1} = -1 < 0$

The number -4 is a solution of the inequality, so the interval A is part of the solution set.

B: Test 0, $\dfrac{1}{0+3} = \dfrac{1}{3} \not< 0$

The number 0 is not a solution of the inequality, so the interval B is not part of the solution set. The solution set is $(-\infty, -3)$, or $\{x | x < -3\}$.

20. $(-4, \infty)$, or $\{x | x > -4\}$

21. $\dfrac{x+1}{x-5} \geq 0$

Solve the related equation.

$$\frac{x+1}{x-5} = 0$$
$$x + 1 = 0$$
$$x = -1$$

Find the values that make the denominator 0.

$$x - 5 = 0$$
$$x = 5$$

Use the numbers -1 and 5 to divide the number line into intervals as shown:

Try test numbers in each interval.

A: Test -2, $\dfrac{-2+1}{-2-5} = \dfrac{-1}{-7} = \dfrac{1}{7} > 0$

The number -2 is a solution of the inequality, so interval A is part of the solution set.

B: Test 0, $\dfrac{0+1}{0-5} = \dfrac{1}{-5} = -\dfrac{1}{5} \not> 0$

The number 0 is not a solution of the inequality, so the interval B is not part of the solution set.

C: Test 6, $\dfrac{6+1}{6-5} = \dfrac{7}{1} = 7 > 0$

The number 6 is a solution of the inequality, so the interval C is part of the solution set.

The solution set includes intervals A and C. The number -1 is also included since the inequality symbol is $\geq$ and -1 is the solution of the related equation. The number 5 is not included since $\dfrac{x+1}{x-5}$ is undefined for $x = 5$. The solution set is $(-\infty, -1] \cup (5, \infty)$, or $\{x | x \leq -1 \ or \ x > 5\}$.

22. $(-5, 2]$, or $\{x | -5 < x \leq 2\}$

23. $\dfrac{3x+2}{2x-4} \leq 0$

Solve the related equation.

$$\frac{3x+2}{2x-4} = 0$$
$$3x + 2 = 0$$
$$3x = -2$$
$$x = -\frac{2}{3}$$

Find the values that make the denominator 0.

$$2x - 4 = 0$$
$$2x = 4$$
$$x = 2$$

Use the numbers $-\dfrac{2}{3}$ and 2 to divide the number line into intervals as shown:

Try test numbers in each interval.

A: Test -1, $\dfrac{3(-1)+2}{2(-1)-4} = \dfrac{-1}{-6} = \dfrac{1}{6} \not\leq 0$

The number -1 is not a solution of the inequality, so the interval A is not part of the solution set.

B: Test 0, $\dfrac{3 \cdot 0 + 2}{2 \cdot 0 - 4} = \dfrac{2}{-4} = -\dfrac{1}{2} \leq 0$

The number 0 is a solution of the inequality, so the interval B is part of the solution set.

C: Test 4, $\dfrac{3 \cdot 4 + 2}{2 \cdot 4 - 4} = \dfrac{14}{4} = \dfrac{7}{2} \not\leq 0$

The number 4 is not a solution of the inequality, so the interval C is not part of the solution set. The solution set includes the interval B. The number $-\dfrac{2}{3}$ is also included since the inequality symbol is $\leq$

and $-\dfrac{2}{3}$ is the solution of the related equation. The number 2 is not included since $\dfrac{3x+2}{2x-4}$ is undefined for $x=2$. The solution set is $\left[-\dfrac{2}{3},2\right)$, or $\left\{x\middle|-\dfrac{2}{3}\le x<2\right\}$.

24. $\left(-\infty,-\dfrac{3}{4}\right)\cup\left[\dfrac{5}{2},\infty\right)$, or $\left\{x\middle|x<-\dfrac{3}{4}\ or\ x\ge\dfrac{5}{2}\right\}$

25. $\dfrac{x+1}{x+6}>1$

Solve the related equation.
$$\dfrac{x+1}{x+6}=1$$
$$x+1=x+6$$
$$1=6$$

The related equation has no solution.

Find the values that make the denominator 0.
$$x+6=0$$
$$x=-6$$

Use the number -6 to divide the number line into two intervals.

Try test numbers in each interval.

A: Test -7, $\dfrac{-7+1}{-7+6}=\dfrac{-6}{-1}=6>1$.

The number -7 is a solution of the inequality, so the interval A is part of the solution set.

B: Test 0, $\dfrac{0+1}{0+6}=\dfrac{1}{6}\not>1$

The number 0 is not a solution of the inequality, so the interval B is not part of the solution set.

The solution set is $(-\infty,-6)$, or $\{x|x<-6\}$.

26. $(-\infty,2)$, or $\{x|x<2\}$

27. $\dfrac{(x-2)(x+1)}{x-5}\le0$

Solve the related equation.
$$\dfrac{(x-2)(x+1)}{x-5}=0$$
$$(x-2)(x+1)=0$$
$$x=2\ or\ x=-1$$

Find the values that make the denominator 0.
$$x-5=0$$
$$x=5$$

Use the numbers 2, -1, and 5 to divide the number line into intervals as shown:

Try test numbers in each interval.

A: Test -2, $\dfrac{(-2-2)(-2+1)}{-2-5}=\dfrac{-4(-1)}{-7}=-\dfrac{4}{7}\le0$

Interval A is part of the solution set.

B: Test 0, $\dfrac{(0-2)(0+1)}{0-5}=\dfrac{-2\cdot1}{-5}=\dfrac{2}{5}\not\le0$

Interval B is not part of the solution set.

C: Test 3, $\dfrac{(3-2)(3+1)}{3-5}=\dfrac{1\cdot4}{-2}=-2\le0$

Interval C is part of the solution set.

D: Test 6, $\dfrac{(6-2)(6+1)}{6-5}=\dfrac{4\cdot7}{1}=28\not\le0$

Interval D is not part of the solution set.

The solution set includes intervals A and C. The numbers -1 and 2 are also included since the inequality symbol is $\le$ and -1 and 2 are the solutions of the related equation. The number 5 is not included since $\dfrac{(x-2)(x+1)}{x-5}$ is undefined for $x=5$.
The solution set is $(-\infty,-1]\cup[2,5)$, or $\{x|x\le-1\ or\ 2\le x<5\}$.

28. $[-4,-3)\cup[1,\infty)$, or $\{x|-4\le x<-3\ or\ x\ge1\}$

29. $\dfrac{x}{x+3}\ge0$

Solve the related equation.
$$\dfrac{x}{x+3}=0$$
$$x=0$$

Find the values that make the denominator 0.
$$x+3=0$$
$$x=-3$$

Use the numbers 0 and -3 to divide the number line into intervals as shown.

Try test numbers in each interval.

A: Test -4, $\dfrac{-4}{-4+3}=\dfrac{-4}{-1}=4\ge0$

Interval A is part of the solution set.

B: Test -1, $\dfrac{-1}{-1+3} = \dfrac{-1}{2} = -\dfrac{1}{2} \not\geq 0$

Interval *B* is not part of the solution set.

C: Test 1, $\dfrac{1}{1+3} = \dfrac{1}{4} \geq 0$

The interval *C* is part of the solution set.

The solution set includes intervals *A* and *C*. The number 0 is also included since the inequality symbol is $\geq$ and 0 is the solution of the related equation. The number -3 is not included since $\dfrac{x}{x+3}$ is undefined for $x = -3$. The solution set is $(-\infty, -3) \cup [0, \infty)$, or $\{x | x < -3 \text{ or } x \geq 0\}$.

30. $(0, 2]$, or $\{x | 0 < x \leq 2\}$

31. $\dfrac{x-5}{x} < 1$

Solve the related equation.

$$\frac{x-5}{x} = 1$$
$$x - 5 = x$$
$$-5 = 0$$

The related equation has no solution.

Find the values that make the denominator 0.

$$x = 0$$

Use the number 0 to divide the number line into two intervals as shown.

Try test numbers in each interval.

A: Test -1, $\dfrac{-1-5}{-1} = \dfrac{-6}{-1} = 6 \not< 1$

Interval *A* is not part of the solution set.

B: Test 1, $\dfrac{1-5}{1} = \dfrac{-4}{1} = -4 < 1$

Interval *B* is part of the solution set.

The solution set is $(0, \infty)$ or $\{x | x > 0\}$.

32. $(1, 2)$, or $\{x | 1 < x < 2\}$

33. $\dfrac{x-1}{(x-3)(x+4)} \leq 0$

Solve the related equation.

$$\frac{x-1}{(x-3)(x+4)} = 0$$
$$x - 1 = 0$$
$$x = 1$$

Find the values that make the denominator 0.

$$(x-3)(x+4) = 0$$

$x = 3 \text{ or } x = -4$

Use the numbers 1, 3, and -4 to divide the number line into intervals as shown:

Try test numbers in each interval.

A: Test -5, $\dfrac{-5-1}{(-5-3)(-5+4)} = \dfrac{-6}{-8(-1)} =$ $-\dfrac{3}{4} < 0$

Interval *A* is part of the solution set.

B: Test 0, $\dfrac{0-1}{(0-3)(0+4)} = \dfrac{-1}{-3 \cdot 4} = \dfrac{1}{12} \not< 0$

Interval *B* is not part of the solution set.

C: Test 2, $\dfrac{2-1}{(2-3)(2+4)} = \dfrac{1}{-1 \cdot 6} = -\dfrac{1}{6} < 0$

Interval *C* is part of the solution set.

D: Test 4, $\dfrac{4-1}{(4-3)(4+4)} = \dfrac{3}{1 \cdot 8} = \dfrac{3}{8} \not< 0$

Interval *D* is not part of the solution set.

The solution set includes intervals *A* and *C*. The number 1 is also included since the inequality symbol is $\leq$ and 1 is the solution of the related equation. The numbers -4 and 3 are not included since $\dfrac{x-1}{(x-3)(x+4)}$ is undefined for $x = -4$ and for $x = 3$. The solution set is $(-\infty, -4) \cup [1, 3)$, or $\{x | x < -4 \text{ or } 1 \leq x < 3\}$.

34. $(-7, -2] \cup (2, \infty)$, or $\{x | -7 < x \leq -2 \text{ or } x > 2\}$

35. $4 < \dfrac{1}{x}$

Solve the related equation.

$$4 = \frac{1}{x}$$
$$x = \frac{1}{4}$$

Find the values that make the denominator 0.

$$x = 0$$

Use the numbers $\dfrac{1}{4}$ and 0 to divide the number line into intervals as shown.

Try test numbers in each interval.

A: Test -1, $\dfrac{1}{-1} = -1 \not> 4$

Interval *A* is not part of the solution set.

B: Test $\dfrac{1}{8}$, $\dfrac{1}{\frac{1}{8}} = 8 > 4$

Interval B is part of the solution set.

C: Test 1, $\dfrac{1}{1} = 1 \not> 4$

Interval C is not part of the solution set.

The solution set is $\left(0, \dfrac{1}{4}\right)$, or $\left\{x \middle| 0 < x < \dfrac{1}{4}\right\}$.

36. $(-\infty, 0) \cup \left[\dfrac{1}{5}, \infty\right)$, or $\left\{x \middle| x < 0 \text{ or } x \geq \dfrac{1}{5}\right\}$

37. Writing exercise

38. Writing exercise

39. $(2a^3b^2c^4)^3 = 2^3(a^3)^3(b^2)^3(c^4)^3 = 8a^{3\cdot3}b^{2\cdot3}c^{4\cdot3} = 8a^9b^6c^{12}$

40. $25a^8b^{14}$

41. $2^{-5} = \dfrac{1}{2^5} = \dfrac{1}{32}$

42. $\dfrac{1}{81}$

43. $f(x) = 3x^2$

$f(a+1) = 3(a+1)^2 = 3(a^2 + 2a + 1) = 3a^2 + 6a + 3$

44. $5a + 7$

45. Writing exercise

46. Writing exercise

47. $x^2 + 2x > 5$

$x^2 + 2x - 5 > 0$

Using the quadratic formula, we find that the solutions of the related equation are $x = -1 \pm \sqrt{6}$. These numbers divide the real-number line into three intervals as shown:

We try test numbers in each interval.

A: Test -4, $f(-4) = (-4)^2 + 2(-4) - 5 = 3$

B: Test 0, $f(0) = 0^2 + 2 \cdot 0 - 5 = -5$

C: Test 2, $f(2) = 2^2 + 2 \cdot 2 - 5 = 3$

The function value will be positive for all numbers in intervals A and C. The solution set is $(-\infty, -1 - \sqrt{6}) \cup (-1 + \sqrt{6}, \infty)$, or $\{x | x < -1 - \sqrt{6} \text{ or } x > -1 + \sqrt{6}\}$.

48. $(-\infty, \infty)$, or the set of all real numbers

49. $x^4 + 3x^2 \leq 0$

$x^2(x^2 + 3) \leq 0$

$x^2 = 0$ for $x = 0$, $x^2 > 0$ for $x \neq 0$, $x^2 + 3 > 0$ for all x

The solution set is $\{0\}$.

50. $\left\{x \middle| x \leq \dfrac{1}{4} \text{ or } x \geq \dfrac{5}{2}\right\}$, or $\left(-\infty, \dfrac{1}{4}\right] \cup \left[\dfrac{5}{2}, \infty\right)$.

51. a) $-3x^2 + 630x - 6000 > 0$

$x^2 - 210x + 2000 < 0$ Multiplying by $-\dfrac{1}{3}$

$(x - 200)(x - 10) < 0$

The solutions of $f(x) = (x - 200)(x - 10) = 0$ are 200 and 10. They divide the number line as shown:

A: Test 0, $f(0) = 0^2 - 210 \cdot 0 + 2000 = 2000$

B: Test 20, $f(20) = 20^2 - 210 \cdot 20 + 2000 = -1800$

C: Test 300, $f(300) = 300^2 - 210 \cdot 300 + 2000 = 29{,}000$

The company makes a profit for values of x such that $10 < x < 200$, or for values of x in the interval $(10, 200)$.

b) See part (a). Keep in mind that x must be nonnegative since negative numbers have no meaning in this application.

The company loses money for values of x such that $0 \leq x < 10$ or $x > 200$, or for values of x in the interval $[0, 10) \cup (200, \infty)$.

52. a) $\{t | 0 \text{ sec} < t < 2 \text{ sec}\}$; b) $\{t | t > 10 \text{ sec}\}$

53. We find values of n such that $N \geq 66$ *and* $N \leq 300$.

For $N \geq 66$:

$\dfrac{n(n-1)}{2} \geq 66$

$n(n-1) \geq 132$

$n^2 - n - 132 \geq 0$

$(n - 12)(n + 11) \geq 0$

The solutions of $f(n) = (n - 12)(n + 11) = 0$ are 12 and -11. They divide the number line as shown:

However, only positive values of n have meaning in this exercise so we need only consider the intervals shown below:

A: Test 1, $f(1) = 1^2 - 1 - 132 = -132$

B: Test 20, $f(20) = 20^2 - 20 - 132 = 248$

Thus, $N \geq 66$ for $\{n | n \geq 12\}$.

For $N \leq 300$:
$$\frac{n(n-1)}{2} \leq 300$$
$$n(n-1) \leq 600$$
$$n^2 - n - 600 \leq 0$$
$$(n - 25)(n + 24) \leq 0$$

The solutions of $f(n) = (n - 25)(n + 24) = 0$ are 25 and -24. They divide the number line as shown:

However, only positive values of n have meaning in this exercise so we need only consider the intervals shown below:

A: Test 1, $f(1) = 1^2 - 1 - 600 = -600$

B: Test 30, $f(30) = 30^2 - 30 - 600 = 270$

Thus, $N \leq 300$ (and $n > 0$) for $\{n | 0 < n \leq 25\}$.

Then $66 \leq N \leq 300$ for $\{n | n$ is an integer and $12 \leq n \leq 25\}$.

54. $\{n | n$ is an integer and $9 \leq n \leq 23\}$

55. From the graph we determine the following:

The solutions of $f(x) = 0$ are -2, 1, and 3.

The solution of $f(x) < 0$ is $(-\infty, -2) \cup (1, 3)$, or $\{x | x < -2 \ or \ 1 < x < 3\}$.

The solution of $f(x) > 0$ is $(-2, 1) \cup (3, \infty)$, or $\{x | -2 < x < 1 \ or \ x > 3\}$.

56. $f(x) = 0$ for $x = -2 \ or \ x = 1$;

$f(x) < 0$ for $(-\infty, -2)$, or $\{x | x < -2\}$;

$f(x) > 0$ for $(-2, 1) \cup (1, \infty)$, or $\{x | -2 < x < 1 \ or \ x > 1\}$

57. From the graph we determine the following:

$f(x)$ has no zeros.

The solutions $f(x) < 0$ are $(-\infty, 0)$, or $\{x | x < 0\}$.

The solutions of $f(x) > 0$ are $(0, \infty)$, or $\{x | x > 0\}$.

58. $f(x) = 0$ for $x = 0 \ or \ x = 1$;

$f(x) < 0$ for $(0, 1)$, or $\{x | 0 < x < 1\}$;

$f(x) > 0$ for $(1, \infty)$, or $\{x | x > 1\}$

59. From the graph we determine the following:

The solutions of $f(x) = 0$ are -1 and 0.

The solutions of $f(x) < 0$ are $(-\infty, -3) \cup (-1, 0)$, or $\{x | x < -3 \ or \ -1 < x < 0\}$.

The solutions of $f(x) > 0$ are $(-3, -1) \cup (0, 2) \cup (2, \infty)$, or $\{x | -3 < x < -1 \ or \ 0 < x < 2 \ or \ x > 2\}$.

60. $f(x) = 0$ for -2, 1, 2, and 3;

$f(x) < 0$ for $(-2, 1) \cup (2, 3)$, or $\{x | -2 < x < 1 \ or \ 2 < x < 3\}$;

$f(x) > 0$ for $(-\infty, -2) \cup (1, 2) \cup (3, \infty)$, or $\{x | x < -2 \ or \ 1 < x < 2 \ or \ x > 3\}$.

61.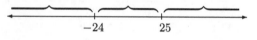

Chapter 12

Exponential and Logarithmic Functions

1. $(f \circ g)(1) = f(g(1)) = f(2 \cdot 1 + 1)$
$\qquad = f(3) = 3^2 + 3$
$\qquad = 9 + 3 = 12$
$(g \circ f)(1) = g(f(1)) = g(1^2 + 3)$
$\qquad = g(4) = 2 \cdot 4 + 1 = 9$
$(f \circ g)(x) = f(g(x)) = f(2x + 1)$
$\qquad = (2x + 1)^2 + 3$
$\qquad = 4x^2 + 4x + 1 + 3$
$\qquad = 4x^2 + 4x + 4$
$(g \circ f)(x) = g(f(x)) = g(x^2 + 3)$
$\qquad = 2(x^2 + 3) + 1$
$\qquad = 2x^2 + 6 + 1$
$\qquad = 2x^2 + 7$

2. $-7; \ 4; \ -9; \ 4x^2 + 4x - 4$

3. $(f \circ g)(x) = f(g(1)) = f(5 \cdot 1^2 + 2)$
$\qquad = f(7) = 3 \cdot 7 - 1$
$\qquad = 21 - 1 = 20$
$(g \circ f)(1) = g(f(1)) = g(3 \cdot 1 - 1)$
$\qquad = g(2) = 5 \cdot 2^2 + 2$
$\qquad = 5 \cdot 4 + 2 = 20 + 2 = 22$
$(f \circ g)(x) = f(g(x)) = f(5x^2 + 2)$
$\qquad = 3(5x^2 + 2) - 1$
$\qquad = 15x^2 + 6 - 1$
$\qquad = 15x^2 + 5$
$(g \circ f)(x) = g(f(x)) = g(3x - 1)$
$\qquad = 5(3x - 1)^2 + 2$
$\qquad = 5(9x^2 - 6x + 1) + 2$
$\qquad = 45x^2 - 30x + 5 + 2$
$\qquad = 45x^2 - 30x + 7$

4. $31; \ 27; \ 48x^2 - 24x + 7; \ 12x^2 + 15$

5. $(f \circ g)(1) = f(g(1)) = f\left(\dfrac{1}{1^2}\right)$
$\qquad = f(1) = 1 + 7 = 8$

$(g \circ f)(1) = g(f(1)) = g(1 + 7)$
$\qquad = g(8) = \dfrac{1}{8^2} = \dfrac{1}{64}$
$(f \circ g)(x) = f(g(x))$
$\qquad = f\left(\dfrac{1}{x^2}\right) = \dfrac{1}{x^2} + 7$
$(g \circ f)(x) = g(f(x))$
$\qquad = g(x + 7) = \dfrac{1}{(x + 7)^2}$

6. $\dfrac{1}{9}; \ 3; \ \dfrac{1}{(x+2)^2}; \ \dfrac{1}{x^2} + 2$

7. $h(x) = (7 + 5x)^2$

This is $7 + 5x$ raised to the second power, so the two most obvious functions are $f(x) = x^2$ and $g(x) = 7 + 5x$.

8. $f(x) = x^2, \ g(x) = 3x - 1$

9. $h(x) = \sqrt{2x + 7}$

We have $2x + 7$ and take the square root of their expression, so the two most obvious functions are $f(x) = \sqrt{x}$ and $g(x) = 2x + 7$.

10. $f(x) = \sqrt{x}, \ g(x) = 5x + 2$

11. $h(x) = \dfrac{2}{x - 3}$

This is 2 divided by $x - 3$, so two functions that can be used are $f(x) = \dfrac{2}{x}$ and $g(x) = x - 3$.

12. $f(x) = x + 4, \ g(x) = \dfrac{3}{x}$

13. $h(x) = \dfrac{1}{\sqrt{7x + 2}}$

This is the reciprocal of the square root of $7x + 2$. Two functions that can be used are $f(x) = \dfrac{1}{\sqrt{x}}$ and $g(x) = 7x + 2$.

14. $f(x) = \sqrt{x} - 3, \ g(x) = x - 7$

15. $h(x) = \dfrac{1}{\sqrt{3x}} + \sqrt{3x}$

This is the reciprocal of the square root of $3x$ plus the square root of $3x$. Two functions that can be used are $f(x) = \dfrac{1}{x} + x$ and $g(x) = \sqrt{3x}$.

16. $f(x) = \dfrac{1}{x} - x$, $g(x) = \sqrt{2x}$

17. The graph of $f(x) = x - 5$ is shown below.

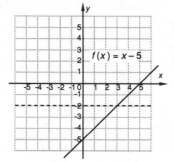

Since there is no horizontal line that crosses the graph more than once, the function is one-to-one.

18. Yes

19. $f(x) = x^2 + 1$

Observe that the graph of this function is a parabola that opens up. Thus, there are many horizontal lines that cross the graph more than once, so the function is not one-to-one. We can also draw the graph as shown below.

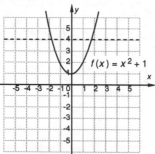

There are many horizontal lines that cross the graph more than once. In particular, the line $y = 4$ crosses the graph more than once. The function is not one-to-one.

20. No

21. The graph of $g(x) = x^3$ is shown below.

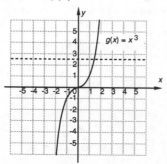

Since no horizontal line crosses the graph more than once, the function is one-to-one.

22. Yes

23. The graph of $g(x) = |x|$ is shown below.

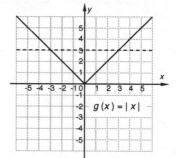

There are many horizontal lines that cross the graph more than once. In particular, the line $y = 3$ crosses the graph more than once. The function is not one-to-one.

24. No

25. a) The function $f(x) = x - 4$ is a linear function that is not constant, so it passes the horizontal-line test. Thus, f is one-to-one.

b) Replace $f(x)$ by y: $y = x - 4$

Interchange x and y: $x = y - 4$

Solve for y: $x + 4 = y$

Replace y by $f^{-1}(x)$: $f^{-1}(x) = x + 4$

26. a) Yes; b) $f^{-1}(x) = x + 2$

27. a) The function $f(x) = 3 + x$ is a linear function that is not constant, so it passes the horizontal-line test. Thus, f is one-to-one.

b) Replace $f(x)$ by y: $y = 3 + x$

Interchange x and y: $x = 3 + y$

Solve for y: $y = x - 3$

Replace y by $f^{-1}(x)$: $f^{-1}(x) = x - 3$

28. a) Yes; b) $f^{-1}(x) = x - 9$

29. a) The function $g(x) = x + 5$ is a linear function that is not constant, so it passes the horizontal-line test. Thus, g is one-to-one.

b) Replace $g(x)$ by y: $y = x + 5$

Interchange x and y: $x = y + 5$

Solve for y: $x - 5 = y$

Replace y by $g^{-1}(x)$: $g^{-1}(x) = x - 5$

30. a) Yes; b) $g^{-1}(x) = x - 8$

31. a) The function $f(x) = 4x$ is a linear function that is not constant, so it passes the horizontal-line test. Thus, f is one-to-one.

  b) Replace $f(x)$ by y: $y = 4x$

  Interchange x and y: $x = 4y$

  Solve for y: $\dfrac{x}{4} = y$

  Replace y by $f^{-1}(x)$: $f^{-1}(x) = \dfrac{x}{4}$

32. a) Yes; b) $f^{-1}(x) = \dfrac{x}{7}$

33. a) The function $g(x) = 4x - 1$ is a linear function that is not constant, so it passes the horizontal-line test. Thus, g is one-to-one.

  b) Replace $g(x)$ by y: $y = 4x - 1$

  Interchange variables: $x = 4y - 1$

  Solve for y: $x + 1 = 4y$

  $\dfrac{x+1}{4} = y$

  Replace y by $g^{-1}(x)$: $g^{-1}(x) = \dfrac{x+1}{4}$

34. a) Yes; b) $g^{-1}(x) = \dfrac{x+6}{4}$

35. a) The graph of $h(x) = 5$ is shown below. The horizontal line $y = 5$ crosses the graph more than once, so the function is not one-to-one.

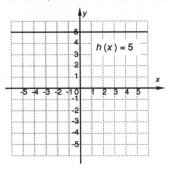

36. a) No

37. a) The graph of $f(x) = \dfrac{1}{x}$ is shown below. It passes the horizontal-line test, so the function is one-to-one.

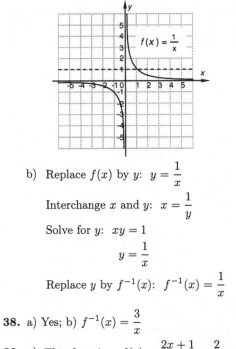

  b) Replace $f(x)$ by y: $y = \dfrac{1}{x}$

  Interchange x and y: $x = \dfrac{1}{y}$

  Solve for y: $xy = 1$

  $y = \dfrac{1}{x}$

  Replace y by $f^{-1}(x)$: $f^{-1}(x) = \dfrac{1}{x}$

38. a) Yes; b) $f^{-1}(x) = \dfrac{3}{x}$

39. a) The function $f(x) = \dfrac{2x+1}{3} = \dfrac{2}{3}x + \dfrac{1}{3}$ is a linear function that is not constant, so it passes the horizontal-line test. Thus, f is one-to-one.

  b) Replace $f(x)$ by y: $y = \dfrac{2x+1}{3}$

  Interchange x and y: $x = \dfrac{2y+1}{3}$

  Solve for y: $3x = 2y + 1$

  $3x - 1 = 2y$

  $\dfrac{3x-1}{2} = y$

  Replace y by $f^{-1}(x)$: $f^{-1}(x) = \dfrac{3x-1}{2}$

40. a) Yes; b) $f^{-1}(x) = \dfrac{5x-2}{3}$

41. a) The graph of $f(x) = x^3 - 5$ is shown below. It passes the horizontal-line test, so the function is one-to-one.

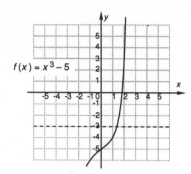

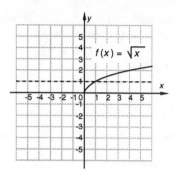

b) Replace $f(x)$ by y: $y = x^3 - 5$
Interchange x and y: $x = y^3 - 5$
Solve for y: $x + 5 = y^3$
$\sqrt[3]{x+5} = y$
Replace y by $f^{-1}(x)$: $f^{-1}(x) = \sqrt[3]{x+5}$

42. a) Yes; b) $f^{-1}(x) = \sqrt[3]{x-2}$

43. a) The graph of $g(x) = (x-2)^3$ is shown below. It passes the horizontal-line test, so the
function is one-to-one.

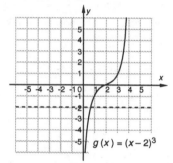

b) Replace $g(x)$ by y: $y = (x-2)^3$
Interchange x and y: $x = (y-2)^3$
Solve for y: $\sqrt[3]{x} = y - 2$
$\sqrt[3]{x} + 2 = y$
Replace y by $g^{-1}(x)$: $g^{-1}(x) = \sqrt[3]{x} + 2$

44. a) Yes; b) $g^{-1}(x) = \sqrt[3]{x} - 7$

45. a) The graph of $f(x) = \sqrt{x}$ is shown below. It
passes the horizontal-line test, so the function
is one-to-one.

b) Replace $f(x)$ by y: $y = \sqrt{x}$ (Note that
$f(x) \geq 0$.)
Interchange x and y: $x = \sqrt{y}$
Solve for y: $x^2 = y$
Replace y by $f^{-1}(x)$: $f^{-1}(x) = x^2, x \geq 0$

46. a) Ys; b) $f^{-1}(x) = x^2 + 1, x \geq 0$

47. a) The graph of $f(x) = 2x^2 + 1$, $x \geq 0$, is shown
below. It passes the horizontal-line test, so
the function is one-to-one.

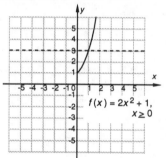

b) Replace $f(x)$ by y: $y = 2x^2 + 1$
Interchange x and y: $x = 2y^2 + 1$
Solve for y: $x - 1 = 2y^2$
$$\frac{x-1}{2} = y^2$$
$$\sqrt{\frac{x-1}{2}} = y$$

(We take the principal square root since
$y \geq 0$.)

Replace y by $f^{-1}(x)$: $f^{-1}(x) = \sqrt{\dfrac{x-1}{2}}$

48. a) Yes; b) $f^{-1}(x) = \sqrt{\dfrac{x+2}{3}}$

49. First graph $f(x) = \dfrac{1}{3}x - 2$. Then graph the inverse
function by reflecting the graph of $f(x) = \dfrac{1}{3}x - 2$
across the line $y = x$. The graph of the inverse function can also be found by first finding a formula for

the inverse, substituting to find function values, and then plotting points.

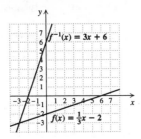

50.

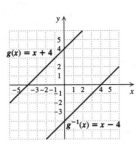

51. Follow the procedure described in Exercise 49 to graph the function and its inverse.

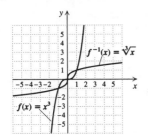

52.

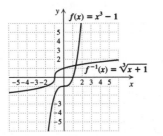

53. Use the procedure described in Exercise 49 to graph the function and its inverse.

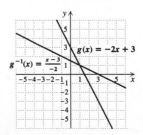

54.

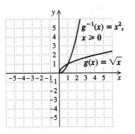

55. Use the procedure described in Exercise 49 to graph the function and its inverse.

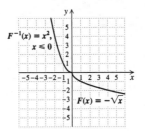

56.

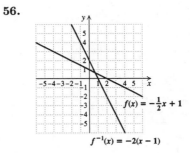

57. Use the procedure described in Exercise 49 to graph the function and its inverse.

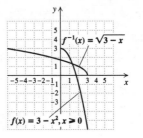

58.

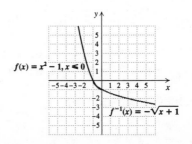

59. We check to see that $f^{-1} \circ f(x) = x$ and $f \circ f^{-1}(x) = x$.

a) $f^{-1} \circ f(x) = f^{-1}(f(x)) = f^{-1}\left(\frac{4}{5}x\right) =$

$\frac{5}{4} \cdot \frac{4}{5}x = x$

b) $f \circ f^{-1}(x) = f(f^{-1}(x)) = f\left(\frac{5}{4}x\right) =$

$\frac{4}{5} \cdot \frac{5}{4}x = x$

60. a) $f^{-1} \circ f(x) = 3\left(\frac{x+7}{3}\right) - 7 = x + 7 - 7 = x$

b) $f \circ f^{-1}(x) = \frac{(3x-7)+7}{3} = \frac{3x}{3} = x$

61. We check to see that $f^{-1} \circ f(x) = x$ and $f \circ f^{-1}(x) = x$.

a) $f^{-1} \circ f(x) = f^{-1}(f(x)) = f^{-1}\left(\frac{1-x}{x}\right) =$

$\frac{1}{\frac{1-x}{x}+1} = \frac{1}{\frac{1-x}{x}+1} \cdot \frac{x}{x} = \frac{x}{1-x+x} =$

$\frac{x}{1} = x$

b) $f \circ f^{-1}(x) = f(f^{-1}(x)) = f\left(\frac{1}{x+1}\right) =$

$\frac{1 - \frac{1}{x+1}}{\frac{1}{x+1}} = \frac{1 - \frac{1}{x+1}}{\frac{1}{x+1}} \cdot \frac{x+1}{x+1} =$

$\frac{x+1-1}{1} = \frac{x}{1} = x$

62. a) $f^{-1} \circ f(x) = \sqrt[3]{x^3 - 5 + 5} = \sqrt[3]{x^3} = x$

b) $f \circ f^{-1}(x) = (\sqrt[3]{x+5})^3 - 5 = x + 5 - 5 = x$

63. a) $f(8) = 8 + 32 = 40$

Size 40 in France corresponds to size 8 in the U.S.

$f(10) = 10 + 32 = 42$

Size 42 in France corresponds to size 10 in the U.S.

$f(14) = 14 + 32 = 46$

Size 46 in France corresponds to size 14 in the U.S.

$f(18) = 18 + 32 = 50$

Size 50 in France corresponds to size 18 in the U.S.

b) The function $f(x) = x + 32$ is a linear function that is not constant, so it passes the horizontal-line test. Thus, f is one-to-one and, hence, has an inverse that is a function. We now find a formula for the inverse.

Replace $f(x)$ by y: $y = x + 32$

Interchange x and y: $x = y + 32$

Solve for y: $x - 32 = y$

Replace y by $f^{-1}(x)$: $f^{-1}(x) = x - 32$

c) $f^{-1}(40) = 40 - 32 = 8$

Size 8 in the U.S. corresponds to size 40 in France.

$f^{-1}(42) = 42 - 32 = 10$

Size 10 in the U.S. corresponds to size 42 in France.

$f^{-1}(46) = 46 - 32 = 14$

Size 14 in the U.S. corresponds to size 46 in France.

$f^{-1}(50) = 50 - 32 = 18$

Size 18 in the U.S. corresponds to size 50 in France.

64. a) 40; 44; 52; 60

b) $f^{-1}(x) = \frac{x-24}{2}$, or $\frac{x}{2} - 12$

c) 8; 10; 14; 18

65. Writing exercise

66. Writing exercise

67. $(a^5b^4)^2(a^3b^5) = (a^5)^2(b^4)^2(a^3b^5)$

$\qquad = a^{5\cdot2}b^{4\cdot2}a^3b^5$

$\qquad = a^{10}b^8a^3b^5$

$\qquad = a^{10+3}b^{8+5}$

$\qquad = a^{13}b^{13}$

68. $x^{10}y^{12}$

69. $27^{4/3} = (3^3)^{4/3} = 3^{3\cdot\frac{4}{3}} = 3^4 = 81$

70. 125

71. $\qquad x = \frac{2}{3}y - 7$

$\qquad x + 7 = \frac{2}{3}y$

$\qquad \frac{3}{2}(x+7) = y$

72. $y = \dfrac{10 - x}{3}$

73. Writing exercise

74. Writing exercise

75. Reflect the graph of f across the line $y = x$.

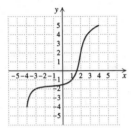

76.

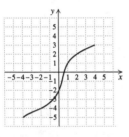

77. From Exercise 64(b), we know that a function that converts dress sizes in Italy to those in the United States is $g(x) = \dfrac{x - 24}{2}$. From Exercise 63, we know that a function that converts dress sizes in the United States to those in France is $f(x) = x + 32$. Then a function that converts dress sizes in Italy to those in France is

$$h(x) = (f \circ g)(x)$$
$$h(x) = f\left(\dfrac{x - 24}{2}\right)$$
$$h(x) = \dfrac{x - 24}{2} + 32$$
$$h(x) = \dfrac{x}{2} - 12 + 32$$
$$h(x) = \dfrac{x}{2} + 20.$$

78. $d(x) = 2(x - 20)$

79. Writing exercise

80. $((f \circ g) \circ h)(x) = (f \circ g)(h(x))$
$$= f(g(h(x))) = f((g \circ h)(x))$$
$$= (f \circ (g \circ h))(x)$$

81. Suppose that $h(x) = (f \circ g)(x)$. First note that for $I(x) = x$, $(f \circ I)(x) = f(I(x))$ for any function f.

i) $((g^{-1} \circ f^{-1}) \circ h)(x) = ((g^{-1} \circ f^{-1}) \circ (f \circ g))(x)$
$$= ((g^{-1} \circ (f^{-1} \circ f)) \circ g)(x)$$
$$= ((g^{-1} \circ I) \circ g)(x)$$
$$= (g^{-1} \circ g)(x) = x$$

ii) $(h \circ (g^{-1} \circ f^{-1}))(x) = ((f \circ g) \circ (g^{-1} \circ f^{-1}))(x)$
$$= ((f \circ (g \circ g^{-1})) \circ f^{-1})(x)$$
$$= ((f \circ I) \circ f^{-1})(x)$$
$$= (f \circ f^{-1})(x) = x$$

Therefore, $(g^{-1} \circ f^{-1})(x) = h^{-1}(x)$.

82. No

83. $(f \circ g)(x) = x$ and $(g \circ f)(x) = x$, so the functions are inverses.

84. Yes

85. $(f \circ g)(x) \neq x$, so the functions are not inverses. (It is also true that $(g \circ f)(x) \neq x$.)

86. Yes

87. (1) C; (2) A; (3) B; (4) D

88. Writing exercise

89. Writing exercise

90. $f(x) = \dfrac{1}{2}x + 3$, $g(x) = 2x - 6$; yes

Exercise Set 12.2

---

1. Graph: $y = 2^x$

We compute some function values, thinking of y as $f(x)$, and keep the results in a table.

$f(0) = 2^0 = 1$

$f(1) = 2^1 = 2$

$f(2) = 2^2 = 4$

$f(-1) = 2^{-1} = \dfrac{1}{2^1} = \dfrac{1}{2}$

$f(-2) = 2^{-2} = \dfrac{1}{2^2} = \dfrac{1}{4}$

x	y, or $f(x)$
0	1
1	2
2	4
-1	$\dfrac{1}{2}$
-2	$\dfrac{1}{4}$

Next we plot these points and connect them with a smooth curve.

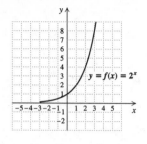

$y = f(x) = 2^x$

2.

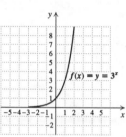

$f(x) = y = 3^x$

3. Graph: $y = 5^x$

We compute some function values, thinking of y as $f(x)$, and keep the results in a table.

$f(0) = 5^0 = 1$

$f(1) = 5^1 = 5$

$f(2) = 5^2 = 25$

$f(-1) = 5^{-1} = \dfrac{1}{5^1} = \dfrac{1}{5}$

$f(-2) = 5^{-2} = \dfrac{1}{5^2} = \dfrac{1}{25}$

x	y, or $f(x)$
0	1
1	5
2	25
-1	$\dfrac{1}{5}$
-2	$\dfrac{1}{25}$

Next we plot these points and connect them with a smooth curve.

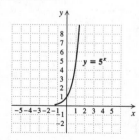

$y = 5^x$

4.

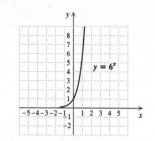

$y = 6^x$

5. Graph: $y = 2^x + 3$

We compute some function values, thinking of y as $f(x)$, and keep the results in a table.

$f(-4) = 2^{-4} + 3 = \dfrac{1}{2^4} + 3 = \dfrac{1}{16} + 3 = 3\dfrac{1}{16}$

$f(-2) = 2^{-2} + 3 = \dfrac{1}{2^2} + 3 = \dfrac{1}{4} + 3 = 3\dfrac{1}{4}$

$f(0) = 2^0 + 3 = 1 + 3 = 4$

$f(1) = 2^1 + 3 = 2 + 3 = 5$

$f(2) = 2^2 + 3 = 4 + 3 = 7$

x	y, or $f(x)$
-4	$3\dfrac{1}{16}$
-2	$3\dfrac{1}{4}$
0	4
1	5
2	7

Next we plot these points and connect them with a smooth curve.

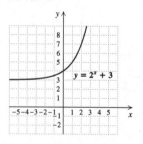

$y = 2^x + 3$

6.

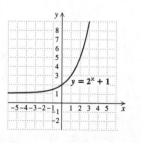

$y = 2^x + 1$

7. Graph: $y = 3^x - 1$

We compute some function values, thinking of y as $f(x)$, and keep the results in a table.

$$f(-3) = 3^{-3} - 1 = \frac{1}{3^3} - 1 = \frac{1}{27} - 1 = -\frac{26}{27}$$

$$f(-1) = 3^{-1} - 1 = \frac{1}{3} - 1 = -\frac{2}{3}$$

$$f(0) = 3^0 - 1 = 1 - 1 = 0$$

$$f(1) = 3^1 - 1 = 3 - 1 = 2$$

$$f(2) = 3^2 - 1 = 9 - 1 = 8$$

x	y, or $f(x)$
-3	$-\dfrac{26}{27}$
-1	$-\dfrac{2}{3}$
0	0
1	2
2	8

Next we plot these points and connect them with a smooth curve.

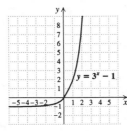

8.

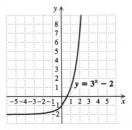

9. Graph: $y = 2^x - 4$

We construct a table of values, thinking of y as $f(x)$. Then we plot the points and connect them with a smooth curve.

$$f(0) = 2^0 - 4 = 1 - 4 = -3$$

$$f(1) = 2^1 - 4 = 2 - 4 = -2$$

$$f(2) = 2^2 - 4 = 4 - 4 = 0$$

$$f(3) = 2^3 - 4 = 8 - 4 = 4$$

$$f(-1) = 2^{-1} - 4 = \frac{1}{2} - 4 = -\frac{7}{2}$$

$$f(-2) = 2^{-2} - 4 = \frac{1}{4} - 4 = -\frac{15}{4}$$

$$f(-4) = 2^{-4} - 4 = \frac{1}{16} - 4 = -\frac{63}{16}$$

x	y, or $f(x)$
0	-3
1	-2
2	0
3	4
-1	$-\dfrac{7}{2}$
-2	$-\dfrac{15}{4}$
-4	$-\dfrac{63}{16}$

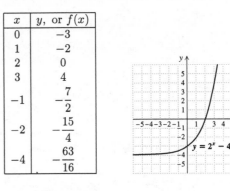

10.

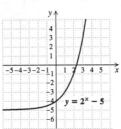

11. Graph: $y = 2^{x-1}$

We construct a table of values, thinking of y as $f(x)$. Then we plot the points and connect them with a smooth curve.

$$f(0) = 2^{0-1} = 2^{-1} = \frac{1}{2}$$

$$f(-1) = 2^{-1-1} = 2^{-2} = \frac{1}{2^2} = \frac{1}{4}$$

$$f(-2) = 2^{-2-1} = 2^{-3} = \frac{1}{2^3} = \frac{1}{8}$$

$$f(1) = 2^{1-1} = 2^0 = 1$$

$$f(2) = 2^{2-1} = 2^1 = 2$$

$$f(3) = 2^{3-1} = 2^2 = 4$$

$$f(4) = 2^{4-1} = 2^3 = 8$$

x	y, or $f(x)$
0	$\dfrac{1}{2}$
-1	$\dfrac{1}{4}$
-2	$\dfrac{1}{8}$
1	1
2	2
3	4
4	8

12.

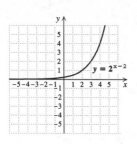

13. Graph: $y = 2^{x+3}$

We construct a table of values, thinking of y as $f(x)$. Then we plot the points and connect them with a smooth curve.

$$f(-4) = 2^{-4+3} = 2^{-1} = \frac{1}{2}$$
$$f(-2) = 2^{-2+3} = 2$$
$$f(-1) = 2^{-1+3} = 2^2 = 4$$
$$f(0) = 2^{0+3} = 2^3 = 8$$

x	y, or $f(x)$
-4	$\dfrac{1}{2}$
-2	2
-1	4
0	8

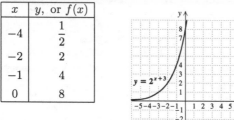

14.

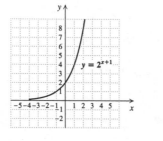

15. Graph: $y = \left(\dfrac{1}{5}\right)^x$

We construct a table of values, thinking of y as $f(x)$. Then we plot the points and connect them with a smooth curve.

$$f(0) = \left(\frac{1}{5}\right)^0 = 1$$
$$f(1) = \left(\frac{1}{5}\right)^1 = \frac{1}{5}$$

$$f(2) = \left(\frac{1}{5}\right)^2 = \frac{1}{25}$$
$$f(-1) = \left(\frac{1}{5}\right)^{-1} = \frac{1}{\frac{1}{5}} = 5$$
$$f(-2) = \left(\frac{1}{5}\right)^{-2} = \frac{1}{\frac{1}{25}} = 25$$

x	y, or $f(x)$
0	1
1	$\dfrac{1}{5}$
2	$\dfrac{1}{25}$
-1	5
-2	25

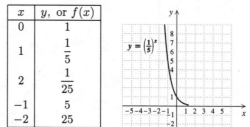

16.

17. Graph: $y = \left(\dfrac{1}{2}\right)^x$

We construct a table of values, thinking of y as $f(x)$. Then we plot the points and connect them with a smooth curve.

$$f(0) = \left(\frac{1}{2}\right)^0 = 1$$
$$f(1) = \left(\frac{1}{2}\right)^1 = \frac{1}{2}$$
$$f(2) = \left(\frac{1}{2}\right)^2 = \frac{1}{4}$$
$$f(3) = \left(\frac{1}{2}\right)^3 = \frac{1}{8}$$
$$f(-1) = \left(\frac{1}{2}\right)^{-1} = \frac{1}{\left(\frac{1}{2}\right)^1} = \frac{1}{\frac{1}{2}} = 2$$
$$f(-2) = \left(\frac{1}{2}\right)^{-2} = \frac{1}{\left(\frac{1}{2}\right)^2} = \frac{1}{\frac{1}{4}} = 4$$
$$f(-3) = \left(\frac{1}{2}\right)^{-3} = \frac{1}{\left(\frac{1}{2}\right)^3} = \frac{1}{\frac{1}{8}} = 8$$

x	y, or $f(x)$
0	1
1	$\dfrac{1}{2}$
2	$\dfrac{1}{4}$
3	$\dfrac{1}{8}$
−1	2
−2	4
−3	8

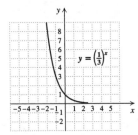

18.

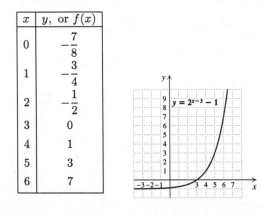

19. Graph: $y = 2^{x-3} - 1$

We construct a table of values, thinking of y as $f(x)$. Then we plot the points and connect them with a smooth curve.

$$f(0) = 2^{0-3} - 1 = 2^{-3} - 1 = \frac{1}{8} - 1 = -\frac{7}{8}$$

$$f(1) = 2^{1-3} - 1 = 2^{-2} - 1 = \frac{1}{4} - 1 = -\frac{3}{4}$$

$$f(2) = 2^{2-3} - 1 = 2^{-1} - 1 = \frac{1}{2} - 1 = -\frac{1}{2}$$

$$f(3) = 2^{3-3} - 1 = 2^{0} - 1 = 1 - 1 = 0$$

$$f(4) = 2^{4-3} - 1 = 2^{1} - 1 = 2 - 1 = 1$$

$$f(5) = 2^{5-3} - 1 = 2^{2} - 1 = 4 - 1 = 3$$

$$f(6) = 2^{6-3} - 1 = 2^{3} - 1 = 8 - 1 = 7$$

x	y, or $f(x)$
0	$-\dfrac{7}{8}$
1	$-\dfrac{3}{4}$
2	$-\dfrac{1}{2}$
3	0
4	1
5	3
6	7

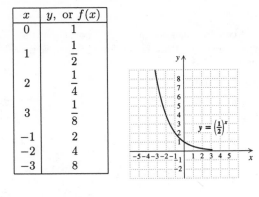

20.

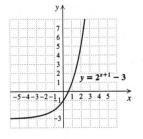

21. Graph: $x = 3^{y}$

We can find ordered pairs by choosing values for y and then computing values for x.

For $y = 0$, $x = 3^{0} = 1$.

For $y = 1$, $x = 3^{1} = 3$.

For $y = 2$, $x = 3^{2} = 9$.

For $y = 3$, $x = 3^{3} = 27$.

For $y = -1$, $x = 3^{-1} = \dfrac{1}{3^{1}} = \dfrac{1}{3}$.

For $y = -2$, $x = 3^{-2} = \dfrac{1}{3^{2}} = \dfrac{1}{9}$.

For $y = -3$, $x = 3^{-3} = \dfrac{1}{3^{3}} = \dfrac{1}{27}$.

x	y
1	0
3	1
9	2
27	3
$\dfrac{1}{3}$	−1
$\dfrac{1}{9}$	−2
$\dfrac{1}{27}$	−3

(1) Choose values for y.

(2) Compute values for x.

We plot the points and connect them with a smooth curve.

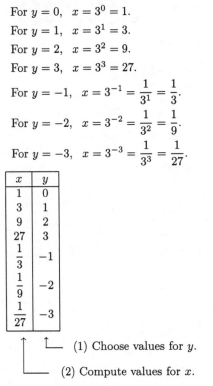

22.

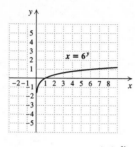

23. Graph: $x = 2^{-y} = \left(\dfrac{1}{2}\right)^{y}$

We can find ordered pairs by choosing values for y and then computing values for x. Then we plot these points and connect them with a smooth curve.

For $y = 0$, $x = \left(\dfrac{1}{2}\right)^{0} = 1$.

For $y = 1$, $x = \left(\dfrac{1}{2}\right)^{1} = \dfrac{1}{2}$.

For $y = 2$, $x = \left(\dfrac{1}{2}\right)^{2} = \dfrac{1}{4}$.

For $y = 3$, $x = \left(\dfrac{1}{2}\right)^{3} = \dfrac{1}{8}$.

For $y = -1$, $x = \left(\dfrac{1}{2}\right)^{-1} = \dfrac{1}{\dfrac{1}{2}} = 2$.

For $y = -2$, $x = \left(\dfrac{1}{2}\right)^{-2} = \dfrac{1}{\dfrac{1}{4}} = 4$.

For $y = -3$, $x = \left(\dfrac{1}{2}\right)^{-3} = \dfrac{1}{\dfrac{1}{8}} = 8$.

x	y
1	0
$\dfrac{1}{2}$	1
$\dfrac{1}{4}$	2
$\dfrac{1}{8}$	3
2	-1
4	-2
8	-3

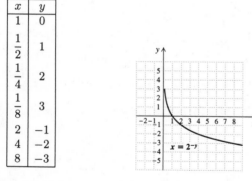

24.

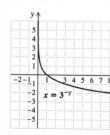

25. Graph: $x = 5^{y}$

We can find ordered pairs by choosing values for y and then computing values for x. Then we plot these points and connect them with a smooth curve.

For $y = 0$, $x = 5^{0} = 1$.

For $y = 1$, $x = 5^{1} = 5$.

For $y = 2$, $x = 5^{2} = 25$.

For $y = -1$, $x = 5^{-1} = \dfrac{1}{5}$.

For $y = -2$, $x = 5^{-2} = \dfrac{1}{25}$.

x	y
1	0
5	1
25	2
$\dfrac{1}{5}$	-1
$\dfrac{1}{25}$	-2

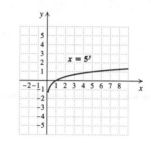

26.

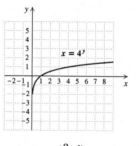

27. Graph: $x = \left(\dfrac{3}{2}\right)^{y}$

We can find ordered pairs by choosing values for y and then computing values for x. Then we plot these points and connect them with a smooth curve.

For $y = 0$, $x = \left(\dfrac{3}{2}\right)^{0} = 1$.

For $y = 1$, $x = \left(\dfrac{3}{2}\right)^{1} = \dfrac{3}{2}$.

For $y = 2$, $x = \left(\dfrac{3}{2}\right)^{2} = \dfrac{9}{4}$.

For $y = 3$, $x = \left(\dfrac{3}{2}\right)^{3} = \dfrac{27}{8}$.

For $y = -1$, $x = \left(\dfrac{3}{2}\right)^{-1} = \dfrac{1}{\dfrac{3}{2}} = \dfrac{2}{3}$.

For $y = -2$, $x = \left(\dfrac{3}{2}\right)^{-2} = \dfrac{1}{\dfrac{9}{4}} = \dfrac{4}{9}$.

For $y = -3$, $\quad x = \left(\dfrac{3}{2}\right)^{-3} = \dfrac{1}{\dfrac{27}{8}} = \dfrac{8}{27}$.

x	y
1	0
$\dfrac{3}{2}$	1
$\dfrac{9}{4}$	2
$\dfrac{27}{8}$	3
$\dfrac{2}{3}$	-1
$\dfrac{4}{9}$	-2
$\dfrac{8}{27}$	-3

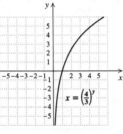

28.

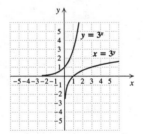

29. Graph $y = 3^x$ (see Exercise 2) and $x = 3^y$ (see Exercise 21) using the same set of axes.

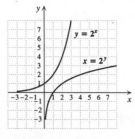

30.

31. Graph $y = \left(\dfrac{1}{2}\right)^x$ (see Exercise 13) and $x = \left(\dfrac{1}{2}\right)^y$ (see Exercise 23) using the same set of axes.

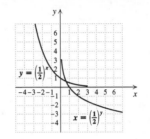

32.

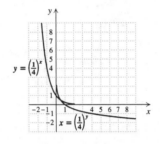

33. a) In 2004, $t = 2004 - 1975 = 29$.

$$P(29) = 4(1.0164)^{29} \approx 6.4$$

The world population will be about 6.4 billion in 2004.
In 2008, $t = 2008 - 1975 = 33$.

$$P(33) = 4(1.0164)^{33} \approx 6.8$$

The world population will be about 6.8 billion in 2008.
In 2012, $t = 2012 - 1975 = 37$.

$$P(37) = 4(1.0164)^{37} \approx 7.3$$

The world population will be about 7.3 billion in 2012.

b)

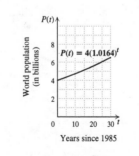

34. a) 4243; 6000; 8485; 12,000; 24,000

b)

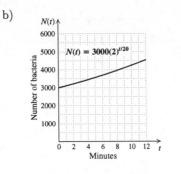

35. a) In 1930, $t = 1930 - 1900 = 30$.
$$P(t) = 150(0.960)^t$$
$$P(30) = 150(0.960)^{30}$$
$$\approx 44.079$$

In 1930, about 44.079 thousand, or 44,079, humpback whales were alive.

In 1960, $t = 1960 - 1900 = 60$.
$$P(t) = 150(0.960)^t$$
$$P(60) = 150(0.960)^{60}$$
$$\approx 12.953$$

In 1960, about 12.953 thousand, or 12,953, humpback whales were alive.

b) Plot the points found in part (a), $(30, 44,079)$ and $(60, 12,953)$ and additional points as needed and graph the function.

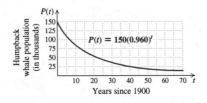

36. a) About 8706 whales; about 13,163 whales

b)

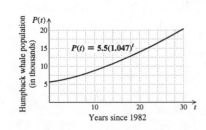

37. a) Substitute for t.
$$N(0) = 250,000\left(\frac{2}{3}\right)^0 = 250,000 \cdot 1 = 250,000;$$
$$N(1) = 250,000\left(\frac{2}{3}\right)^1 = 250,000 \cdot \frac{2}{3} = 166,667;$$
$$N(4) = 250,000\left(\frac{2}{3}\right)^4 = 250,000 \cdot \frac{16}{81} \approx 49,383;$$
$$N(10) = 250,000\left(\frac{2}{3}\right)^{10} = 250,000 \cdot \frac{1024}{59,049} \approx 4335$$

b) We use the function values computed in part (a) to draw the graph of the function. Note that the axes are scaled differently because of the large function values.

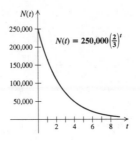

38. a) $5200; $4160; $3328; $1703.94; $558.35

b)

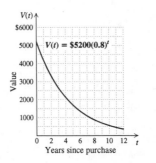

39. a) In 1985, $t = 1985 - 1985 = 0$.

$N(0) = 0.3(1.4477)^0 = 0.3(1) = 0.3$ million, or 300,000

In 1995, $t = 1995 - 1985 = 10$.

$N(10) = 0.3(1.4477)^{10} \approx 12.1$ million

In 2005, $t = 2005 - 1985 = 20$.

$N(20) = 0.3(1.4477)^{20} \approx 490.6$ million

In 2010, $t = 2010 - 1985 = 25$.

$N(25) = 0.3(1.4477)^{25} \approx 3119.5$ million, or 3.1195 billion

b) We use the function values computed in part (a) to draw the graph of the function. Note that the axes are scaled differently because of the large function values.

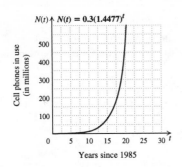

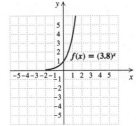

x	y
0	1
1	3.8
2	14.44
3	54.872
-1	0.26
-2	0.7

40. a) 454,354,240 cm^2; 525,233,501,400 cm^2

b)

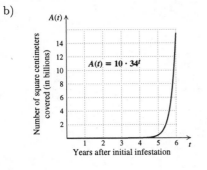

54.

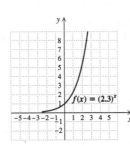

41. Writing exercise

42. Writing exercise

43. $5^{-2} = \dfrac{1}{5^2} = \dfrac{1}{25}$

44. $\dfrac{1}{32}$

45. $1000^{2/3} = (10^3)^{2/3} = 10^{3 \cdot \frac{2}{3}} = 10^2 = 100$

46. 125

47. $\dfrac{10a^8b^7}{2a^2b^4} = \dfrac{10}{2}a^{8-2}b^{7-4} = 5a^6b^3$

48. $6x^4y$

49. Writing exercise

50. Writing exercise

51. Since the bases are the same, the one with the larger exponent is the larger number. Thus $\pi^{2.4}$ is larger.

52. $8^{\sqrt{3}}$

53. Graph: $f(x) = 3.8^x$

Use a calculator with a power key to construct a table of values. (We will round values of $f(x)$ to the nearest hundredth.) Then plot these points and connect them with a smooth curve.

55. Graph: $y = 2^x + 2^{-x}$

Construct a table of values, thinking of y as $f(x)$. Then plot these points and connect them with a curve.

$f(0) = 2^0 + 2^{-0} = 1 + 1 = 2$

$f(1) = 2^1 + 2^{-1} = 2 + \dfrac{1}{2} = 2\dfrac{1}{2}$

$f(2) = 2^2 + 2^{-2} = 4 + \dfrac{1}{4} = 4\dfrac{1}{4}$

$f(3) = 2^3 + 2^{-3} = 8 + \dfrac{1}{8} = 8\dfrac{1}{8}$

$f(-1) = 2^{-1} + 2^{-(-1)} = \dfrac{1}{2} + 2 = 2\dfrac{1}{2}$

$f(-2) = 2^{-2} + 2^{-(-2)} = \dfrac{1}{4} + 4 = 4\dfrac{1}{4}$

$f(-3) = 2^{-3} + 2^{-(-3)} = \dfrac{1}{8} + 8 = 8\dfrac{1}{8}$

x	y, or $f(x)$
0	2
1	$2\dfrac{1}{2}$
2	$4\dfrac{1}{4}$
3	$8\dfrac{1}{8}$
-1	$2\dfrac{1}{2}$
-2	$4\dfrac{1}{4}$
-3	$8\dfrac{1}{8}$

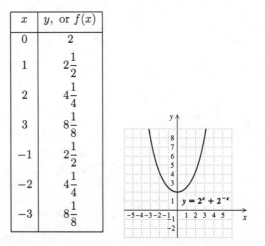

56.

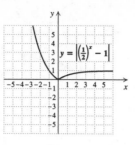

57. Graph: $y = |2^x - 2|$

We construct a table of values, thinking of y as $f(x)$. Then plot these points and connect them with a curve.

$$f(0) = |2^0 - 2| = |1 - 2| = |-1| = 1$$
$$f(1) = |2^1 - 2| = |2 - 2| = |0| = 0$$
$$f(2) = |2^2 - 2| = |4 - 2| = |2| = 2$$
$$f(3) = |2^3 - 2| = |8 - 2| = |6| = 6$$
$$f(-1) = |2^{-1} - 2| = \left|\frac{1}{2} - 2\right| = \left|-\frac{3}{2}\right| = \frac{3}{2}$$
$$f(-3) = |2^{-3} - 2| = \left|\frac{1}{8} - 2\right| = \left|-\frac{15}{8}\right| = \frac{15}{8}$$
$$f(-5) = |2^{-5} - 2| = \left|\frac{1}{32} - 2\right| = \left|-\frac{63}{32}\right| = \frac{63}{32}$$

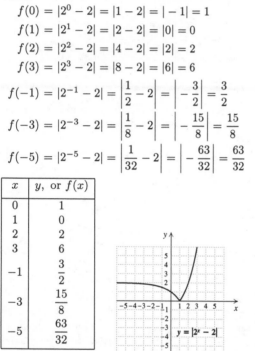

x	y, or $f(x)$
0	1
1	0
2	2
3	6
-1	$\frac{3}{2}$
-3	$\frac{15}{8}$
-5	$\frac{63}{32}$

58.

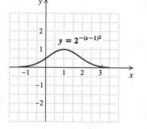

59. Graph: $y = |2x^2 - 1|$

We construct a table of values, thinking of y as $f(x)$. Then we plot these points and connect them with a curve.

$$f(0) = |2^{0^2} - 1| = |1 - 1| = 0$$
$$f(1) = |2^{1^2} - 1| = |2 - 1| = 1$$
$$f(2) = |2^{2^2} - 1| = |16 - 1| = 15$$
$$f(-1) = |2^{(-1)^2} - 1| = |2 - 1| = 1$$
$$f(-2) = |2^{(-2)^2} - 1| = |16 - 1| = 15$$

x	y, or $f(x)$
0	0
1	1
2	15
-1	1
-2	15

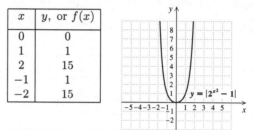

60.

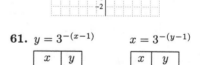

61. $y = 3^{-(x-1)}$ $x = 3^{-(y-1)}$

x	y
0	3
1	1
2	$\frac{1}{3}$
3	$\frac{1}{9}$
-1	9

x	y
3	0
1	1
$\frac{1}{3}$	2
$\frac{1}{9}$	3
9	-1

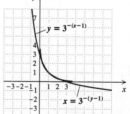

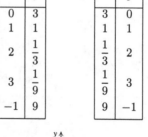

62.

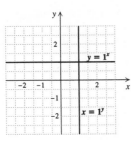

63. Enter the data points $(0, 171)$, $(1, 421)$, and $(2, 1099)$ and then use the exponential regression feature of the grapher to find an exponential function that models the data;

$A(t) = 169.339338(2.535133248)^t$, where $A(t)$ is total U.S. sales, in millions of dollars, t years after 1997.

In 2005, $t = 2005 - 1997 = 8$.

$A(8) = 169.3393318(2.535133248)^8 \approx \$288,911.0615$ million, or $\$288,911,061,500$

64. $N(t) = 9047.179795(1.393338213)^t$, where t is the number of years after 1985; 17,564; 484,386

65. Enter the function on the grapher and then use the graph or a table of values set in ASK mode.

$S(10) \approx 19$ words per minute

$S(40) \approx 66$ words per minute

$S(80) \approx 110$ words per minute

66. Writing exercise

Exercise Set 12.3

1. $\log_{10} 100$ is the power to which we raise 10 to get 100. Since $10^2 = 100$, $\log_{10} 100 = 2$.

2. 3

3. $\log_2 8$ is the power to which we raise 2 to get 8. Since $2^3 = 8$, $\log_2 8 = 3$.

4. 4

5. $\log_3 81$ is the power to which we raise 3 to get 81. Since $3^4 = 81$, $\log_3 81 = 4$.

6. 3

7. $\log_4 \dfrac{1}{16}$ is the power to which we raise 4 to get $\dfrac{1}{16}$. Since $4^{-2} = \dfrac{1}{16}$, $\log_4 \dfrac{1}{16} = -2$.

8. -1

9. Since $7^{-1} = \dfrac{1}{7}$, $\log_7 \dfrac{1}{7} = -1$.

10. -2

11. Since $5^4 = 625$, $\log_5 625 = 4$.

12. 3

13. Since $6^1 = 6$, $\log_6 6 = 1$.

14. 0

15. Since $8^0 = 1$, $\log_8 1 = 0$.

16. 1

17. $\log_9 9^7$ is the power to which we raise 9 to get 9^7. Clearly, this power is 7, so $\log_9 9^7 = 7$.

18. 10

19. Since $10^{-1} = \dfrac{1}{10} = 0.1$, $\log_{10} 0.1 = -1$.

20. -2

21. Since $9^{1/2} = 3$, $\log_9 3 = \dfrac{1}{2}$.

22. $\dfrac{1}{2}$

23. Since $9 = 3^2$ and $(3^2)^{3/2} = 3^3 = 27$, $\log_9 27 = \dfrac{3}{2}$.

24. $\dfrac{3}{2}$

25. Since $1000 = 10^3$ and $(10^3)^{2/3} = 10^2 = 100$, $\log_{1000} 100 = \dfrac{2}{3}$.

26. $\dfrac{2}{3}$

27. Since $\log_5 7$ is the power to which we raise 5 to get 7, then 5 raised to this power is 7. That is, $5^{\log_5 7} = 7$.

28. 13

29. Graph: $y = \log_{10} x$

The equation $y = \log_{10} x$ is equivalent to $10^y = x$. We can find ordered pairs by choosing values for y and computing the corresponding x-values.

For $y = 0$, $x = 10^0 = 1$.

For $y = 1$, $x = 10^1 = 10$.

For $y = 2$, $x = 10^2 = 100$.

For $y = -1$, $x = 10^{-1} = \dfrac{1}{10}$.

For $y = -2$, $x = 10^{-2} = \dfrac{1}{100}$.

x, or 10^y	y
1	0
10	1
100	2
$\dfrac{1}{10}$	-1
$\dfrac{1}{100}$	-2

            (1) Select y.

            (2) Compute x.

We plot the set of ordered pairs and connect the points with a smooth curve.

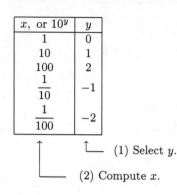

30.

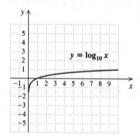

31. Graph: $y = \log_3 x$

The equation $y = \log_3 x$ is equivalent to $3^y = x$. We can find ordered pairs by choosing values for y and computing the corresponding x-values.

For $y = 0,\quad x = 3^0 = 1.$

For $y = 1,\quad x = 3^1 = 3.$

For $y = 2,\quad x = 3^2 = 9.$

For $y = -1,\quad x = 3^{-1} = \dfrac{1}{3}.$

For $y = -2,\quad x = 3^{-2} = \dfrac{1}{9}.$

x, or 3^y	y
1	0
3	1
9	2
$\dfrac{1}{3}$	-1
$\dfrac{1}{9}$	-2

We plot the set of ordered pairs and connect the points with a smooth curve.

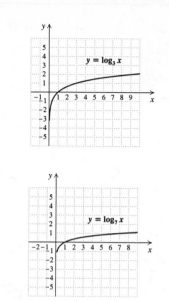

32.

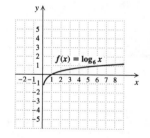

33. Graph: $f(x) = \log_6 x$

Think of $f(x)$ as y. Then $y = \log_6 x$ is equivalent to $6^y = x$. We find ordered pairs by choosing values for y and computing the corresponding x-values. Then we plot the points and connect them with a smooth curve.

For $y = 0,\quad x = 6^0 = 1.$

For $y = 1,\quad x = 6^1 = 6.$

For $y = 2,\quad x = 6^2 = 36.$

For $y = -1,\quad x = 6^{-1} = \dfrac{1}{6}.$

For $y = -2,\quad x = 6^{-2} = \dfrac{1}{36}.$

x, or 6^y	y
1	0
6	1
36	2
$\dfrac{1}{6}$	-1
$\dfrac{1}{36}$	-2

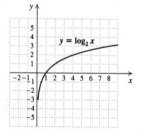

34.

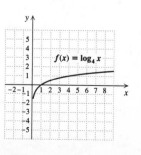

35. Graph: $f(x) = \log_{2.5} x$

Think of $f(x)$ as y. Then $y = \log_{2.5} x$ is equivalent to $2.5^y = x$. We construct a table of values, plot these points and connect them with a smooth curve.

For $y = 0, x = 2.5^0 = 1$.

For $y = 1, x = 2.5^1 = 2.5$.

For $y = 2, x = 2.5^2 = 6.25$.

For $y = 3, x = 2.5^3 = 15.625$.

For $y = -1, x = 2.5^{-1} = 0.4$.

For $y = -2, x = 2.5^{-2} = 0.16$.

x, or 2.5^y	y
1	0
2.5	1
6.25	2
15.625	3
0.4	-1
0.16	-2

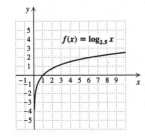

36.

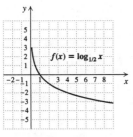

37. Graph $f(x) = 3^x$ (see Exercise Set 12.2, Exercise 2) and $f^{-1}(x) = \log_3 x$ (see Exercise 31 above) on the same set of axes.

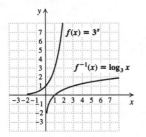

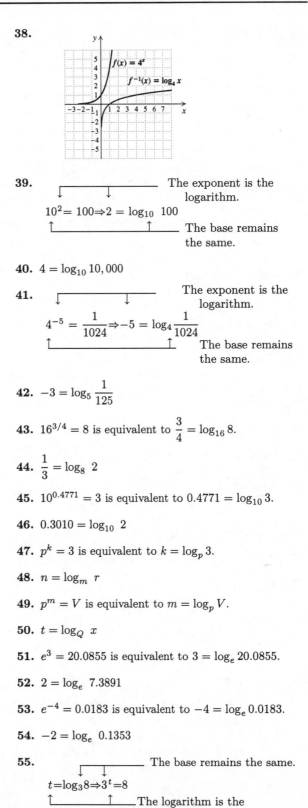

38.

39. The exponent is the logarithm.

$10^2 = 100 \Rightarrow 2 = \log_{10} 100$

The base remains the same.

40. $4 = \log_{10} 10,000$

41. The exponent is the logarithm.

$4^{-5} = \dfrac{1}{1024} \Rightarrow -5 = \log_4 \dfrac{1}{1024}$

The base remains the same.

42. $-3 = \log_5 \dfrac{1}{125}$

43. $16^{3/4} = 8$ is equivalent to $\dfrac{3}{4} = \log_{16} 8$.

44. $\dfrac{1}{3} = \log_8 2$

45. $10^{0.4771} = 3$ is equivalent to $0.4771 = \log_{10} 3$.

46. $0.3010 = \log_{10} 2$

47. $p^k = 3$ is equivalent to $k = \log_p 3$.

48. $n = \log_m r$

49. $p^m = V$ is equivalent to $m = \log_p V$.

50. $t = \log_Q x$

51. $e^3 = 20.0855$ is equivalent to $3 = \log_e 20.0855$.

52. $2 = \log_e 7.3891$

53. $e^{-4} = 0.0183$ is equivalent to $-4 = \log_e 0.0183$.

54. $-2 = \log_e 0.1353$

55. The base remains the same.

$t = \log_3 8 \Rightarrow 3^t = 8$

The logarithm is the exponent.

56. $7^h = 10$

57.

$$\log_5 25 = 2 \Rightarrow 5^2 = 25$$

The logarithm is the exponent.

The base remains the same.

58. $6^1 = 6$

59. $\log_{10} 0.1 = -1$ is equivalent to $10^{-1} = 0.1$.

60. $10^{-2} = 0.01$

61. $\log_{10} 7 = 0.845$ is equivalent to $10^{0.845} = 7$.

62. $10^{0.4771} = 3$

63. $\log_c m = 8$ is equivalent to $c^8 = m$.

64. $b^{23} = n$

65. $\log_t Q = r$ is equivalent to $t^r = Q$.

66. $m^a = P$

67. $\log_e 0.25 = -1.3863$ is equivalent to $e^{-1.3863} = 0.25$.

68. $e^{-0.0111} = 0.989$

69. $\log_r T = -x$ is equivalent to $r^{-x} = T$.

70. $c^{-w} = M$

71. $\log_3 x = 2$

$\qquad 3^2 = x$ Converting to an exponential equation

$\qquad 9 = x$ Computing 3^2

72. 64

73. $\log_x 64 = 3$

$\qquad x^3 = 64$ Converting to an exponential equation

$\qquad x = 4$ Taking cube roots

74. 5

75. $\log_5 25 = x$

$\qquad 5^x = 25$ Converting to an exponential equation

$\qquad 5^x = 5^2$

$\qquad x = 2$ The exponents must be the same.

76. 4

77. $\log_4 16 = x$

$\qquad 4^x = 16$ Converting to an exponential equation

$\qquad 4^x = 4^2$

$\qquad x = 2$ The exponents must be the same.

78. 3

79. $\log_x 7 = 1$

$\qquad x^1 = 7$ Converting to an exponential equation

$\qquad x = 7$ Siimplifying x^1

80. 8

81. $\log_9 x = 1$

$\qquad 9^1 = x$ Converting to an exponential equation

$\qquad 9 = x$ Simplifying 9^1

82. 1

83. $\log_3 x = -2$

$\qquad 3^{-2} = x$ Converting to an exponential equation

$\qquad \dfrac{1}{9} = x$ Simplifying

84. $\dfrac{1}{2}$

85. $\log_{32} x = \dfrac{2}{5}$

$\qquad 32^{2/5} = x$ Converting to an exponential equation

$\qquad (2^5)^{2/5} = x$

$\qquad 4 = x$

86. 4

87. Writing exercise

88. Writing exercise

89. $\dfrac{x^{12}}{x^4} = x^{12-4} = x^8$

90. a^{12}

91. $(a^4 b^6)(a^3 b^2) = a^{4+3} b^{6+2} = a^7 b^8$

92. $x^5 y^{12}$

93. $\dfrac{\dfrac{3}{x} - \dfrac{2}{xy}}{\dfrac{2}{x^2} + \dfrac{1}{xy}}$

The LCD of all the denominators is x^2y. We multiply numerator and denominator by the LCD.

$$\frac{\dfrac{3}{x} - \dfrac{2}{xy}}{\dfrac{2}{x^2} + \dfrac{1}{xy}} \cdot \frac{x^2y}{x^2y} = \frac{\left(\dfrac{3}{x} - \dfrac{2}{xy}\right)x^2y}{\left(\dfrac{2}{x^2} + \dfrac{1}{xy}\right)x^2y}$$

$$= \frac{\dfrac{3}{x} \cdot x^2y - \dfrac{2}{xy} \cdot x^2y}{\dfrac{2}{x^2} \cdot x^2y + \dfrac{1}{xy} \cdot x^2y}$$

$$= \frac{3xy - 2x}{2y + x}, \text{ or}$$

$$\frac{x(3y - 2)}{2y + x}$$

94. $\dfrac{x + 2}{x + 1}$

95. Writing exercise

96. Writing exercise

97. Graph: $y = \left(\dfrac{3}{2}\right)^x$ 　　Graph: $y = \log_{3/2} x$, or

$$x = \left(\dfrac{3}{2}\right)^y$$

x	y, or $\left(\dfrac{3}{2}\right)^x$
0	1
1	$\dfrac{3}{2}$
2	$\dfrac{9}{4}$
3	$\dfrac{27}{8}$
-1	$\dfrac{2}{3}$
-2	$\dfrac{4}{9}$

x, or $\left(\dfrac{3}{2}\right)^y$	y
1	0
$\dfrac{3}{2}$	1
$\dfrac{9}{4}$	2
$\dfrac{27}{8}$	3
$\dfrac{2}{3}$	-1
$\dfrac{4}{9}$	-2

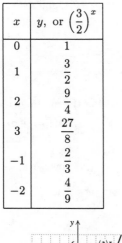

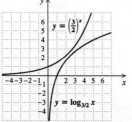

98.

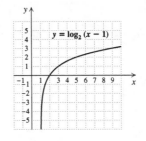

99. Graph: $y = \log_3 |x + 1|$

x	y
0	0
2	1
8	2
-2	0
-4	1
-9	2

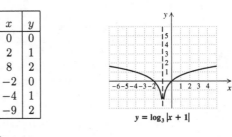

$y = \log_3 |x + 1|$

100. $\dfrac{1}{9}$, 9

101. $\log_{125} x = \dfrac{2}{3}$

$$125^{2/3} = x$$
$$(5^3)^{2/3} = x$$
$$5^2 = x$$
$$25 = x$$

102. 6

103. $\log_8(2x + 1) = -1$

$$8^{-1} = 2x + 1$$
$$\frac{1}{8} = 2x + 1$$
$$1 = 16x + 8 \quad \text{Multiplying by 8}$$
$$-7 = 16x$$
$$-\frac{7}{16} = x$$

104. -25, 4

105. Let $\log_{1/4} \dfrac{1}{64} = x$. Then

$$\left(\frac{1}{4}\right)^x = \frac{1}{64}$$
$$\left(\frac{1}{4}\right)^x = \left(\frac{1}{4}\right)^3$$
$$x = 3.$$

Thus, $\log_{1/4} \dfrac{1}{64} = 3$.

106. -2

107. $\log_{81} 3 \cdot \log_3 81$

$\quad = \dfrac{1}{4} \cdot 4 \qquad \left(\log_{81} 3 = \dfrac{1}{4}, \ \log_3 81 = 4 \right)$

$\quad = 1$

108. 0

109. $\log_2(\log_2(\log_4 256))$

$\quad = \log_2(\log_2 4) \qquad (\log_4 256 = 4)$

$\quad = \log_2 2 \qquad\qquad (\log_2 4 = 2)$

$\quad = 1$

110. Let $b = 0$, $x = 1$, and $y = 2$. Then $0^1 = 0^2$, but $1 \neq 2$. Let $b = 1$, $x = 1$, and $y = 2$. Then $1^1 = 1^2$, but $1 \neq 2$.

Exercise Set 12.4

1. $\log_3 (81 \cdot 27) = \log_3 81 + \log_3 27$ Using the product rule

2. $\log_2 16 + \log_2 32$

3. $\log_4 (64 \cdot 16) = \log_4 64 + \log_4 16$ Using the product rule

4. $\log_5 25 + \log_5 125$

5. $\log_c rst$

$\quad = \log_c r + \log_c s + \log_c t$ Using the product rule

6. $\log_t 3 + \log_t a + \log_t b$

7. $\log_a 5 + \log_a 14 = \log_a (5 \cdot 14)$ Using the product rule

The result can also be expressed as $\log_a 70$.

8. $\log_b (65 \cdot 2)$, or $\log_b 130$

9. $\log_c t + \log_c y = \log_c (t \cdot y)$ Using the product rule

10. $\log_t (H \cdot M)$

11. $\log_a r^8 = 8 \log_a r$ Using the power rule

12. $5 \log_b t$

13. $\log_c y^6 = 6 \log_c y$ Using the power rule

14. $7 \log_{10} y$

15. $\log_b C^{-3} = -3 \log_b C$ Using the power rule

16. $-5 \log_c M$

17. $\log_2 \dfrac{53}{17} = \log_2 53 - \log_2 17$ Using the quotient rule

18. $\log_3 23 - \log_3 9$

19. $\log_b \dfrac{m}{n} = \log_b m - \log_b n$ Using the quotient rule

20. $\log_a y - \log_a x$

21. $\log_a 15 - \log_a 3$

$\quad = \log_a \dfrac{15}{3}, \qquad$ Using the quotient rule

$\quad$ or $\log_a 5$

22. $\log_b \dfrac{42}{7}$, or $\log_b 6$

23. $\log_b 36 - \log_b 4$

$\quad = \log_b \dfrac{36}{4}, \qquad$ Using the quotient rule

$\quad$ or $\log_b 9$

24. $\log_a \dfrac{26}{2}$, or $\log_a 13$

25. $\log_a 7 - \log_z 18 = \log_a \dfrac{7}{18}$ Using the quotient rule

26. $\log_b \dfrac{5}{13}$

27. $\log_a x^5 y^7 z^6$

$\quad = \log_a x^5 + \log_a y^7 + \log_a z^6$ Using the product rule

$\quad = 5 \log_a x + 7 \log_a y + 6 \log_a z$ Using the power rule

28. $\log_a x + 41 \log_a y + 3 \log_a z$

29. $\log_b \dfrac{xy^2}{z^3}$

$\quad = \log_b xy^2 - \log_b z^3$ Using the quotient rule

$\quad = \log_b x + \log_b y^2 - \log_b z^3$ Using the product rule

$\quad = \log_b x + 2 \log_b y - 3 \log_b z$ Using the power rule

30. $2 \log_b x + 5 \log_b y - 4 \log_b w - 7 \log_b z$

31. $\log_a \dfrac{x^4}{y^3 z}$

$\quad = \log_a x^4 - \log_a y^3 z$ Using the quotient rule

$\quad = \log_a x^4 - (\log_a y^3 + \log_a z)$ Using the product rule

$\quad = \log_a x^4 - \log_a y^3 - \log_a z$ Removing parentheses

$\quad = 4 \log_a x - 3 \log_a y - \log_a z$ Using the power rule

32. $4 \log_a x - \log_a y - 2 \log_a z$

33. $\log_b \dfrac{xy^2}{wz^3}$

$= \log_b xy^2 - \log_b wz^3$ Using the quotient rule

$= \log_b x + \log_b y^2 - (\log_b w + \log_b z^3)$
Using the product rule

$= \log_b x + \log_b y^2 - \log_b w - \log_b z^3$
Removing parentheses

$= \log_b x + 2 \log_b y - \log_b w - 3 \log_b z$
Using the power rule

34. $2 \log_b w + \log_b x - 3 \log_b y - \log_b z$

35. $\log_a \sqrt{\dfrac{x^7}{y^5 z^8}}$

$= \log_a \left(\dfrac{x^7}{y^5 z^8} \right)^{1/2}$

$= \dfrac{1}{2} \log_a \dfrac{x^7}{y^5 z^8}$ Using the power rule

$= \dfrac{1}{2} (\log_a x^7 - \log_a y^5 z^8)$ Using the quotient rule

$= \dfrac{1}{2} \left[\log_a x^7 - (\log_a y^5 + \log_a z^8) \right]$
Using the product rule

$= \dfrac{1}{2} (\log_a x^7 - \log_a y^5 - \log_a z^8)$
Removing parentheses

$= \dfrac{1}{2} (7 \log_a x - 5 \log_a y - 8 \log_a z)$
Using the power rule

36. $\dfrac{1}{3} (4 \log_c x - 3 \log_c y - 2 \log_c z)$

37. $\log_a \sqrt[3]{\dfrac{x^6 y^3}{a^2 z^7}}$

$= \log_a \left(\dfrac{x^6 y^3}{a^2 z^7} \right)^{1/3}$

$= \dfrac{1}{3} \log_a \dfrac{x^6 y^3}{a^2 z^7}$ Using the power rule

$= \dfrac{1}{3} (\log_a x^6 y^3 - \log_a a^2 z^7)$ Using the quotient rule

$= \dfrac{1}{3} [\log_a x^6 + \log_a y^3 - (\log_a a^2 + \log_a z^7)]$
Using the product rule

$= \dfrac{1}{3} (\log_a x^6 + \log_a y^3 - \log_a a^2 - \log_a z^7)$
Removing parentheses

$= \dfrac{1}{3} (\log_a x^6 + \log_a y^3 - 2 - \log_a z^7)$
2 is the number to which
we raise a to get a^2.

$= \dfrac{1}{3} (6 \log_a x + 3 \log_a y - 2 - 7 \log_a z)$
Using the power rule

38. $\dfrac{1}{4} (8 \log_a x + 12 \log_a y - 3 - 5 \log_a z)$

39. $7 \log_a x + 3 \log_a z$

$= \log_a x^7 + \log_a z^3$ Using the power rule

$= \log_a x^7 z^3$ Using the product rule

40. $\log_b m^2 n^{1/2}$, or $\log_b m^2 \sqrt{n}$

41. $\log_a x^2 - 2 \log_a \sqrt{x}$

$= \log_a x^2 - \log_a (\sqrt{x})^2$ Using the power rule

$= \log_a x^2 - \log_a x$ $(\sqrt{x})^2 = x$

$= \log_a \dfrac{x^2}{x}$ Using the quotient rule

$= \log_a x$ Simplifying

42. $\log_a \dfrac{\sqrt{a}}{x}$

43. $\dfrac{1}{2} \log_a x + 5 \log_a y - 2 \log_a x$

$= \log_a x^{1/2} + \log_a y^5 - \log_a x^2$ Using the power rule

$= \log_a x^{1/2} y^5 - \log_a x^2$ Using the product rule

$= \log_a \dfrac{x^{1/2} y^5}{x^2}$ Using the quotient rule

The result can also be expressed as $\log_a \dfrac{\sqrt{x} y^5}{x^2}$ or as $\log_a \dfrac{y^5}{x^{3/2}}$.

44. $\log_a \dfrac{2x^4}{y^3}$

45.

$$\log_a(x^2 - 4) - \log_a(x + 2)$$

$$= \log_a \frac{x^2 - 4}{x + 2} \qquad \text{Using the quotient rule}$$

$$= \log_a \frac{(x + 2)(x - 2)}{x + 2}$$

$$= \log_a \frac{(x\!\!\!/\!+2)(x - 2)}{x\!\!\!/\!+2} \qquad \text{Simplifying}$$

$$= \log_a(x - 2)$$

46. $\log_a \dfrac{2}{x - 5}$

47.

$$\log_b 15 = \log_b(3 \cdot 5)$$

$$= \log_b 3 + \log_b 5 \qquad \text{Using the product rule}$$

$$= 0.792 + 1.161$$

$$= 1.953$$

48. 0.369

49.

$$\log_b \frac{3}{5} = \log_b 3 - \log_b 5 \qquad \text{Using the quotient rule}$$

$$= 0.792 - 1.161$$

$$= -0.369$$

50. -0.792

51.

$$\log_b \frac{1}{5} = \log_b 1 - \log_b 5 \qquad \text{Using the quotient rule}$$

$$= 0 - 1.161 \qquad (\log_b 1 = 0)$$

$$= -1.161$$

52. $\dfrac{1}{2}$

53. $\log_b \sqrt{b^3} = \log_b b^{3/2} = \dfrac{3}{2}$ 3/2 is the number to which we raise b to get $b^{3/2}$.

54. 1.792

55. $\log_b 6$

Since 6 cannot be expressed using the numbers 1, 3, and 5, we cannot find $\log_b 6$ using the given information.

56. 2.745

57.

$$\log_b 75$$

$$= \log_b(3 \cdot 5^2)$$

$$= \log_b 3 + \log_b 5^2 \qquad \text{Using the product rule}$$

$$= \log_b 3 + 2\log_b 5 \qquad \text{Using the power rule}$$

$$= 0.792 + 2(1.161)$$

$$= 3.114$$

58. Cannot be found

59. $\log_t t^9 = 9$ 9 is the power to which we raise t to get t^9.

60. 4

61. $\log_e e^m = m$ m is the power to which we raise e to get e^m.

62. -2

63. Writing exercise

64. Writing exercise

65. Graph $f(x) = \sqrt{x} - 3$.

We construct a table of values, plot points, and connect them with a smooth curve. Note that we must choose nonnegative values of x in order for $\sqrt{x}$ to be a real number.

x	$f(x)$
0	-3
1	-2
4	-1
9	0

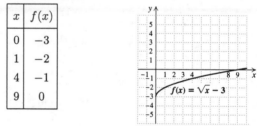

$f(x) = \sqrt{x} - 3$

66.

$f(x) = \sqrt{x} + 2$

67. Graph $g(x) = \sqrt[3]{x} + 1$.

We construct a table of values, plot points, and connect them with a smooth curve.

x	$g(x)$
-8	-1
-1	0
0	1
1	2
8	3

$g(x) = \sqrt[3]{x} + 1$

68.

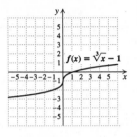

$f(x) = \sqrt[3]{x} - 1$

69. $(a^3b^2)^5(a^2b^7) = (a^{3 \cdot 5}b^{2 \cdot 5})(a^2b^7) =$
$a^{15}b^{10}a^2b^7 = a^{15+2}b^{10+7} = a^{17}b^{17}$

70. $x^{11}y^6z^8$

71. Writing exercise

72. Writing exercise

73. $\log_a (x^8 - y^8) - \log_a (x^2 + y^2)$

$= \log_a \dfrac{x^8 - y^8}{x^2 + y^2}$

$= \log_a \dfrac{(x^4 + y^4)(x^2 + y^2)(x + y)(x - y)}{x^2 + y^2}$

$= \log_a [(x^4 + y^4)(x^2 - y^2)]$ Simplifying

$= \log_a (x^6 - x^4y^2 + x^2y^4 - y^6)$

74. $\log_a (x^3 + y^3)$

75. $\log_a \sqrt{1 - s^2}$

$= \log_a (1 - s^2)^{1/2}$

$= \dfrac{1}{2} \log_a (1 - s^2)$

$= \dfrac{1}{2} \log_a [(1 - s)(1 + s)]$

$= \dfrac{1}{2} \log_a (1 - s) + \dfrac{1}{2} \log_a (1 + s)$

76. $\dfrac{1}{2} \log_a (c - d) - \dfrac{1}{2} \log_a (c + d)$

77. $\log_a \dfrac{\sqrt[3]{x^2 z}}{\sqrt[3]{y^2 z^{-2}}}$

$= \log_a \left(\dfrac{x^2 z^3}{y^2}\right)^{1/3}$

$= \dfrac{1}{3}(\log_a x^2 z^3 - \log_a y^2)$

$= \dfrac{1}{3}(2 \log_a x + 3 \log_a z - 2 \log_a y)$

$= \dfrac{1}{3}[2 \cdot 2 + 3 \cdot 4 - 2 \cdot 3]$

$= \dfrac{1}{3}(10)$

$= \dfrac{10}{3}$

78. -2

79. $\log_a x = 2$, so $a^2 = x$.

Let $\log_{1/a} x = n$ and solve for n.

$\log_{1/a} a^2 = n$ Substituting a^2 for x

$\left(\dfrac{1}{a}\right)^n = a^2$

$(a^{-1})^n = a^2$

$a^{-n} = a^2$

$-n = 2$

$n = -2$

Thus, $\log_{1/a} x = -2$ when $\log_a x = 2$.

80. False

81. True; $\log_a(Q + Q^2) = \log_a[Q(1 + Q)] = \log_a Q + \log_a(1 + Q) = \log_a Q + \log_a(Q + 1)$.

82. Graph $y_1 = \log x^2$ and $y_2 = \log x \cdot \log x$ and observe that the graphs do not coincide.

Exercise Set 12.5

1. 0.7782

2. 0.6990

3. 1.8621

4. 1.8686

5. Since $10^3 = 1000$, $\log 1000 = 3$.

6. 2

7. -0.2782

8. -0.3072

9. 1.7986

10. 1.9219

11. 199.5262

12. 2511.8864

13. 1.4894

14. 1.7660

15. 0.0011

16. 79,104.2833

17. 1.6094

18. 0.6931

19. 4.0431

20. 3.4012

21. -5.0832

22. -7.2225

23. 96.7583

24. 107.8516

25. 15.0293

26. 21.3276

27. 0.0305

28. 0.0714

29. 109.9472

30. 3.4212

31. We will use common logarithms for the conversion. Let $a = 10$, $b = 6$, and $M = 92$ and substitute in the change-of-base formula.

$$\log_b M = \frac{\log_a M}{\log_a b}$$

$$\log_6 92 = \frac{\log_{10} 92}{\log_{10} 6}$$

$$\approx \frac{1.963787827}{0.7781512504}$$

$$\approx 2.5237$$

32. 3.9656

33. We will use common logarithms for the conversion. Let $a = 10$, $b = 2$, and $M = 100$ and substitute in the change-of-base formula.

$$\log_2 100 = \frac{\log_{10} 100}{\log_{10} 2}$$

$$\approx \frac{2}{0.3010}$$

$$\approx 6.6439$$

34. 2.3666

35. We will use natural logarithms for the conversion. Let $a = e$, $b = 7$, and $M = 65$ and substitute in the change-of-base formula.

$$\log_7 65 = \frac{\ln 65}{\ln 7}$$

$$\approx \frac{4.1744}{1.9459}$$

$$\approx 2.1452$$

36. 2.3223

37. We will use natural logarithms for the conversion. Let $a = e$, $b = 0.5$, and $M = 5$ and substitute in the change-of-base formula.

$$\log_{0.5} 5 = \frac{\ln 5}{\ln 0.5}$$

$$\approx \frac{1.6094}{-0.6931}$$

$$\approx -2.3219$$

38. -0.4771

39. We will use common logarithms for the conversion. Let $a = 10$, $b = 2$, and $M = 0.2$ and substitute in the change-of-base formula.

$$\log_2 0.2 = \frac{\log_{10} 0.2}{\log_{10} 2}$$

$$\approx \frac{-0.6990}{0.3010}$$

$$\approx -2.3219$$

40. -3.6439

41. We will use natural logarithms for the conversion. Let $a = e$, $b = \pi$, and $M = 58$ and substitute in the change-of-base formula.

$$\log_\pi 58 = \frac{\ln 58}{\ln \pi}$$

$$\approx \frac{4.0604}{1.1447}$$

$$\approx 3.5471$$

42. 4.6284

43. Graph: $f(x) = e^x$

We find some function values with a calculator. We use these values to plot points and draw the graph.

x	e^x
0	1
1	2.7
2	7.4
3	20.1
-1	0.4
-2	0.1

The domain is the set of real numbers and the range is $(0, \infty)$.

44.

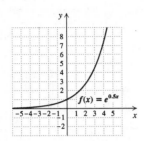

The domain is the set of real numbers and the range is $(0, \infty)$.

45. Graph: $f(x) = e^{-0.4x}$

We find some function values, plot points, and draw the graph.

x	$e^{-0.4x}$
0	1
1	0.67
2	0.45
−1	1.49
−2	2.23
−3	3.32
−4	4.95

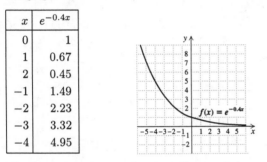

The domain is the set of real numbers and the range is $(0, \infty)$.

46.

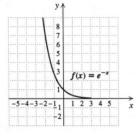

The domain is the set of real numbers and the range is $(0, \infty)$.

47. Graph: $f(x) = e^x + 1$

We find some function values, plot points, and draw the graph.

x	$e^x + 1$
0	2
1	3.72
2	8.39
−1	1.37
−2	1.14

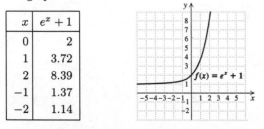

The domain is the set of real numbers and the range is $(1, \infty)$.

48.

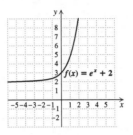

The domain is the set of real numbers and the range is $(2, \infty)$.

49. Graph: $f(x) = e^x - 2$

We find some function values, plot points, and draw the graph.

x	$e^x - 2$
0	−1
1	0.72
2	5.4
−1	−1.6
−2	−1.9

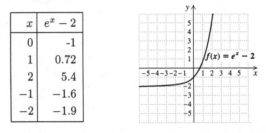

The domain is the set of real numbers and the range is $(-2, \infty)$.

50.

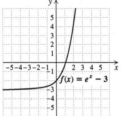

The domain is the set of real numbers and the range is $(-3, \infty)$.

51. Graph: $f(x) = 0.5e^x$

We find some function values, plot points, and draw the graph.

x	$0.5e^x$
0	0.5
1	1.36
2	3.69
−1	0.18
−2	0.07

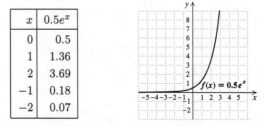

The domain is the set of real numbers and the range is $(0, \infty)$.

52.

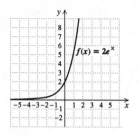

The domain is the set of real numbers and the range is $(0, \infty)$.

53. Graph: $f(x) = 2e^{-0.5x}$

We find some function values, plot points, and draw the graph.

x	$2e^{-0.5x}$
0	2
1	1.21
2	0.74
3	0.45
-1	3.30
-2	5.44
-3	8.96

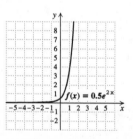

The domain is the set of real numbers and the range is $(0, \infty)$.

54.

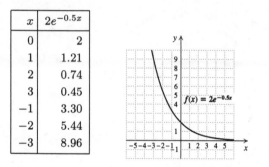

The domain is the set of real numbers and the range is $(0, \infty)$.

55. Graph: $f(x) = e^{x-2}$

We find some function values, plot points, and draw the graph.

x	e^{x-2}
0	0.14
2	1
4	7.39
-1	0.05
-2	0.02

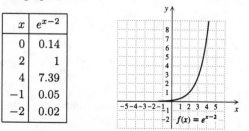

The domain is the set of real numbers and the range is $(0, \infty)$.

56.

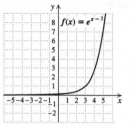

The domain is the set of real numbers and the range is $(0, \infty)$.

57. Graph: $f(x) = e^{x+3}$

We find some function values, plot points, and draw the graph.

x	e^{x+3}
0	20.09
1	54.60
-1	7.39
-3	1
-4	0.37

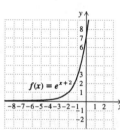

The domain is the set of real numbers and the range is $(0, \infty)$.

58.

The domain is the set of real numbers and the range is $(0, \infty)$.

59. Graph: $f(x) = 2 \ln x$

x	$2 \ln x$
0.5	-1.4
1	0
2	1.4
3	2.2
4	2.8
5	3.2
6	3.6

The domain is $(0, \infty)$ and the range is the set of real numbers.

60.

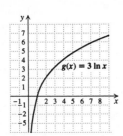

The domain is $(0, \infty)$ and the range is the set of real numbers.

61. Graph: $f(x) = 0.5 \ln x$

x	$0.5 \ln x$
0.5	−0.35
1	0
2	0.35
3	0.55
4	0.69
5	0.80

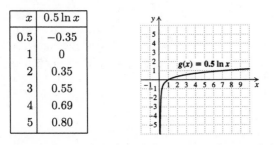

The domain is $(0, \infty)$ and the range is the set of real numbers.

62.

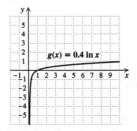

The domain is $(0, \infty)$ and the range is the set of real numbers.

63. Graph: $g(x) = \ln x + 3$

x	$\ln x + 3$
1	3
2	3.69
3	4.10
4	4.39
5	4.61

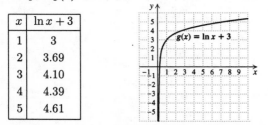

The domain is $(0, \infty)$ and the range is the set of real numbers.

64.

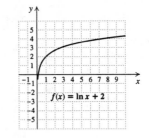

The domain is $(0, \infty)$ and the range is the set of real numbers.

65. Graph: $g(x) = \ln x - 2$

x	$\ln x - 2$
1	−2
2	−1.31
3	−0.90
4	−0.61
5	−0.39

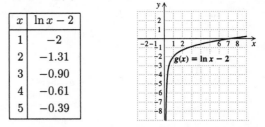

The domain is $(0, \infty)$ and the range is the set of real numbers.

66.

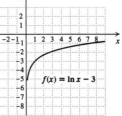

The domain is $(0, \infty)$ and the range is the set of real numbers.

67. Graph: $f(x) = \ln(x + 1)$

We find some function values, plot points, and draw the graph.

x	$\ln(x + 1)$
0	0
1	0.69
2	1.10
4	1.61
6	1.95
−1	Undefined

The domain is $(-1, \infty)$ and the range is the set of real numbers.

68.

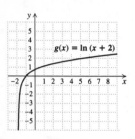

The domain is $(-2, \infty)$ and the range is the set of real numbers.

69. Graph: $g(x) = \ln(x - 3)$

We find some function values, plot points, and draw the graph.

x	$\ln(x-3)$
3.1	-2.30
4	0
5	0.69
6	1.10
7	1.39

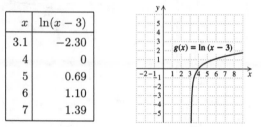

The domain is $(3, \infty)$ and the range is the set of real numbers.

70.

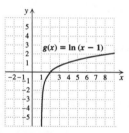

The domain is $(1, \infty)$ and the range is the set of real numbers.

71. Writing exercise

72. Writing exercise

73.
$$4x^2 - 25 = 0$$
$$(2x+5)(2x-5) = 0$$
$$2x+5 = 0 \quad \text{or} \quad 2x-5 = 0$$
$$2x = -5 \quad \text{or} \quad 2x = 5$$
$$x = -\frac{5}{2} \quad \text{or} \quad x = \frac{5}{2}$$
The solutions are $-\frac{5}{2}$ and $\frac{5}{2}$.

74. $0, \dfrac{7}{5}$

75.
$$17x - 15 = 0$$
$$17x = 15$$
$$x = \frac{15}{17}$$
The solution is $\dfrac{15}{17}$.

76. $\dfrac{9}{13}$

77. $x^{1/2} - 6x^{1/4} + 8 = 0$

Let $u = x^{1/4}$.
$$u^2 - 6u + 8 = 0 \quad \text{Substituting}$$
$$(u-4)(u-2) = 0$$
$$u = 4 \quad \text{or} \quad u = 2$$
$$x^{1/4} = 4 \quad \text{or} \quad x^{1/4} = 2$$
$$x = 256 \quad \text{or} \quad x = 16 \quad \text{Raising both sides to the fourth power}$$
Both numbers check. The solutions are 256 and 16.

78. $\dfrac{1}{4}$, 9

79. Writing exercise

80. Writing exercise

81. We use the change-of-base formula.
$$\log_6 81 = \frac{\log 81}{\log 6}$$
$$= \frac{\log 3^4}{\log(2 \cdot 3)}$$
$$= \frac{4 \log 3}{\log 2 + \log 3}$$
$$\approx \frac{4(0.477)}{0.301 + 0.477}$$
$$\approx 2.452$$

82. 1.262

83. We use the change-of-base formula.
$$\log_{12} 36 = \frac{\log 36}{\log 12}$$
$$= \frac{\log(2 \cdot 3)^2}{\log(2^2 \cdot 3)}$$
$$= \frac{2\log(2 \cdot 3)}{\log 2^2 + \log 3}$$
$$= \frac{2(\log 2 + \log 3)}{2\log 2 + \log 3}$$
$$\approx \frac{2(0.301 + 0.477)}{2(0.301) + 0.477}$$
$$\approx 1.442$$

84. $\ln M = \dfrac{\log M}{\log e}$

85. Use the change-of-base formula with $a = e$ and $b = 10$. We obtain

$$\log M = \frac{\ln M}{\ln 10}.$$

86. $\pm 6.0302 \times 10^{17}$

87. $\log(492x) = 5.728$

$10^{5.728} = 492x$

$\dfrac{10^{5.728}}{492} = x$

$1086.5129 \approx x$

88. 1.5893

89. $\log 692 + \log x = \log 3450$

$\log x = \log 3450 - \log 692$

$\log x = \log \dfrac{3450}{692}$

$x = \dfrac{3450}{692}$

$x \approx 4.9855$

90. (a) Domain: $\{x | x > 0\}$, or $(0, \infty)$; range: the set of real numbers;

(b) $[-3, 10, -100, 1000]$, Xscl $= 1$, Yscl $= 100$;

(c)

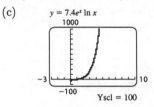
$y = 7.4e^x \ln x$

91. (a) Domain: $\{x | x > 0\}$, or $(0, \infty)$; range: $\{y | y < 0.5135\}$, or $(-\infty, 0.5135)$;

(b) $[-1, 5, -10, 5]$;

(c)

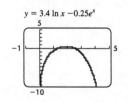
$y = 3.4 \ln x - 0.25e^x$

92. (a) Domain: $\{x | x > 2.1\}$, or $(2.1, \infty)$; range: the set of real numbers;

(b) $[-1, 10, -10, 20]$, Xscl $= 1$, Yscl $= 5$;

(c) $y = x \ln(x - 2.1)$

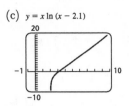

93. (a) Domain $\{x | x > 0\}$, or $(0, \infty)$; range: $\{y | y > -0.2453\}$, or $(-0.2453, \infty)$

(b) $[-1, 5, -1, 10]$;

(c)

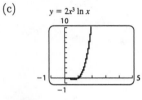
$y = 2x^3 \ln x$

94.

95.

96. Writing exercise

Exercise Set 12.6

1. $2^x = 16$

$2^x = 2^4$

$x = 4$ The exponents must be the same.

The solution is 4.

2. 3

3. $3^x = 27$

$3^x = 3^3$

$x = 3$ The exponents must be the same.

The solution is 3.

4. 3

5. $2^{x+3} = 32$

$2^{x+3} = 2^5$

$x + 3 = 5$

$x = 2$

The solution is 2.

6. 5

7. $5^{3x} = 625$

$5^{3x} = 5^4$

$3x = 4$

$x = \dfrac{4}{3}$

The solution is $\dfrac{4}{3}$.

8. $\dfrac{3}{2}$

9. $7^{4x} = 1$

Since $a^0 = 1 \; a \neq 0$, then $4x = 0$ and thus $x = 0$. The solution is 0.

10. 0

11. $4^{2x-1} = 64$

$4^{2x-1} = 4^3$

$2x - 1 = 3$

$2x = 4$

$x = 2$

The solution is 2.

12. $\dfrac{5}{2}$

13. $3^{x^2} \cdot 3^{3x} = 81$

$3^{x^2+3x} = 3^4$

$x^2 + 3x = 4$

$x^2 + 3x - 4 = 0$

$(x+4)(x-1) = 0$

$x = -4 \;\; or \;\; x = 1$

The solutions are -4 and 1.

14. $-3, \; -1$

15. $2^x = 15$

$\log 2^x = \log 15$

$x \log 2 = \log 15$

$x = \dfrac{\log 15}{\log 2}$

$x \approx 3.907$

The solution is $\log 15 / \log 2$, or approximately 3.907.

16. $\dfrac{\log 19}{\log 2} \approx 4.248$

17. $4^{x+1} = 13$

$\log 4^{x+1} = \log 13$

$(x + 1) \log 4 = \log 13$

$x + 1 = \dfrac{\log 13}{\log 4}$

$x = \dfrac{\log 13}{\log 4} - 1$

$x \approx 0.850$

The solution is $\log 13 / \log 4 - 1$ or approximately 0.850.

18. $\dfrac{\log 17}{\log 8} + 1 \approx 2.362$

19. $e^t = 100$

$\ln e^t = \ln 100$ Taking ln on both sides

$t = \ln 100$ Finding the logarithm of the base to a power

$t \approx 4.605$ Using a calculator

20. $\ln 1000 \approx 6.908$

21. $e^{-0.07t} + 3 = 3.08$

$e^{-0.07t} = 0.08$

$\ln e^{-0.07t} = \ln 0.08$ Taking ln on both sides

$-0.07t = \ln 0.08$ Finding the logarithm of the base to a power

$t = \dfrac{\ln 0.08}{-0.07}$

$t \approx 36.082$

22. $\dfrac{\ln 5}{0.03} \approx 53.648$

23. $2^x = 3^{x-1}$

$\log 2^x = \log 3^{x-1}$

$x \log 2 = (x - 1) \; \log 3$

$x \log 2 = x \log 3 - \log 3$

$\log 3 = x \log 3 - x \log 2$

$\log 3 = x(\log 3 - \log 2)$

$\dfrac{\log 3}{\log 3 - \log 2} = x$

$2.710 \approx x$

24. $\dfrac{\log 3}{\log 5 - \log 3} \approx 2.151$

25.
$$4^{x+1} = 5^x$$
$$\log 4^{x+1} = \log 5^x$$
$$(x+1)\log 4 = x \log 5$$
$$x \log 4 + \log 4 = x \log 5$$
$$\log 4 = x \log 5 - x \log 4$$
$$\log 4 = x(\log 5 - \log 4)$$
$$\frac{\log 4}{\log 5 - \log 4} = x$$
$$6.213 \approx x$$

26. $\dfrac{3 \log 2}{\log 7 - \log 2} \approx 1.660$

27. $7.2^x - 65 = 0$
$$7.2^x = 65$$
$$\log 7.2^x = \log 65$$
$$x \log 7.2 = \log 65$$
$$x = \frac{\log 65}{\log 7.2}$$
$$x \approx 2.115$$

28. $\dfrac{\log 87}{\log 4.9} \approx 2.810$

29. $\log_5 x = 3$
$$x = 5^3 \quad \text{Writing an equivalent} \atop \text{exponential equation}$$
$$x = 125$$

30. 81

31. $\log_4 x = \dfrac{1}{2}$
$$x = 4^{1/2} \quad \text{Writing an equivalent} \atop \text{exponential equation}$$
$$x = 2$$

32. $\dfrac{1}{8}$

33. $\log x = 3 \qquad$ The base is 10.
$$x = 10^3$$
$$x = 1000$$

34. 10

35. $2 \log x = -8$
$$\log x = -4 \qquad \text{The base is 10.}$$
$$x = 10^{-4}$$
$$x = \frac{1}{10,000}, \text{ or } 0.0001$$

36. $\dfrac{1}{10,000}$

37. $\ln x = 1$
$$x = e \approx 2.718$$

38. $e^2 \approx 7.389$

39. $5 \ln x = -15$
$$\ln x = -3$$
$$x = e^{-3} \approx 0.050$$

40. $e^{-1} \approx 0.368$

41. $\log_2(8 - 6x) = 5$
$$8 - 6x = 2^5$$
$$8 - 6x = 32$$
$$-6x = 24$$
$$x = -4$$
The answer checks. The solution is -4.

42. 66

43. $\log(x - 9) + \log x = 1 \qquad$ The base is 10.
$$\log_{10}[(x - 9)(x)] = 1 \quad \text{Using the product rule}$$
$$x(x - 9) = 10^1$$
$$x^2 - 9x = 10$$
$$x^2 - 9x - 10 = 0$$
$$(x + 1)(x - 10) = 0$$
$$x = -1 \text{ or } x = 10$$
Check: For -1:
$$\frac{\log(x - 9) + \log x = 1}{\log(-1 + 9) + \log(-1) \; ? \; 1} \qquad \text{FALSE}$$

For 10:
$$\frac{\log(x - 9) + \log x = 1}{\log(10 - 9) + \log(10) \; ? \; 1}$$
$$\log 1 + \log 10 \qquad\qquad$$
$$0 + 1 \qquad\qquad$$
$$1 \quad \Big| \quad 1 \qquad \text{TRUE}$$

The number -1 does not check, because negative numbers do not have logarithms. The solution is 10.

44. 1

45. $\log x - \log(x + 3) = 1$ The base is 10.

$$\log_{10} \frac{x}{x+3} = 1 \quad \text{Using the quotient rule}$$

$$\frac{x}{x+3} = 10^1$$

$$x = 10(x + 3)$$

$$x = 10x + 30$$

$$-9x = 30$$

$$x = -\frac{10}{3}$$

The number $-\dfrac{10}{3}$ does not check. The equation has no solution.

46. $\dfrac{7}{9}$

47. $\log_4(x + 3) - \log_4(x - 5) = 2$

$$\log_4 \frac{x+3}{x-5} = 2 \quad \text{Using the}$$
$$\text{quotient rule}$$

$$\frac{x+3}{x-5} = 4^2$$

$$\frac{x+3}{x-5} = 16$$

$$x + 3 = 16(x - 5)$$

$$x + 3 = 16x - 80$$

$$83 = 15x$$

$$\frac{83}{15} = x$$

The number $\dfrac{83}{15}$ checks. It is the solution.

48. 5

49. $\log_7(x + 2) + \log_7(x + 1) = \log_7 6$

$$\log_7[(x + 2)(x + 1)] = \log_7 6 \quad \text{Using the}$$
$$\text{product rule}$$

$$\log_7(x^2 + 3x + 2) = \log_7 6$$

$$x^2 + 3x + 2 = 6 \quad \text{Using the}$$
$$\text{property of logarithmic equality}$$

$$x^2 + 3x - 4 = 0$$

$$(x + 4)(x - 1) = 0$$

$$x = -4 \quad or \quad x = 1$$

The number 1 checks, but -4 does not. The solution is 1.

50. 2

51. $\log_3(x + 4) + \log_3(x - 4) = 2$

$$\log_3[(x + 4)(x - 4)] = 2$$

$$(x + 4)(x - 4) = 3^2$$

$$x^2 - 16 = 9$$

$$x^2 = 25$$

$$x = \pm 5$$

The number 5 checks, but -5 does not. The solution is 5.

52. 4

53. $\log_{12}(x + 5) - \log_{12}(x - 4) = \log_{12} 3$

$$\log_{12} \frac{x+5}{x-4} = \log_{12} 3$$

$$\frac{x+5}{x-4} = 3 \quad \text{Using the prop-}$$
$$\text{erty of logarithmic}$$
$$\text{equality}$$

$$x + 5 = 3(x - 4)$$

$$x + 5 = 3x - 12$$

$$17 = 2x$$

$$\frac{17}{2} = x$$

The number $\dfrac{17}{2}$ checks and is the solution.

54. $\dfrac{17}{4}$

55. $\log_2(x - 2) + \log_2 x = 3$

$$\log_2[(x - 2)(x)] = 3$$

$$x(x - 2) = 2^3$$

$$x^2 - 2x = 8$$

$$x^2 - 2x - 8 = 0$$

$$(x - 4)(x + 2) = 0$$

$$x = 4 \quad or \quad x = -2$$

The number 4 checks, but -2 does not. The solution is 4.

56. $\dfrac{2}{5}$

57. Writing exercise

58. Writing exercise

59. $y = kx$

$$7.2 = k(0.8) \quad \text{Substituting}$$

$$9 = k \quad \text{Variation constant}$$

$$y = 9x \quad \text{Equation of variation}$$

60. $y = \dfrac{21.35}{x}$

61.

$$T = 2\pi \sqrt{\frac{L}{32}}$$

$$\frac{T}{2\pi} = \sqrt{\frac{L}{32}}$$

$$\left(\frac{T}{2\pi}\right)^2 = \left(\sqrt{\frac{L}{32}}\right)^2$$

$$\frac{T^2}{4\pi^2} = \frac{L}{32}$$

$$32 \cdot \frac{T^2}{4\pi^2} = L$$

$$\frac{8T^2}{\pi^2} = L$$

62. $c = \sqrt{\dfrac{E}{m}}$

63. *Familiarize*. Let $t =$ the time, in hours, it takes Joni and Miles to key in the score, working together. Then in t hours Joni does $\dfrac{t}{2}$ of the job, Miles does $\dfrac{t}{3}$, and together they do 1 entire job.

Translate.

$$\frac{t}{2} + \frac{t}{3} = 1$$

Carry out. We solve the equation. First we multiply by the LCD, 6.

$$6\left(\frac{t}{2} + \frac{t}{3}\right) = 6 \cdot 1$$

$$6 \cdot \frac{t}{2} + 6 \cdot \frac{t}{3} = 6$$

$$3t + 2t = 6$$

$$5t = 6$$

$$t = \frac{6}{5}$$

Check. In $\dfrac{6}{5}$ hr Joni does $\dfrac{6/5}{2}$, or $\dfrac{3}{5}$ of the job, and Miles does $\dfrac{6/5}{3}$, or $\dfrac{2}{5}$ of the job. Together they do $\dfrac{3}{5} + \dfrac{2}{5}$ or 1 entire job. The answer checks.

State. It takes Joni and Miles $\dfrac{6}{5}$ hr, or $1\dfrac{1}{5}$ hr, to do the job, working together.

64. $9\dfrac{3}{8}$ min

65. Writing exercise

66. Writing exercise

67.

$$100^{3x} = 1000^{2x+1}$$

$$(10^2)^{3x} = (10^3)^{2x+1}$$

$$10^{6x} = 10^{6x+1}$$

$$6x = 6x + 1$$

$$0 = 1$$

We get a false equation, so the equation has no solution.

68. $\dfrac{12}{5}$

69.

$$8^x = 16^{3x+9}$$

$$(2^3)^x = (2^4)^{3x+9}$$

$$2^{3x} = 2^{12x+36}$$

$$3x = 12x + 36$$

$$-36 = 9x$$

$$-4 = x$$

The solution is -4.

70. $\sqrt[3]{3}$

71.

$$\log_6 (\log_2 x) = 0$$

$$\log_2 x = 6^0$$

$$\log_2 x = 1$$

$$x = 2^1$$

$$x = 2$$

The solution is 2.

72. -1

73.

$$\log_5 \sqrt{x^2 - 9} = 1$$

$$\sqrt{x^2 - 9} = 5^1$$

$$(\sqrt{x^2 - 9})^2 = 5^2$$

$$x^2 - 9 = 25$$

$$x^2 = 34$$

$$x = \pm\sqrt{34}$$

The solutions are $\pm\sqrt{34}$.

74. $-3, -1$

75.

$$\log (\log x) = 5$$

$$\log x = 10^5$$

$$\log x = 100,000$$

$$x = 10^{100,000}$$

The solution is $10^{100,000}$.

76. $-625, 625$

77. $\log x^2 = (\log x)^2$

$2 \log x = (\log x)^2$

$0 = (\log x)^2 - 2 \log x$

Let $u = \log x$.

$0 = u^2 - 2u$

$0 = u(u - 2)$

$u = 0 \quad or \quad u = 2$

$\log x = 0 \quad or \quad \log x = 2$ Replacing u with $\log x$

$x = 10^0 \quad or \quad x = 10^2$

$x = 1 \quad or \quad x = 100$

Both numbers check. The solutions are 1 and 100.

78. $\dfrac{1}{2}$, 5000

79. $\log x^{\log x} = 25$

$\log x (\log x) = 25$ Using the power rule

$(\log x)^2 = 25$

$\log x = \pm 5$

$x = 10^5 \qquad or \quad x = 10^{-5}$

$x = 100{,}000 \quad or \quad x = \dfrac{1}{100{,}000}$

Both numbers check. The solutions are 100,000 and $\dfrac{1}{100{,}000}$.

80. 1, $\dfrac{\log 5}{\log 3} \approx 1.465$

81. $(81^{x-2})(27^{x+1}) = 9^{2x-3}$

$[(3^4)^{x-2}][(3^3)^{x+1}] = (3^2)^{2x-3}$

$(3^{4x-8})(3^{3x+3}) = 3^{4x-6}$

$3^{7x-5} = 3^{4x-6}$

$7x - 5 = 4x - 6$

$3x = -1$

$x = -\dfrac{1}{3}$

The solution is $-\dfrac{1}{3}$.

82. $\dfrac{3}{2}$

83. $2^y = 16^{x-3} \quad$ and $\quad 3^{y+2} = 27^x$

$2^y = (2^4)^{x-3} \quad$ and $\quad 3^{y+2} = (3^3)^x$

$y = 4x - 12 \quad$ and $\quad y + 2 = 3x$

$12 = 4x - y \quad$ and $\qquad 2 = 3x - y$

Solving this system of equations we get $x = 10$ and $y = 28$. Then $x + y = 10 + 28 = 38$.

84. -3

85. Find the first coordinate of the point of intersection of $y_1 = \ln x$ and $y_2 = \log x$. The value of x for which the natural logarithm of x is the same as the common logarithm of x is 1.

86.

Exercise Set 12.7

1. a) Replace $N(t)$ with 200 and solve for t.

$N(t) = 112(1.37)^t$

$200 = 112(1.37)^t$

$1.7857 \approx (1.37)^t \qquad$ Dividing by 112

$\ln 1.7857 \approx \ln(1.37)^t \quad$ Taking the natural
logarithm on both sides

$\ln 1.7857 \approx t \ln 1.37$

$\dfrac{\ln 1.7857}{\ln 1.37} \approx t$

$2 \approx t$

200 million cellular phones would be in use 2 yr after 2001, or in 2003.

b) Replace $N(t)$ with 2(112), or 224, and solve for t.

$224 = 112(1.37)^t$

$2 = (1.37)^t$

$\ln 2 = \ln(1.37)^t$

$\ln 2 = t \ln(1.37)$

$\dfrac{\ln 2}{\ln 1.37} = t$

$2.2 \approx t$

The doubling time is about 2.2 years.

2. a) 13.5 yr; **b)** 5.7 yr

3. a) Replace $A(t)$ with 40,000 and solve for t.

$A(t) = 29{,}000(1.08)^t$

$40{,}000 = 29{,}000(1.08)^t$

$1.379 \approx (1.08)^t$

$\log 1.379 \approx \log(1.08)^t$

$\log 1.379 \approx t \log 1.08$

$\dfrac{\log 1.379}{\log 1.08} \approx t$

$4.2 \approx t$

The amount due will reach \$40,000 after about 4.2 years.

b) Replace $A(t)$ with $2(29{,}000)$, or $58{,}000$, and solve for t.

$$58{,}000 = 29{,}000(1.08)^t$$
$$2 = (1.08)^t$$
$$\log 2 = \log(1.08)^t$$
$$\log 2 = t \log 1.08$$
$$\frac{\log 2}{\log 1.08} = t$$
$$9.0 \approx t$$

The doubling time is about 9.0 years.

4. a) 3.6 days; b) 0.6 days

5. a) Find $N(41)$.

$$N(x) = 600(0.873)^{x-16}$$

$$N(41) = 600(0.873)^{41-16}$$
$$= 600(0.873)^{25}$$
$$\approx 20.114$$

There are about 20.114 thousand, or 20,114, 41-yr-old skateboarders.

b) Substitute 2 for $N(x)$ and solve for x. (Remember that $N(x)$ is in thousands.)

$$2 = 600(0.873)^{x-16}$$
$$0.0033 \approx (0.873)^{x-16}$$
$$\log 0.0033 \approx (x-16)\log 0.873$$
$$\frac{\log 0.0033}{\log 0.873} \approx x - 16$$
$$\frac{\log 0.0033}{\log 0.873} + 16 \approx x$$
$$58 \approx x$$

There are only 2000 skateboarders at age 58.

6. a) 3.5 yr; b) 13.6 yr

7. $\text{pH} = -\log[H^+]$
$$= -\log[1.3 \times 10^{-5}]$$
$$\approx -(-4.886057) \quad \text{Using a calculator}$$
$$\approx 4.9$$

The pH of fresh-brewed coffee is about 4.9.

8. 6.8

9. $\text{pH} = -\log[H^+]$
$$7.0 = -\log[H^+]$$
$$-7.0 = \log[H^+]$$
$$10^{-7.0} = [H^+] \quad \text{Converting to an exponential equation}$$

The hydrogen ion concentration is 10^{-7} moles per liter.

10. 1.58×10^{-8} moles per liter

11. $L = 10 \cdot \log \dfrac{I}{I_0}$
$$= 10 \cdot \log \frac{3.2 \times 10^{-6}}{10^{-12}}$$
$$= 10 \cdot \log(3.2 \times 10^6)$$
$$\approx 10(6.5)$$
$$\approx 65$$

The intensity of sound in normal conversation is about 65 decibels.

12. 95 dB

13.
$$L = 10 \cdot \log \frac{I}{I_0}$$
$$105 = 10 \cdot \log \frac{I}{10^{-12}}$$
$$10.5 = \log \frac{I}{10^{-12}}$$
$$10.5 = \log I - \log 10^{-12} \quad \text{Using the quotient rule}$$
$$10.5 = \log I - (-12) \quad (\log 10^a = a)$$
$$10.5 = \log I + 12$$
$$-1.5 = \log I$$
$$10^{-1.5} = I \quad \text{Converting to an exponential equation}$$
$$3.2 \times 10^{-2} \approx I$$

The intensity of the sound is $10^{-1.5}$ W/m^2, or about 3.2×10^{-2} W/m^2.

14. $10^{-0.9}$ W/m^2, or 1.3×10^{-1} W/m^2

15. a) Substitute 0.06 for k:

$$P(t) = P_0 e^{0.06t}$$

b) To find the balance after one year, replace P_0 with 5000 and t with 1. We find $P(1)$:

$$P(1) = 5000 \, e^{0.06(1)} = 5000 \, e^{0.06} \approx$$
$$5000(1.061836547) \approx \$5309.18$$

To find the balance after 2 years, replace P_0 with 5000 and t with 2. We find $P(2)$:

$$P(2) = 5000 \, e^{0.06(2)} = 5000 \, e^{0.12} \approx$$
$$5000(1.127496852) \approx \$5637.48$$

c) To find the doubling time, replace P_0 with 5000 and $P(t)$ with 10,000 and solve for t.

$10,000 = 5000 \, e^{0.06t}$

$2 = e^{0.06t}$

$\ln 2 = \ln \, e^{0.06t}$ Taking the natural loga-
                               rithm on both sides

$\ln 2 = 0.06t$ Finding the logarithm of
                    the base to a power

$\dfrac{\ln 2}{0.06} = t$

$11.6 \approx t$

The investment will double in about 11.6 years.

16. a) $P(t) = P_0 e^{0.05t}$

b) $1051.27; $1105.17

c) 13.9 years

17. a) $P(t) = 283.75e^{0.013t}$, where $P(t)$ is in millions
and t is the number of years after 2000.

b) In 2005, $t = 2005 - 2000 = 5$. Replace t with 5
and compute $P(5)$.

$P(5) = 283.75e^{0.013(5)}$

$= 283.75e^{0.065}$

≈ 302.81

The U.S. population in 2005 will be about 302.81
million.

c) Replace $P(t)$ with 325 and solve for t.

$325 = 283.75e^{0.013t}$

$1.1454 \approx e^{0.031t}$

$\ln 1.1454 \approx \ln e^{0.013t}$

$\ln 1.1454 \approx 0.013t$

$\dfrac{\ln 1.1454}{0.013} \approx t$

$10 \approx t$

The U.S. population will reach 325 million about
10 years after 2000, or in 2010.

18. a) $P(t) = 6.1e^{0.014t}$, where $P(t)$ is in billions and t
is the numbers of years after 2001.

b) 6.5 billion

c) 2020

19. a) Replace $N(t)$ with 60,000 and solve for t.

$60,000 = 3000(2)^{t/20}$

$20 = (2)^{t/20}$

$\log 20 = \log(2)^{t/20}$

$\log 20 = \dfrac{t}{20} \log 2$

$20 \log 20 = t \log 2$

$\dfrac{20 \log 20}{\log 2} = t$

$86.4 \approx t$

There will be 60,000 bacteria after about 86.4
minutes.

b) Replace $N(t)$ with 100,000,000 and solve for t.

$100,000,000 = 3000(2)^{t/20}$

$33,333.333 = (2)^{t/20}$

$\log 33,333.333 = \log(2)^{t/20}$

$\log 33,333.333 = \dfrac{t}{20} \log 2$

$20 \log 33,333.333 = t \log 2$

$\dfrac{20 \log 33,333.333}{\log 2} = t$

$300.5 \approx t$

About 300.5 minutes would have to pass in order
for a possible infection to occur.

c) Replace $P(t)$ with 6000 and solve for t.

$6000 = 3000(2)^{t/20}$

$2 = (2)^{t/20}$

$1 = \dfrac{t}{20}$ The exponents must
                        be the same.

$t = 20$

The doubling time is 20 minutes.

20. 19.8 years

21. a) Replace a with 1 and compute $N(1)$.

$N(a) = 2000 + 500 \log a$

$N(1) = 2000 + 500 \log 1$

$N(1) = 2000 + 500 \cdot 0$

$N(1) = 2000$

2000 units were sold after $1000 was spent.

b) Find $N(8)$.

$N(8) = 2000 + 500 \log 8$

$N(8) \approx 2451.5$

About 2452 units were sold after $8000 was
spent.

c) Using the values we computed in parts (a) and (b) and any others we wish to calculate, we sketch the graph:

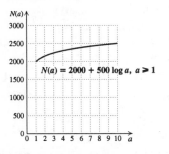

$N(a) = 2000 + 500 \log a, \ a \geqslant 1$

d) Replace $N(a)$ with 5000 and solve for a.

$$5000 = 2000 + 500 \log a$$
$$3000 = 500 \log a$$
$$6 = \log a$$
$$a = 10^6 = 1{,}000{,}000$$

$1{,}000{,}000 thousand, or \$1{,}000{,}000{,}000$ would have to be spent.

22. a) 68%; b) 54%, 40%;

c)

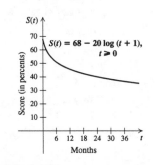

$S(t) = 68 - 20 \log (t + 1)$,
$t \geqslant 0$

d) 6.9 months

23. a) We use the growth equation $N(t) = N_0 e^{kt}$, where t is the number of years since 1995. In 1995, at $t = 0$, 17 people were infected. We substitute 17 for N_0:

$$N(t) = 17 e^{kt}.$$

To find the exponential growth rate k, observe that 1 year later 29 people were infected.

$$N(1) = 17 e^{k \cdot 1} \quad \text{Substituting 1 for } t$$
$$29 = 17 e^{k} \quad \text{Substituting 29 for } N(1)$$
$$1.706 \approx e^{k}$$
$$\ln 1.706 \approx \ln e^{k}$$
$$\ln 1.706 \approx k$$
$$0.534 \approx k$$

The exponential function is $N(t) = 17 e^{0.534t}$, where t is the number of years since 1995.

b) In 2001, $t = 2001 - 1995$, or 6. Find $N(6)$.

$$N(6) = 17 e^{0.534(6)}$$
$$= 17 e^{3.204}$$
$$\approx 418.7$$

Approximately 419 people will be infected in 2001.

24. a) $N(t) = 1418 \, e^{0.036t}$, where t is the number of years after 1987;

b) about 3488

25. We start with the exponential growth equation

$$D(t) = D_0 \, e^{kt}, \text{ where } t \text{ is the number of}$$

years after 1995.

Substitute $2 D_0$ for $D(t)$ and 0.1 for k and solve for t.

$$2 D_0 = D_0 \, e^{0.1t}$$
$$2 = e^{0.1t}$$
$$\ln 2 = 0.1t$$
$$\ln 2 = \ln e^{0.1t}$$
$$\frac{\ln 2}{0.1} = t$$
$$6.9 \approx t$$

The demand will be double that of 1995 in $1995 + 7$, or 2002.

26. 2013

27. a) We use the exponential decay equation $W(t) = W_0 e^{-kt}$, where t is the number of years after 1996 and $W(t)$ is in millions of tons. In 1996, at $t = 0$, 17.5 million tons of yard waste were discarded. We substitute 17.5 for W_0.

$$W(t) = 17.5 e^{-kt}.$$

To find the exponential decay rate k, observe that 2 years after 1996, in 1998, 14.5 million tons of yard waste were discarded. We substitute 2 for t and 14.5 for $W(t)$.

$$14.5 = 17.5 e^{-k \cdot 2}$$
$$0.8286 \approx e^{-2k}$$
$$\ln 0.8286 \approx \ln e^{-2k}$$
$$\ln 0.8286 \approx -2k$$
$$\frac{\ln 0.8286}{-2} \approx k$$
$$0.094 \approx k$$

Then we have $W(t) = 17.5 e^{-0.094t}$, where t is the number of years after 1996 and $W(t)$ is in millions of tons.

b) In 2006, $t = 2006 - 1996 = 10$.
$$W(10) = 17.5e^{-0.094(10)}$$
$$= 17.5e^{-0.94}$$
$$\approx 6.8$$

In 2006, about 6.8 million tons of yard waste were discarded.

c) 1 ton is equivalent to 0.000001 million tons.
$$0.000001 = 17.5e^{-0.094t}$$
$$5.71 \times 10^{-8} \approx e^{-0.094t}$$
$$\ln(5.71 \times 10^{-8}) \approx \ln e^{-0.094t}$$
$$\ln(5.71 \times 10^{-8}) \approx -0.094t$$
$$\frac{\ln(5.71 \times 10^{-8})}{-0.094} \approx t$$
$$177 \approx t$$

Only one ton of yard waste will be discarded about 177 years after 1996, or in 2173.

28. a) $k \approx 0.315$; $M(t) = 5300e^{-0.315t}$, where t is the number of years after 1990

b) 64

c) 2017

29. We will use the function derived in Example 7:
$$P(t) = P_0 e^{-0.00012t}$$

If the scrolls had lost 22.3% of their carbon-14 from an initial amount P_0, then 77.7%(P_0) is the amount present. To find the age t of the scrolls, we substitute 77.7%(P_0), or $0.777P_0$, for $P(t)$ in the function above and solve for t.
$$0.777P_0 = P_0 e^{-0.00012t}$$
$$0.777 = e^{-0.00012t}$$
$$\ln 0.777 = \ln e^{-0.00012t}$$
$$-0.2523 \approx -0.00012t$$
$$t \approx \frac{-0.2523}{-0.00012} \approx 2103$$
The scrolls are about 2103 years old.

30. 1654 yr

31. The function $P(t) = P_0 e^{-kt}$, $k > 0$, can be used to model decay. For iodine-131, $k = 9.6\%$, or 0.096. To find the half-life we substitute 0.096 for k and $\frac{1}{2} P_0$ for $P(t)$, and solve for t.

$$\frac{1}{2} P_0 = P_0 e^{-0.096t}, \text{ or } \frac{1}{2} = e^{-0.096t}$$
$$\ln \frac{1}{2} = \ln e^{-0.096t} = -0.096t$$
$$t = \frac{\ln 0.5}{-0.096} \approx \frac{-0.6931}{-0.096} \approx 7.2 \text{ days}$$

32. 11 years

33. The function $P(t) = P_0 e^{-kt}$, $k > 0$, can be used to model decay. We substitute $\frac{1}{2} P_0$ for $P(t)$ and 1 for t and solve for the decay rate k.
$$\frac{1}{2} P_0 = P_0 e^{-k \cdot 1}$$
$$\frac{1}{2} = e^{-k}$$
$$\ln \frac{1}{2} = \ln e^{-k}$$
$$-0.693 \approx -k$$
$$0.693 \approx k$$

The decay rate is 0.693, or 69.3% per year.

34. 3.15% per year

35. a) We start with the exponential growth equation
$$V(t) = V_0 e^{kt}, \text{ where } t \text{ is the number}$$
of years after 1996.

Substituting 640,500 for V_0, we have
$$V(t) = 640,500 e^{kt}.$$

To find the exponential growth rate k, observe that the card sold for $1.1 million, or $1,100,000 in 2000, or 4 years after 1996. We substitute and solve for k.
$$V(5) = 640,500 e^{k \cdot 4}$$
$$1,100,000 = 640,500 e^{4k}$$
$$1.7174 \approx e^{4k}$$
$$\ln 1.7174 \approx \ln e^{4k}$$
$$\ln 1.7174 \approx 4k$$
$$\frac{\ln 1.7174}{4} \approx k$$
$$0.135 \approx k$$

Thus, the exponential growth function is $V(t) = 640,500e^{0.135t}$, where t is the number of years after 1996.

b) In 2006, $t = 2006 - 1996 = 10$
$$V(10) = 640,500e^{0.135(10)} \approx 2,470,681$$

The card's value in 2006 will be about $2.47 million

c) Substitute 2($640,500), or $1,281,000 for $V(t)$ and solve for t.

$$1,281,000 = 640,500\,e^{0.135t}$$
$$2 = e^{0.135t}$$
$$\ln 2 = \ln e^{0.135t}$$
$$\ln 2 = 0.135t$$
$$\frac{\ln 2}{0.135} = t$$
$$5.1 \approx t$$

The doubling time is about 5.1 years.

d) Substitute $2,000,000 for $V(t)$ and solve for t.

$$2,000,000 = 640,500\,e^{0.135t}$$
$$3.1226 \approx e^{0.135t}$$
$$\ln 3.1226 \approx \ln e^{0.135t}$$
$$\ln 3.1226 \approx 0.135t$$
$$\frac{\ln 3.1226}{0.135} \approx t$$
$$8 \approx t$$

The value of the card will first exceed $2,000,000 about 8 years after 1996, or in 2004.

36. a) $k \approx 0.117$; $V(t) = 58e^{0.117t}$, where t is the number of years after 1987 and $V(t)$ is in millions of dollars

b) $602.1 million

c) 5.9 yr

d) 24.3 yr

37. Writing exercise

38. Writing exercise

39. Graph $y = x^2 - 8x$.

First we find the vertex.

$$-\frac{b}{2a} = -\frac{-8}{2 \cdot 1} = 4$$

When $x = 4$, $y = 4^2 - 8 \cdot 4 = 16 - 32 = -16$.

The vertex is $(4, -16)$ and the axis of symmetry is $x = 4$. We plot a few points on either side of the vertex and graph the parabola.

x	y
4	-16
0	0
2	-12
5	-15
6	-12

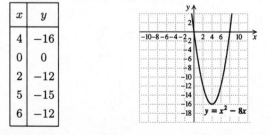

40.

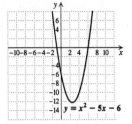

41. Graph $f(x) = 3x^2 - 5x - 1$

First we find the vertex.

$$-\frac{b}{2a} = -\frac{-5}{2 \cdot 3} = \frac{5}{6}$$

$$f\left(\frac{5}{6}\right) = 3\left(\frac{5}{6}\right)^2 - 5 \cdot \frac{5}{6} - 1 = -\frac{37}{12}$$

The vertex is $\left(\frac{5}{6}, -\frac{37}{12}\right)$ and the axis of symmetry is $x = \frac{5}{6}$. We plot a few points on either side of the vertex and graph the parabola.

x	$f(x)$
$\dfrac{5}{6}$	$-\dfrac{37}{12}$
0	-1
-1	7
2	1
3	11

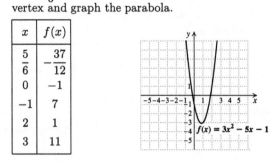

42.

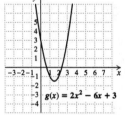

43.
$$x^2 - 8x = 7$$
$$x^2 - 8x + 16 = 7 + 16 \qquad \text{Adding } \left[\frac{1}{2}(-8)\right]^2$$
$$(x - 4)^2 = 23$$
$$x - 4 = \pm\sqrt{23}$$
$$x = 4 \pm \sqrt{23}$$

The solutions are $4 \pm \sqrt{23}$.

44. $-5 \pm \sqrt{31}$

45. Writing exercise

46. Writing exercise

47. We will use the exponential growth equation $V(t) = V_0 e^{kt}$, where t is the number of years after 2001 and $V(t)$ is in millions of dollars. We substitute 21 for $V(t)$, 0.05 for k, and 9 for t and solve for V_0.

$$21 = V_0 e^{0.05(9)}$$
$$21 = V_0 e^{0.45}$$
$$\frac{21}{e^{0.45}} = V_0$$
$$13.4 \approx V_0$$

George Steinbrenner needs to invest \$13.4 million at 5% interest compounded continuously in order to have \$21 million to pay Derek Jeter in 2010.

48. $(6, \$403)$

49. Use the function in Exercise 1 to find the number of cellular phones in use in 2000:

$$N(-1) = 112(1.37)^{-1} \approx 81.75182482 \text{ million}$$

Use the function found in Exercise 17 to find the U.S. population in 2001:

$$P(1) = 283.75 e^{0.013(1)} \approx 287.4628311 \text{ million}$$

Find the percentage of U.S. residents owning a cellular phone in 2000:

$$\frac{81.75182482}{238.75} \approx 0.288112158 \approx 28.8112158\%$$

Find the percentage of U.S. residents owning a cellular phone in 2001:

$$\frac{112}{287.4628311} \approx 0.3896155881 \approx 38.96155881\%$$

Now find an equation that models the percentage of U.S. residents owning a cellular phone t years after 2001:

$$28.8112158 = 38.96155881 e^{k(-1)}$$
$$0.7394780055 \approx e^{-k}$$
$$\ln 0.7394780055 \approx \ln e^{-k}$$
$$\ln 0.7394780055 \approx -k$$
$$-\ln 0.7394780055 \approx k$$
$$0.302 \approx k$$

A function that can be used is

$$P(t) = 39 e^{0.302t},$$

where t is the number of years after 2001 and $P(t)$ is a percent. (This assumes that each resident owns no more than one cellular phone.)

50. Writing exercise

51. First we find k. When $t = 24{,}360$, $P(t) = 0.5P_0$.

$$0.5P_0 = P_0 e^{-k \cdot 24,360}$$
$$0.5 = e^{-24,360k}$$
$$\ln 0.5 = \ln e^{-24,360k}$$
$$\ln 0.5 = -24,360k$$
$$\frac{\ln 0.5}{-24,360} = k$$
$$0.0000285 \approx k$$

Now we have a function for the decay of plutonium-239.

$$P(t) = P_0 e^{-0.0000285t}$$

If a fuel rod has lost 90% of its plutonium, then 10% of the initial amount is still present. We substitute and solve for t.

$$0.1P_0 = P_0 e^{-0.0000285t}$$
$$0.1 = e^{-0.0000285t}$$
$$\ln 0.1 = \ln e^{-0.0000285t}$$
$$\ln 0.1 = -0.0000285t$$
$$\frac{\ln 0.1}{-0.0000285} = t$$
$$80,792 \approx t$$

It will take about 80,792 yr for the fuel rod of plutonium -239 to lose 90% of its radioactivity.

Chapter 13

Conic Sections

Exercise Set 13.1

1. $y = -x^2$

   a) This is equivalent to $y = -(x - 0)^2 + 0$. The vertex is $(0, 0)$.

   b) We choose some x-values on both sides of the vertex and compute the corresponding values of y. The graph opens down, because the coefficient of x^2, -1, is negative.

x	y
0	0
1	-1
2	-4
-1	-1
-2	-4

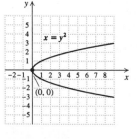

2.

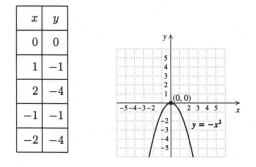

3. $y = -x^2 + 4x - 5$

   a) We can find the vertex by computing the first coordinate, $x = -b/2a$, and then substituting to find the second coordinate:

$$x = -\frac{b}{2a} = -\frac{4}{2(-1)} = 2$$

$$y = -x^2 + 4x - 5 = -(2)^2 + 4(2) - 5 = -1$$

The vertex is $(2, -1)$.

b) We choose some x-values and compute the corresponding values for y. The graph opens downward because the coefficient of x^2, -1, is negative.

x	y
2	-1
3	-2
4	-5
1	-2
0	-5

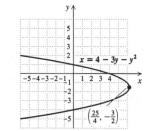

4.

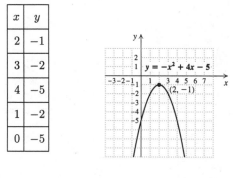

5. $x = y^2 - 4y + 1$

   a) We find the vertex by completing the square.

$$x = (y^2 - 4y + 4) + 1 - 4$$

$$x = (y - 2)^2 - 3$$

The vertex is $(-3, 2)$.

b) To find ordered pairs, we choose values for y and compute the corresponding values of x. The graph opens to the right, because the coefficient of y^2, 1, is positive.

x	y
6	-1
1	0
-2	1
-3	2
-2	3

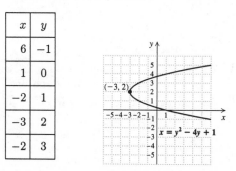

6.

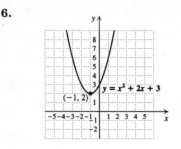

7. $x = y^2 + 1$

a) $x = (y - 0)^2 + 1$

The vertex is $(1, 0)$.

b) To find the ordered pairs, we choose y-values and compute the corresponding values for x. The graph opens to the right, because the coefficient of y^2, 1, is positive.

x	y
1	0
2	1
5	2
2	−1
5	−2

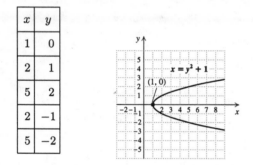

8.

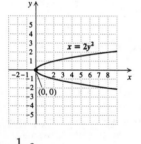

9. $x = -\dfrac{1}{2}y^2$

a) $x = -\dfrac{1}{2}(y - 0)^2 + 0$

The vertex is $(0, 0)$.

b) We choose y-values and compute the corresponding values for x. The graph opens to the left, because the coefficient of y^2, $-\dfrac{1}{2}$, is negative.

x	y
0	0
−2	2
−8	4
−2	−2
−8	−4

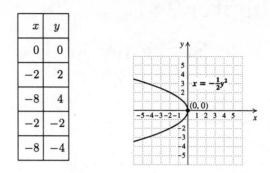

10.

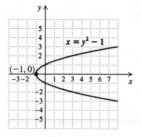

11. $x = -y^2 - 4y$

a) We find the vertex by computing the second coordinate, $y = -b/2a$, and then substituting to find the first coordinate:

$$y = -\frac{b}{2a} = -\frac{-4}{2(-1)} = -2$$
$$x = -y^2 - 4y = (-2)^2 - 4(-2) = 4$$

The vertex is $(4, -2)$.

b) We choose y-values and compute the corresponding values for x. The graph opens to the left, because the coefficient of y^2, -1, is negative.

x	y
4	−2
−5	1
0	0
3	−1
3	−3

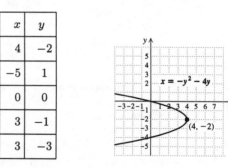

12.

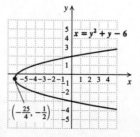

13. $x = 8 - y - y^2$

   a) We find the vertex by completing the square.

$$x = -(y^2 + y) + 8$$

$$x = -\left(y^2 + y + \frac{1}{4}\right) + 8 + \frac{1}{4}$$

$$x = -\left(y + \frac{1}{2}\right)^2 + \frac{33}{4}$$

The vertex is $\left(\dfrac{33}{4}, -\dfrac{1}{2}\right)$.

   b) We choose y-values and compute the corresponding values for x. The graph opens to the left, because the coefficient of y^2, -1, is negative.

x	y
$\dfrac{33}{4}$	$-\dfrac{1}{2}$
8	0
6	1
2	2
8	-1
6	-2
2	-3

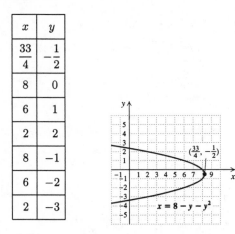

14.

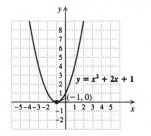

15. $y = x^2 - 2x + 1$

   a) $y = (x - 1)^2 + 0$

The vertex is $(1, 0)$.

   b) We choose x-values and compute the corresponding values for y. The graph opens upward, because the coefficient of x^2, 1, is positive.

x	y
1	0
0	1
-1	4
2	1
3	4

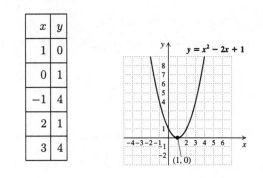

16.

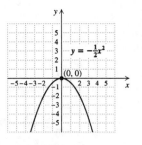

17. $x = -y^2 + 2y - 1$

   a) We find the vertex by computing the second coordinate, $y = -b/2a$, and then substituting to find the first coordinate.

$$y = -\frac{b}{2a} = -\frac{2}{2(-1)} = 1$$

$$x = -y^2 + 2y - 1 = -(1)^2 + 2(1) - 1 = 0$$

The vertex is $(0, 1)$.

   b) We choose y-values and compute the corresponding values for x. The graph opens to the left, because the coefficient of y^2, -1, is negative.

x	y
-4	3
-1	2
-1	0
-4	-1
-4	3

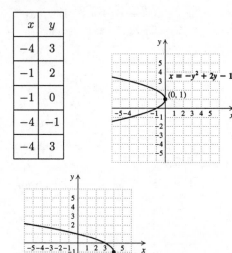

18.

19. $x = -2y^2 - 4y + 1$

   a) We find the vertex by completing the square.

$$x = -2(y^2 + 2y) + 1$$
$$x = -2(y^2 + 2y + 1) + 1 + 2$$
$$x = -2(y + 1)^2 + 3$$

    The vertex is $(-3, -1)$.

   b) We choose y-values and compute the corresponding values for x. The graph opens to the left, because the coefficient of y^2, -2, is negative.

x	y
3	-1
1	-2
-5	-3
1	0
-5	1

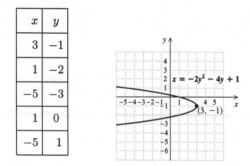

20.

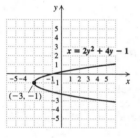

21. $\begin{aligned} d &= \sqrt{(x_2 - x_1)^2 + (y_2 + y_1)^2} & \text{Distance formula} \\ &= \sqrt{(5 - 1)^2 + (9 - 6)^2} & \text{Substituting} \\ &= \sqrt{4^2 + 3^2} \\ &= \sqrt{25} = 5 \end{aligned}$

22. 10

23. $\begin{aligned} d &= \sqrt{(x_2 - x_1)^2 + (y_2 - y_1)^2} & \text{Distance formula} \\ &= \sqrt{(3 - 0)^2 + [-4 - (-7)]^2} & \text{Substituting} \\ &= \sqrt{3^2 + 3^2} \\ &= \sqrt{18} \approx 4.243 & \text{Simplifying and approximating} \end{aligned}$

24. 10

25. $\begin{aligned} d &= \sqrt{(x_2 - x_1)^2 + (y_2 - y_1)^2} \\ &= \sqrt{[6 - (-4)]^2 + (-6 - 4)^2} \\ &= \sqrt{200} \approx 14.142 \end{aligned}$

26. $\sqrt{464} \approx 21.541$

27. $\begin{aligned} d &= \sqrt{(x_2 - x_1)^2 + (y_2 - y_1)^2} \\ &= \sqrt{(-9.2 - 8.6)^2 + [-3.4 - (-3.4)]^2} \\ &= \sqrt{(-17.8)^2 + 0^2} \\ &= \sqrt{316.84} = 17.8 \end{aligned}$

(Since these points are on a horizontal line, we could have found the distance between them by finding $|x_2 - x_1| = |-9.2 - 8.6| = |-17.8| = 17.8$.)

28. $\sqrt{98.93} \approx 9.946$

29. $\begin{aligned} d &= \sqrt{(x_2 - x_1)^2 + (y_2 - y_1)^2} \\ d &= \sqrt{\left(\frac{5}{7} - \frac{1}{7}\right)^2 + \left(\frac{1}{14} - \frac{11}{14}\right)^2} \\ &= \sqrt{\left(\frac{4}{7}\right)^2 + \left(-\frac{5}{7}\right)^2} \\ &= \sqrt{\frac{16}{49} + \frac{25}{49}} \\ &= \sqrt{\frac{41}{49}} \\ &= \frac{\sqrt{41}}{7} \approx 0.915 \end{aligned}$

30. $\sqrt{13} \approx 3.606$

31. $\begin{aligned} d &= \sqrt{(x_2 - x_1)^2 + (y_2 - y_1)^2} \\ d &= \sqrt{[0 - (-\sqrt{6})]^2 + (0 - \sqrt{2})^2} \\ &= \sqrt{6 + 2} \\ &= \sqrt{8} \approx 2.828 \end{aligned}$

32. $\sqrt{8} \approx 2.828$

33. $\begin{aligned} d &= \sqrt{(x_2 - x_1)^2 + (y_2 - y_1)^2} \\ &= \sqrt{(-\sqrt{7} - \sqrt{2})^2 + [\sqrt{5} - (-\sqrt{3})]^2} \\ &= \sqrt{7 + 2\sqrt{14} + 2 + 5 + 2\sqrt{15} + 3} \\ &= \sqrt{17 + 2\sqrt{14} + 2\sqrt{15}} \approx 5.677 \end{aligned}$

34. $\sqrt{22 + 2\sqrt{40} + 2\sqrt{18}} \approx 6.568$

35. $\begin{aligned} d &= \sqrt{(x_2 - x_1)^2 + (y_2 - y_1)^2} \\ d &= \sqrt{(s - 0)^2 + (t - 0)^2} \\ &= \sqrt{s^2 + t^2} \end{aligned}$

36. $\sqrt{p^2 + q^2}$

37. We use the midpoint formula:

$$\left(\frac{x_1 + x_2}{2}, \frac{y_1 + y_2}{2}\right) = \left(\frac{-7 + 9}{2}, \frac{6 + 2}{2}\right), \text{ or }$$

$$\left(\frac{2}{2}, \frac{8}{2}\right), \text{ or } (1, 4)$$

38. $\left(\frac{13}{2}, -1\right)$

39. We use the midpoint formula:
$$\left(\frac{x_1 + x_2}{2}, \frac{y_1 + y_2}{2}\right) = \left(\frac{2 + 5}{2}, \frac{-1 + 8}{2}\right), \text{ or }$$
$$\left(\frac{7}{2}, \frac{7}{2}\right)$$

40. $\left(0, -\frac{1}{2}\right)$

41. We use the midpoint formula:
$$\left(\frac{x_1 + x_2}{2}, \frac{y_1 + y_2}{2}\right) = \left(\frac{-8 + 6}{2}, \frac{-5 + (-1)}{2}\right), \text{ or }$$
$$\left(\frac{-2}{2}, \frac{-6}{2}\right), \text{ or } (-1, -3)$$

42. $\left(\frac{5}{2}, 1\right)$

43. We use the midpoint formula:
$$\left(\frac{x_1 + x_2}{2}, \frac{y_1 + y_2}{2}\right) = \left(\frac{-3.4 + 2.9}{2}, \frac{8.1 + (-8.7)}{2}\right),$$
$$\text{or } \left(\frac{-0.5}{2}, \frac{-0.6}{2}\right), \text{ or } (-0.25, -0.3)$$

44. $(4.65, 0)$

45. We use the midpoint formula:
$$\left(\frac{x_1 + x_2}{2}, \frac{y_1 + y_2}{2}\right) = \left(\frac{\frac{1}{6} + \left(-\frac{1}{3}\right)}{2}, \frac{-\frac{3}{4} + \frac{5}{6}}{2}\right),$$
$$\text{or } \left(\frac{-\frac{1}{6}}{2}, \frac{\frac{1}{12}}{2}\right), \text{ or } \left(-\frac{1}{12}, \frac{1}{24}\right)$$

46. $\left(-\frac{27}{80}, \frac{1}{24}\right)$

47. We use the midpoint formula:
$$\left(\frac{x_1 + x_2}{2}, \frac{y_1 + y_2}{2}\right) = \left(\frac{\sqrt{2} + \sqrt{3}}{2}, \frac{-1 + 4}{2}\right), \text{ or }$$
$$\left(\frac{\sqrt{2} + \sqrt{3}}{2}, \frac{3}{2}\right)$$

48. $\left(\frac{5}{2}, \frac{7\sqrt{3}}{2}\right)$

49. $\quad (x - h)^2 + (y - k)^2 = r^2 \quad$ Standard form
$$(x - 0)^2 + (y - 0)^2 = 6^2 \quad \text{Substituting}$$
$$x^2 + y^2 = 36 \quad \text{Simplifying}$$

50. $x^2 + y^2 = 25$

51. $\quad (x - h)^2 + (y - k)^2 = r^2 \quad$ Standard form
$$(x - 7)^2 + (y - 3)^2 = (\sqrt{5})^2 \quad \text{Substituting}$$
$$(x - 7)^2 + (y - 3)^2 = 5$$

52. $(x - 5)^2 + (y - 6)^2 = 2$

53. $\quad (x - h)^2 + (y - k)^2 = r^2 \quad$ Standard form
$$[x - (-4)]^2 + (y - 3)^2 = (4\sqrt{3})^2 \quad \text{Substituting}$$
$$(x + 4)^2 + (y - 3)^2 = 48$$
$$[(4\sqrt{3})^2 = 16 \cdot 3 = 48]$$

54. $(x + 2)^2 + (y - 7)^2 = 20$

55. $\quad (x - h)^2 + (y - k)^2 = r^2$
$$[x - (-7)]^2 + [y - (-2)]^2 = (5\sqrt{2})^2$$
$$(x + 7)^2 + (y + 2)^2 = 50$$

56. $(x + 5)^2 + (y + 8)^2 = 18$

57. Since the center is $(0, 0)$, we have
$$(x - 0)^2 + (y - 0)^2 = r^2 \text{ or } x^2 + y^2 = r^2$$
The circle passes through $(-3, 4)$. We find r^2 by substituting -3 for x and 4 for y.
$$(-3)^2 + 4^2 = r^2$$
$$9 + 16 = r^2$$
$$25 = r^2$$
Then $x^2 + y^2 = 25$ is an equation of the circle.

58. $(x - 3)^2 + (y + 2)^2 = 64$

59. Since the center is $(-4, 1)$, we have
$$[x - (-4)]^2 + (y - 1)^2 = r^2, \text{ or }$$
$$(x + 4)^2 + (y - 1)^2 = r^2.$$
The circle passes through $(-2, 5)$. We find r^2 by substituting -2 for x and 5 for y.
$$(-2 + 4)^2 + (5 - 1)^2 = r^2$$
$$4 + 16 = r^2$$
$$20 = r^2$$
Then $(x + 4)^2 + (y - 1)^2 = 20$ is an equation of the circle.

60. $(x + 1)^2 + (y + 3)^2 = 34$

61. We write standard form.

$(x - 0)^2 + (y - 0)^2 = 7^2$

The center is $(0, 0)$, and the radius is 7.

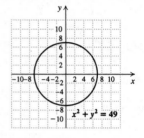

62. Center: $(0, 0)$; radius: 6

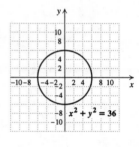

63. $(x + 1)^2 + (y + 3)^2 = 4$

$[x - (-1)]^2 + [y - (-3)]^2 = 2^2$ Standard form

The center is $(-1, -3)$, and the radius is 2.

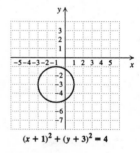

64. Center: $(2, -3)$

Radius: 1

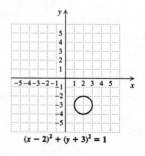

65. $(x - 4)^2 + (y + 3)^2 = 10$

$(x - 4)^2 + [y - (-3)]^2 = (\sqrt{10})^2$

The center is $(4, -3)$, and the radius is $\sqrt{10}$.

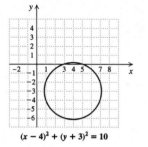

66. Center: $(-5, 1)$

Radius: $\sqrt{15}$

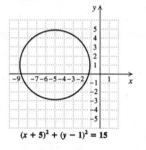

67. $x^2 + y^2 = 7$

$(x - 0)^2 + (y - 0)^2 = (\sqrt{7})^2$ Standard form

The center is $(0, 0)$, and the radius is $\sqrt{7}$.

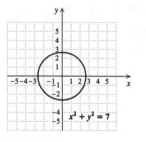

68. Center: $(0, 0)$

Radius: $\sqrt{8}$

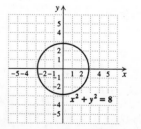

69.
$$(x-5)^2 + y^2 = \frac{1}{4}$$

$$(x-5)^2 + (y-0)^2 = \left(\frac{1}{2}\right)^2 \quad \text{Standard form}$$

The center is $(5,0)$, and the radius is $\frac{1}{2}$.

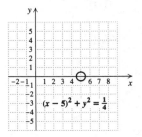

70. Center: $(0,1)$

Radius: $\frac{1}{5}$

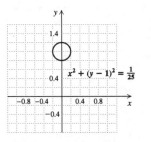

71.
$$x^2 + y^2 + 8x - 6y - 15 = 0$$
$$x^2 + 8x + y^2 - 6y = 15$$
$$(x^2 + 8x + 16) + (y^2 - 6y + 9) = 15 + 16 + 9$$
$$\text{Completing the square twice}$$
$$(x+4)^2 + (y-3)^2 = 40$$
$$[x - (-4)]^2 + (y-3)^2 = (\sqrt{40})^2$$
$$\text{Standard form}$$

The center is $(-4, 3)$, and the radius is $\sqrt{40}$, or $2\sqrt{10}$.

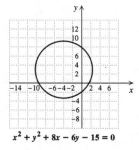

72. Center: $(-3, 2)$

Radius: $\sqrt{28}$, or $2\sqrt{7}$

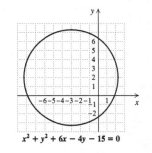

73.
$$x^2 + y^2 - 8x + 2y + 13 = 0$$
$$x^2 - 8x + y^2 + 2y = -13$$
$$(x^2 - 8x + 16) + (y^2 + 2y + 1) = -13 + 16 + 1$$
$$\text{Completing the square twice}$$
$$(x-4)^2 + (y+1)^2 = 4$$
$$(x-4)^2 + [y - (-1)]^2 = 2^2$$
$$\text{Standard form}$$

The center is $(4, -1)$, and the radius is 2.

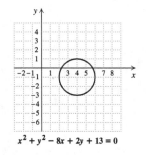

74. Center: $(-3, -2)$

Radius: 1

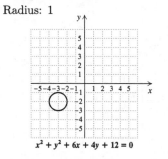

75.
$$x^2 + y^2 + 10y - 75 = 0$$
$$x^2 + y^2 + 10y = 75$$
$$x^2 + (y^2 + 10y + 25) = 75 + 25$$
$$(x-0)^2 + (y+5)^2 = 100$$
$$(x-0)^2 + [y - (-5)]^2 = 10^2$$

The center is $(0, -5)$, and the radius is 10.

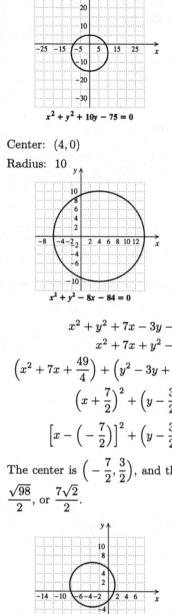

$$x^2 + y^2 + 10y - 75 = 0$$

76. Center: $(4, 0)$

Radius: 10

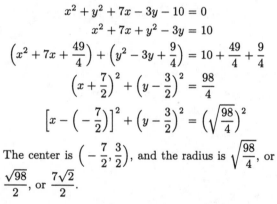

$$x^2 + y^2 - 8x - 84 = 0$$

77.
$$x^2 + y^2 + 7x - 3y - 10 = 0$$
$$x^2 + 7x + y^2 - 3y = 10$$
$$\left(x^2 + 7x + \frac{49}{4}\right) + \left(y^2 - 3y + \frac{9}{4}\right) = 10 + \frac{49}{4} + \frac{9}{4}$$
$$\left(x + \frac{7}{2}\right)^2 + \left(y - \frac{3}{2}\right)^2 = \frac{98}{4}$$
$$\left[x - \left(-\frac{7}{2}\right)\right]^2 + \left(y - \frac{3}{2}\right)^2 = \left(\sqrt{\frac{98}{4}}\right)^2$$

The center is $\left(-\frac{7}{2}, \frac{3}{2}\right)$, and the radius is $\sqrt{\frac{98}{4}}$, or $\frac{\sqrt{98}}{2}$, or $\frac{7\sqrt{2}}{2}$.

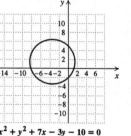

$$x^2 + y^2 + 7x - 3y - 10 = 0$$

78. Center: $\left(\frac{21}{2}, \frac{33}{2}\right)$; radius: $\frac{\sqrt{1462}}{2}$

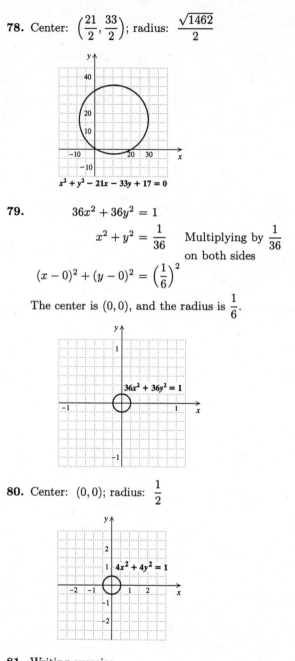

$$x^2 + y^2 - 21x - 33y + 17 = 0$$

79.
$$36x^2 + 36y^2 = 1$$
$$x^2 + y^2 = \frac{1}{36} \quad \text{Multiplying by } \frac{1}{36}$$
$$\text{on both sides}$$
$$(x - 0)^2 + (y - 0)^2 = \left(\frac{1}{6}\right)^2$$

The center is $(0, 0)$, and the radius is $\frac{1}{6}$.

$$36x^2 + 36y^2 = 1$$

80. Center: $(0, 0)$; radius: $\frac{1}{2}$

$$4x^2 + 4y^2 = 1$$

81. Writing exercise

82. Writing exercise

83. $\dfrac{x}{4} + \dfrac{5}{6} = \dfrac{2}{3}$, LCD is 12

$$12\left(\dfrac{x}{4} + \dfrac{5}{6}\right) = 12 \cdot \dfrac{2}{3}$$

$$12 \cdot \dfrac{x}{4} + 12 \cdot \dfrac{5}{6} = 8$$

$$3x + 10 = 8$$

$$3x = -2$$

$$x = -\dfrac{2}{3}$$

The solution is $-\dfrac{2}{3}$.

84. $\dfrac{25}{6}$

85. *Familiarize*. We make a drawing and label it. Let x represent the width of the border.

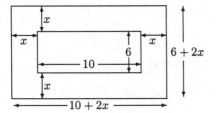

The perimeter of the larger rectangle is

$$2(10 + 2x) + 2(6 + 2x), \text{ or } 8x + 32.$$

The perimeter of the smaller rectangle is

$$2(10) + 2(6), \text{ or } 32.$$

Translate. The perimeter of the larger rectangle is twice the perimeter of the smaller rectangle.

$$8x + 32 = 2 \cdot 32$$

Carry out. We solve the equation.

$$8x + 32 = 64$$

$$8x = 32$$

$$x = 4$$

Check. If the width of the border is 4 in., then the length and width of the larger rectangle are 18 in. and 14 in. Thus its perimeter is $2(18) + 2(14)$, or 64 in. The perimeter of the smaller rectangle is 32 in. The perimeter of the larger rectangle is twice the perimeter of the smaller rectangle.

State. The width of the border is 4 in.

86. 2640 mi

87. $3x - 8y = 5$, (1)

$2x + 6y = 5$ (2)

Multiply Equation (1) by 3, multiply Equation (2) by 4, and add.

$$\begin{array}{r} 9x - 24y = 15 \\ 8x + 24y = 20 \\ \hline 17x \qquad = 35 \end{array}$$

$$x = \dfrac{35}{17}$$

Now substitute $\dfrac{35}{17}$ for x in one of the original equations and solve for y. We use Equation (2).

$$2x + 6y = 5$$

$$2\left(\dfrac{35}{17}\right) + 6y = 5$$

$$\dfrac{70}{17} + 6y = 5$$

$$6y = \dfrac{15}{17}$$

$$y = \dfrac{5}{34}$$

The solution is $\left(\dfrac{35}{17}, \dfrac{5}{34}\right)$.

88. $\left(0, -\dfrac{9}{5}\right)$

89. Writing exercise

90. Writing exercise

91. We make a drawing of the circle with center $(3, -5)$ and tangent to the y-axis.

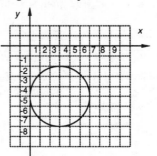

We see that the circle touches the y-axis at $(0, -5)$. Hence the radius is the distance between $(0, -5)$ and $(3, -5)$, or $\sqrt{(3 - 0)^2 + [-5 - (-5)]^2}$, or 3. Now we write the equation of the circle.

$$(x - h)^2 + (y - k)^2 = r^2$$

$$(x - 3)^2 + [y - (-5)]^2 = 3^2$$

$$(x - 3)^2 + (y + 5)^2 = 9$$

92. $(x + 7)^2 + (y + 4)^2 = 16$

93. First we use the midpoint formula to find the center:

$$\left(\frac{7+(-1)}{2}, \frac{3+(-3)}{2}\right), \text{ or } \left(\frac{6}{2}, \frac{0}{2}\right), \text{ or } (3,0)$$

The length of the radius is the distance between the center $(3,0)$ and either endpoint of a diameter. We will use endpoint $(7,3)$ in the distance formula:

$$r = \sqrt{(7-3)^2 + (3-0)^2} = \sqrt{25} = 5$$

Now we write the equation of the circle:

$$(x-h)^2 + (y-k)^2 = r^2$$
$$(x-3)^2 + (y-0)^2 = 5^2$$
$$(x-3)^2 + y^2 = 25$$

94. $(x+3)^2 + (y-5)^2 = 16$

95. Let $(0,y)$ be the point on the y-axis that is equidistant from $(2,10)$ and $(6,2)$. Then the distance between $(2,10)$ and $(0,y)$ is the same as the distance between $(6,2)$ and $(0,y)$.

$$\sqrt{(0-2)^2 + (y-10)^2} = \sqrt{(0-6)^2 + (y-2)^2}$$
$$(-2)^2 + (y-10)^2 = (-6)^2 + (y-2)^2$$

Squaring both sides

$$4 + y^2 - 20y + 100 = 36 + y^2 - 4y + 4$$
$$64 = 16y$$
$$4 = y$$

This number checks. The point is $(0,4)$.

96. $(-5,0)$

97. a) Use the fact that the center of the circle $(0,k)$ is equidistant from the points $(-575,0)$ and $(0,19.5)$.

$$\sqrt{(-575-0)^2 + (0-k)^2} = \sqrt{(0-0)^2 + (19.5-k)^2}$$
$$\sqrt{330,625 + k^2} = \sqrt{380.25 - 39k + k^2}$$
$$330,625 + k^2 = 380.25 - 39k + k^2$$

Squaring both sides

$$330,244.75 = -39k$$
$$-8467.8 \approx k$$

Then the center of the circle is about $(0, -8467.8)$.

b) To find the radius we find the distance from the center, $(0, -8467.8)$ to any one of the points $(-575,0)$, $(0,19.5)$, or $(575,0)$. We use $(0,19.5)$.

$$r = \sqrt{(0-0)^2 + [19.5 - (-8467.8)]^2} \approx 8487.3 \text{ mm}$$

98. 8186.6 mm

99. a) Use the fact that the center of the circle, $(0,k)$ is equidistant from the points $(0,2.1)$ and $(80,0)$.

$$\sqrt{(80-0)^2 + (0-k)^2} = \sqrt{(0-0)^2 + (2.1-k)^2}$$
$$\sqrt{6400 + k^2} = \sqrt{4.41 - 4.2k + k^2}$$
$$6400 + k^2 = 4.41 - 4.2k + k^2$$

Squaring both sides

$$6395.59 = -4.2k$$
$$-1522.8 \approx k$$

Then the center of the circle is about $(0, -1522.8)$.

b) To find the radius we find the distance from the center, $(0, -1522.8)$, to either of the points $(0,2.1)$ or $(80,0)$. We use $(0,2.1)$.

$$r = \sqrt{(0-0)^2 + [2.1 - (-1522.8)]^2} \approx 1524.9 \text{ cm}$$

100. a) $(0,-3)$; b) 5 ft

101. Position a coordinate system as shown below so that the center of the top of the bowl is at the origin. Let r represent the radius of the circle.

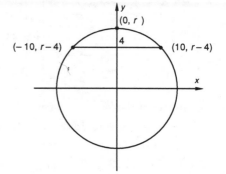

The equation of the circle is $x^2 + y^2 = r^2$, and one point on the circle is $(10, r-4)$. We substitute 10 for x and $r-4$ for y and solve for r.

$$10^2 + (r-4)^2 = r^2$$
$$100 + r^2 - 8r + 16 = r^2$$
$$r^2 - 8r + 116 = r^2$$
$$116 = 8r$$
$$14.5 = r$$

Then the original diameter of the bowl is $2r = 2(14.5)$, or 29 cm.

102. $x^2 + (y-30.6)^2 = 590.49$

103. First we graph $x = y^2 - y - 6$, $x = 2$, and $x = -3$ on the same set of axes.

$$x = y^2 - y - 6$$
$$x = \left(y^2 - y + \frac{1}{4}\right) - 6 - \frac{1}{4}$$
$$x = \left(y - \frac{1}{2}\right)^2 - \frac{25}{4}$$

The vertex is $\left(-\frac{25}{4}, \frac{1}{2}\right)$.

x	y
$-\dfrac{25}{4}$	$\dfrac{1}{2}$
-6	1
-4	2
0	3
-6	0
-4	-1
0	-2

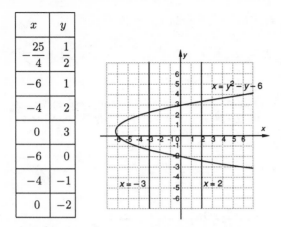

a) Graph $x = 2$ on the same set of axes as $x = y^2 - y - 6$ and approximate the y-coordinates of the points of intersection. (See the graph above.) The solutions are approximately 3.4 and -2.4.

b) Graph $x = -3$ on the same set of axes as $x = y^2 - y - 6$ and approximate the y-coordinates of the points of intersection. (See the graph above.) The solutions are approximately 2.3 and -1.3.

104.

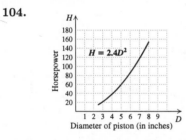

105. Let $P_1 = (x_1, y_1)$, $P_2 = (x_2, y_2)$, and $M = \left(\dfrac{x_1 + x_2}{2}, \dfrac{y_1 + y_2}{2}\right)$. Let $d(AB)$ denote the distance from point A to point B.

i) $d(P_1M)$

$= \sqrt{\left(\dfrac{x_1 + x_2}{2} - x_1\right)^2 + \left(\dfrac{y_1 + y_2}{2} - y_1\right)^2}$

$= \dfrac{1}{2}\sqrt{(x_2 - x_1)^2 + (y_2 - y_1)^2};$

$d(P_2M)$

$= \sqrt{\left(\dfrac{x_1 + x_2}{2} - x_2\right)^2 + \left(\dfrac{y_1 + y_2}{2} - y_2\right)^2}$

$= \dfrac{1}{2}\sqrt{(x_1 - x_2)^2 + (y_1 - y_2)^2}$

$= \dfrac{1}{2}\sqrt{(x_2 - x_1)^2 + (y_2 - y_1)^2} = d(P_1M).$

ii) $d(P_1M) + d(P_2M)$

$= \dfrac{1}{2}\sqrt{(x_2 - x_1)^2 + (y_2 - y_1)^2} +$

$\quad \dfrac{1}{2}\sqrt{(x_2 - x_1)^2 + (y_2 - y_1)^2}$

$= \sqrt{(x_2 - x_1)^2 + (y_2 - y_1)^2}$

$= d(P_1P_2)$

106. a) $y = -1 \pm \sqrt{-x^2 + 6x + 7}$

b)

107. Writing exercise

108. Writing exercise

Exercise Set 13.2

1. $\dfrac{x^2}{1} + \dfrac{y^2}{4} = 1$

$\dfrac{x^2}{1^2} + \dfrac{y^2}{2^2} = 1$

The x-intercepts are $(1, 0)$ and $(-1, 0)$, and the y-intercepts are $(0, 2)$ and $(0, -2)$. We plot these points and connect them with an oval-shaped curve.

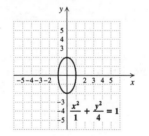

2.

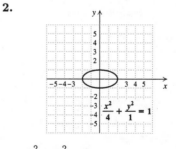

3. $\dfrac{x^2}{25} + \dfrac{y^2}{9} = 1$

$\dfrac{x^2}{5^2} + \dfrac{y^2}{3^2} = 1$

The x-intercepts are $(5, 0)$ and $(-5, 0)$, and the y-intercepts are $(0, 3)$ and $(0, -3)$. We plot these points and connect them with an oval-shaped curve.

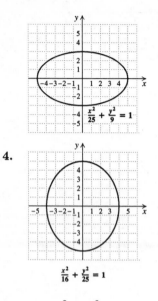

4.

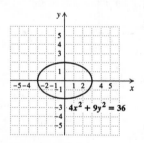

5. $\qquad 4x^2 + 9y^2 = 36$

$$\frac{1}{36}(4x^2 + 9y^2) = \frac{1}{36}(36) \qquad \text{Multiplying by } \frac{1}{36}$$

$$\frac{x^2}{9} + \frac{y^2}{4} = 1$$

$$\frac{x^2}{3^2} + \frac{y^2}{2^2} = 1$$

The x-intercepts are $(-3, 0)$ and $(3, 0)$, and the y-intercepts are $(0, -2)$ and $(0, 2)$. We plot these points and connect them with an oval-shaped curve.

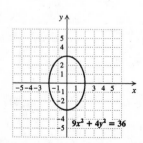

6.

7. $\quad 16x^2 + 9y^2 = 144$

$$\frac{x^2}{9} + \frac{y^2}{16} = 1 \quad \text{Multiplying by } \frac{1}{144}$$

$$\frac{x^2}{3^2} + \frac{y^2}{4^2} = 1$$

The x-intercepts are $(3, 0)$ and $(-3, 0)$, and the y-intercepts are $(0, 4)$ and $(0, -4)$. We plot these points and connect them with an oval-shaped curve.

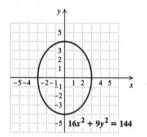

8.

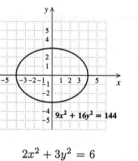

9. $\qquad 2x^2 + 3y^2 = 6$

$$\frac{x^2}{3} + \frac{y^2}{2} = 1 \quad \text{Multiplying by } \frac{1}{6}$$

$$\frac{x^2}{(\sqrt{3})^2} + \frac{y^2}{(\sqrt{2})^2} = 1$$

The x-intercepts are $(\sqrt{3}, 0)$ and $(-\sqrt{3}, 0)$, and the y-intercepts are $(0, \sqrt{2})$ and $(0, -\sqrt{2})$. We plot these points and connect them with an oval-shaped curve.

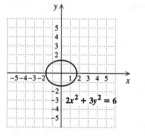

10.

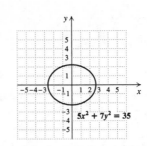

11. $5x^2 + 5y^2 = 125$

Observe that the x^2- and y^2-terms have the same coefficient. We divide both sides of the equation by 5 to obtain $x^2 + y^2 = 25$. This is the equation of a circle with center $(0,0)$ and radius 5.

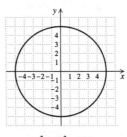

$$5x^2 + 5y^2 = 125$$

12.

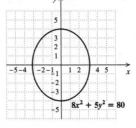

13. $\quad 3x^2 + 7y^2 - 63 = 0$

$$3x^2 + 7y^2 = 63$$

$$\frac{x^2}{21} + \frac{y^2}{9} = 1 \quad \text{Multiplying by } \frac{1}{63}$$

$$\frac{x^2}{(\sqrt{21})^2} + \frac{y^2}{3^2} = 1$$

The x-intercepts are $(\sqrt{21}, 0)$ and $(-\sqrt{21}, 0)$, or about $(4.583, 0)$ and $(-4.583, 0)$. The y-intercepts are $(0, 3)$ and $(0, -3)$. We plot these points and connect them with an oval-shaped curve.

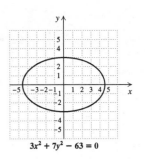

$$3x^2 + 7y^2 - 63 = 0$$

14.

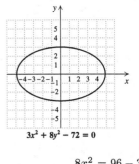

$$3x^2 + 8y^2 - 72 = 0$$

15. $\qquad\qquad 8x^2 = 96 - 3y^2$

$$8x^2 + 3y^2 = 96$$

$$\frac{x^2}{12} + \frac{y^2}{32} = 1$$

$$\frac{x^2}{(\sqrt{12})^2} + \frac{y^2}{(\sqrt{32})^2} = 1$$

The x-intercepts are $(\sqrt{12}, 0)$ and $(-\sqrt{12}, 0)$, or about $(3.464, 0)$ and $(-3.464, 0)$. The y-intercepts are $(0, \sqrt{32})$ and $(0, -\sqrt{32})$, or about $(0, 5.657)$ and $(0, -5.657)$. We plot these points and connect them with an oval-shaped curve.

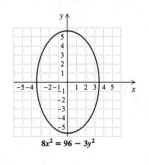

$$8x^2 = 96 - 3y^2$$

16.

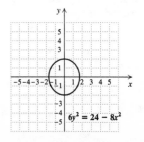

$$6y^2 = 24 - 8x^2$$

17. $16x^2 + 25y^2 = 1$

Note that $16 = \dfrac{1}{\dfrac{1}{16}}$ and $25 = \dfrac{1}{\dfrac{1}{25}}$. Thus, we can rewrite the equation:

$$\frac{x^2}{\dfrac{1}{16}} + \frac{y^2}{\dfrac{1}{25}} = 1$$

$$\frac{x^2}{\left(\dfrac{1}{4}\right)^2} + \frac{y^2}{\left(\dfrac{1}{5}\right)^2} = 1$$

The x-intercepts are $\left(\dfrac{1}{4}, 0\right)$ and $\left(-\dfrac{1}{4}, 0\right)$, and the y-intercepts are $\left(0, \dfrac{1}{5}\right)$ and $\left(0, -\dfrac{1}{5}\right)$. We plot these points and connect them with an oval-shaped curve.

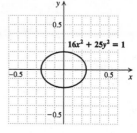

18.

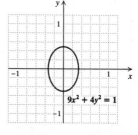

19. $\dfrac{(x-2)^2}{9} + \dfrac{(y-1)^2}{25} = 1$

$$\frac{(x-2)^2}{3^2} + \frac{(y-1)^2}{5^2} = 1$$

The center of the ellipse is $(2, 1)$. Note that $a = 3$ and $b = 5$. We locate the center and then plot the points $(2+3, 1)$ $(2-3, 1)$, $(2, 1+5)$, and $(2, 1-5)$, or $(5, 1)$, $(-1, 1)$, $(2, 6)$, and $(2, -4)$. Connect these points with an oval-shaped curve.

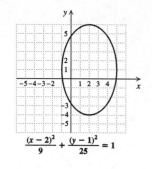

$$\frac{(x-2)^2}{9} + \frac{(y-1)^2}{25} = 1$$

20.

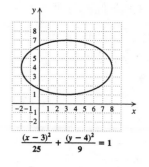

$$\frac{(x-3)^2}{25} + \frac{(y-4)^2}{9} = 1$$

21. $\dfrac{(x+4)^2}{16} + \dfrac{(y-3)^2}{49} = 1$

$$\frac{(x-(-4))^2}{4^2} + \frac{(y-3)^2}{7^2} = 1$$

The center of the ellipse is $(-4, 3)$. Note that $a = 4$ and $b = 7$. We locate the center and then plot the points $(-4+4, 3)$, $(-4-4, 3)$, $(-4, 3+7)$, and $(-4, 3-7)$, or $(0, 3)$, $(-8, 3)$, $(-4, 10)$, and $(-4, -4)$. Connect these points with an oval-shaped curve.

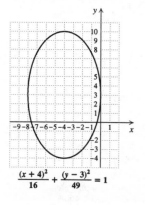

$$\frac{(x+4)^2}{16} + \frac{(y-3)^2}{49} = 1$$

22.

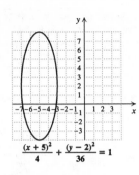

$$\frac{(x+5)^2}{4} + \frac{(y-2)^2}{36} = 1$$

23. $12(x-1)^2 + 3(y+4)^2 = 48$

$$\frac{(x-1)^2}{4} + \frac{(y+4)^2}{16} = 1$$

$$\frac{(x-1)^2}{2^2} + \frac{(y-(-4))^2}{4^2} = 1$$

The center of the ellipse is $(1, -4)$. Note that $a = 2$ and $b = 4$. We locate the center and then plot the points $(1+2, -4)$, $(1-2, -4)$, $(1, -4+4)$, and $(1, -4-4)$, or $(3, -4)$, $(-1, -4)$, $(1, 0)$, and $(1, -8)$. Connect these points with an oval-shaped curve.

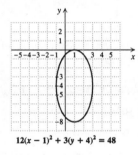

$$12(x-1)^2 + 3(y+4)^2 = 48$$

24.

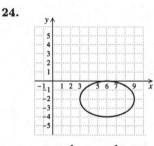

$$4(x-6)^2 + 9(y+2)^2 = 36$$

25. $4(x+3)^2 + 4(y+1)^2 - 10 = 90$

$$4(x+3)^2 + 4(y+1)^2 = 100$$

Observe that the x^2- and y^2-terms have the some coefficient. Dividing both sides by 4, we have

$$(x+3)^2 + (y+1)^2 = 25.$$

This is the equation of a circle with center $(-3, -1)$ and radius 5.

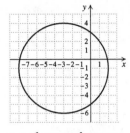

$$4(x+3)^2 + 4(y+1)^2 - 10 = 90$$

26.

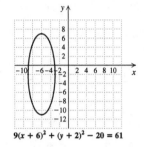

$$9(x+6)^2 + (y+2)^2 - 20 = 61$$

27. Writing exercise

28. Writing exercise

29. $\dfrac{3}{x-2} - \dfrac{5}{x-2} = 9$

Note that the denominators are 0 when $x = 2$, so 2 cannot be a solution. We multiply by the LCD, $x - 2$.

$$x - 2\left(\frac{3}{x-2} - \frac{5}{x-2}\right) = (x-2)9$$

$$(x-2) \cdot \frac{3}{x-2} - (x-2) \cdot \frac{5}{x-2} = 9x - 18$$

$$3 - 5 = 9x - 18$$

$$-2 = 9x - 18$$

$$16 = 9x$$

$$\frac{16}{9} = x$$

The number $\dfrac{16}{9}$ checks and is the solution.

30. $-\dfrac{19}{8}$

31. $\dfrac{x}{x-4} - \dfrac{3}{x-5} = \dfrac{2}{x-4}$

Note that $x - 4$ is 0 when $x = 4$ and $x - 5$ is 0 when x is 5, so 4 and 5 cannot be solutions. We multiply by the LCD, $(x-4)(x-5)$.

$$(x-4)(x-5)\left(\frac{x}{x-4}-\frac{3}{x-5}\right) =$$

$$(x-4)(x-5)\cdot\frac{2}{x-4}$$

$$(x-4)(x-5)\cdot\frac{x}{x-4}-(x-4)(x-5)\frac{3}{x-5}=2(x-5)$$

$$x(x-5)-3(x-4)=2(x-5)$$

$$x^2-5x-3x+12=2x-10$$

$$x^2-8x+12=2x-10$$

$$x^2-10x+22=0$$

We use the quadratic formula with $a=1$, $b=-10$, and $c=22$.

$$x=\frac{-b\pm\sqrt{b^2-4ac}}{2a}$$

$$x=\frac{-(-10)\pm\sqrt{(-10)^2-4\cdot1\cdot22}}{2\cdot1}$$

$$x=\frac{10\pm\sqrt{12}}{2}=\frac{10\pm2\sqrt{3}}{2}$$

$$x=\frac{\cancel{2}(5\pm\sqrt{3})}{\cancel{2}\cdot1}=5\pm\sqrt{3}$$

Both numbers check. The solutions are $5\pm\sqrt{3}$.

32. $3\pm\sqrt{7}$

33. $9-\sqrt{2x+1}=7$

$-\sqrt{2x+1}=-2$ Isolating the radical

$(-\sqrt{2x+1})^2=(-2)^2$

$2x+1=4$

$2x=3$

$x=\frac{3}{2}$

The number $\frac{3}{2}$ checks and is the solution.

34. No solution

35. Writing exercise

36. Writing exercise

37. Plot the given points.

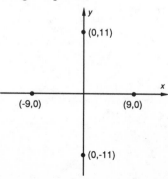

From the location of these points, we see that the ellipse that contains them is centered at the origin with $a=9$ and $b=11$. We write the equation of the ellipse:

$$\frac{x^2}{9^2}+\frac{y^2}{11^2}=1$$

$$\frac{x^2}{81}+\frac{y^2}{121}=1$$

38. $\frac{x^2}{49}+\frac{y^2}{25}=1$

39. Plot the given points.

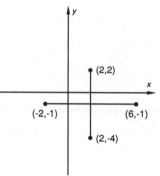

The midpoint of the segment from $(-2,-1)$ to $(6,-1)$ is $\left(\frac{-2+6}{2},\frac{-1-1}{2}\right)$, or $(2,-1)$. The midpoint of the segment from $(2,-4)$ to $(2,2)$ is $\left(\frac{2+2}{2},\frac{-4+2}{2}\right)$, or $(2,-1)$. Thus, we can conclude that $(2,-1)$ is the center of the ellipse. The distance from $(-2,-1)$ to $(2,-1)$ is $\sqrt{[2-(-2)]^2+[-1-(-1)]^2}=\sqrt{16}=4$, so $a=4$. The distance from $(2,2)$ to $(2,-1)$ is $\sqrt{(2-2)^2+(-1-2)^2}=\sqrt{9}=3$, so $b=3$. We write the equation of the ellipse.

$$\frac{(x-2)^2}{4^2}+\frac{(y-(-1))^2}{3^2}=1$$

$$\frac{(x-2)^2}{16}+\frac{(y+1)^2}{9}=1$$

40. $\frac{(x+1)^2}{25}+\frac{(y-3)^2}{16}=1$

41. We make a drawing.

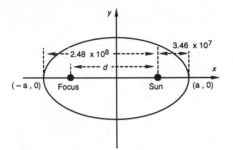

The distance between vertex $(a, 0)$ and the sun is the same as the distance between vertex $(-a, 0)$ and the other focus. Then

$$d = 2.48 \times 10^8 - 3.46 \times 10^7 =$$
$$2.48 \times 10^8 - 0.346 \times 10^8 = 2.134 \times 10^8 \text{ mi.}$$

42. a) Let $F_1 = (-c, 0)$ and $F_2 = (c, 0)$. Then the sum of the distances from the foci to P is $2a$. By the distance formula,

$$\sqrt{(x+c)^2 + y^2} + \sqrt{(x-c)^2 + y^2} = 2a, \text{ or}$$
$$\sqrt{(x+c)^2 + y^2} = 2a - \sqrt{(x-c)^2 + y^2}.$$

Squaring, we get

$$(x+c)^2 + y^2 = 4a^2 - 4a\sqrt{(x-c)^2 + y^2} + (x-c)^2 + y^2,$$

or $x^2 + 2cx + c^2 + y^2$

$$= 4a^2 - 4a\sqrt{(x-c)^2 + y^2} + x^2 - 2cx + c^2 + y^2.$$

Thus

$$-4a^2 + 4cx = -4a\sqrt{(x-c)^2 + y^2}$$
$$a^2 - cx = a\sqrt{(x-c)^2 + y^2}.$$

Squaring again, we get

$$a^4 - 2a^2cx + c^2x^2 = a^2(x^2 - 2cx + c^2 + y^2)$$
$$a^4 - 2a^2cx + c^2x^2 = a^2x^2 - 2a^2cx + a^2c^2 + a^2y^2,$$

or

$$x^2(a^2 - c^2) + a^2y^2 = a^2(a^2 - c^2)$$
$$\frac{x^2}{a^2} + \frac{y^2}{a^2 - c^2} = 1.$$

b) When P is at $(0, b)$, it follows that $b^2 = a^2 - c^2$. Substituting, we have

$$\frac{x^2}{a^2} + \frac{y^2}{b^2} = 1.$$

43. Position the ellipse on a coordinate system as shown below.

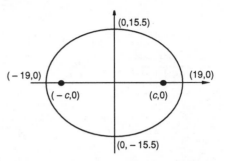

In order to best use the room's acoustics, the President and the advisor should be seated at the foci of the ellipse, or at $(-c, 0)$ and $(c, 0)$. We use the equation relating the coordinates of the foci and the intercepts to find c:

$$b^2 = a^2 - c^2$$
$$(15.5)^2 = (19)^2 - c^2$$
$$240.25 = 361 - c^2$$
$$c^2 = 120.75$$
$$c \approx 11$$

We make a sketch.

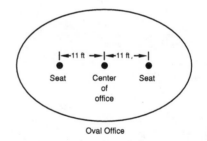

Oval Office

44. 5.66 ft

45.
$$\frac{x^2}{40,000} + \frac{y^2}{10,000} = 1, \text{ or}$$
$$\frac{x^2}{(200)^2} + \frac{y^2}{(100)^2} = 1, \text{ and}$$
$$\frac{x^2}{250,000} + \frac{y^2}{10,000} = 1, \text{ or}$$
$$\frac{x^2}{(500)^2} + \frac{y^2}{(100)^2} = 1$$

For each ellipse $b = 100$ yd and the width of the fire is twice this length, or 200 yd.

For the smaller ellipse, $a = 200$ yd, and for the larger ellipse, $a = 500$ yd. The length of the fire is the sum of these lengths, or 700 yd.

46. $\dfrac{(x-2)^2}{16} + \dfrac{(y+1)^2}{4} = 1$

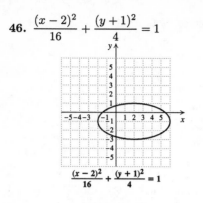

$$\dfrac{(x-2)^2}{16} + \dfrac{(y+1)^2}{4} = 1$$

47.

$$4x^2 + 24x + y^2 - 2y - 63 = 0$$
$$4(x^2 + 6x) + y^2 - 2y = 63$$
$$4(x^2+6x+9-9)+(y^2-2y+1-1) = 63$$
$$4(x^2 + 6x + 9) + (y^2 - 2y + 1) = 63 + 4\cdot 9 + 1$$
$$4(x + 3)^2 + (y - 1)^2 = 100$$
$$\dfrac{(x+3)^2}{25} + \dfrac{(y-1)^2}{100} = 1$$

The center of the ellipse is $(-3,1)$. Note that $a = 5$ and $b = 10$. Locate the center and then plot the points $(-3 + 5, 1)$, $(-3 - 5, 1)$, $(-3, 1 + 10)$, and $(-3, 1-10)$, or $(2, 1)$, $(-8, 1)$, $(-3, 11)$, and $(-3, -9)$. Connect these points with an oval-shaped curve.

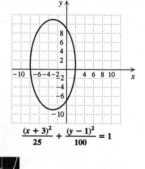

$$\dfrac{(x+3)^2}{25} + \dfrac{(y-1)^2}{100} = 1$$

48.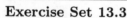

Exercise Set 13.3

1. $\dfrac{y^2}{9} - \dfrac{x^2}{9} = 1$

$$\dfrac{y^2}{3^2} - \dfrac{x^2}{3^2} = 1$$

$a = 3$ and $b = 3$, so the asymptotes are $y = \dfrac{3}{3}x$ and $y = -\dfrac{3}{3}x$, or $y = x$ and $y = -x$. We sketch them.

Replacing x with 0 and solving for y, we get $y = \pm 3$, so the intercepts are $(0, 3)$ and $(0, -3)$.

We plot the intercepts and draw smooth curves through them that approach the asymptotes.

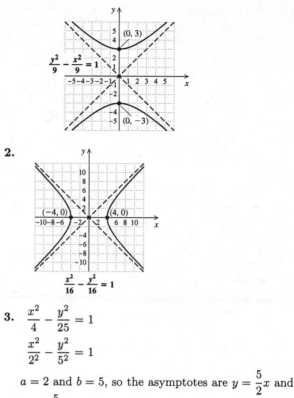

$$\dfrac{y^2}{9} - \dfrac{x^2}{9} = 1$$

2.

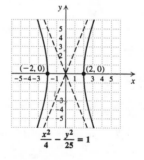

$$\dfrac{x^2}{16} - \dfrac{y^2}{16} = 1$$

3. $\dfrac{x^2}{4} - \dfrac{y^2}{25} = 1$

$$\dfrac{x^2}{2^2} - \dfrac{y^2}{5^2} = 1$$

$a = 2$ and $b = 5$, so the asymptotes are $y = \dfrac{5}{2}x$ and $y = -\dfrac{5}{2}x$. We sketch them.

Replacing y with 0 and solving for x, we get $x = \pm 2$, so the intercepts are $(2, 0)$ and $(-2, 0)$.

We plot the intercepts and draw smooth curves through them that approach the asymptotes.

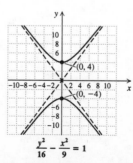

$$\dfrac{x^2}{4} - \dfrac{y^2}{25} = 1$$

4.

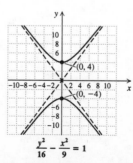

$$\dfrac{y^2}{16} - \dfrac{x^2}{9} = 1$$

5. $\dfrac{y^2}{36} - \dfrac{x^2}{9} = 1$

$\dfrac{y^2}{6^2} - \dfrac{x^2}{3^2} = 1$

$a = 3$ and $b = 6$, so the asymptotes are $y = \dfrac{6}{3}x$ and $y = -\dfrac{6}{3}x$, or $y = 2x$ and $y = -2x$. We sketch them.

Replacing x with 0 and solving for y, we get $y = \pm 6$, so the intercepts are $(0, 6)$ and $(0, -6)$.

We plot the intercepts and draw smooth curves through them that approach the asymptotes.

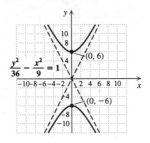

6.

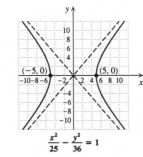

7. $y^2 - x^2 = 25$

$\dfrac{y^2}{25} - \dfrac{x^2}{25} = 1$

$\dfrac{y^2}{5^2} - \dfrac{x^2}{5^2} = 1$

$a = 5$ and $b = 5$, so the asymptotes are $y = \dfrac{5}{5}x$ and $y = -\dfrac{5}{5}x$, or $y = x$ and $y = -x$. We sketch them.

Replacing x with 0 and solving for y, we get $y = \pm 5$, so the intercepts are $(0, 5)$ and $(0, -5)$.

We plot the intercepts and draw smooth curves through them that approach the asymptotes.

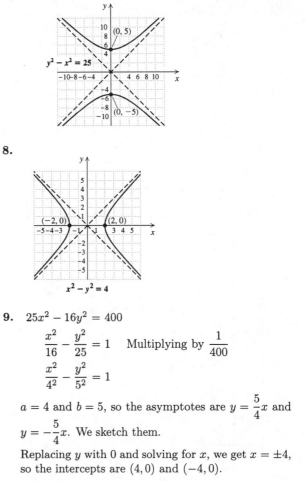

8.

9. $25x^2 - 16y^2 = 400$

$\dfrac{x^2}{16} - \dfrac{y^2}{25} = 1$ Multiplying by $\dfrac{1}{400}$

$\dfrac{x^2}{4^2} - \dfrac{y^2}{5^2} = 1$

$a = 4$ and $b = 5$, so the asymptotes are $y = \dfrac{5}{4}x$ and $y = -\dfrac{5}{4}x$. We sketch them.

Replacing y with 0 and solving for x, we get $x = \pm 4$, so the intercepts are $(4, 0)$ and $(-4, 0)$.

We plot the intercepts and draw smooth curves through them that approach the asymptotes.

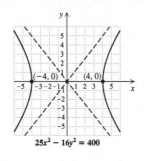

10.

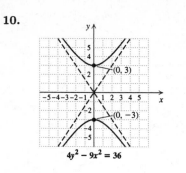

$4y^2 - 9x^2 = 36$

11. $xy = -6$

$y = -\dfrac{6}{x}$ Solving for y

We find some solutions, keeping the results in a table.

x	y
$\dfrac{1}{2}$	-12
1	-6
2	-3
4	$-\dfrac{3}{2}$
8	$-\dfrac{3}{4}$
$-\dfrac{1}{2}$	12
-1	6
-2	3
-8	$\dfrac{3}{4}$

Note that we cannot use 0 for x. The x-axis and the y-axis are the asymptotes.

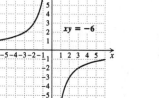

12.

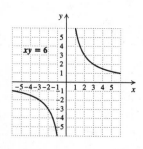

13. $xy = 4$

$y = \dfrac{4}{x}$ Solving for y

We find some solutions, keeping the results in a table.

x	y
$\dfrac{1}{2}$	8
1	4
4	1
8	$\dfrac{1}{2}$
$-\dfrac{1}{2}$	-8
-1	-4
-2	-2
-4	-1

Note that we cannot use 0 for x. The x-axis and the y-axis are the asymptotes.

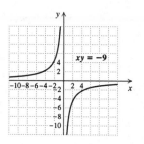

14.

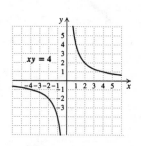

15. $xy = -2$

$y = -\dfrac{2}{x}$ Solving for y

x	y
$\dfrac{1}{2}$	-4
1	-2
2	-1
4	$-\dfrac{1}{2}$
$-\dfrac{1}{2}$	4
-1	2
-2	1
-4	$\dfrac{1}{2}$

Note that we cannot use 0 for x. The x-axis and the y-axis are the asymptotes.

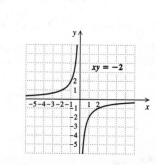

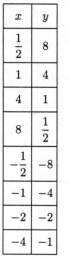
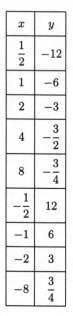

16.

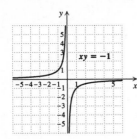

$xy = -1$

17. $xy = 1$

$$y = \frac{1}{x} \qquad \text{Solving for } y$$

x	y
$\frac{1}{4}$	4
$\frac{1}{2}$	2
1	1
2	$\frac{1}{2}$
4	$\frac{1}{4}$
$-\frac{1}{4}$	-4
$-\frac{1}{2}$	-2
-1	-1
-2	$-\frac{1}{2}$
-4	$-\frac{1}{4}$

Note that we cannot use 0 for x. The x-axis and the y-axis are the asymptotes.

$xy = 1$

18.

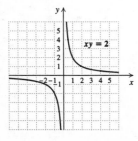

$xy = 2$

19. $x^2 + y^2 - 10x + 8y - 40 = 0$

Completing the square twice, we obtain an equivalent equation:

$$(x^2 - 10x) + (y^2 + 8y) = 40$$
$$(x^2 - 10x + 25) + (y^2 + 8y + 16) = 40 + 25 + 16$$
$$(x - 5)^2 + (y + 4)^2 = 81$$

The graph is a circle.

20. Parabola

21. $9x^2 + 4y^2 - 36 = 0$

$$9x^2 + 4y^2 = 36$$
$$\frac{x^2}{4} + \frac{y^2}{9} = 1$$

The graph is an ellipse.

22. Parabola

23. $4x^2 - 9y^2 - 72 = 0$

$$4x^2 - 9y^2 = 72$$
$$\frac{x^2}{18} - \frac{y^2}{8} = 1$$

The graph is a hyperbola.

24. Circle

25. $$x^2 + y^2 = 2x + 4y + 4$$
$$x^2 - 2x + y^2 - 4y = 4$$
$$(x^2 - 2x + 1) + (y^2 - 4y + 4) = 4 + 1 + 4$$
$$(x - 1)^2 + (y - 2)^2 = 9$$

The graph is a circle.

26. Circle

27. $$4x^2 = 64 - y^2$$
$$4x^2 + y^2 = 64$$
$$\frac{x^2}{16} + \frac{y^2}{64} = 1$$

The graph is an ellipse.

28. Hyperbola

29. $x - \dfrac{3}{y} = 0$

$$x = \frac{3}{y}$$
$$xy = 3$$

The graph is a hyperbola.

30. Parabola

31. $y + 6x = x^2 + 5$

$$y = x^2 - 6x + 5$$

The graph is a parabola.

32. Hyperbola

33.
$$9y^2 = 36 + 4x^2$$
$$9y^2 - 4x^2 = 36$$
$$\frac{y^2}{4} - \frac{x^2}{9} = 1$$
The graph is a hyperbola.

34. Circle

35.
$$3x^2 + y^2 - x = 2x^2 - 9x + 10y + 40$$
$$x^2 + y^2 + 8x - 10y = 40$$
Both variables are squared, so the graph is not a parabola. The plus sign between x^2 and y^2 indicates that we have either a circle or an ellipse. Since the coefficients of x^2 and y^2 are the same, the graph is a circle.

36. Ellipse

37.
$$16x^2 + 5y^2 - 12x^2 + 8y^2 - 3x + 4y = 568$$
$$4x^2 + 13y^2 - 3x + 4y = 568$$
Both variables are squared, so the graph is not a parabola. The plus sign between x^2 and y^2 indicates that we have either a circle or an ellipse. Since the coefficients of x^2 and y^2 are different, the graph is an ellipse.

38. Ellipse

39. Writing exercise

40. Writing exercise

41.
$$5x + 6y = -12, \quad (1)$$
$$3x + 9y = 15 \quad\quad (2)$$
We will use the elimination method. First multiply equation (1) by 3 and equation (2) by -2 and add.
$$15x + 18y = -36$$
$$\underline{-6x - 18y = -30}$$
$$9x \quad\quad\quad = -66$$
$$x = -\frac{22}{3}$$
Now substitute $-\frac{22}{3}$ for x in one of the original equations and solve for y.
$$5x + 6y = -12 \quad (1)$$
$$5\left(-\frac{22}{3}\right) + 6y = -12$$
$$-\frac{110}{3} + 6y = -12$$
$$6y = \frac{74}{3}$$
$$y = \frac{37}{9}$$

The solution is $\left(-\dfrac{22}{3}, \dfrac{37}{9}\right)$.

42. $(9, -4)$

43.
$$y^2 - 3 = 6$$
$$y^2 = 9$$
$y = 3 \ \ or \ \ y = -3$ Principle of square roots
The solutions are 3 and -3.

44. ± 1

45. ***Familiarize.*** Let $p = $ the price of the radio before the tax was added. Then the total price is $p + 5\%p$, or $p + 0.05p$, or $1.05p$.

Translate.

$\underbrace{\text{The total price}}$ is \$36.75.

$\quad\quad\quad \downarrow \quad\quad\quad\quad \downarrow \quad \downarrow$

$\quad\quad 1.05p \quad\quad = \quad 36.75$

Carry out. We solve the equation.
$$1.05p = 36.75$$
$$p = \frac{36.75}{1.05}$$
$$p = 35$$

Check. 5% of \$35 is \$1.75 and \$35 + \$1.75 = \$36.75. The answer checks.

State. The price before tax was \$35.

46. 69

47. Writing exercise

48. Writing exercise

49. Since the intercepts are $(0, 6)$ and $(0, -6)$, we know that the hyperbola is of the form $\dfrac{y^2}{b^2} - \dfrac{x^2}{a^2} = 1$ and that $b = 6$. The equations of the asymptotes tell us that $b/a = 3$, so
$$\frac{6}{a} = 3$$
$$a = 2.$$
The equation is $\dfrac{y^2}{6^2} - \dfrac{x^2}{2^2} = 1$, or $\dfrac{y^2}{36} - \dfrac{x^2}{4} = 1$.

50. $\dfrac{x^2}{64} - \dfrac{y^2}{1024} = 1$

51.
$$\frac{(x-5)^2}{36} - \frac{(y-2)^2}{25} = 1$$
$$\frac{(x-5)^2}{6^2} - \frac{(y-2)^2}{5^2} = 1$$
$h = 5, \ k = 2, \ a = 6, \ b = 5$
Center: $(5, 2)$

Vertices: $(-1, 2)$ and $(11, 2)$

Asymptotes: $y - 2 = \dfrac{5}{6}(x - 5)$ and $y - 2 = -\dfrac{5}{6}(x - 5)$

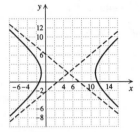

52. Center: $(2, 1)$

Vertices: $(-1, 1)$ and $(5, 1)$

Asymptotes: $y - 1 = \dfrac{2}{3}(x - 2)$ and $y - 1 = -\dfrac{2}{3}(x - 2)$

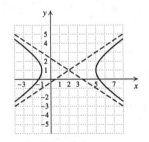

53.
$$8(y + 3)^2 - 2(x - 4)^2 = 32$$
$$\frac{(y + 3)^2}{4} - \frac{(x - 4)^2}{16} = 1$$
$$\frac{(y - (-3))^2}{2^2} - \frac{(x - 4)^2}{4^2} = 1$$
$h = 4,\ k = -3,\ a = 4,\ b = 2$

Center: $(4, -3)$

Vertices: $(4, -3 + 2)$ and $(4, -3 - 2)$, or $(4, -1)$ and $(4, -5)$

Asymptotes: $y - (-3) = \dfrac{2}{4}(x - 4)$ and

$y - (-3) = -\dfrac{2}{4}(x - 4)$, or $y + 3 = \dfrac{1}{2}(x - 4)$ and

$y + 3 = -\dfrac{1}{2}(x - 4)$

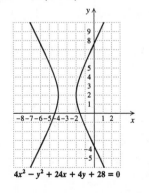

54. Center: $(4, -5)$

Vertices: $(2, -5)$ and $(6, -5)$

Asymptotes: $y + 5 = \dfrac{5}{2}(x - 4)$ and $y + 5 = -\dfrac{5}{2}(x - 4)$

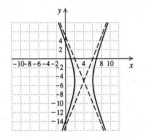

55.
$$4x^2 - y^2 + 24x + 4y + 28 = 0$$
$$4(x^2 + 6x) - (y^2 - 4y) = -28$$
$$4(x^2 + 6x + 9 - 9) - (y^2 - 4y + 4 - 4) = -28$$
$$4(x^2 + 6x + 9) - (y^2 - 4y + 4) = -28 + 4 \cdot 9 - 4$$
$$4(x + 3)^2 - (y - 2)^2 = 4$$
$$\frac{(x + 3)^2}{1} - \frac{(y - 2)^2}{4} = 1$$
$$\frac{(x - (-3))^2}{1^2} - \frac{(y - 2)^2}{2^2} = 1$$
$h = -3,\ k = 2,\ a = 1,\ b = 2$

Center: $(-3, 2)$

Vertices: $(-3 - 1, 2)$, and $(-3 + 1, 2)$, or $(-4, 2)$ and $(-2, 2)$

Asymptotes: $y - 2 = \dfrac{2}{1}(x - (-3))$ and

$y - 2 = -\dfrac{2}{1}(x - (-3))$, or $y - 2 = 2(x + 3)$ and

$y - 2 = -2(x + 3)$

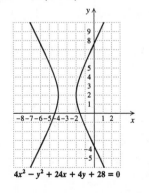

$4x^2 - y^2 + 24x + 4y + 28 = 0$

56. Center: $(-2, 1)$

Vertices: $(-2, 6)$ and $(-2, -4)$

Asymptotes: $y - 1 = \dfrac{5}{2}(x + 2)$ and

$y - 1 = -\dfrac{5}{2}(x + 2)$

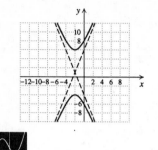

57. ▨

Exercise Set 13.4

---

1. $x^2 + y^2 = 25,$ (1)

$y - x = 1$ (2)

First solve Eq. (2) for y.

$y = x + 1$ (3)

Then substitute $x + 1$ for y in Eq. (1) and solve for x.

$$x^2 + y^2 = 25$$
$$x^2 + (x + 1)^2 = 25$$
$$x^2 + x^2 + 2x + 1 = 25$$
$$2x^2 + 2x - 24 = 0$$
$$x^2 + x - 12 = 0$$
$$(x + 4)(x - 3) = 0$$

$x + 4 = 0$ *or* $x - 3 = 0$ Principle of zero products

$x = -4$ *or* $x = 3$

Now substitute these numbers in Eq. (3) and solve for y.

$y = -4 + 1 = -3$

$y = 3 + 1 = 4$

The pairs $(-4, -3)$ and $(3, 4)$ check, so they are the solutions.

2. $(-8, -6),\ (6, 8)$

3. $9x^2 + 4y^2 = 36,$ (1)

$3x + 2y = 6$ (2)

First solve Eq. (2) for x.

$3x = 6 - 2y$

$x = 2 - \dfrac{2}{3}y$ (3)

Then substitute $2 - \dfrac{2}{3}y$ for x in Eq. (1) and solve for y.

$$9x^2 + 4y^2 = 36$$
$$9\left(2 - \frac{2}{3}y\right)^2 + 4y^2 = 36$$
$$9\left(4 - \frac{8}{3}y + \frac{4}{9}y^2\right) + 4y^2 = 36$$
$$36 - 24y + 4y^2 + 4y^2 = 36$$
$$8y^2 - 24y = 0$$
$$y^2 - 3y = 0$$
$$y(y - 3) = 0$$

$y = 0$ *or* $y = 3$

Now substitute these numbers in Eq. (3) and solve for x.

$x = 2 - \dfrac{2}{3}(0) = 2$

$x = 2 - \dfrac{2}{3}(3) = 0$

The pairs $(2, 0)$ and $(0, 3)$ check, so they are the solutions.

4. $(0, 2),\ (3, 0)$

5. $y = x^2,$ (1)

$3x = y + 2$ (2)

First solve Eq. (2) for y.

$y = 3x - 2$ (3)

Then substitute $3x - 2$ for y in Eq. (1) and solve for x.

$$y = x^2$$
$$3x - 2 = x^2$$
$$0 = x^2 - 3x + 2$$
$$0 = (x - 2)(x - 1)$$

$x = 2$ *or* $x = 1$

Now substitute these numbers in Eq. (3) and solve for y.

$y = 3 \cdot 2 - 2 = 4$

$y = 3 \cdot 1 - 2 = 1$

The pairs $(2, 4)$ and $(1, 1)$ check, so they are the solutions.

6. $(-2, 1)$

7. $2y^2 + xy + x^2 = 7,$ (1)

$x - 2y = 5$ (2)

First solve Eq. (2) for x.

$x = 2y + 5$ (3)

Then substitute $2y + 5$ for x in Eq. (1) and solve for y.

$$2y^2 + xy + x^2 = 7$$
$$2y^2 + (2y + 5)y + (2y + 5)^2 = 7$$
$$2y^2 + 2y^2 + 5y + 4y^2 + 20y + 25 = 7$$
$$8y^2 + 25y + 18 = 0$$
$$(8y + 9)(y + 2) = 0$$

$$y = -\frac{9}{8} \text{ or } y = -2$$

Now substitute these numbers in Eq. (3) and solve for x.

$$x = 2\left(-\frac{9}{8}\right) + 5 = \frac{11}{4}$$

$$x = 2(-2) + 5 = 1$$

The pairs $\left(\frac{11}{4}, -\frac{9}{8}\right)$ and $(1, -2)$ check, so they are the solutions.

8. $\left(\frac{5 + \sqrt{70}}{3}, \frac{-1 + \sqrt{70}}{3}\right)$, $\left(\frac{5 - \sqrt{70}}{3}, \frac{-1 - \sqrt{70}}{3}\right)$

9. $x^2 - y^2 = 16$, (1)

$\quad x - 2y = 1$ (2)

First solve Eq. (2) for x.

$$x = 2y + 1 \quad (3)$$

Then substitute $2y + 1$ for x in Eq. (1) and solve for y.

$$x^2 - y^2 = 16$$
$$(2y + 1)^2 - y^2 = 16$$
$$4y^2 + 4y + 1 - y^2 = 16$$
$$3y^2 + 4y - 15 = 0$$
$$(3y - 5)(y + 3) = 0$$

$$y = \frac{5}{3} \text{ or } y = -3$$

Now substitute these numbers in Eq. (3) and find x.

$$x = 2\left(\frac{5}{3}\right) + 1 = \frac{13}{3}$$

$$x = 2(-3) + 1 = -5$$

The pairs $\left(\frac{13}{3}, \frac{5}{3}\right)$ and $(-5, -3)$ check, so they are the solutions.

10. $\left(4, \frac{3}{2}\right)$, $(3, 2)$

11. $m^2 + 3n^2 = 10$, (1)

$\quad m - n = 2$ (2)

First solve Eq. (2) for m.

$$m = n + 2 \quad (3)$$

Then substitute $n + 2$ for m in Eq. (1) and solve for n.

$$m^2 + 3n^2 = 10$$
$$(n + 2)^2 + 3n^2 = 10$$
$$n^2 + 4n + 4 + 3n^2 = 10$$
$$4n^2 + 4n - 6 = 0$$
$$2n^2 + 2n - 3 = 0$$

$$n = \frac{-2 \pm \sqrt{2^2 - 4(2)(-3)}}{2 \cdot 2} = \frac{-1 \pm \sqrt{7}}{2}$$

Now substitute these numbers in Eq. (3) and solve for m.

$$m = \frac{-1 + \sqrt{7}}{2} + 2 = \frac{3 + \sqrt{7}}{2}$$

$$m = \frac{-1 - \sqrt{7}}{2} + 2 = \frac{3 - \sqrt{7}}{2}$$

The pairs $\left(\frac{3 + \sqrt{7}}{2}, \frac{-1 + \sqrt{7}}{2}\right)$ and

$\left(\frac{3 - \sqrt{7}}{2}, \frac{-1 - \sqrt{7}}{2}\right)$ check, so they are the solutions.

12. $\left(\frac{7}{3}, \frac{1}{3}\right)$, $(1, -1)$

13. $2y^2 + xy = 5$, (1)

$\quad 4y + x = 7$ (2)

First solve Eq. (2) for x.

$$x = -4y + 7 \quad (3)$$

Then substitute $-4y + 7$ for x in Eq. (3) and solve for y.

$$2y^2 + xy = 5$$
$$2y^2 + (-4y + 7)y = 5$$
$$2y^2 - 4y^2 + 7y = 5$$
$$0 = 2y^2 - 7y + 5$$
$$0 = (2y - 5)(y - 1)$$

$$y = \frac{5}{2} \text{ or } y = 1$$

Now substitute these numbers in Eq. (3) and solve for x.

$$x = -4\left(\frac{5}{2}\right) + 7 = -3$$

$$x = -4(1) + 7 = 3$$

The pairs $\left(-3, \frac{5}{2}\right)$ and $(3, 1)$ check, so they are the solutions.

14. $\left(\frac{11}{4}, -\frac{5}{4}\right)$, $(1, 4)$

15. $p + q = -6$, (1)

$\quad pq = -7$ (2)

First solve Eq. (1) for p.

$$p = -q - 6 \quad (3)$$

Then substitute $-q - 6$ for p in Eq. (2) and solve for q.

$$pq = -7$$
$$(-q - 6)q = -7$$
$$-q^2 - 6q = -7$$
$$0 = q^2 + 6q - 7$$
$$0 = (q + 7)(q - 1)$$

$q = -7 \ or \ q = 1$

Now substitute these numbers in Eq. (3) and solve for p.

$$p = -(-7) - 6 = 1$$
$$p = -1 - 6 = -7$$

The pairs $(1, -7)$ and $(-7, 1)$ check, so they are the solutions.

16. $\left(\dfrac{7 - \sqrt{33}}{2}, \dfrac{7 + \sqrt{33}}{2} \right), \left(\dfrac{7 + \sqrt{33}}{2}, \dfrac{7 - \sqrt{33}}{2} \right)$

17. $4x^2 + 9y^2 = 36, \quad (1)$

   $x + 3y = 3 \qquad (2)$

First solve Eq. (1) for x.

$x = -3y + 3 \qquad (3)$

Then substitute $-3y + 3$ for x in Eq. (1) and solve for y.

$$4x^2 + 9y^2 = 36$$
$$4(-3y + 3)^2 + 9y^2 = 36$$
$$4(9y^2 - 18y + 9) + 9y^2 = 36$$
$$36y^2 - 72y + 36 + 9y^2 = 36$$
$$45y^2 - 72y = 0$$
$$5y^2 - 8y = 0$$
$$y(5y - 8) = 0$$

$y = 0 \ or \ y = \dfrac{8}{5}$

Now substitute these numbers in Eq. (3) and solve for x.

$x = -3 \cdot 0 + 3 = 3$

$x = -3\left(\dfrac{8}{5}\right) + 3 = -\dfrac{9}{5}$

The pairs $(3, 0)$ and $\left(-\dfrac{9}{5}, \dfrac{8}{5} \right)$ check, so they are the solutions.

18. $(3, -5), (-1, 3)$

19. $xy = 4, \qquad (1)$

   $x + y = 5 \qquad (2)$

First solve Eq. (2) for x.

$x = -y + 5 \quad (3)$

Substitute $-y + 5$ for x in Eq. (1) and solve for y.

$$xy = 4$$
$$(-y + 5)y = 4$$
$$-y^2 + 5y = 4$$
$$0 = y^2 - 5y + 4$$
$$0 = (y - 4)(y - 1)$$

$y = 4 \ or \ y = 1$

Then substitute these numbers in Eq. (3) and solve for x.

$x = -4 + 5 = 1$

$x = -1 + 5 = 4$

The pairs $(1, 4)$ and $(4, 1)$ check, so they are the solutions.

20. $(-5, -8), (8, 5)$

21. $y = x^2, \quad (1)$

   $x = y^2 \quad (2)$

Eq. (1) is already solved for y. Substitute x^2 for y in Eq. (2) and solve for x.

$x = y^2$
$x = (x^2)^2$
$x = x^4$
$0 = x^4 - x$
$0 = x(x^3 - 1)$
$0 = x(x - 1)(x^2 + x + 1)$

$x = 0 \quad or \quad x = 1 \quad or \quad x = \dfrac{-1 \pm \sqrt{1^2 - 4 \cdot 1 \cdot 1}}{2}$

$x = 0 \quad or \quad x = 1 \quad or \quad x = -\dfrac{1}{2} \pm \dfrac{\sqrt{3}}{2}i$

Substitute these numbers in Eq. (1) and solve for y.

$y = 0^2 = 0$

$y = 1^2 = 1$

$y = \left(-\dfrac{1}{2} + \dfrac{\sqrt{3}}{2}i \right)^2 = -\dfrac{1}{2} - \dfrac{\sqrt{3}}{2}i$

$y = \left(-\dfrac{1}{2} - \dfrac{\sqrt{3}}{2}i \right)^2 = -\dfrac{1}{2} + \dfrac{\sqrt{3}}{2}i$

The pairs $(0, 0)$, $(1, 1)$, $\left(-\dfrac{1}{2} + \dfrac{\sqrt{3}}{2}i, -\dfrac{1}{2} - \dfrac{\sqrt{3}}{2}i \right)$, and $\left(-\dfrac{1}{2} - \dfrac{\sqrt{3}}{2}i, -\dfrac{1}{2} + \dfrac{\sqrt{3}}{2}i \right)$ check, so they are the solutions.

22. $(-5, 0), (4, 3), (4, -3)$

23. $x^2 + y^2 = 9, \qquad (1)$

   $x^2 - y^2 = 9 \qquad (2)$

Here we use the elimination method.

$$x^2 + y^2 = 9 \quad (1)$$
$$\underline{x^2 - y^2 = 9 \quad (2)}$$
$$2x^2 \quad\quad = 18 \quad \text{Adding}$$
$$x^2 = 9$$
$$x = \pm 3$$

If $x = 3$, $x^2 = 9$, and if $x = -3$, $x^2 = 9$, so substituting 3 or -3 in Eq. (1) gives us

$$x^2 + y^2 = 9$$
$$9 + y^2 = 9$$
$$y^2 = 0$$
$$y = 0.$$

The pairs $(3, 0)$ and $(-3, 0)$ check. They are the solutions.

24. $(0, 2)$, $(0, -2)$

25. $x^2 + y^2 = 25$, $\quad$ (1)

$\quad xy = 12 \quad\quad$ (2)

First we solve Eq. (2) for y.

$$xy = 12$$
$$y = \frac{12}{x}$$

Then we substitute $\frac{12}{x}$ for y in Eq. (1) and solve for x.

$$x^2 + y^2 = 25$$
$$x^2 + \left(\frac{12}{x}\right)^2 = 25$$
$$x^2 + \frac{144}{x^2} = 25$$
$$x^4 + 144 = 25x^2 \quad \text{Multiplying by } x^2$$
$$x^4 - 25x^2 + 144 = 0$$
$$u^2 - 25u + 144 = 0 \quad\quad \text{Letting } u = x^2$$
$$(u - 9)(u - 16) = 0$$
$$u = 9 \ \text{ or } \ u = 16$$

We now substitute x^2 for u and solve for x.

$$x^2 = 9 \quad \text{or} \quad x^2 = 16$$
$$x = \pm 3 \quad \text{or} \quad x = \pm 4$$

Since $y = 12/x$, if $x = 3$, $y = 4$; if $x = -3$, $y = -4$; if $x = 4$, $y = 3$; and if $x = -4$, $y = -3$. The pairs $(3, 4)$, $(-3, -4)$, $(4, 3)$, and $(-4, -3)$ check. They are the solutions.

26. $(-5, 3)$, $(-5, -3)$, $(4, 0)$

27. $x^2 + y^2 = 4$, $\quad\quad$ (1)

$\quad 9x^2 + 16y^2 = 144 \quad$ (2)

$$-9x^2 - 9y^2 = -36 \quad \text{Multiplying (1) by } -9$$
$$\underline{9x^2 + 16y^2 = 144}$$
$$7y^2 = 108 \quad \text{Adding}$$
$$y^2 = \frac{108}{7}$$
$$y = \pm\sqrt{\frac{108}{7}} = \pm 6\sqrt{\frac{3}{7}}$$
$$y = \pm\frac{6\sqrt{21}}{7} \quad \text{Rationalizing the denominator}$$

Substituting $\frac{6\sqrt{21}}{7}$ or $-\frac{6\sqrt{21}}{7}$ for y in Eq. (1) gives us

$$x^2 + \frac{36 \cdot 21}{49} = 4$$
$$x^2 = 4 - \frac{108}{7}$$
$$x^2 = -\frac{80}{7}$$
$$x = \pm\sqrt{-\frac{80}{7}} = \pm 4i\sqrt{\frac{5}{7}}$$
$$x = \pm\frac{4i\sqrt{35}}{7}. \quad \text{Rationalizing the denominator}$$

The pairs $\left(\frac{4i\sqrt{35}}{7}, \frac{6\sqrt{21}}{7}\right)$, $\left(-\frac{4i\sqrt{35}}{7}, \frac{6\sqrt{21}}{7}\right)$, $\left(\frac{4i\sqrt{35}}{7}, -\frac{6\sqrt{21}}{7}\right)$, and $\left(-\frac{4i\sqrt{35}}{7}, -\frac{6\sqrt{21}}{7}\right)$ check. They are the solutions.

28. $\left(\frac{16}{3}, \frac{5\sqrt{7}}{3}i\right)$, $\left(\frac{16}{3}, -\frac{5\sqrt{7}}{3}i\right)$, $\left(-\frac{16}{3}, \frac{5\sqrt{7}}{3}i\right)$, $\left(-\frac{16}{3}, -\frac{5\sqrt{7}}{3}i\right)$

29. $\quad x^2 + y^2 = 16$, $\quad\quad\quad x^2 + y^2 = 16$, $\quad$ (1)

$\quad\quad\quad\quad\quad\quad\quad\quad$ or

$\quad y^2 - 2x^2 = 10 \quad\quad -2x^2 + y^2 = 10 \quad$ (2)

Here we use the elimination method.

$$2x^2 + 2y^2 = 32 \quad \text{Multiplying (1) by 2}$$
$$\underline{-2x^2 + y^2 = 10}$$
$$3y^2 = 42 \quad \text{Adding}$$
$$y^2 = 14$$
$$y = \pm\sqrt{14}$$

Substituting $\sqrt{14}$ or $-\sqrt{14}$ for y in Eq. (1) gives us

$$x^2 + 14 = 16$$
$$x^2 = 2$$
$$x = \pm\sqrt{2}$$

The pairs $(-\sqrt{2}, -\sqrt{14})$, $(-\sqrt{2}, \sqrt{14})$, $(\sqrt{2}, -\sqrt{14})$, and $(\sqrt{2}, \sqrt{14})$ check. They are the solutions.

30. $(-3, -\sqrt{5})$, $(-3, \sqrt{5})$, $(3, -\sqrt{5})$, $(3, \sqrt{5})$

31. $x^2 + y^2 = 5$,   (1)
$xy = 2$ (2)

First we solve Eq. (2) for y.

$xy = 2$

$y = \dfrac{2}{x}$

Then we substitute $\dfrac{2}{x}$ for y in Eq. (1) and solve for x.

$$x^2 + y^2 = 5$$
$$x^2 + \left(\frac{2}{x}\right)^2 = 5$$
$$x^2 + \frac{4}{x^2} = 5$$
$$x^4 + 4 = 5x^2 \quad \text{Multiplying by } x^2$$
$$x^4 - 5x^2 + 4 = 0$$
$$u^2 - 5u + 4 = 0 \quad \text{Letting } u = x^2$$
$$(u - 4)(u - 1) = 0$$
$$u = 4 \ or \ u = 1$$

We now substitute x^2 for u and solve for x.
$x^2 = 4 \quad or \quad x^2 = 1$
$x = \pm 2 \quad or \quad x = \pm 1$

Since $y = 2/x$, if $x = 2$, $y = 1$; if $x = -2$, $y = -1$; if $x = 1$, $y = 2$; and if $x = -1$, $y = -2$. The pairs $(2, 1)$, $(-2, -1)$, $(1, 2)$, and $(-1, -2)$ check. They are the solutions.

32. $(4, 2)$, $(-4, -2)$, $(2, 4)$, $(-2, -4)$

33. $x^2 + y^2 = 13$,   (1)

$xy = 6$ (2)

First we solve Eq. (2) for y.

$xy = 6$

$y = \dfrac{6}{x}$

Then we substitute $\dfrac{6}{x}$ for y in Eq. (1) and solve for x.

$$x^2 + y^2 = 13$$
$$x^2 + \left(\frac{6}{x}\right)^2 = 13$$
$$x^2 + \frac{36}{x^2} = 13$$
$$x^4 + 36 = 13x^2 \quad \text{Multiplying by } x^2$$
$$x^4 - 13x^2 + 36 = 0$$
$$u^2 - 13u + 36 = 0 \qquad \text{Letting } u = x^2$$
$$(u - 9)(u - 4) = 0$$
$$u = 9 \quad or \quad u = 4$$

We now substitute x^2 for u and solve for x.
$x^2 = 9 \quad or \quad x^2 = 4$
$x = \pm 3 \quad or \quad x = \pm 2$

Since $y = 6/x$, if $x = 3$, $y = 2$; if $x = -3$, $y = -2$; if $x = 2$, $y = 3$; and if $x = -2$, $y = -3$. The pairs $(3, 2)$, $(-3, -2)$, $(2, 3)$, and $(-2, -3)$ check. They are the solutions.

34. $(4, 1)$, $(-4, -1)$, $(2, 2)$, $(-2, -2)$

35. $3xy + x^2 = 34$,   (1)
$2xy - 3x^2 = 8$ (2)

$$
\begin{array}{ll}
6xy + \ 2x^2 = \ 68 & \text{Multiplying (1) by 2} \\
\underline{-6xy + \ 9x^2 = -24} & \text{Multiplying (2) by } -3 \\
11x^2 = \ 44 & \text{Adding} \\
x^2 = \ 4 & \\
x = \ \pm 2 &
\end{array}
$$

Substitute for x in Eq. (1) and solve for y.

When $x = 2$: $3 \cdot 2 \cdot y + 2^2 = 34$
$$6y + 4 = 34$$
$$6y = 30$$
$$y = 5$$

When $x = -2$: $3(-2)(y) + (-2)^2 = 34$
$$-6y + 4 = 34$$
$$-6y = 30$$
$$y = -5$$

The pairs $(2, 5)$ and $(-2, -5)$ check. They are the solutions.

36. $(2, 1)$, $(-2, -1)$

37. $xy - y^2 = 2$,   (1)
$2xy - 3y^2 = 0$ (2)

$$
\begin{array}{ll}
-2xy + \ 2y^2 = -4 & \text{Multiplying (1) by } -2 \\
\underline{2xy - \ 3y^2 = \ 0} & \\
-y^2 = -4 & \text{Adding} \\
y^2 = \ 4 & \\
y = \ \pm 2 &
\end{array}
$$

We substitute for y in Eq. (1) and solve for x.

When $y = 2$: $x \cdot 2 - 2^2 = 2$
$$2x - 4 = 2$$
$$2x = 6$$
$$x = 3$$

When $y = -2$: $x(-2) - (-2)^2 = 2$
$$-2x - 4 = 2$$
$$-2x = 6$$
$$x = -3$$

The pairs $(3, 2)$ and $(-3, -2)$ check. They are the solutions.

38. $\left(2, -\frac{4}{5}\right)$, $\left(-2, -\frac{4}{5}\right)$, $(5, 2)$, $(-5, 2)$

39. $x^2 - y = 5$, (1)

$x^2 + y^2 = 25$ (2)

We solve Eq. (1) for y.

$x^2 - 5 = y$ (3)

Substitute $x^2 - 5$ for y in Eq. (2) and solve for x.

$$x^2 + (x^2 - 5)^2 = 25$$
$$x^2 + x^4 - 10x^2 + 25 = 25$$
$$x^4 - 9x^2 = 0$$
$$u^2 - 9u = 0 \quad \text{Letting } u = x^2$$
$$u(u - 9) = 0$$

$u = 0 \quad or \quad u = 9$

$x^2 = 0 \quad or \quad x^2 = 9$

$x = 0 \quad or \quad x = \pm 3$

Substitute in Eq. (3) and solve for y.

When $x = 0$: $y = 0^2 - 5 = -5$

When $x = 3$ or -3: $y = 9 - 5 = 4$

The pairs $(0, -5)$, $(3, 4)$, and $(-3, 4)$ check. They are the solutions.

(This exercise could also be solved using the elimination method.)

40. $(-\sqrt{2}, \sqrt{2})$, $(\sqrt{2}, -\sqrt{2})$

41. *Familiarize*. We first make a drawing. We let l and w represent the length and width, respectively.

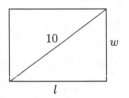

Translate. The perimeter is 28 cm.

$2l + 2w = 28$, or $l + w = 14$

Using the Pythagorean theorem we have another equation.

$l^2 + w^2 = 10^2$, or $l^2 + w^2 = 100$

Carry out. We solve the system:

$l + w = 14$, (1)

$l^2 + w^2 = 100$ (2)

First solve Eq. (1) for w.

$w = 14 - l$ (3)

Then substitute $14 - l$ for w in Eq. (2) and solve for l.

$$l^2 + w^2 = 100$$
$$l^2 + (14 - l)^2 = 100$$
$$l^2 + 196 - 28l + l^2 = 100$$
$$2l^2 - 28l + 96 = 0$$
$$l^2 - 14l + 48 = 0$$
$$(l - 8)(l - 6) = 0$$

$l = 8 \quad or \quad l = 6$

If $l = 8$, then $w = 14 - 8$, or 6. If $l = 6$, then $w = 14 - 6$, or 8. Since the length is usually considered to be longer than the width, we have the solution $l = 8$ and $w = 6$, or $(8, 6)$.

Check. If $l = 8$ and $w = 6$, then the perimeter is $2 \cdot 8 + 2 \cdot 6$, or 28. The length of a diagonal is $\sqrt{8^2 + 6^2}$, or $\sqrt{100}$, or 10. The numbers check.

State. The length is 8 cm, and the width is 6 cm.

42. Length: 2 yd; width: 1 yd

43. *Familiarize*. We first make a drawing. Let $l =$ the length and $w =$ the width of the rectangle.

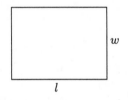

Translate.

Area: $lw = 20$

Perimeter: $2l + 2w = 18$, or $l + w = 9$

Carry out. We solve the system:

Solve the second equation for l: $l = 9 - w$

Substitute $9 - w$ for l in the first equation and solve for w.

$$(9 - w)w = 20$$
$$9w - w^2 = 20$$
$$0 = w^2 - 9w + 20$$
$$0 = (w - 5)(w - 4)$$

$w = 5 \quad or \quad w = 4$

If $w = 5$, then $l = 9 - w$, or 4. If $w = 4$, then $l = 9 - 4$, or 5. Since length is usually considered to be longer than width, we have the solution $l = 5$ and $w = 4$, or $(5, 4)$.

Check. If $l = 5$ and $w = 4$, the area is $5 \cdot 4$, or 20. The perimeter is $2 \cdot 5 + 2 \cdot 4$, or 18. The numbers check.

State. The length is 5 in. and the width is 4 in.

44. Length is 2 in.; width: 1 in.

45. Familiarize. We first make a drawing. Let l = the length and w = the width of the cargo area, in feet.

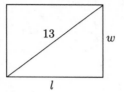

Translate. The cargo area must be 60 ft^2, so we have one equation:

$$lw = 60$$

The Pythagorean equation gives us another equation:

$$l^2 + w^2 = 13^2, \text{ or } l^2 + w^2 = 169$$

Carry out. We solve the system of equations.

$$lw = 60, \qquad (1)$$
$$l^2 + w^2 = 169 \quad (2)$$

First solve Eq. (1) for w:

$$lw = 60$$
$$w = \frac{60}{l} \quad (3)$$

Then substitute $60/l$ for w in Eq. (2) and solve for l.

$$l^2 + w^2 = 169$$
$$l^2 + \left(\frac{60}{l}\right)^2 = 169$$
$$l^2 + \frac{3600}{l^2} = 169$$
$$l^4 + 3600 = 169l^2$$
$$l^4 - 169l^2 + 3600 = 0$$

Let $u = l^2$ and $u^2 = l^4$ and substitute.

$$u^2 - 169u + 3600 = 0$$
$$(u - 144)(u - 25) = 0$$
$$u = 144 \quad or \quad u = 25$$
$$l^2 = 144 \quad or \quad l^2 = 25 \quad \text{Replacing } u \text{ with } l^2$$
$$l = \pm 12 \quad or \quad l = \pm 5$$

Since the length cannot be negative, we consider only 12 and 5. We substitute in Eq. (3) to find w. When $l = 12$, $w = 60/12 = 5$; when $l = 5$, $w = 60/5 = 12$. Since we usually consider length to be longer than width, we check the pair (12.5).

Check. If the length is 12 ft and the width is 5 ft, then the area is $12 \cdot 5$, or 60 ft^2. Also $12^2 + 5^2 = 144 + 25 = 169 = 13^2$. The answer checks.

State. The length is 12 ft and the width is 5 ft.

46. Length: 20 ft; widt: 15 ft

47. Familiarize. We make a drawing and label it. Let x and y represent the lengths of the legs of the triangle.

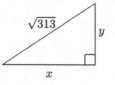

Translate. The product of the lengths of the legs is 156, so we have:

$$xy = 156$$

We use the Pythagorean theorem to get a second equation:

$$x^2 + y^2 = (\sqrt{313})^2, \text{ or } x^2 + y^2 = 313$$

Carry out. We solve the system of equations.

$$xy = 156, \qquad (1)$$
$$x^2 + y^2 = 313 \qquad (2)$$

First solve Equation (1) for y.

$$xy = 156$$
$$y = \frac{156}{x}$$

Then we substitute $\frac{156}{x}$ for y in Eq. (2) and solve for x.

$$x^2 + y^2 = 313 \qquad (2)$$
$$x^2 + \left(\frac{156}{x}\right)^2 = 313$$
$$x^2 + \frac{24,336}{x^2} = 313$$
$$x^4 + 24,336 = 313x^2$$
$$x^4 - 313x^2 + 24,336 = 0$$
$$u^2 - 313u + 24,336 = 0 \qquad \text{Letting } u = x^2$$
$$(u - 169)(u - 144) = 0$$
$$u - 169 = 0 \quad or \quad u - 144 = 0$$
$$u = 169 \quad or \qquad u = 144$$

We now substitute x^2 for u and solve for x.

$$x = \pm 13 \quad or \quad x = \pm 12$$

Since $y = 156/x$, if $x = 13$, $y = 12$; if $x = -13$, $y = -12$; if $x = 12$, $y = 13$; and if $x = -12$, $y = -13$. The possible solutions are $(13, 12)$, $(-13, -12)$, $(12, 13)$, and $(-12, -13)$.

Check. Since measurements cannot be negative, we consider only $(13, 12)$ and $(12, 13)$. Since both possible solutions give the same pair of legs, we only need to check $(13, 12)$. If $x = 13$ and $y = 12$, their product is 156. Also, $\sqrt{13^2 + 12^2} = \sqrt{313}$. The numbers check.

State. The lengths of the legs are 13 and 12.

48. 6 and 10 or −6 and −10

49. *Familiarize*. Let p = the principal and r = the interest rate. If \$750 more had been invested, then the principle would have been $p + 750$. If the interest rate had been 1% less, it would have been $r - 0.01$. Recall that Interest = Principal × Rate.

***Translate*.** With principal p and interest rate r, the interest is \$225, so we have

$$pr = 225. \quad (1)$$

With principal $p + 750$ and interest rate $r - 0.01$, the interest is also \$225, so we have

$$(p + 750)(r - 0.01) = 225 \quad (2)$$

***Carry out*.** We solve the system of equations. First solve Eq. (1) for r.

$$pr = 225$$
$$r = \frac{225}{p} \quad (3)$$

Now substitute $\dfrac{225}{p}$ for r in Eq. (2) and solve for p.

$$(p + 750)\left(\frac{225}{p} - 0.01\right) = 225$$
$$225 - 0.01p + \frac{168,750}{p} - 7.5 = 225$$
$$-0.01p + \frac{168,750}{p} - 7.5 = 0$$

$-0.01p^2 + 168,750 - 7.5p = 0$ Multiplying by p

$p^2 + 750p + 16,875,000 = 0$ Multiplying by -100 and rearranging

$$(p - 3750)(p + 4500) = 0$$
$$p = 3750 \ \ or \ \ p = -4500$$

Since the principal cannot be negative, we consider only 3750. Substitute 3750 for p in Eq. (3) and find r.

$$r = \frac{225}{3750} = 0.06$$

***Check*.** \$3750 × 0.06 = \$225. Also, \$3750 + \$750 = \$4500, 0.06 − 0.01 = 0.05, and \$4500 × 0.05 = \$225. The answer checks.

***State*.** The principal was \$3750, and the interest rate was 0.06, or 6%.

50. 24 ft, 16 ft

51. *Familiarize*. We first make a drawing. Let l = the length and w = the width.

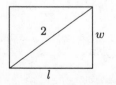

***Translate*.**

Area: $lw = \sqrt{3}$ (1)

From the Pythagorean theorem: $l^2 + w^2 = 2^2$ (2)

***Carry out*.** We solve the system of equations.

We first solve Eq. (1) for w.

$$lw = \sqrt{3}$$
$$w = \frac{\sqrt{3}}{l}$$

Then we substitute $\dfrac{\sqrt{3}}{l}$ for w in Eq. 2 and solve for l.

$$l^2 + \left(\frac{\sqrt{3}}{l}\right)^2 = 4$$
$$l^2 + \frac{3}{l^2} = 4$$
$$l^4 + 3 = 4l^2$$
$$l^4 - 4l^2 + 3 = 0$$
$$u^2 - 4u + 3 = 0 \quad \text{Letting } u = l^2$$
$$(u - 3)(u - 1) = 0$$
$$u = 3 \ or \ u = 1$$

We now substitute l^2 for u and solve for l.

$$l^2 = 3 \quad or \quad l^2 = 1$$
$$l = \pm\sqrt{3} \quad or \quad l = \pm 1$$

Measurements cannot be negative, so we only need to consider $l = \sqrt{3}$ and $l = 1$. Since $w = \sqrt{3}/l$, if $l = \sqrt{3}$, $w = 1$ and if $l = 1$, $w = \sqrt{3}$. Length is usually considered to be longer than width, so we have the solution $l = \sqrt{3}$ and $w = 1$, or $(\sqrt{3}, 1)$.

***Check*.** If $l = \sqrt{3}$ and $w = 1$, the area is $\sqrt{3} \cdot 1 = \sqrt{3}$. Also $(\sqrt{3})^2 + 1^2 = 3 + 1 = 4 = 2^2$. The numbers check.

***State*.** The length is $\sqrt{3}$ m, and the width is 1 m.

52. Length: $\sqrt{2}$ m; width: 1 m

53. Writing exercise

54. Writing exercise

55. $(-1)^9(-2)^4 = -1 \cdot 16 = -16$

56. −32

57. $\dfrac{(-1)^k}{k - 5} = \dfrac{(-1)^6}{6 - 5} = \dfrac{1}{1} = 1$

58. $-\dfrac{1}{4}$

59. $\dfrac{n}{2}(3 + n) = \dfrac{8}{2}(3 + 8) = 4 \cdot 11 = 44$

60. 28

61. Writing exercise

62. Writing exercise

63. Let $x =$ the length of the longer piece and $y =$ the length of the shorter piece. Then the lengths of the sides of the squares are $\frac{x}{4}$ and $\frac{y}{4}$. Solve the system:

$x + y = 100,$

$\left(\frac{x}{4}\right)^2 = \left(\frac{y}{4}\right)^2 + 144$

The solution is $(61.52, 38.48)$. One piece should be 61.52 cm long, and then the other will be 38.48 cm long.

64. $(x + 2)^2 + (y - 1)^2 = 4$

65. $\dfrac{x^2}{a^2} + \dfrac{y^2}{b^2} = 1$ Standard form

Substitute the coordinates of the given points:

$\dfrac{2^2}{a^2} + \dfrac{(-3)^2}{b^2} = 1,$

$\dfrac{1^2}{a^2} + \dfrac{(\sqrt{13})^2}{b^2} = 1,$ or

$\dfrac{4}{a^2} + \dfrac{9}{b^2} = 1,$ (1)

$\dfrac{1}{a^2} + \dfrac{13}{b^2} = 1$ (2)

Solve Eq. (2) for $\dfrac{1}{a^2}$:

$\dfrac{1}{a^2} = 1 - \dfrac{13}{b^2}$

$\dfrac{1}{a^2} = \dfrac{b^2 - 13}{b^2}$ (3)

Substitute $\dfrac{b^2 - 13}{b^2}$ for $\dfrac{1}{a^2}$ in Eq. (1) and solve for b^2.

$4\left(\dfrac{b^2 - 13}{b^2}\right) + \dfrac{9}{b^2} = 1$

$\dfrac{4b^2 - 52}{b^2} + \dfrac{9}{b^2} = 1$

$4b^2 - 52 + 9 = b^2$

$3b^2 = 43$

$b^2 = \dfrac{43}{3}$

Substitute $\dfrac{43}{3}$ for b^2 in Eq. (3) and solve for a^2.

$\dfrac{1}{a^2} = \dfrac{\frac{43}{3} - 13}{\frac{43}{3}} = \dfrac{\frac{43}{3} - 13}{\frac{43}{3}} \cdot \dfrac{3}{3}$

$\dfrac{1}{a^2} = \dfrac{43 - 3 \cdot 13}{43} = \dfrac{43 - 39}{43}$

$\dfrac{1}{a^2} = \dfrac{4}{43}$

$a^2 = \dfrac{43}{4}$

The equation of the ellipse is

$\dfrac{x^2}{\frac{43}{4}} + \dfrac{y^2}{\frac{43}{3}} = 1,$ or

$\dfrac{4x^2}{43} + \dfrac{3y^2}{43} = 1,$ or

$4x^2 + 3y^2 = 43.$

66. $(-2, 3),\ (2, -3),\ (-3, 2),\ (3, -2)$

67. $a + b = \dfrac{5}{6},$ (1)

   $\dfrac{a}{b} + \dfrac{b}{a} = \dfrac{13}{6}$ (2)

$b = \dfrac{5}{6} - a = \dfrac{5 - 6a}{6}$ Solving Eq. (1) for b

$\dfrac{a}{\frac{5 - 6a}{6}} + \dfrac{\frac{5 - 6a}{6}}{a} = \dfrac{13}{6}$ Substituting for b in Eq. (2)

$\dfrac{6a}{5 - 6a} + \dfrac{5 - 6a}{6a} = \dfrac{13}{6}$

$36a^2 + 25 - 60a + 36a^2 = 65a - 78a^2$

$150a^2 - 125a + 25 = 0$

$6a^2 - 5a + 1 = 0$

$(3a - 1)(2a - 1) = 0$

$a = \dfrac{1}{3}\ or\ a = \dfrac{1}{2}$

Substitute for a and solve for b.

When $a = \dfrac{1}{3},\ b = \dfrac{5 - 6\left(\frac{1}{3}\right)}{6} = \dfrac{1}{2}.$

When $a = \dfrac{1}{2},\ b = \dfrac{5 - 6\left(\frac{1}{2}\right)}{6} = \dfrac{1}{3}.$

The pairs $\left(\dfrac{1}{3}, \dfrac{1}{2}\right)$ and $\left(\dfrac{1}{2}, \dfrac{1}{3}\right)$ check. They are the solutions.

68. 10 in. by 7 in. by 5 in.

69. *Familiarize*. Let l = the length and h = the height, in cm.

Translate. Since the ratio of the length to the height is 4 to 3, we have one equation:

$$\frac{l}{h} = \frac{4}{3}$$

The Pythagorean equation gives us a second equation:

$$l^2 + h^2 = 31^2, \text{ or } l^2 + h^2 = 961$$

We have a system of equations.

$$\frac{l}{h} = \frac{4}{3}, \qquad (1)$$
$$l^2 + h^2 = 961 \quad (2)$$

Carry out. We solve the system of equations. First we solve Eq. (1) for l:

$$\frac{l}{h} = \frac{4}{3}$$

$$l = \frac{4}{3}h \quad (3)$$

Now substitute $\frac{4}{3}h$ for w in Eq. (2) and solve for h.

$$l^2 + h^2 = 961$$

$$\left(\frac{4}{3}h\right)^2 + h^2 = 961$$

$$\frac{16}{9}h^2 + h^2 = 961$$

$$\frac{25}{9}h^2 = 961$$

$$h^2 = \frac{961 \cdot 9}{25} = \frac{8649}{25}$$

$$h = \frac{93}{5} \quad or \quad h = -\frac{93}{5}$$

Since the height cannot be negative, we consider only $\frac{93}{5}$. Substitute $\frac{93}{5}$ for h in Eq. (3) and find l.

$$l = \frac{4}{3} \cdot \frac{93}{5} = \frac{124}{5}$$

Check. The ratio of 124/5 to 93/5 is $\dfrac{124/5}{93/5} = \dfrac{124}{93} = \dfrac{4}{3}$. Also $\left(\dfrac{124}{5}\right)^2 + \left(\dfrac{93}{5}\right)^2 = 961 = 31^2$. The answer checks.

State. The length is $\dfrac{124}{5}$ cm, or 24.8 cm, and the height is $\dfrac{93}{5}$ cm, or 18.6 cm.

70. Length: 61.02 in.; height: 34.32 in.

71.
$$R = C$$
$$100x + x^2 = 80x + 1500$$
$$x^2 + 20x - 1500 = 0$$
$$(x - 30)(x + 50) = 0$$

$x = 30 \text{ } or \text{ } x = -50$

Since the number of units cannot be negative, the solution of the problem is 30. Thus, 30 units must be sold in order to break even.

72.

Chapter 14

Sequences, Series, and Combinatorics

1. $a_n = 5n - 2$

$a_1 = 5 \cdot 1 - 2 = 3,$

$a_2 = 5 \cdot 2 - 2 = 8,$

$a_3 = 5 \cdot 3 - 2 = 13,$

$a_4 = 5 \cdot 4 - 2 = 18;$

$a_{10} = 5 \cdot 10 - 2 = 48;$

$a_{15} = 5 \cdot 15 - 2 = 73$

2. $5, 7, 9, 11; 23; 33$

3. $a_n = \dfrac{n}{n+1}$

$a_1 = \dfrac{1}{1+1} = \dfrac{1}{2},$

$a_2 = \dfrac{2}{2+1} = \dfrac{2}{3},$

$a_3 = \dfrac{3}{3+1} = \dfrac{3}{4},$

$a_4 = \dfrac{4}{4+1} = \dfrac{4}{5};$

$a_{10} = \dfrac{10}{10+1} = \dfrac{10}{11};$

$a_{15} = \dfrac{15}{15+1} = \dfrac{15}{16}$

4. $3, 6, 11, 18; 102; 227$

5. $a_n = n^2 - 2n$

$a_1 = 1^2 - 2 \cdot 1 = -1,$

$a_2 = 2^2 - 2 \cdot 2 = 0,$

$a_3 = 3^2 - 2 \cdot 3 = 3,$

$a_4 = 4^2 - 2 \cdot 4 = 8;$

$a_{10} = 10^2 - 2 \cdot 10 = 80;$

$a_{15} = 15^2 - 2 \cdot 15 = 195$

6. $0, \dfrac{3}{5}, \dfrac{4}{5}, \dfrac{15}{17}; \dfrac{99}{101}; \dfrac{112}{113}$

7. $a_n = n + \dfrac{1}{n}$

$a_1 = 1 + \dfrac{1}{1} = 2,$

$a_2 = 2 + \dfrac{1}{2} = 2\dfrac{1}{2},$

$a_3 = 3 + \dfrac{1}{3} = 3\dfrac{1}{3},$

$a_4 = 4 + \dfrac{1}{4} = 4\dfrac{1}{4};$

$a_{10} = 10 + \dfrac{1}{10} = 10\dfrac{1}{10};$

$a_{15} = 15 + \dfrac{1}{15} = 15\dfrac{1}{15}$

8. $1, -\dfrac{1}{2}, \dfrac{1}{4}, -\dfrac{1}{8}; -\dfrac{1}{512}; \dfrac{1}{16,384}$

9. $a_n = (-1)^n n^2$

$a_1 = (-1)^1 1^2 = -1,$

$a_2 = (-1)^2 2^2 = 4,$

$a_3 = (-1)^3 3^2 = -9,$

$a_4 = (-1)^4 4^2 = 16;$

$a_{10} = (-1)^{10} 10^2 = 100;$

$a_{15} = (-1)^{15} 15^2 = -225$

10. $-4, 5, -6, 7; 13; -18$

11. $a_n = (-1)^{n+1}(3n - 5)$

$a_1 = (-1)^{1+1}(3 \cdot 1 - 5) = -2,$

$a_2 = (-1)^{2+1}(3 \cdot 2 - 5) = -1,$

$a_3 = (-1)^{3+1}(3 \cdot 3 - 5) = 4,$

$a_4 = (-1)^{4+1}(3 \cdot 4 - 5) = -7;$

$a_{10} = (-1)^{10+1}(3 \cdot 10 - 5) = -25;$

$a_{15} = (-1)^{15+1}(3 \cdot 15 - 5) = 40$

12. $0, 7, -26, 63; 999; -3374$

13. $a_n = 2n - 5$

$a_7 = 2 \cdot 7 - 5 = 14 - 5 = 9$

14. 26

15. $a_n = (3n + 1)(2n - 5)$

$a_9 = (3 \cdot 9 + 1)(2 \cdot 9 - 5) = 28 \cdot 13 = 364$

16. 400

17. $a_n = (-1)^{n-1}(3.4n - 17.3)$

$a_{12} = (-1)^{12-1}[3.4(12) - 17.3] = -23.5$

18. $-37,916,508.16$

19. $a_n = 3n^2(9n - 100)$

$a_{11} = 3 \cdot 11^2(9 \cdot 11 - 100) = 3 \cdot 121(-1) = -363$

20. 9680

21. $a_n = \left(1 + \dfrac{1}{n}\right)^2$

$a_{20} = \left(1 + \dfrac{1}{20}\right)^2 = \left(\dfrac{21}{20}\right)^2 = \dfrac{441}{400}$

22. $\dfrac{2744}{3375}$

23. 1, 3, 5, 7, 9, . . .

These are odd integers, so the general term could be $2n - 1$.

24. $2n$

25. 1, −1, 1, −1, . . .

1 and −1 alternate, beginning with 1, so the general term could be $(-1)^{n+1}$.

26. $(-1)^n$

27. −1, 2, −3, 4, . . .

These are the first four natural numbers, but with alternating signs, beginning with a negative number. The general term could be $(-1)^n \cdot n$.

28. $(-1)^{n+1} \cdot n$

29. −2, 6, −18, 54, . . .

We can see a pattern if we write the sequence as

$-1 \cdot 2 \cdot 1,\ 1 \cdot 2 \cdot 3,\ -1 \cdot 2 \cdot 9,\ 1 \cdot 2 \cdot 27, \ldots$

The general term could be $(-1)^n 2(3)^{n-1}$.

30. $5n - 7$

31. $\dfrac{1}{2}, \dfrac{2}{3}, \dfrac{3}{4}, \dfrac{4}{5}, \dfrac{5}{6}, \ldots$

These are fractions in which the denominator is 1 greater than the numerator. Also, each numerator is 1 greater than the preceding numerator. The general term could be $\dfrac{n}{n+1}$.

32. $n(n + 1)$

33. 5, 25, 125, 625, . . .

This is powers of 5, so the general term could be 5^n.

34. 4^n

35. −1, 4, −9, 16, . . .

This is the squares of the first four natural numbers, but with alternating signs, beginning with a negative number. The general term could be $(-1)^n \cdot n^2$.

36. $(-1)^{n+1} \cdot n^2$

37. 1, −2, 3, −4, 5, −6, . . .

$S_7 = 1 - 2 + 3 - 4 + 5 - 6 + 7 = 4$

38. −8

39. 2, 4, 6, 8, . . .

$S_5 = 2 + 4 + 6 + 8 + 10 = 30$

40. $\dfrac{5269}{3600}$

41. $\displaystyle\sum_{k=1}^{5} \dfrac{1}{2k} = \dfrac{1}{2 \cdot 1} + \dfrac{1}{2 \cdot 2} + \dfrac{1}{2 \cdot 3} + \dfrac{1}{2 \cdot 4} + \dfrac{1}{2 \cdot 5}$

$= \dfrac{1}{2} + \dfrac{1}{4} + \dfrac{1}{6} + \dfrac{1}{8} + \dfrac{1}{10}$

$= \dfrac{60}{120} + \dfrac{30}{120} + \dfrac{20}{120} + \dfrac{15}{120} + \dfrac{12}{120}$

$= \dfrac{137}{120}$

42. $\dfrac{6508}{3465}$

43. $\displaystyle\sum_{k=0}^{4} 3^k = 3^0 + 3^1 + 3^2 + 3^3 + 3^4$

$= 1 + 3 + 9 + 27 + 81$

$= 121$

44. 13.7952

45. $\displaystyle\sum_{k=1}^{8} \dfrac{k}{k+1} = \dfrac{1}{1+1} + \dfrac{2}{2+1} + \dfrac{3}{3+1} + \dfrac{4}{4+1} +$

$\dfrac{5}{5+1} + \dfrac{6}{6+1} + \dfrac{7}{7+1} + \dfrac{8}{8+1}$

$= \dfrac{1}{2} + \dfrac{2}{3} + \dfrac{3}{4} + \dfrac{4}{5} + \dfrac{5}{6} + \dfrac{6}{7} + \dfrac{7}{8} + \dfrac{8}{9}$

$= \dfrac{15{,}551}{2520}$

46. $\dfrac{17}{84}$

47. $\displaystyle\sum_{k=1}^{8} (-1)^{k+1} 2^k = (-1)^2 2^1 + (-1)^3 2^2 + (-1)^4 2^3 +$

$(-1)^5 2^4 + (-1)^6 2^5 + (-1)^7 2^6 +$

$(-1)^8 2^7 + (-1)^9 2^8$

$= 2 - 4 + 8 - 16 + 32 - 64 +$

$128 - 256$

$= -170$

48. −52,432

49.
$$\sum_{k=0}^{5} (k^2 - 2k + 3)$$
$$= (0^2 - 2 \cdot 0 + 3) + (1^2 - 2 \cdot 1 + 3) +$$
$$(2^2 - 2 \cdot 2 + 3) + (3^2 - 2 \cdot 3 + 3) +$$
$$(4^2 - 2 \cdot 4 + 3) + (5^2 - 2 \cdot 5 + 3)$$
$$= 3 + 2 + 3 + 6 + 11 + 18$$
$$= 43$$

50. 34

51.
$$\sum_{k=3}^{5} \frac{(-1)^k}{k(k+1)} = \frac{(-1)^3}{3(3+1)} + \frac{(-1)^4}{4(4+1)} + \frac{(-1)^5}{5(5+1)}$$
$$= \frac{-1}{3 \cdot 4} + \frac{1}{4 \cdot 5} + \frac{-1}{5 \cdot 6}$$
$$= -\frac{1}{12} + \frac{1}{20} - \frac{1}{30}$$
$$= -\frac{4}{60} = -\frac{1}{15}$$

52. $\dfrac{119}{128}$

53. $\dfrac{2}{3} + \dfrac{3}{4} + \dfrac{4}{5} + \dfrac{5}{6} + \dfrac{6}{7}$

This is a sum of fractions in which the denominator is one greater than the numerator. Also, each numerator is 1 greater than the preceding numerator. Sigma notation is
$$\sum_{k=1}^{5} \frac{k+1}{k+2}.$$

54. $\displaystyle\sum_{k=1}^{5} 3k$

55. $1 + 4 + 9 + 16 + 25 + 36$

This is the sum of the squares of the first six natural numbers. Sigma notation is
$$\sum_{k=1}^{6} k^2.$$

56. $\displaystyle\sum_{k=1}^{5} \frac{1}{k^2}$

57. $4 - 9 + 16 - 25 + \ldots + (-1)^n n^2$

This is a sum of terms of the form $(-1)^k k^2$, beginning with $k = 2$ and continuing through $k = n$. Sigma notation is
$$\sum_{k=2}^{n} (-1)^k k^2.$$

58. $\displaystyle\sum_{k=3}^{n} (-1)^{k+1} k^2$

59. $5 + 10 + 15 + 20 + 25 + \ldots$

This is a sum of multiples of 5, and it is an infinite series. Sigma notation is
$$\sum_{k=1}^{\infty} 5k.$$

60. $\displaystyle\sum_{k=1}^{\infty} 7k$

61. $\dfrac{1}{1 \cdot 2} + \dfrac{1}{2 \cdot 3} + \dfrac{1}{3 \cdot 4} + \dfrac{1}{4 \cdot 5} + \ldots$

This is a sum of fractions in which the numerator is 1 and the denominator is a product of two consecutive integers. The larger integer in each product is the smaller integer in the succeeding product. It is an infinite series. Sigma notation is
$$\sum_{k=1}^{\infty} \frac{1}{k(k+1)}.$$

62. $\displaystyle\sum_{k=1}^{\infty} \frac{1}{k(k+1)^2}$

63. Writing exercise

64. Writing exercise

65. $\dfrac{7}{2}(a_1 + a_7) = \dfrac{7}{2}(8 + 14) = \dfrac{7}{2} \cdot 22 = 77$

66. 23

67.
$$(x+y)^3$$
$$= (x+y)(x+y)^2$$
$$= (x+y)(x^2 + 2xy + y^2)$$
$$= x(x^2 + 2xy + y^2) + y(x^2 + 2xy + y^2)$$
$$= x^3 + 2x^2 y + xy^2 + x^2 y + 2xy^2 + y^3$$
$$= x^3 + 3x^2 y + 3xy^2 + y^3$$

68. $a^3 - 3a^2 b + 3ab^2 - b^3$

69.
$$(2a - b)^3$$
$$= (2a - b)(2a - b)^2$$
$$= (2a - b)(4a^2 - 4ab + b^2)$$
$$= 2a(4a^2 - 4ab + b^2) - b(4a^2 - 4ab + b^2)$$
$$= 8a^3 - 8a^2 b + 2ab^2 - 4a^2 b + 4ab^2 - b^3$$
$$= 8a^3 - 12a^2 b + 6ab^2 - b^3$$

70. $8x^3 + 12x^2 y + 6xy^2 + y^3$

71. Writing exercise

72. Writing exercise

73. $a_1 = 1$, $a_{n+1} = 5a_n - 2$

$a_1 = 1$

$a_2 = 5 \cdot 1 - 2 = 3$

$a_3 = 5 \cdot 3 - 2 = 13$

$a_4 = 5 \cdot 13 - 2 = 63$

$a_5 = 5 \cdot 63 - 2 = 313$

$a_6 = 5 \cdot 313 - 2 = 1563$

74. 0, 3, 12, 147, 21,612, 467,078,547

75. Find each term by multiplying the preceding term by 2:

1, 2, 4, 8, 16, 32, 64, 128, 256, 512, 1024,

2048, 4096, 8192, 16,384, 32,768, 65,536

76. \$5200, \$3900, \$2925, \$2193.75, \$1645.31, \$1233.98, \$925.49, \$694.12, \$520.59, \$390.44

77. $a_n = (-1)^n$

This sequence is of the form $-1, 1, -1, 1, \ldots$. Each pair of terms adds to 0. S_{100} has 50 such pairs, so $S_{100} = 0$. S_{101} consists of the 50 pairs in S_{100} that add to 0 as well as a_{101}, or -1, so $S_{101} = -1$.

78. $\dfrac{3}{2}, \dfrac{3}{2}, \dfrac{9}{8}, \dfrac{3}{4}, \dfrac{15}{32}, \dfrac{171}{32}$

79. $a_n = i^n$

$a_1 = i^1 = i$

$a_2 = i^2 = -1$

$a_3 = i^3 = i^2 \cdot i = -1 \cdot i = -i$

$a_4 = i^4 = (i^2)^2 = (-1)^2 = 1$

$a_5 = i^5 = (i^2)^2 \cdot i = (-1)^2 \cdot i = 1 \cdot i = i$

$S_5 = i - 1 - i + 1 + i = i$

80. $\{x | x = 4n - 1,$ where n is a natural number$\}$

81. Enter $y_1 = 14x^4 + 6x^3 + 416x^2 - 655x - 1050$. Then scroll through a table of values. We see that $y_1 = 6144$ when $x = 11$, so the 11th term of the sequence is 6144.

82. 1225

Exercise Set 14.2

1. 2, 6, 10, 14, . . .

$a_1 = 2$

$d = 4 \qquad (6 - 2 = 4,\ 10 - 6 = 4,\ 14 - 10 = 4)$

2. $a_1 = 1.06$, $d = 0.06$

3. 6, 2, -2, -6, . . .

$a_1 = 6$

$d = -4 \qquad (2 - 6 = -4, -2 - 2 = -4,$
$-6 - (-2) = -4)$

4. $a_1 = -9$, $d = 3$

5. $\dfrac{3}{2}, \dfrac{9}{4}, 3, \dfrac{15}{4}, \ldots$

$a_1 = \dfrac{3}{2}$

$d = \dfrac{3}{4} \qquad \left(\dfrac{9}{4} - \dfrac{3}{2} = \dfrac{3}{4},\ 3 - \dfrac{9}{4} = \dfrac{3}{4} \right)$

6. $a_1 = \dfrac{3}{5}$, $d = -\dfrac{1}{2}$

7. \$5.12, \$5.24, \$5.36, \$5.48, . . .

$a_1 = \$5.12$

$d = \$0.12 \quad (\$5.24 - \$5.12 = \$0.12,\ \$5.36 -$
$\$5.24 = \$0.12,\ \$5.48 - \$5.36 =$
$\$0.12)$

8. $a_1 = \$214$, $d = -\$3$

9. 3, 7, 11, . . .

$a_1 = 3$, $d = 4$, and $n = 12$

$a_n = a_1 + (n - 1)d$

$a_{12} = 3 + (12 - 1)4 = 3 + 11 \cdot 4 = 3 + 44 = 47$

10. 0.57

11. 7, 4, 1, . . .

$a_1 = 7$, $d = -3$, and $n = 17$

$a_n = a_1 + (n - 1)d$

$a_{17} = 7 + (17 - 1)(-3) = 7 + 16(-3) =$
$7 - 48 = -41$

12. $-\dfrac{17}{3}$

13. \$1200, \$964.32, \$728.64, . . .

$a_1 = \$1200$, $d = \$964.32 - \$1200 = -\$235.68$,

and $n = 13$

$a_n = a_1 + (n - 1)d$

$a_{13} = \$1200 + (13 - 1)(-\$235.68) =$
$\$1200 + 12(-\$235.68) = \$1200 - \$2828.16 =$
$-\$1628.16$

14. \$7941.62

15. $a_1 = 3$, $d = 4$

$a_n = a_1 + (n - 1)d$

Let $a_n = 107$, and solve for n.

$107 = 3 + (n - 1)(4)$

$107 = 3 + 4n - 4$

$107 = 4n - 1$

$108 = 4n$

$27 = n$

The 27th term is 107.

16. 33rd

17. $a_1 = 7$, $d = -3$

$a_n = a_1 + (n - 1)d$

$-296 = 7 + (n - 1)(-3)$

$-296 = 7 - 3n + 3$

$-306 = -3n$

$102 = n$

The 102nd term is -296.

18. 46th

19. $a_n = a_1 + (n - 1)d$

$a_{17} = 2 + (17 - 1)5$ Substituting 17 for n, 2 for a_1, and 5 for d

$= 2 + 16 \cdot 5$

$= 2 + 80$

$= 82$

20. -43

21. $a_n = a_1 + (n - 1)d$

$33 = a_1 + (8 - 1)4$ Substituting 33 for a_8, 8 for n, and 4 for d

$33 = a_1 + 28$

$5 = a_1$

(Note that this procedure is equivalent to subtracting d from a_8 seven times to get a_1: $33 - 7(4) = 33 - 28 = 5$)

22. -54

23. $a_n = a_1 + (n - 1)d$

$-76 = 5 + (n - 1)(-3)$ Substituting -76 for a_n, 5 for a_1, and -3 for d

$-76 = 5 - 3n + 3$

$-76 = 8 - 3n$

$-84 = -3n$

$28 = n$

24. 39

25. We know that $a_{17} = -40$ and $a_{28} = -73$. We would have to add d eleven times to get from a_{17} to a_{28}. That is,

$-40 + 11d = -73$

$11d = -33$

$d = -3.$

Since $a_{17} = -40$, we subtract d sixteen times to get to a_1.

$a_1 = -40 - 16(-3) = -40 + 48 = 8$

We write the first five terms of the sequence:

$8, 5, 2, -1, -4$

26. $a_1 = \dfrac{1}{3}$; $d = \dfrac{1}{2}$; $\dfrac{1}{3}, \dfrac{5}{6}, \dfrac{4}{3}, \dfrac{11}{6}, \dfrac{7}{3}$

27. $a_{13} = 13$ and $a_{54} = 54$

Observe that for this to be true, $a_1 = 1$ and $d = 1$.

28. $a_1 = 2$, $d = 2$

29. $1 + 5 + 9 + 13 + \ldots$

Note that $a_1 = 1$, $d = 4$, and $n = 20$. Before using the formula for S_n, we find a_{20}:

$a_{20} = 1 + (20 - 1)4$ Substituting into the formula for a_n

$= 1 + 19 \cdot 4$

$= 77$

Then

$S_{20} = \dfrac{20}{2}(1 + 77)$ Using the formula for S_n

$= 10(78)$

$= 780.$

30. -210

31. The sum is $1 + 2 + 3 + \ldots + 249 + 250$. This is the sum of the arithmetic sequence for which $a_1 = 1$, $a_n = 250$, and $n = 250$. We use the formula for S_n.

$S_n = \dfrac{n}{2}(a_1 + a_n)$

$S_{300} = \dfrac{250}{2}(1 + 250) = 125(251) = 31,375$

32. 80,200

33. The sum is $2 + 4 + 6 + \ldots + 98 + 100$. This is the sum of the arithmetic sequence for which $a_1 = 2$, $a_n = 100$, and $n = 50$. We use the formula for S_n.

$S_n = \dfrac{n}{2}(a_1 + a_n)$

$S_{50} = \dfrac{50}{2}(2 + 100) = 25(102) = 2550$

34. 2500

35. The sum is $6 + 12 + 18 + \ldots + 96 + 102$. This is the sum of the arithmetic sequence for which $a_1 = 6$, $a_n = 102$, and $n = 17$. We use the formula for S_n.

$$S_n = \frac{n}{2}(a_1 + a_n)$$

$$S_{17} = \frac{17}{2}(6 + 102) = \frac{17}{2}(108) = 918$$

36. 34,036

37. Before using the formula for S_n, we find a_{20}:

$$a_{20} = 4 + (20 - 1)5 \quad \text{Substituting into the formula for } a_n$$

$$= 4 + 19 \cdot 5 = 99$$

Then

$$S_{20} = \frac{20}{2}(4 + 99) \quad \text{Using the formula for } S_n$$

$$= 10(103) = 1030.$$

38. -1200

39. *Familiarize*. We want to find the fifteenth term and the sum of an arithmetic sequence with $a_1 = 14$, $d = 2$, and $n = 15$. We will first use the formula for a_n to find a_{15}. This result is the number of marchers in the last row. Then we will use the formula for S_n to find S_{15}. This is the total number of marchers.

Translate. Substituting into the formula for a_n, we have

$$a_{15} = 14 + (15 - 1)2.$$

Carry out. We first find a_{15}.

$$a_{15} = 14 + 14 \cdot 2 = 42$$

Then use the formula for S_n to find S_{15}.

$$S_{15} = \frac{15}{2}(14 + 42) = \frac{15}{2}(56) = 420$$

Check. We can do the calculations again. We can also do the entire addition.

$$14 + 16 + 18 + \cdots + 42.$$

State. There are 42 marchers in the last row, and there are 420 marchers altogether.

40. 3; 210

41. *Familiarize*. We go from 50 poles in a row, down to six poles in the top row, so there must be 45 rows. We want the sum $50 + 49 + 48 + \ldots + 6$. Thus we want the sum of an arithmetic sequence. We will use the formula $S_n = \frac{n}{2}(a_1 + a_n)$.

Translate. We want to find the sum of the first 45 terms of an arithmetic sequence with $a_1 = 50$ and $a_{45} = 6$.

Carry out. Substituting into the formula for S_n, we have

$$S_{45} = \frac{45}{2}(50 + 6)$$

$$= \frac{45}{2} \cdot 56 = 1260$$

Check. We can do the calculation again, or we can do the entire addition:

$$50 + 49 + 48 + \ldots + 6.$$

State. There will be 1260 poles in the pile.

42. $49.60

43. *Familiarize*. We want to find the sum of an arithmetic sequence with $a_1 = \$600$, $d = \$100$, and $n = 20$. We will use the formula for a_n to find a_{20}, and then we will use the formula for S_n to find S_{20}.

Translate. Substituting into the formula for a_n, we have

$$a_{20} = 600 + (20 - 1)(100).$$

Carry out. We first find a_{20}.

$$a_{20} = 600 + 19 \cdot 100 = 600 + 1900 = 2500$$

Then we use the formula for S_n to find S_{20}.

$$S_{20} = \frac{20}{2}(600 + 2500) = 10(3100) = 31,000$$

Check. We can do the calculation again.

State. They save $31,000 (disregarding interest).

44. $10,230

45. *Familiarize*. We want to find the sum of an arithmetic sequence with $a_1 = 20$, $d = 2$, and $n = 19$. We will use the formula for a_n to find a_{19}, and then we will use the formula for S_n to find S_{19}.

Translate. Substituting into the formula for a_n, we have

$$a_{19} = 20 + (19 - 1)(2).$$

Carry out. We find a_{19}.

$$a_{19} = 20 + 18 \cdot 2 = 56$$

Then we use the formula for S_n to find S_{19}.

$$S_{19} = \frac{19}{2}(20 + 56) = 722$$

Check. We can do the calculation again.

State. There are 722 seats.

46. $462,500

47. Writing exercise

48. Writing exercise

49. $\dfrac{3}{10x} + \dfrac{2}{15x}$, LCD is $30x$

$$= \dfrac{3}{10x} \cdot \dfrac{3}{3} + \dfrac{2}{15x} \cdot \dfrac{2}{2}$$

$$= \dfrac{9}{30x} + \dfrac{4}{30x}$$

$$= \dfrac{13}{30x}$$

50. $\dfrac{23}{36t}$

51.

$\log_a P = k \qquad a^k = P$

The logarithm is the exponent.

The base does not change.

52. $e^a = t$

53. Standard form for the equation of a circle with center (h, k) and radius r is

$$(x - h)^2 + (y - k)^2 = r^2.$$

We substitute 0 for h, 0 for k, and 9 for r:

$$(x - 0)^2 + (y - 0)^2 = 9^2$$
$$x^2 + y^2 = 81$$

54. $(x + 2)^2 + (y - 5)^2 = 18$

55. Writing exercise

56. Writing exercise

57. $a_1 = 1$, $d = 2$, $n = n$

$$a_n = 1 + (n - 1)2 = 1 + 2n - 2 = 2n - 1$$

$$S_n = \dfrac{n}{2}[1 + (2n - 1)] = \dfrac{n}{2} \cdot 2n = n^2$$

Thus, the formula $S_n = n^2$ can be used to find the sum of the first n consecutive odd numbers starting with 1.

58. 3, 5, 7

59.

$a_1 = \$8760$

$a_2 = \$8760 + (-\$798.23) = \$7961.77$

$a_3 = \$8760 + 2(-\$798.23) = \$7163.54$

$a_4 = \$8760 + 3(-\$798.23) = \$6365.31$

$a_5 = \$8760 + 4(-\$798.23) = \$5567.08$

$a_6 = \$8760 + 5(-\$798.23) = \$4768.85$

$a_7 = \$8760 + 6(-\$798.23) = \$3970.62$

$a_8 = \$8760 + 7(-\$798.23) = \$3172.39$

$a_9 = \$8760 + 8(-\$798.23) = \$2374.16$

$a_{10} = \$8760 + 9(-\$798.23) = \$1575.93$

60. \$51,679.65

61. See the answer section in the text.

62. a) $a_t = \$5200 - \$512.50t$

b) \$5200, \$4687.50, \$4175, \$3662.50, \$3150, \$1612.50, \$1100

c) $a_0 = \$5200$, $a_t = a_{t-1} - \$512.50$

63. Each integer from 501 through 750 is 500 more than the corresponding integer from 1 through 250. There are 250 integers from 501 through 750, so their sum is the sum of the integers from 1 to 250 plus $250 \cdot 500$. From Exercise 31, we know that the sum of the integers from 1 through 250 is 31,375. Thus, we have

$$31,375 + 250 \cdot 500, \text{ or } 156,375.$$

Exercise Set 14.3

1. 7, 14, 28, 56, . . .

$$\dfrac{14}{7} = 2, \quad \dfrac{28}{14} = 2, \quad \dfrac{56}{28} = 2$$

$$r = 2$$

2. 3

3. 5, −5, 5, −5, . . .

$$\dfrac{-5}{5} = -1, \quad \dfrac{5}{-5} = -1, \quad \dfrac{-5}{5} = -1$$

$$r = -1$$

4. 0.1

5. $\dfrac{1}{2}, -\dfrac{1}{4}, \dfrac{1}{8}, -\dfrac{1}{16}, \cdots$

$$\dfrac{-\frac{1}{4}}{\frac{1}{2}} = -\dfrac{1}{4} \cdot \dfrac{2}{1} = -\dfrac{2}{4} = -\dfrac{1}{2}$$

$$\dfrac{\frac{1}{8}}{-\frac{1}{4}} = \dfrac{1}{8} \cdot \left(-\dfrac{4}{1}\right) = -\dfrac{4}{8} = -\dfrac{1}{2}$$

$$\dfrac{-\frac{1}{16}}{\frac{1}{8}} = -\dfrac{1}{16} \cdot \dfrac{8}{1} = -\dfrac{8}{16} = -\dfrac{1}{2}$$

$$r = -\dfrac{1}{2}$$

6. −2

7. $75, 15, 3, \frac{3}{5}, \ldots$

$$\frac{15}{75} = \frac{1}{5}, \quad \frac{3}{15} = \frac{1}{5}, \quad \frac{\frac{3}{5}}{3} = \frac{3}{5} \cdot \frac{1}{3} = \frac{1}{5}$$

$$r = \frac{1}{5}$$

8. $-\frac{1}{3}$

9. $\frac{1}{m}, \frac{3}{m^2}, \frac{9}{m^3}, \frac{27}{m^4}, \ldots$

$$\frac{\frac{3}{m^2}}{\frac{1}{m}} = \frac{3}{m^2} \cdot \frac{m}{1} = \frac{3}{m}$$

$$\frac{\frac{9}{m^3}}{\frac{3}{m^2}} = \frac{9}{m^3} \cdot \frac{m^2}{3} = \frac{3}{m}$$

$$\frac{\frac{27}{m^4}}{\frac{9}{m^3}} = \frac{27}{m^4} \cdot \frac{m^3}{9} = \frac{3}{m}$$

$$r = \frac{3}{m}$$

10. $\frac{m}{5}$

11. $3, 6, 12, \ldots$

$a_1 = 3$, $n = 7$, and $r = \frac{6}{3} = 2$

We use the formula $a_n = a_1 r^{n-1}$.

$a_7 = 3 \cdot 2^{7-1} = 3 \cdot 2^6 = 3 \cdot 64 = 192$

12. $131,072$

13. $5, 5\sqrt{2}, 10, \ldots$

$a_1 = 5$, $n = 9$, and $r = \frac{5\sqrt{2}}{5} = \sqrt{2}$

$a_n = a_1 r^{n-1}$

$a_9 = 5(\sqrt{2})^{9-1} = 5(\sqrt{2})^8 = 5 \cdot 16 = 80$

14. $108\sqrt{3}$

15. $-\frac{8}{243}, \frac{8}{81}, -\frac{8}{27}, \ldots$

$a_1 = -\frac{8}{243}$, $n = 10$, and $r = \dfrac{\frac{8}{81}}{-\frac{8}{243}} = $

$$\frac{8}{81}\left(-\frac{243}{8}\right) = -3$$

$a_n = a_1 r^{n-1}$

$$a_{10} = -\frac{8}{243}(-3)^{10-1} = -\frac{8}{243}(-3)^9 =$$

$$-\frac{8}{243}(-19,683) = 648$$

16. $2,734,375$

17. $\$1000, \$1080, \$1166.40, \ldots$

$a_1 = \$1000$, $n = 12$, and $r = \dfrac{\$1080}{\$1000} = 1.08$

$a_n = a_1 r^{n-1}$

$a_{12} = \$1000(1.08)^{12-1} \approx \$1000(2.331638997) \approx$
$\$2331.64$

18. $\$1967.15$

19. $1, 3, 9, \ldots$

$a_1 = 1$ and $r = \frac{3}{1}$, or 3

$a_n = a_1 r^{n-1}$

$a_n = 1(3)^{n-1} = 3^{n-1}$

20. 5^{3-n}

21. $1, -1, 1, -1, \ldots$

$a_1 = 1$ and $r = \frac{-1}{1} = -1$

$a_n = a_1 r^{n-1}$

$a_n = 1(-1)^{n-1} = (-1)^{n-1}$

22. 2^n

23. $\frac{1}{x}, \frac{1}{x^2}, \frac{1}{x^2}, \ldots$

$a_1 = \frac{1}{x}$ and $r = \dfrac{\frac{1}{x^2}}{\frac{1}{x}} = \frac{1}{x^2} \cdot \frac{x}{1} = \frac{1}{x}$

$a_n = a_1 r^{n-1}$

$a_n = \frac{1}{x}\left(\frac{1}{x}\right)^{n-1} = \frac{1}{x} \cdot \frac{1}{x^{n-1}} = \frac{1}{x^{1+n-1}} = \frac{1}{x^n}$

24. $5\left(\dfrac{m}{2}\right)^{n-1}$

25. $6 + 12 + 24 + \ldots$

$a_1 = 6$, $n = 7$, and $r = \frac{12}{6} = 2$

$S_n = \frac{a_1(1 - r^n)}{1 - r}$

$S_7 = \frac{6(1 - 2^7)}{1 - 2} = \frac{6(1 - 128)}{-1} = \frac{6(-127)}{-1} = 762$

26. 10.5

27. $\dfrac{1}{18} - \dfrac{1}{6} + \dfrac{1}{2} - \ldots$

$$a_1 = \frac{1}{18}, \ n = 7, \text{ and } r = \frac{-\dfrac{1}{6}}{\dfrac{1}{18}} = -\frac{1}{6} \cdot \frac{18}{1} = -3$$

$$S_n = \frac{a_1(1 - r^n)}{1 - r}$$

$$S_7 = \frac{\dfrac{1}{18}\left[1 - (-3)^7\right]}{1 - (-3)} = \frac{\dfrac{1}{18}(1 + 2187)}{4} = \frac{\dfrac{1}{18}(2188)}{4} =$$

$$\frac{1}{18}(2188)\left(\frac{1}{4}\right) = \frac{547}{18}$$

28. 7.7777

29. $1 + x + x^2 + x^3 + \ldots$

$$a_1 = 1, \ n = 8, \text{ and } r = \frac{x}{1}, \text{ or } x$$

$$S_n = \frac{a_1(1 - r^n)}{1 - r}$$

$$S_8 = \frac{1(x - x^8)}{1 - x} = \frac{(1 + x^4)(1 - x^4)}{1 - x} =$$

$$\frac{(1 + x^4)(1 + x^2)(1 - x^2)}{1 - x} =$$

$$\frac{(1 + x^4)(1 + x^2)(1 + x)(1 - x)}{1 - x} =$$

$$(1 + x^4)(1 + x^2)(1 + x)$$

30. $\dfrac{1 - x^{20}}{1 - x^2}$

31. $200, \ 200(1.06), \ 200(1.06)^2, \ldots$

$$a_1 = \$200, \ n = 16, \text{ and } r = \frac{\$200(1.06)}{\$200} = 1.06$$

$$S_n = \frac{a_1(1 - r^n)}{1 - r}$$

$$S_{16} = \frac{\$200[1 - (1.06)^{16}]}{1 - 1.06} \approx$$

$$\frac{\$200(1 - 2.540351685)}{-0.06} \approx \$5134.51$$

32. \$60,893.30

33. $16 + 4 + 1 + \ldots$

$$|r| = \left|\frac{4}{16}\right| = \left|\frac{1}{4}\right| = \frac{1}{4}, \text{ and since } |r| < 1, \text{ the series}$$
does have a sum.

$$S_\infty = \frac{a_1}{1 - r} = \frac{16}{1 - \dfrac{1}{4}} = \frac{16}{\dfrac{3}{4}} = 16 \cdot \frac{4}{3} = \frac{64}{3}$$

34. 16

35. $7 + 3 + \dfrac{9}{7} + \ldots$

$$|r| = \left|\frac{3}{7}\right| = \frac{3}{7}, \text{ and since } |r| < 1, \text{ the series}$$
does have a sum.

$$S_\infty = \frac{a_1}{1 - r} = \frac{7}{1 - \dfrac{3}{7}} = \frac{7}{\dfrac{4}{7}} = 7 \cdot \frac{7}{4} = \frac{49}{4}$$

36. 48

37. $3 + 15 + 75 + \ldots$

$$|r| = \left|\frac{15}{3}\right| = |5| = 5, \text{ and since } |r| \not< 1 \text{ the series does}$$
not have a sum.

38. No

39. $4 - 6 + 9 - \dfrac{27}{2} + \ldots$

$$|r| = \left|\frac{-6}{4}\right| = \left|-\frac{3}{2}\right| = \frac{3}{2}, \text{ and since } |r| \not< 1 \text{ the series}$$
does not have a sum.

40. -4

41. $0.43 + 0.0043 + 0.000043 + \ldots$

$$|r| = \left|\frac{0.0043}{0.43}\right| = |0.01| = 0.01, \text{ and since } |r| < 1,$$
the series does have a sum.

$$S_\infty = \frac{a_1}{1 - r} = \frac{0.43}{1 - 0.01} = \frac{0.43}{0.99} = \frac{43}{99}$$

42. $\dfrac{37}{99}$

43. $\$500(1.02)^{-1} + \$500(1.02)^{-2} + \$500(1.02)^{-3} + \ldots$

$$|r| = \left|\frac{\$500(1.02)^{-2}}{\$500(1.02)^{-1}}\right| = |(1.02)^{-1}| = (1.02)^{-1}, \text{ or}$$

$$\frac{1}{1.02}, \text{ and since } |r| < 1, \text{ the series does have a sum.}$$

$$S_\infty = \frac{a_1}{1 - r} = \frac{\$500(1.02)^{-1}}{1 - \left(\dfrac{1}{1.02}\right)} = \frac{\dfrac{\$500}{1.02}}{\dfrac{0.02}{1.02}} =$$

$$\frac{\$500}{1.02} \cdot \frac{1.02}{0.02} = \$25,000$$

44. \$12,500

45. $0.7777\ldots = 0.7 + 0.07 + 0.007 + 0.0007 + \ldots$

This is an infinite geometric series with $a_1 = 0.7$.

$$|r| = \left|\frac{0.07}{0.7}\right| = |0.1| = 0.1 < 1, \text{ so the series has a}$$
sum.

$$S_\infty = \frac{a_1}{1 - r} = \frac{0.7}{1 - 0.1} = \frac{0.7}{0.9} = \frac{7}{9}$$

Fractional notation for $0.7777\ldots$ is $\dfrac{7}{9}$.

46. $\dfrac{2}{9}$

47. $8.3838\ldots = 8.3 + 0.083 + 0.00083 + \ldots$

This is an infinite geometric series with $a_1 = 8.3$.

$|r| = \left|\dfrac{0.083}{8.3}\right| = |0.01| = 0.01 < 1$, so the series has a sum.

$$S_\infty = \frac{a_1}{1-r} = \frac{8.3}{1-0.01} = \frac{8.3}{0.99} = \frac{830}{99}$$

Fractional notation for $8.3838\ldots$ is $\dfrac{830}{99}$.

48. $\dfrac{740}{99}$

49. $0.15151515\ldots = 0.15 + 0.0015 + 0.000015 + \ldots$

This is an infinite geometric series with $a_1 = 0.15$.

$|r| = \left|\dfrac{0.0015}{0.15}\right| = |0.01| = 0.01 < 1$, so the series has a sum.

$$S_\infty = \frac{a_1}{1-r} = \frac{0.15}{1-0.01} = \frac{0.15}{0.99} = \frac{15}{99} = \frac{5}{33}$$

Fractional notation for $0.15151515\ldots$ is $\dfrac{5}{33}$.

50. $\dfrac{4}{33}$

51. *Familiarize*. The rebound distances form a geometric sequence:

$$\frac{1}{4} \times 20, \quad \left(\frac{1}{4}\right)^2 \times 20, \quad \left(\frac{1}{4}\right)^3 \times 20, \ldots,$$

or $5, \quad \dfrac{1}{4} \times 5, \quad \left(\dfrac{1}{4}\right)^2 \times 5, \ldots$

The height of the 6th rebound is the 6th term of the sequence.

Translate. We will use the formula $a_n = a_1 r^{n-1}$, with $a_1 = 5$, $r = \dfrac{1}{4}$, and $n = 6$:

$$a_6 = 5\left(\frac{1}{4}\right)^{6-1}$$

Carry out. We calculate to obtain $a_6 = \dfrac{5}{1024}$.

Check. We can do the calculation again.

State. It rebounds $\dfrac{5}{1024}$ ft the 6th time.

52. $6\dfrac{2}{3}$ ft

53. *Familiarize*. In one year, the population will be $100,000 + 0.03(100,000)$, or $(1.03)100,000$. In two years, the population will be $(1.03)100,000 + 0.03(1.03)100,000$, or $(1.03)^2 100,000$. Thus the populations form a geometric sequence:

$$100,000, \quad (1.03)100,000, \quad (1.03)^2 100,000, \ldots$$

The population in 15 years will be the 16th term of the sequence.

Translate. We will use the formula $a_n = a_1 r^{n-1}$ with $a_1 = 100,000$, $r = 1.03$, and $n = 16$:

$$a_{16} = 100,000(1.03)^{16-1}$$

Carry out. We calculate to obtain $a_{16} \approx 155,797$.

Check. We can do the calculation again.

State. In 15 years the population will be about 155,797.

54. About 24 years

55. *Familiarize*. The amounts owed at the beginning of successive years form a geometric sequence:

$$\$15,000, \quad (1.085)\$15,000, \quad (1.085)^2\$15,000,$$
$$(1.085)^3\$15,000, \ldots$$

The amount to be repaid at the end of 13 years is the amount owed at the beginning of the 14th year.

Translate. We use the formula $a_n = a_1 r^{n-1}$ with $a_1 = 15,000$, $r = 1.085$, and $n = 14$:

$$a_{14} = 15,000(1.085)^{14-1}$$

Carry out. We calculate to obtain $a_{14} \approx 43,318.94$.

Check. We can do the calculation again.

State. At the end of 13 years, $43,318.94 will be repaid.

56. 2710

57. We have a geometric sequence

$$5000, \quad 5000(0.96), \quad 5000(0.96)^2, \ldots$$

where the general term $5000(0.96)^n$ represents the number of fruit flies remaining alive after n minutes. We find the value of n for which the general term is 1800.

$$1800 = 5000(0.96)^n$$
$$0.36 = (0.96)^n$$
$$\log 0.36 = \log(0.96)^n$$
$$\log 0.36 = n \log 0.96$$
$$\frac{\log 0.36}{\log 0.96} = n$$
$$25 \approx n$$

It will take about 25 minutes for only 1800 fruit flies to remain alive.

58. $213,609.57

59. *Familiarize*. The lengths of the falls form a geometric sequence:

$$556, \quad \left(\frac{3}{4}\right)556, \quad \left(\frac{3}{4}\right)^2 556, \quad \left(\frac{3}{4}\right)^3 556, \ldots$$

The total length of the first 6 falls is the sum of the first six terms of this sequence. The heights of the rebounds also form a geometric sequence:

$$\left(\frac{3}{4}\right)556, \quad \left(\frac{3}{4}\right)^2 556, \quad \left(\frac{3}{4}\right)^3 556, \ldots, \quad \text{or}$$

$$417, \quad \left(\frac{3}{4}\right)417, \quad \left(\frac{3}{4}\right)^2 417, \ldots$$

When the ball hits the ground for the 6th time, it will have rebounded 5 times. Thus the total length of the rebounds is the sum of the first five terms of this sequence.

Translate. We use the formula $S_n = \dfrac{a_1(1 - r^n)}{1 - r}$

twice, once with $a_1 = 556$, $r = \dfrac{3}{4}$, and $n = 6$ and

a second time with $a_1 = 417$, $r = \dfrac{3}{4}$, and $n = 5$.

D = Length of falls + length of rebounds

$$= \frac{556\left[1 - \left(\frac{3}{4}\right)^6\right]}{1 - \frac{3}{4}} + \frac{417\left[1 - \left(\frac{3}{4}\right)^5\right]}{1 - \frac{3}{4}}.$$

Carry out. We use a calculator to obtain $D \approx 3100.35$.

Check. We can do the calculations again.

State. The ball will have traveled about 3100.35 ft.

60. 3892 ft

61. Familiarize. The heights of the stack form a geometric sequence:

$0.02, 0.02(2), 0.02(2^2), \ldots$

The height of the stack after it is doubled 10 times is given by the 11th term of this sequence.

Translate. We have a geometric sequence with $a_1 = 0.02$, $r = 2$, and $n = 11$. We use the formula

$a_n = a_1 r^{n-1}$.

Carry out. We substitute and calculate.

$a_{11} = 0.02(2^{11-1})$

$a_{11} = 0.02(1024) = 20.48$

Check. We can do the calculation again.

State. The final stack will be 20.48 in. high.

62. \$2,684,354.55

63. Writing exercise

64. Writing exercise

65. $\quad (x + y)(x^2 + 2xy + y^2)$

$= x(x^2 + 2xy + y^2) + y(x^2 + 2xy + y^2)$

$= x^3 + 2x^2 y + xy^2 + x^2 y + 2xy^2 + y^3$

$= x^3 + 3x^2 y + 3xy^2 + y^3$

66. $a^3 - 3a^2 b + 3ab^2 - b^3$

67. $\quad 5x - 2y = -3, \quad (1)$

$2x + 5y = -24 \quad (2)$

Multiply Eq. (1) by 5 and Eq. (2) by 2 and add.

$25x - 10y = -15$

$\underline{4x + 10y = -48}$

$29x \qquad\quad = -63$

$x = -\dfrac{63}{29}$

Substitute $-\dfrac{63}{29}$ for x in the second equation and solve for y.

$2\left(-\dfrac{63}{29}\right) + 5y = -24$

$-\dfrac{126}{29} + 5y = -24$

$5y = -\dfrac{570}{29}$

$y = -\dfrac{114}{29}$

The solution is $\left(-\dfrac{63}{29}, -\dfrac{114}{29}\right)$.

68. $(-1, 2, 3)$

69. Writing exercise

70. Writing exercise

71. $x^2 - x^3 + x^4 + x^5 + \ldots$

This is a geometric series with $a_1 = x^2$ and $r = -x$.

$S_n = \dfrac{a_1(1 - r^n)}{1 - r} = \dfrac{x^2[1 - (-x)^n]}{1 - (-x)} = \dfrac{x^2[1 - (-x)^n]}{1 + x}$

72. $\dfrac{1 - x^n}{1 - x}$

73. The length of a side of the first square is 16 cm. The length of a side of the next square is the length of the hypotenuse of a right triangle with legs 8 cm and 8 cm, or $8\sqrt{2}$ cm. The length of a side of the next square is the length of the hypotenuse of a right triangle with legs $4\sqrt{2}$ cm and $4\sqrt{2}$ cm, or 8 cm. The areas of the squares form a sequence:

$(16)^2, \quad (8\sqrt{2})^2, \quad (8)^2, \ldots, \quad \text{or}$

$256, \quad 128, \quad 64, \ldots.$

This is a geometric sequence with $a_1 = 256$ and $r = \dfrac{1}{2}$.

We find the sum of the infinite geometric series $256 + 128 + 64 + \ldots.$

$S_\infty = \dfrac{256}{1 - \dfrac{1}{2}} = \dfrac{256}{\dfrac{1}{2}} = 512 \text{ cm}^2$

74. $0.999\ldots = 0.9 + 0.09 + 0.009 + \ldots$

$|r| = \left| \dfrac{0.09}{0.9} \right| = |0.1| = 0.1 < 1$, so the series has a sum.

$S_\infty = \dfrac{0.9}{1 - 0.1} = \dfrac{0.9}{0.9} = 1$

Thus, $0.999\ldots = 1$.

75. Writing exercise

76. Writing exercise

Exercise Set 14.4

1. $9! = 9 \cdot 8 \cdot 7 \cdot 6 \cdot 5 \cdot 4 \cdot 3 \cdot 2 \cdot 1 = 362,880$

2. $3,628,800$

3. $11! = 11 \cdot 10 \cdot 9 \cdot 8 \cdot 7 \cdot 6 \cdot 5 \cdot 4 \cdot 3 \cdot 2 \cdot 1 = 39,916,800$

4. $479,001,600$

5. $0!$ is defined to be 1.

6. 1

7. $\dfrac{7!}{4!} = \dfrac{7 \cdot 6 \cdot 5 \cdot 4!}{4!} = 7 \cdot 6 \cdot 5 = 210$

8. 56

9. $\dfrac{9!}{5!} = \dfrac{9 \cdot 8 \cdot 7 \cdot 6 \cdot 5!}{5!} = 9 \cdot 8 \cdot 7 \cdot 6 = 3024$

10. 720

11. $(8-3)! = 5! = 5 \cdot 4 \cdot 3 \cdot 2 \cdot 1 = 120$

12. 24

13. $8! - 3! = (8 \cdot 7 \cdot 6 \cdot 5 \cdot 4 \cdot 3 \cdot 2 \cdot 1) - (3 \cdot 2 \cdot 1) =$
$40,320 - 6 = 40,314$

14. $362,760$

15. $_6P_6 = 6! = 6 \cdot 5 \cdot 4 \cdot 3 \cdot 2 \cdot 1 = 720$

16. 120

17. Using formula (1), we have $_4P_3 = 4 \cdot 3 \cdot 2 = 24$.

Using formula (2), we have

$_4P_3 = \dfrac{4!}{(4-3)!} = \dfrac{4!}{1!} = \dfrac{4 \cdot 3 \cdot 2 \cdot 1}{1} = 24.$

18. 2520

19. Using formula (1), we have

$_{10}P_7 = 10 \cdot 9 \cdot 8 \cdot 7 \cdot 6 \cdot 5 \cdot 4 = 604,800.$

Using formula (2), we have

$_{10}P_7 = \dfrac{10!}{(10-7)!} = \dfrac{10!}{3!} = \dfrac{10 \cdot 9 \cdot 8 \cdot 7 \cdot 6 \cdot 5 \cdot 4 \cdot 3!}{3!} =$
$604,800.$

20. 720

21. Using formula (1), we have $_6P_1 = 6$.

Using formula (2), we have

$_6P_1 = \dfrac{6!}{(6-1)!} = \dfrac{6!}{5!} = \dfrac{6 \cdot 5!}{5!} = 6.$

22. 12

23. Using formula (1), we have $_6P_5 = 6 \cdot 5 \cdot 4 \cdot 3 \cdot 2 = 720$.

Using formula (2), we have

$_6P_5 = \dfrac{6!}{(6-5)!} = \dfrac{6!}{1!} = \dfrac{6 \cdot 5 \cdot 4 \cdot 3 \cdot 2 \cdot 1}{1} = 720.$

24. $479,001,600$

25. The box can be chosen in 3 ways, the wrapping paper can be chosen in 10 ways, the shipping company can be chosen in 5 ways, and the insurance can be chosen in 3 ways. By the fundamental counting principle, the total number of different ways the candles can be packed and shipped is $3 \cdot 10 \cdot 5 \cdot 3$, or 450.

26. $4 \cdot 33 \cdot 5 \cdot 2$, or 1320

27. There are 3 choices for the first letter, 2 for the second, and 1 for the third. The number of permutations is $3 \cdot 2 \cdot 1$, or $_3P_3$, or $3!$, or 6.

28. $2!$, or 2

29. There are 7 choices for the first letter, 6 for the second, 5 for the third, 4 for the fourth, 3 for the fifth, 2 for the sixth, and 1 for the seventh. The number of permutations is $7 \cdot 6 \cdot 5 \cdot 4 \cdot 3 \cdot 2 \cdot 1$, or $_7P_7$, or $7!$, or 5040.

30. $5!$, or 120

31. $_7P_4 = \dfrac{7!}{4!} = \dfrac{7!}{(7-4)!} = \dfrac{7!}{3!} = \dfrac{7 \cdot 6 \cdot 5 \cdot 4 \cdot 3!}{3!} = 840$

32. $_5P_3$, or 60

33. Without repetition, the total is the number of permutations of 4 objects taken 4 at a time:

$_4P_4 = 4! = 24$

With repetition each of the 4 digits can be chosen in 4 ways:

$4 \cdot 4 \cdot 4 \cdot 4 = 256$

34. 5!, or 120; 5^5, or 3125

35. The number of arrangements is the number of permutations of 5 objects taken 5 at a time:
$$_5P_5 = 5! = 120$$

36. 7!, or 5040

37. There are only 8 choices for the first digit since 0 and 1 are excluded. There are 9 choices for the second digit since 0 and 1 can be included and the first digit cannot be repeated. Because no digit is used more than once there are only 8 choices for the third digit, 7 for the fourth, 6 for the fifth, 5 for the sixth, and 4 for the seventh. By the fundamental counting principle the total number of permutations is

$8 \cdot 9 \cdot 8 \cdot 7 \cdot 6 \cdot 5 \cdot 4$, or 483,840.

Thus 483,840 7-digit phone numbers can be formed.

38. $12 \cdot 11 \cdot 10 \cdot 9$, or 11,880

39. a) The number of ways in which the coins can be lined up is $_5P_5 = 5! = 120$.

b) There are 5 choices for the first coin and 2 possibilities (head or tail) for each choice. This results in a total of 10 choices for the first selection.

There are 4 choices (no coin can be used more than once) for the second coin and 2 possibilities (head or tail) for each choice. This results in a total of 8 choices for the second selection.

There are 3 choices for the third coin and 2 possibilities (head or tail) for each choice. This results in a total of 6 choices for the third selection.

Likewise there are 4 choices for the fourth selection and 2 choices for the fifth selection.
Using the fundamental counting principle we know there are

$$10 \cdot 8 \cdot 6 \cdot 4 \cdot 2, \text{ or } 3840$$

ways the coins can be lined up.

40. a) 4!, or 24; b) $4! \cdot 2^4$, or 384

41. $_{52}P_4 = \dfrac{52!}{(52-4)!} = \dfrac{52!}{48!} = \dfrac{52 \cdot 51 \cdot 50 \cdot 49 \cdot 48!}{48!} =$
6,497,400

42. $50 \cdot 49 \cdot 48 \cdot 47 \cdot 46$, or 254,251,200

43. There are 80 choices for the number of the county, 26 choices for the letter of the alphabet, and 9999 choices for the number that follows the letter. By the fundamental counting principle we know there are $80 \cdot 26 \cdot 9999$, or 20,797,920 possible license plates.

44. a) $26 \cdot 10 \cdot 26 \cdot 10 \cdot 26 \cdot 10$, or 17,576,600; b) no

45. a) Since repetition is allowed, each of the 5 digits can be chosen in 10 ways. The number of zip-codes possible is $10 \cdot 10 \cdot 10 \cdot 10 \cdot 10$, or 100,000.

b) Since there are 100,000 possible zip-codes, there could be 100,000 post offices.

46. a) 10^9, or 1,000,000,000; b) yes

47. a) Since repetition is allowed, each digit can be chosen in 10 ways. There can be
$10 \cdot 10 \cdot 10 \cdot 10 \cdot 10 \cdot 10 \cdot 10 \cdot 10 \cdot 10$, or 1,000,000,000 social security numbers.

b) Since more than 266 million social security numbers are possible, each person can have a unique social security number.

48. About 41,466 yr

49. Writing exercise

50. Writing exercise

51. $\log_b 35 = \log_b (5 \cdot 7)$
$= \log_b 5 + \log_b 7$
$= 1.609 + 1.946$
$= 3.555$

52. -0.337

53. $\log_b 49 = \log_b 7^2$
$= 2 \log_b 7$
$= 2(1.946)$
$= 3.892$

54. 10

55. $\log_b b^{-5} = -5$
(The power to which you raise b in order to get b^{-5} is -5).

56. m

57. Writing exercise

58. Writing exercise

59. $_nP_5 = 7 \cdot {}_nP_4$
$\dfrac{n!}{(n-5)!} = 7 \cdot \dfrac{n!}{(n-4)!}$ Formula (2)
$\dfrac{n!}{7(n-5)!} = \dfrac{n!}{(n-4)!}$ Dividing by 7
$7(n-5)! = (n-4)!$ The denominators must be the same.
$7(n-5)! = (n-4)(n-5)!$
$7 = n-4$ Dividing by $(n-5)!$
$11 = n$

60. 8

61.
$$_nP_5 = 9 \cdot {}_{n-1}P_4$$

$$\frac{n!}{(n-5)!} = 9 \cdot \frac{(n-1)!}{(n-1-4)!} \qquad \text{Formula (2)}$$

$$\frac{n!}{(n-5)!} = 9 \cdot \frac{(n-1)!}{(n-5)!}$$

$$\qquad n! = 9(n-1)! \qquad \text{Multiplying by } (n-5)!$$

$$n(n-1)! = 9(n-1)!$$

$$\qquad n = 9 \qquad \text{Dividing by } (n-1)!$$

62. 11

63. a) $_6P_6 = 6! = 720$

b) If an adult is placed in the first seat, we have
$$_3P_3 \cdot {}_3P_3 = 3! \cdot 3!.$$
Similarly, if a child is placed in the first seat, we have $_3P_3 \cdot {}_3P_3 = 3! \cdot 3!.$
The total number of arrangements is $2 \cdot 3! \cdot 3!$, or 72.

c) The adult and child who must sit together can be seated in 5 different pairs of chairs (the first and second chairs, or the second and third chairs, and so on to the fifth and sixth chairs). In addition, they can be seated in $_2P_2$, or $2!$, or 2 ways and the 4 remaining people can be seated in $_4P_4$, or $4!$, or 24 ways.
Then the total number of arrangements is $5 \cdot 2! \cdot 4!$, or 240.

d) The number of arrangements if a particular adult and child must not sit together is the difference between the number of arrangements with no seating restrictions and the number of arrangements when a particular adult and child must sit together. Using the results of parts (a) and (c), we have $720 - 240$, or 480 arrangements.

64. a) $_6P_6$, or $6!$, or 720

b) $4 \cdot 2! \cdot 4!$, or 192

c) $8 \cdot 2! \cdot 2! \cdot 2!$, or 64

65. We will only consider factorizations in which the factors of a are both positive. Since b and c are both positive, we will also consider only positive factors of c. Now 6 has 2 such factorizations ($1 \cdot 6$ and $2 \cdot 3$), and 12 has 3 such factorizations ($1 \cdot 12$, $2 \cdot 6$, and $3 \cdot 4$). For each trial factorization $(px+\ )(rx+\ )$ there are 6 arrangements for the second terms of the factors. These are the 3 pairs of factors of 12 in the order shown above and the 3 pairs formed by taking these factors in the opposite order. Since there are 2 possible pairs of choices for p and r, the number of possible trial factorizations is $2 \cdot 6$, or 12. (Note

that if we also reverse the order of p and r to form 2 additional sets of possibilities we only repeat the 12 original factorizations.)

66. 2

Exercise Set 14.5

---

1. $_{13}C_2 = \dfrac{13!}{(13-2)!2!}$

$$= \frac{13!}{11!2!} = \frac{13 \cdot 12 \cdot 11!}{11! \cdot 2 \cdot 1}$$

$$= \frac{13 \cdot 12}{2 \cdot 1} = \frac{13 \cdot 6 \cdot 2}{2 \cdot 1}$$

$$= 78$$

2. 84

3. $\dbinom{13}{11} = \dfrac{13!}{(13-11)!11!}$

$$= \frac{13!}{2!11!}$$

$$= 78 \qquad \text{(See Exercise 1.)}$$

4. 84

5. $\dbinom{7}{1} = \dfrac{7!}{(7-1)!1!}$

$$= \frac{7!}{6!1!} = \frac{7 \cdot 6!}{6! \cdot 1}$$

$$= 7$$

6. 1

7. $\dfrac{_5P_3}{3!} = \dfrac{5 \cdot 4 \cdot 3}{3!}$

$$= \frac{5 \cdot 4 \cdot 3}{3 \cdot 2 \cdot 1} = \frac{5 \cdot 2 \cdot 2 \cdot 3}{3 \cdot 2 \cdot 1}$$

$$= 5 \cdot 2 = 10$$

8. 252

9. $\dbinom{6}{0} = \dfrac{6!}{(6-0)!0!}$

$$= \frac{6!}{6!0!} = \frac{6!}{6! \cdot 1}$$

$$= 1$$

10. 20

11. $_{12}C_{11} = \dfrac{12!}{(12-11)!11!}$

$$= \frac{12!}{1!11!} = \frac{12 \cdot 11!}{1 \cdot 11!}$$

$$= 12$$

12. 66

13. $_{20}C_{18} = \dfrac{20!}{(20-18)!18!}$

$= \dfrac{20!}{2!18!} = \dfrac{20 \cdot 19 \cdot 18!}{2 \cdot 1 \cdot 18!}$

$= \dfrac{20 \cdot 19}{2 \cdot 1} = \dfrac{2 \cdot 10 \cdot 19}{2 \cdot 1}$

$= 190$

14. 4060

15. $\dbinom{35}{2} = \dfrac{35!}{(35-2)!2!}$

$= \dfrac{35!}{33!2!} = \dfrac{35 \cdot 34 \cdot 33!}{33! \cdot 2 \cdot 1}$

$= \dfrac{35 \cdot 34}{2 \cdot 1} = \dfrac{35 \cdot 2 \cdot 17}{2 \cdot 1}$

$= 595$

16. 780

17. $_{10}C_5 = \dfrac{10!}{(10-5)!5!}$

$= \dfrac{10!}{5!5!} = \dfrac{10 \cdot 9 \cdot 8 \cdot 7 \cdot 6 \cdot 5!}{5 \cdot 4 \cdot 3 \cdot 2 \cdot 1 \cdot 5!}$

$= \dfrac{10 \cdot 9 \cdot 8 \cdot 7 \cdot 6}{5 \cdot 4 \cdot 3 \cdot 2 \cdot 1} = \dfrac{5 \cdot 2 \cdot 3 \cdot 3 \cdot 4 \cdot 2 \cdot 7 \cdot 6}{5 \cdot 4 \cdot 3 \cdot 2 \cdot 1}$

$= 252$

18. 1365

19. $_{23}C_4 = \dfrac{23!}{(23-4)!4!}$

$= \dfrac{23!}{19!4!} = \dfrac{23 \cdot 22 \cdot 21 \cdot 20 \cdot 19!}{19! \cdot 4 \cdot 3 \cdot 2 \cdot 1}$

$= \dfrac{23 \cdot 22 \cdot 21 \cdot 20}{4 \cdot 3 \cdot 2 \cdot 1} = \dfrac{23 \cdot 2 \cdot 11 \cdot 3 \cdot 7 \cdot 4 \cdot 5}{4 \cdot 3 \cdot 2 \cdot 1}$

$= 8855$

20. $_9C_2$, or 36; $2 \cdot_9 C_2$, or 72

21. $_{10}C_6 = \dfrac{10!}{(10-6)!6!}$

$= \dfrac{10!}{4!6!} = \dfrac{10 \cdot 9 \cdot 8 \cdot 7 \cdot 6!}{4 \cdot 3 \cdot 2 \cdot 1 \cdot 6!}$

$= \dfrac{10 \cdot 9 \cdot 8 \cdot 7}{4 \cdot 3 \cdot 2 \cdot 1} = \dfrac{10 \cdot 3 \cdot 3 \cdot 4 \cdot 2 \cdot 7}{4 \cdot 3 \cdot 2 \cdot 1}$

$= 210$

22. $_{10}C_7 \cdot_5 C_3$, or 1200

23. Since two points determine a line and no three of these 8 points are collinear, we need to find the number of combinations of 8 points taken 2 at a time, $_8C_2$.

$_8C_2 = \dbinom{8}{2} = \dfrac{8!}{2!(8-2)!}$

$= \dfrac{8 \cdot 7 \cdot 6!}{2 \cdot 1 \cdot 6!} = \dfrac{4 \cdot 2 \cdot 7}{2 \cdot 1}$

$= 28$

Thus 28 lines are determined.

Since three noncolinear points determine a triangle, we need to find the number of combinations of 8 points taken 3 at a time, $_8C_3$.

$_8C_3 = \dbinom{8}{3} = \dfrac{8!}{3!(8-3)!}$

$= \dfrac{8 \cdot 7 \cdot 6 \cdot 5!}{3 \cdot 2 \cdot 1 \cdot 5!} = \dfrac{8 \cdot 7 \cdot 3 \cdot 2}{3 \cdot 2 \cdot 1}$

$= 56$

Thus 56 triangles are determined.

24. a) $_7C_2$, or 21; b) $_7C_3$, or 35

25. Using the fundamental counting principle, we have $_{58}C_6 \cdot_{42} C_4$.

26. $_{63}C_8 \cdot_{37} C_{12}$

27. We use the fundamental counting principle. The blue die can fall in 6 ways and the red die can also fall in 6 ways.

$6 \cdot 6 = 36$

28. $6 \cdot 9 \cdot 7$, or 378

29. $_{52}C_5$

30. $_{52}C_{13}$

31. We use the fundamental counting principle.

$_5C_2 \cdot_6 C_3 \cdot_3 C_1$

$= \dfrac{5!}{(5-2)!2!} \cdot \dfrac{6!}{(6-3)!3!} \cdot \dfrac{3!}{(3-1)!1!}$

$= \dfrac{5!}{3!2!} \cdot \dfrac{6!}{3!3!} \cdot \dfrac{3!}{2!1!}$

$= \dfrac{5 \cdot 4 \cdot 3!}{3! \cdot 2!} \cdot \dfrac{6 \cdot 5 \cdot 4 \cdot 3!}{3! \cdot 3 \cdot 2 \cdot 1} \cdot \dfrac{3 \cdot 2!}{2! \cdot 1}$

$= \dfrac{5 \cdot 4}{2} \cdot \dfrac{6 \cdot 5 \cdot 4}{3 \cdot 2} \cdot \dfrac{3}{1} = 10 \cdot 20 \cdot 3$

$= 600$

32. $_8C_2 \cdot_6 C_5 \cdot_1 C_1$, or 168

33. The pizza can have no toppings or 1 topping or 2 or 3 or 4 or 5 or 6 or 7 or 8 or 9 or 10 toppings. We add these combinations to find the total number possible.

$$\binom{10}{0} + \binom{10}{1} + \binom{10}{2} + \binom{10}{3} + \binom{10}{4} + \binom{10}{5} +$$

$$\binom{10}{6} + \binom{10}{7} + \binom{10}{8} + \binom{10}{9} + \binom{10}{10} = 1024$$

34. $2 \cdot 3 \cdot 1024$, or 6144

35. In a 52-card deck there are 4 aces and 48 cards that are not aces. We use the fundamental counting principle.

$$\binom{4}{3} \cdot \binom{48}{2} = \frac{4!}{(4-3)!3!} \cdot \frac{48!}{(48-2)!2!}$$

$$= \frac{4!}{1!3!} \cdot \frac{48!}{46!2!}$$

$$= \frac{4 \cdot 3!}{1 \cdot 3!} \cdot \frac{48 \cdot 47 \cdot 46!}{46! \cdot 2 \cdot 1}$$

$$= \frac{4}{1} \cdot \frac{48 \cdot 47}{2 \cdot 1}$$

$$= 4512$$

36. $\binom{4}{2} \cdot \binom{48}{3}$, or 103,776

37. a) If order is considered and repetition is not allowed, we have

$$_{33}P_3 = 33 \cdot 32 \cdot 31 = 32,736.$$

b) If order is considered and repetition is allowed, each scoop of ice cream can be chosen in 33 ways, so we have

$$33 \cdot 33 \cdot 33 = 35,937.$$

c) If order is not considered and repetition is not allowed, we have

$$_{33}C_3 = \frac{33!}{(33-3)!3!}$$

$$= \frac{33!}{30!3!} = \frac{33 \cdot 32 \cdot 31 \cdot 30!}{30!3 \cdot 2 \cdot 1}$$

$$= 5456.$$

38. a) $_{31}P_2$, or 930; b) $31 \cdot 31$, or 961; c) $_{31}C_2$, or 465

39. Writing exercise

40. Writing exercise

41. $2^x = \dfrac{1}{4}$

$2^x = \dfrac{1}{2^2}$

$2^x = 2^{-2}$

$x = -2$ The exponents must be the same.

The solution is -2.

42. $\dfrac{3}{2}$

43. $\log_5 (x+1) = 2$

$\qquad 5^2 = x + 1$ Writing an equivalent exponential equation

$\qquad 25 = x + 1$

$\qquad 24 = x$

The number 24 checks and is the solution.

44. 5

45. $\log_x 5 = 1$

$\qquad 5 = x^1$

$\qquad 5 = x$

The solution is 5.

46. 1.861

47. Writing exercise

48. Writing exercise

49. $\binom{m}{1} = \dfrac{m!}{(m-1)!1!}$

$\qquad = \dfrac{m(m-1)!}{(m-1)! \cdot 1}$

$\qquad = m$

50. m

51. $\binom{m}{0} = \dfrac{m!}{(m-0)!0!}$

$\qquad = \dfrac{m!}{m! \cdot 1}$

$\qquad = 1$

52. $\dfrac{m(m-1)}{2}$

53.
$$\binom{n+1}{3} = 2 \cdot \binom{n}{2}$$

$$\frac{(n+1)!}{(n+1-3)!3!} = 2 \cdot \frac{n!}{(n-2)!2!}$$

$$\frac{(n+1)!}{(n-2)!3!} = 2 \cdot \frac{n!}{(n-2)!2!}$$

$$\frac{(n+1)(n)(n-1)(n-2)!}{(n-2)!3 \cdot 2 \cdot 1} = 2 \cdot \frac{n(n-1)(n-2)!}{(n-2)! \cdot 2 \cdot 1}$$

$$\frac{(n+1)(n)(n-1)}{6} = n(n-1)$$

$$\frac{n^3 - n}{6} = n^2 - n$$

$$n^3 - n = 6n^2 - 6n$$

$$n^3 - 6n^2 + 5n = 0$$

$$n(n^2 - 6n + 5) = 0$$

$$n(n-5)(n-1) = 0$$

$$n = 0 \text{ or } n - 5 = 0 \text{ or } n - 1 = 0$$

$$n = 0 \text{ or } \qquad n = 5 \text{ or } \qquad n = 1$$

Only 5 checks. The solution is 5.

54. 4

55.
$$\binom{n+2}{4} = 6 \cdot \binom{n}{2}$$

$$\frac{(n+2)!}{(n+2-4)!4!} = 6 \cdot \frac{n!}{(n-2)!2!}$$

$$\frac{(n+2)!}{(n-2)!4!} = 6 \cdot \frac{n!}{(n-2)!2!}$$

$$\frac{(n+2)!}{4!} = 6 \cdot \frac{n!}{2!} \quad \text{Multiplying by } (n{-}2)!$$

$$4! \cdot \frac{(n+2)!}{4!} = 4! \cdot 6 \cdot \frac{n!}{2!}$$

$$(n+2)! = 72 \cdot n!$$

$$(n+2)(n+1)n! = 72 \cdot n!$$

$$(n+2)(n+1) = 72 \quad \text{Dividing by } n!$$

$$n^2 + 3n + 2 = 72$$

$$n^2 + 3n - 70 = 0$$

$$(n+10)(n-7) = 0$$

$$n + 10 = 0 \quad \text{ or } \quad n - 7 = 0$$

$$n = -10 \text{ or } \qquad n = 7$$

Only 7 checks. The solution is 7.

56. 6

57. There is one losing team per game. In order to leave one tournament winner there must be $n - 1$ losers produced in $n - 1$ games.

58. $2n - 1$

59.
$$\binom{m}{3} = \frac{m!}{(m-3)!3!}$$

$$= \frac{m(m-1)(m-2)(m-3)!}{(m-3)! \cdot 3 \cdot 2 \cdot 1}$$

$$= \frac{m(m-1)(m-2)}{6}$$

60. $\binom{m}{2}\binom{n}{2}$, or $\dfrac{mn(m-1)(n-1)}{4}$

61. See the answer section in the text.

Exercise Set 14.6

1. Expand $(m + n)^5$.

Form 1: The expansion of $(m + n)^5$ has $5 + 1$, or 6 terms. The sum of the exponents in each term is 5. The exponents of m start with 5 and decrease to 0. The last term has no factor of m. The first term has no factor of n. The exponents of n start in the second term with 1 and increase to 5. We get the coefficients from the 6th row of Pascal's triangle.

$$1$$
$$1 \qquad 1$$
$$1 \qquad 2 \qquad 1$$
$$1 \qquad 3 \qquad 3 \qquad 1$$
$$1 \qquad 4 \qquad 6 \qquad 4 \qquad 1$$
$$1 \qquad 5 \qquad 10 \qquad 10 \qquad 5 \qquad 1$$

$$(m+n)^5 = 1 \cdot m^5 + 5 \cdot m^4 n^1 + 10 \cdot m^3 \cdot n^2 +$$
$$10 \cdot m^2 \cdot n^3 + 5 \cdot m \cdot n^4 + 1 \cdot n^5$$
$$= m^5 + 5m^4 n + 10m^3 n^2 + 10m^2 n^3 +$$
$$5mn^4 + n^5$$

Form 2: We have $a = m$, $b = n$, and $n = 5$.

$$(m+n)^5 = \binom{5}{0}m^5 + \binom{5}{1}m^4 n + \binom{5}{2}m^3 n^2 +$$
$$\binom{5}{3}m^2 n^3 + \binom{5}{4}mn^4 + \binom{5}{5}n^5$$

$$= \frac{5!}{5!0!}m^5 + \frac{5!}{4!1!}m^4 n + \frac{5!}{3!2!}m^3 n^2 +$$
$$\frac{5!}{2!3!}m^2 n^3 + \frac{5!}{1!4!}mn^4 + \frac{5!}{0!5!}m^5$$

$$= m^5 + 5m^4 n + 10m^3 n^2 + 10m^2 n^3 +$$
$$5mn^4 + n^5$$

2. $a^4 - 4a^3 b + 6a^2 b^2 - 4ab^3 + b^4$

3. Expand $(x - y)^6$.

Form 1: The expansion of $(x - y)^6$ has $6 + 1$, or 7 terms. The sum of the exponents in each term is 6.

The exponents of x start with 6 and decrease to 0. The last term has no factor of x. The first term has no factor of $-y$. The exponents of $-y$ start in the second term with 1 and increase to 6. We get the coefficients from the 7th row of Pascal's triangle.

$$
\begin{array}{ccccccccccccc}
& & & & & & 1 & & & & & & \\
& & & & & 1 & & 1 & & & & & \\
& & & & 1 & & 2 & & 1 & & & & \\
& & & 1 & & 3 & & 3 & & 1 & & & \\
& & 1 & & 4 & & 6 & & 4 & & 1 & & \\
& 1 & & 5 & & 10 & & 10 & & 5 & & 1 & \\
1 & & 6 & & 15 & & 20 & & 15 & & 6 & & 1
\end{array}
$$

$$
\begin{aligned}
(x-y)^6 &= 1 \cdot x^6 + 6 \cdot x^5 \cdot (-y) + 15 \cdot x^4 \cdot (-y)^2 + \\
&\quad 20 \cdot x^3 \cdot (-y)^3 + 15 \cdot x^2 \cdot (-y)^4 + \\
&\quad 6 \cdot x \cdot (-y)^5 + 1 \cdot (-y)^6 \\
&= x^6 - 6x^5 y + 15x^4 y^2 - 20x^3 y^3 + \\
&\quad 15x^2 y^4 - 6xy^5 + y^6
\end{aligned}
$$

Form 2: We have $a = x$, $b = -y$, and $n = 6$.

$$
\begin{aligned}
(x-y)^6 &= \binom{6}{0} x^6 + \binom{6}{1} x^5(-y) + \binom{6}{2} x^4(-y)^2 + \\
&\quad \binom{6}{3} x^3(-y)^3 + \binom{6}{4} x^2(-y)^4 + \\
&\quad \binom{6}{5} x(-y)^5 + \binom{6}{6} (-y)^6 \\
&= \frac{6!}{6!0!} x^6 + \frac{6!}{5!1!} x^5(-y) + \frac{6!}{4!2!} x^4 y^2 + \\
&\quad \frac{6!}{3!3!} x^3(-y^3) + \frac{6!}{2!4!} x^2 y^4 + \frac{6!}{1!5!} x(-y^5) + \\
&\quad \frac{6!}{0!6!} y^6 \\
&= x^6 - 6x^5 y + 15x^4 y^2 - 20x^3 y^3 + \\
&\quad 15x^2 y^4 - 6xy^5 + y^6
\end{aligned}
$$

4. $p^7 + 7p^6 q + 21p^5 q^2 + 35p^4 q^3 + 35p^3 q^4 + 21p^2 q^5 + 7pq^6 + q^7$

5. Expand $(x^2 - 3y)^5$.

We have $a = x^2$, $b = -3y$, and $n = 5$.

Form 1: We get the coefficients from the 6th row of Pascal's triangle. From Exercise 17 we know that the coefficients are

$$
\begin{array}{cccccc}
1 & 5 & 10 & 10 & 5 & 1.
\end{array}
$$

$$
\begin{aligned}
(x^2 - 3y)^5 &= 1 \cdot (x^2)^5 + 5 \cdot (x^2)^4 \cdot (-3y) + \\
&\quad 10 \cdot (x^2)^3 \cdot (-3y)^2 + 10 \cdot (x^2)^2 \cdot (-3y)^3 + \\
&\quad 5 \cdot (x^2) \cdot (-3y)^4 + 1 \cdot (-3y)^5 \\
&= x^{10} - 15x^8 y + 90x^6 y^2 - 270x^4 y^3 + \\
&\quad 405x^2 y^4 - 243y^5
\end{aligned}
$$

Form 2:

$$
\begin{aligned}
(x^2 + 3y)^5 &= \binom{5}{0}(x^2)^5 + \binom{5}{1}(x^2)^4(-3y) + \\
&\quad \binom{5}{2}(x^2)^3(-3y)^2 + \binom{5}{3}(x^2)^2(-3y)^3 + \\
&\quad \binom{5}{4} x^2(-3y)^4 + \binom{5}{5}(-3y)^5 \\
&= \frac{5!}{5!0!} x^{10} + \frac{5!}{4!1!} x^8(-3y) + \frac{5!}{3!2!} x^6(9y^2) + \\
&\quad \frac{5!}{2!3!} x^4(-27y^3) + \frac{5!}{1!4!} x^2(81y^4) + \\
&\quad \frac{5!}{0!5!}(-243y^5) \\
&= x^{10} - 15x^8 y + 90x^6 y^2 - 270x^4 y^3 + \\
&\quad 405x^2 y^4 - 243y^5
\end{aligned}
$$

6. $2187c^7 - 5103c^6 d + 5103c^5 d^2 - 2835c^4 d^3 + 945c^3 d^4 - 189c^2 d^5 + 21cd^6 - d^7$

7. Expand $(3c - d)^6$.

We have $a = 3c$, $b = -d$, and $n = 6$.

Form 1: We get the coefficients from the 7th row of Pascal's triangle. From Exercise 19 we know that the coefficients are

$$
\begin{array}{cccccc}
1 & 6 & 15 & 20 & 15 & 6 & 1.
\end{array}
$$

$$
\begin{aligned}
(3c - d)^6 &= 1 \cdot (3c)^6 + 6 \cdot (3c)^5 \cdot (-d) + \\
&\quad 15 \cdot (3c)^4 \cdot (-d)^2 + 20 \cdot (3c)^3 \cdot (-d)^3 + \\
&\quad 15 \cdot (3c)^2 \cdot (-d)^4 + 6 \cdot (3c) \cdot (-d)^5 + \\
&\quad 1 \cdot (-d)^6 \\
&= 3^6 c^6 - 6 \cdot 3^5 c^5 d + 15 \cdot 3^4 c^4 d^2 - \\
&\quad 20 \cdot 3^3 c^3 d^3 + 15 \cdot 3^2 c^2 d^4 - 6 \cdot 3cd^5 + d^6 \\
&= 729c^6 - 6 \cdot 243c^5 d + 15 \cdot 81c^4 d^2 - \\
&\quad 20 \cdot 27c^3 d^3 + 15 \cdot 9c^2 d^4 - 6 \cdot 3cd^5 + d^6 \\
&= 729c^6 - 1458c^5 d + 1215c^4 d^2 - 540c^3 d^3 + \\
&\quad 135c^2 d^4 - 18cd^5 + d^6
\end{aligned}
$$

Form 2:

$$
\begin{aligned}
(3c - d)^6 &= \binom{6}{0}(3c)^6 + \binom{6}{1}(3c)^5(-d) + \\
&\quad \binom{6}{2}(3c)^4(-d)^2 + \binom{6}{3}(3c)^3(-d)^3 + \\
&\quad \binom{6}{4}(3c)^2(-d)^4 + \binom{6}{5}(3c)(-d)^5 + \\
&\quad \binom{6}{6}(-d)^6
\end{aligned}
$$

$$= \frac{6!}{6!0!}(729c^6) + \frac{6!}{5!1!}(243c^5)(-d) +$$

$$\frac{6!}{4!2!}(81c^4)(d^2) + \frac{6!}{3!3!}(27c^3)(-d^3) +$$

$$\frac{6!}{2!4!}(9c^2)(d^4) + \frac{6!}{1!5!}(3c)(-d^5) +$$

$$\frac{6!}{0!6!}d^6$$

$$= 729c^6 - 1458c^5d + 1215c^4d^2 - 540c^3d^3 +$$
$$135c^2d^4 - 18cd^5 + d^6$$

8. $t^{-12} + 12t^{-10} + 60t^{-8} + 160t^{-6} + 240t^{-4} + 192t^{-2} + 64$

9. Expand $(x - y)^3$.

We have $a = x$, $b = -y$, and $n = 3$.

Form 1: We get the coefficients from the 4th row of Pascal's triangle.

$$1$$
$$1 \quad 1$$
$$1 \quad 2 \quad 1$$
$$1 \quad 3 \quad 3 \quad 1$$

$$(x - y)^3$$
$$= 1 \cdot x^3 + 3x^2(-y) + 3x(-y)^2 + 1 \cdot (-y)^3$$
$$= x^3 - 3x^2y + 3xy^2 - y^3$$

Form 2:
$$(x - y)^3$$

$$= \binom{3}{0}x^3 + \binom{3}{1}x^2(-y) + \binom{3}{2}x(-y)^2 +$$

$$\binom{3}{3}(-y)^3$$

$$= \frac{3!}{3!0!}x^3 + \frac{3!}{2!1!}x^2(-y) + \frac{3!}{1!2!}xy^2 +$$

$$\frac{3!}{0!3!}(-y^3)$$

$$= x^3 - 3x^2y + 3xy^2 - y^3$$

10. $x^5 - 5x^4y + 10x^3y^2 - 10x^2y^3 + 5xy^4 - y^5$

11. Expand $\left(x + \dfrac{2}{y}\right)^9$.

We have $a = x$, $b = \dfrac{2}{y}$, and $n = 9$.

Form 1: We get the coefficients from the 10th row of Pascal's triangle.

$$1$$
$$1 \quad 1$$
$$1 \quad 2 \quad 1$$
$$1 \quad 3 \quad 3 \quad 1$$
$$1 \quad 4 \quad 6 \quad 4 \quad 1$$
$$1 \quad 5 \quad 10 \quad 10 \quad 5 \quad 1$$
$$1 \quad 6 \quad 15 \quad 20 \quad 15 \quad 6 \quad 1$$
$$1 \quad 7 \quad 21 \quad 35 \quad 35 \quad 21 \quad 7 \quad 1$$
$$1 \quad 8 \quad 28 \quad 56 \quad 70 \quad 56 \quad 28 \quad 8 \quad 1$$
$$1 \quad 9 \quad 36 \quad 84 \quad 126 \quad 126 \quad 84 \quad 36 \quad 9 \quad 1$$

$$\left(x + \frac{2}{y}\right)^9 = 1 \cdot x^9 + 9x^8\left(\frac{2}{y}\right) + 36x^7\left(\frac{2}{y}\right)^2 +$$

$$84x^6\left(\frac{2}{y}\right)^3 + 126x^5\left(\frac{2}{y}\right)^4 +$$

$$126x^4\left(\frac{2}{y}\right)^5 + 84x^3\left(\frac{2}{y}\right)^6 +$$

$$36x^2\left(\frac{2}{y}\right)^7 + 9x\left(\frac{2}{y}\right)^8 + 1 \cdot \left(\frac{2}{y}\right)^9$$

$$= x^9 + \frac{18x^8}{y} + \frac{144x^7}{y^2} + \frac{672x^6}{y^3} +$$

$$\frac{2016x^5}{y^4} + \frac{4032x^4}{y^5} + \frac{5376x^3}{y^6} +$$

$$\frac{4608x^2}{y^7} + \frac{2304x}{y^8} + \frac{512}{y^9}$$

Form 2:

$$\left(x - \frac{2}{y}\right)^9$$

$$= \binom{9}{0}x^9 + \binom{9}{1}x^8\left(\frac{2}{y}\right) + \binom{9}{2}x^7\left(\frac{2}{y}\right)^2 +$$

$$\binom{9}{3}x^6\left(\frac{2}{y}\right)^3 + \binom{9}{4}x^5\left(\frac{2}{y}\right)^4 +$$

$$\binom{9}{5}x^4\left(\frac{2}{y}\right)^5 + \binom{9}{6}x^3\left(\frac{2}{y}\right)^6 +$$

$$\binom{9}{7}x^2\left(\frac{2}{y}\right)^7 + \binom{9}{8}x\left(\frac{2}{y}\right)^8 +$$

$$\binom{9}{9}\left(\frac{2}{y}\right)^9$$

$$= \frac{9!}{9!0!}x^9 + \frac{9!}{8!1!}x^8\left(\frac{2}{y}\right) + \frac{9!}{7!2!}x^7\left(\frac{4}{y^2}\right) +$$

$$\frac{9!}{6!3!}x^6\left(\frac{8}{y^3}\right) + \frac{9!}{5!4!}x^5\left(\frac{16}{y^4}\right) +$$

$$\frac{9!}{4!5!}x^4\left(\frac{32}{y^5}\right) + \frac{9!}{3!6!}x^3\left(\frac{64}{y^6}\right) +$$

$$\frac{9!}{2!7!}x^2\left(\frac{128}{y^7}\right) + \frac{9!}{1!8!}x\left(\frac{256}{y^8}\right) +$$

$$\frac{9!}{0!9!}\left(\frac{512}{y^9}\right)$$

$$= x^9 + 9x^8\left(\frac{2}{y}\right) + 36x^7\left(\frac{4}{y^2}\right) + 84x^6\left(\frac{8}{y^3}\right) +$$

$$126x^5\left(\frac{16}{y^4}\right) + 126x^4\left(\frac{32}{y^5}\right) + 84x^3\left(\frac{64}{y^6}\right) +$$

$$36x^2\left(\frac{128}{y^7}\right) + 9x\left(\frac{256}{y^8}\right) + \frac{512}{y^9}$$

$$= x^9 + \frac{18x^8}{y} + \frac{144x^7}{y^2} + \frac{672x^6}{y^3} +$$

$$\frac{2016x^5}{y^4} + \frac{4032x^4}{y^5} + \frac{5376x^3}{y^6} +$$

$$\frac{4608x^2}{y^7} + \frac{2304x}{y^8} + \frac{512}{y^9}$$

12. $19,683s^9 + \dfrac{59,049s^8}{t} + \dfrac{78,732s^7}{t^2} + \dfrac{61,236s^6}{t^3} +$
$\dfrac{30,618s^5}{t^4} + \dfrac{10,206s^4}{t^5} + \dfrac{2268s^3}{t^6} + \dfrac{324s^2}{t^7} + \dfrac{27s}{t^8} + \dfrac{1}{t^9}$

13. Expand $(a^2 - b^3)^5$.

We have $a = a^2$, $b = -b^3$, and $n = 5$.

Form 1: We get the coefficient from the 6th row of Pascal's triangle. From Exercise 17 we know that the coefficients are

$$1 \quad 5 \quad 10 \quad 10 \quad 5 \quad 1.$$

$$(a^2 - b^3)^5$$
$$= 1 \cdot (a^2)^5 + 5(a^2)^4(-b^3) + 10(a^2)^3(-b^3)^2 +$$
$$10(a^2)^2(-b^3)^3 + 5(a^2)(-b^3)^4 + 1 \cdot (-b^3)^5$$
$$= a^{10} - 5a^8b^3 + 10a^6b^6 - 10a^4b^9 +$$
$$5a^2b^{12} - b^{15}$$

Form 2:
$$(a^2 - b^3)^5$$
$$= \binom{5}{0}(a^2)^5 + \binom{5}{1}(a^2)^4(-b^3) +$$
$$\binom{5}{2}(a^2)^3(-b^3)^2 + \binom{5}{3}(a^2)^2(-b^3)^3 +$$
$$\binom{5}{4}(a^2)(-b^3)^4 + \binom{5}{5}(-b^3)^5$$
$$= \frac{5!}{5!0!}a^{10} + \frac{5!}{4!1!}a^8(-b^3) + \frac{5!}{3!2!}a^6(b^6) +$$
$$\frac{5!}{2!3!}a^4(-b^9) + \frac{5!}{1!4!}a^2(b^{12}) + \frac{5!}{0!5!}(-b^{15})$$
$$= a^{10} - 5a^8b^3 + 10a^6b^6 - 10a^4b^9 +$$
$$5a^2b^{12} - b^{15}$$

14. $x^{15} - 10x^{12}y + 40x^9y^2 - 80x^6y^3 + 80x^3y^4 - 32y^5$

15. Expand $(\sqrt{3} - t)^4$.

We have $a = \sqrt{3}$, $b = -t$, and $n = 4$.

Form 1: We get the coefficients from the 5th row of Pascal's triangle.

$$1$$
$$1 \quad 1$$
$$1 \quad 2 \quad 1$$
$$1 \quad 3 \quad 3 \quad 1$$
$$1 \quad 4 \quad 6 \quad 4 \quad 1$$

$$(\sqrt{3} - t)^4 = 1 \cdot (\sqrt{3})^4 + 4(\sqrt{3})^3(-t) +$$
$$6(\sqrt{3})^2(-t)^2 + 4(\sqrt{3})(-t)^3 + 1 \cdot (-t)^4$$
$$= 9 - 12\sqrt{3}t + 18t^2 - 4\sqrt{3}t^3 + t^4$$

Form 2:
$$(\sqrt{3} - t)^4 = \binom{4}{0}(\sqrt{3})^4 + \binom{4}{1}(\sqrt{3})^3(-t) +$$
$$\binom{4}{2}(\sqrt{3})^2(-t)^2 + \binom{4}{3}(\sqrt{3})(-t)^3 +$$
$$\binom{4}{4}(-t)^4$$
$$= \frac{4!}{4!0!}(9) + \frac{4!}{3!1!}(3\sqrt{3})(-t) +$$
$$\frac{4!}{2!2!}(3)(t^2) + \frac{4!}{1!3!}(\sqrt{3})(-t^3) +$$
$$\frac{4!}{0!4!}(t^4)$$
$$= 9 - 12\sqrt{3}t + 18t^2 - 4\sqrt{3}t^3 + t^4$$

16. $125 + 150\sqrt{5}\,t + 375t^2 + 100\sqrt{5}\,t^3 + 75t^4 + 6\sqrt{5}\,t^5 + t^6$

17. Expand $(x^{-2} + x^2)^4$.

We have $a = x^{-2}$, $b = x^2$, and $n = 4$.

Form 1: We get the coefficients from the fifth row of Pascal's triangle. From Exercise 31 we know that the coefficients are

$$1 \quad 4 \quad 6 \quad 4 \quad 1.$$
$$(x^{-2} + x^2)^4$$
$$= 1 \cdot (x^{-2})^4 + 4(x^{-2})^3(x^2) + 6(x^{-2})^2(x^2)^2 +$$
$$4(x^{-2})(x^2)^3 + 1 \cdot (x^2)^4$$
$$= x^{-8} + 4x^{-4} + 6 + 4x^4 + x^8$$

Form 2:
$$(x^{-2} + x^2)^4$$

$$= \binom{4}{0}(x^{-2})^4 + \binom{4}{1}(x^{-2})^3(x^2) +$$

$$\binom{4}{2}(x^{-2})^2(x^2)^2 + \binom{4}{3}(x^{-2})(x^2)^3 +$$

$$\binom{4}{4}(x^2)^4$$

$$= \frac{4!}{4!0!}(x^{-8}) + \frac{4!}{3!1!}(x^{-6})(x^2) + \frac{4!}{2!2!}(x^{-4})(x^4) +$$

$$\frac{4!}{1!3!}(x^{-2})(x^6) + \frac{4!}{0!4!}(x^8)$$

$$= x^{-8} + 4x^{-4} + 6 + 4x^4 + x^8$$

18. $x^{-3} - 6x^{-2} + 15x^{-1} - 20 + 15x - 6x^2 + x^3$

19. Find the 3rd term of $(a + b)^6$.

First, we note that $3 = 2 + 1$, $a = a$, $b = b$, and $n = 6$. Then the 3rd term of the expansion of $(a + b)^6$ is

$$\binom{6}{2}a^{6-2}b^2, \text{ or } \frac{6!}{4!2!}a^4b^2, \text{ or } 15a^4b^2.$$

20. $21x^2y^5$

21. Find the 12th term of $(a - 3)^{14}$.

First, we note that $12 = 11 + 1$, $a = a$, $b = -3$, and $n = 14$. Then the 12th term of the expansion of $(a - 3)^{14}$ is

$$\binom{14}{11}a^{14-11} \cdot (-3)^{11} = \frac{14!}{3!11!}a^3(-177,147)$$
$$= 364a^3(-177,147)$$
$$= -64,481,508a^3$$

22. $67,584x^2$

23. Find the 5th term of $(2x^3 - \sqrt{y})^8$.

First, we note that $5 = 4 + 1$, $a = 2x^3$, $b = -\sqrt{y}$, and $n = 8$. Then the 5th term of the expansion of $(2x^3 - \sqrt{y})^8$ is

$$\binom{8}{4}(2x^3)^{8-4}(-\sqrt{y})^4$$
$$= \frac{8!}{4!4!}(2x^3)^4(-\sqrt{y})^4$$
$$= 70(16x^{12})(y^2)$$
$$= 1120x^{12}y^2$$

24. $\dfrac{35c^3}{b^8}$

25. The expansion of $(2u - 3v^2)^{10}$ has 11 terms so the 6th term is the middle term. Note that $6 = 5 + 1$, $a = 2u$, $b = -3v^2$, and $n = 10$. Then the 6th term of the expansion of $(2u - 3v^2)^{10}$ is

$$\binom{10}{5}(2u)^{10-5}(-3v^2)^5$$
$$= \frac{10!}{5!5!}(2u)^5(-3v^2)^5$$
$$= 252(32u^5)(-243v^{10})$$
$$= -1,959,552u^5v^{10}$$

26. $30x\sqrt{x}$; $30x\sqrt{3}$

27. The 9th term of $(x - y)^8$ is the last term, y^8.

28. $-b^9$

29. Writing exercise

30. Writing exercise

31.
$$\log_2 x + \log_2(x - 2) = 3$$
$$\log_2 x(x - 2) = 3$$
$$x(x - 2) = 2^3$$
$$x^2 - 2x = 8$$
$$x^2 - 2x - 8 = 0$$
$$(x - 4)(x + 2) = 0$$
$$x = 4 \text{ or } x = -2$$

Only 4 checks. It is the solution.

32. $\dfrac{5}{2}$

33.
$$e^t = 280$$
$$\ln e^t = \ln 280$$
$$t = \ln 280$$
$$t \approx 5.6348$$

34. ± 5

35. Writing exercise

36. Writing exercise

37. Find the third term of $(0.325 + 0.675)^5$.

$\binom{5}{n}(0.325)^{5-2}(0.675)^2 = \dfrac{5!}{3!2!}(0.325)^3(0.675)^2 \approx$
0.156

38. $\binom{8}{5}(0.15)^{8-5}(0.85)^5 \approx 0.084$

39. Find and add the 3rd through the 6th terms of $(0.325 + 0.675)^5$:

$\binom{5}{2}(0.325)^3(0.675)^2 + \binom{5}{3}(0.325)^2(0.675)^3 +$

$\binom{5}{4}(0.325)(0.675)^4 + \binom{5}{5}(0.675)^5 \approx 0.959$

40. $\binom{8}{6}(0.15)^2(0.85)^6 + \binom{8}{7}(0.15)(0.85)^7 +$

$\binom{8}{8}(0.85)^8 \approx 0.89$

41. The $(r+1)$st term of $\left(\dfrac{3x^2}{2} - \dfrac{1}{3x}\right)^{12}$ is

$\binom{12}{r}\left(\dfrac{3x^2}{2}\right)^{12-r}\left(-\dfrac{1}{3x}\right)^r$. In the term which does
not contain x, the exponent of x in the numerator is
equal to the exponent of x in the denominator.

$$2(12 - r) = r$$
$$24 - 2r = r$$
$$24 = 3r$$
$$8 = r$$

Find the $(8+1)$st, or 9th term:

$\binom{12}{8}\left(\dfrac{3x^2}{2}\right)^4\left(-\dfrac{1}{3x}\right)^8 = \dfrac{12!}{4!8!}\left(\dfrac{3^4x^8}{2^4}\right)\left(\dfrac{1}{3^8x^8}\right) = \dfrac{55}{144}$

42. $-4320x^6y^{9/2}$

43.

$\dfrac{\binom{5}{3}(p^2)^2\left(-\frac{1}{2}p\sqrt[3]{q}\right)^3}{\binom{5}{2}(p^2)^3\left(-\frac{1}{2}p\sqrt[3]{q}\right)^2} = \dfrac{-\frac{1}{8}p^7q}{\frac{1}{4}p^8\sqrt[3]{q^2}} =$

$\dfrac{-\frac{1}{8}p^7q}{\frac{1}{4}p^8q^{2/3}} = -\dfrac{1}{8}\cdot\dfrac{4}{1}\cdot p^{7-8}\cdot q^{1-2/3} =$

$-\dfrac{1}{2}p^{-1}q^{1/3} = -\dfrac{\sqrt[3]{q}}{2p}$

44. $-\dfrac{35}{x^{1/6}}$

45. The degree of $(x^2 + 3)^4$ is the degree of $(x^2)^4 = x^8$, or 8.

46. $x^7 + 7x^6y + 21x^5y^2 + 35x^4y^3 + 35x^3y^4 + 21x^2y^5 + 7xy^6 + y^7$

Exercise Set 14.7

1. 100 people were surveyed. 57 wore either glasses or contacts. $100 - 57$, or 43 wore neither glasses nor contacts.

We use Principle P.

The probability that a person wears either glasses or contacts is P where

$P = \dfrac{57}{100}$, or 0.57.

The probability that a person wears neither glasses nor contacts is P where

$P = \dfrac{43}{100}$, or 0.43.

2. a) 0.18, 0.24, 0.23, 0.23, 0.12;

b) People tend not to choose the first or last numbers.

3. There was a total of 9136 letters.
   A occurred 853 times.

   E occurred 1229 times.

   I occurred 539 times.

   O occurred 705 times.

   U occurred 240 times.

We use Principle P.

The probability of the occurrence of an A is
$\dfrac{853}{9136} \approx 0.093$.

The probability of the occurrence of an E is
$\dfrac{1229}{9136} \approx 0.135$.

The probability of the occurrence of an I is
$\dfrac{539}{9136} \approx 0.059$.

The probability of the occurrence of an O is
$\dfrac{705}{9136} \approx 0.077$.

The probability of the occurrence of a U is
$\dfrac{240}{9136} \approx 0.026$.

4. 0.390

5. There was a total of 9136 letters.

The total number of vowels was $853 + 1229 + 539 + 705 + 240$, or 3566.

The total number of consonants was $9136 - 3566$, or 5570.

The probability of a consonant occurring is
$\dfrac{5570}{9136} \approx 0.610$.

6. Q; 1.000 (If the probability were rounded to 4 decimal places, it would be given as 0.9996.)

7. a) T is the consonant with the largest number of occurrences, so it is the consonant with the greatest probability of occurring.

b) E is the vowel with the largest number of occurrences, so it is the vowel with the greatest probability of occurring.

c) The contestant should choose the five consonants with the largest numbers of occurrences among the consonants and the vowel with the largest number of occurrences among the vowels. From part (a) we know that one consonant should be T. The four consonants with the next highest numbers of occurrences are S, R, N, and L. From part (b) we know that the vowel should be E. Thus, the contestant should choose T, S, R, N, L, and E, based solely on the results of the given experiment.

8. 52

9. Since there are 52 equally likely outcomes and there are 4 ways to obtain a queen, by Principle P we have
$$P(\text{drawing a queen}) = \frac{4}{52}, \text{ or } \frac{1}{13}.$$

10. $\frac{1}{4}$

11. Since there are 52 equally likely outcomes and there are 26 ways to obtain a spade or a club (13 spades and 13 clubs), by Principle P we have
$$P(\text{drawing a spade or club}) = \frac{26}{52}, \text{ or } \frac{1}{2}.$$

12. $\frac{1}{2}$

13. Since there are 52 equally likely outcomes and there are 8 ways to obtain a 9 or a king (four 9's and four kings), we have, by Principle P,
$$P(\text{drawing a 9 or a king}) = \frac{8}{52} = \frac{2}{13}.$$

14. $\frac{1}{26}$

15. Since there are 14 equally likely ways of selecting a marble from a bag containing 4 red marbles and 10 green marbles, we have, by Principle P,
$$P(\text{selecting a red marble}) = \frac{4}{14} = \frac{2}{7}.$$

16. $\frac{5}{7}$

17. There are 14 equally likely ways of selecting any marble from a bag containing 4 red marbles and 10 green marbles. Since the bag does not contain any purple marbles, there are 0 ways of selecting a purple marble. By Principle P, we have
$$P(\text{selecting a purple marble}) = \frac{0}{14} = 0.$$

18. 0

19. The number of ways of drawing 4 cards from a deck of 52 cards is $_{52}C_4$. Now 13 of the 52 cards are spades, so the number of ways of drawing 4 spades is $_{13}C_4$. Thus,
$$P(\text{getting 4 spades}) = \frac{_{13}C_4}{_{52}C_4}, \text{ or } \frac{11}{4165}.$$

20. $\frac{_{13}C_4}{_{52}C_4}$, or $\frac{11}{4165}$

21. The number of ways of selecting 6 coins from a group of 20 is $_{20}C_6$. Three nickels can be selected in $_6C_3$ ways, two dimes can be selected in $_{10}C_2$ ways, and one quarter can be selected in $_4C_1$ ways. By the fundamental counting principle, the number of ways of selecting 3 nickels, 2 dimes, and 1 quarter is $_6C_3 \cdot _{10}C_2 \cdot _4C_1$. Thus,
$$P(\text{3 nickels, 2 dimes, and 1 quarter}) =$$
$$\frac{_6C_3 \cdot _{10}C_2 \cdot _4C_1}{_{20}C_6}, \text{ or } \frac{30}{323}.$$

22. $\frac{_8C_2 \cdot _7C_2}{_{15}C_4}$, or $\frac{28}{65}$

23. On each die there are 6 possible outcomes. The outcomes are paired so there are $6 \cdot 6$, or 36 possible ways in which the two can fall. The pairs that total 6 are (1,5), (2,4), (3,3), (4,2), and (5,1). Thus there are 5 possible ways of getting a total of 6, so the probability is $\frac{5}{36}$.

24. $\frac{1}{36}$

25. On each die there are 6 possible outcomes. The outcomes are paired so there are $6 \cdot 6$, or 36 possible ways in which the two can fall. There are 6 possible doubles: (1,1), (2,2), (3,3), (4,4), (5,5), and (6,6), so the probability of rolling doubles is $\frac{6}{36}$, or $\frac{1}{6}$.

26. $\frac{5}{12}$

27. THe bottle contains 7·4, or 28 vitamins. The number of ways of selecting 4 vitamins from a group of 28 is $_{28}C_4$. The number of ways of selecting 1 vitamin A tablet from a group of 7 is $_7C_1$. The same is true for selecting 1 vitamin C, E, or B-12 tablet from a group of 7 each. Thus

P(selecting 1 each of vitamins A, C, E, and B-12) =

$$\frac{_7C_1 \cdot _7C_1 \cdot _7C_1 \cdot _7C_1}{_{28}C_4}, \text{ or } \frac{343}{2925}.$$

28. $\dfrac{_4C_1 \cdot _{10}C_1 \cdot _6C_1 \cdot _5C_1}{_{25}C_4}$, or $\dfrac{24}{253}$

29. The roulette wheel contains 38 equally likely slots. Eighteen of the 38 slots are colored black. Thus, by Principle P,

P(the ball falls in a black slot) $= \dfrac{18}{38} = \dfrac{9}{19}$.

30. $\dfrac{18}{19}$

31. The roulette wheel contains 38 equally likely slots. Only 1 slot is numbered 0. Then, by Principle P,

P(the ball falls in the 0 slot) $= \dfrac{1}{38}$.

32. $\dfrac{1}{19}$

33. The one red region on the left-hand side of the dartboard occupies $\dfrac{1}{3}$ of that side or $\dfrac{1}{3} \cdot \dfrac{1}{2}$, or $\dfrac{1}{6}$ of the dartboard. The two red regions on the right-hand side of the dartboard each occupy $\dfrac{1}{4}$ of that side, so together they occupy $2 \cdot \dfrac{1}{4} \cdot \dfrac{1}{2}$, or $\dfrac{1}{4}$ of the dartboard. Then all of the red regions together occupy $\dfrac{1}{6} + \dfrac{1}{4}$, or $\dfrac{5}{12}$ of the dartboard, so

$$P(\text{red}) = \frac{5}{12}.$$

34. $\dfrac{1}{6}$

35. The blue regions occupy the same amount of space as the red regions, or $\dfrac{5}{12}$ of the dartboard. (See Exercise 33.) The yellow region occupies $\dfrac{1}{3}$ of $\dfrac{1}{2}$, or $\dfrac{1}{3} \cdot \dfrac{1}{2}$, or $\dfrac{1}{6}$ of the dartboard. Thus,

$$P(\text{blue or yellow}) = \frac{5}{12} + \frac{1}{6} = \frac{7}{12}.$$

36. $\dfrac{5}{6}$

37. Since the dartboard contains only red, blue, and yellow regions, a dart that hits the dartboard must hit one of these colors. Thus,

$$P(\text{red or blue or yellow}) = 1.$$

38. 0

39. Writing exercise

40. Writing exercise.

41. $2x + 5y = 7$,   (1)
$3x + 2y = 16$ (2)

Multiply Equation (1) by 2, multiply Equation (2) by -5, and add.

$$
\begin{array}{r}
4x + 10y = 14 \\
-15x - 10y = -80 \\
\hline
-11x = -66 \\
x = 6
\end{array}
$$

Substitute 6 for x in Equation (2) and solve for y.

$3 \cdot 6 + 2y = 16$
$18 + 2y = 16$
$2y = -2$
$y = -1$

The solution is $(6, -1)$.

42. -1

43.
$$\log_a \frac{x^2 y}{z^3}$$
$= \log_a x^2 y - \log_a z^3$ Quotient rule
$= \log_a x^2 + \log_a y - \log_a z^3$ Product rule
$= 2\log_a x + \log_a y - 3\log_a z$ Power rule

44. $x^2 + (y + 3)^2 = 12$

45. $3^4 = x$

46. $\log_4 10 = y$

47. Writing exercise

48. Writing exercise

49. $_{52}C_5 = \dfrac{52!}{47!5!} = \dfrac{52 \cdot 51 \cdot 50 \cdot 49 \cdot 48 \cdot 47!}{47! \cdot 5 \cdot 4 \cdot 3 \cdot 2 \cdot 1}$
$= 26 \cdot 17 \cdot 10 \cdot 49 \cdot 12$
$= 2{,}598{,}960$

50. a) 4;   b) $\dfrac{4}{_{52}C_5} \approx 0.0000015$

51. Consider a suit

A K Q J 10 9 8 7 6 5 4 3 2

A straight flush can be any of the following combinations in the same suit.

K	Q	J	10	9
Q	J	10	9	8
J	10	9	8	7
10	9	8	7	6
9	8	7	6	5
8	7	6	5	4
7	6	5	4	3
6	5	4	3	2
5	4	3	2	A

Remember a straight flush does not include A K Q J 10 which is a royal flush.

a) Since there are 9 straight flushes per suit, there are $9 \cdot 4$, or 36 straight flushes in all 4 suits.

b) Since 2,598,960, or $_{52}C_5$, poker hands can be dealt from a standard 52-card deck and 36 of those hands are straight flushes, the probability of getting a straight flush is $\dfrac{36}{2,598,960}$, or 0.0000139.

52. a) $13 \cdot 48$, or 624; b) $\dfrac{624}{_{52}C_5} \approx 0.00024$

53. a) There are 13 ways to select a denomination. Then from that denomination there are $_4C_3$ ways to pick 3 of the 4 cards in that denomination. Now there are 12 ways to select any one of the remaining 12 denominations and $_4C_2$ ways to pick 2 cards from the 4 cards in that denomination. Thus the number of full houses is $(13 \cdot_4 C_3) \cdot (12 \cdot_4 C_2)$ or 3744.

b) $\dfrac{3744}{_{52}C_5} = \dfrac{3744}{2,598,960} \approx 0.00144$

54. a) $13 \cdot \dbinom{4}{2}\dbinom{12}{3}\dbinom{4}{1}\dbinom{4}{1}\dbinom{4}{1}$, or 1,098,240

b) $\dfrac{1,098,240}{_{52}C_5} \approx 0.423$

55. a) There are 13 ways to select a denomination and then $\dbinom{4}{3}$ ways to choose 3 of the 4 cards in that denomination. Now there are $\dbinom{48}{2}$ ways to choose 2 cards from the 12 remaining denominations $(4 \cdot 12$, or 48 cards). But these combinations include the 3744 hands in a full house like Q-Q-Q-4-4 (Exercise 53), so these must be subtracted. Thus the number of three of a kind hands is $13 \cdot \dbinom{4}{3} \cdot \dbinom{48}{2} - 3744$, or 54,912.

b) $\dfrac{54,912}{_{52}C_5} = \dfrac{54,912}{2,598,960} \approx 0.0211$

56. a) $4 \cdot \dbinom{13}{5} - 4 - 36$, or 5108

b) $\dfrac{5108}{_{52}C_5} \approx 0.00197$

57. a) There are $\dbinom{13}{2}$ ways to select 2 denominations from the 13 denominations. Then in each denomination there are $\dbinom{4}{2}$ ways to choose 2 of the 4 cards. Finally there are $\dbinom{44}{1}$ ways to choose the fifth card from the 11 remaining denominations $(4 \cdot 11$, or 44 cards). Thus the number of two pairs hands is $\dbinom{13}{2} \cdot \dbinom{4}{2} \cdot \dbinom{4}{2} \cdot \dbinom{44}{1}$, or 123,552.

b) $\dfrac{123,552}{_{52}C_5} = \dfrac{123,552}{2,598,960} \approx 0.0475$

58. a) $10 \cdot 4^5 - 4 - 36$, or 10,200

b) $\dfrac{10,200}{_{52}C_5} \approx 0.00392$